World and United States Aviation and Space Records

(as of December 31, 2003)

Publisher
Donald J. Koranda

Editor
A. W. Greenfield

Assistant Editor
Michael R. Pablo

1815 North Fort Myer Drive, Suite 500
Arlington, VA 22209

www.naa-usa.org

Cover: Reproduction of the Wright Brothers' 1902 glider, modified for flight testing
Photo by Paul Glenshaw, courtesy of The Wright Experience, Inc.

ISBN 0-9709897-3-3 21.95

Table of Contents

Most Memorable Aviation Records of 2003

- Fred Coon and Mark Stolzberg broke the record for piston-engine aircraft in a Grumman Cheetah on October 16. They flew Santa Ana, Calif., to the site of the Wrights' first flight, Kitty Hawk, N.C. in 14 hours 53 minutes and 32 seconds, averaging a speed of 159 miles per hour during the trip.

- Steve Fossett and Douglas Travis set a record for a transcontinental flight for jet aircraft in a Cessna Citation X. On February 5, the pair flew from San Diego to Charleston, S.C. in 2 hours 56 minutes 20 seconds, averaging a speed of 726 mph.

- The next day Fossett flew along with Joseph Ritchie in a Piaggio Avanti for a transcontinental record for turboprop aircraft. They also flew from San Diego to Charleston, covering the distance in 3 hours 51 minutes 52 seconds, an average speed of 546 mph.

- Randolph Pentel and Mark Anderson set a different national record on December 16 and 17--the Fastest Time to Fly Around the Border of the Continental U.S. The pair took off in a Cessna Citation Ultra from International Falls, Minn., and returned 45 hours and 27 minutes later, taking a route that traced the border of the contiguous 48 states.

- Also in December, William Watters, Raymond Wellington and Ahmed Ragheb set a record for Distance Without Landing in a Gulfstream G550, the 2003 Collier Trophy winner. On December 6, Watters and crew flew from Savannah, Ga., to Dubai, United Arab Emirates, a distance of 7,546 miles. The jet averaged a speed of 549 mph and took under 14 hours to make the trip.

- Bruce Bohannon continued to push the envelope for aircraft going up into the sky--he set another record for altitude last year. Bohannon took off from Angleton, Texas, on November 15 and climbed to 47,067 feet in his piston-engine homebuilt "Flyin' Tiger."

- Jon Jacobs set a record for Distance in a Straight Line with Limited Fuel for ultralight aircraft on November 30. He covered 170 miles with just 7.5 kilograms of fuel--or about 2.7 gallons--in his Mitchell Wing b-10 to break a 15-year-old record.

- Finally, an aeromodel called TAM-5 became the first aircraft of its type to cross the Atlantic Ocean, piloted by Maynard Hill, Barrett Foster and David Brown. The radio-controlled model set a record for Distance in a Straight Line, traveling 1,882 miles before landing on August 11. The flight from Cape Spear, Newfoundland, to Mannin Beach, Ireland, took just under 39 hours.

CHECKLIST FOR SETTING A RECORD

- ☐ Use this book to determine what the existing record is, if any, for the record you would like to set.
- ☐ Contact NAA regarding the flight you are considering. We will check to see if anyone has set a record since the date of this publication. We will also send you a "Record Attempt Kit," or you can download online it at: www.naa-usa.org
- ☐ Carefully review the record attempt kit and the various forms in it to get an understanding of the documentation requirements.
- ☐ Send NAA a completed sanction application along with the applicable fee. If you are not already a member of NAA, you must join at this time. You will also need a valid FAI Sporting License, which can be issued by NAA.
- ☐ Once the sanction request is approved, you should discuss with NAA who will serve as directing official or designated observers. If the flight is a city-to-city record, usually Air Traffic Control personnel will serve as designated observers for the purpose of certifying the times of start and finish.
- ☐ You should get in touch with your directing official or designated observers to discuss the flight and procedures for documenting the record.
- ☐ The fun part—you make the flight.
- ☐ *Within 72 hours of the flight* you must notify NAA in writing (via fax, E-mail, or overnight letter) that the flight was successful, and provide details of the flight. Follow-up your notification with a phone call, to confirm that your claim was received.
- ☐ Within 30 days you must send NAA complete details of the flight on the forms prescribed. Please include any photographs that NAA may wish to use for promotional purposes. The registration fee must be paid in full at this time.
- ☐ After NAA receives and reviews the documentation file, NAA may certify the record as a United States National Record.
- ☐ For world records, NAA will then forward a complete dossier to FAI in Switzerland requesting their approval.

- ☐ It usually takes FAI 30–60 days to approve the record, or to request additional documentation.
- ☐ NAA can present your certificates at one of its many awards ceremonies held throughout the year. If you prefer, the certificates can also be mailed to you.
- ☐ Your record will be published in the next edition of this book. Congratulations, you have become an official part of aviation history!

National Aeronautic Association
1815 N. Ft. Myer Dr., Ste. 500
Arlington, VA 22209

Phone: (703) 527–0226
Fax: (703) 527–0229
E-mail: records@naa-usa.org
Web site: www.naa-usa.org

Official Records

To establish uniformity in the recognition of the many possible aircraft performances, a standard record classification has been adopted by the FAI for "Absolute World" and "World Class" records. U.S. National records are also recognized following these FAI classifications.

Absolute World Records

Absolute World Records are defined as the maximum performances of all vehicles in Classes C, H, M, and N (see below). The following achievements are recognized:

Distance without Landing
Distance Over a Closed Circuit without Landing
Altitude
Altitude in Horizontal Flight
Speed Over a 15/25 Kilometer Straight Course
Speed Over a Closed Circuit
Speed Around the World, Non-stop, Non-refueled

World Class Records

All other records which are international in scope and necessarily the best world performances for a particular type of aviation are termed "World Class" Records to differentiate from the foregoing maximum achievements. World Class Records, therefore, are defined as the best international performances for the classes and categories recognized by the FAI. The classes which have been adopted by the FAI for the purpose of providing a standard classification are:

Class A: Free Balloons
Class B: Airships
Class C: Aeroplanes (C-1), Seaplanes (C-2), Amphibians (C-3) (These classes are further divided into weight subclasses, see page 6)
Class D: Gliders
Class E: Rotorcraft (This class is further divided into subclasses, see page 7)
Class F: Model Aircraft
Class G: Parachutes
Class H: Vertical Takeoff and Landing (VTOL) Aircraft
Class I: Human Powered Aircraft
Class K: Spacecraft
Class M: Tilt Wing / Tilt Engine Aircraft
Class N: Short Take-off and Landing (STOL) Aircraft
Class O: Hang Gliders / Paragliders
Class P: Aerospacecraft

Class R: Microlights
Class S: Space Models
Class U: Unmanned Aerial Vehicles

Records in classes C and E may be classified by weight as follows:

C-1 (Landplanes)

Class C-1.a/0	Takeoff weight less than 661 lbs
Class C-1.a	Takeoff weight from 661 to less than 1,102 lbs
Class C-1.b	Takeoff weight from 1,102 to less than 2,205 lbs
Class C-1.c	Takeoff weight from 2,205 to less than 3,858 lbs
Class C-1.d	Takeoff weight from 3,858 to less than 6,614 lbs
Class C-1.e	Takeoff weight from 6,614 to less than 13,228 lbs
Class C-1.f	Takeoff weight from 13,228 to less than 19,842 lbs
Class C-1.g	Takeoff weight from 19,842 to less than 26,455 lbs
Class C-1.h	Takeoff weight from 26,455 to less than 35,274 lbs
Class C-1.i	Takeoff weight from 35,274 to less than 44,092 lbs
Class C-1.j	Takeoff weight from 44,092 to less than 55,116 lbs
Class C-1.k	Takeoff weight from 55,116 to less than 77,162 lbs
Class C-1.l	Takeoff weight from 77,162 to less than 99,208 lbs
Class C-1.m	Takeoff weight from 99,208 to less than 132,277 lbs
Class C-1.n	Takeoff weight from 132,277 to less than 176,370 lbs
Class C-1.o	Takeoff weight from 176,370 to less than 220,462 lbs
Class C-1.p	Takeoff weight from 220,462 to less than 330,693 lbs
Class C-1.q	Takeoff weight from 330,693 to less than 440,924 lbs
Class C-1.r	Takeoff weight from 440,924 to less than 551,155 lbs
Class C-1.s	Takeoff weight from 551,155 to less than 661,386 lbs
Class C-1.t	Takeoff weight 661,386 lbs or greater

C-2 (Seaplanes), and C-3 (Amphibians)

Classes C-2.a/0 and C-3.a/0	Takeoff weight less than 661 lbs
Classes C-2.a and C-3.a	Takeoff weight from 661 to less than 1,323 lbs
Classes C-2.b and C-3.b	Takeoff weight from 1,323 to less than 2,646 lbs
Classes C-2.c and C-3.c	Takeoff weight from 2,646 to less than 4,630 lbs
Classes C-2.d and C-3.d	Takeoff weight from 4,630 to less than 7,496 lbs
Classes C-2.e and C-3.e	Takeoff weight from 7,496 to less than 13,228 lbs
Classes C-2.f and C-3.f	Takeoff weight from 13,228 to less than 22,046 lbs
Classes C-2.g and C-3.g	Takeoff weight from 22,046 to less than 44,092 lbs
Classes C-2.h and C-3.h	Takeoff weight from 44,092 to less than 55,116 lbs
Classes C-2.i and C-3.i	Takeoff weight from 55,116 to less than 77,162 lbs
Classes C-2.j and C-3.j	Takeoff weight from 77,162 to less than 99,208 lbs
Classes C-2.k and C-3.k	Takeoff weight from 99,208 to less than 132,277 lbs
Classes C-2.l and C-3.l	Takeoff weight 132,277 lbs or greater

E-1 (Helicopters)

Class E-1.a	Takeoff weight less than 1,102 lbs
Class E-1.b	Takeoff weight from 1,102 to less than 2,205 lbs
Class E-1.c	Takeoff weight from 2,205 to less than 3,858 lbs
Class E-1.d	Takeoff weight from 3,858 to less than 6,614 lbs
Class E-1.e	Takeoff weight from 6,614 to less than 9,921 lbs
Class E-1.f	Takeoff weight from 9,921 to less than 13,228 lbs
Class E-1.g	Takeoff weight from 13,228 to less than 22,046 lbs
Class E-1.h	Takeoff weight from 22,046 to less than 44,092 lbs
Class E-1.i	Takeoff weight from 44,092 to less than 66,139 lbs
Class E-1.j	Takeoff weight from 66,139 to less than 88,185 lbs
Class E-1.k	Takeoff weight from 88,185 to less than 110,231 lbs
Class E-1.l	Takeoff weight 110,231 lbs or greater

Class E-3 (Autogyros)

Class E-3.a	Takeoff weight less than 1,102 lbs
Class E-3.b	Takeoff weight from 1,102 to less than 2,205 lbs
Class E-3.c	Takeoff weight from 2,204 to less than 3,858 lbs
Class E-3.d	Takeoff weight 3,858 lbs or greater

These classes are further categorized by type of powerplant:

Group	I	Piston
Group	II	Turboprop
Group	III	Jet
Group	IV	Rocket

Guide to Record Listings

The following examples are intended for the reader's use in understanding and interpreting the record data listed in this publication.

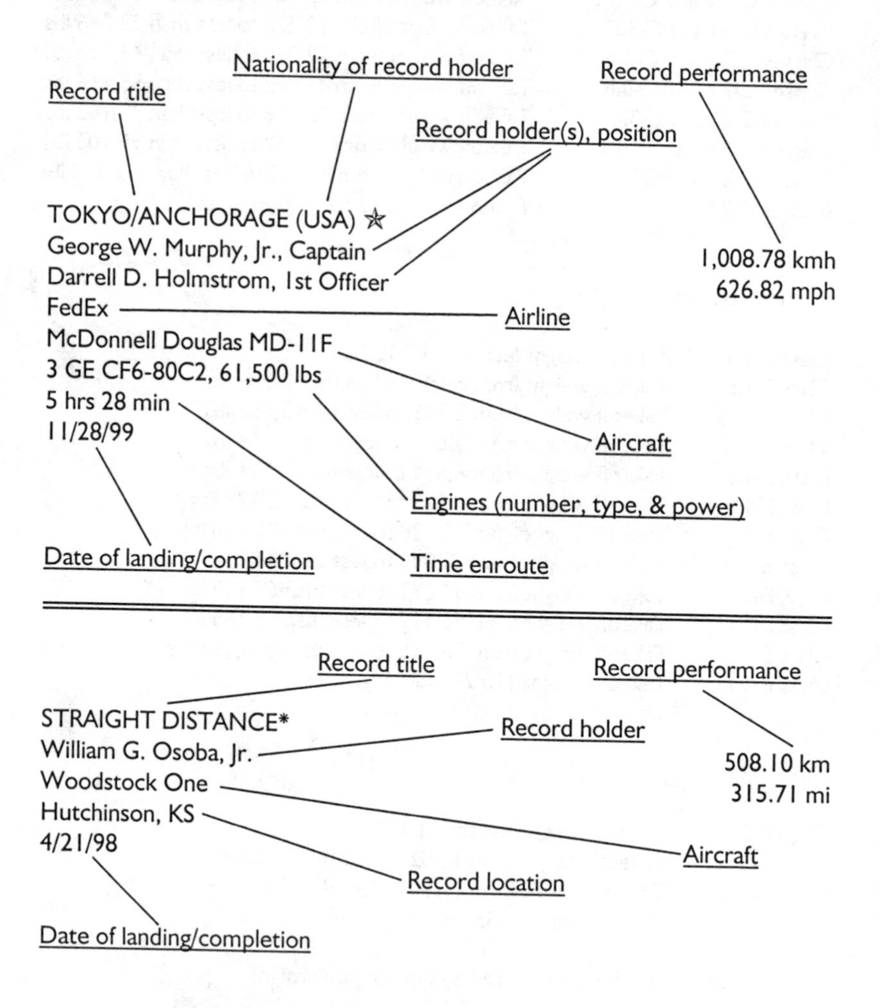

Guide to symbols used with record titles:

- * U.S. National record only
- § Speed Over a Commercial Air Route records in excess of 6,500 km
- ✯ NAA "Most Memorable" record

Absolute World Records

Absolute World Records are the supreme achievements of the hundreds of records open to flying machines. All types of airplanes are eligible for these few very special records. Airplanes may be powered by piston, turboprop, turbojet, rocket engines, or a combination. They may be landplanes, seaplanes, or amphibians; they may be light planes, business planes, military, or commercial airplanes.

Generally, the cost of developing high-performance aircraft has been so great that only airplanes created for military purposes have held these records. However, there have been two exceptions.

Most recently and most dramatically, the Rutan-designed "Voyager" shattered the theory that only a complicated military behemoth could hold an Absolute World Record. The Voyager team and its non-stop, non-refueled flight around the world proved that the dreams of dedicated individuals, combined with creative engineering, new technology, and hard work, could conquer the world.

The other exception was the X-15 rocket-powered research airplane. It was used for both civilian and military research during its highly productive lifetime.

Within the seven basic categories of absolute records for airplanes, the Voyager stands out. This homebuilt, experimental aircraft deserves special recognition. The others were all designed and built by major government facilities. They are: the little-known Soviet E-266M, a rocket-boosted version of the MiG-25 "Foxbat" interceptor, which holds the absolute altitude record, the North American X-15, which holds the record for carrier-launched altitude, and the Lockheed SR-71 "Blackbird" photo-reconnaissance airplane, which holds three records for speed and altitude.

As evidence of the exceptional status of these records, all are more than 15 years old. Since the performance of many new government-backed high performance airplanes is wrapped in a blanket of national security, the breaking of some of these records will depend as much on political considerations as on technical ones.

CLASSES C, H, M, AND N

DISTANCE WITHOUT LANDING (USA)
Richard Rutan, Pilot
Jeana Yeager, Pilot
Voyager, N269VA
Front Engine O-240 Air Cooled, 130 bhp
Rear Engine 10L Liquid Cooled, 110 bhp
Edwards AFB, CA - Edwards AFB, CA
12/14-23/86

40,212.14 km
24,986.73 mi

DISTANCE OVER A CLOSED CIRCUIT WITHOUT LANDING (USA)
Richard Rutan, Pilot
Jeana Yeager, Pilot
Voyager, N269VA
Front Engine O-240 Air Cooled, 130 bhp
Rear Engine 10L Liquid Cooled, 110 bhp
Edwards AFB, CA - Edwards AFB, CA
12/14-23/86

40,212.14 km
24,986.73 mi

ALTITUDE (USSR)
Alexander Fedotov
MIG-25, E-266M; "Foxbat"
RD, 30,865 lbs
Podmoscovnoe, USSR
8/31/77

37,650 m
123,524 ft

ALTITUDE IN HORIZONTAL FLIGHT (USA)
CAPT Robert C. Helt, USAF
Lockheed SR-71A "Blackbird"
2 P&W, 30,000 lbs
Beale AFB, CA
7/28/76

25,929 m
85,069 ft

SPEED OVER A 15/25 KILOMETER STRAIGHT COURSE (USA)
CAPT Eldon W. Joersz, USAF
Lockheed SR-71A "Blackbird"
2 P&W, 30,000 lbs
Beale AFB, CA
7/28/76

3,529.56 kmh
2,193.16 mph

SPEED OVER A CLOSED CIRCUIT (USA)
MAJ Adolphus H. Bledsoe
Lockheed SR-71A "Blackbird"
2 P&W, 30,000 lbs
Beale AFB, CA
7/27/76

3,367.22 kmh
2,092.29 mph

SPEED AROUND THE WORLD, NON-STOP, NON-REFUELED (USA)
Richard Rutan, Pilot
Jeana Yeager, Pilot
Voyager, N269VA
Front Engine O-240 Air Cooled, 130 bhp
Rear Engine 10L, Liquid Cooled, 110 bhp
Elapsed Time: 216 hrs 3 min 44 sec
Edwards AFB, CA - Edwards AFB, CA
12/14-23/86

186.11 kmh
115.65 mph

SPACECRAFT

The Space Age began on April 12, 1961, when Soviet Cosmonaut Yuri Gagarin made the first orbit of the Earth by a human in his Vostok I spacecraft. In just over an hour, he shattered all the existing records for speed, altitude, and distance.

His speed of 18,000 mph was more than 11 times faster than that achieved by an airplane. The airplane record at that time was 1,526 mph, set by U.S. Air Force Maj. Joseph Rogers flying a Convair F-106 Delta Dagger jet fighter.

His altitude of 203 miles (slightly over one million feet) was 10 times the highest level yet reached. That record was 103,389 feet, set by U.S. Air Force Capt. Joseph Jordan in a Lockheed F-104C Starfighter.

His distance of about 24,000 miles was double the longest non-stop flight of any airplane, which was 11,234 miles from Australia to Ohio in a U.S. Navy Lockheed P2V Neptune patrol bomber in 1946. Within a few months, fellow Cosmonaut Gehrman Titov stretched the distance record to about 400,000 miles with a 17-orbit flight.

The true magnitude of achievement in this first manned space flight is seen when compared to the first powered airplane flight. That first powered airplane flight took place only 58 years earlier, at a speed of less than 10 mph, an altitude of a couple of yards, and covered a distance of 120 feet.

Space records are kept separate from those of aircraft, as the operating conditions are simply not comparable. With the advent of NASA's Space Shuttle, a further division has been established between true "spacecraft" which operate exclusively in space, and "aerospace craft" like the Shuttle which function both in space and in the atmosphere.

ABSOLUTE RECORDS

DURATION (RUSSIA)
Valeri V. Poliakov — 437 days
Mir Orbital Complex — 17 hrs 58 min
3/22/95 — 17 sec

ALTITUDE (USA)
Frank Borman — 377,668.9 km
James A. Lovell, Jr. — 234,672.5 mi
William Anders
Apollo 8
12/21-27/68

DISTANCE TRAVELED (USSR)
A. N. Beresovoy — 140,800,000 km
V. V. Lebedev — 87,436,800 mi
Salyut 7, Soyuz T-5, Soyuz T-7
5/13-12/10/82

GREATEST MASS LIFTED TO ALTITUDE (USA)
Frank Borman — 127,980 kg
James A. Lovell, Jr. — 282,197 lb
William Anders
Apollo 8
12/21-27/68

EXTRAVEHICULAR DURATION IN SPACE (USA) ✯
Thomas D. Akers, Mission Specialist — 8 hrs
Richard J. Hieb, Mission Specialist — 20 min
Pierre J. Thuot, Mission Specialist
STS-49, Space Shuttle Orbiter "Endeavour," OV-105
John F. Kennedy Space Center, FL
5/13/92

ACCUMULATED SPACE FLIGHT TIME, GENERAL (RUSSIA)
Serguei Avdeyev — 747 days
8/28/99 — 11 hrs

ACCUMULATED SPACE FLIGHT TIME, WOMAN'S RECORD (USA) ✯
Shannon W. Lucid — 223 days
Space Shuttle Orbiters — 2 hrs 53 min
and Mir Orbital Complex — 17 sec
6/17/85-9/26/96

RECORDS WITH TWO OR MORE SPACECRAFT AND CREWS OF TWO OR MORE NATIONS

ASSEMBLED MASS OF SPACECRAFT IN LINKED FLIGHT (RUSSIA/USA)
STS-112:
Jeffrey S. Ashby, Commander — 264,433 kg
Pamela A. Melroy, Pilot — 582,975 lbs
Sandra H. Magnus, Mission Specialist
Piers J. Sellers, Mission Specialist
David A. Wolf, Mission Specialist
Fyodor N. Yurchikhin, Cosmonaut
Expedition 5:
Valery G. Korzun, Commander
Peggy A. Whitson, Flight Engineer
Sergei Y. Treschev, Flight Engineer
STS-112, Space Shuttle Orbiter "Atlantis," OV-104
& International Space Station
10/18/02

DURATION IN LINKED FLIGHT (USA/RUSSIA)
STS-102:
James D. Wetherbee, Commander — 8 days
James M. Kelly, Pilot — 21 hrs 53 min
Andrew S.W. Thomas, Mission Specialist — 25 sec
Paul W. Richards, Mission Specialist
ISS Expedition One:
William M. Shepherd, Commander
Yuri P. Gidzenko, Flight Engineer
Sergei K. Krikalev, Flight Engineer
ISS Expedition Two:
Yury V. Usachev., Commander
James S. Voss, Flight Engineer
Susan J. Helms, Flight Engineer
STS-102, Space Shuttle Orbiter "Discovery," OV-103 & International Space Station
3/19/01

ALTITUDE IN LINKED FLIGHT (RUSSIA/USA)
STS-71:
Robert L. Gibson, Commander — 416 km
Charles J. Precourt, Pilot — 258 mi
Ellen S. Baker, Mission Specialist
Bonnie J. Dunbar, Mission Specialist
Gregory J. Harbaugh, Mission Specialist
Mir-18:
Vladimir N. Dezhurov, Commander
Gennadiy M. Strekalov, Engineer
Norman E. Thagard, Mission Specialist
Mir-19:
Anatoly Y. Solovyev, Commander
Nikolai M. Budarin, Engineer
STS-71, Space Shuttle Orbiter "Atlantis," OV-104
& Mir-18/19, Orbital Station
6/29/95

DISTANCE TRAVELLED IN LINKED FLIGHT (USA/USSR)
Thomas P. Stafford (USA) — 1,309,974 km
Vance D. Brand — 813,980 mi
Donald K. Slayton
Aleksey Leonov (USSR)
Valeriy Kubasov
Spacecraft: Apollo & Soyuz
7/15-24/75

CATEGORY RECORDS

CLASS K, SPACECRAFT

K-2 ORBITAL MISSIONS, ONE ASTRONAUT

DURATION (USSR)
V. F. Bikovsky — 118 hrs 56 min 41 sec
Vostok 5
6/14-19/63

ALTITUDE (USSR)
Yuri Gagarin — 327 km / 203 mi
Vostok 1
4/12/61

GREATEST MASS LIFTED TO ALTITUDE (USSR)
Georgiy Beregovoi — 6,575 kg / 14,495 lb
Soyuz 3
10/26-30/68

K-2 ORBITAL MISSIONS, TWO OR MORE ASTRONAUTS

DURATION *General Category* (RUSSIA)
Valeri V. Poliakov — 437 days 17 hrs 58 min 17 sec
Mir Orbital Complex
3/22/95

DURATION *Feminine Category* (RUSSIA)
Helena V. Kondakova — 169 days 5 hrs 21 min 20 sec
Mir Orbital Complex
3/22/95

ALTITUDE (USA)
Frank Borman — 377,688.9 km / 234,685.0 mi
James A. Lovell, Jr.
William Anders
Apollo 8
12/21-27/68

GREATEST MASS LIFTED TO ALTITUDE (USA)
Frank Borman — 127,980 kg / 282,197 lb
James A. Lovell, Jr.
William Anders
Apollo 8
12/21-27/68

K-3 LUNAR AND PLANETARY MISSIONS, ONE ASTRONAUT

DURATION IN LUNAR ORBIT (USA)
Ronald E. Evans — 147 hrs 41 min 13 sec
Apollo 17
12/10-16/72

K-3 LUNAR AND PLANETARY MISSIONS, TWO OR MORE ASTRONAUTS

DURATION OF STAY ON THE LUNAR SURFACE (USA)
John W. Young — 71 hrs 2 min 13 sec
Charles M. Duke
Apollo 16 & LM Orion
4/21-24/72

EXTRAVEHICULAR DURATION ON THE MOON (USA)
Eugene A. Cernan — 21 hrs 31 min 44 sec
Apollo 17 & LM Challenger
12/12-24/72

DURATION OF STAY IN ORBIT AROUND A CELESTIAL BODY
(No record registered)

GREATEST MASS LANDED ON THE MOON (USA)
John W. Young — 8,257.6 kg / 18,208.0 lb
Charles M. Duke
Apollo 16 & LM Orion
4/21/72

DISTANCE TRAVELED ON THE LUNAR SURFACE (USA)
Eugene A. Cernan — 7,370 m / 24,180 ft
Harrison H. Schmitt
Apollo 17, LM Challenger & Lunar Rover
12/13/72

DURATION OF LUNAR MISSION (USA)
Eugene A. Cernan — 301 hrs 51 min 57 sec
Harrison H. Schmitt
Ronald E. Evans
Apollo 17
12/7-19/72

GREATEST MASS LIFTED FROM THE MOON (USA)
John W. Young — 4,965.5 kg / 10,949.0 lb
Charles M. Duke
Apollo 16 & LM Orion
4/24/72

CLASS P AEROSPACECRAFT

CLASS P-1 SUBORBITAL MISSIONS

(No records registered)

CLASS P-2 ORBITAL MISSIONS

DURATION (USA) ★
Kenneth D. Cockrell, Commander — 17 days 15 hrs 53 min 17 sec
Kent V. Rominger, Pilot
Tamara E. Jernigan, Thomas D. Jones, Mission Specialists
F. Story Musgrave, Payload Specialist
STS-80, Space Shuttle Orbiter "Columbia," OV-102
John F. Kennedy Space Center, FL
11/19-12/7/96

ALTITUDE IN ELLIPTICAL ORBIT (USA)
Kenneth D. Bowersox, Commander — 622.77 km / 386.97 mi
Scott J. Horowitz, Pilot
Gregory J. Harbaugh, Steven A. Hawley, Mark C. Lee, Steven L. Smith, Joseph R. Tanner, Mission Specialists
STS-82, Space Shuttle Orbiter "Discovery," OV-103
John F. Kennedy Space Center, FL
2/11-21/97

ALTITUDE IN PARACIRCULAR ORBIT (USA) ★
Kenneth D. Bowersox, Commander — 623.64 km / 387.51 mi
Scott J. Horowitz, Pilot
Gregory J. Harbaugh, Steven A. Hawley, Mark C. Lee, Steven L. Smith, Joseph R. Tanner, Mission Specialists
STS-82, Space Shuttle Orbiter "Discovery," OV-103
John F. Kennedy Space Center, FL
2/11-21/97

GREATEST MASS LIFTED TO ALTITUDE (USA)
John E. Blaha, Commander — 114,969.8 kg / 253,465.0 lb
Michael A. Baker, Pilot
G. David Low, Shannon W. Lucid, James C. Adamson, Mission Specialists
STS-43, Space Shuttle Orbiter "Atlantis," OV-104
John F. Kennedy Space Center, FL
8/2-11/91

GREATEST PAYLOAD MASS LIFTED TO GREATEST ALTITUDE (USA)
Paul J. Weitz, Commander — 20,477.0 kg / 45,049.4 lb to 291.73 km / 181.28 mi
Karol J. Bobko, Pilot
Donald H. Peterson, Mission Specialist
F. Story Musgrave, Mission Specialist
STS-6, Space Shuttle Orbiter "Challenger," OV-099
Edwards AFB, CA
4/4-9/83

Piston Engine Aircraft

Rapid development of the piston-engine airplane in the first decade of the 20th century made record setting and documentation a major part of the FAI's business. With steady design improvements of airframes and engines, the performance of airplanes increased so quickly that surpassing previous performances became an almost daily occurrence.

SPEED

When Santos-Dumont set the first official airplane records in 1906, he was also making the first official flights in all of Europe. Prior to this, the best anyone had done there was "hop" for a hundred feet or so, without any sign of flying under control or being able to sustain a flight beyond the initial surge that may have been the result of the wind.

By the time the Aero Club of France recognized the first speed and distance records, the Wright brothers had already flown considerably faster, farther, and longer than the French pilots. On October 5, 1905, the Wrights flew more than 24 miles, at a speed in excess of 38 mph, but they were intent on perfecting their airplane and showed little interest in setting records. Since the Aero Club of America did not yet exist, there would have been no one to make their records official, even if they had been interested.

The speed record section of the official record book became the private preserve of the French until long after World War I, as a series of pilots in Blériot, Antoinette, and Deperdussin airplanes broke record after record. The 100 mph "barrier" was broken in 1912, and 125 mph was achieved the following year. After a pause of more than six years during World War I, the FAI record book "reopened" in 1920 and the French went back to work, quickly boosting the speed record to 171 mph. Within a year and a half, the 200 mph milestone was reached.

In late 1922, the U.S. recaptured the speed crown with the first in a series of Curtiss military racers. General Billy Mitchell flew 223 mph, and then, after a French pilot surpassed that speed, other Americans pushed the mark to over 265 mph by late 1923.

The French came back in 1924 and the spotlight shifted to racing seaplanes, inspired by the prestigious Schneider Trophy Races. While the landplane record languished below 300 mph, seaplanes from Italy pushed the absolute speed record to over 300 mph. The British topped 400 mph with a Supermarine S.6B in 1931, and finally, an Italian seaplane racer was flown at 441 mph in 1934.

Without the stimulus of the Schneider Trophy, the record for landplanes eased slowly upward. In 1933, American racing pilot Jimmy Weddell topped 300 mph, and in 1935 Howard Hughes flew faster than 350 mph. It was not until the German Luftwaffe sprung into action that a landplane again flew faster than any seaplane. In 1939, a specially built Heinkel upped the record to 464 mph and then a highly modified Messerschmitt raised it to 469 mph.

Until 1945, all airplane records were set with piston engines, since they were the only powerplants available for record purposes. The jet engine was invented in 1939, but World War II precluded any record setting until the mid-1940's. Soon after the end of the war, the British set the first jet speed record, using a Gloster Meteor fighter to break both the 500 mph and 600 mph "barriers" on a single flight. This marked the end of the airplane's 39-year reign as the King of Speed.

Not until pylon air racing experienced a re-birth in 1964 was any serious thought given to breaking the 1939 German piston engine record. In 1969, racing pilot Darryl Greenamyer was clocked at 482 mph in a highly modified Grumman F8F-2 Bearcat, the last American production-line, piston-engine fighter plane. His record lasted for 10 years until it was broken by another racing pilot, Steve Hinton, at 499 mph. Hinton's airplane was a P-51D Mustang "Red Baron" modified with a Rolls Royce Griffon engine and counter-rotating propellers. In 1989, racing pilot Lyle Shelton moved the record to 528 mph in his modified F8F Bearcat "Rare Bear."

DISTANCE

The initial distance record set by Santos-Dumont in 1906 lasted less than a year and was broken by a far greater margin than his speed record. By 1907, it was up to a half mile, by 1908 to 77 miles, and by 1909 to 150 miles. An airplane had flown non-stop for more than 1,000 miles by the time World War I began.

The 1920's saw a flurry of long-distance flying. In 1926 the distance record was raised from just under 2,000 miles to more than 3,300 miles by the French. Charles Lindbergh pushed this to 3,600 miles in 1927, and was then up-staged two weeks later by a flight of 3,900 miles by Clarence Chamberlin. The next spring, a pair of Italian pilots flew almost 4,500 miles without stopping. A year later, the record was increased to 4,900 miles by a French crew.

The early 1930's saw a continued assault on the distance record. By 1933, it was raised to more than 5,600 miles. In 1937, a Soviet crew flew a specially-built airplane from the USSR to California, a distance of almost 6,300 miles. The British then took over, flying a Vickers Welleseley bomber more than 7,150 miles in 1938.

The distance race resumed after World War II when an American Boeing B-29 Superfortress was flown more than 7,900 miles in 1945. The record for piston-engine airplanes was increased the following year by a U.S. Navy patrol bomber which flew from Australia to Ohio, a distance of 11,236 miles.

1986 witnessed the first real distance record in many years; some had called it the final aviation record yet to be set. It was attained by the team of Burt Rutan, Dick Rutan, and Jeana Yeager. The Voyager, an aptly-named design by Burt Rutan, was built and tested in a modest hangar in Mojave, California. Then, the Rutan/Yeager pilot team delivered an incomparable act of daring and skill: the first non-stop, non-refueled, flight around the world. With a total distance of 24,986 miles and a duration of over 216 hours, the pilots and their hand-picked volunteers proved that determination and teamwork can triumph over seemingly insurmountable odds.

ALTITUDE

The altitude record moved along a somewhat different path from the speed and distance records, as the climbing ability of airplanes improved much faster than did their ability to fly fast or far. Hubert Latham set the first official altitude mark of 509 feet in an Antoinette monoplane in August, 1909. It was raised to almost 1,500 feet by the end of the year. In 1910, it climbed by leaps and bounds, reaching more than 10,000 feet. By the start of World War I, George Legagneux, of France, had flown a Nieuport airplane above 20,000 feet.

The United States got into the game as World War I ended and the FAI was again actively recognizing records. Army Lt. R.W. Schroeder flew a LePere biplane to 33,000 feet, thanks to technical developments made during the war. From then on, advancements were made at the rate of about 1,000 feet at a time. The 40,000-foot "barrier" was broken in 1929 by a German pilot flying an all-metal Junkers monoplane. Not until 1937 did any pilot officially top 50,000 feet when Mario Pezzi, of Italy, flew to 51,364 feet in a Caproni biplane.

The current record for piston-engine airplanes, of just over 56,000 feet, was set in 1938 by Pezzi in a similar Caproni. He used a partial pressure suit filled with water, while his engine continued to run in that rarified atmosphere despite the absence of a supercharger. This is one of the oldest records still in the book. After World War II, jet airplanes took over the altitude record. A British deHavilland Vampire fighter, flown by John Cunningham, surpassed Pezzi's record by several thousand feet.

All types of light airplanes—from ultralights to Formula One racers, factory-built touring planes to twin-engine business airplanes—are eligible to set records. They have been regularly setting new speed records on 3 kilometer and 15/25 kilometer straight courses and closed circuit courses from 100 to 2,000 kilometers; altitude; altitude in horizontal flight, and time-to-climb to various altitudes.

Activity has been increasing in many of these categories since the 1980s. The creation of subclasses of aircraft (by weight and propulsion system) has permitted hundreds of pilots in the U.S. and worldwide to set world records in their own or rented aircraft, and has stimulated demands for improved performance through use of new stronger and lighter materials, better power/weight ratios, improved fuel specifics, and precise navigation and weather avoidance equipment.

CLASS C-1 (UNLIMITED)
GROUP I (PISTON ENGINE)

DISTANCE

DISTANCE WITHOUT LANDING (USA)
Richard Rutan, Pilot 40,212.14 km
Jeana Yeager, Pilot 24,986.73 mi
Voyager, N269VA
Front Engine O-240 Air Cooled, 130 bhp
Rear Engine 10L Liquid Cooled, 110 bhp
Edwards AFB, CA - Edwards, AFB, CA
12/14-23/86

DISTANCE OVER A CLOSED CIRCUIT WITHOUT LANDING (USA)
Richard Rutan, Pilot 40,212.14 km
Jeana Yeager, Pilot 24,986.73 mi
Voyager, N269VA
Front Engine O-240 Air Cooled, 130 bhp
Rear Engine 10L Liquid Cooled, 110 bhp
Edwards AFB, CA - Edwards, AFB, CA
12/14-23/86

ALTITUDE

ALTITUDE (ITALY)
Mario Pezzi 17,083 m
Caproni 161 56,046 ft
Piaggio XI R.C.
Montecelio, Italy
10/22/38

ALTITUDE IN HORIZONTAL FLIGHT (USA)
Bruce Bohannon 14,301 m
Bohannon B-1 46,919 ft
1 Mattituck/Lycoming IO-540, 350 bhp
Angleton, TX
11/15/03

Superseded
ALTITUDE IN HORIZONTAL FLIGHT (USA)
Bruce Bohannon 12,983 m
Bohannon B-1 42,595 ft
1 Mattituck/Lycoming IO-540, 350 bhp
Angleton, TX
10/18/03

ALTITUDE WITH 1,000 KG PAYLOAD (USA)
MAJ Finley F. Ross, USAAF 14,603 m
Boeing B-29 Superfortress 47,910 ft
4 Wright R-3350-23A
Harmon Field, Guam, M.I.
5/16/46

ALTITUDE WITH 2,000 KG PAYLOAD (USA)
COL E.D. Reynolds, USAAF 14,180 m
Boeing B-29 Superfortress 46,522 ft
4 Wright R-3350-23
Harmon Field, Guam, M.I.
5/13/46

ALTITUDE WITH 5,000 KG PAYLOAD (USA)
LT J.P. Tobison, USAAF 13,793 m
Boeing B-29 Superfortress 45,253 ft
4 Wright, 2,000 bhp
Harmon Field, Guam, M.I.
5/4/46

ALTITUDE WITH 10,000 KG PAYLOAD (USA)
CAPT A.A. Pearson, USAAF 12,668 m
Boeing B-29 Superfortress 41,562 ft
4 Wright R-3350-23
Harmon AFB, Newfoundland
5/8/46

ALTITUDE WITH 15,000 KG PAYLOAD (USA)
COL J.B. Warren, USAAF 12,046 m
Boeing B-29 Superfortress 39,521 ft
4 Wright R-3350-23
Guam, M.I.
5/11/46

GREATEST PAYLOAD CARRIED TO 2,000 METERS (USA)
COL J.B. Warren, USAAF 15,166.00 kg
Boeing B-29 Superfortress 33,435.30 lb
4 Wright R-3350-23
Guam, M.I.
5/11/46

SPEED

SPEED OVER A 3 KM STRAIGHT COURSE (USA)
Lyle T. Shelton 850.25 kmh
Grumman F8F Bearcat, N777L 528.33 mph
1 Wright R3350, 3800 bhp
Las Vegas, NM
08/21/89

SPEED OVER A 15/25 KM STRAIGHT COURSE (USA)
Frank Taylor 832.12 kmh
P-51D "Dago Red" 517.06 mph
1 R/R Merlin V-1650-9, 3,000 bhp
Mojave, CA
7/30/83

SPEED OVER A 100 KM CLOSED CIRCUIT WITHOUT PAYLOAD (USA)
Jacqueline Cochran 755.67 kmh
North American P-51C Mustang 469.55 mph
Packard RR Merlin V-1650
Coachella Valley, CA
12/10/47

SPEED OVER A 500 KM CLOSED CIRCUIT WITHOUT PAYLOAD (USA)
Jacqueline Cochran 703.38 kmh
North American P-51C Mustang 433.99 mph
Packard RR Merlin V-1650
Desert Center - Mount Wilson, CA
12/29/49

SPEED OVER A 1,000 KM CLOSED CIRCUIT WITHOUT PAYLOAD (USA)
Jacqueline Cochran 693.78 kmh
North American P-51C Mustang 431.09 mph
Packard RR Merlin V-1650
Santa Rosa Summit, CA -
Santa Fe, NM- Flagstaff, AZ
5/24/48

SPEED OVER A 1,000 KM CLOSED CIRCUIT WITH 1,000 KG PAYLOAD (USA)
LT E.M. Grabowski, USAAF 594.96 kmh
Boeing B-29 Superfortress 369.69 mph
4 Wright R-3350-23
Dayton, OH
5/16/46

SPEED OVER A 1,000 KM CLOSED CIRCUIT WITH 2,000 KG PAYLOAD (USA)
LT E.M. Grabowski, USAAF 594.96 kmh
Boeing B-29 Superfortress 369.69 mph
4 Wright R-3350-23
Dayton, OH
5/17/46

SPEED OVER A 1,000 KM CLOSED CIRCUIT WITH 5,000 KG PAYLOAD (USA)
LT E.M. Grabowski, USAAF 594.96 kmh
Boeing B-29 Superfortress 369.69 mph
4 Wright R-3350-23
5/17/46

SPEED OVER A 1,000 KM CLOSED CIRCUIT WITH 10,000 KG PAYLOAD (USA)
CAPT J.D. Bartlett, USAAF — 575.71 kmh
Boeing B-29 Superfortress — 357.73 mph
4 Wright R-3350-23
Dayton, OH
5/19/46

SPEED OVER A 2,000 KM CLOSED CIRCUIT WITHOUT PAYLOAD (USA)
Jacqueline Cochran — 720.13 kmh
North American P-51C Mustang — 447.47 mph
Packard RR Merlin V-1650
Santa Rosa Summit, CA
5/22/48

SPEED OVER A 2,000 KM CLOSED CIRCUIT WITH 1,000 KG PAYLOAD (USA)
LT E.M. Grabowski, USAAF — 588.46 kmh
Boeing B-29 Superfortress — 365.65 mph
4 Wright R-3350-23
Dayton, OH
5/17/46

SPEED OVER A 2,000 KM CLOSED CIRCUIT WITH 2,000 KG PAYLOAD (USA)
LT E.M. Grabowski, USAAF — 588.46 kmh
Boeing B-29 Superfortress — 365.65 mph
4 Wright R-3350-23
Dayton, OH
5/17/46

SPEED OVER A 2,000 KM CLOSED CIRCUIT WITH 5,000 KG PAYLOAD (USA)
LT E. M. Grabowski, USAAF — 588.46 kmh
Boeing B-29 Superfortress — 365.65 mph
4 Wright R-3350-23
Dayton, OH
5/17/46

SPEED OVER A 2,000 KM CLOSED CIRCUIT WITH 10,000 KG PAYLOAD (USA)
CAPT J.D. Bartlett, USAAF — 574.59 kmh
Boeing B-29 Superfortress — 357.04 mph
4 Wright R-3350-23
Dayton, OH
5/19/46

SPEED OVER A 5,000 KM CLOSED CIRCUIT WITHOUT PAYLOAD (USA)
CAPT James Bauer, USAAF — 544.59 kmh
Boeing B-29 Superfortress — 338.39 mph
4 Wright R-3350-23
Dayton, OH
6/28/46

SPEED OVER A 5,000 KM CLOSED CIRCUIT WITH 1,000 KG PAYLOAD (USA)
CAPT James Bauer, USAAF — 544.59 kmh
Boeing B-29 Superfortress — 338.39 mph
4 Wright R-3350-23
Dayton, OH
6/28/46

SPEED OVER A 5,000 KM CLOSED CIRCUIT WITH 2,000 KG PAYLOAD (USA)
CAPT James Bauer, USAAF — 544.59 kmh
Boeing B-29 Superfortress — 338.39 mph
4 Wright R-3350-23
Dayton, OH
6/28/46

SPEED OVER A 5,000 KM CLOSED CIRCUIT WITH 5,000 KG PAYLOAD (USA)
LTC R.G. Ruegg, USAAF — 428.12 kmh
Boeing B-29 Superfortress — 266.02 mph
4 Wright R-3350-23
Dayton, OH
6/21/46

SPEED OVER A 5,000 KM CLOSED CIRCUIT WITH 10,000 KG PAYLOAD (USA)
LTC R.G. Ruegg, USAAF — 428.12 kmh
Boeing B-29 Superfortress — 266.02 mph
4 Wright R-3350-23
Dayton, OH
6/21/46

SPEED OVER A 10,000 KM CLOSED CIRCUIT WITHOUT PAYLOAD (USA)
LTC O.F. Lassiter, USAF — 439.67 kmh
Boeing B-29 Superfortress — 273.19 mph
4 Wright R-3350-23
Dayton, OH
7/29-30/47

SPEED AROUND THE WORLD, EASTBOUND (USA)
Philander P. Claxton, Pilot — 353.57 kmh
John L. Cink, Copilot — 219.70 mph
Aerostar 601P
2 Lycoming IO-540, 290 bhp
Los Angeles, Peoria, Gander, Frankfurt, Tehran, Madras, Kota Kinabalu, Manila, Guam, Kwajalein Island, Wake Island, Honolulu, Los Angeles
Distance: 36,803.95 km (22,868.91 mi)
104 hrs 5 min 30 sec
11/4-9/77

SPEED AROUND THE WORLD, WESTBOUND (USA)
Max Conrad — 198.27 kmh
Piper Aztec — 123.19 mph
2 Lycoming O-540, 250 bhp
Miami, Long Beach, Honolulu, Wake Island, Guam, Manila, Singapore, Bombay, Nairobi, Lagos, Dakar, Amapa, Atkinson Field, Port of Spain, Miami
3/8/61

SPEED AROUND THE WORLD OVER BOTH THE EARTH'S POLES (USA)
Elgen M. Long — 93.05 kmh
Piper Navajo, N9097Y — 57.82 mph
2 Lycoming TIO-540, 310 bhp
San Francisco, Fairbanks, Pt. Barrow, North Pole, Tromso, Stockholm, London, Accra, Equator, Recife, Rio de Janeiro, Punta Arenas, South Pole, McMurdo, Sydney, Nandi, Equator, Wake Island, Tokyo, Wake Island, Honolulu, San Francisco
62,597.04 km (38,896.03 mi)
28 days 00 hrs 43 min
11/5-12/3/71

TIME TO CLIMB

TIME TO 3,000 METERS (USA)
Lyle T. Shelton — 1 min
Grumman F8F2 Bearcat, N777L — 31 sec
1 Wright R3350
Thermal, CA
2/6/72

TIME TO 6,000 METERS (USA) ✯
Bruce Bohannon — 6 min
Bohannon B-1 — 40 sec
1 Lycoming IO-540, 280 bhp
Angleton, TX
11/5/00

TIME TO 9,000 METERS (USA)
Bruce Bohannon — 15 min
Bohannon B-1 — 32 sec
1 Mattituck/Lycoming IO-540, 350 bhp
Angleton, TX
10/18/03

TIME TO 12,000 METERS (USA)
Bruce Bohannon — 23 min
Bohannon B-1 — 41 sec
1 Mattituck/Lycoming IO-540, 350 bhp
Angleton, TX
10/18/03

CLASS C-1.A/O (LESS THAN 300 KG/661 LBS) GROUP I (PISTON ENGINE)

DISTANCE

DISTANCE WITHOUT LANDING (AUSTRIA)
Wilhem Lischak 1,527.78 km
Experimental LW 949.36 mi
Voslau - Brest
6/8/88

DISTANCE OVER A CLOSED CIRCUIT WITHOUT LANDING (AUSTRIA)
Wilhelm Lischak 2,702.16 km
Experimental LW02 1,679.12 mi
Wels
6/18/88

ALTITUDE

ALTITUDE (USA)
Richard J. Rowley 7,907 m
Mitchell U2 Superwing, 25,940 ft
1 Cuyuna 430D, 30 HP
Colorado Springs, CO
9/17/83

ALTITUDE IN HORIZONTAL FLIGHT (AUSTRALIA)
Eric S. Winton 9,189 m
Facet Opal 30,148 ft
1 Rotax 447, 40 bhp
Tyagarah, Australia
4/8/89

SPEED

SPEED OVER A 3 KM STRAIGHT COURSE (AUSTRIA)
Peter Scheichenberger 351.39 kmh
Bede BD-5B 218.34 mph
Zeltweg, Austria
8/28/99

SPEED OVER A 3 KM STRAIGHT COURSE* ✭
Michael S. Arnold 213.18 mph
Arnold AR-5
1 Rotax 582, 63 bhp
Davis, CA
8/30/92

SPEED OVER A 15/25 KM STRAIGHT COURSE (USA)
Brian Dempsey 292.15 kmh
Sonerai, N8FV 181.54 mph
1 VW 1600
Easton, MD
02/19/89

SPEED OVER A 100 KM CLOSED CIRCUIT WITHOUT PAYLOAD (USA)
Charles T. Andrews 297.72 kmh
Monnett "Monex" 185.00 mph
Monnett Aero Vee 70 bhp
Fond du Lac, WI
8/3/82

SPEED OVER A 500 KM CLOSED CIRCUIT WITHOUT PAYLOAD (USA)
Charles T. Andrews 293.04 kmh
Monnett "Monex" 182.10 mph
Monnett Aero Vee, 70 bhp
Fond du Lac, WI
8/3/82

SPEED OVER A 1,000 KM CLOSED CIRCUIT WITHOUT PAYLOAD (AUSTRIA)
Wilhem Lischak 193.83 km
LW 02 120.44 mi
Steyr-Puch mod. 36PS
Vöslau, Austria
5/20/93

SPEED OVER A 2,000 KM CLOSED CIRCUIT WITHOUT PAYLOAD (AUSTRIA)
Wilhelm Lischak 178.19 kmh
Experimental LW02 110.73 mph
Wels
6/18/88

TIME TO CLIMB

TIME TO 3,000 METERS (USSR)
Mikhail Markov 13 min
May-89 39.2 sec
1 Rotax 532, 64 bhp
8/6/90

TIME TO 3,000 METERS*
Brian Dempsey 9 min
Sonerai, N8FV 36.29 sec
1 VW 1600
Easton, MD
02/19/89

CLASS C-1.A (300-500 KG / 661-1,102 LBS) GROUP I, PISTON ENGINE

DISTANCE

DISTANCE WITHOUT LANDING (USA)
F. Gary Hertzler 3,563.02 km
Rutan VariEze 2,214.06 mi
1 Continental A-80, 80 bhp
Mojave, CA to Martinsburg, WV
7/15/84

DISTANCE OVER A CLOSED CIRCUIT WITHOUT LANDING (USA) ✭
F. Gary Hertzler 4,007.76 km
VariEze 2,490.32 mi
1 Continental A-80, 80 bhp
Chandler, AZ–Perris, CA
10/31/93

ALTITUDE

ALTITUDE (USA)
James A. Price 10,676 m
LongEz 35,027 ft
1 Lycoming O-320, 150 bhp
Minden, NV
5/5/96

ALTITUDE IN HORIZONTAL FLIGHT (USA)
James A. Price 10,645 m
LongEz 34,926 ft
1 Lycoming O-320, 150 bhp
Minden, NV
5/5/96

SPEED

SPEED OVER A 3 KM STRAIGHT COURSE (USA)
Jon M. Sharp 466.83 kmh
Sharp DR90 290.08 mph
1 Continental O-200, 100 bhp
Mojave, CA
11/15/98

SPEED OVER A 15/25 KM STRAIGHT COURSE (USA)
Jon M. Sharp 454.77 kmh
Sharp DR90 282.58 mph
1 Continental O-200, 100 bhp
Mojave, CA
10/31/98

SPEED OVER A 100 KM CLOSED CIRCUIT WITHOUT PAYLOAD (USA)
James W. Miller 369.20 kmh
Miller GEM 260, N177M 229.41 mph
1 Continental O-200, 100 bhp
Fond du Lac, WI
7/31/84

SPEED OVER A 500 KM CLOSED CIRCUIT WITHOUT PAYLOAD (USA)
James W. Miller — 367.17 kmh
Miller GEM 260, N177M — 228.15 mph
1 Continental O-200, 100 bhp
Fond du Lac, WI
7/31/84

SPEED OVER 1,000 KM CLOSED CIRCUIT WITHOUT PAYLOAD (FRG)
Klaus H. Savier — 327.78 kmh
Vari-Eze — 203.67 mph
1 Continental, 120 bhp
Homeland, CA - Chandler, AZ
3/14/92

SPEED OVER A 2,000 KM CLOSED CIRCUIT WITHOUT PAYLOAD (FRG)
Klaus H. Savier — 322.06 kmh
Vari-Eze — 200.12 mph
1 Continental, 120 bhp
Homeland, CA - Chandler, AZ
3/14/92

SPEED AROUND THE WORLD, EASTBOUND (UK)
Colin Bodill — 16.53 kmh
Blade 912 — 10.27 mph
1 Rotax 912 ULS
Brooklands, UK
9/6/00

TIME TO CLIMB

TIME TO 3,000 METERS (USA)
Bruce Bohannon — 3 min
Miller Special, M-105 — 8 sec
Continental O-200, 100 bhp
Oshkosh, WI
8/3/96

TIME TO 6,000 METERS (USA) ★
Bruce Bohannon — 12 min
Miller Special, M-105 — 50 sec
Continental O-200, 100 bhp
Oshkosh, WI
7/30/94

TIME TO 9,000 METERS (USA)
Bruce Bohannon — 41 min
Miller Special, M-105 — 35 sec
Continental O-200, 100 bhp
Galveston, TX
1/23/96

CLASS C-1.B (500-1,000 KG / 1,102-2,205 LBS) GROUP I (PISTON ENGINE)

DISTANCE

DISTANCE WITHOUT LANDING (USA)
Richard Rutan — 7,344.56 km
Rutan Long EZ — 4,563.70 mi
1 Lycoming O-320-DZG, 160 bhp
Anchorage, AK to Grand Turk, British West Indies
6/5-6/81

DISTANCE OVER A CLOSED CIRCUIT WITHOUT LANDING (USA)
Richard Rutan — 7,727.96 km
Rutan Long EZ 001 — 4,800.03 mi
1 Lycoming O-235, 115 bhp
Mojave/Bishop, CA
12/15-16/79

ALTITUDE

ALTITUDE (USA) ★
Bruce Bohannon — 14,346 m
Bohannon B-1 — 47,067 ft
1 Mattituck/Lycoming IO-540, 350 bhp
Angleton, TX
11/15/03

Superseded
ALTITUDE (USA)
Bruce Bohannon — 13,343 m
Bohannon B-1 — 43,778 ft
1 Mattituck/Lycoming IO-540, 350 bhp
Angleton, TX
10/18/03

ALTITUDE IN HORIZONTAL FLIGHT (USA)
Bruce Bohannon — 14,301 m
Bohannon B-1 — 46,919 ft
1 Mattituck/Lycoming IO-540, 350 bhp
Angleton, TX
11/15/03

Superseded
ALTITUDE IN HORIZONTAL FLIGHT (USA)
Bruce Bohannon — 12,983 m
Bohannon B-1 — 42,595 ft
1 Mattituck/Lycoming IO-540, 350 bhp
Angleton, TX
10/18/03

SPEED

SPEED OVER A 3 KM STRAIGHT COURSE (USA)
Richard J. Gritter — 533.88 kmh
MayCay Beeler — 331.75 mph
Questair Venture, N62V
1 Continental IO-550C, 280 bhp
Kitty Hawk, NC
6/8/89

SPEED OVER A 15/25 KM STRAIGHT COURSE (USA)
MayCay Beeler — 487.08 kmh
Richard J. Gritter — 302.67 kmh
Questair Venture, N62V
1 Continental IO-550C, 280 bhp
Kitty Hawk, NC
6/8/89

SPEED OVER A 100 KM CLOSED CIRCUIT WITHOUT PAYLOAD (USA)
Richard J. Gritter — 491.08 kmh
MayCay Beeler — 305.15 mph
Questair Venture, N62V
1 Continental IO-550C, 280 bhp
Kitty Hawk, NC
6/7/89

SPEED OVER A 500 KM CLOSED CIRCUIT WITHOUT PAYLOAD (USA) ★
Richard C. Keyt — 488.26 kmh
Polen Special II — 303.39 mph
1 Lycoming TIO-360, 180 bhp
Oshkosh, WI
7/26/01

SPEED OVER A 1,000 KM CLOSED CIRCUIT WITHOUT PAYLOAD (USA)
MayCay Beeler — 456.84 kmh
Richard G. Gritter — 283.88 mph
Questair Venture, N62V
1 Continental IO-550C, 280 bhp
Greensboro, NC
6/11/89

SPEED OVER A 2,000 KM CLOSED CIRCUIT WITHOUT PAYLOAD (USA) ✯
Richard Rutan 401.46 kmh
Rutan Catbird 249.45 mph
1 Lycoming TIO-360, 210 bhp
Lake Hughes, CA - Boise, ID
1/29/94

SPEED AROUND THE WORLD, EASTBOUND (SWITZERLAND)
Hans G. Schmid 69.87 kmh
Long-EZ 43.41 mph
1 Lycoming O-320, 160 bhp
3/28/00

SPEED AROUND THE WORLD, WESTBOUND (SWITZERLAND)
Hans G. Schmid 69.99 kmh
Long-EZ 43.49 mph
1 Lycoming O-320, 160 bhp
4/29/00

TIME TO CLIMB

TIME TO 3,000 METERS (USA)
Bruce Bohannon 2 min
Bohannon B-1 20 sec
1 Lycoming IO-540, 280 bhp
Oshkosh, WI
7/31/99

TIME TO 6,000 METERS (USA) ✯
Bruce Bohannon 6 min
Bohannon B-1 40 sec
1 Lycoming IO-540, 280 bhp
Angleton, TX
11/5/00

TIME TO 9,000 METERS (USA)
Bruce Bohannon 15 min
Bohannon B-1 32 sec
1 Mattituck/Lycoming IO-540, 350 bhp
Angleton, TX
10/18/03

TIME TO 12,000 METERS (USA)
Bruce Bohannon 23 min
Bohannon B-1 41 sec
1 Mattituck/Lycoming IO-540, 350 bhp
Angleton, TX
10/18/03

CLASS C-1.C (1,000-1,750 KG / 2,205-3,858 LBS) GROUP I (PISTON ENGINE)

DISTANCE

DISTANCE WITHOUT LANDING (USA)
Max Conrad 11,211.83 km
Piper Comanche PA24-180 6,966.75 mi
1 Lycoming O-360, 180 bhp
Casablanca to El Paso, TX
11/24/59

DISTANCE OVER A CLOSED CIRCUIT WITHOUT LANDING (USA)
Max Conrad 11,138.72 km
Piper Comanche PA24-180 6,921.28 mi
1 Lycoming O-360, 180 bhp
Minneapolis, Chicago,
Des Moines, Minneapolis
7/14/60

ALTITUDE

ALTITUDE (USA)
Walter D. Cable 12,906 m
Cessna T-210G 42,344 ft
1 Continental TSIO-520-C, 285 bhp
Upland, CA
5/13/67

ALTITUDE IN HORIZONTAL FLIGHT (W. GERMANY)
Wilhelm Heller 10,889 m
Cessna P 210N 35,725 ft
1 Continental TSIO-520, 310 bhp
2/12/84

SPEED

SPEED OVER A 3 KM STRAIGHT COURSE (USA)
Fred E. Schrameck II 555.33 kmh
Lancair Legacy 345.06 mph
1 Continental GTSIO-550, 550 bhp
Reno, NV
9/13/02

SPEED OVER A 15/25 KM STRAIGHT COURSE (USA)
Richard A. Vandam 483.34 kmh
Glasair III 300.33 mph
1 Lycoming IO-540, 300 bhp
Reno, NV
12/31/99

SPEED OVER A 100 KM CLOSED CIRCUIT WITHOUT PAYLOAD (USA)
John P. Harris 455.23 kmh
Bellanca Skyrocket II 282.87 mph
1 Continental GTSIO-520F
Huntington, WV
7/28/75

SPEED OVER A 500 KM CLOSED CIRCUIT WITHOUT PAYLOAD (USA)
John P. Harris 477.15 kmh
Bellanca Skyrocket II 296.49 mph
1 Continental GTSIO-520
Huntington, WV
7/22/75

SPEED OVER A 1,000 KM CLOSED CIRCUIT WITHOUT PAYLOAD (USA)
John P. Harris 484.62 kmh
Bellanca Skyrocket II 301.13 mph
1 Continental GTSIO-520F
Huntington, WV
7/22/75

SPEED OVER A 2,000 KM CLOSED CIRCUIT WITHOUT PAYLOAD (USA) ✯
Michael W. Melvill 413.78 kmh
Rutan Catbird 257.11 mph
1 Lycoming TIO-360, 210 bhp
Lake Hughes, CA - Boise, ID
3/2/94

SPEED AROUND THE WORLD, EASTBOUND (CANADA)
Donald F. Muir, pilot 243.75 kmh
Andre Daemen 151.46 mph
Cessna 210N
Continental IO-520-L, 300 bhp
Montreal, St. John's, Shannon, Naples, Cairo, Bahrain, Madras, Kuala Lumpur, Manila, Agana, Majuro, Honolulu, Hilo, San Francisco, Denver, Lansing, Montreal
8/1-7/82

SPEED AROUND THE WORLD, WESTBOUND (USA)
Charles Classen, Pilot 87.50 kmh
Phillip Greth, Copilot 54.37 mph
Beech Bonanza G-35, N4493D
1 Continental, E0225, 225 bhp
Waukegan, San Jose, Honolulu, Majuro, Guadalcanal, Cairns, Brisbane, Adelaide, Yulara, Alice Springs, Darwin, Bali, Singapore, Madras, Bombay, Dubai, Cairo, Palma de Mallorca, Santa Maria, St. Johns, Waukegan
5/27-6/6/88

TIME TO CLIMB

TIME TO 3,000 METERS (W. GERMANY)
Wilhelm Heller — 3 min
Cessna P 210N — 9 sec
Continental TSIO 520 310 bhp
2/12/84

TIME TO 6,000 METERS (W. GERMANY)
Wilhelm Heller — 8 min
Cessna P 210N — 52 sec
Continental TSIO 520 310 bhp
2/12/84

TIME TO 9,000 METERS (W. GERMANY)
Wilhelm Heller — 20 min
Cessna P 210N — 54 sec
Continental TSIO 520 310 bhp
2/12/84

CLASS C-1.D (1,750-3,000 KG / 3,858-6,614 LBS)
GROUP I (PISTON ENGINE)

DISTANCE

DISTANCE WITHOUT LANDING (AUSTRALIA)
Peter Wilkins — 12,760.00 km
Piper Malibu, N9221M — 7,929.71 mi
1 Continental TSIO-BE, 310 bhp
Sydney, Australia to Phoenix, AZ
3/30-4/1/87

DISTANCE OVER A CLOSED CIRCUIT WITHOUT LANDING (USA)
Richard Rutan, Pilot — 18,658.16 km
Jeana Yeager, Pilot — 11,593.68 mi
Voyager, N269VA
Front Engine O-240 Air Cooled, 130 bhp
Rear Engine 10L Liquid Cooled, 110 bhp
Vandenberg AFB, CA - Mojave Airport, CA
7/10-15/86

ALTITUDE

ALTITUDE (USA)
Frances Bera — 12,239 m
Piper Aztec N5978Y — 40,154 ft
2 Lycoming IO-540
Long Beach, CA
7/16/66

ALTITUDE IN HORIZONTAL FLIGHT (USA)
C. Lee Merritt — 11,613 m
Beech Baron AR6TC — 38,100 ft
2 Lycoming TIO-541-E1B4, 380 bhp
Denver, CO
2/16/80

SPEED

SPEED OVER A 3 KM STRAIGHT COURSE (USA)
James L. Wright — 489.35 kmh
homebuilt Hughes H-1 replica — 304.07 mph
1 P&W R-1535, 750 bhp
Reno, NV
9/13/02

SPEED OVER A 15/25 KM STRAIGHT COURSE (USA)
Moton H. Crockett, Pilot — 467.44 kmh
Charles I. Fitzsimmons, Copilot — 290.46 mph
Aerostar 601P, N5444P
2 Lycoming IO-540, 340 hp ea
Austin, TX
6/6/87

SPEED OVER A 100 KM CLOSED CIRCUIT WITHOUT PAYLOAD (UK)
Ms. R. M. Sharpe — 519.48 kmh
Vickers Supermarine — 322.78 mph
Spitfire 5B
Rolls Royce Merlin 55M
Wolverhampton, England
6/17/50

SPEED OVER A 500 KM CLOSED CIRCUIT WITHOUT PAYLOAD (USA)
John P. Harris — 525.45 kmh
Bellanca Skyrocket II — 326.50 mph
1 Continental GTSIO-250F
Huntington, WV
8/1/75

SPEED OVER A 1,000 KM CLOSED CIRCUIT WITHOUT PAYLOAD (USA)
John P. Harris — 505.31 kmh
Bellanca Skyrocket II — 313.99 mph
1 Continental GTSIO-520F
Huntington, WV
9/3/75

SPEED OVER A 2,000 KM CLOSED CIRCUIT WITHOUT PAYLOAD (USA)
Jack F. Chrysler — 439.07 kmh
Aerostar 601 — 272.85 mph
2 Lycoming TSIO-540
Van Nuys, CA
8/6/75

SPEED AROUND THE WORLD, EASTBOUND (AUSTRALIA)
Trevor K. Brougham — 318.28 kmh
Beechcraft Baron B55 — 197.77 mph
2 Rolls Royce Continental IO-470-L
Darwin, Rabaul, Bucholz, Honolulu, San Francisco, Toronto, Gander, London, Athens, Bahrain, Bombay, Singapore, Darwin
Distance: 39,911 km (24,800 mi)
8/4-10/71

SPEED AROUND THE WORLD, WESTBOUND (USA)
Max Conrad — 198.27 kmh
Piper Aztec — 123.19 mph
2 Lycoming O-540, 250 bhp
Miami, Long Beach, Honolulu, Wake Island, Guam, Manila, Singapore, Bombay, Nairobi, Lagos, Dakar, Amapa, Atkinson Field, Port of Spain, Miami
3/8/61

SPEED AROUND THE WORLD OVER BOTH THE EARTH'S POLES (USA)
Richard Norton, Pilot — 14.04 kmh
Calin Rosetti, Copilot — 8.72 mph
Piper Malibu, N26033
1 Continental TSIO-520-B, 310 bhp
Paris, Dakar, Rio de Janeiro, Commodore Rivadiva, Rio Grande, Maramio, South Pole, Teniente Marsh, Marambio, Santiago, Easter Island, Tahiti, Honolulu, Fairbanks, Mould Bay, North Pole, Spitzbergen, Oslo, Paris
1/21-6/15/87

TIME TO CLIMB

TIME TO 3,000 METERS (USA)
Russel H. Hancock — 3 min
Piper Panther Navajo — 33 sec
2 Lycoming TIO-540 J2BD, 350 bhp
Nashville, TN
10/6/78

TIME TO 6,000 METERS (USA)
Russell H. Hancock — 7 min
Piper Panther Navajo — 39 sec
2 Lycoming TIO-540 J2BD, 350 bhp
Nashville, TN
10/6/78

TIME TO 9,000 METERS (UK)
David Long — 19 min
Piper Panther Navajo — 6 sec
2 Lycoming TIO-540-J2BD 350 hp
Cranfield, England
9/8/79

CLASS C-1.E (3,000-6,000 KG / 6,614-13,228 LBS)
GROUP I (PISTON ENGINE)

DISTANCE

DISTANCE WITHOUT LANDING (USA)
Richard Rutan, Pilot — 40,212.14 km
Jeana Yeager, Pilot — 24,986.73 mi
Voyager, N269VA
Front Engine O-240 Air Cooled, 130 bhp
Rear Engine 10L Liquid Cooled, 110 bhp
Edward AFB, CA - Edwards AFB, CA
12/14-23/86

DISTANCE OVER A CLOSED CIRCUIT WITHOUT LANDING (USA)
Richard Rutan, Pilot — 40,212.14 km
Jeana Yeager, Pilot — 24,986.73 mi
Voyager, N269VA
Front Engine O-240 Air Cooled, 130 bhp
Rear Engine 10L Liquid Cooled, 110 bhp
Edwards AFB, CA - Edwards AFB, CA
12/14-23/86

ALTITUDE

ALTITUDE (USSR)
V. Kalinin — 11,248 m
Ourkrania — 36,904 ft
1 ACH, 1,000 bhp
6/9/54

ALTITUDE IN HORIZONTAL FLIGHT (USA)
James L. Badgett, pilot — 10,625 m
Aero Commander 685 — 34,860 ft
2 Continental GTSIO-520
Santa Barbara, CA
1/23/73

ALTITUDE WITH 1,000 KG PAYLOAD (AUSTRALIA)
M. F. Tuytjens — 8,624 m
Dornier Do 28D-1 Skyservant — 28,294 ft
2 Lycoming, 380 bhp
3/15/72

GREATEST PAYLOAD CARRIED TO 2,000 METERS (AUSTRALIA)
M. F. Tuytjens — 1,000 kg
Dornier Do 28D-1 Skyservant — 2,205 lb
2 Lycoming, 380 bhp
3/15/72

SPEED

SPEED OVER A 15/25 KM STRAIGHT COURSE (USA)
Charles I. Fitzsimmons, CoCaptain — 484.91 kmh
James Schwertner, CoCaptain — 301.32 mph
Robert Howell, Flight Engineer
Piper Navajo, N27946
2 Lycoming TIO-540, 350 bhp
San Antonio, TX
11/21/87

SPEED OVER A 100 KM CLOSED CIRCUIT WITHOUT PAYLOAD (USA)
Jack Chrysler — 408.96 kmh
Mr. RPM Turbo 800 — 254.12 mph
2 Lycoming IO-720-B1B
Van Nuys, CA
1/20/76

SPEED OVER A 500 KM CLOSED CIRCUIT WITHOUT PAYLOAD (AUSTRIA)
Franz Achleitner — 407.98 kmh
Cessna 421C — 253.52 mph
2 Continental GTSIO-520-N, 375 bhp
Salzburg, Austria
9/11/81

SPEED OVER A 1,000 KM CLOSED CIRCUIT WITHOUT PAYLOAD (AUSTRIA)
Franz Achleitner — 407.75 kmh
Cessna 421C — 253.37 mph
2 Continental GTSIO-520-N, 375 bhp
Salzburg, Austria
9/11/81

SPEED AROUND THE WORLD, EASTBOUND (USA)
Philander P. Claxton, Pilot — 353.57 kmh
John L. Cink, Copilot — 219.70 mph
Aerostar 601P
2 Lycoming IO-540, 290 bhp
Los Angeles, Peoria, Gander, Frankfurt, Tehran, Madras, Kota Kinabalu, Manila, Guam, Kwajalein Island, Wake Island, Honolulu, Los Angeles
Distance: 36,803.95 km (22,868.91 mi)
104 hrs 5 min 30 sec
11/4-9/77

SPEED AROUND THE WORLD, WESTBOUND (QATAR) ★
Hamad Al-Thani — 132.02 kmh
Piper Aerostar 601P — 82.03 mph
2 Lycoming IO-540, 290 bhp
San Jose, Honolulu, Majuro, Yap, Bandar Seri Begawan, Phuket, Bombay, Doha, Bahrain, Larnaca, Iraklion, Faro, Santa Maria, Halifax, Washington, San Jose
4/17-29/92

SPEED AROUND THE WORLD OVER BOTH THE EARTH'S POLES (USA)
Elgen M. Long — 93.05 kmh
Piper Navajo, N9097Y — 57.82 mph
2 Lycoming TIO-540, 310 bhp
San Francisco, Fairbanks, Pt. Barrow, North Pole, Tromso, Stockholm, London, Accra, Equator, Recife, Rio de Janeiro, Punta Arenas, South Pole, McMurdo, Sydney, Nandi, Equator, Wake Island, Tokyo, Wake Island, Honolulu, San Francisco
28 days 00 hrs 43 min
11/5-12/3/71

TIME TO CLIMB

TIME TO 3,000 METERS (AUSTRIA)
Franz Achleitner — 5 min
Cessna 421C — 14 sec
2 Continental GTSIO-520-N, 375 bhp
Salzburg, Austria
10/25/81

TIME TO 6,000 METERS (AUSTRIA)
Franz Achleitner — 11 min
Cessna 421C — 1 sec
2 Continental GTSIO-520-N, 375 bhp
Salzburg, Austria
10/25/81

TIME TO 9,000 METERS (AUSTRIA)
Franz Achleitner — 24 min
Cessna 421C — 30 sec
Continental GTSIO-520-N, 375 bhp
Salzburg, Austria
10/25/81

TURBOPROP

Turboprop airplanes are part propeller and part jet, with performance that falls somewhere between the two. They offer economy and acceleration not ordinarily possible with a true jet engine, and considerably greater power (relative to engine weight) than any piston engine.

The first airplane to fly with a propeller driven by a gas turbine engine was a jet fighter that had been modified for experimental purposes. A British Gloster Meteor, the only jet fighter other than German to see action in World War II, flew on the power of two Rolls Royce Trent turboprop engines on September 20, 1945.

While the rapid development of the pure turbojet engine eliminated some of the advantages of the jet-with-a-propeller, it remained an excellent powerplant for short and medium-range airliners, and long-range patrol and heavy-duty cargo airplanes for the military.

In the United States, Lockheed Aircraft Co. produced the greatest number of turboprop airplanes, with its Electra airliner, Hercules cargoplane, and Orion maritime patrol version of the Electra. In Great Britain, Vickers created the Viscount and Vanguard airliners. In the USSR, there are still many turboprop--powered long-range military surveillance, bomber, and transport airplanes, as well as civil airliners—many of them using contra-rotating propellers to absorb the great power of their engines.

Today, there is a marked trend toward fitting twin-engine business airplanes with more powerful, yet lighter, turboprop engines, and these are setting many of the latest records for speed.

CLASS C-1 (UNLIMITED)
GROUP II (TURBOPROP)

DISTANCE

DISTANCE WITHOUT LANDING (USA)
LTC Edgar L. Allison, USAF 14,052.95 km
Lockheed HC-130 Hercules 8,732.09 mi
4 T-56A15
Ching-Chuan Kang, Taiwan to
Scott AFB, IL
2/20/72

DISTANCE OVER A CLOSED CIRCUIT WITHOUT LANDING (USA)
CDR Philip R. Hite, USN 10,103.51 km
Lockheed RP-3D Orion 6,278.05 mi
4 T56A14
NAS Patuxent River, MD
11/4/72

ALTITUDE

ALTITUDE (USA)
Einar Envoldson 16,329 m
Egrett-1, N14ES 53,574 ft
1 Garrett TPE331-14A, 750 shp
Greenville, TX
9/1/88

ALTITUDE IN HORIZONTAL FLIGHT (USA)
Einar Envoldson 16,239 m
Egrett-1, N14ES 53,276 ft
1 Garrett TPE331-14A, 750 shp
Greenville, TX
9/1/88

ALTITUDE WITH 1,000 KG PAYLOAD (GERMANY)
Werner Kraut 15,552 m
Burkhart Grob G-520 Egrett 51,023 ft
1 Garrett TPE331, 750 shp
Mindelheim, Germany
3/31/94

ALTITUDE WITH 1,000 KG PAYLOAD* ✭
LCDR Eric Hinger, USN, Pilot 41,254 ft
LCDR Matt Klunder, USN, Copilot
Grumman E-2C Plus
2 Allison T56-A-427, 5,250 shp
NAS Patuxent River, MD
12/19/91

ALTITUDE WITH 2,000 KG PAYLOAD (USSR)
V. E. Egorov, Pilot 12,265 m
M. M. Bachkirov, Copilot 40,240 ft
S. N. Martyanov, Navigator
Tupolev Tu-142
4 Kuznetsov NK-12MA, 15,000 shp
10/05/89

ALTITUDE WITH 5,000 KG PAYLOAD (USSR)
V. E. Egorov, Pilot 12,265 m
M. M. Bachkirov, Copilot 40,240 ft
S. N. Martyanov, Navigator
Tupolev Tu-142
4 Kuznetsov NK-12MA, 15,000 shp
10/05/89

ALTITUDE WITH 10,000 KG PAYLOAD (USSR)
Vladimir Kokkinaki 13,154 m
Ilyushin IL-18 43,156 ft
4 AI-20
Moscow, USSR
11/15/58

ALTITUDE WITH 15,000 KG PAYLOAD (USSR)
Vladimir Kokkinaki 12,471 m
Ilyushin IL-18 40,915 ft
4 AI-20
Aerodrome of Bikovo, Moscow
11/14/58

ALTITUDE WITH 20,000 KG PAYLOAD (USSR)
Vladimir Kokkinaki 12,118 m
Ilyushin IL-18 39,747 ft
4 AI-20
Podmoskovnoe, USSR
11/25/59

ALTITUDE WITH 25,000 KG PAYLOAD (USSR)
Ivan Soukhomline 12,073 m
Tupolev Tu-114 39,610 ft
4 Kuznetsov NK-12M
Vnoukovo Airport, USSR
7/12/61

ALTITUDE WITH 30,000 KG PAYLOAD (USSR)
Ivan Soukhomline 12,073 m
Tupolev Tu-114 39,610 ft
4 Kuznetsov NK-12M
Vnoukovo Airport, USSR
7/12/61

ALTITUDE WITH 35,000 KG PAYLOAD (USSR)
Ivan Davidov 7,848 m
Antonov An-22 25,748 ft
4 Kuznetsov NK-12MA
10/26/67

ALTITUDE WITH 40,000 KG PAYLOAD (USSR)
Ivan Davidov 7,848 m
Antonov An-22 25,748 ft
4 Kuznetsov NK-12MA
10/26/67

ALTITUDE WITH 45,000 KG PAYLOAD (USSR)
Ivan Davidov 7,848 m
Antonov An-22 25,748 ft
4 Kuznetsov NK-12MA
10/26/67

ALTITUDE WITH 50,000 KG PAYLOAD (USSR)
Ivan Davidov 7,848 m
Antonov An-22 25,748 ft
4 Kuznetsov NK-12MA
10/26/67

ALTITUDE WITH 55,000 KG PAYLOAD (USSR)
Ivan Davidov 7,848 m
Antonov An-22 25,748 ft
4 Kuznetsov NK-12MA
10/26/67

ALTITUDE WITH 60,000 KG PAYLOAD (USSR)
Ivan Davidov 7,848 m
Antonov An-22 25,748 ft
4 Kuznetsov NK-12MA
10/26/67

ALTITUDE WITH 65,000 KG PAYLOAD (USSR)
Ivan Davidov 7,848 m
Antonov An-22 25,748 ft
4 Kuznetsov NK-12MA
10/26/67

ALTITUDE WITH 70,000 KG PAYLOAD (USSR)
Ivan Davidov 7,848 m
Antonov An-22 25,748 ft
4 Kuznetsov NK-12MA
10/26/67

ALTITUDE WITH 75,000 KG PAYLOAD (USSR)
Ivan Davidov 7,848 m
Antonov An-22 25,748 ft
4 Kuznetsov NK-12MA
10/26/67

ALTITUDE WITH 80,000 KG PAYLOAD (USSR)
Ivan Davidov 7,848 m
Antonov An-22 25,748 ft
4 Kuznetsov NK-12MA
10/26/67

ALTITUDE WITH 85,000 KG PAYLOAD (USSR)
Ivan Davidov 7,848 m
Antonov An-22 25,748 ft
4 Kuznetsov NK-12MA
10/26/67

ALTITUDE WITH 90,000 KG PAYLOAD (USSR)
Ivan Davidov 7,848 m
Antonov An-22 25,748 ft
4 Kuznetsov NK-12MA
10/26/67

ALTITUDE WITH 95,000 KG PAYLOAD (USSR)
Ivan Davidov 7,848 m
Antonov An-22 25,748 ft
4 Kuznetsov NK-12MA
10/26/67

ALTITUDE WITH 100,000 KG PAYLOAD (USSR)
Ivan Davidov 7,848 m
Antonov An-22 25,748 ft
4 Kuznetsov NK-12MA
10/26/67

GREATEST PAYLOAD CARRIED TO 2,000 METERS (USSR)
Ivan Davidov 100,444.60 kg
Antonov An-22 221,442.00 lbs
4 Kuznetsov NK-12MA
10/26/67

SPEED

SPEED OVER A 15/25 KM STRAIGHT COURSE (USA)
CDR D.H. Lilienthal, USN 806.10 kmh
Lockheed P3C Orion 501.44 mph
4 Allison T56-A014
NAS Patuxent River, MD
1/27/71

SPEED OVER A 100 KM CLOSED CIRCUIT WITHOUT PAYLOAD (USSR)
B. Konstantinov 706.00 kmh
Ilyushin Il-18 438.68 mph
4 Ivchenko AI-20
5/16/68

SPEED OVER A 500 KM CLOSED CIRCUIT WITHOUT PAYLOAD (USSR)
Alexandre Metronie 730.61 kmh
Antonov An-10 453.90 mph
4 TV-12
Sternberg, USSR
4/29/61

SPEED OVER A 1,000 KM CLOSED CIRCUIT WITHOUT PAYLOAD (USSR)
Ivan Soukhomline 871.38 kmh
Tupolev Tu-114 541.45 mph
4 TV-12
Sternberg, USSR
3/24/60

SPEED OVER A 1,000 KM CLOSED CIRCUIT WITH 1,000 KG PAYLOAD (USSR)
Ivan Soukhomline 871.38 kmh
Tupolev Tu-114 541.45 mph
4 TV-12
Sternberg, USSR
3/24/60

SPEED OVER A 1,000 KM CLOSED CIRCUIT WITH 2,000 KG PAYLOAD (USSR)
Ivan Soukhomline 871.38 kmh
Tupolev Tu-114 541.45 mph
4 TV-12
Sternberg, USSR
3/24/60

SPEED OVER A 1,000 KM CLOSED CIRCUIT WITH 5,000 KG PAYLOAD (USSR)
Ivan Soukhomline 871.38 kmh
Tupolev Tu-114 541.45 mph
4 TV-12
Sternberg, USSR
3/24/60

SPEED OVER A 1,000 KM CLOSED CIRCUIT WITH 10,000 KG PAYLOAD (USSR)
Ivan Soukhomline 871.38 kmh
Tupolev Tu-114 541.45 mph
4 TV-12
Sternberg, USSR
3/24/60

SPEED OVER A 1,000 KM CLOSED CIRCUIT WITH 15,000 KG PAYLOAD (USSR)
Ivan Soukhomline 871.38 kmh
Tupolev Tu-114 541.45 mph
4 TV-12
Sternberg, USSR
3/24/60

SPEED OVER A 1,000 KM CLOSED CIRCUIT WITH 20,000 KG PAYLOAD (USSR)
Ivan Soukhomline 871.38 kmh
Tupolev Tu-114 541.45 mph
4 TV-12
Sternberg, USSR
3/24/60

SPEED OVER A 1,000 KM CLOSED CIRCUIT WITH 25,000 KG PAYLOAD (USSR)
Ivan Soukhomline 871.38 kmh
Tupolev Tu-114 541.45 mph
4 TV-12
Sternberg, USSR
3/24/60

SPEED OVER A 1,000 KM CLOSED CIRCUIT WITH 30,000 KG PAYLOAD (USSR)
L. V. Kozlov 816.25 kmh
VP-021 507.19 mph
4 Kuznetsov NK-12MA, 15,000 shp
9/26/89

SPEED OVER A 1,000 KM CLOSED CIRCUIT WITH 35,000 KG PAYLOAD (USSR)
M. Popovitch 608.45 kmh
A. Timofeev 378.07 mph
Antonov An-22
4 Kuznetsov NK-12MA
2/21/72

SPEED OVER A 1,000 KM CLOSED CIRCUIT WITH 40,000 KG PAYLOAD (USSR)
M. Popovitch 608.45 kmh
A. Timofeev 378.07 mph
Antonov An-22
4 Kuznetsov NK-12MA
2/21/72

SPEED OVER A 1,000 KM CLOSED CIRCUIT WITH 45,000 KG PAYLOAD (USSR)
M. Popovitch 608.45 kmh
A. Timofeev 378.07 mph
Antonov An-22
4 Kuznetsov NK-12MA
2/21/72

SPEED OVER A 1,000 KM CLOSED CIRCUIT WITH 50,000 KG PAYLOAD (USSR)
M. Popovitch 608.45 kmh
A. Timofeev 378.07 mph
Antonov An-22
4 Kuznetsov NK-12MA
2/21/72

SPEED OVER A 2,000 KM CLOSED CIRCUIT WITHOUT PAYLOAD (USSR)
Ivan Soukhomline 857.28 kmh
Tupolev Tu-114 532.69 mph
4 TV-12
Sternberg, USSR
4/1/60

SPEED OVER A 2,000 KM CLOSED CIRCUIT WITH 1,000 KG PAYLOAD (USSR)
Ivan Soukhomline 857.28 kmh
Tupolev Tu-114 532.69 mph
4 TV-12
Sternberg, USSR
4/1/60

SPEED OVER A 2,000 KM CLOSED CIRCUIT WITH 2,000 KG PAYLOAD (USSR)
Ivan Soukhomline 857.28 kmh
Tupolev Tu-114 532.69 mph
4 TV-12
Sternberg, USSR
4/1/60

SPEED OVER A 2,000 KM CLOSED CIRCUIT WITH 5,000 KG PAYLOAD (USSR)
Ivan Soukhomline 857.28 kmh
Tupolev Tu-114 532.69 mph
4 TV-12
Sternberg, USSR
4/1/60

SPEED OVER A 2,000 KM CLOSED CIRCUIT WITH 10,000 KG PAYLOAD (USSR)
Ivan Soukhomline 857.28 kmh
Tupolev Tu-114 532.69 mph
4 TV-12
Sternberg, USSR
4/1/60

SPEED OVER A 2,000 KM CLOSED CIRCUIT WITH 15,000 KG PAYLOAD (USSR)
Ivan Soukhomline 857.28 kmh
Tupolev Tu-114 532.69 mph
4 TV-12
Sternberg, USSR
4/1/60

SPEED OVER A 2,000 KM CLOSED CIRCUIT WITH 20,000 KG PAYLOAD (USSR)
Ivan Soukhomline 857.28 kmh
Tupolev Tu-114 532.69 mph
4 TV-12
Sternberg, USSR
4/1/60

SPEED OVER A 2,000 KM CLOSED CIRCUIT WITH 25,000 KG PAYLOAD (USSR)
Ivan Soukhomline 857.28 kmh
Tupolev Tu-114 532.69 mph
4 TV-12
Sternberg, USSR
4/1/60

SPEED OVER A 2,000 KM CLOSED CIRCUIT WITH 30,000 KG PAYLOAD (USSR)
Viktor Naimushin 834.82 kmh
Sergei Osipov 518.73 mph
Fedor Ivlev
Tupolev Tu-142
9/28/89

SPEED OVER A 2,000 KM CLOSED CIRCUIT WITH 35,000 KG PAYLOAD (USSR)
M. Popovitch 593.32 kmh
A. Timofeev 368.67 mph
Antonov An-22
4 Kuznetsov NK-12MA
2/19/72

SPEED OVER A 2,000 KM CLOSED CIRCUIT WITH 40,000 KG PAYLOAD (USSR)
M. Popovitch 593.32 kmh
A. Timofeev 368.07 mph
Antonov An-22
4 Kuznetsov NK-12MA
2/19/72

SPEED OVER A 2,000 KM CLOSED CIRCUIT WITH 45,000 KG PAYLOAD (USSR)
M. Popovitch 593.32 kmh
A. Timofeev 368.67 mph
Antonov An-22
4 Kuznetsov NK-12MA
2/19/72

SPEED OVER A 2,000 KM CLOSED CIRCUIT WITH 50,000 KG PAYLOAD (USSR)
M. Popovitch 593.32 kmh
A. Timofeev 368.67 mph
Antonov An-22
4 Kuznetsov NK-12MA
2/19/72

SPEED OVER A 5,000 KM CLOSED CIRCUIT WITHOUT PAYLOAD (USSR)
Ivan Soukhomline 877.21 kmh
Tupolev Tu-114 545.07 mph
4 TV-12
Sternberg, USSR
4/9/60

SPEED OVER A 5,000 KM CLOSED CIRCUIT WITH 1,000 KG PAYLOAD (USSR)
Ivan Soukhomline 877.21 kmh
Tupolev Tu-114 545.07 mph
4 TV-12
Sternberg, USSR
4/9/60

SPEED OVER A 5,000 KM CLOSED CIRCUIT WITH 2,000 KG PAYLOAD (USSR)
Ivan Soukhomline 877.21 kmh
Tupolev Tu-114 545.07 mph
4 TV-12
Sternberg, USSR
4/9/60

SPEED OVER A 5,000 KM CLOSED CIRCUIT WITH 5,000 KG PAYLOAD (USSR)
Ivan Soukhomline 877.21 kmh
Tupolev Tu-114 545.07 mph
4 TV-12
Sternberg, USSR
4/9/60

SPEED OVER A 5,000 KM CLOSED CIRCUIT WITH 10,000 KG PAYLOAD (USSR)
Ivan Soukhomline 877.21 kmh
Tupolev Tu-114 545.07 mph
4 TV-12
Sternberg, USSR
4/9/60

SPEED OVER A 5,000 KM CLOSED CIRCUIT WITH 15,000 KG PAYLOAD (USSR)
Ivan Soukhomline 877.21 kmh
Tupolev Tu-114 545.07 mph
4 TV-12
Sternberg, USSR
4/9/60

SPEED OVER A 5,000 KM CLOSED CIRCUIT WITH 20,000 KG PAYLOAD (USSR)
Ivan Soukhomline 877.21 kmh
Tupolev Tu-114 545.07 mph
4 TV-12
Sternberg, USSR
4/9/60

SPEED OVER A 5,000 KM CLOSED CIRCUIT WITH 25,000 KG PAYLOAD (USSR)
Ivan Soukhomline 877.21 kmh
Tupolev Tu-114 545.07 mph
4 TV-12
Sternberg, USSR
4/9/60

SPEED OVER A 5,000 KM CLOSED CIRCUIT WITH 30,000 KG PAYLOAD (USSR)
S. Dedoukh 597.28 kmh
Antonov An-22 371.13 mph
4 Kuznetsov NK-12MA
10/21/74

SPEED OVER A 5,000 KM CLOSED CIRCUIT WITH 35,000 KG PAYLOAD (USSR)
Yuri Romanov 588.64 kmh
Antonov An-22 365.76 mph
4 Kuznetsov NK-12MA
10/24/74

SPEED OVER A 5,000 KM CLOSED CIRCUIT WITH 40,000 KG PAYLOAD (USSR)
George Pakilev 584.04 kmh
Antonov An-22 362.90 mph
4 Kuznetsov NK-12MA
4/17/75

SPEED OVER A 10,000 KM CLOSED CIRCUIT WITHOUT PAYLOAD (USSR)
Ivan Soukhomline 737.35 kmh
P. Soldatov 458.17 mph
Tupolev Tu-114
4 Kuznetsov NK-12M, 12,000 shp
Sternberg, USSR
4/21/62

SPEED OVER A 10,000 KM CLOSED CIRCUIT WITH 1,000 KG PAYLOAD (USSR)
Ivan Soukhomline 737.35 kmh
P. Soldatov 458.17 mph
Tupolev Tu-114
4 Kuznetsov NK-12M, 12,000 shp
Sternberg, USSR
4/21/62

SPEED OVER A 10,000 KM CLOSED CIRCUIT WITH 2,000 KG PAYLOAD (USSR)
I. Soukhomline 737.35 kmh
P. Soldatov 458.17 mph
Tupolev Tu-114
4 Kuznetsov NK-12M, 12,000 shp
Sternberg, USSR
4/21/62

SPEED OVER A 10,000 KM CLOSED CIRCUIT WITH 5,000 KG PAYLOAD (USSR)
Ivan Soukhomline 737.35 kmh
P. Soldatov 458.17 mph
Tupolev Tu-114
4 Kuznetsov NK-12M, 12,000 shp
Sternberg, USSR
4/21/62

SPEED OVER A 10,000 KM CLOSED CIRCUIT WITH 10,000 KG PAYLOAD (USSR)
Ivan Soukhomline 737.35 kmh
P. Soldatov 458.17 mph
Tupolev Tu-114
4 Kuznetsov NK-12M, 12,000 shp
Sternberg, USSR
4/21/62

SPEED AROUND THE WORLD, EASTBOUND (USA)
Joe Harnish, Pilot 490.51 kmh
David B. Webster, Pilot 304.80 mph
Gulfstream Commander 695A
2 Garrett TPE-331-501K, 820 shp
Elkhart, Goose Bay, Keflavik, Vienna, Cairo, Luxor, Sharjah, Colombo, Singapore, Manila, Agana, Wake Island, Midway Island, Honolulu, San Francisco, Elkhart
3/21-24/83

TIME TO CLIMB

TIME TO 3,000 METERS (USA) ✭
Wayne E. Handley 1 min
Giles G-750 9 sec
1 P&W PT6A-25C, 750 shp
Salinas, CA
1/20/99

TIME TO 6,000 METERS (USA)
Wayne E. Handley 3 min
Giles G-750 6 sec
1 P&W PT6A-25C, 750 shp
Oshkosh, WI
7/30/99

TIME TO 9,000 METERS (USA)
Charles "Chuck" Yeager, Pilot 6 min
Renald Davenport, Copilot 34 sec
Piper Cheyenne 400LS
Portland, OR
4/16/85

TIME TO 12,000 METERS (USA)
Charles "Chuck" Yeager, Pilot 11 min
Renald Davenport, Copilot 8 sec
Piper Cheyenne 400LS
Portland, OR
4/16/85

TIME TO 15,000 METERS (USA)
Einar Envoldson 40 min
Burkhart Grob Egrett-1 47 sec
1 Garrett TPE331-14A, 750 shp
Greenville, TX
9/1/88

CLASS C-1.B (500-1,000 KG / 1,102-2,205 LBS) GROUP II (TURBOPROP)

ALTITUDE

ALTITUDE (USA)
William D. Thompson, Jr. 11,297 m
Cessna XL-19B 37,063 ft
502-B LX-50-BL-1
7/16/53

TIME TO CLIMB

TIME TO 3,000 METERS (USA) ✭
Wayne E. Handley 1 min
Giles G-750 9 sec
1 P&W PT6A-25C, 750 shp
Salinas, CA
1/20/99

TIME TO 6,000 METERS (USA)
Wayne E. Handley 3 min
Giles G-750 6 sec
1 P&W PT6A-25C, 750 shp
Oshkosh, WI
7/30/99

CLASS C-1.C (1,000-1,750 KG / 2,205-3,858 LBS)
GROUP II (TURBOPROP)

ALTITUDE

ALTITUDE (USA)
Donald R. Wilson — 15,549 m
LTV L450F — 51,014 ft
1 PT6A-34
Greenville, TX
3/27/72

ALTITUDE IN HORIZONTAL FLIGHT (USA)
Donald R. Wilson — 15,456 m
LTV L450F — 50,708 ft
1 PT6A-34
Greenville, TX
3/27/72

SPEED

SPEED OVER A 3 KM STRAIGHT COURSE (USA)
Wesley E. Behel, Jr. — 570.33 kmh
Lancair Turbine IV-P — 354.38 mph
1 Walters M 601D, 650 shp
Reno, NV
9/11/02

SPEED OVER A 15/25 KM STRAIGHT COURSE (CHILE)
Jaime Acosta Herrera — 382.43 kmh
Felipe Fernandez Mesa — 237.63 mph
ENAER T-35DT Turbo Pillán
Allison 250 B17D, 429 shp
Santiago, Chile
3/10/00

SPEED OVER A 100 KM CLOSED CIRCUIT WITHOUT PAYLOAD (FRANCE)
Pierre Bonneau — 436.32 kmh
Sipa Antelope — 271.11 mph
1 Astazou
Istres, France
4/4/65

SPEED OVER A 500 KM CLOSED CIRCUIT WITHOUT PAYLOAD (FRANCE)
Pierre Bonneau — 421.15 kmh
M. Lesieur — 261.69 mph
Sipa Antelope
1 Astazou
Istres, France
12/9/64

SPEED OVER A 1,000 KM CLOSED CIRCUIT WITHOUT PAYLOAD (FRANCE)
Pierre Bonneau — 406.35 kmh
Sipa Antelope — 252.49 mph
1 Astazou
Istres, France
10/5/64

SPEED OVER A 2,000 KM CLOSED CIRCUIT WITHOUT PAYLOAD (FRANCE)
Pierre Bonneau — 379.46 kmh
Sipa Antelope — 235.77 mph
1 Astazou
Istres, France
10/6/64

TIME TO CLIMB

TIME TO 3,000 METERS (USA)
Donald R. Wilson — 3 min
LTV L450F — 30 sec
1 PT6A-34
Greenville, TX
3/27/72

TIME TO 6,000 METERS (USA)
Donald R. Wilson — 7 min
LTV L450F — 16 sec
1 PT6A-34
Greenville, TX
3/27/72

TIME TO 9,000 METERS (USA)
Donald R. Wilson — 12 min
LTV L450F — 18 sec
1 PT6A-34
Greenville, TX
3/27/72

TIME TO 12,000 METERS (USA)
Donald R. Wilson — 20 min
LTV L450F — 15 sec
1 PT6A-34
Greenville, TX
3/27/72

TIME TO 15,000 METERS (USA)
Donald R. Wilson — 46 min
LTV L450F — 29 sec
1 PT6A-34
Greenville, TX
3/27/72

CLASS C-1.D (1,750-3,000 KG / 3,858-6,614 LBS)
GROUP II (TURBOPROP)

DISTANCE

DISTANCE WITHOUT LANDING (AUSTRALIA)
MAJ D.B. Coffey — 3,893.56 km
Pilatus Porter PC-6 — 2,419.46 mi
1 PTA6A-20, 550 shp
Carnavon - Brisbane
8/8/83

ALTITUDE

ALTITUDE (USA)
Donald R. Wilson — 13,783 m
LTV L450F — 45,222 ft
1 PT6A-34
Greenville, TX
3/23/72

ALTITUDE IN HORIZONTAL FLIGHT (USA)
Donald R. Wilson — 13,779 m
LTV L450F — 45,207 ft
1 PT6A-34
Greenville, TX
3/23/72

SPEED

SPEED OVER A 15/25 KM STRAIGHT COURSE (USA)
Robert C. Hayes, Pilot — 354.78 kmh
Jack Schweibold, Copilot — 220.46 mph
Beech Bonanza
1 Allison 250 B-17D, 420 shp
Reno, NV
9/10/86

SPEED OVER A 100 KM CLOSED CIRCUIT WITHOUT PAYLOAD (FRANCE)
Pierre Bonneau — 521.50 kmh
M. Duberger — 324.04 mph
Marquis F-BJSI
2 Astazoux
Istres, France
5/12/65

SPEED OVER A 500 KM CLOSED CIRCUIT WITHOUT PAYLOAD (FRANCE)
Pierre Bonneau 522.68 kmh
M. Duberger 324.77 mph
Marquis F-BJSI
2 Astazoux
Istres, France
5/12/65

SPEED OVER A 1,000 KM CLOSED CIRCUIT WITHOUT PAYLOAD (FRANCE)
Pierre Bonneau 524.01 kmh
M. Duberger 325.60 mph
Marquis F-BJSI
2 Astazoux
Istres, France
5/14/65

SPEED OVER A 2,000 KM CLOSED CIRCUIT WITHOUT PAYLOAD (FRANCE)
Pierre Bonneau 501.42 kmh
M. Duberger 311.56 mph
Marquis F-BJSI
2 Astazoux
Istres, France
5/13/65

SPEED AROUND THE WORLD, EASTBOUND (FRANCE)
Jacques Lemaigre du Breuil 403.03 kmh
Nicolas Gorodiche 250.43 mph
Olivier Waisblat
Socata TBM 700
6/13-17/93

TIME TO CLIMB

TIME TO 3,000 METERS (USA)
William C. Brodbeck 2 min
Ayres S2R-T65 45 sec
1 P&W PT6A-65R, 1,230 shp
Albany, Georgia
11/27/85

TIME TO 6,000 METERS (USA)
William C. Brodbeck 6 min
Ayres S2R-T65 47 sec
1 P&W PT6A-65R, 1,230 shp
Albany, Georgia
11/27/85

TIME TO 9,000 METERS (USA)
William C. Brodbeck 14 min
Ayres S2R-T65 8 sec
1 P&W PT6A-65R, 1,230 shp
Albany, Georgia
11/27/85

TIME TO 12,000 METERS (USA)
Donald R. Wilson 27 min
LTV L450F 39 sec
1 PT6A-34
Greenville, TX
3/23/72

CLASS C-1.E (3,000-6,000 KG / 6,614-13,228 LBS) GROUP II (TURBOPROP)

DISTANCE

DISTANCE WITHOUT LANDING (W. GERMANY)
Egon Evertz, Pilot 6,171.35 km
Stefan Evertz, Navigator 3,834.88 mi
Gulfstream Commander 980
Toronto - Dusseldorf
8/13/88

ALTITUDE

ALTITUDE (USA)
Einar Envoldson 16,329 m
Egrett-1, N14ES 53,574 ft
1 Garrett TPE331-14A, 750 shp
Greenville, TX
9/1/88

ALTITUDE IN HORIZONTAL FLIGHT (USA)
Einar Envoldson 16,239 m
Egrett-1, N14ES 53,276 ft
1 Garrett TPE331-14A, 750 shp
Greenville, TX
9/1/88

ALTITUDE WITH 1,000 KG PAYLOAD (GERMANY)
Werner Kraut 15,552 m
Burkhart Grob G-520 Egrett 51,023 ft
1 Garrett TPE331, 750 shp
Mindelheim, Germany
3/31/94

ALTITUDE WITH 2,000 KG PAYLOAD (USSR)
Sergi Gorbik 6,150 m
Antonov An-3 20,177 ft
1 TBD-20, 1,450 shp
Podkievskoie
12/13/85

GREATEST PAYLOAD CARRIED TO 2,000 METERS (USSR)
Sergei Gorbik 2,375.00 kg
Antonov An-3 5,235.92 lbs
1 TBD-20, 1,450 shp
Podkievskoie
12/13/85

SPEED

SPEED OVER A 100 KM CLOSED CIRCUIT WITHOUT PAYLOAD (W. GERMANY)
Joachim H. Blumschein 571.43 kmh
Gulfstream Commander 680 355.07 mph
2 Garrett TPE331-10-510K, 710 shp
Leine
5/22/82

SPEED OVER A 500 KM CLOSED CIRCUIT WITHOUT PAYLOAD (W. GERMANY)
Joachim H. Blumschein 571.43 kmh
Gulfstream Commander 680 355.07 mph
2 Garrett TPE331-10-510K, 710 shp
Leine
5/22/82

SPEED OVER A 1,000 KM CLOSED CIRCUIT WITHOUT PAYLOAD (W. GERMANY)
Joachim H. Blumschein 572.08 kmh
Gulfstream Commander 680 355.47 mph
2 Garret TPE331-10-510K, 710 shp
5/22/82

SPEED OVER A 2,000 KM CLOSED CIRCUIT WITHOUT PAYLOAD (W. GERMANY)
Joachim H. Blumschein 569.85 kmh
Gulfstream Commander 680 354.09 mph
2 Garrett TPE331-10-510K, 710 shp
Leine
5/22/82

SPEED AROUND THE WORLD, EASTBOUND (USA)
Joe Harnish, Pilot 490.95 kmh
David B. Webster, Pilot 304.80 mph
Gulfstream Commander 695A
2 Garrett TPE-331-501K, 820 shp
Elkhart, Goose Bay, Keflavik, Vienna, Cairo, Luxor, Sharjah, Colombo, Singapore, Manila, Agana, Wake Island, Midway Island, Honolulu, San Francisco, Elkhart
3/21-24/83

TIME TO CLIMB

TIME TO 3,000 METERS (USA)
Charles "Chuck" Yeager, Pilot — 1 min 48 sec
Renald Davenport, Copilot
Piper Cheyenne 400LS
Portland, OR
4/16/85

TIME TO 6,000 METERS (USA)
Charles "Chuck" Yeager, Pilot — 3 min 43 sec
Renald Davenport, Copilot
Piper Cheyenne 400LS
Portland, OR
4/16/85

TIME TO 9,000 METERS (USA)
Charles "Chuck" Yeager, Pilot — 6 min 34 sec
Renald Davenport, Copilot
Piper Cheyenne 400LS
Portland, OR
4/16/85

TIME TO 12,000 METERS (USA)
Charles "Chuck" Yeager, Pilot — 11 min 8 sec
Renald Davenport, Copilot
Piper Cheyenne 400LS
Portland, OR
4/16/85

TIME TO 15,000 METERS (USA)
Einar Envoldson — 40 min 47 sec
Egrett-1, N14ES
1 Garrett TPE331-14A, 750 shp
Greenville, TX
9/1/88

CLASS C-1.F (6,000-9,000 KG / 13,228-19,842 LBS) GROUP II (TURBOPROP)

DISTANCE

DISTANCE WITHOUT LANDING (USA)
Donald T. Holmes, Pilot — 6,625.45 km
Erdoğan Menekse, Copilot — 4,117.05 mi
Piper Cheyenne IIIA
2 P&W PT6A-41, 720 shp
Gander - Ankara
12/17-18/85

ALTITUDE

ALTITUDE IN HORIZONTAL FLIGHT (USA)
James F. Peters — 9,753 m
Grumman OV-1 Mohawk — 32,000 ft
2 Lycoming T-53-L-7
Peconic River, Long Island, NY
6/16/66

ALTITUDE WITH 1,000 KG PAYLOAD (USA)
L. Michael Billow — 10,516 m
M. Gunner Kyle — 34,501 ft
Beech C-12T3
2 P&W PT6A, 850 shp
Reno, NV
9/15/02

ALTITUDE WITH 2,000 KG PAYLOAD (USSR)
Vladimir Lysenko — 6,100 m
Antonov An-3 — 20,013 ft
1 Glushenkov TVD-20, 1,450 shp
Podkievskoie, USSR
12/12/85

GREATEST PAYLOAD CARRIED TO 2,000 METERS (USSR)
Vladimir Lysenko — 2,583.00 kg
Antonov An-3 — 5,694.48 lbs
1 TBD-20, 1,450 shp
Podkievskoie, USSR
12/12/85

SPEED

SPEED OVER A 100 KM CLOSED CIRCUIT WITHOUT PAYLOAD (USA)
LTC E. L. Nielsen, USA — 472.21 kmh
Grumman OV-1 Mohawk — 293.41 mph
2 Lycoming T-53-L-7
Peconic River, Long Island, NY
6/17/66

TIME TO CLIMB

TIME TO 3,000 METERS (USA)
James F. Peters — 3 min 45 sec
Grumman OV-1 Mohawk
2 Lycoming T-53-L-7
Peconic River, Long Island, NY
6/16/66

TIME TO 6,000 METERS (USA)
James F. Peters — 9 min 9 sec
Grumman OV-1 Mohawk
2 Lycoming T-53-L-7
Peconic River, Long Island, NY
6/16/66

CLASS C-1.G (9,000-12,000 KG / 19,842-26,455 LBS) GROUP II (TURBOPROP)

ALTITUDE

ALTITUDE (USA) ★
William G. Walker — 10,892 m
Wyatt C. Ingram — 35,735 ft
Marsh S-2F3T Turbotracker
2 Garrett TPE331, 1,645 shp
8/31/93

SPEED

SPEED OVER A 100 KM CLOSED CIRCUIT (USA) ★
William G. Walker — 454.53 kmh
Wyatt C. Ingram — 282.43 mph
Marsh S-2F3T Turbotracker
2 Garrett TPE331, 1,645 shp
8/31/93

TIME TO CLIMB

TIME TO 3,000 METERS (USA) ★
William G. Walker — 3 min 40 sec
Wyatt C. Ingram
Marsh S-2F3T Turbotracker
2 Garrett TPE331, 1,645 shp
8/30/93

CLASS C-1.H (12,000-16,000 KG / 26,455-35,274 LBS) GROUP II (TURBOPROP)

ALTITUDE

ALTITUDE (USSR)
Marina Popovich, Pilot — 11,050 m
Galina Kortchuganova, Copilot — 36,255 ft
Antonov An-24
2 AI-24, 2,820 shp
Podmoscovnoe, USSR
5/7/82

ALTITUDE IN HORIZONTAL FLIGHT (USSR)
Marina Popovich, Pilot 10,920 m
Galina Kortchuganova, Copilot 35,829 ft
Antonov An-24
2 AI-24, 2,820 shp
Podmoscovnoe, USSR
5/11/82

SPEED

SPEED OVER A 100 KM CLOSED CIRCUIT (USSR)
Marina Popovich, Pilot 502 kmh
Galina Kortchuganova, Copilot 312 mph
Antonov An-24
2 AI-24, 2,820 shp
Podmoscovnoe, USSR
7/14/82

TIME TO CLIMB

TIME TO 3,000 METERS (CANADA)
T. E. Appleton 2 min
de Havilland DHC-5D Buffalo 12 sec
2 GE CT64-820-4
Downsview, Ontario
2/16/76

TIME TO 6,000 METERS (CANADA)
T. E. Appleton 4 min
de Havilland DHC-5D Buffalo 27 sec
2 GE CT64-820-4
Downsview, Ontario
2/16/76

TIME TO 9,000 METERS (CANADA)
T. E. Appleton 8 min
de Havilland DHC-5D Buffalo 3 sec
2 GE CT64-820-4
Downsview, Ontario
2/16/76

CLASS C-1.1 (16,000-20,000 KG / 35,274-44,092 LBS) GROUP II (TURBOPROP)

ALTITUDE

ALTITUDE (USA) ✯
LT Steve Schmeiser, USN, Pilot 12,517 m
LCDR Matt Klunder, USN, Copilot 41,068 ft
Grumman E-2C Plus
2 Allison T56-A-427, 5,250 shp
NAS Patuxent River, MD
12/19/91

ALTITUDE IN HORIZONTAL FLIGHT (USA)
LT Steve Schmeiser, USN, Pilot 12,150 m
LCDR Matt Klunder, USN, Copilot 39,861 ft
Grumman E-2C Plus
2 Allison T56-A-427, 5,250 shp
NAS Patuxent River, MD
12/19/91

ALTITUDE WITH 1,000 KG PAYLOAD (USA)
LT Steve Schmeiser, USN, Pilot 12,517 m
LCDR Matt Klunder, USN, Copilot 41,068 ft
Grumman E-2C Plus
2 Allison T56-A-427, 5,250 shp
NAS Patuxent River, MD
12/19/91

ALTITUDE WITH 2,000 KG PAYLOAD (USA)
LCDR Matt Klunder, USN, Pilot 12,107 m
LT Steve Schmeiser, USN, Copilot 39,722 ft
Grumman E-2C Plus
2 Allison T56-A-427, 5,250 shp
NAS Patuxent River, MD
12/18/91

SPEED

SPEED OVER A 100 KM CLOSED CIRCUIT WITHOUT PAYLOAD (USA)
LCDR Matt Klunder, USN, Pilot 600.00 kmh
LT Pete Tomczak, USN, Copilot 372.82 mph
Grumman E-2C Plus
2 Allison T56-A-427, 5,250 shp
NAS Patuxent River, MD
12/17/91

SPEED OVER A 100 KM CLOSED CIRCUIT WITH 1,000 KG PAYLOAD (USA)
LCDR Matt Klunder, USN, Pilot 600.00 kmh
LT Pete Tomczak, USN, Copilot 372.82 mph
Grumman E-2C Plus
2 Allison T56-A-427, 5,250 shp
NAS Patuxent River, MD
12/17/91

SPEED OVER A 100 KM CLOSED CIRCUIT WITH 2,000 KG PAYLOAD (USA)
LCDR Matt Klunder, USN, Pilot 600.00 kmh
LT Pete Tomczak, USN, Copilot 372.82 mph
Grumman E-2C Plus
2 Allison T56-A-427, 5,250 shp
NAS Patuxent River, MD
12/17/91

SPEED OVER A 500 KM CLOSED CIRCUIT WITHOUT PAYLOAD (USSR)
Galina Kortchuganova, Pilot 516.90 kmh
Marina Popovich, Copilot 321.20 mph
Antonov An-24
2 AI-24, 2,820 shp
Podmoscovnoe, USSR
6/12/82

SPEED OVER A 1,000 KM CLOSED CIRCUIT WITHOUT PAYLOAD (USSR)
Galina Kortchuganova, Pilot 517.20 kmh
Marina Popovich, Copilot 321.40 mph
Antonov An-24
2 AI-24, 2,820 shp
Podmoscovnoe, USSR
6/17/82

SPEED OVER A 2,000 KM CLOSED CIRCUIT WITHOUT PAYLOAD (USSR)
Marina Popovich, Pilot 490 kmh
Galina Kortchuganova, Copilot 304 mph
Antonov An-24
2 AI-24, 2,820 shp
Podmoscovnoe, USSR
7/13/82

TIME TO CLIMB

TIME TO 3,000 METERS (SWEDEN)
Gideon Singer 2 min
Kjell Nordstrom 25.8 sec
Saab 2000
5/19/93

TIME TO 6,000 METERS (SWEDEN)
Gideon Singer 4 min
Kjell Nordstrom 45.2 sec
Saab 2000
5/19/93

TIME TO 9,000 METERS (SWEDEN)
Gideon Singer 8 min
Kjell Nordstrom 00.8 sec
Saab 2000
5/19/93

CLASS C-1.J (20,000-25,000 KG / 44,092-55,116 LBS)
GROUP II (TURBOPROP)

DISTANCE

DISTANCE OVER A CLOSED CIRCUIT WITHOUT LANDING (USSR)
Lev Vasilyevich Kozlov 2,250.71 km
Antonov An-26 1,398.52 mi
2 Ivchenko AI-24, 2,820 shp
Akhtubinsk, Russia
10/8/91

ALTITUDE

ALTITUDE (USA)
LCDR Eric Hinger, USN, Pilot 12,574 m
LCDR Matt Klunder, USN, Copilot 41,254 ft
Grumman E-2C Plus
2 Allison T56-A-427, 5,250 shp
NAS Patuxent River, MD
12/19/91

ALTITUDE IN HORIZONTAL FLIGHT (USA)
LCDR Eric Hinger, USN, Pilot 12,347 m
LCDR Matt Klunder, USN, Copilot 40,510 ft
Grumman E-2C Plus
2 Allison T56-A-427, 5,250 shp
NAS Patuxent River, MD
12/19/91

ALTITUDE WITH 1,000 KG PAYLOAD (USA)
LCDR Eric Hinger, USN, Pilot 12,574 m
LCDR Matt Klunder, USN, Copilot 41,254 ft
Grumman E-2C Plus
2 Allison T56-A-427, 5,250 shp
NAS Patuxent River, MD
12/19/91

ALTITUDE WITH 2,000 KG PAYLOAD (USA)
LT Eric Hinger, USN, Pilot 12,178 m
LT Steve Schmeiser, USN, Copilot 39,954 ft
Grumman E-2C Plus
2 Allison T56-A-427, 5,250 shp
NAS Patuxent River, MD
12/18/91

ALTITUDE WITH 5,000 KG PAYLOAD(USSR)
Petr Kiritchuk, Pilot 11,230 m
Alexandre Tkachenko, Copilot 36,844 ft
Antonov An-32
2 Ivchenko AI-20DM, 5,180 shp
Podkievskoie, USSR
11/5/85

GREATEST PAYLOAD CARRIED TO 2,000 METERS (USSR)
Marina Popovich, Pilot 8,096 kg
Galina Kortchuganova, Copilot 17,844 lbs
Antonov An-24
2 AI-24, 2,820 shp
Podmoscovnoe, USSR
7/7/82

SPEED

SPEED OVER A 100 KM CLOSED CIRCUIT WITHOUT PAYLOAD (USA)
LCDR Eric Hinger, USN, Pilot 595.04 kmh
LCDR Matt Klunder, USN, Copilot 369.74 mph
Grumman E-2C Plus
2 Allison T56-A-427, 5,250 shp
NAS Patuxent River, MD
12/19/91

SPEED OVER A 100 KM CLOSED CIRCUIT WITH 1,000 KG PAYLOAD (USA)
LCDR Eric Hinger, USN, Pilot 595.04 kmh
LCDR Matt Klunder, USN, Copilot 369.74 mph
Grumman E-2C Plus
2 Allison T56-A-427, 5,250 shp
NAS Patuxent River, MD
12/19/91

SPEED OVER A 100 KM CLOSED CIRCUIT WITH 2,000 KG PAYLOAD (USA)
LT Eric Hinger, USN, Pilot 593.08 kmh
LT Steve Schmeiser, USN, Copilot 368.52 mph
Grumman E-2C Plus
2 Allison T56-A-427, 5,250 shp
NAS Patuxent River, MD
12/18/91

SPEED OVER A 500 KM CLOSED CIRCUIT WITHOUT PAYLOAD (USSR)
Galina Kortchuganova, Pilot 519.20 kmh
Marina Popovich, Copilot 322.60 mph
Antonov An-24
2 Ivchenko AI-24, 2,820 shp
Podmoscovnoe, USSR
6/11/82

SPEED OVER A 1,000 KM CLOSED CIRCUIT WITHOUT PAYLOAD (RUSSIA)
Alexander Butakov 384.58 kmh
Antonov An-26 238.96 mph
2 Ivchenko AI-24VT, 2,820 shp
12/13/91

SPEED OVER A 2,000 KM CLOSED CIRCUIT WITHOUT PAYLOAD (USSR)
Galina Kortchuganova, Pilot 492.80 kmh
Marina Popovich, Copilot 306.20 mph
Antonov An-24
2 Ivchenko AI-24, 2,820 shp
Podmoscovnoe, USSR
7/9/82

TIME TO CLIMB

TIME TO 3,000 METERS (USA)
LT Eric Hinger, USN, Pilot 3 min
LT Steve Schmeiser, USN, Copilot 1.2 sec
Grumman E-2C Plus
2 Allison T56-A-427, 5,250 shp
NAS Patuxent River, MD
12/18/91

TIME TO 6,000 METERS (USA)
LT Eric Hinger, USN, Pilot 6 min
LT Steve Schmeiser, USN, Copilot 16.0 sec
Grumman E-2C Plus
2 Allison T56-A-427, 5,250 shp
NAS Patuxent River, MD
12/18/91

TIME TO 9,000 METERS (USA)
LT Eric Hinger, USN, Pilot 11 min
LT Steve Schmeiser, USN, Copilot 24.8 sec
Grumman E-2C Plus
2 Allison T56-A-427, 5,250 shp
NAS Patuxent River, MD
12/18/91

CLASS C-1.K (25,000-35,000 KG / 55,116-77,162 LBS)
GROUP II (TURBOPROP)

ALTITUDE

ALTITUDE (USSR)
Youri Kourline, Pilot — 10,940 m
Gueorgi Pobol, Copilot — 35,894 ft
Antonov An-32
2 Ivchenko AI-20DM, 5,180 shp
Podkievskoie, USSR
10/24/85

ALTITUDE IN HORIZONTAL FLIGHT (USSR)
Youri Kourline, Pilot — 10,420 m
Petr Kiritchuk, Copilot — 34,188 ft
Antonov An-32
2 Ivchenko AI-20DM, 5,180 shp
Podkievskoie, USSR
10/21/85

ALTITUDE WITH 1,000 KG PAYLOAD (USSR)
Alexandre Tkachenko, Pilot — 11,120 m
Vladimir Lysenko, Copilot — 36,485 ft
Antonov An-32
2 Ivchenko AI-20DM, 5,180 shp
Podkievskoie, USSR
10/28/85

ALTITUDE WITH 2,000 KG PAYLOAD (USSR)
Alexandre Tkachenko, Pilot — 10,890 m
Vladimir Lysenko, Copilot — 35,730 ft
Antonov An-32
2 Ivchenko AI-20DM, 5,180 shp
Podkievskoie, USSR
10/28/85

ALTITUDE WITH 5,000 KG PAYLOAD (USSR)
Petr Kiritchuk, Pilot — 10,510 m
Alexandre Tkachenko, Copilot — 34,483 ft
Antonov An-32
2 Ivchenko AI-20DM, 5,180 shp
Podkievskoie, USSR
11/4/85

GREATEST PAYLOAD CARRIED TO 2,000 METERS (USSR)
Petr Kiritchuk, Pilot — 7,256.00 kg
Alexandre Tkachenko, Copilot — 15,992.95 lbs
Antonov An-32
2 Ivchenko AI-20DM, 5,180 shp
Podkievskoie, USSR
11/4/85

SPEED

(No records registered.)

CLASS C-1.L (35,000-45,000 KG / 77,162-99,208 LBS)
GROUP II (TURBOPROP)

ALTITUDE

ALTITUDE (RUSSIA)
Victor Krasilnikov — 11,590 m
Antonov An-12 — 38,025 ft
10/2/91

ALTITUDE IN HORIZONTAL FLIGHT (RUSSIA)
Oleg Koudryashov — 10,870 m
Antonov An-12 — 35,663 ft
3/25/92

SPEED

SPEED OVER A 100 KM CLOSED CIRCUIT WITHOUT PAYLOAD (RUSSIA)
Alexander A. Glyadkov — 546.39 kmh
Antonov An-12 — 339.51 mph
11/20/91

SPEED OVER A 500 KM CLOSED CIRCUIT WITHOUT PAYLOAD (RUSSIA)
Lev V. Kozlov — 574.15 kmh
Antonov An-12 — 356.76 mph
10/9/91

SPEED OVER A 1,000 KM CLOSED CIRCUIT WITHOUT PAYLOAD (USSR)
Valery V. Kunesky — 593.17 kmh
Antonov An-12 — 368.57 mph
7/22/91

CLASS C-1.M (45,000-60,000 KG / 99,208-132,277 LBS)
GROUP II (TURBOPROP)

ALTITUDE

ALTITUDE (USSR)
Anatoly Rudj, Pilot — 10,720 m
Alexander A. Gliadkov, Copilot — 35,171 ft
Antonov An-12
3/5/91

ALTITUDE IN HORIZONTAL FLIGHT (RUSSIA)
Victor Krasilnikov — 10,070 m
Antonov An-12 — 33,038 ft
2/5/92

ALTITUDE WITH 1,000 KG PAYLOAD (USSR)
Anatoly Rudj, Pilot — 10,720 m
Alexander A. Gliadkov, Copilot — 35,171 ft
Antonov An-12
3/5/91

ALTITUDE WITH 2,000 KG PAYLOAD (USSR)
Anatoly Rudj, Pilot — 10,720 m
Alexander A. Gliadkov, Copilot — 35,171 ft
Antonov An-12
3/5/91

ALTITUDE WITH 5,000 KG PAYLOAD (USSR)
Anatoly Rudj, Pilot — 10,720 m
Alexander A. Gliadkov, Copilot — 35,171 ft
Antonov An-12
3/5/91

ALTITUDE WITH 10,000 KG PAYLOAD (USSR)
Anatoly Rudj, Pilot — 10,720 m
Alexander A. Gliadkov, Copilot — 35,171 ft
Antonov An-12
3/5/91

SPEED

SPEED OVER A 500 KM CLOSED CIRCUIT WITHOUT PAYLOAD (RUSSIA)
Anatoli Rudj — 560.23 kmh
Antonov An-12 — 348.11 mph
12/13/91

SPEED OVER A 1,000 KM CLOSED CIRCUIT WITHOUT PAYLOAD (USSR)
Evgenij Bistrov, Pilot — 587.53 kmh
Alexei Marenkov, Copilot — 365.07 mph
Antonov An-12
3/22/91

SPEED OVER A 1,000 KM CLOSED CIRCUIT WITH 1,000 KG PAYLOAD (USSR)
Evgenij Bistrov, Pilot 587.53 kmh
Alexei Marenkov, Copilot 365.07 mph
Antonov An-12
3/22/91

SPEED OVER A 1,000 KM CLOSED CIRCUIT WITH 2,000 KG PAYLOAD (USSR)
Evgenij Bistrov, Pilot 587.53 kmh
Alexei Marenkov, Copilot 365.07 mph
Antonov An-12
3/22/91

SPEED OVER A 1,000 KM CLOSED CIRCUIT WITH 5,000 KG PAYLOAD (USSR)
Evgenij Bistrov, Pilot 587.53 kmh
Alexei Marenkov, Copilot 365.07 mph
Antonov An-12
3/22/91

SPEED OVER A 1,000 KM CLOSED CIRCUIT WITH 10,000 KG PAYLOAD (USSR)
Evgenij Bistrov, Pilot 587.53 kmh
Alexei Marenkov, Copilot 365.07 mph
Antonov An-12
3/22/91

SPEED OVER A 2,000 KM CLOSED CIRCUIT WITHOUT PAYLOAD (USSR)
Alexander Taradin, Pilot 543.57 kmh
Vjatcheslav Fatkulin, Copilot 337.75 mph
Antonov An-12
3/27/91

SPEED OVER A 2,000 KM CLOSED CIRCUIT WITH 1,000 KG PAYLOAD (USSR)
Alexander Taradin, Pilot 543.57 kmh
Vjatcheslav Fatkulin, Copilot 337.75 mph
Antonov An-12
3/27/91

SPEED OVER A 2,000 KM CLOSED CIRCUIT WITH 2,000 KG PAYLOAD (USSR)
Alexander Taradin, Pilot 543.57 kmh
Vjatcheslav Fatkulin, Copilot 337.75 mph
Antonov An-12
3/27/91

SPEED OVER A 2,000 KM CLOSED CIRCUIT WITH 5,000 KG PAYLOAD (USSR)
Alexander Taradin, Pilot 543.57 kmh
Vjatcheslav Fatkulin, Copilot 337.75 mph
Antonov An-12
3/27/91

SPEED OVER A 2,000 KM CLOSED CIRCUIT WITH 10,000 KG PAYLOAD (USSR)
Alexander Taradin, Pilot 543.57 kmh
Vjatcheslav Fatkulin, Copilot 337.75 mph
Antonov An-12
3/27/91

CLASS C-1.N (60,000-80,000 KG / 132,277-176,370 LBS) GROUP II (TURBOPROP)

ALTITUDE

ALTITUDE (USA)
Lyle H. Schaefer, Pilot 11,143 m
Arlen D. Rens, Copilot 36,560 ft
Timothy L. Gomez, Aircraft Systems Specialist
Lockheed Martin C-130J
2 RR Allison AE 2100, 4,591 shp
Marietta, GA
4/20/99

ALTITUDE WITH 1,000 KG PAYLOAD (USA)
Lyle H. Schaefer, Pilot 11,143 m
Arlen D. Rens, Copilot 36,560 ft
Timothy L. Gomez, Aircraft Systems Specialist
Lockheed Martin C-130J
2 RR Allison AE 2100, 4,591 shp
Marietta, GA
4/20/99

ALTITUDE WITH 2,000 KG PAYLOAD (USA)
Lyle H. Schaefer, Pilot 11,143 m
Arlen D. Rens, Copilot 36,560 ft
Timothy L. Gomez, Aircraft Systems Specialist
Lockheed Martin C-130J
2 RR Allison AE 2100, 4,591 shp
Marietta, GA
4/20/99

ALTITUDE WITH 5,000 KG PAYLOAD (USA)
Lyle H. Schaefer, Pilot 11,143 m
Arlen D. Rens, Copilot 36,560 ft
Timothy L. Gomez, Aircraft Systems Specialist
Lockheed Martin C-130J
2 RR Allison AE 2100, 4,591 shp
Marietta, GA
4/20/99

ALTITUDE WITH 10,000 KG PAYLOAD (USA)
Lyle H. Schaefer, Pilot 11,143 m
Arlen D. Rens, Copilot 36,560 ft
Timothy L. Gomez, Aircraft Systems Specialist
Lockheed Martin C-130J
2 RR Allison AE 2100, 4,591 shp
Marietta, GA
4/20/99

ALTITUDE WITH 15,000 KG PAYLOAD (USA)
Lyle H. Schaefer, Pilot 11,143 m
Arlen D. Rens, Copilot 36,560 ft
Timothy L. Gomez, Aircraft Systems Specialist
Lockheed Martin C-130J
2 RR Allison AE 2100, 4,591 shp
Marietta, GA
4/20/99

ALTITUDE WITH 20,000 KG PAYLOAD (USA)
Lyle H. Schaefer, Pilot 11,143 m
Arlen D. Rens, Copilot 36,560 ft
Timothy L. Gomez, Aircraft Systems Specialist
Lockheed Martin C-130J
2 RR Allison AE 2100, 4,591 shp
Marietta, GA
4/20/99

SPEED

SPEED OVER A 1,000 KM CLOSED CIRCUIT WITHOUT PAYLOAD (USA)
Arlen D. Rens, Pilot 637.58 kmh
Lyle H. Schaefer, Copilot 396.17 mph
Timothy L. Gomez, Aircraft Systems Specialist
Lockheed Martin C-130J
2 RR Allison AE 2100, 4,591 shp
Marietta, GA
4/20/99

SPEED OVER A 1,000 KM CLOSED CIRCUIT WITH 1,000 KG PAYLOAD (USA)
Arlen D. Rens, Pilot 637.58 kmh
Lyle H. Schaefer, Copilot 396.17 mph
Timothy L. Gomez, Aircraft Systems Specialist
Lockheed Martin C-130J
2 RR Allison AE 2100, 4,591 shp
Marietta, GA
4/20/99

SPEED OVER A 1,000 KM CLOSED CIRCUIT WITH 2,000 KG PAYLOAD (USA)
Arlen D. Rens, Pilot 637.58 kmh
Lyle H. Schaefer, Copilot 396.17 mph
Timothy L. Gomez, Aircraft Systems Specialist
Lockheed Martin C-130J
2 RR Allison AE 2100, 4,591 shp
Marietta, GA
4/20/99

SPEED OVER A 1,000 KM CLOSED CIRCUIT WITH 5,000 KG PAYLOAD (USA)
Arlen D. Rens, Pilot 637.58 kmh
Lyle H. Schaefer, Copilot 396.17 mph
Timothy L. Gomez, Aircraft Systems Specialist
Lockheed Martin C-130J
2 RR Allison AE 2100, 4,591 shp
Marietta, GA
4/20/99

SPEED OVER A 1,000 KM CLOSED CIRCUIT WITH 10,000 KG PAYLOAD (USA)
Arlen D. Rens, Pilot 637.58 kmh
Lyle H. Schaefer, Copilot 396.17 mph
Timothy L. Gomez, Aircraft Systems Specialist
Lockheed Martin C-130J
2 RR Allison AE 2100, 4,591 shp
Marietta, GA
4/20/99

SPEED OVER A 1,000 KM CLOSED CIRCUIT WITH 15,000 KG PAYLOAD (USA)
Arlen D. Rens, Pilot 637.58 kmh
Lyle H. Schaefer, Copilot 396.17 mph
Timothy L. Gomez, Aircraft Systems Specialist
Lockheed Martin C-130J
2 RR Allison AE 2100, 4,591 shp
Marietta, GA
4/20/99

SPEED OVER A 1,000 KM CLOSED CIRCUIT WITH 20,000 KG PAYLOAD (USA)
Arlen D. Rens, Pilot 637.58 kmh
Lyle H. Schaefer, Copilot 396.17 mph
Timothy L. Gomez, Aircraft Systems Specialist
Lockheed Martin C-130J
2 RR Allison AE 2100, 4,591 shp
Marietta, GA
4/20/99

SPEED OVER A 2,000 KM CLOSED CIRCUIT WITHOUT PAYLOAD (USA)
Arlen D. Rens, Pilot 635.49 kmh
Lyle H. Schaefer, Copilot 394.87 mph
Timothy L. Gomez, Aircraft Systems Specialist
Lockheed Martin C-130J
2 RR Allison AE 2100, 4,591 shp
Marietta, GA
4/20/99

SPEED OVER A 2,000 KM CLOSED CIRCUIT WITH 1,000 KG PAYLOAD (USA)
Arlen D. Rens, Pilot 635.49 kmh
Lyle H. Schaefer, Copilot 394.87 mph
Timothy L. Gomez, Aircraft Systems Specialist
Lockheed Martin C-130J
2 RR Allison AE 2100, 4,591 shp
Marietta, GA
4/20/99

SPEED OVER A 2,000 KM CLOSED CIRCUIT WITH 2,000 KG PAYLOAD (USA)
Arlen D. Rens, Pilot 635.49 kmh
Lyle H. Schaefer, Copilot 394.87 mph
Timothy L. Gomez, Aircraft Systems Specialist
Lockheed Martin C-130J
2 RR Allison AE 2100, 4,591 shp
Marietta, GA
4/20/99

SPEED OVER A 2,000 KM CLOSED CIRCUIT WITH 5,000 KG PAYLOAD (USA)
Arlen D. Rens, Pilot 635.49 kmh
Lyle H. Schaefer, Copilot 394.87 mph
Timothy L. Gomez, Aircraft Systems Specialist
Lockheed Martin C-130J
2 RR Allison AE 2100, 4,591 shp
Marietta, GA
4/20/99

SPEED OVER A 2,000 KM CLOSED CIRCUIT WITH 10,000 KG PAYLOAD (USA)
Arlen D. Rens, Pilot 635.49 kmh
Lyle H. Schaefer, Copilot 394.87 mph
Timothy L. Gomez, Aircraft Systems Specialist
Lockheed Martin C-130J
2 RR Allison AE 2100, 4,591 shp
Marietta, GA
4/20/99

SPEED OVER A 2,000 KM CLOSED CIRCUIT WITH 15,000 KG PAYLOAD (USA)
Arlen D. Rens, Pilot 635.49 kmh
Lyle H. Schaefer, Copilot 394.87 mph
Timothy L. Gomez, Aircraft Systems Specialist
Lockheed Martin C-130J
2 RR Allison AE 2100, 4,591 shp
Marietta, GA
4/20/99

SPEED OVER A 2,000 KM CLOSED CIRCUIT WITH 20,000 KG PAYLOAD (USA)
Arlen D. Rens, Pilot 635.49 kmh
Lyle H. Schaefer, Copilot 394.87 mph
Timothy L. Gomez, Aircraft Systems Specialist
Lockheed Martin C-130J
2 RR Allison AE 2100, 4,591 shp
Marietta, GA
4/20/99

CLASS C-1.P (100,000-150,000 KG / 220,462-330,693 LBS) GROUP II (TURBOPROP)

ALTITUDE

ALTITUDE (USSR)
V. E. Egorov, Pilot 12,265 m
M. M. Bachkirov, Copilot 40,240 ft
S.N. Martyanov, Navigator
Tupolev Tu-142
4 Kuznetsov NK-12MA, 15,000 shp
10/05/89

ALTITUDE WITH 1,000 KG PAYLOAD (USSR)
V. E. Egorov, Pilot 12,265 m
M. M. Bachkirov, Copilot 40,240 ft
S.N. Martyanov, Navigator
Tupolev Tu-142
4 Kuznetsov NK-12MA, 15,000 shp
10/05/89

ALTITUDE WITH 2,000 KG PAYLOAD (USSR)
V. E. Egorov, Pilot 12,265 m
M. M. Bachkirov, Copilot 40,240 ft
S.N. Martyanov, Navigator
Tupolev Tu-142
4 Kuznetsov NK-12MA, 15,000 shp
10/05/89

ALTITUDE WITH 5,000 KG PAYLOAD (USSR)
V. E. Egorov, Pilot 12,265 m
M. M. Bachkirov, Copilot 40,240 ft
S.N. Martyanov, Navigator
Tupolev Tu-142
4 Kuznetsov NK-12MA, 15,000 shp
10/05/89

ALTITUDE WITH 10,000 KG PAYLOAD (USSR)
Valery V. Alfeurov, Pilot 12,240 m
Viktor K. Nikolaev, Copilot 40,157 ft
Tupolev Tu-142 (No. 173)
11/28/90

ALTITUDE WITH 15,000 KG PAYLOAD (USSR)
Valery V. Alfeurov, Pilot 12,240 m
Viktor K. Nikolaev, Copilot 40,157 ft
Tupolev Tu-142 (No. 173)
11/28/90

ALTITUDE WITH 20,000 KG PAYLOAD (USSR)
Valery V. Alfeurov, Pilot 12,240 m
Viktor K. Nikolaev, Copilot 40,157 ft
Tupolev Tu-142 (No. 173)
11/28/90

ALTITUDE WITH 25,000 KG PAYLOAD (USSR)
Mikhail M. Bachirov, Pilot 11,410 m
Viktor V. Samorodov, Copilot 37,434 ft
Tupolev Tu-142M
11/22/90

ALTITUDE WITH 30,000 KG PAYLOAD (USSR)
Mikhail M. Bachirov, Pilot 11,410 m
Viktor V. Samorodov, Copilot 37,434 ft
Tupolev Tu-142M
11/22/90

ALTITUDE WITH 35,000 KG PAYLOAD (UKRAINE)
Vitalii Horovienko 7,356 m
Igor Kurzantsev 24,133 ft
Antonov An-70
4 ZMKB Progress/Ivchenko D-27, 13,800 shp
Kiev, Ukraine
11/10/03

ALTITUDE WITH 40,000 KG PAYLOAD (UKRAINE)
Vitalii Horovienko 7,356 m
Igor Kurzantsev 24,133 ft
Antonov An-70
4 ZMKB Progress/Ivchenko D-27, 13,800 shp
Kiev, Ukraine
11/10/03

ALTITUDE WITH 45,000 KG PAYLOAD (UKRAINE)
Vitalii Horovienko 7,356 m
Igor Kurzantsev 24,133 ft
Antonov An-70
4 ZMKB Progress/Ivchenko D-27, 13,800 shp
Kiev, Ukraine
11/10/03

ALTITUDE WITH 50,000 KG PAYLOAD (UKRAINE)
Vitalii Horovienko 7,356 m
Igor Kurzantsev 24,133 ft
Antonov An-70
4 ZMKB Progress/Ivchenko D-27, 13,800 shp
Kiev, Ukraine
11/10/03

ALTITUDE WITH 55,000 KG PAYLOAD (UKRAINE)
Vitalii Horovienko 7,356 m
Igor Kurzantsev 24,133 ft
Antonov An-70
4 ZMKB Progress/Ivchenko D-27, 13,800 shp
Kiev, Ukraine
11/10/03

GREATEST PAYLOAD CARRIED TO 2,000 METERS (UKRAINE)
Vitalii Horovienko 55,063 kg
Igor Kurzantsev 121,393 lbs
Antonov An-70
4 ZMKB Progress/Ivchenko D-27, 13,800 shp
Kiev, Ukraine
11/10/03

SPEED

SPEED OVER A 1,000 KM CLOSED CIRCUIT WITHOUT PAYLOAD (USSR)
V. E. Mossolov, Pilot 807.37 kmh
I. A. Tchalov, Copilot 501.67 mph
VP-021 (Tu 95)
9/26/89

SPEED OVER A 1,000 KM CLOSED CIRCUIT WITH 1,000 KG PAYLOAD (USSR)
V. E. Mossolov, Pilot 807.37 kmh
I. A. Tchalov, Copilot 501.67 mph
VP-021 (Tu 95)
9/26/89

SPEED OVER A 1,000 KM CLOSED CIRCUIT WITH 2,000 KG PAYLOAD (USSR)
V. E. Mossolov, Pilot 807.37 kmh
I. A. Tchalov, Copilot 501.67 mph
VP-021 (Tu 95)
9/26/89

SPEED OVER A 1,000 KM CLOSED CIRCUIT WITH 5,000 KG PAYLOAD (USSR)
V. E. Mossolov, Pilot 807.37 kmh
I. A. Tchalov, Copilot 501.67 mph
VP-021 (Tu 95)
9/26/89

SPEED OVER A 1,000 KM CLOSED CIRCUIT WITH 10,000 KG PAYLOAD (USSR)
V. E. Mossolov, Pilot 807.37 kmh
I. A. Tchalov, Copilot 501.67 mph
VP-021 (Tu 95)
9/26/89

SPEED OVER A 1,000 KM CLOSED CIRCUIT WITH 15,000 KG PAYLOAD (USSR)
V. E. Mossolov, Pilot 807.37 kmh
I. A. Tchalov, Copilot 501.67 mph
VP-021 (Tu 95)
9/26/89

SPEED OVER A 1,000 KM CLOSED CIRCUIT WITH 20,000 KG PAYLOAD (USSR)
V. E. Mossolov, Pilot 807.37 kmh
I. A. Tchalov, Copilot 501.67 mph
VP-021 (Tu 95)
9/26/89

SPEED OVER A 1,000 KM CLOSED CIRCUIT WITH 25,000 KG PAYLOAD (USSR)
V. E. Mossolov, Pilot 807.37 kmh
I. A. Tchalov, Copilot 501.67 mph
VP-021 (Tu 95)
9/26/89

SPEED OVER A 1,000 KM CLOSED CIRCUIT WITH 30,000 KG PAYLOAD (USSR)
V. E. Mossolov, Pilot 807.37 kmh
I. A. Tchalov, Copilot 501.67 mph
VP-021 (Tu 95)
9/26/89

SPEED OVER A 2,000 KM CLOSED CIRCUIT WITHOUT PAYLOAD (USSR)
Viktor Naimuhin, Pilot 834.82 kmh
Sergei Osipov, Copilot 518.73 mph
Fedor Ivlev, Navigator
Tupolev Tu-142
9/28/89

SPEED OVER A 2,000 KM CLOSED CIRCUIT WITH 1,000 KG PAYLOAD (USSR)
Viktor Naimuhin, Pilot 834.82 kmh
Sergei Osipov, Copilot 518.73 mph
Fedor Ivlev, Navigator
Tupolev Tu-142
9/28/89

SPEED OVER A 2,000 KM CLOSED CIRCUIT WITH 2,000 KG PAYLOAD (USSR)
Viktor Naimuhin, Pilot 834.82 kmh
Sergei Osipov, Copilot 518.73 mph
Fedor Ivlev, Navigator
Tupolev Tu-142
9/28/89

SPEED OVER A 2,000 KM CLOSED CIRCUIT WITH 5,000 KG PAYLOAD (USSR)
Viktor Naimuhin, Pilot 834.82 kmh
Sergei Osipov, Copilot 518.73 mph
Fedor Ivlev, Navigator
Tupolev Tu-142
9/28/89

SPEED OVER A 2,000 KM CLOSED CIRCUIT WITH 10,000 KG PAYLOAD (USSR)
Viktor Naimuhin, Pilot 834.82 kmh
Sergei Osipov, Copilot 518.73 mph
Fedor Ivlev, Navigator
Tupolev Tu-142
9/28/89

SPEED OVER A 2,000 KM CLOSED CIRCUIT WITH 15,000 KG PAYLOAD (USSR)
Viktor Naimuhin, Pilot 834.82 kmh
Sergei Osipov, Copilot 518.73 mph
Fedor Ivlev, Navigator
Tupolev Tu-142
9/28/89

SPEED OVER A 2,000 KM CLOSED CIRCUIT WITH 20,000 KG PAYLOAD (USSR)
Viktor Naimuhin, Pilot 834.82 kmh
Sergei Osipov, Copilot 518.73 mph
Fedor Ivlev, Navigator
Tupolev Tu-142
9/28/89

SPEED OVER A 2,000 KM CLOSED CIRCUIT WITH 25,000 KG PAYLOAD (USSR)
Viktor Naimuhin, Pilot 834.82 kmh
Sergei Osipov, Copilot 518.73 mph
Fedor Ivlev, Navigator
Tupolev Tu-142
9/28/89

SPEED OVER A 2,000 KM CLOSED CIRCUIT WITH 30,000 KG PAYLOAD (USSR)
Viktor Naimuhin, Pilot 834.82 kmh
Sergei Osipov, Copilot 518.73 mph
Fedor Ivlev, Navigator
Tupolev Tu-142
9/28/89

SPEED OVER A 5,000 KM CLOSED CIRCUIT WITHOUT PAYLOAD (USSR)
V. R. Smelov, Pilot 786.10 kmh
V. A. Vaciliev, Navigator 488.46 mph
V. V. Ketov, Navigator
Tupolev Tu-142
4 Kuznetsov NK-12MA, 15,000 shp
10/11/89

SPEED OVER A 5,000 KM CLOSED CIRCUIT WITH 1,000 KG PAYLOAD (USSR)
V. R. Smelov, Pilot 786.10 kmh
V. A. Vaciliev, Navigator 488.46 mph
V. V. Ketov, Navigator
Tupolev Tu-142
4 Kuznetsov NK-12MA, 15,000 shp
10/11/89

SPEED OVER A 5,000 KM CLOSED CIRCUIT WITH 2,000 KG PAYLOAD (USSR)
V. R. Smelov, Pilot 786.10 kmh
V. A. Vaciliev, Navigator 488.46 mph
V. V. Ketov, Navigator
Tupolev Tu-142
4 Kuznetsov NK-12MA, 15,000 shp
10/11/89

SPEED OVER A 5,000 KM CLOSED CIRCUIT WITH 5,000 KG PAYLOAD (USSR)
V. R. Smelov, Pilot 786.10 kmh
V. A. Vaciliev, Navigator 488.46 mph
V. V. Ketov, Navigator
Tupolev Tu-142
4 Kuznetsov NK-12MA, 15,000 shp
10/11/89

TIME TO CLIMB

TIME TO 6,000 METERS (USSR)
Valery Vanshin 4 min
Alexandre Artiukhin 23 sec
Tupolev Tu-142
5/3/90

TIME TO 9,000 METERS (USSR)
Valery Vanshin 6 min
Alexandre Artiukhin 4 sec
Tupolev Tu-142
5/3/90

CLASS C-1.Q (150,000-200,000 KG / 330,693-440,924 LBS) GROUP II (TURBOPROP)

ALTITUDE

ALTITUDE (USSR)
Yuri Kabanov, Pilot 10,823 m
V. V. Alferov, Copilot 35,509 ft
V. A. Vaciliev and V. V. Ketov, Navigators
Tupolev Tu-142
4 Kuznetsov NK-12MA, 15,000 shp
10/05/89

ALTITUDE WITH 1,000 KG PAYLOAD (USSR)
Yuri Kabanov, Pilot 10,823 m
V. V. Alferov, Copilot 35,509 ft
V. A. Vaciliev and V. V. Ketov, Navigators
Tupolev Tu-142
4 Kuznetsov NK-12MA, 15,000 shp
10/05/89

ALTITUDE WITH 2,000 KG PAYLOAD (USSR)
Yuri Kabanov, Pilot 10,823 m
V. V. Alferov, Copilot 35,509 ft
V. A. Vaciliev and V. V. Ketov, Navigators
Tupolev Tu-142
4 Kuznetsov NK-12MA, 15,000 shp
10/05/89

ALTITUDE WITH 5,000 KG PAYLOAD (USSR)
Yuri Kabanov, Pilot 10,823 m
V. V. Alferov, Copilot 35,509 ft
V. A. Vaciliev and V. V. Ketov, Navigators
Tupolev Tu-142
4 Kuznetsov NK-12MA, 15,000 shp
10/05/89

ALTITUDE WITH 10,000 KG PAYLOAD (USSR)
Vladimir Bobileuv, Pilot 11,100 m
Yuri P. Makarov, Copilot 36,417 ft
Tupolev Tu-142 (No. 173)
11/28/90

ALTITUDE WITH 15,000 KG PAYLOAD (USSR)
Vladimir Bobileuv, Pilot 11,100 m
Yuri P. Makarov, Copilot 36,417 ft
Tupolev Tu-142 (No. 173)
11/28/90

ALTITUDE WITH 20,000 KG PAYLOAD (USSR)
Vladimir Bobileuv, Pilot 11,100 m
Yuri P. Makarov, Copilot 36,417 ft
Tupolev Tu-142 (No. 173)
11/28/90

ALTITUDE WITH 25,000 KG PAYLOAD (USSR)
Viktor K. Nikolaev, Pilot 10,110 m
Evgeni S. Kobyakov, Copilot 33,169 ft
Tupolev Tu-142 (No. 173)
11/16/90

ALTITUDE WITH 30,000 KG PAYLOAD (USSR)
Viktor K. Nikolaev, Pilot 10,110 m
Evgeni S. Kobyakov, Copilot 33,169 ft
Tupolev Tu-142 (No. 173)
11/16/90

SPEED

SPEED OVER A 1,000 KM CLOSED CIRCUIT WITHOUT PAYLOAD (USSR)
L. V. Kozlov, Pilot 816.25 kmh
S. S. Popov, Copilot 507.19 mph
G. P. Malzev, Navigator
Tupolev Tu-142
4 Kuznetsov NK-12MA, 15,000 shp
9/26/89

SPEED OVER A 1,000 KM CLOSED CIRCUIT WITH 1,000 KG PAYLOAD (USSR)
L. V. Kozlov, Pilot 816.25 kmh
S. S. Popov, Copilot 507.19 mph
G. P. Malzev, Navigator
Tupolev Tu-142
4 Kuznetsov NK-12MA, 15,000 shp
9/26/89

SPEED OVER A 1,000 KM CLOSED CIRCUIT WITH 2,000 KG PAYLOAD (USSR)
L. V. Kozlov, Pilot 816.25 kmh
S. S. Popov, Copilot 507.19 mph
G. P. Malzev, Navigator
Tupolev Tu-142
4 Kuznetsov NK-12MA, 15,000 shp
9/26/89

SPEED OVER A 1,000 KM CLOSED CIRCUIT WITH 5,000 KG PAYLOAD (USSR)
L. V. Kozlov, Pilot 816.25 kmh
S. S. Popov, Copilot 507.19 mph
G. P. Malzev, Navigator
Tupolev Tu-142
4 Kuznetsov NK-12MA, 15,000 shp
9/26/89

SPEED OVER A 1,000 KM CLOSED CIRCUIT WITH 10,000 KG PAYLOAD (USSR)
L. V. Kozlov, Pilot 816.25 kmh
S. S. Popov, Copilot 507.19 mph
G. P. Malzev, Navigator
Tupolev Tu-142
4 Kuznetsov NK-12MA, 15,000 shp
9/26/89

SPEED OVER A 1,000 KM CLOSED CIRCUIT WITH 15,000 KG PAYLOAD (USSR)
L. V. Kozlov, Pilot 816.25 kmh
S. S. Popov, Copilot 507.19 mph
G. P. Malzev, Navigator
Tupolev Tu-142
4 Kuznetsov NK-12MA, 15,000 shp
9/26/89

SPEED OVER A 1,000 KM CLOSED CIRCUIT WITH 20,000 KG PAYLOAD (USSR)
L. V. Kozlov, Pilot 816.25 kmh
S. S. Popov, Copilot 507.19 mph
G. P. Malzev, Navigator
Tupolev Tu-142
4 Kuznetsov NK-12MA, 15,000 shp
9/26/89

SPEED OVER A 1,000 KM CLOSED CIRCUIT WITH 25,000 KG PAYLOAD (USSR)
L. V. Kozlov, Pilot 816.25 kmh
S. S. Popov, Copilot 507.19 mph
G. P. Malzev, Navigator
Tupolev Tu-142
4 Kuznetsov NK-12MA, 15,000 shp
9/26/89

SPEED OVER A 1,000 KM CLOSED CIRCUIT WITH 30,000 KG PAYLOAD (USSR)
L. V. Kozlov, Pilot 816.25 kmh
S. S. Popov, Copilot 507.19 mph
G. P. Malzev, Navigator
Tupolev Tu-142
4 Kuznetsov NK-12MA, 15,000 shp
9/26/89

SPEED OVER A 2,000 KM CLOSED CIRCUIT WITHOUT PAYLOAD (USSR)
Viacheslav N. Gorelov, Pilot 813.13 kmh
Michael I. Pozdnyakov, Copilot 505.26 mph
Tupolev Tu-95
9/27/89

SPEED OVER A 2,000 KM CLOSED CIRCUIT WITH 1,000 KG PAYLOAD (USSR)
Viacheslav N. Gorelov, Pilot 813.13 kmh
Michael I. Pozdnyakov, Copilot 505.26 mph
Tupolev Tu-95
9/27/89

SPEED OVER A 2,000 KM CLOSED CIRCUIT WITH 2,000 KG PAYLOAD (USSR)
Viacheslav N. Gorelov, Pilot 813.13 kmh
Michael I. Pozdnyakov, Copilot 505.26 mph
Tupolev Tu-95
9/27/89

SPEED OVER A 2,000 KM CLOSED CIRCUIT WITH 5,000 KG PAYLOAD (USSR)
Viacheslav N. Gorelov, Pilot 813.13 kmh
Michael I. Pozdnyakov, Copilot 505.26 mph
Tupolev Tu-95
9/27/89

SPEED OVER A 2,000 KM CLOSED CIRCUIT WITH 10,000 KG PAYLOAD (USSR)
Viacheslav N. Gorelov, Pilot 813.13 kmh
Michael I. Pozdnyakov, Copilot 505.26 mph
Tupolev Tu-95
9/27/89

SPEED OVER A 2,000 KM CLOSED CIRCUIT WITH 15,000 KG PAYLOAD (USSR)
Viacheslav N. Gorelov, Pilot 813.13 kmh
Michael I. Pozdnyakov, Copilot 505.26 mph
Tupolev Tu-95
9/27/89

SPEED OVER A 2,000 KM CLOSED CIRCUIT WITH 20,000 KG PAYLOAD (USSR)
Viacheslav N. Gorelov, Pilot 813.13 kmh
Michael I. Pozdnyakov, Copilot 505.26 mph
Tupolev Tu-95
9/27/89

SPEED OVER A 2,000 KM CLOSED CIRCUIT WITH 25,000 KG PAYLOAD (USSR)
Viacheslav N. Gorelov, Pilot 813.13 kmh
Michael I. Pozdnyakov, Copilot 505.26 mph
Tupolev Tu-95
9/27/89

SPEED OVER A 2,000 KM CLOSED CIRCUIT WITH 30,000 KG PAYLOAD (USSR)
Viacheslav N. Gorelov, Pilot 813.13 kmh
Michael I. Pozdnyakov, Copilot 505.26 mph
Tupolev Tu-95
9/27/89

SPEED OVER A 5,000 KM CLOSED CIRCUIT WITHOUT PAYLOAD (USSR)
K. I. Prispouskov, Pilot 785.30 kmh
V. D. Baskarov, Copilot 487.96 mph
V. N. Cedov, Navigator
Tupolev Tu-142
4 Kuznetsov NK-12MA, 15,000 shp
10/10/89

SPEED OVER A 5,000 KM CLOSED CIRCUIT WITH 1,000 KG PAYLOAD (USSR)
K. I. Prispouskov, Pilot 785.30 kmh
V. D. Baskarov, Copilot 487.96 mph
V. N. Cedov, Navigator
Tupolev Tu-142
4 Kuznetsov NK-12MA, 15,000 shp
10/10/89

SPEED OVER A 5,000 KM CLOSED CIRCUIT WITH 2,000 KG PAYLOAD (USSR)
K. I. Prispouskov, Pilot 785.30 kmh
V. D. Baskarov, Copilot 487.96 mph
V. N. Cedov, Navigator
Tupolev Tu-142
4 Kuznetsov NK-12MA, 15,000 shp
10/10/89

SPEED OVER A 5,000 KM CLOSED CIRCUIT WITH 5,000 KG PAYLOAD (USSR)
K. I. Prispouskov, Pilot 785.30 kmh
V. D. Baskarov, Copilot 487.96 mph
V. N. Cedov, Navigator
Tupolev Tu-142
4 Kuznetsov NK-12MA, 15,000 shp
10/10/89

SPEED OVER A 5,000 KM CLOSED CIRCUIT WITH 10,000 KG PAYLOAD (USSR)
K. I. Prispouskov, Pilot 785.30 kmh
V. D. Baskarov, Copilot 487.96 mph
V. N. Cedov, Navigator
Tupolev Tu-142
4 Kuznetsov NK-12MA, 15,000 shp
10/10/89

SPEED OVER A 5,000 KM CLOSED CIRCUIT WITH 15,000 KG PAYLOAD (USSR)
K. I. Prispouskov, Pilot 785.30 kmh
V. D. Baskarov, Copilot 487.96 mph
V. N. Cedov, Navigator
Tupolev Tu-142
4 Kuznetsov NK-12MA, 15,000 shp
10/10/89

SPEED OVER A 5,000 KM CLOSED CIRCUIT WITH 20,000 KG PAYLOAD (USSR)
K. I. Prispouskov, Pilot 785.30 kmh
V. D. Baskarov, Copilot 487.96 mph
V. N. Cedov, Navigator
Tupolev Tu-142
4 Kuznetsov NK-12MA, 15,000 shp
10/10/89

SPEED OVER A 10,000 KM CLOSED CIRCUIT WITHOUT PAYLOAD (USSR)
Valery I. Pavlov, Pilot 647.89 kmh
N. Sattarov, Copilot 402.58 mph
Tupolev Tu-142 (VP-021 N 31)
10/31-11/01/89

SPEED OVER A 10,000 KM CLOSED CIRCUIT WITH 1,000 KG PAYLOAD (USSR)
Valery I. Pavlov, Pilot 647.89 kmh
N. Sattarov, Copilot 402.58 mph
Tupolev Tu-142 (VP-021 N 31)
10/31-11/01/89

SPEED OVER A 10,000 KM CLOSED CIRCUIT WITH 2,000 KG PAYLOAD (USSR)
Valery I. Pavlov, Pilot 647.89 kmh
N. Sattarov, Copilot 402.58 mph
Tupolev Tu-142 (VP-021 N 31)
10/31-11/01/89

JET ENGINE

The idea of propelling a vehicle by the reaction effect of a blast of hot gas goes back into antiquity, but its realization is relatively recent. Not until 1937 did an airplane fly on rocket power, followed two years later by the first flight of a true jet-propelled airplane.

Credit for these milestones goes to the Germans, who were engaged in preparation for World War II. They flew a Heinkel He-112 propeller-driven fighterplane with an auxiliary rocket motor in the summer of 1937, with at least one flight that included brief periods of rocket-only propulsion.

On June 20, 1939, the Germans flew the first airplane designed expressly for rocket power, the Heinkel He-176, and on August 7, 1939, the first true jet airplane, the Heinkel He-178. By 1941, the British and Italians had begun testing jet airplanes, and in 1942 the U.S.A. joined this rapidly growing movement.

Not until the spring of 1944 did any turbojet-powered airplane enter service, due to the technical complexity of this completely new type of engine and to the obstruction of tradition-minded governments. By the end of World War II, hardly a year and a half later, it was clear that no airplane powered by a piston engine could hope to match the speed of a jet airplane, even though the latter's poor acceleration and generally delicate nature limited its usefulness.

On November 7, 1945, Group Captain Hugh Wilson, of the Royal Air Force, set the world's first official record by a jet propelled airplane. He was clocked at a speed of 606.38 mph in a Gloster Meteor Mk. 4 fighter. The existing record by a propeller airplane was 469 mph, set in 1939 by a special Messerschmitt Me-209.

From then on, jet airplanes dominated speed records, with a North American F-100 Super Saber exceeding the speed of sound on an 822 mph record run at Edwards Air Force Base, CA, in 1955. The 1,000 mph "barrier" was broken in 1956 when Peter Twiss flew a British Fairey Delta 2 at 1,132 mph.

The altitude record was taken over by jet airplanes in 1948 when John Cunningham flew a British deHavilland Vampire fighter to 59,446 feet. In 1959, the record went over the 100,000-foot mark during a flight by Joe Jordan in a Lockheed F-104 Starfighter.

Distance records remained the province of slower, but more efficient, propeller airplanes. Soon, they too, had to temporarily bow to the increasingly sophisticated jets. In 1962, a Boeing B-52 Stratofortress was flown more than 12,500 miles from Okinawa to Spain. In 1986, the Voyager team retook this record with the most efficient propeller aircraft ever designed.

CLASS C-1 (UNLIMITED)
GROUP III (JET ENGINE)

DISTANCE

DISTANCE WITHOUT LANDING (USA)
MAJ Clyde P. Evely, USAF 20,168.78 km
Boeing B-52H Stratofortress 12,532.28 mi
8 P&W TF-330-3
Kadena, Okinawa to Madrid, Spain
1/10-11/62

DISTANCE OVER A CLOSED CIRCUIT WITHOUT LANDING (USSR)
Vladimir Tersky, Pilot 20,150.92 km
Yuri Resnitsky, Copilot 12,521.78 mi
AN-124
4 D-18T, 23,400 kg
Podmoskovnoye, USSR
5/6-7/87

ALTITUDE

ALTITUDE (USSR)
Alexander Fedotov 37,650 m
E-266M 123,524 ft
2 RD
Podmoscovnoe, USSR
8/31/77

ALTITUDE IN HORIZONTAL FLIGHT (USA)
CAPT Robert C. Helt, USAF 25,929 m
Lockheed SR-71 Blackbird 85,069 ft
2 P&W, 30,000 lbs
7/28/76

ALTITUDE WITH 1,000 KG PAYLOAD (USSR)
Alexander Fedotov 37,080 m
E-266 M 121,654 ft
2 RD, 30,865 lbs
Podmoscovnoe, USSR
7/22/77

ALTITUDE WITH 2,000 KG PAYLOAD (USSR)
Alexander Fedotov 37,080 m
E-266 M 121,654 ft
2 RD, 30,865 lbs
7/22/77

ALTITUDE WITH 5,000 KG PAYLOAD (USA)
MAJ F.L. Fulton, USAF 26,018 m
Convair B-58 HUSTLER 85,361 ft
4 GE J-79-GE-5B
Edwards AFB, CA
9/14/62

ALTITUDE WITH 10,000 KG PAYLOAD (USSR)
S. Agapov, Pilot 18,200 m
B. Veremey, Copilot 59,696 ft
Tupolev Tu-144
4 "57," 20,000 kg
Podmoscovnoe, USSR
7/20/83

ALTITUDE WITH 15,000 KG PAYLOAD (USSR)
S. Agapov, Pilot 18,200 m
B. Veremey, Copilot 59,696 ft
Tupolev Tu-144
4 "57," 20,000 kg
Podmoscovnoe, USSR
7/20/83

ALTITUDE WITH 20,000 KG PAYLOAD (USSR)
S. Agapov, Pilot 18,200 m
B. Veremey, Copilot 59,696 ft
Tupolev Tu-144
4 "57," 20,000 kg
Podmoscovnoe, USSR
7/20/83

ALTITUDE WITH 25,000 KG PAYLOAD (USSR)
S. Agapov, Pilot 18,200 m
B. Veremey, Copilot 59,696 ft
Tupolev Tu-144
4 "57," 20,000 kg
Podmoscovnoe, USSR
7/20/83

ALTITUDE WITH 30,000 KG PAYLOAD (USSR)
S. Agapov, Pilot 18,200 m
B. Veremey, Copilot 59,696 ft
Tupolev Tu-144
4 "57," 20,000 kg
Podmoscovnoe, USSR
7/20/83

ALTITUDE WITH 35,000 KG PAYLOAD (USSR)
Boris Stepanov 13,121 m
102M 43,048 ft
4 D-15
Podmoscovnoe, USSR
10/29/59

ALTITUDE WITH 40,000 KG PAYLOAD (USSR)
Boris Stepanov 13,121 m
102M 43,048 ft
4 D-15
Podmoscovnoe, USSR
10/29/59

ALTITUDE WITH 45,000 KG PAYLOAD (USSR)
Boris Stepanov 13,121 m
102M 43,048 ft
4 D-15
Podmoscovnoe, USSR
10/29/59

ALTITUDE WITH 50,000 KG PAYLOAD (USSR)
Boris Stepanov 13,121 m
102M 43,048 ft
4 D-15
Podmoscovnoe, USSR
10/29/59

ALTITUDE WITH 55,000 KG PAYLOAD (USSR)
Boris Stepanov 13,121 m
102M 43,048 ft
4 D-15
Podmoscovnoe, USSR
10/29/59

ALTITUDE WITH 60,000 KG PAYLOAD (USSR)
Alexandr V. Galounenko, Captain 12,340 m
Sergei A. Borbik, Copilot 40,486 ft
Sergei F. Netchaev, Navigator
Antonov 225 "Mriya"
3/22/89

ALTITUDE WITH 65,000 KG PAYLOAD (USSR)
Alexandr V. Galounenko, Captain 12,340 m
Sergei A. Borbik, Copilot 40,486 ft
Sergei F. Netchaev, Navigator
Antonov 225 "Mriya"
3/22/89

ALTITUDE WITH 70,000 KG PAYLOAD (USSR)
Alexandr V. Galounenko, Captain 12,340 m
Sergei A. Borbik, Copilot 40,486 ft
Sergei F. Netchaev, Navigator
Antonov 225 "Mriya"
3/22/89

ALTITUDE WITH 75,000 KG PAYLOAD (USSR)
Alexandr V. Galounenko, Captain 12,340 m
Sergei A. Borbik, Copilot 40,486 ft
Sergei F. Netchaev, Navigator
Antonov 225 "Mriya"
3/22/89

ALTITUDE WITH 80,000 KG PAYLOAD (USSR)
Alexandr V. Galounenko, Captain 12,340 m
Sergei A. Borbik, Copilot 40,486 ft
Sergei F. Netchaev, Navigator
Antonov 225 "Mriya"
3/22/89

ALTITUDE WITH 85,000 KG PAYLOAD (USSR)
Alexandr V. Galounenko, Captain 12,340 m
Sergei A. Borbik, Copilot 40,486 ft
Sergei F. Netchaev, Navigator
Antonov 225 "Mriya"
3/22/89

ALTITUDE WITH 90,000 KG PAYLOAD (USSR)
Alexandr V. Galounenko, Captain 12,340 m
Sergei A. Borbik, Copilot 40,486 ft
Sergei F. Netchaev, Navigator
Antonov 225 "Mriya"
3/22/89

ALTITUDE WITH 95,000 KG PAYLOAD (USSR)
Alexandr V. Galounenko, Captain 12,340 m
Sergei A. Borbik, Copilot 40,486 ft
Sergei F. Netchaev, Navigator
Antonov 225 "Mriya"
3/22/89

ALTITUDE WITH 100,000 KG PAYLOAD (USSR)
Alexandr V. Galounenko, Captain 12,340 m
Sergei A. Borbik, Copilot 40,486 ft
Sergei F. Netchaev, Navigator
Antonov 225 "Mriya"
3/22/89

ALTITUDE WITH 105,000 KG PAYLOAD (USSR)
Alexandr V. Galounenko, Captain 12,340 m
Sergei A. Borbik, Copilot 40,486 ft
Sergei F. Netchaev, Navigator
Antonov 225 "Mriya"
3/22/89

ALTITUDE WITH 110,000 KG PAYLOAD (USSR)
Alexandr V. Galounenko, Captain 12,340 m
Sergei A. Borbik, Copilot 40,486 ft
Sergei F. Netchaev, Navigator
Antonov 225 "Mriya"
3/22/89

ALTITUDE WITH 115,000 KG PAYLOAD (USSR)
Alexandr V. Galounenko, Captain 12,340 m
Sergei A. Borbik, Copilot 40,486 ft
Sergei F. Netchaev, Navigator
Antonov 225 "Mriya"
3/22/89

ALTITUDE WITH 120,000 KG PAYLOAD (USSR)
Alexandr V. Galounenko, Captain 12,340 m
Sergei A. Borbik, Copilot 40,486 ft
Sergei F. Netchaev, Navigator
Antonov 225 "Mriya"
3/22/89

ALTITUDE WITH 125,000 KG PAYLOAD (USSR)
Alexandr V. Galounenko, Captain 12,340 m
Sergei A. Borbik, Copilot 40,486 ft
Sergei F. Netchaev, Navigator
Antonov 225 "Mriya"
3/22/89

ALTITUDE WITH 130,000 KG PAYLOAD (USSR)
Alexandr V. Galounenko, Captain 12,340 m
Sergei A. Borbik, Copilot 40,486 ft
Sergei F. Netchaev, Navigator
Antonov 225 "Mriya"
3/22/89

ALTITUDE WITH 135,000 KG PAYLOAD (USSR)
Alexandr V. Galounenko, Captain 12,340 m
Sergei A. Borbik, Copilot 40,486 ft
Sergei F. Netchaev, Navigator
Antonov 225 "Mriya"
3/22/89

ALTITUDE WITH 140,000 KG PAYLOAD (USSR)
Alexandr V. Galounenko, Captain 12,340 m
Sergei A. Borbik, Copilot 40,486 ft
Sergei F. Netchaev, Navigator
Antonov 225 "Mriya"
3/22/89

ALTITUDE WITH 145,000 KG PAYLOAD (USSR)
Alexandr V. Galounenko, Captain 12,340 m
Sergei A. Borbik, Copilot 40,486 ft
Sergei F. Netchaev, Navigator
Antonov 225 "Mriya"
3/22/89

ALTITUDE WITH 150,000 KG PAYLOAD (USSR)
Alexandr V. Galounenko, Captain 12,340 m
Sergei A. Borbik, Copilot 40,486 ft
Sergei F. Netchaev, Navigator
Antonov 225 "Mriya"
3/22/89

ALTITUDE WITH 155,000 KG PAYLOAD (USSR)
Alexandr V. Galounenko, Captain 12,340 m
Sergei A. Borbik, Copilot 40,486 ft
Sergei F. Netchaev, Navigator
Antonov 225 "Mriya"
3/22/89

ALTITUDE WITH 160,000 KG PAYLOAD (USSR)
Vladimir Tersky, Pilot 10,750 m
Alexandre Galounenko, Copilot 35,269 ft
Antonov An-124
4 D-18T, 23,400 kg
Podmoscovnoe, USSR
7/26/85

ALTITUDE WITH 165,000 KG PAYLOAD (USSR)
Vladimir Tersky, Pilot 10,750 m
Alexandre Galounenko, Copilot 35,269 ft
Antonov An-124
4 D-18T, 23,400 kg
Podmoscovnoe, USSR
7/26/85

ALTITUDE WITH 170,000 KG PAYLOAD (USSR)
Vladimir Tersky, Pilot 10,750 m
Alexandre Galounenko, Copilot 35,269 ft
Antonov An-124
4 D-18T, 23,400 kg
Podmoscovnoe, USSR
7/26/85

ALTITUDE WITH 175,000 KG PAYLOAD (UKRAINE)
Olexander Halunenko 10,750 m
Anatolii Moisseiev 35,269 ft
Antonov An-225 Mriya
6 Progress D-18T, 51,590 lbs
Kyiv, Ukraine
9/11/01

ALTITUDE WITH 180,000 KG PAYLOAD (UKRAINE)
Olexander Halunenko 10,750 m
Anatolii Moisseiev 35,269 ft
Antonov An-225 Mriya
6 Progress D-18T, 51,590 lbs
Kyiv, Ukraine
9/11/01

ALTITUDE WITH 185,000 KG PAYLOAD (UKRAINE)
Olexander Halunenko 10,750 m
Anatolii Moisseiev 35,269 ft
Antonov An-225 Mriya
6 Progress D-18T, 51,590 lbs
Kyiv, Ukraine
9/11/01

ALTITUDE WITH 190,000 KG PAYLOAD (UKRAINE)
Olexander Halunenko 10,750 m
Anatolii Moisseiev 35,269 ft
Antonov An-225 Mriya
6 Progress D-18T, 51,590 lbs
Kyiv, Ukraine
9/11/01

ALTITUDE WITH 195,000 KG PAYLOAD (UKRAINE)
Olexander Halunenko 10,750 m
Anatolii Moisseiev 35,269 ft
Antonov An-225 Mriya
6 Progress D-18T, 51,590 lbs
Kyiv, Ukraine
9/11/01

ALTITUDE WITH 200,000 KG PAYLOAD (UKRAINE)
Olexander Halunenko 10,750 m
Anatolii Moisseiev 35,269 ft
Antonov An-225 Mriya
6 Progress D-18T, 51,590 lbs
Kyiv, Ukraine
9/11/01

ALTITUDE WITH 205,000 KG PAYLOAD (UKRAINE)
Olexander Halunenko 10,750 m
Anatolii Moisseiev 35,269 ft
Antonov An-225 Mriya
6 Progress D-18T, 51,590 lbs
Kyiv, Ukraine
9/11/01

ALTITUDE WITH 210,000 KG PAYLOAD (UKRAINE)
Olexander Halunenko 10,750 m
Anatolii Moisseiev 35,269 ft
Antonov An-225 Mriya
6 Progress D-18T, 51,590 lbs
Kyiv, Ukraine
9/11/01

ALTITUDE WITH 215,000 KG PAYLOAD (UKRAINE)
Olexander Halunenko 10,750 m
Anatolii Moisseiev 35,269 ft
Antonov An-225 Mriya
6 Progress D-18T, 51,590 lbs
Kyiv, Ukraine
9/11/01

ALTITUDE WITH 220,000 KG PAYLOAD (UKRAINE)
Olexander Halunenko 10,750 m
Anatolii Moisseiev 35,269 ft
Antonov An-225 Mriya
6 Progress D-18T, 51,590 lbs
Kyiv, Ukraine
9/11/01

ALTITUDE WITH 225,000 KG PAYLOAD (UKRAINE)
Olexander Halunenko 10,750 m
Anatolii Moisseiev 35,269 ft
Antonov An-225 Mriya
6 Progress D-18T, 51,590 lbs
Kyiv, Ukraine
9/11/01

ALTITUDE WITH 230,000 KG PAYLOAD (UKRAINE)
Olexander Halunenko 10,750 m
Anatolii Moisseiev 35,269 ft
Antonov An-225 Mriya
6 Progress D-18T, 51,590 lbs
Kyiv, Ukraine
9/11/01

ALTITUDE WITH 235,000 KG PAYLOAD (UKRAINE)
Olexander Halunenko 10,750 m
Anatolii Moisseiev 35,269 ft
Antonov An-225 Mriya
6 Progress D-18T, 51,590 lbs
Kyiv, Ukraine
9/11/01

ALTITUDE WITH 240,000 KG PAYLOAD (UKRAINE)
Olexander Halunenko 10,750 m
Anatolii Moisseiev 35,269 ft
Antonov An-225 Mriya
6 Progress D-18T, 51,590 lbs
Kyiv, Ukraine
9/11/01

ALTITUDE WITH 245,000 KG PAYLOAD (UKRAINE)
Olexander Halunenko 10,750 m
Anatolii Moisseiev 35,269 ft
Antonov An-225 Mriya
6 Progress D-18T, 51,590 lbs
Kyiv, Ukraine
9/11/01

ALTITUDE WITH 250,000 KG PAYLOAD (UKRAINE)
Olexander Halunenko 10,750 m
Anatolii Moisseiev 35,269 ft
Antonov An-225 Mriya
6 Progress D-18T, 51,590 lbs
Kyiv, Ukraine
9/11/01

GREATEST PAYLOAD CARRIED TO 2,000 METERS (UKRAINE)
Olexander Halunenko 253,820 kg
Anatolii Moisseiev 559,576 lbs
Antonov An-225 Mriya
6 Progress D-18T, 51,590 lbs
Kyiv, Ukraine
9/11/01

GREATEST PAYLOAD CARRIED TO 15,000 METERS (USA)
Lt Col Bryan Galbreath 1,503 kg
Lockheed Martin U-2S 3,314 lbs
1 GE F118-101, 16,000 lbs
Palmdale, CA
11/18/98

SPEED

SPEED OVER A 3 KM STRAIGHT COURSE (USA)
Darryl G. Greenamyer 1,590.45 kmh
Lockheed F-104 Starfighter 988.26 mph
1 GE J-79, 18,000 lbs
Tonopah, NV
10/24/77

SPEED OVER A 15/25 KM STRAIGHT COURSE (USA)
CAPT Eldon W. Joersz, USAF 3,529.56 kmh
Lockheed SR-71 Blackbird 2,193.16 mph
2 P&W, 30,000 lbs
Beale AFB, CA
7/28/76

SPEED OVER A 100 KM CLOSED CIRCUIT WITHOUT PAYLOAD (USSR)
Alexander Fedotov 2,605.10 kmh
E-266 1,618.70 mph
2 RD
4/8/73

SPEED OVER A 500 KM CLOSED CIRCUIT WITHOUT PAYLOAD (USSR)
Mikhail Komarov 2,981.50 kmh
E-266 1,852.61 mph
2 RD
10/5/67

SPEED OVER A 1,000 KM CLOSED CIRCUIT WITHOUT PAYLOAD (USA)
MAJ Adolphus H. Bledsoe 3,367.22 kmh
Lockheed SR-71 Blackbird 2,092.29 mph
2 P&W, 30,000 lbs
Beale AFB, CA
7/27/76

SPEED OVER A 1,000 KM CLOSED CIRCUIT WITH 1,000 KG PAYLOAD (USA)
MAJ Adolphus H. Bledsoe 3,367.22 kmh
Lockheed SR-71 Blackbird 2,092.29 mph
2 P&W, 30,000 lbs
Edwards AFB, CA
7/27/76

SPEED OVER A 1,000 KM CLOSED CIRCUIT WITH 2,000 KG PAYLOAD (USSR)
Piotor Ostapenko 2,920.67 kmh
E-266 M 1,814.81 mph
2 RD turbojets of 30,865 lbs
10/27/67

SPEED OVER A 1,000 KM CLOSED CIRCUIT WITH 5,000 KG PAYLOAD (USSR)
Serguei Agapov, Pilot 2012.26 kmh
Boris Veremey, Copilot 1,250.42 mph
Tupolev Tu-144
4 "57", 20,000 kg
Podmoscovnoe, USSR
7/13/83

SPEED OVER A 1,000 KM CLOSED CIRCUIT WITH 10,000 KG PAYLOAD (USSR)
Serguei Agapov, Pilot 2,031.55 kmh
Boris Veremey, Copilot 1,262.40 mph
Tupolev Tu-144
4 "57", 20,000 kg
Podmoscovnoe, USSR
7/13/83

SPEED OVER A 1,000 KM CLOSED CIRCUIT WITH 15,000 KG PAYLOAD (USSR)
L. V. Kozlov, Pilot 1,731.40 kmh
M. I. Pozdnyakov, Copilot 1,075.84 mph
Tupolev Tu-160
10/31/89

SPEED OVER A 1,000 KM CLOSED CIRCUIT WITH 20,000 KG PAYLOAD (USSR)
Sergei Agopov, Pilot 2,031.55 kmh
Boris Veremey, Copilot 1,262.40 mph
Tupolev Tu-144
4 "57", 20,000 kg
Podmoscovoe, USSR
7/13/83

SPEED OVER A 1,000 KM CLOSED CIRCUIT WITH 25,000 KG PAYLOAD (USSR)
L. V. Kozlov, Pilot 1,731.40 kmh
M. I. Pozdnyakov, Copilot 1,075.84 mph
Tupolev Tu-160
10/31/89

SPEED OVER A 1,000 KM CLOSED CIRCUIT WITH 30,000 KG PAYLOAD (USSR)
Serguei Agapov, Pilot 2,031.55 kmh
Boris Veremey, Copilot 1,262.40 mph
Tupolev Tu-144
4 "57", 20,000 kg
Podmoscovnoe, USSR
7/13/83

SPEED OVER A 1,000 KM CLOSED CIRCUIT WITH 35,000 KG PAYLOAD (USSR)
G. Volokhov, Pilot 962.00 kmh
A. Turumine, Copilot 597.78 mph
Ilyushin IL-86
4 TBL DHK-86, 13,000 kg
Podmoscovnoe, USSR
9/24/81

SPEED OVER A 1,000 KM CLOSED CIRCUIT WITH 40,000 KG PAYLOAD (USSR)
G. Volokhov, Pilot 962.00 kmh
A. Turumine, Copilot 597.78 mph
Ilyushin IL-86
4 TBL DHK-86, 13,000 kg
Podmoscovnoe, USSR
9/24/81

SPEED OVER A 1,000 KM CLOSED CIRCUIT WITH 45,000 KG PAYLOAD (USSR)
G. Volokhov, Pilot 962.00 kmh
A. Turumine, Copilot 597.78 mph
Ilyushin IL-86
4 TBL DHK-86, 13,000 kg
Podmoscovnoe, USSR
9/24/81

SPEED OVER A 1,000 KM CLOSED CIRCUIT WITH 50,000 KG PAYLOAD (USSR)
G. Volokhov, Pilot 962.00 kmh
A. Turumine, Copilot 597.78 mph
Ilyushin IL-86
4 TBL DHK-86, 13,000 kg
Podmoscovnoe, USSR
9/24/81

SPEED OVER A 1,000 KM CLOSED CIRCUIT WITH 55,000 KG PAYLOAD (USSR)
G. Volokhov, Pilot 962.00 kmh
A. Turumine, Copilot 597.78 mph
Ilyushin IL-86
4 TBL DHK-86, 13,000 kg
Podmoscovnoe, USSR
9/24/81

SPEED OVER A 1,000 KM CLOSED CIRCUIT WITH 60,000 KG PAYLOAD (USSR)
G. Volokhov, Pilot 962.00 kmh
A. Turumine, Copilot 597.78 mph
Ilyushin IL-86
4 TBL DHK-86, 13,000 kg
Podmoscovnoe, USSR
9/24/81

SPEED OVER A 1,000 KM CLOSED CIRCUIT WITH 65,000 KG PAYLOAD (USSR)
G. Volokhov 962.00 kmh
Ilyushin IL-86 597.78 mph
4 TBL DHK-86, 13,000 kg
Podmoscovnoe, USSR
9/24/81

SPEED OVER A 1,000 KM CLOSED CIRCUIT WITH 70,000 KG PAYLOAD (USSR)
G. Volokhov, Pilot 962.00 kmh
A. Turumine, Copilot 597.78 mph
Ilyushin IL-86
4 TBL DHK-86, 13,000 kg
Podmoscovnoe, USSR
9/24/81

SPEED OVER A 1,000 KM CLOSED CIRCUIT WITH 75,000 KG PAYLOAD (USSR)
G. Volokhov, Pilot 962.00 kmh
A. Turumine, Copilot 597.78 mph
Ilyushin IL-86
4 TBL DHK-86, 13,000 kg
Podmoscovnoe, USSR
9/24/81

SPEED OVER A 1,000 KM CLOSED CIRCUIT WITH 80,000 KG PAYLOAD (USSR)
G. Volokhov, Pilot 962.00 kmh
A. Turumine, Copilot 597.78 mph
Ilyushin IL-86
4 TBL DHK-86, 13,000 kg
Podmoscovnoe, USSR
9/24/81

SPEED OVER A 1,000 KM CLOSED CIRCUIT WITH 85,000 KG PAYLOAD (UKRAINE)
Olexander Halunenko 763.20 kmh
Anatolii Moisseiev 474.23 mph
Antonov An-225 Mriya
6 Progress D-18T, 51,590 lbs
Kyiv, Ukraine
9/11/01

SPEED OVER A 1,000 KM CLOSED CIRCUIT WITH 90,000 KG PAYLOAD (UKRAINE)
Olexander Halunenko 763.20 kmh
Anatolii Moisseiev 474.23 mph
Antonov An-225 Mriya
6 Progress D-18T, 51,590 lbs
Kyiv, Ukraine
9/11/01

SPEED OVER A 1,000 KM CLOSED CIRCUIT WITH 95,000 KG PAYLOAD (UKRAINE)
Olexander Halunenko 763.20 kmh
Anatolii Moisseiev 474.23 mph
Antonov An-225 Mriya
6 Progress D-18T, 51,590 lbs
Kyiv, Ukraine
9/11/01

SPEED OVER A 1,000 KM CLOSED CIRCUIT WITH 100,000 KG PAYLOAD (UKRAINE)
Olexander Halunenko 763.20 kmh
Anatolii Moisseiev 474.23 mph
Antonov An-225 Mriya
6 Progress D-18T, 51,590 lbs
Kyiv, Ukraine
9/11/01

SPEED OVER A 1,000 KM CLOSED CIRCUIT WITH 105,000 KG PAYLOAD (UKRAINE)
Olexander Halunenko 763.20 kmh
Anatolii Moisseiev 474.23 mph
Antonov An-225 Mriya
6 Progress D-18T, 51,590 lbs
Kyiv, Ukraine
9/11/01

SPEED OVER A 1,000 KM CLOSED CIRCUIT WITH 110,000 KG PAYLOAD (UKRAINE)
Olexander Halunenko 763.20 kmh
Anatolii Moisseiev 474.23 mph
Antonov An-225 Mriya
6 Progress D-18T, 51,590 lbs
Kyiv, Ukraine
9/11/01

SPEED OVER A 1,000 KM CLOSED CIRCUIT WITH 115,000 KG PAYLOAD (UKRAINE)
Olexander Halunenko 763.20 kmh
Anatolii Moisseiev 474.23 mph
Antonov An-225 Mriya
6 Progress D-18T, 51,590 lbs
Kyiv, Ukraine
9/11/01

SPEED OVER A 1,000 KM CLOSED CIRCUIT WITH 120,000 KG PAYLOAD (UKRAINE)
Olexander Halunenko 763.20 kmh
Anatolii Moisseiev 474.23 mph
Antonov An-225 Mriya
6 Progress D-18T, 51,590 lbs
Kyiv, Ukraine
9/11/01

SPEED OVER A 1,000 KM CLOSED CIRCUIT WITH 125,000 KG PAYLOAD (UKRAINE)
Olexander Halunenko 763.20 kmh
Anatolii Moisseiev 474.23 mph
Antonov An-225 Mriya
6 Progress D-18T, 51,590 lbs
Kyiv, Ukraine
9/11/01

SPEED OVER A 1,000 KM CLOSED CIRCUIT WITH 130,000 KG PAYLOAD (UKRAINE)
Olexander Halunenko 763.20 kmh
Anatolii Moisseiev 474.23 mph
Antonov An-225 Mriya
6 Progress D-18T, 51,590 lbs
Kyiv, Ukraine
9/11/01

SPEED OVER A 1,000 KM CLOSED CIRCUIT WITH 135,000 KG PAYLOAD (UKRAINE)
Olexander Halunenko 763.20 kmh
Anatolii Moisseiev 474.23 mph
Antonov An-225 Mriya
6 Progress D-18T, 51,590 lbs
Kyiv, Ukraine
9/11/01

SPEED OVER A 1,000 KM CLOSED CIRCUIT WITH 140,000 KG PAYLOAD (UKRAINE)
Olexander Halunenko 763.20 kmh
Anatolii Moisseiev 474.23 mph
Antonov An-225 Mriya
6 Progress D-18T, 51,590 lbs
Kyiv, Ukraine
9/11/01

SPEED OVER A 1,000 KM CLOSED CIRCUIT WITH 145,000 KG PAYLOAD (UKRAINE)
Olexander Halunenko 763.20 kmh
Anatolii Moisseiev 474.23 mph
Antonov An-225 Mriya
6 Progress D-18T, 51,590 lbs
Kyiv, Ukraine
9/11/01

SPEED OVER A 1,000 KM CLOSED CIRCUIT WITH 150,000 KG PAYLOAD (UKRAINE)
Olexander Halunenko 763.20 kmh
Anatolii Moisseiev 474.23 mph
Antonov An-225 Mriya
6 Progress D-18T, 51,590 lbs
Kyiv, Ukraine
9/11/01

SPEED OVER A 1,000 KM CLOSED CIRCUIT WITH 155,000 KG PAYLOAD (UKRAINE)
Olexander Halunenko 763.20 kmh
Anatolii Moisseiev 474.23 mph
Antonov An-225 Mriya
6 Progress D-18T, 51,590 lbs
Kyiv, Ukraine
9/11/01

SPEED OVER A 1,000 KM CLOSED CIRCUIT WITH 160,000 KG PAYLOAD (UKRAINE)
Olexander Halunenko 763.20 kmh
Anatolii Moisseiev 474.23 mph
Antonov An-225 Mriya
6 Progress D-18T, 51,590 lbs
Kyiv, Ukraine
9/11/01

SPEED OVER A 1,000 KM CLOSED CIRCUIT WITH 165,000 KG PAYLOAD (UKRAINE)
Olexander Halunenko 763.20 kmh
Anatolii Moisseiev 474.23 mph
Antonov An-225 Mriya
6 Progress D-18T, 51,590 lbs
Kyiv, Ukraine
9/11/01

SPEED OVER A 1,000 KM CLOSED CIRCUIT WITH 170,000 KG PAYLOAD (UKRAINE)
Olexander Halunenko 763.20 kmh
Anatolii Moisseiev 474.23 mph
Antonov An-225 Mriya
6 Progress D-18T, 51,590 lbs
Kyiv, Ukraine
9/11/01

SPEED OVER A 1,000 KM CLOSED CIRCUIT WITH 175,000 KG PAYLOAD (UKRAINE)
Olexander Halunenko 763.20 kmh
Anatolii Moisseiev 474.23 mph
Antonov An-225 Mriya
6 Progress D-18T, 51,590 lbs
Kyiv, Ukraine
9/11/01

SPEED OVER A 1,000 KM CLOSED CIRCUIT WITH 180,000 KG PAYLOAD (UKRAINE)
Olexander Halunenko 763.20 kmh
Anatolii Moisseiev 474.23 mph
Antonov An-225 Mriya
6 Progress D-18T, 51,590 lbs
Kyiv, Ukraine
9/11/01

SPEED OVER A 1,000 KM CLOSED CIRCUIT WITH 185,000 KG PAYLOAD (UKRAINE)
Olexander Halunenko 763.20 kmh
Anatolii Moisseiev 474.23 mph
Antonov An-225 Mriya
6 Progress D-18T, 51,590 lbs
Kyiv, Ukraine
9/11/01

SPEED OVER A 1,000 KM CLOSED CIRCUIT WITH 190,000 KG PAYLOAD (UKRAINE)
Olexander Halunenko 763.20 kmh
Anatolii Moisseiev 474.23 mph
Antonov An-225 Mriya
6 Progress D-18T, 51,590 lbs
Kyiv, Ukraine
9/11/01

SPEED OVER A 1,000 KM CLOSED CIRCUIT WITH 195,000 KG PAYLOAD (UKRAINE)
Olexander Halunenko 763.20 kmh
Anatolii Moisseiev 474.23 mph
Antonov An-225 Mriya
6 Progress D-18T, 51,590 lbs
Kyiv, Ukraine
9/11/01

SPEED OVER A 1,000 KM CLOSED CIRCUIT WITH 200,000 KG PAYLOAD (UKRAINE)
Olexander Halunenko 763.20 kmh
Anatolii Moisseiev 474.23 mph
Antonov An-225 Mriya
6 Progress D-18T, 51,590 lbs
Kyiv, Ukraine
9/11/01

SPEED OVER A 1,000 KM CLOSED CIRCUIT WITH 205,000 KG PAYLOAD (UKRAINE)
Olexander Halunenko 763.20 kmh
Anatolii Moisseiev 474.23 mph
Antonov An-225 Mriya
6 Progress D-18T, 51,590 lbs
Kyiv, Ukraine
9/11/01

SPEED OVER A 1,000 KM CLOSED CIRCUIT WITH 210,000 KG PAYLOAD (UKRAINE)
Olexander Halunenko 763.20 kmh
Anatolii Moisseiev 474.23 mph
Antonov An-225 Mriya
6 Progress D-18T, 51,590 lbs
Kyiv, Ukraine
9/11/01

SPEED OVER A 1,000 KM CLOSED CIRCUIT WITH 215,000 KG PAYLOAD (UKRAINE)
Olexander Halunenko 763.20 kmh
Anatolii Moisseiev 474.23 mph
Antonov An-225 Mriya
6 Progress D-18T, 51,590 lbs
Kyiv, Ukraine
9/11/01

SPEED OVER A 1,000 KM CLOSED CIRCUIT WITH 220,000 KG PAYLOAD (UKRAINE)
Olexander Halunenko 763.20 kmh
Anatolii Moisseiev 474.23 mph
Antonov An-225 Mriya
6 Progress D-18T, 51,590 lbs
Kyiv, Ukraine
9/11/01

SPEED OVER A 1,000 KM CLOSED CIRCUIT WITH 225,000 KG PAYLOAD (UKRAINE)
Olexander Halunenko 763.20 kmh
Anatolii Moisseiev 474.23 mph
Antonov An-225 Mriya
6 Progress D-18T, 51,590 lbs
Kyiv, Ukraine
9/11/01

SPEED OVER A 1,000 KM CLOSED CIRCUIT WITH 230,000 KG PAYLOAD (UKRAINE)
Olexander Halunenko 763.20 kmh
Anatolii Moisseiev 474.23 mph
Antonov An-225 Mriya
6 Progress D-18T, 51,590 lbs
Kyiv, Ukraine
9/11/01

SPEED OVER A 1,000 KM CLOSED CIRCUIT WITH 235,000 KG PAYLOAD (UKRAINE)
Olexander Halunenko 763.20 kmh
Anatolii Moisseiev 474.23 mph
Antonov An-225 Mriya
6 Progress D-18T, 51,590 lbs
Kyiv, Ukraine
9/11/01

SPEED OVER A 1,000 KM CLOSED CIRCUIT WITH 240,000 KG PAYLOAD (UKRAINE)
Olexander Halunenko 763.20 kmh
Anatolii Moisseiev 474.23 mph
Antonov An-225 Mriya
6 Progress D-18T, 51,590 lbs
Kyiv, Ukraine
9/11/01

SPEED OVER A 1,000 KM CLOSED CIRCUIT WITH 245,000 KG PAYLOAD (UKRAINE)
Olexander Halunenko 763.20 kmh
Anatolii Moisseiev 474.23 mph
Antonov An-225 Mriya
6 Progress D-18T, 51,590 lbs
Kyiv, Ukraine
9/11/01

SPEED OVER A 1,000 KM CLOSED CIRCUIT WITH 250,000 KG PAYLOAD (UKRAINE)
Olexander Halunenko 763.20 kmh
Anatolii Moisseiev 474.23 mph
Antonov An-225 Mriya
6 Progress D-18T, 51,590 lbs
Kyiv, Ukraine
9/11/01

SPEED OVER A 2,000 KM CLOSED CIRCUIT WITHOUT PAYLOAD (USSR)
S. Agapov, Pilot 2,012.26 kmh
B. Veremey, Copilot 1,250.42 mph
Tupolev Tu-144
4 "57" 20,000 kg
Podmoscovnoe, USSR
7/20/83

SPEED OVER A 2,000 KM CLOSED CIRCUIT WITH 1,000 KG PAYLOAD (USA)
MAJ H.J. Deutschendorf, Jr. 1,708.82 kmh
Convair B-58 Hustler 1,061.81 mph
4 GE J-79-GE-5B
Edwards AFB, CA
1/12/61

SPEED OVER A 2,000 KM CLOSED CIRCUIT WITH
2,000 KG PAYLOAD (USA)
MAJ H.J. Deutschendorf, Jr. 1,708.82 kmh
Convair B-58 Hustler 1,061.81 mph
4 GE J-79-GE-5B
Edwards AFB, CA
1/12/61

SPEED OVER A 2,000 KM CLOSED CIRCUIT WITH
5,000 KG PAYLOAD (USSR)
S. Agapov, Pilot 2,012.26 kmh
B. Veremey, Copilot 1,250.42 mph
Tupolev Tu-144
4 "57", 20,000 kg
Podmoscovnoe, USSR
7/20/83

SPEED OVER A 2,000 KM CLOSED CIRCUIT WITH
10,000 KG PAYLOAD (USSR)
S. Agapov, Pilot 2,012.26 kmh
B. Veremey, Copilot 1,250.42 mph
Tupolev Tu-144
4 "57" 20,000 kg
Podmoscovnoe, USSR
7/20/83

SPEED OVER A 2,000 KM CLOSED CIRCUIT WITH
15,000 KG PAYLOAD (USSR)
B. Veremey, Pilot 1,678.00 kmh
G.N. Chapoval, Copilot 1,042.66 mph
Tupolev Tu-160
11/03/89

SPEED OVER A 2,000 KM CLOSED CIRCUIT WITH
20,000 KG PAYLOAD (USSR)
S. Agapov, Pilot 2,012.26 kmh
B. Veremey, Co-pilot 1,250.42 mph
Tupolev Tu-144
4 "57"; 20,000 kg
Podmoscovnoe, USSR
7/20/83

SPEED OVER A 2,000 KM CLOSED CIRCUIT WITH
25,000 KG PAYLOAD (USSR)
B. Veremey, Pilot 1,678.00 kmh
G.N. Chapoval, Copilot 1,042.66 mph
Tupolev Tu-160
11/03/89

SPEED OVER A 2,000 KM CLOSED CIRCUIT WITH
30,000 KG PAYLOAD (USSR)
S. Agapov, Pilot 2,012.26 kmh
B. Veremey, Copilot 1,250.42 mph
Tupolev Tu-144
4 "57" 20,000 kg
Podmoscovnoe, USSR
7/20/83

SPEED OVER A 2,000 KM CLOSED CIRCUIT WITH
35,000 KG PAYLOAD (USSR)
G. Volokhov, Pilot 975.30 kmh
A. Turumine, Copilot 606.05 mph
Ilyushin IL-86
4 TBL DHK, 13000 kg
Podmoscovnoe, USSR
9/22/81

SPEED OVER A 2,000 KM CLOSED CIRCUIT WITH
40,000 KG PAYLOAD (USSR)
G. Volokhov 975.30 kmh
Ilyushin IL-86 606.05 mph
4 TBL DHK-86, 13,000 kg
Podmoscovnoe, USSR
9/22/81

SPEED OVER A 2,000 KM CLOSED CIRCUIT WITH
45,000 KG PAYLOAD (USSR)
G. Volokhov, Pilot 975.30 mph
A. Turumine, Copilot 606.05 mph
Ilyushin IL-86
4 TBL DHK-86, 13,000 kg
Podmoscovnoe, USSR
9/22/81

SPEED OVER A 2,000 KM CLOSED CIRCUIT WITH
50,000 KG PAYLOAD (USSR)
G. Volokhov 975.30 kmh
Ilyushin IL-86 606.05 mph
4 TBL DHK-86, 13,000 kg
Podmoscovnoe, USSR
9/22/81

SPEED OVER A 2,000 KM CLOSED CIRCUIT WITH
55,000 KG PAYLOAD (USSR)
G. Volokhov 975.30 kmh
Ilyushin IL-86 606.05 mph
4 TBL DHK-86, 13,000 kg
Podmoscovnoe, USSR
9/22/81

SPEED OVER A 2,000 KM CLOSED CIRCUIT WITH
60,000 KG PAYLOAD (USSR)
G. Volokhov, Pilot 975.30 kmh
A. Turumine, Copilot 606.05 mph
Ilyushin IL-86
4 TBL DHK-86, 13,000 kg
Podmoscovnoe, USSR
9/22/81

SPEED OVER A 2,000 KM CLOSED CIRCUIT WITH
65,000 KG PAYLOAD (USSR)
G. Volokhov, Pilot 975.30 kmh
A. Turumine, Copilot 606.05 mph
Ilyushin IL-86
4 TBL DHK-86, 13,000 kg
Podmoscovnoe, USSR
9/24/81

SPEED OVER A 2,000 KM CLOSED CIRCUIT WITH
70,000 KG PAYLOAD (USSR)
Alexandr V. Galounenko, Captain 813.09 kmh
Sergei A. Borbik, Copilot 505.25 mph
Sergei F. Netchaev, Navigator
Antonov 225 "Mriya"
3/22/89

SPEED OVER A 2,000 KM CLOSED CIRCUIT WITH
75,000 KG PAYLOAD (USSR)
Alexandr V. Galounenko, Captain 813.09 kmh
Sergei A. Borbik, Copilot 505.25 mph
Sergei F. Netchaev, Navigator
Antonov 225 "Mriya"
3/22/89

SPEED OVER A 2,000 KM CLOSED CIRCUIT WITH
80,000 KG PAYLOAD (USSR)
Alexandr V. Galounenko, Captain 813.09 kmh
Sergei A. Borbik, Copilot 505.25 mph
Sergei F. Netchaev, Navigator
Antonov 225 "Mriya"
3/22/89

SPEED OVER A 2,000 KM CLOSED CIRCUIT WITH
85,000 KG PAYLOAD (USSR)
Alexandr V. Galounenko, Captain 813.09 kmh
Sergei A. Borbik, Copilot 505.25 mph
Sergei F. Netchaev, Navigator
Antonov 225 "Mriya"
3/22/89

SPEED OVER A 2,000 KM CLOSED CIRCUIT WITH 90,000 KG PAYLOAD (USSR)
Alexandr V. Galounenko, Captain 813.09 kmh
Sergei A. Borbik, Copilot 505.25 mph
Sergei F. Netchaev, Navigator
Antonov 225 "Mriya"
3/22/89

SPEED OVER A 2,000 KM CLOSED CIRCUIT WITH 95,000 KG PAYLOAD (USSR)
Alexandr V. Galounenko, Captain 813.09 kmh
Sergei A. Borbik, Copilot 505.25 mph
Sergei F. Netchaev, Navigator
Antonov 225 "Mriya"
3/22/89

SPEED OVER A 2,000 KM CLOSED CIRCUIT WITH 100,000 KG PAYLOAD (USSR)
Alexandr V. Galounenko, Captain 813.09 kmh
Sergei A. Borbik, Copilot 505.25 mph
Sergei F. Netchaev, Navigator
Antonov 225 "Mriya"
3/22/89

SPEED OVER A 2,000 KM CLOSED CIRCUIT WITH 105,000 KG PAYLOAD (USSR)
Alexandr V. Galounenko, Captain 813.09 kmh
Sergei A. Borbik, Copilot 505.25 mph
Sergei F. Netchaev, Navigator
Antonov 225 "Mriya"
3/22/89

SPEED OVER A 2,000 KM CLOSED CIRCUIT WITH 110,000 KG PAYLOAD (USSR)
Alexandr V. Galounenko, Captain 813.09 kmh
Sergei A. Borbik, Copilot 505.25 mph
Sergei F. Netchaev, Navigator
Antonov 225 "Mriya"
3/22/89

SPEED OVER A 2,000 KM CLOSED CIRCUIT WITH 115,000 KG PAYLOAD (USSR)
Alexandr V. Galounenko, Captain 813.09 kmh
Sergei A. Borbik, Copilot 505.25 mph
Sergei F. Netchaev, Navigator
Antonov 225 "Mriya"
3/22/89

SPEED OVER A 2,000 KM CLOSED CIRCUIT WITH 120,000 KG PAYLOAD (USSR)
Alexandr V. Galounenko, Captain 813.09 kmh
Sergei A. Borbik, Copilot 505.25 mph
Sergei F. Netchaev, Navigator
Antonov 225 "Mriya"
3/22/89

SPEED OVER A 2,000 KM CLOSED CIRCUIT WITH 125,000 KG PAYLOAD (USSR)
Alexandr V. Galounenko, Captain 813.09 kmh
Sergei A. Borbik, Copilot 505.25 mph
Sergei F. Netchaev, Navigator
Antonov 225 "Mriya"
3/22/89

SPEED OVER A 2,000 KM CLOSED CIRCUIT WITH 130,000 KG PAYLOAD (USSR)
Alexandr V. Galounenko, Captain 813.09 kmh
Sergei A. Borbik, Copilot 505.25 mph
Sergei F. Netchaev, Navigator
Antonov 225 "Mriya"
3/22/89

SPEED OVER A 2,000 KM CLOSED CIRCUIT WITH 135,000 KG PAYLOAD (USSR)
Alexandr V. Galounenko, Captain 813.09 kmh
Sergei A. Borbik, Copilot 505.25 mph
Sergei F. Netchaev, Navigator
Antonov 225 "Mriya"
3/22/89

SPEED OVER A 2,000 KM CLOSED CIRCUIT WITH 140,000 KG PAYLOAD (USSR)
Alexandr V. Galounenko, Captain 813.09 kmh
Sergei A. Borbik, Copilot 505.25 mph
Sergei F. Netchaev, Navigator
Antonov 225 "Mriya"
3/22/89

SPEED OVER A 2,000 KM CLOSED CIRCUIT WITH 145,000 KG PAYLOAD (USSR)
Alexandr V. Galounenko, Captain 813.09 kmh
Sergei A. Borbik, Copilot 505.25 mph
Sergei F. Netchaev, Navigator
Antonov 225 "Mriya"
3/22/89

SPEED OVER A 2,000 KM CLOSED CIRCUIT WITH 150,000 KG PAYLOAD (USSR)
Alexandr V. Galounenko, Captain 813.09 kmh
Sergei A. Borbik, Copilot 505.25 mph
Sergei F. Netchaev, Navigator
Antonov 225
3/22/89

SPEED OVER A 2,000 KM CLOSED CIRCUIT WITH 155,000 KG PAYLOAD (USSR)
Alexandr V. Galounenko, Captain 813.09 kmh
Sergei A. Borbik, Copilot 505.25 mph
Sergei F. Netchaev, Navigator
Antonov 225 "Mriya"
3/22/89

SPEED OVER A 5,000 KM CLOSED CIRCUIT WITHOUT PAYLOAD (USA)
MAJ H. Hedgpeth, Pilot 1,054.21 kmh
LTC Robt Chamberlain, Copilot 655.09 mph
CPT Alexander Ivanchishin, Crew Member
CPT Daniel Novick, Crew Member
B-1B, Heavy Bomber, S/N 70
4 GE F101-GE-102, 30,780 lbs
Palmdale, CA
9/17/87

SPEED OVER A 5,000 KM CLOSED CIRCUIT WITH 1,000 KG PAYLOAD (USA)
MAJ H. Hedgpeth, Pilot 1,054.21 kmh
LTC Robt Chamberlain, Copilot 655.09 mph
CPT Alexander Ivanchishin, Crew Member
CPT Daniel Novick, Crew Member
B-1B, Heavy Bomber, S/N 70
4 GE F101-GE-102, 30,780 lbs
Palmdale, CA
9/17/87

SPEED OVER A 5,000 KM CLOSED CIRCUIT WITH 2,000 KG PAYLOAD (USA)
MAJ H. Hedgpeth, Pilot 1,054.21 kmh
LTC Robt Chamberlain, Copilot 655.09 mph
CPT Alexander Ivanchishin, Crew Member
CPT Daniel Novick, Crew Member
B-1B, Heavy Bomber, S/N 70
4 GE F101-GE-102, 30,780 lbs
Palmdale, CA
9/17/87

SPEED OVER A 5,000 KM CLOSED CIRCUIT WITH 5,000 KG PAYLOAD (USA)
MAJ H. Hedgpeth, Pilot 1,054.21 kmh
LTC Robt Chamberlain, Copilot 655.09 mph
CPT Alexander Ivanchishin, Crew Member
CPT Daniel Novick, Crew Member
B-1B, Heavy Bomber, S/N 70
4 GE F101-GE-102, 30,780 lbs
Palmdale, CA
9/17/87

SPEED OVER A 5,000 KM CLOSED CIRCUIT WITH 10,000 KG PAYLOAD (USA)
MAJ H. Hedgpeth, Pilot 1,054.21 kmh
LTC Robt Chamberlain, Copilot 655.09 mph
CPT Alexander Ivanchishin, Crew Member
CPT Daniel Novick, Crew Member
B-1B, Heavy Bomber, S/N 70
4 GE F101-GE-102, 30,780 lbs
Palmdale, CA
9/17/87

SPEED OVER A 5,000 KM CLOSED CIRCUIT WITH 15,000 KG PAYLOAD (USA)
MAJ H. Hedgpeth, Pilot 1,054.21 kmh
LTC Robt Chamberlain, Copilot 655.09 mph
CPT Alexander Ivanchishin, Crew Member
CPT Daniel Novick, Crew Member
B-1B, Heavy Bomber, S/N 70
4 GE F101-GE-102, 30,780 lbs
Palmdale, CA
9/17/87

SPEED OVER A 5,000 KM CLOSED CIRCUIT WITH 20,000 KG PAYLOAD (USA)
MAJ H. Hedgpeth, Pilot 1,054.21 kmh
LTC Robt Chamberlain, Copilot 655.09 mph
CPT Alexander Ivanchishin, Crew Member
CPT Daniel Novick, Crew Member
B-1B, Heavy Bomber, S/N 70
4 GE F101-GE-102, 30,780 lbs
Palmdale, CA
9/17/87

SPEED OVER A 5,000 KM CLOSED CIRCUIT WITH 25,000 KG PAYLOAD (USA)
MAJ H. Hedgpeth, Pilot 1,054.21 kmh
LTC Robt Chamberlain, Copilot 655.09 mph
CPT Alexander Ivanchishin, Crew Member
CPT Daniel Novick, Crew Member
B-1B, Heavy Bomber, S/N 70
4 GE F101-GE-102, 30,780 lbs
Palmdale, CA
9/17/87

SPEED OVER A 5,000 KM CLOSED CIRCUIT WITH 30,000 KG PAYLOAD (USA)
MAJ H. Hedgpeth, Pilot 1,054.21 kmh
LTC Robt Chamberlain, Copilot 655.09 mph
CPT Alexander Ivanchishin, Crew Member
CPT Daniel Novick, Crew Member
B-1B, Heavy Bomber, S/N 70
4 GE F101-GE-102, 30,780 lbs
Palmdale, CA
9/17/87

SPEED OVER A 5,000 KM CLOSED CIRCUIT WITH 35,000 KG PAYLOAD (USSR)
Alexandre Turumine 815.97 kmh
Ilyushin IL-76 507.02 mph
4 Sol. D-30KP TF, 26,400 lbs
Bikovo
7/10/75

SPEED OVER A 5,000 KM CLOSED CIRCUIT WITH 40,000 KG PAYLOAD (USSR)
Alexandre Turumine 815.97 kmh
Ilyushin IL-76 507.02 mph
4 D-30KP, 26,400 lbs
Bikovo
7/10/75

SPEED OVER A 10,000 KM CLOSED CIRCUIT WITHOUT PAYLOAD (USA)
Capt. Michael S. Menser, Commander 964.95 kmh
Capt. Brian P. Gallagher, Pilot 599.59 mph
Capt. Robert P. Boman, OSO
Capt. Matthew E. Grant, WSO
Rockwell International B-1B
4 GE F101, 12,000 lbs
Grand Forks, ND, Monroeville, AL, Mullan, ID
4/7-8/94

SPEED OVER A 10,000 KM CLOSED CIRCUIT WITH 1,000 KG PAYLOAD (USA)
Capt. Russell F. Mathers 884.26 kmh
Capt. Daniel G. Manuel, Jr. 549.45 mph
Capt. Henry C. Jenkins, Jr.
1st Lt. Ralph DeLatour
Capt. Allen D. Patton
Boeing B-52H
8 P&W TF-33, 17,100 lbs
Edwards AFB, CA
8/26/95

SPEED OVER A 10,000 KM CLOSED CIRCUIT WITH 2,000 KG PAYLOAD (USA)
Capt. Russell F. Mathers 884.26 kmh
Capt. Daniel G. Manuel, Jr. 549.45 mph
Capt. Henry C. Jenkins, Jr.
1st Lt. Ralph DeLatour
Capt. Allen D. Patton
Boeing B-52H
8 P&W TF-33, 17,100 lbs
Edwards AFB, CA
8/26/95

SPEED OVER A 10,000 KM CLOSED CIRCUIT WITH 5,000 KG PAYLOAD (USA) ★
Capt. Russell F. Mathers 884.26 kmh
Capt. Daniel G. Manuel, Jr. 549.45 mph
Capt. Henry C. Jenkins, Jr.
1st Lt. Ralph DeLatour
Capt. Allen D. Patton
Boeing B-52H
8 P&W TF-33, 17,100 lbs
Edwards AFB, CA
8/26/95

SPEED AROUND THE WORLD, EASTBOUND (FRANCE)
Michel Dupont 1,305.93 kmh
Claude Hetru 811.46 mph
BAe/Aérospatiale Concorde
4 RR Olympus, 38,000 lbs
New York, Toulouse, Dubai, Bangkok, Guam, Honolulu, Acapulco, New York
8/16/95

SPEED AROUND THE WORLD, EASTBOUND (WITH REFUELING IN FLIGHT) (USA) ★
Col. Douglas L. Raaberg 1,015.76 kmh
Capt. Ricky W. Carver 631.16 mph
Capt. Gerald V. Goodfellow
Capt. Kevin D. Clotfelter
Rockwell International B-1B
4 GE F101-GE-102, 30,780 lbs
Elapsed Time: 36 hrs 13 min 36 sec
Dyess AFB, Abilene, TX
6/3/95

SPEED AROUND THE WORLD, WESTBOUND (FRANCE)
Claude Delorme 1,231.12 kmh
Jean Boye 764.98 mph
BAe/Aérospatiale Concorde
4 RR Olympus, 38,000 lbs
Lisbon, Santo Domingo, Acapulco, Honolulu, Guam, Bangkok, Bahrain, Lisbon
Elapsed Time: 32 hrs 49 min 3 sec
10/12/92

SPEED AROUND THE WORLD OVER BOTH THE EARTH'S POLES (USA)
Capt. W. H. Mullikin 784.31 kmh
Capt. A. A. Frink 487.31 mph
S. Beckett, F. Cassaniti, E. Shields
Pan American World Airways
Boeing 747 SP, N533PA
4 P&W JT9D-7A
San Francisco, North Pole, London, Capetown, South Pole, Auckland, San Francisco
Distance: 42,458.92 km (26,382.755 mi)
54 hrs 7 min 12 sec
10/28-31/77

TIME TO CLIMB

TIME TO 3,000 METERS (USSR)
Victor Pugachev 25 sec
Sukhoi P-42
2 RR-32, 30,000 lbs
Podmoscovnoe, USSR
10/27/86

TIME TO 6,000 METERS (USSR)
Victor Pugachev 37 sec
Sukhoi P-42
2 RR-32, 30,000 lbs
Podmoscovnoe, USSR
11/15/86

TIME TO 9,000 METERS (USSR)
Nikolai Sadovnikov 44 sec
Sukhoi P-42
2 P-32, 13,600 kg
Podmoskovnoye, USSR
3/10/87

TIME TO 12,000 METERS (USSR)
Nikolai Sadovnikov 56 sec
Sukhoi P-42
2 P-32, 13,600 kg
Podmoskovnoye, USSR
3/10/87

TIME TO 15,000 METERS (USSR)
Nikolai Sadovnikov 1 min
Sukhoi P-42 10 sec
2 P-32, 13,600 kg
Podmoskovnoye, USSR
3/23/88

TIME TO 20,000 METERS (USA)
MAJ Roger J. Smith, USAF 2 min
McDonnell Douglas F-15 Eagle 3 sec
2 P&W F-100
Grand Forks, ND
1/19/75

TIME TO 25,000 METERS (USSR)
Alexandre Fedotov 2 min
E-266 M 34 sec
2 RD-F, 14,000 kg
Podmoscovnoe, USSR
5/17/75

TIME TO 30,000 METERS (USSR)
Piotr Ostapenko 3 min
E-266 M 10 sec
2 RD-F, 14,000 kg
Podmoscovnoe, USSR
5/17/75

TIME TO 35,000 METERS (USSR)
Alexandre Fedotov 4 min
E-266 M 12 sec
2 RD-F, 14,000 kg
Podmoscovnoe, USSR
5/17/75

CLASS C-1.A (300-500 KG / 661-1,102 LBS) GROUP III (JET ENGINE)

DISTANCE

DISTANCE OVER A CLOSED CIRCUIT WITHOUT LANDING (USA)
J.G. Mercer 502.60 km
Bede MJ-90 312.60 mi
1 TRS-018, 218 lbs
Pontiac, MI
3/16/79

ALTITUDE

ALTITUDE (USA)
J.G. Mercer 8,047 m
Bede MJ-90 26,400 ft
1 TRS-018, 218 lbs
Pontiac, MI
3/22/79

ALTITUDE IN HORIZONTAL FLIGHT (USA)
J.G. Mercer 7,955 m
Bede MJ-90 26,100 ft
1 TRS-018, 218 lbs
Pontiac, MI
3/22/79

SPEED

SPEED OVER A 100 KM CLOSED CIRCUIT WITHOUT PAYLOAD (USA)
J.G. Mercer 450.12 kmh
Bede MJ-90 279.69 mph
1 TRS-018, 218 lbs
Pontiac, MI
3/8/79

SPEED OVER A 500 KM CLOSED CIRCUIT WITHOUT PAYLOAD (USA)
J.G. Mercer 342.00 kmh
Bede MJ-90 211.85 mph
1 TRS-018, 218 lbs
Pontiac, MI
3/16/79

TIME TO CLIMB

TIME TO 3,000 METERS (USA)
J.G. Mercer 5 min
Bede MJ-90 11 sec
1 TRS-018, 218 lbs
Pontiac, MI
3/12/79

CLASS C-1.B (500-1,000 KG / 1,102-2,205 LBS) GROUP III (JET ENGINE)

ALTITUDE

ALTITUDE (ITALY)
Col. Adriano Mantelli 9,366 m
"Canguro" MM 1000028 30,728 ft
1 Palas
Guidonia Airport, Italy
9/24/64

CLASS C-1.C (1,000-1,750 KG / 2,205-3,858 LBS) GROUP III (JET ENGINE)

SPEED

SPEED OVER A 15/25 KM STRAIGHT COURSE (USSR)
Rozalia Chikhina 755.00 kmh
YAK 32 469.13 mph
1 TRD29
Podmoscovnoe, USSR
2/19/65

SPEED OVER A 100 KM CLOSED CIRCUIT WITHOUT PAYLOAD (USSR)
Galina Kortchuganova — 724.43 kmh
YAK 30 — 441.17 mph
1/14/65

CLASS C-1.D (1,750-3,000 KG / 3,858-6,614 LBS) GROUP III (JET ENGINE)

DISTANCE

DISTANCE WITHOUT LANDING (ITALY)
Massino Ralli — 970.01 km
Aermacchi MB-326 — 602.67 mi
1 BS Viper II
7/18/67

DISTANCE OVER A CLOSED CIRCUIT WITHOUT LANDING (ITALY)
Massino Ralli — 773.557 km
Aermacchi MB-326 — 480.665 mi
1 BS Viper II
12/9/67

ALTITUDE

ALTITUDE (ITALY)
Massino Ralli — 17,315 m
Aermacchi MB-326 — 56,776 ft
1 BS Viper II
3/18/66

ALTITUDE IN HORIZONTAL FLIGHT (ITALY)
Massino Ralli — 15,668 m
Aermacchi MB-326 — 51,305 ft
1 BS Viper II
3/18/66

SPEED

SPEED OVER A 3 KM STRAIGHT COURSE (ITALY)
Massino Ralli — 871.80 kmh
Aermacchi MB-326 — 541.60 mph
1 BS Viper II
8/2/67

SPEED OVER A 15/25 KM STRAIGHT COURSE (ITALY)
Massino Ralli — 880.00 kmh
Aermacchi MB-326 — 546.80 mph
1 BS Viper II
12/1/67

SPEED OVER A 100 KM CLOSED CIRCUIT (ITALY)
Massino Ralli — 831.01 kmh
Aermacchi MB-326 — 516.40 mph
1 BS Viper II
12/3/67

SPEED OVER A 500 KM CLOSED CIRCUIT (ITALY)
Massino Ralli — 777.67 kmh
Aermacchi MB-326 — 483.20 mph
1 BS Viper II
12/6/67

TIME TO CLIMB

TIME TO 3,000 METERS (USA)
Richard B. Hunt — 1 min
Super Pinto T-610, N7754A — 50 sec
1 GE CJ610-4
Washington, DC
5/27/72

TIME TO 6,000 METERS (USA)
Richard B. Hunt — 3 min
Super Pinto T-610, N7754A — 40 sec
1 GE CJ610-4
Washington, DC
5/27/72

TIME TO 9,000 METERS (USA)
Richard B. Hunt — 6 min
Super Pinto T-610, N7754A — 6 sec
GE CJ610-4
Washington, DC
5/27/72

TIME TO 12,000 METERS (USA)
Richard B. Hunt — 9 min
Super Pinto T-610, N7754A — 11 sec
1 GE CJ610-4
Washington, DC
5/27/72

CLASS C-1.E (3,000-6,000 KG / 6,614-13,228 LBS) GROUP III (JET ENGINE)

ALTITUDE

ALTITUDE (USA) ✯
Michael W. Melvill — 19,277 m
Robert J. Waldmiller — 63,245 ft
Scaled Composites Proteus
2 Williams-Rolls FJ44-2A, 2,200 lbs
Mojave, CA
10/25/00

ALTITUDE IN HORIZONTAL FLIGHT (USA)
Michael W. Melvill — 19,015 m
Robert J. Waldmiller — 62,385 ft
Scaled Composites Proteus
2 Williams-Rolls FJ44-2A, 2,200 lbs
Mojave, CA
10/25/00

ALTITUDE WITH 1,000 KG PAYLOAD (USA)
Michael W. Melvill — 17,067 m
Robert J. Waldmiller — 55,994 ft
Scaled Composites Proteus
2 Williams-Rolls FJ44-2A, 2,200 lbs
Mojave, CA
10/27/00

SPEED

SPEED OVER A 3 KM STRAIGHT COURSE (USA)
Jack De Boer, Pilot — 774.45 kmh
Craig Tylski, Copilot — 481.24 mph
Learjet 31, N984JD
Las Vegas, NM
8/19/89

SPEED OVER A 15/25 KM STRAIGHT COURSE (USA)
Richard A. Vandam — 860.76 kmh
William H. Strom — 534.85 mph
MiG 15 UTI
1 Klimov VK-1, 5,600 lbs
Reno, NV
9/19/97

SPEED OVER A 100 KM CLOSED CIRCUIT WITHOUT PAYLOAD (USA)
Richard A. Vandam — 834.64 kmh
Howard A. Torman — 518.62 mph
MiG 15 UTI
1 Klimov VK-1, 5,600 lbs
Reno, NV
11/29/96

SPEED OVER A 500 KM CLOSED CIRCUIT WITHOUT PAYLOAD (USA)
Richard A. Vandam — 830.21 kmh
Steven M. Korcheck — 515.87 mph
MiG 15 UTI
1 Klimov VK-1, 5,600 lbs
Reno, NV
9/10/97

SPEED AROUND THE WORLD, EASTBOUND (UK)
M. Naviede 448.22 kmh
Cessna 550 278.51 mph
5/10-13/91

TIME TO CLIMB

TIME TO 3,000 METERS (USA)
Edward H. Wachs, Pilot 52 sec
Betsy R. Benton, Copilot
Learjet 24, N500SW
2 CJ610-6, 2,950 lbs
Kenosha, WI
2/6/89

TIME TO 6,000 METERS (USA)
Edward H. Wachs, Pilot 1 min
Betsy R. Benton, Copilot 43 sec
Learjet 24, N500SW
2 CJ610-6, 2,950 lbs
Kenosha, WI
2/6/89

TIME TO 9,000 METERS (USA)
Edward H. Wachs, Pilot 2 min
Betsy R. Benton, Copilot 42 sec
Learjet 24, N500SW
2 CJ610-6, 2,950 lbs
Kenosha, WI
2/6/89

TIME TO 12,000 METERS (USA)
Edward H. Wachs, Pilot 6 min
Betsy R. Benton, Copilot 5 sec
Learjet 24, N500SW
2 CJ610-6, 2,950 lbs
Duluth, MN
6/27/88

TIME TO 15,000 METERS (USA)
Edward H. Wachs, Pilot 11 min
William Benton, Copilot 52 sec
Learjet 28, N500LG
2 CJ610-8A, 2,950 lbs
Bangor, ME
6/28/88

CLASS C-1.F (6,000-9,000 KG / 13,228-19,842 LBS)
GROUP III (JET ENGINE)

ALTITUDE

ALTITUDE (USA)
Jerry Hoyt 22482 m
Lockheed U2C, 56-6682 73,761 ft
1 P&W J75 P-138
Edwards AFB, CA
4/17/89

ALTITUDE IN HORIZONTAL FLIGHT (USA)
Jerry Hoyt 22,475 m
Lockheed U2C, 56-6682 73,739 ft
1 P&W J75 P-138
Edwards AFB, CA
4/17/89

ALTITUDE WITH 1,000 KG PAYLOAD (USSR)
Vasily M. Titarenko 14,670 m
MiG-21 48,130 ft
7/11/91

GREATEST PAYLOAD CARRIED TO 2,000 METERS (RUSSIA)
Vasily M. Titarenko 1,450.0 kg
MiG-21 3,196.7 lb
10/10/91

SPEED

SPEED OVER A 100 KM CLOSED CIRCUIT WITHOUT PAYLOAD (RUSSIA)
Vladimir Jachanov 1,031.00 kmh
MiG-21 640.63 mph
10/1/91

SPEED OVER A 500 KM CLOSED CIRCUIT WITHOUT PAYLOAD (RUSSIA)
Alexander Dronov 1,023.52 kmh
MiG-21 635.98 mph
10/2/91

SPEED OVER A 1,000 KM CLOSED CIRCUIT WITHOUT PAYLOAD (FRANCE)
H. Leprince-Ringuet 930.40 kmh
Falcon 10 578.10 mph
2 GE CJ-610-6
6/1/71

SPEED OVER A 2,000 KM CLOSED CIRCUIT WITHOUT PAYLOAD (FRANCE)
Paul Albert 917.02 kmh
Falcon 10 569.80 mph
2 Air TFE731-2
Yvelines-Bonifacio,
Cape Pertusato, France
5/29/73

SPEED AROUND THE WORLD, EASTBOUND (USA)
Mark E. Calkins 752.53 kmh
Charles Conrad, Jr. 467.60 mph
Paul Thayer
Daniel Miller
Learjet 35A
2 Garrett TFE731, 3,500 lbs
Denver, CO
2/12-14/96

TIME TO CLIMB

TIME TO 3,000 METERS (USA)
Jerry Hoyt 55 sec
Lockheed U2C, 56-6682
1 P&W J75 P-138
Edwards AFB, CA
4/17/89

TIME TO 6,000 METERS (USA)
Jerry Hoyt 1 min
Lockheed U2C, 56-6682 50 sec
1 P&W J75 P-138
Edwards AFB, CA
4/17/89

TIME TO 9,000 METERS (USA)
Jerry Hoyt 2 min
Lockheed U2C, 56-6682 51 sec
1 P&W J75 P-138
Edwards AFB, CA
4/17/89

TIME TO 12,000 METERS (USA)
Jerry Hoyt 4 min
Lockheed U2C, 56-6682 17 sec
1 P&W J75 P-138
Edwards AFB, CA
4/17/89

TIME TO 15,000 METERS (USA)
Jerry Hoyt 6 min
Lockheed U2C, 56-6682 16 sec
1 P&W J75 P-138
Edwards AFB, CA
4/17/89

TIME TO 20,000 METERS (USA)
Jerry Hoyt — 12 min 14 sec
Lockheed U2C, 56-6682
1 P&W J75 P-138
Edwards AFB, CA
4/17/89

CLASS C-1.G (9,000-12,000 KG / 19,842-26,455 LBS)
GROUP III (JET ENGINE)

ALTITUDE

ALTITUDE (USA)
Ronald R. Williams — 22,198 m / 72,829 ft
Lockheed U-2C, 56-6682
1 P&W J75 P-138
Edwards AFB, CA
4/18/89

ALTITUDE IN HORIZONTAL FLIGHT (USA)
Ronald R. Williams — 22,198 m / 72,829 ft
Lockheed U-2C, 56-6682
1 P&W J75 P-138
Edwards AFB, CA
4/18/89

SPEED

SPEED OVER A 1,000 KM CLOSED CIRCUIT WITHOUT PAYLOAD (FRANCE)
Jacqueline Auriol — 859.51 kmh / 534.07 mph
Mystere 20
2 GE CF700-2B
Istres, Cazaux, Istres, France
6/10/65

SPEED OVER A 2,000 KM CLOSED CIRCUIT WITHOUT PAYLOAD (FRANCE)
Jacqueline Auriol — 819.13 kmh / 508.98 mph
Mystere 20
2 GE CF700-2B
Istres, Cazaux, Istres, France
6/15/65

SPEED AROUND THE WORLD, EASTBOUND (USA)
Arnold Palmer — 644.11 kmh
James E. Bir — 400.23 mph
Lewis L. Purkey
Learjet 36
2 Garrett TFE-731-2-2B
Denver, Boston, Cardiff, Paris, Tehran, Columbo, Ceylon, Jakarta, Manila, Wake Island, Honolulu, Denver
57 hrs 25 min 42 sec
5/17-19/76

TIME TO CLIMB

TIME TO 3,000 METERS (USA)
Ronald R. Williams — 1 min 9 sec
Lockheed U-2C, 56-6682
1 P&W J75 P-138
Edwards AFB, CA
4/18/89

TIME TO 6,000 METERS (USA)
Ronald R. Williams — 2 min 13 sec
Lockheed U-2C, 56-6682
1 P&W J75 P-138
Edwards AFB, CA
4/18/89

TIME TO 9,000 METERS (USA)
Ronald R. Williams — 3 min 30 sec
Lockheed U-2C, 56-6682
1 P&W J75 P-138
Edwards AFB, CA
4/18/89

TIME TO 12,000 METERS (USA)
Ronald R. Williams — 5 min 10 sec
Lockheed U-2C, 56-6682
1 P&W J75 P-138
Edwards AFB, CA
4/18/89

TIME TO 15,000 METERS (USA)
Ronald R. Williams — 8 min 9 sec
Lockheed U-2C, 56-6682
1 P&W J75 P-138
Edwards AFB, CA
4/18/89

TIME TO 20,000 METERS (USA)
Ronald R. Williams — 19 min 37 sec
Lockheed U-2C, 56-6682
1 P&W J75 P-138
Edwards AFB, CA
4/18/89

CLASS C-1.H (12,000-16,000 KG / 26,455-35,274 LBS)
GROUP III (JET ENGINE)

DISTANCE

DISTANCE WITHOUT LANDING (USA)
Randolph M. Kennedy — 6,099.91 km / 3,790.30 mi
Falcon 50
3 Garrett 731-3, 3,700 lbs
Bordeaux, France to Washington, DC
3/31/79

ALTITUDE

ALTITUDE (RUSSIA)
Roman Taskaev — 27,460 m / 90,092 ft
Mikoyan MiG-29
2 RD-33, 18,300 lbs
Akhtubinsk, Russia
4/26/95

ALTITUDE IN HORIZONTAL FLIGHT (USA)
James L. Barrilleaux — 20,479 m / 67,188 ft
Lockheed Martin ER-2
1 GE F118-101, 16,000 lbs
Edwards AFB, CA
11/19/98

ALTITUDE WITH 1,000 KG PAYLOAD (RUSSIA)
Oleg Antonovich — 25,150 m / 82,513 ft
Mikoyan MiG-29
2 RD-33, 18,300 lbs
Akhtubinsk, Russia
7/5/95

ALTITUDE WITH 2,000 KG PAYLOAD (USA)
Maj Alan W. Zwick — 20,399 m / 66,925 ft
Lockheed Martin U-2S
1 GE F118-101, 16,000 lbs
Beale AFB, CA
12/12/98

GREATEST PAYLOAD CARRIED TO 2,000 METERS (USSR)
Svetlana Savitskaya — 4,084 kg / 9,001 lbs
Galina Kortchuganova
YAK 40
3 AI-25, 1,500 kg
Podmoscovnoe, USSR
4/24/81

GREATEST PAYLOAD CARRIED TO 15,000 METERS (USA)
Lt Col Bryan Galbreath — 1,503 kg / 3,314 lbs
Lockheed Martin U-2S
1 GE F118-101, 16,000 lbs
Palmdale, CA
11/18/98

SPEED

SPEED OVER A 2,000 KM CLOSED CIRCUIT WITHOUT PAYLOAD (USA)
Steve Fossett, Pilot — 962.81 kmh
Harrison S. Holland, Copilot — 598.26 mph
Cessna 750 Citation X
2 R-R AE 3007C, 6,442 lbs
Boulder City, NV
9/26/99

SPEED OVER A 5,000 KM CLOSED CIRCUIT WITHOUT PAYLOAD (USA)
Steve Fossett, Pilot — 921.02 kmh
Darrin L. Adkins, Copilot — 572.29 mph
Cessna 750 Citation X
2 R-R AE 3007C, 6,442 lbs
Alamosa, CO
7/14/00

SPEED AROUND THE WORLD, EASTBOUND (USA) ✯
Steve Fossett, Pilot — 901.07 kmh
Darrin L. Adkins, Copilot — 559.89 mph
Alexander M. Tai, Copilot
Cessna 750 Citation X
2 R-R AE 3007C, 6,442 lbs
Los Angeles, Hamilton, Agadir, Luxor, Calcutta, Nagasaki, Midway Island, Los Angeles
2/16/00

SPEED AROUND THE WORLD, WESTBOUND (USA)
Steve Fossett, Pilot — 805.59 kmh
Alexander M. Tai, Copilot — 500.56 mph
Pierre F. d'Avenas, Copilot
Cessna 750 Citation X
2 R-R AE 3007C, 6,442 lbs
San José del Cabo, Kona, Majuro, Palau, Singapore, Male, Nairobi, Abidjan, Fortaleza, Barranquilla, San José del Cabo
11/24/00

TIME TO CLIMB

TIME TO 3,000 METERS (USSR)
Victor Pugachev — 25 sec
Sukhoi P-42
2 RR-32, 30,000 lbs
Podmoscovnoe, USSR
10/27/86

TIME TO 6,000 METERS (USSR)
Victor Pugachev — 37 sec
Sukhoi P-42
2 RR-32, 30,000 lbs
Podmoscovnoe, USSR
11/15/86

TIME TO 9,000 METERS (USSR)
Nikolai Sadovnikov — 44 sec
Sukhoi P-42
2 P-32, 13,600 kg
Podmoskovnoye, USSR
3/10/87

TIME TO 12,000 METERS (USSR)
Nikolai Sadovnikov — 55 sec
Sukhoi P-42
2 P-32, 13,600 kg
Podmoskovnoye, USSR
3/10/87

TIME TO 15,000 METERS (USSR)
Victor Pugachev — 82 sec
Sukhoi P-42
3/29/90

TIME TO 15,000 METERS WITH 1,000 KG PAYLOAD (USSR)
Victor Pugachev — 82 sec
Sukhoi P-42
3/29/90

CLASS C-1.I (16,000-20,000 KG / 35,274-44,092 LBS) GROUP III (JET ENGINE)

DISTANCE

DISTANCE WITHOUT LANDING (USA) ✯
Gene "Ed" Allen, Captain — 8,067.23 km
Herve Leprince-Ringuet, Co-Captain — 5,012.74 mi
Guy Mitaux Maurouard, 1st Officer
Dassault Falcon 900B
3 Garrett TFE731, 4,750 lbs
Paris, France - Houston, TX
7/16/91

DISTANCE OVER A CLOSED CIRCUIT WITHOUT LANDING (USA)
William S. Dirks — 5,075.88 km
Jeffrey C. Brollier — 3,154.00 mi
Cessna Citation X
2 Allison AE 3007C, 6,400 lbs
Reno, NV - Galveston, TX
9/13/97

ALTITUDE

ALTITUDE (USSR)
Vladmir V. Arkipenko — 21,830 m
Myasishchev M-17 — 71,621 ft
3/28/90

ALTITUDE IN HORIZONTAL FLIGHT (USSR)
Vladmir V. Arkipenko — 21,830 m
Myasishchev M-17 — 71,621 ft
3/28/90

ALTITUDE WITH 1,000 KG PAYLOAD (RUSSIA)
Victor Pugachev — 22,250 m
Sukhoi P-42 — 72,999 ft
Podomoscovnoe, Russia
5/20/93

ALTITUDE WITH 2,000 KG PAYLOAD (USSR)
Nicolas N. Generalov — 21,540 m
Myasishchev M-17 — 70,669 ft
4/17/90

GREATEST PAYLOAD CARRIED TO 2,000 METERS (USSR)
Svetlana Savitskaya — 5,012.00 kg
Galina Kortchuganova — 11,046.45 lbs
YAK 40
3 AI-25, 1,500 kg
Podmoscovnoe, USSR
4/23/81

GREATEST PAYLOAD CARRIED TO 15,000 METERS (RUSSIA)
Victor Pugachev — 1,015 kg
Sukhoi P-42 — 2,238 lbs
Podomoscovnoe, Russia
5/20/93

SPEED

SPEED OVER A 15/25 KM STRAIGHT COURSE (USA)
Maj. Richard A. Vandam — 1,704.04 kmh
Lt. Aaron M. Zeff — 1,058.84 mph
McDonnell Douglas RF-4C
2 GE J79-GE-15E, 17,800 lbs
Reno, NV
9/12/95

SPEED OVER A 100 KM CLOSED CIRCUIT WITHOUT PAYLOAD (USA)
Maj. Richard A. Vandam 1,264.47 kmh
Lt. Col. Bradley N. Wilkerson 785.70 mph
McDonnell Douglas RF-4C
2 GE J79-GE-15E, 17,800 lbs
Reno, NV
9/9/95

SPEED OVER A 500 KM CLOSED CIRCUIT WITHOUT PAYLOAD (USA)
Maj. Richard A. Vandam 1,273.96 kmh
Lt. Col. Bradley N. Wilkerson 791.60 mph
McDonnell Douglas RF-4C
2 GE J79-GE-15E, 17,800 lbs
Reno, NV
9/9/95

SPEED OVER A 1,000 KM CLOSED CIRCUIT WITHOUT PAYLOAD (USA)
Capt. Keith S. Ernst 986.59 kmh
Lt. Allan S. Renwick 613.04 mph
McDonnell Douglas RF-4C
2 GE J79-GE-15E, 17,800 lbs
Reno, NV
9/9/95

SPEED OVER A 1,000 KM CLOSED CIRCUIT WITH 1,000 KG PAYLOAD (USSR)
Oleg G. Smirnov 645.80 kmh
Myasishchev M-17 401.28 mph
4/20/90

SPEED OVER A 1,000 KM CLOSED CIRCUIT WITH 2,000 KG PAYLOAD (USSR)
Oleg G. Smirnov 645.80 kmh
Myasishchev M-17 401.28 mph
4/20/90

SPEED OVER A 5,000 KM CLOSED CIRCUIT WITHOUT PAYLOAD (USA)
William S. Dirks 876.66 kmh
Jeffrey C. Brollier 544.73 mph
Cessna Citation X
2 Allison AE 3007C, 6,400 lbs
Reno, NV - Galveston, TX
9/13/97

SPEED AROUND THE WORLD, EASTBOUND (SAUDI ARABIA)
Aziz Ojjeh 751.70 kmh
Challenger 601 467.11 mph
Nice to Nice
7/22-24/84

TIME TO CLIMB

TIME TO 3,000 METERS (USA)
Gary M. Freeman 1 min
Edward D. Mendenhall 32 sec
William M. Osborne
Gulfstream IV SP
2 RR Tay Mk 611-8, 13,850 lbs
Savannah, GA
5/21/95

TIME TO 3,000 METERS WITH 1,000 KG PAYLOAD (USA)
Gary M. Freeman 1 min
Edward D. Mendenhall 32 sec
William M. Osborne
Gulfstream IV SP
2 RR Tay Mk 611-8, 13,850 lbs
Savannah, GA
5/21/95

TIME TO 3,000 METERS WITH 2,000 KG PAYLOAD (USSR)
Oleg G. Smirnov 2 min
Myasishchev M-17 51.48 sec
4/19/90

TIME TO 6,000 METERS (USA)
Edward D. Mendenhall 3 min
Gary M. Freeman 40 sec
William M. Osborne
Gulfstream IV SP
2 RR Tay Mk 611-8, 13,850 lbs
Savannah, GA
5/21/95

TIME TO 6,000 METERS WITH 1,000 KG PAYLOAD (USA)
Edward D. Mendenhall 3 min
Gary M. Freeman 40 sec
William M. Osborne
Gulfstream IV SP
2 RR Tay Mk 611-8, 13,850 lbs
Savannah, GA
5/21/95

TIME TO 6,000 METERS WITH 2,000 KG PAYLOAD (USSR)
Oleg G. Smirnov 4 min
Myasishchev M-17 43.36 sec
4/19/90

TIME TO 9,000 METERS (USA)
Edward D. Mendenhall 5 min
Gary M. Freeman 28 sec
William M. Osborne
Gulfstream IV SP
2 RR Tay Mk 611-8, 13,850 lbs
Savannah, GA
5/21/95

TIME TO 9,000 METERS WITH 1,000 KG PAYLOAD (USA)
Edward D. Mendenhall 5 min
Gary M. Freeman 28 sec
William M. Osborne
Gulfstream IV SP
2 RR Tay Mk 611-8, 13,850 lbs
Savannah, GA
5/21/95

TIME TO 9,000 METERS WITH 2,000 KG PAYLOAD (USSR)
Oleg G. Smirnov 6 min
Myasishchev M-17 25.15 sec
4/19/90

TIME TO 12,000 METERS (USSR)
Vladmir V. Arkipenko 7 min
Myasishchev M-17 41 sec
4/6/90

TIME TO 12,000 METERS WITH 1,000 KG PAYLOAD (USSR)
Oleg G. Smirnov 8 min
Myasishchev M-17 44.35 sec
4/19/90

TIME TO 12,000 METERS WITH 2,000 KG PAYLOAD (USSR)
Oleg G. Smirnov 8 min
Myasishchev M-17 44.35 sec
4/19/90

TIME TO 15,000 METERS (RUSSIA)
Victor Pugachev 2 min
Sukhoi P-42 6 sec
Podomoscovnoe, Russia
5/20/93

TIME TO 15,000 METERS WITH 1,000 KG PAYLOAD (RUSSIA)
Victor Pugachev 2 min
Sukhoi P-42 6 sec
Podomoscovnoe, Russia
5/20/93

TIME TO 15,000 METERS WITH 2,000 KG PAYLOAD (USSR)
Nicolas N. Generalov — 12 min 49.10 sec
Myasishchev M-17
5/3/90

TIME TO 20,000 METERS (USSR)
Vladmir V. Arkipenko — 21 min 58 sec
Myasishchev M-17
4/6/90

TIME TO 20,000 METERS WITH 1,000 KG PAYLOAD (USSR)
Nicolas N. Generalov — 25 min 5.20 sec
Myasishchev M-17
5/14/90

TIME TO 20,000 METERS WITH 2,000 KG PAYLOAD (USSR)
Nicolas N. Generalov — 25 min 5.20 sec
Myasishchev M-17
5/14/90

CLASS C-1.J (20,000-25,000 KG / 44,092-55,116 LBS)
GROUP III (JET ENGINE)

DISTANCE

DISTANCE WITHOUT LANDING (FRANCE)
Jean Bongiraud — 8,873.75 km
Daniel Acton — 5,513.89 mi
Dassault Falcon 900 EX
3 Allied Signal TFE-731, 4,250 lbs
Palm Springs, CA - Amsterdam, The Netherlands
2/1/97

ALTITUDE

ALTITUDE (RUSSIA)
Victor Vasenkov — 21,360 m
M-55 — 70,079 ft
2 D-30812, 5,000 kg
Akhtubinsk, Russia
9/21/93

ALTITUDE* ✯
Gary M. Freeman — 57,260 ft
Edward D. Mendenhall
William M. Osborne
Gulfstream IV SP
2 RR Tay Mk 611-8, 13,850 lbs
Savannah, GA
5/21/95

ALTITUDE IN HORIZONTAL FLIGHT (USSR)
Victor Vasenkov — 21,340 m
M-55 — 70,013 ft
2 D-30812, 5,000 kg
Akhtubinsk, Russia
9/21/93

ALTITUDE IN HORIZONTAL FLIGHT*
Gary M. Freeman — 57,260 ft
Edward D. Mendenhall
William M. Osborne
Gulfstream IV SP
2 RR Tay Mk 611-8, 13,850 lbs
Savannah, GA
5/21/95

ALTITUDE WITH 1,000 KG PAYLOAD (USA)
Gary M. Freeman — 17,453 m
Edward D. Mendenhall — 57,260 ft
William M. Osborne
Gulfstream IV SP
2 RR Tay Mk 611-8, 13,850 lbs
Savannah, GA
5/21/95

ALTITUDE WITH 2,000 PAYLOAD (USSR)
Victor Vasenkov — 21,360 m
M-55 — 70,079 ft
2 D-30812, 5,000 kg
Akhtubinsk, Russia
9/21/93

ALTITUDE WITH 2,000 KG PAYLOAD*
Gary M. Freeman — 57,260 ft
Edward D. Mendenhall
William M. Osborne
Gulfstream IV SP
2 RR Tay Mk 611-8, 13,850 lbs
Savannah, GA
5/21/95

GREATEST PAYLOAD CARRIED TO 2,000 METERS (USSR)
Marina Popovich, Pilot — 3,528.00 kg
Sergei Maksimov, Copilot — 7,761.60 lbs
Antonov AN-72
2 D-36, 6,000 kg
Podmoscovnoe, USSR
11/17/83

GREATEST PAYLOAD CARRIED TO 15,000 METERS (USA)
Gary M. Freeman — 2,006 kg
Edward D. Mendenhall — 4,424 lbs
William M. Osborne
Gulfstream IV SP
2 RR Tay Mk 611-8, 13,850 lbs
Savannah, GA
5/21/95

SPEED

SPEED OVER A 15/25 KM STRAIGHT COURSE (USA)
Col. Stephen L. Vonderheide — 1,692.05 kmh
Capt. Matthew R. Speth — 1,051.39 mph
McDonnell Douglas RF-4C
2 GE J79-GE-15E, 17,800 lbs
Reno, NV
9/12/95

SPEED OVER A 100 KM CLOSED CIRCUIT WITHOUT PAYLOAD (USA)
Lt. Col. Wesley E. Behel, Jr. — 1,014.64 kmh
Lt. Col. Laurence D. Matlock — 630.46 mph
McDonnell Douglas RF-4C
2 GE J79-GE-15E, 17,800 lbs
Reno, NV
9/9/95

SPEED OVER A 500 KM CLOSED CIRCUIT WITHOUT PAYLOAD (USA)
Lt. Col. Wesley E. Behel, Jr. — 1,241.94 kmh
Lt. Col. Laurence D. Matlock — 771.70 mph
McDonnell Douglas RF-4C
2 GE J79-GE-15E, 17,800 lbs
Reno, NV
9/9/95

SPEED OVER A 1,000 KM CLOSED CIRCUIT WITHOUT PAYLOAD (USA)
Lt. Col. Charles M. Hanson — 993.16 kmh
Maj. Michael L. Caraker — 617.12 mph
McDonnell Douglas RF-4C
2 GE J79-GE-15E, 17,800 lbs
Reno, NV
9/9/95

SPEED OVER A 2,000 KM CLOSED CIRCUIT WITHOUT PAYLOAD (USSR)
Georgy Pobol, Pilot — 673.03 kmh
Vladimir Tkachenko, Copilot — 418.22 mph
Antonov AN-72
2 D-36, 6,000 kg
Podkievskoie, USSR
1/3/86

TIME TO CLIMB

TIME TO 3,000 METERS (USA)
Edward D. Mendenhall — 1 min 31 sec
Gary M. Freeman
William M. Osborne
Gulfstream IV SP
2 RR Tay Mk 611-8, 13,850 lbs
Savannah, GA
5/21/95

TIME TO 3,000 METERS WITH 1,000 KG PAYLOAD (USA)
Edward D. Mendenhall — 1 min 31 sec
Gary M. Freeman
William M. Osborne
Gulfstream IV SP
2 RR Tay Mk 611-8, 13,850 lbs
Savannah, GA
5/21/95

TIME TO 6,000 METERS (USA)
James R. McClellan — 3 min 46 sec
Gary M. Freeman
William M. Osborne
Gulfstream IV SP
2 RR Tay Mk 611-8, 13,850 lbs
Savannah, GA
5/21/95

TIME TO 6,000 METERS WITH 1,000 KG PAYLOAD (USA)
James R. McClellan — 3 min 46 sec
Gary M. Freeman
William M. Osborne
Gulfstream IV SP
2 RR Tay Mk 611-8, 13,850 lbs
Savannah, GA
5/21/95

TIME TO 9,000 METERS (RUSSIA)
Oleg Schepetkov — 5 min 56 sec
M-55
2 D-30812, 5,000 kg
Akhtubinsk, Russia
9/24/93

TIME TO 9,000 METERS*
James R. McClellan — 6 min 32 sec
Gary M. Freeman
William M. Osborne
Gulfstream IV SP
2 RR Tay Mk 611-8, 13,850 lbs
Savannah, GA
5/21/95

TIME TO 9,000 METERS WITH 1,000 KG PAYLOAD (RUSSIA)
Oleg Schepetkov — 5 min 56 sec
M-55
2 D-30812, 5,000 kg
Akhtubinsk, Russia
9/24/93

TIME TO 9,000 METERS WITH 1,000 KG PAYLOAD*
James R. McClellan — 6 min 32 sec
Gary M. Freeman
William M. Osborne
Gulfstream IV SP
2 RR Tay Mk 611-8, 13,850 lbs
Savannah, GA
5/21/95

TIME TO 9,000 METERS WITH 2,000 KG PAYLOAD (RUSSIA)
Oleg Schepetkov — 5 min 56 sec
M-55
2 D-30812, 5,000 kg
Akhtubinsk, Russia
9/24/93

TIME TO 12,000 METERS (RUSSIA)
Oleg Schepetkov — 8 min 13 sec
M-55
2 D-30812, 5,000 kg
Akhtubinsk, Russia
9/24/93

TIME TO 12,000 METERS*
James R. McClellan — 8 min 46 sec
Gary M. Freeman
William M. Osborne
Gulfstream IV SP
2 RR Tay Mk 611-8, 13,850 lbs
Savannah, GA
5/21/95

TIME TO 12,000 METERS WITH 1,000 KG PAYLOAD (RUSSIA)
Oleg Schepetkov — 8 min 13 sec
M-55
2 D-30812, 5,000 kg
Akhtubinsk, Russia
9/24/93

TIME TO 12,000 METERS WITH 1,000 KG PAYLOAD*
James R. McClellan — 8 min 46 sec
Gary M. Freeman
William M. Osborne
Gulfstream IV SP
2 RR Tay Mk 611-8, 13,850 lbs
Savannah, GA
5/21/95

TIME TO 12,000 METERS WITH 2,000 KG PAYLOAD (RUSSIA)
Oleg Schepetkov — 8 min 13 sec
M-55
2 D-30812, 5,000 kg
Akhtubinsk, Russia
9/24/93

TIME TO 15,000 METERS (RUSSIA)
Vladimir Bukhtoyarov — 12 min 6 sec
M-55
2 D-30812, 5,000 kg
Akhtubinsk, Russia
10/4/93

TIME TO 15,000 METERS*
Gary M. Freeman — 15 min 25 sec
Edward D. Mendenhall
William M. Osborne
Gulfstream IV SP
2 RR Tay Mk 611-8, 13,850 lbs
Savannah, GA
5/21/95

TIME TO 15,000 METERS WITH 1,000 KG PAYLOAD (RUSSIA)
Vladimir Bukhtoyarov — 12 min 6 sec
M-55
2 D-30812, 5,000 kg
Akhtubinsk, Russia
10/4/93

TIME TO 15,000 METERS WITH 1,000 KG PAYLOAD*
Gary M. Freeman — 15 min 25 sec
Edward D. Mendenhall
William M. Osborne
Gulfstream IV SP
2 RR Tay Mk 611-8, 13,850 lbs
Savannah, GA
5/21/95

TIME TO 15,000 METERS WITH 2,000 KG PAYLOAD (RUSSIA)
Vladimir Bukhtoyarov 12 min
M-55 6 sec
2 D-30812, 5,000 kg
Akhtubinsk, Russia
10/4/93

TIME TO 15,000 METERS WITH 2,000 KG PAYLOAD*
Gary M. Freeman 15 min
Edward D. Mendenhall 25 sec
William M. Osborne
Gulfstream IV SP
2 RR Tay Mk 611-8, 13,850 lbs
Savannah, GA
5/21/95

TIME TO 20,000 METERS (RUSSIA)
Victor Vasenkov 22 min
M-55 14 sec
2 D-30812, 5,000 kg
Akhtubinsk, Russia
9/21/93

TIME TO 20,000 METERS WITH 1,000 KG PAYLOAD (RUSSIA)
Victor Vasenkov 22 min
M-55 14 sec
2 D-30812, 5,000 kg
Akhtubinsk, Russia
9/21/93

TIME TO 20,000 METERS WITH 2,000 KG PAYLOAD (RUSSIA)
Victor Vasenkov 22 min
M-55 14 sec
2 D-30812, 5,000 kg
Akhtubinsk, Russia
9/21/93

CLASS C-1.K (25,000-35,000 KG / 55,116-77,162 LBS)
GROUP III (JET ENGINE)

DISTANCE

DISTANCE WITHOUT LANDING (USA)
Wayne Altman, Pilot 9,524.60 km
Robert Kirksey, Copilot 5,918.31 mi
Gulfstream IV
2 RR Tay Mk 611-8, 13,850 lbs
Tokyo - Albuquerque
3/3/93

DISTANCE OVER A CLOSED CIRCUIT WITHOUT LANDING (USSR)
Vladimir A. Stepanov 1,021.00 km
SU-24 634.42 mi
7/11/91

ALTITUDE

ALTITUDE (RUSSIA)
Ivan Pyshny 19,780 m
Ural Sultanov 64,895 ft
MiG-25 PU
2 P15 B-300, 11,300 kg
Ramenskoye, Russia
7/3/97

ALTITUDE IN HORIZONTAL FLIGHT (RUSSIA)
Ivan Pyshny 19,720 m
Ural Sultanov 64,698 ft
MiG-25 PU
2 P15 B-300, 11,300 kg
Ramenskoye, Russia
7/3/97

ALTITUDE WITH 1,000 KG PAYLOAD (USA)
Gary M. Freeman 16,784 m
Edward D. Mendenhall 55,067 ft
William M. Osborne
Gulfstream V
2 BMW Rolls-Royce BR710, 14,845 lbs
Savannah, GA
9/14/97

ALTITUDE WITH 2,000 KG PAYLOAD (USA)
Gary M. Freeman 16,784 m
Edward D. Mendenhall 55,067 ft
William M. Osborne
Gulfstream V
2 BMW Rolls-Royce BR710, 14,845 lbs
Savannah, GA
9/14/97

ALTITUDE WITH 5,000 KG PAYLOAD (RUSSIA)
Vyacheslav Petrucha 15,063 m
Alexandre Oshchepkov 49,419 ft
Sukhoi Su-32FN
2 Saturn AL-31F, 27,557 lbs
Aktyubinsk, Kazakhstan
8/3/99

GREATEST PAYLOAD CARRIED TO 2,000 METERS (USSR)
Sergei Maksimov, Pilot 8,064.00 kg
Marina Popovich, Copilot 17,740.08 lbs
Antonov AN-72
2 D-36, 6,000 kg
Podkievskoie, USSR
11/18/83

SPEED

SPEED OVER A 100 KM CLOSED CIRCUIT WITHOUT PAYLOAD (USSR)
Mikhail Kosarev 606.20 kmh
Sergei Sabaev 376.67 mph
Antonov An-72
2 Zaporozhye/Lotarev D-36, 13,227 lbs
Akhtubinsk, Russia
7/25/91

SPEED OVER A 500 KM CLOSED CIRCUIT WITHOUT PAYLOAD (USSR)
Lev Vasilyevich Koslov 643.43 kmh
Yuri Klishin 399.74 mph
Antonov An-72
2 Zaporozhye/Lotarev D-36, 13,227 lbs
Akhtubinsk, Russia
7/25/91

SPEED OVER A 1,000 KM CLOSED CIRCUIT WITHOUT PAYLOAD (USSR)
Alexandre Tkachenko 667.99 kmh
Georgy Pobol 415.07 mph
Antonov An-72
2 Zaporozhye/Lotarev D-36, 13,227 lbs
Akhtubinsk, Russia
11/27/85

SPEED OVER A 1,000 KM CLOSED CIRCUIT WITH 1,000 KG PAYLOAD (USSR)
Mikhail Kosarev 597.48 kmh
Alexander Butakov 371.26 mph
Antonov An-72
2 Zaporozhye/Lotarev D-36, 13,227 lbs
Akhtubinsk, Russia
3/20/91

SPEED OVER A 1,000 KM CLOSED CIRCUIT WITH 2,000 KG PAYLOAD (USSR)
Mikhail Kosarev 597.48 kmh
Alexander Butakov 371.26 mph
Antonov An-72
2 Zaporozhye/Lotarev D-36, 13,227 lbs
Akhtubinsk, Russia
3/20/91

SPEED OVER A 1,000 KM CLOSED CIRCUIT WITH 5,000 KG PAYLOAD (USSR)
Mikhail Konov 618.76 kmh
Alexander Butakov 384.47 mph
Antonov An-72
2 Zaporozhye/Lotarev D-36, 13,227 lbs
Akhtubinsk, Russia
3/22/91

SPEED OVER A 2,000 KM CLOSED CIRCUIT WITHOUT PAYLOAD (USSR)
Sergei Gorbik, Pilot 681.80 kmh
Georgy Pobol, Copilot 423.67 mph
Antonov AN-72
2 D-36, 6,000 kg
Podkieveskoie, USSR
11/28/85

SPEED OVER A 2,000 KM CLOSED CIRCUIT WITH 1,000 KG PAYLOAD (USSR)
Valery Savchuk 591.63 kmh
Alexander Kuznetsov 367.62 mph
Antonov An-72
2 Zaporozhye/Lotarev D-36, 13,227 lbs
Akhtubinsk, Russia
3/26/91

SPEED OVER A 2,000 KM CLOSED CIRCUIT WITH 2,000 KG PAYLOAD (USSR)
Valery Savchuk 591.63 kmh
Alexander Kuznetsov 367.62 mph
Antonov An-72
2 Zaporozhye/Lotarev D-36, 13,227 lbs
Akhtubinsk, Russia
3/26/91

SPEED AROUND THE WORLD, EASTBOUND (USA)
Allen E. Paulson, Captain 1,026.26 kmh
Robert K. Smyth, CoCaptain 637.71 mph
John Salamankas, CoCaptain
Jeff Bailey, CoCaptain
Gulfstream IV, N400GA
2 RR TAY MK611-8
Houston, Lake Charles, Shannon, Dubai, Hong Kong, Taipei, Honolulu, Maui, Houston, Lake Charles, Houston
Elapsed Time: 36 hrs 8 min 34 sec
2/26-28/88

SPEED AROUND THE WORLD, WESTBOUND (USA)
Allen E. Paulson, Captain 810.93 kmh
K.C. Edgecomb, 503.91 mph
Jefferson Bailey, John Salamankas, Pilots
Colin B. Allen, Flight Engineer
Gulfstream IV, N440GA
2 RR TAY MK610-8
Elapsed Time: 45 hrs 25 min 10 sec
6/12-14/87

SPEED AROUND THE WORLD OVER BOTH THE EARTH'S POLES (USA)
Brooke Knapp 538.83 kmh
Gulfstream III 334.83 mph
2 RR Spey MK511-8, 11,400 lbs
Los Angeles, Honolulu, Pago Pago, Christchurch, McMurdo Station, Punta Arenas, Recife, Tenerife, Trondheim, Fairbanks, Los Angeles
11/15-11/18/83

TIME TO CLIMB

TIME TO 3,000 METERS (USA)
Edward D. Mendenhall 1 min
Gary M. Freeman 54 sec
William M. Osborne
Gulfstream IV SP
2 RR Tay Mk 611-8, 13,850 lbs
Savannah, GA
5/21/95

TIME TO 3,000 METERS WITH 1,000 KG PAYLOAD (USA)
Edward D. Mendenhall 1 min
Gary M. Freeman 54 sec
William M. Osborne
Gulfstream IV SP
2 RR Tay Mk 611-8, 13,850 lbs
Savannah, GA
5/21/95

TIME TO 6,000 METERS (USA)
Gary M. Freeman 3 min
Edward D. Mendenhall 58 sec
William M. Osborne
Gulfstream IV SP
2 RR Tay Mk 611-8, 13,850 lbs
Savannah, GA
5/21/95

TIME TO 6,000 METERS WITH 1,000 KG PAYLOAD (USA)
Gary M. Freeman 3 min
Edward D. Mendenhall 58 sec
William M. Osborne
Gulfstream IV SP
2 RR Tay Mk 611-8, 13,850 lbs
Savannah, GA
5/21/95

TIME TO 9,000 METERS (USA)
Gary M. Freeman 5 min
Edward D. Mendenhall 54.4 sec
William M. Osborne
Gulfstream V
2 BMW Rolls-Royce BR710, 14,845 lbs
Savannah, GA
9/14/97

TIME TO 9,000 METERS WITH 1,000 KG PAYLOAD (USA)
Gary M. Freeman 5 min
Edward D. Mendenhall 54.4 sec
William M. Osborne
Gulfstream V
2 BMW Rolls-Royce BR710, 14,845 lbs
Savannah, GA
9/14/97

TIME TO 9,000 METERS WITH 2,000 KG PAYLOAD (USA)
Gary M. Freeman 5 min
Edward D. Mendenhall 54.4 sec
William M. Osborne
Gulfstream V
2 BMW Rolls-Royce BR710, 14,845 lbs
Savannah, GA
9/14/97

TIME TO 12,000 METERS (USA)
Edward D. Mendenhall 9 min
Gary M. Freeman 37.3 sec
William M. Osborne
Gulfstream V
2 BMW Rolls-Royce BR710, 14,845 lbs
Savannah, GA
9/14/97

TIME TO 12,000 METERS WITH 1,000 KG PAYLOAD (USA)
Edward D. Mendenhall 9 min
Gary M. Freeman 37.3 sec
William M. Osborne
Gulfstream V
2 BMW Rolls-Royce BR710, 14,845 lbs
Savannah, GA
9/14/97

TIME TO 12,000 METERS WITH 2,000 KG PAYLOAD (USA)
Edward D. Mendenhall 9 min
Gary M. Freeman 37.3 sec
William M. Osborne
Gulfstream V
2 BMW Rolls-Royce BR710, 14,845 lbs
Savannah, GA
9/14/97

TIME TO 15,000 METERS (USA)
Gary M. Freeman 16 min
Edward D. Mendenhall 5.3 sec
William M. Osborne
Gulfstream V
2 BMW Rolls-Royce BR710, 14,845 lbs
Savannah, GA
9/14/97

TIME TO 15,000 METERS WITH 1,000 KG PAYLOAD (USA)
Gary M. Freeman 16 min
Edward D. Mendenhall 5.3 sec
William M. Osborne
Gulfstream V
2 BMW Rolls-Royce BR710, 14,845 lbs
Savannah, GA
9/14/97

TIME TO 15,000 METERS WITH 2,000 KG PAYLOAD (USA)
Gary M. Freeman 16 min
Edward D. Mendenhall 5.3 sec
William M. Osborne
Gulfstream V
2 BMW Rolls-Royce BR710, 14,845 lbs
Savannah, GA
9/14/97

CLASS C-1.L (35,000-45,000 KG / 77,162-99,208 LBS)
GROUP III (JET ENGINE)

DISTANCE

DISTANCE WITHOUT LANDING (USA) ✯
William J. Watters 12,145.39 km
Raymond A. Wellington 7,546.79 mi
Ahmed M. Ragheb
Gulfstream G550
2 BMW R-R BR710, 15,385 lbs
Savannah, GA to Dubai, UAE
12/6/03

DISTANCE OVER A CLOSED CIRCUIT WITHOUT LANDING (RUSSIA)
Valery Kobalsko 2,018.59 km
Sukhoi Su-24 1,254.29 mi
10/3/91

ALTITUDE

ALTITUDE (RUSSIA)
Ivan Pyshny 22,150 m
Ural Sultanov 72,670 ft
MiG-25 PU
Ramenskoye, Russia
8/21/97

ALTITUDE IN HORIZONTAL FLIGHT (RUSSIA)
Ivan Pyshny 18,610 m
Ural Sultanov 61,056 ft
MiG-25 PU
Ramenskoye, Russia
8/21/97

ALTITUDE WITH 1,000 KG PAYLOAD (RUSSIA)
Vladimir Gurkin 21,695 m
Alexander Kozachenko 71,177 ft
MiG-31
Akhtubinsk, Russia
8/1/03

ALTITUDE WITH 1,000 KG PAYLOAD*
James R. McClellan 45,198 ft
Gary M. Freeman
William M. Osborne
Gulfstream V
2 BMW Rolls-Royce BR710, 14,845 lbs
Savannah, GA
9/14/97

ALTITUDE WITH 2,000 KG PAYLOAD (RUSSIA)
Vladimir Gurkin 21,695 m
Alexander Kozachenko 71,177 ft
MiG-31
Akhtubinsk, Russia
8/1/03

ALTITUDE WITH 2,000 KG PAYLOAD*
James R. McClellan 45,198 ft
Gary M. Freeman
William M. Osborne
Gulfstream V
2 BMW Rolls-Royce BR710, 14,845 lbs
Savannah, GA
9/14/97

ALTITUDE WITH 5,000 KG PAYLOAD (RUSSIA)
Igor Votinzev 14,727 m
Alexandre Gaivoronsky 48,317 ft
Sukhoi Su-32FN
Akhtubinsk, Russia
7/28/99

ALTITUDE WITH 10,000 KG PAYLOAD (USSR)
Vladimir Tkachenko, Pilot 10,960 m
Anatoly Moisseev, Copilot 35,958 ft
AN-74
2 D-36, 6,500 kg
Podmoskovnoye, USSR
4/30/87

ALTITUDE WITH 15,000 KG PAYLOAD (USSR)
Vladimir Tkachenko, Pilot 10,960 m
Anatoly Moisseev, Copilot 35,958 ft
AN-74
2 D-36, 6,500 kg
Podmoskovnoye, USSR
4/30/87

GREATEST PAYLOAD CARRIED TO 2,000 METERS (USSR)
Vladimir Tkachenko, Pilot 15,256.00 kg
Anatoly Moisseev, Copilot 33,563.20 lb
AN-74
2 D-36, 6,500 kg
Podmoskovnoye, USSR
4/30/87

GREATEST PAYLOAD CARRIED TO 15,000 METERS (RUSSIA)
Igor Solovjev 2,330 kg
Vladimir Shendrik 5,136 lb
Sukhoi Su-32FN
2 Saturn AL-31F, 27,557 lbs
Zhukovskiy, Russia
8/19/99

SPEED

SPEED OVER A 2,000 KM CLOSED CIRCUIT WITHOUT PAYLOAD (USSR)
Valery A. Kobalsko 778.93 kmh
SU-24 484.00 mph
7/18/91

TIME TO CLIMB

TIME TO 3,000 METERS (RUSSIA)
Vladimir Gurkin 1 min
Alexander Kozachenko 22 sec
MiG-31
Akhtubinsk, Russia
8/1/03

TIME TO 3,000 METERS WITH 1,000 KG PAYLOAD (RUSSIA)
Vladimir Gurkin 1 min
Alexander Kozachenko 22 sec
MiG-31
Akhtubinsk, Russia
8/1/03

TIME TO 3,000 METERS WITH 1,000 KG PAYLOAD*
Edward D. Mendenhall 2 min
Gary M. Freeman 38.8 sec
William M. Osborne
Gulfstream V
2 BMW Rolls-Royce BR710, 14,845 lbs
Savannah, GA
9/14/97

TIME TO 3,000 METERS WITH 2,000 KG PAYLOAD (RUSSIA)
Vladimir Gurkin 1 min
Alexander Kozachenko 22 sec
MiG-31
Akhtubinsk, Russia
8/1/03

TIME TO 3,000 METERS WITH 2,000 KG PAYLOAD*
Edward D. Mendenhall 2 min
Gary M. Freeman 38.8 sec
William M. Osborne
Gulfstream V
2 BMW Rolls-Royce BR710, 14,845 lbs
Savannah, GA
9/14/97

TIME TO 6,000 METERS (RUSSIA)
Vladimir Gurkin 1 min
Alexander Kozachenko 50 sec
MiG-31
Akhtubinsk, Russia
8/1/03

TIME TO 6,000 METERS WITH 1,000 KG PAYLOAD (RUSSIA)
Vladimir Gurkin 1 min
Alexander Kozachenko 50 sec
MiG-31
Akhtubinsk, Russia
8/1/03

TIME TO 6,000 METERS WITH 1,000 KG PAYLOAD*
Gary M. Freeman 5 min
James R. McClellan 22.8 sec
William M. Osborne
Gulfstream V
2 BMW Rolls-Royce BR710, 14,845 lbs
Savannah, GA
9/14/97

TIME TO 6,000 METERS WITH 2,000 KG PAYLOAD (RUSSIA)
Vladimir Gurkin 1 min
Alexander Kozachenko 50 sec
MiG-31
Akhtubinsk, Russia
8/1/03

TIME TO 6,000 METERS WITH 2,000 KG PAYLOAD*
Gary M. Freeman 5 min
James R. McClellan 22.8 sec
William M. Osborne
Gulfstream V
2 BMW Rolls-Royce BR710, 14,845 lbs
Savannah, GA
9/14/97

TIME TO 9,000 METERS (RUSSIA)
Vladimir Gurkin 2 min
Alexander Kozachenko 22 sec
MiG-31
Akhtubinsk, Russia
8/1/03

TIME TO 9,000 METERS WITH 1,000 KG PAYLOAD (RUSSIA)
Vladimir Gurkin 2 min
Alexander Kozachenko 22 sec
MiG-31
Akhtubinsk, Russia
8/1/03

TIME TO 9,000 METERS WITH 1,000 KG PAYLOAD*
Edward D. Mendenhall 9 min
Gary M. Freeman 11.6 sec
William M. Osborne
Gulfstream V
2 BMW Rolls-Royce BR710, 14,845 lbs
Savannah, GA
9/14/97

TIME TO 9,000 METERS WITH 2,000 KG PAYLOAD (RUSSIA)
Vladimir Gurkin 2 min
Alexander Kozachenko 22 sec
MiG-31
Akhtubinsk, Russia
8/1/03

TIME TO 9,000 METERS WITH 2,000 KG PAYLOAD*
Edward D. Mendenhall 9 min
Gary M. Freeman 11.6 sec
William M. Osborne
Gulfstream V
2 BMW Rolls-Royce BR710, 14,845 lbs
Savannah, GA
9/14/97

TIME TO 12,000 METERS (RUSSIA)
Ivan Pyshny 5 min
Ural Sultanov 42 sec
MiG-25 PU
Ramenskoye, Russia
8/21/97

TIME TO 12,000 METERS WITH 1,000 KG PAYLOAD (RUSSIA)
Vladimir Gurkin 5 min
Alexander Kozachenko 52 sec
MiG-31
Akhtubinsk, Russia
8/1/03

TIME TO 12,000 METERS WITH 1,000 KG PAYLOAD*
James R. McClellan 16 min
Gary M. Freeman 31.8 sec
William M. Osborne
Gulfstream V
2 BMW Rolls-Royce BR710, 14,845 lbs
Savannah, GA
9/14/97

TIME TO 12,000 METERS WITH 2,000 KG PAYLOAD (RUSSIA)
Vladimir Gurkin — 5 min
Alexander Kozachenko — 52 sec
MiG-31
Akhtubinsk, Russia
8/1/03

TIME TO 12,000 METERS WITH 2,000 KG PAYLOAD*
James R. McClellan — 16 min
Gary M. Freeman — 31.8 sec
William M. Osborne
Gulfstream V
2 BMW Rolls-Royce BR710, 14,845 lbs
Savannah, GA
9/14/97

TIME TO 15,000 METERS (RUSSIA)
Vladimir Gurkin — 7 min
Alexander Kozachenko — 37 sec
MiG-31
Akhtubinsk, Russia
8/1/03

TIME TO 15,000 METERS WITH 1,000 KG PAYLOAD (RUSSIA)
Vladimir Gurkin — 7 min
Alexander Kozachenko — 37 sec
MiG-31
Akhtubinsk, Russia
8/1/03

TIME TO 15,000 METERS WITH 2,000 KG PAYLOAD (RUSSIA)
Vladimir Gurkin — 7 min
Alexander Kozachenko — 37 sec
MiG-31
Akhtubinsk, Russia
8/1/03

TIME TO 20,000 METERS (RUSSIA)
Vladimir Gurkin — 8 min
Alexander Kozachenko — 23 sec
MiG-31
Akhtubinsk, Russia
8/1/03

TIME TO 20,000 METERS WITH 1,000 KG PAYLOAD (RUSSIA)
Vladimir Gurkin — 8 min
Alexander Kozachenko — 23 sec
MiG-31
Akhtubinsk, Russia
8/1/03

TIME TO 20,000 METERS WITH 2,000 KG PAYLOAD (RUSSIA)
Vladimir Gurkin — 8 min
Alexander Kozachenko — 23 sec
MiG-31
Akhtubinsk, Russia
8/1/03

CLASS C-1.M (45,000-60,000 KG / 99,208-132,277 LBS) GROUP III (JET ENGINE)

DISTANCE

DISTANCE WITHOUT LANDING (AUSTRALIA)
Ian Haigh — 6,262.25 km
Boeing 737-400 — 3,891.18 mi
Hilo, HI to Noumea, New Caledonia
7/23/92

ALTITUDE

ALTITUDE (RUSSIA)
Nicolas A. Krupenko — 14,120 m
Tupolev Tu-16 — 46,325 ft
9/27/91

ALTITUDE IN HORIZONTAL FLIGHT (USSR)
Alexander A. Ivlev, Pilot — 14,590 m
Igor L. Kozlov, Copilot — 47,867 ft
Tupolev Tu-16
2/26/91

ALTITUDE WITH 1,000 KG PAYLOAD (RUSSIA)
Igor Shishkin — 14,180 m
Tupolev Tu-16 — 46,522 ft
9/24/91

ALTITUDE WITH 2,000 KG PAYLOAD (RUSSIA)
Igor Shishkin — 14,180 m
Tupolev Tu-16 — 46,522 ft
9/24/91

ALTITUDE WITH 5,000 KG PAYLOAD (RUSSIA)
Igor Shishkin — 14,180 m
Tupolev Tu-16 — 46,522 ft
9/24/91

GREATEST PAYLOAD CARRIED TO 2,000 METERS (USSR)
Valentin Moukhin, Pilot — 20,186 kg
Eduard Kniaginichev — 44,490 lbs
Yak 42
3 D-36, 14,330 lbs
Podmoscovnoe, USSR
1/29/81

SPEED

SPEED OVER A 1,000 KM CLOSED CIRCUIT WITH 2,000 KG PAYLOAD (USSR)
Lev V. Kozlov — 913.30 kmh
Tupolev Tu-16 — 567.49 mph
7/22/91

SPEED OVER A 1,000 KM CLOSED CIRCUIT WITH 5,000 KG PAYLOAD (RUSSIA)
Valery Lobovikov — 881.45 kmh
Tupolev Tu-16 — 547.70 mph
10/1/91

TIME TO CLIMB

TIME TO 3,000 METERS (USA)
LTC Ira S. Paul, Command Pilot — 1 min
CPT David G. Glisson, Arcft Cmdr — 39 sec
1Lt Scott A. Neumann, Pilot
CPT Mark T. Moss, Navigator
SSG Randy S. Seip, Boom Operator
Boeing KC-135R, 62-3554
4 CFM F108(56-2B-1), 22,000 lbs
Robins AFB, GA
11/19/88

TIME TO 6,000 METERS (USA)
LTC Ira S. Paul, Command Pilot — 2 min
CPT David G. Glisson, Arcft Cmdr — 57 sec
1Lt Scott A. Neumann, Pilot
CPT Mark T. Moss, Navigator
SSG Randy S. Seip, Boom Operator
Boeing KC-135R, 62-3554
4 CFM F108(56-2B-1), 22,000 lbs
Robins AFB, GA
11/19/88

TIME TO 9,000 METERS (USA)
LTC Ira S. Paul, Command Pilot — 4 min 23 sec
CPT David G. Glisson, Arcft Cmdr
1Lt Scott A. Neumann, Pilot
CPT Mark T. Moss, Navigator
SSG Randy S. Seip, Boom Operator
Boeing KC-135R, 62-3554
4 CFM F108(56-2B-1), 22,000 lb
Robins AFB, GA
11/19/88

TIME TO 12,000 METERS (USA)
LTC Ira S. Paul, Command Pilot — 5 min 51 sec
CPT David G. Glisson, Arcft Cmdr
1Lt Scott A. Neumann, Pilot
CPT Mark T. Moss, Navigator
SSG Randy S. Seip, Boom Operator
Boeing KC-135R, 62-3554
4 CFM F108(56-2B-1), 22,000 lbs
Robins AFB, GA
11/19/88

TIME TO 15,000 METERS (USA)
LTC Ira S. Paul, Command Pilot — 8 min 15 sec
CPT David G. Glisson, Arcft Cmdr
1Lt Scott A. Neumann, Pilot
CPT Mark T. Moss, Navigator
SSG Randy S. Seip, Boom Operator
Boeing KC-135R, 62-3554
4 CFM F108(56-2B-1), 22,000 lbs
Robins AFB, GA
11/19/88

CLASS C-1.N (60,000-80,000 KG / 132,277-176,36 LBS)
GROUP III (JET ENGINE)

DISTANCE

DISTANCE WITHOUT LANDING (GERMANY)
Thomas Scheel — 8,116.83 km
Boeing 737-700 — 5,043.56 mi
Seattle, WA to Berlin, Germany
3/11/98

ALTITUDE

ALTITUDE (USA)
Robert E. Siman — 11,247 m
Hassan A. Al-Kandry — 36,900 ft
McDonnell Douglas MD-83
2 P&W JT8D-219, 21,700 lbs
5/23/93

ALTITUDE IN HORIZONTAL FLIGHT (USA)
Robert E. Siman — 11,247 m
Hassan A. Al-Kandry — 36,900 ft
McDonnell Douglas MD-83
2 P&W JT8D-219, 21,700 lbs
5/23/93

ALTITUDE WITH 1,000 KG PAYLOAD (USSR)
Valery V. Pavlov, Pilot — 9,546 m
Vladmir Sevankaev, Copilot — 31,319 ft
Tupolev Tu-155
8/4/89

ALTITUDE WITH 2,000 KG PAYLOAD (USA)
Robert E. Siman — 11,247 m
Hassan A. Al-Kandry — 36,900 ft
McDonnell Douglas MD-83
2 P&W JT8D-219, 21,700 lbs
5/23/93

ALTITUDE WITH 5,000 KG PAYLOAD (USSR)
Valery Lobovikov, Pilot — 14,010 m
Vladimir A. Volodkovitch, Copilot — 45,965 ft
Tupolev Tu-16
2/28/91

GREATEST PAYLOAD CARRIED TO 2,000 METERS (USSR)
Sergei Zhuchkov, Pilot — 8,406.00 kg
Valery V. Lobovikobv, Copilot — 18,532.00 lbs
Tupolev Tu-16
3/22/91

SPEED

SPEED OVER A 1,000 KM CLOSED CIRCUIT WITHOUT PAYLOAD (USSR)
Alexander Eboldov, Pilot — 897.16 kmh
Oleg V. Redine, Copilot — 557.46 mph
Tupolev Tu-16
3/5/91

SPEED OVER A 1,000 KM CLOSED CIRCUIT WITH 1,000 KG PAYLOAD (USSR)
Alexander Eboldov, Pilot — 897.16 kmh
Oleg V. Redine, Copilot — 557.46 mph
Tupolev Tu-16
3/5/91

SPEED OVER A 1,000 KM CLOSED CIRCUIT WITH 2,000 KG PAYLOAD (USSR)
Alexander Eboldov, Pilot — 897.16 kmh
Oleg V. Redine, Copilot — 557.46 mph
Tupolev Tu-16
3/5/91

SPEED OVER A 1,000 KM CLOSED CIRCUIT WITH 5,000 KG PAYLOAD (USSR)
Alexander Eboldov, Pilot — 897.16 kmh
Oleg V. Redine, Copilot — 557.46 mph
Tupolev Tu-16
3/5/91

SPEED OVER A 2,000 KM CLOSED CIRCUIT WITHOUT PAYLOAD (USSR)
Nicolas A. Krupenko, Pilot — 889.42 kmh
Alexander N. Mazapine, Copilot — 552.66 mph
Tupolev Tu-16
3/20/91

SPEED OVER A 2,000 KM CLOSED CIRCUIT WITH 1,000 KG PAYLOAD (USSR)
Nicolas A. Krupenko, Pilot — 889.42 kmh
Alexander N. Mazapine, Copilot — 552.66 mph
Tupolev Tu-16
3/20/91

SPEED OVER A 2,000 KM CLOSED CIRCUIT WITH 2,000 KG PAYLOAD (USSR)
Nicolas A. Krupenko, Pilot — 889.42 kmh
Alexander N. Mazapine, Copilot — 552.66 mph
Tupolev Tu-16
3/20/91

SPEED OVER A 2,000 KM CLOSED CIRCUIT WITH 5,000 KG PAYLOAD (USSR)
Nicolas A. Krupenko, Pilot — 889.42 kmh
Alexander N. Mazapine, Copilot — 552.66 mph
Tupolev Tu-16
3/20/91

TIME TO CLIMB

TIME TO 3,000 METERS (USA)
LTC Ira S. Paul, Command Pilot — 1 min 43 sec
MAJ Thomas Yarbrough, Arcft Cmdr
1LT Donald Cobacchini, Pilot
1LT David Whisenand, Navigator
SSG David J. Passey, Boom Operator
Boeing KC-135R, 62-3554
4 CFM F108(56-2B-1), 22,000 lbs
Robins AFB, GA
11/19/88

TIME TO 6,000 METERS (USA)
LTC Ira S. Paul, Command Pilot — 2 min 58 sec
MAJ Thomas Yarbrough, Arcft Cmdr
1LT Donald Cobacchini, Pilot
1LT David Whisenand, Navigator
SSG David J. Passey, Boom Operator
Boeing KC-135R, 62-3554
4 CFM F108(56-2B-1), 22,000 lbs
Robins AFB, GA
11/19/88

TIME TO 9,000 METERS (USA)
LTC Ira S. Paul, Command Pilot — 4 min 29 sec
MAJ Thomas Yarbrough, Arcft Cmdr
1LT Donald Cobacchini, Pilot
1LT David Whisenand, Navigator
SSG David J. Passey, Boom Operator
Boeing KC-135R, 62-3554
4 CFM F108(56-2B-1), 22,000 lbs
Robins AFB, GA
11/19/88

TIME TO 12,000 METERS (USA)
LTC Ira S. Paul, Command Pilot — 5 min 44 sec
MAJ Thomas Yarbrough, Arcft Cmdr
1LT Donald Cobacchini, Pilot
1LT David Whisenand, Navigator
SSG David J. Passey, Boom Operator
Boeing KC-135R, 62-3554
4 CFM F108(56-2B-1), 22,000 lbs
Robins AFB, GA
11/19/88

CLASS C-1.O (80,000-100,000 KG / 176,370-220,462 LBS)
GROUP III (JET ENGINE)

ALTITUDE

ALTITUDE (USSR)
Vladmir Sevankaev, Pilot — 12,170 m
Valery V. Pavlov, Copilot — 39,928 ft
Tupolev Tu-155 (No. 85035)
12/4/90

ALTITUDE IN HORIZONTAL FLIGHT (USSR)
Vladmir Sevankaev, Pilot — 12,130 m
Valery V. Pavlov, Copilot — 39,797 ft
Tupolev Tu-155 (No. 85035)
12/4/90

ALTITUDE WITH 1,000 KG PAYLOAD (USSR)
Vladmir Sevankaev, Pilot — 12,170 m
Valery V. Pavlov, Copilot — 39,928 ft
Tupolev Tu-155 (No. 85035)
12/4/90

ALTITUDE WITH 2,000 KG PAYLOAD (USSR)
Vladmir Sevankaev, Pilot — 12,170 m
Valery V. Pavlov, Copilot — 39,928 ft
Tupolev Tu-155 (No. 85035)
12/4/90

GREATEST PAYLOAD CARRIED TO 2,000 METERS (USSR)
Vladmir Sevankaev, Pilot — 2,308.2 kg
Valery V. Pavlov, Copilot — 5,088.71 lbs
Tupolev Tu-155
3 Kuznetsov NK-88, 10,500 ehp
8/7/89

TIME TO CLIMB

TIME TO 3,000 METERS (USA)
Lt Col William J. Moran Jr. — 1 min 13.06 sec
Capt Mark L. Eby
Capt Richard M. Nehls
Rockwell International B-1B
4 GE F101-GE-102, 30,780 lbs
Grand Forks AFB, ND
2/28/92

TIME TO 6,000 METERS (USA)
Lt Col William J. Moran Jr. — 1 min 42.23 sec
Capt Mark L. Eby
Capt Richard M. Nehls
Rockwell International B-1B
4 GE F101-GE-102, 30,780 lbs
Grand Forks AFB, ND
2/28/92

TIME TO 9,000 METERS (USA)
Lt Col William J. Moran Jr. — 2 min 10.98 sec
Capt Mark L. Eby
Capt Richard M. Nehls
Rockwell International B-1B
4 GE F101-GE-102, 30,780 lbs
Grand Forks AFB, ND
2/28/92

TIME TO 12,000 METERS (USA)
Lt Col James P. Robinson — 5 min 1.63 sec
Capt Scott A. Neumann
Capt Dennis J. Murphy, II
Rockwell International B-1B
4 GE F101-GE-102, 30,780 lbs
Grand Forks AFB, ND
2/28/92

CLASS C-1.P (100,000-150,000 KG / 220,462-330,693 LBS)
GROUP III (JET ENGINE)

DISTANCE

DISTANCE WITHOUT LANDING (USA)
Capt. Jeff Kennedy — 16,227.19 km
Capt. Robert Kilgore — 10,083.11 mi
1st Lt. John Isakson
Capt. Mark Hostetter
1st Lt. Ronald Fischer
MSgt. Temur Ablay
SMSgt. Dan Deloy
Boeing KC-135R
4 CFM F108-CF-100, 22,000 lbs
Okinawa, Japan to McGuire AFB, NJ
12/19/92

ALTITUDE

ALTITUDE (USA)
Capt. R. F. Olszewski — 13,136 m
USAir Boeing 767-201ER — 43,100 ft
San Juan, PR - Charlotte, NC
7/25/91

ALTITUDE IN HORIZONTAL FLIGHT (USA)
Capt. R. F. Olszewski — 13,136 m
USAir Boeing 767-201ER — 43,100 ft
San Juan, PR - Charlotte, NC
7/25/91

ALTITUDE WITH 1,000 KG PAYLOAD (USA)
Capt. R. F. Olszewski — 13,136 m
USAir Boeing 767-201ER — 43,100 ft
San Juan, PR - Charlotte, NC
7/25/91

ALTITUDE WITH 2,000 KG PAYLOAD (USA)
Capt. R. F. Olszewski — 13,136 m
USAir Boeing 767-201ER — 43,100 ft
San Juan, PR - Charlotte, NC
7/25/91

ALTITUDE WITH 5,000 KG PAYLOAD (USA)
Capt. R. F. Olszewski — 13,136 m
USAir Boeing 767-201ER — 43,100 ft
San Juan, PR - Charlotte, NC
7/25/91

ALTITUDE WITH 10,000 KG PAYLOAD (USA)
Capt. R. F. Olszewski 13,136 m
USAir Boeing 767-201ER 43,100 ft
San Juan, PR - Charlotte, NC
7/25/91

ALTITUDE WITH 15,000 KG PAYLOAD (USA)
Capt. R. F. Olszewski 13,136 m
USAir Boeing 767-201ER 43,100 ft
San Juan, PR - Charlotte, NC
7/25/91

SPEED

SPEED OVER A 15/25 KM STRAIGHT COURSE (USA)
Troy A. Asher 1,300.47 kmh
Richard C. Recker 808.87 mph
Jeffery C. Wharton
Rodney F. Todaro
Rockwell B-1B
4 GE F101, 30,780 lbs
Edwards AFB, CA
10/26/03

SPEED OVER A 100 KM CLOSED CIRCUIT WITHOUT PAYLOAD (USA)
Troy A. Asher 1,045.65 kmh
Richard C. Recker 649.74 mph
Jeffery C. Wharton
Rodney F. Todaro
Rockwell B-1B
4 GE F101, 30,780 lbs
Edwards AFB, CA
10/26/03

SPEED OVER A 100 KM CLOSED CIRCUIT WITH 1,000 KG PAYLOAD (USA)
Troy A. Asher 1,045.65 kmh
Richard C. Recker 649.74 mph
Jeffery C. Wharton
Rodney F. Todaro
Rockwell B-1B
4 GE F101, 30,780 lbs
Edwards AFB, CA
10/26/03

SPEED OVER A 100 KM CLOSED CIRCUIT WITH 2,000 KG PAYLOAD (USA)
Troy A. Asher 1,045.65 kmh
Richard C. Recker 649.74 mph
Jeffery C. Wharton
Rodney F. Todaro
Rockwell B-1B
4 GE F101, 30,780 lbs
Edwards AFB, CA
10/26/03

SPEED OVER A 100 KM CLOSED CIRCUIT WITH 5,000 KG PAYLOAD (USA)
Troy A. Asher 1,045.65 kmh
Richard C. Recker 649.74 mph
Jeffery C. Wharton
Rodney F. Todaro
Rockwell B-1B
4 GE F101, 30,780 lbs
Edwards AFB, CA
10/26/03

SPEED OVER A 100 KM CLOSED CIRCUIT WITH 10,000 KG PAYLOAD (USA)
Troy A. Asher 1,045.65 kmh
Richard C. Recker 649.74 mph
Jeffery C. Wharton
Rodney F. Todaro
Rockwell B-1B
4 GE F101, 30,780 lbs
Edwards AFB, CA
10/26/03

SPEED OVER A 100 KM CLOSED CIRCUIT WITH 15,000 KG PAYLOAD (USA)
Troy A. Asher 1,045.65 kmh
Richard C. Recker 649.74 mph
Jeffery C. Wharton
Rodney F. Todaro
Rockwell B-1B
4 GE F101, 30,780 lbs
Edwards AFB, CA
10/26/03

SPEED OVER A 100 KM CLOSED CIRCUIT WITH 20,000 KG PAYLOAD (USA)
Troy A. Asher 1,045.65 kmh
Richard C. Recker 649.74 mph
Jeffery C. Wharton
Rodney F. Todaro
Rockwell B-1B
4 GE F101, 30,780 lbs
Edwards AFB, CA
10/26/03

SPEED OVER A 100 KM CLOSED CIRCUIT WITH 25,000 KG PAYLOAD (USA)
Troy A. Asher 1,045.65 kmh
Richard C. Recker 649.74 mph
Jeffery C. Wharton
Rodney F. Todaro
Rockwell B-1B
4 GE F101, 30,780 lbs
Edwards AFB, CA
10/26/03

SPEED OVER A 500 KM CLOSED CIRCUIT WITHOUT PAYLOAD (USA)
Troy A. Asher 1,036.71 kmh
Richard C. Recker 644.18 mph
Jeffery C. Wharton
Rodney F. Todaro
Rockwell B-1B
4 GE F101, 30,780 lbs
Edwards AFB, CA
10/26/03

SPEED OVER A 500 KM CLOSED CIRCUIT WITH 1,000 KG PAYLOAD (USA)
Troy A. Asher 1,036.71 kmh
Richard C. Recker 644.18 mph
Jeffery C. Wharton
Rodney F. Todaro
Rockwell B-1B
4 GE F101, 30,780 lbs
Edwards AFB, CA
10/26/03

SPEED OVER A 500 KM CLOSED CIRCUIT WITH 2,000 KG PAYLOAD (USA)
Troy A. Asher 1,036.71 kmh
Richard C. Recker 644.18 mph
Jeffery C. Wharton
Rodney F. Todaro
Rockwell B-1B
4 GE F101, 30,780 lbs
Edwards AFB, CA
10/26/03

SPEED OVER A 500 KM CLOSED CIRCUIT WITH 5,000 KG PAYLOAD (USA)
Troy A. Asher 1,036.71 kmh
Richard C. Recker 644.18 mph
Jeffery C. Wharton
Rodney F. Todaro
Rockwell B-1B
4 GE F101, 30,780 lbs
Edwards AFB, CA
10/26/03

SPEED OVER A 500 KM CLOSED CIRCUIT WITH 10,000 KG PAYLOAD (USA)
Troy A. Asher 1,036.71 kmh
Richard C. Recker 644.18 mph
Jeffery C. Wharton
Rodney F. Todaro
Rockwell B-1B
4 GE F101, 30,780 lbs
Edwards AFB, CA
10/26/03

SPEED OVER A 500 KM CLOSED CIRCUIT WITH 15,000 KG PAYLOAD (USA)
Troy A. Asher 1,036.71 kmh
Richard C. Recker 644.18 mph
Jeffery C. Wharton
Rodney F. Todaro
Rockwell B-1B
4 GE F101, 30,780 lbs
Edwards AFB, CA
10/26/03

SPEED OVER A 500 KM CLOSED CIRCUIT WITH 20,000 KG PAYLOAD (USA)
Troy A. Asher 1,036.71 kmh
Richard C. Recker 644.18 mph
Jeffery C. Wharton
Rodney F. Todaro
Rockwell B-1B
4 GE F101, 30,780 lbs
Edwards AFB, CA
10/26/03

SPEED OVER A 500 KM CLOSED CIRCUIT WITH 25,000 KG PAYLOAD (USA)
Troy A. Asher 1,036.71 kmh
Richard C. Recker 644.18 mph
Jeffery C. Wharton
Rodney F. Todaro
Rockwell B-1B
4 GE F101, 30,780 lbs
Edwards AFB, CA
10/26/03

SPEED OVER A 1,000 KM CLOSED CIRCUIT WITHOUT PAYLOAD (USA)
Troy A. Asher 1,016.20 kmh
Richard C. Recker 631.44 mph
Jeffery C. Wharton
Rodney F. Todaro
Rockwell B-1B
4 GE F101, 30,780 lbs
Edwards AFB, CA
10/26/03

SPEED OVER A 1,000 KM CLOSED CIRCUIT WITH 1,000 KG PAYLOAD (USA)
Troy A. Asher 1,016.20 kmh
Richard C. Recker 631.44 mph
Jeffery C. Wharton
Rodney F. Todaro
Rockwell B-1B
4 GE F101, 30,780 lbs
Edwards AFB, CA
10/26/03

SPEED OVER A 1,000 KM CLOSED CIRCUIT WITH 2,000 KG PAYLOAD (USA)
Troy A. Asher 1,016.20 kmh
Richard C. Recker 631.44 mph
Jeffery C. Wharton
Rodney F. Todaro
Rockwell B-1B
4 GE F101, 30,780 lbs
Edwards AFB, CA
10/26/03

SPEED OVER A 1,000 KM CLOSED CIRCUIT WITH 5,000 KG PAYLOAD (USA)
Troy A. Asher 1,016.20 kmh
Richard C. Recker 631.44 mph
Jeffery C. Wharton
Rodney F. Todaro
Rockwell B-1B
4 GE F101, 30,780 lbs
Edwards AFB, CA
10/26/03

SPEED OVER A 1,000 KM CLOSED CIRCUIT WITH 10,000 KG PAYLOAD (USA)
Troy A. Asher 1,016.20 kmh
Richard C. Recker 631.44 mph
Jeffery C. Wharton
Rodney F. Todaro
Rockwell B-1B
4 GE F101, 30,780 lbs
Edwards AFB, CA
10/26/03

SPEED OVER A 1,000 KM CLOSED CIRCUIT WITH 15,000 KG PAYLOAD (USA)
Troy A. Asher 1,016.20 kmh
Richard C. Recker 631.44 mph
Jeffery C. Wharton
Rodney F. Todaro
Rockwell B-1B
4 GE F101, 30,780 lbs
Edwards AFB, CA
10/26/03

SPEED OVER A 1,000 KM CLOSED CIRCUIT WITH 20,000 KG PAYLOAD (USA)
Troy A. Asher 1,016.20 kmh
Richard C. Recker 631.44 mph
Jeffery C. Wharton
Rodney F. Todaro
Rockwell B-1B
4 GE F101, 30,780 lbs
Edwards AFB, CA
10/26/03

SPEED OVER A 1,000 KM CLOSED CIRCUIT WITH 25,000 KG PAYLOAD (USA)
Troy A. Asher 1,016.20 kmh
Richard C. Recker 631.44 mph
Jeffery C. Wharton
Rodney F. Todaro
Rockwell B-1B
4 GE F101, 30,780 lbs
Edwards AFB, CA
10/26/03

TIME TO CLIMB

TIME TO 3,000 METERS (USA)
Lt Col James P. Robinson 1 min
Capt Scott A. Neumann 19.28 sec
Capt Dennis J. Murphy, II
Rockwell International B-1B
4 GE F101-GE-102, 30,780 lbs
Grand Forks AFB, ND
2/29/92

TIME TO 6,000 METERS (USA)
Lt Col James P. Robinson 1 min
Capt Scott A. Neumann 54.95 sec
Capt Dennis J. Murphy, II
Rockwell International B-1B
4 GE F101-GE-102, 30,780 lbs
Grand Forks AFB, ND
2/29/92

TIME TO 9,000 METERS (USA) ★
Lt Col James P. Robinson — 2 min 22.77 sec
Capt Scott A. Neumann
Capt Dennis J. Murphy, II
Rockwell International B-1B
4 GE F101-GE-102, 30,780 lbs
Grand Forks AFB, ND
2/29/92

TIME TO 12,000 METERS (USA)
Lt Col James P. Robinson — 6 min 9.35 sec
Capt Scott A. Neumann
Capt Dennis J. Murphy, II
Rockwell International B-1B
4 GE F101-GE-102, 30,780 lbs
Grand Forks AFB, ND
2/29/92

CLASS C-1.Q (150,000-200,000 KG / 330,693-440,924 LBS)
GROUP III (JET ENGINE)

DISTANCE

DISTANCE WITHOUT LANDING (USA) ★
Michael Fox, Captain — 14,856.19 km
John Lloyd, Co-Captain — 9,231.21 mi
Orme Dockins, Co-Captain
Khalidkhan Asmakhan, 1st Officer
Royal Brunei Airlines Boeing 767-200ER
2 P&W 4050
Seattle, WA to Mombasa, Kenya
06/09/90

ALTITUDE

ALTITUDE (USA)
Norman E. Howell, Pilot — 14,012 m
Maj. Todd M. Markwald, Copilot — 45,971 ft
Gary R. Briscoe, Loadmaster
Boeing C-17A Globemaster III
4 P&W F117-PW-100, 40,700 lbs
Edwards AFB, CA
11/28/01

ALTITUDE IN HORIZONTAL FLIGHT (USA)
Norman E. Howell, Pilot — 13,701 m
Maj. Todd M. Markwald, Copilot — 44,952 ft
Gary R. Briscoe, Loadmaster
Boeing C-17A Globemaster III
4 P&W F117-PW-100, 40,700 lbs
Edwards AFB, CA
11/28/01

ALTITUDE WITH 1,000 KG PAYLOAD (USA)
Norman E. Howell, Pilot — 14,012 m
Maj. Todd M. Markwald, Copilot — 45,971 ft
Gary R. Briscoe, Loadmaster
Boeing C-17A Globemaster III
4 P&W F117-PW-100, 40,700 lbs
Edwards AFB, CA
11/28/01

ALTITUDE WITH 2,000 KG PAYLOAD (USA)
Norman E. Howell, Pilot — 14,012 m
Maj. Todd M. Markwald, Copilot — 45,971 ft
Gary R. Briscoe, Loadmaster
Boeing C-17A Globemaster III
4 P&W F117-PW-100, 40,700 lbs
Edwards AFB, CA
11/28/01

ALTITUDE WITH 5,000 KG PAYLOAD (USA)
Norman E. Howell, Pilot — 14,012 m
Maj. Todd M. Markwald, Copilot — 45,971 ft
Gary R. Briscoe, Loadmaster
Boeing C-17A Globemaster III
4 P&W F117-PW-100, 40,700 lbs
Edwards AFB, CA
11/28/01

ALTITUDE WITH 10,000 KG PAYLOAD (USA)
Norman E. Howell, Pilot — 14,012 m
Maj. Todd M. Markwald, Copilot — 45,971 ft
Gary R. Briscoe, Loadmaster
Boeing C-17A Globemaster III
4 P&W F117-PW-100, 40,700 lbs
Edwards AFB, CA
11/28/01

ALTITUDE WITH 15,000 KG PAYLOAD (USA)
Capt. Christopher S. Morgan, Pilot — 13,404 m
Maj. Christopher J. Lindell, Copilot — 43,978 ft
TSgt. Thomas E. Fields, Jr., Loadmaster
Boeing C-17A Globemaster III
4 P&W F117-PW-100, 40,700 lbs
Edwards AFB, CA
11/27/01

ALTITUDE WITH 20,000 KG PAYLOAD (USA)
Capt. Christopher S. Morgan, Pilot — 13,404 m
Maj. Christopher J. Lindell, Copilot — 43,978 ft
TSgt. Thomas E. Fields, Jr., Loadmaster
Boeing C-17A Globemaster III
4 P&W F117-PW-100, 40,700 lbs
Edwards AFB, CA
11/27/01

ALTITUDE WITH 25,000 KG PAYLOAD (USA)
Capt. Christopher S. Morgan, Pilot — 13,404 m
Maj. Christopher J. Lindell, Copilot — 43,978 ft
TSgt. Thomas E. Fields, Jr., Loadmaster
Boeing C-17A Globemaster III
4 P&W F117-PW-100, 40,700 lbs
Edwards AFB, CA
11/27/01

ALTITUDE WITH 30,000 KG PAYLOAD (USA)
Capt. Christopher S. Morgan, Pilot — 13,404 m
Maj. Christopher J. Lindell, Copilot — 43,978 ft
TSgt. Thomas E. Fields, Jr., Loadmaster
Boeing C-17A Globemaster III
4 P&W F117-PW-100, 40,700 lbs
Edwards AFB, CA
11/27/01

ALTITUDE WITH 35,000 KG PAYLOAD (USA)
Capt. Christopher S. Morgan, Pilot — 13,404 m
Maj. Christopher J. Lindell, Copilot — 43,978 ft
TSgt. Thomas E. Fields, Jr., Loadmaster
Boeing C-17A Globemaster III
4 P&W F117-PW-100, 40,700 lbs
Edwards AFB, CA
11/27/01

ALTITUDE WITH 40,000 KG PAYLOAD (USA)
Capt. Christopher S. Morgan, Pilot — 13,404 m
Maj. Christopher J. Lindell, Copilot — 43,978 ft
TSgt. Thomas E. Fields, Jr., Loadmaster
Boeing C-17A Globemaster III
4 P&W F117-PW-100, 40,700 lbs
Edwards AFB, CA
11/27/01

ALTITUDE WITH 100,000 KG PAYLOAD (USSR)
Alexandr V. Galounenko, Captain — 12,340 m
Sergei A. Borbik, Copilot — 40,486 ft
Sergei F. Netchaev, Navigator
Antonov 225 "Mriya"
3/22/89

GREATEST PAYLOAD CARRIED TO 2,000 METERS (USA)
Capt. Christopher S. Morgan, Pilot — 40,632 kg
Maj. Christopher J. Lindell, Copilot — 89,580 lb
TSgt. Thomas E. Fields, Jr., Loadmaster
Boeing C-17A Globemaster III
4 P&W F117-PW-100, 40,700 lbs
Edwards AFB, CA
11/27/01

SPEED

SPEED OVER A 15/25 KM STRAIGHT COURSE (USA)
William A. Libby 1,333.90 kmh
Hans H. Miller 828.84 mph
Brian A. Tom
Jeremy S. Agte
Rockwell B-1B
4 GE F101, 30,780 lbs
Edwards AFB, CA
10/25/03

SPEED OVER A 100 KM CLOSED CIRCUIT WITHOUT PAYLOAD (USA)
William A. Libby 1,037.87 kmh
Hans H. Miller 644.90 mph
Brian A. Tom
Jeremy S. Agte
Rockwell B-1B
4 GE F101, 30,780 lbs
Edwards AFB, CA
10/25/03

SPEED OVER A 100 KM CLOSED CIRCUIT WITH 1,000 KG PAYLOAD (USA)
William A. Libby 1,037.87 kmh
Hans H. Miller 644.90 mph
Brian A. Tom
Jeremy S. Agte
Rockwell B-1B
4 GE F101, 30,780 lbs
Edwards AFB, CA
10/25/03

SPEED OVER A 100 KM CLOSED CIRCUIT WITH 2,000 KG PAYLOAD (USA)
William A. Libby 1,037.87 kmh
Hans H. Miller 644.90 mph
Brian A. Tom
Jeremy S. Agte
Rockwell B-1B
4 GE F101, 30,780 lbs
Edwards AFB, CA
10/25/03

SPEED OVER A 100 KM CLOSED CIRCUIT WITH 5,000 KG PAYLOAD (USA)
William A. Libby 1,037.87 kmh
Hans H. Miller 644.90 mph
Brian A. Tom
Jeremy S. Agte
Rockwell B-1B
4 GE F101, 30,780 lbs
Edwards AFB, CA
10/25/03

SPEED OVER A 100 KM CLOSED CIRCUIT WITH 10,000 KG PAYLOAD (USA)
William A. Libby 1,037.87 kmh
Hans H. Miller 644.90 mph
Brian A. Tom
Jeremy S. Agte
Rockwell B-1B
4 GE F101, 30,780 lbs
Edwards AFB, CA
10/25/03

SPEED OVER A 100 KM CLOSED CIRCUIT WITH 15,000 KG PAYLOAD (USA)
William A. Libby 1,037.87 kmh
Hans H. Miller 644.90 mph
Brian A. Tom
Jeremy S. Agte
Rockwell B-1B
4 GE F101, 30,780 lbs
Edwards AFB, CA
10/25/03

SPEED OVER A 100 KM CLOSED CIRCUIT WITH 20,000 KG PAYLOAD (USA)
William A. Libby 1,037.87 kmh
Hans H. Miller 644.90 mph
Brian A. Tom
Jeremy S. Agte
Rockwell B-1B
4 GE F101, 30,780 lbs
Edwards AFB, CA
10/25/03

SPEED OVER A 100 KM CLOSED CIRCUIT WITH 25,000 KG PAYLOAD (USA)
William A. Libby 1,037.87 kmh
Hans H. Miller 644.90 mph
Brian A. Tom
Jeremy S. Agte
Rockwell B-1B
4 GE F101, 30,780 lbs
Edwards AFB, CA
10/25/03

SPEED OVER A 500 KM CLOSED CIRCUIT WITHOUT PAYLOAD (USA)
William A. Libby 1,123.63 kmh
Hans H. Miller 698.19 mph
Brian A. Tom
Jeremy S. Agte
Rockwell B-1B
4 GE F101, 30,780 lbs
Edwards AFB, CA
10/25/03

SPEED OVER A 500 KM CLOSED CIRCUIT WITH 1,000 KG PAYLOAD (USA)
William A. Libby 1,123.63 kmh
Hans H. Miller 698.19 mph
Brian A. Tom
Jeremy S. Agte
Rockwell B-1B
4 GE F101, 30,780 lbs
Edwards AFB, CA
10/25/03

SPEED OVER A 500 KM CLOSED CIRCUIT WITH 2,000 KG PAYLOAD (USA)
William A. Libby 1,123.63 kmh
Hans H. Miller 698.19 mph
Brian A. Tom
Jeremy S. Agte
Rockwell B-1B
4 GE F101, 30,780 lbs
Edwards AFB, CA
10/25/03

SPEED OVER A 500 KM CLOSED CIRCUIT WITH 5,000 KG PAYLOAD (USA)
William A. Libby 1,123.63 kmh
Hans H. Miller 698.19 mph
Brian A. Tom
Jeremy S. Agte
Rockwell B-1B
4 GE F101, 30,780 lbs
Edwards AFB, CA
10/25/03

SPEED OVER A 500 KM CLOSED CIRCUIT WITH 10,000 KG PAYLOAD (USA)
William A. Libby 1,123.63 kmh
Hans H. Miller 698.19 mph
Brian A. Tom
Jeremy S. Agte
Rockwell B-1B
4 GE F101, 30,780 lbs
Edwards AFB, CA
10/25/03

SPEED OVER A 500 KM CLOSED CIRCUIT WITH 15,000 KG PAYLOAD (USA)
William A. Libby 1,123.63 kmh
Hans H. Miller 698.19 mph
Brian A. Tom
Jeremy S. Agte
Rockwell B-1B
4 GE F101, 30,780 lbs
Edwards AFB, CA
10/25/03

SPEED OVER A 500 KM CLOSED CIRCUIT WITH 20,000 KG PAYLOAD (USA)
William A. Libby 1,123.63 kmh
Hans H. Miller 698.19 mph
Brian A. Tom
Jeremy S. Agte
Rockwell B-1B
4 GE F101, 30,780 lbs
Edwards AFB, CA
10/25/03

SPEED OVER A 500 KM CLOSED CIRCUIT WITH 25,000 KG PAYLOAD (USA)
William A. Libby 1,123.63 kmh
Hans H. Miller 698.19 mph
Brian A. Tom
Jeremy S. Agte
Rockwell B-1B
4 GE F101, 30,780 lbs
Edwards AFB, CA
10/25/03

SPEED OVER A 1,000 KM CLOSED CIRCUIT WITHOUT PAYLOAD (USA)
William A. Libby 1,105.68 kmh
Hans H. Miller 687.04 mph
Brian A. Tom
Jeremy S. Agte
Rockwell B-1B
4 GE F101, 30,780 lbs
Edwards AFB, CA
10/25/03

SPEED OVER A 1,000 KM CLOSED CIRCUIT WITH 1,000 KG PAYLOAD (USA)

William A. Libby 1,105.68 kmh
Hans H. Miller 687.04 mph
Brian A. Tom
Jeremy S. Agte
Rockwell B-1B
4 GE F101, 30,780 lbs
Edwards AFB, CA
10/25/03

SPEED OVER A 1,000 KM CLOSED CIRCUIT WITH 2,000 KG PAYLOAD (USA)
William A. Libby 1,105.68 kmh
Hans H. Miller 687.04 mph
Brian A. Tom
Jeremy S. Agte
Rockwell B-1B
4 GE F101, 30,780 lbs
Edwards AFB, CA
10/25/03

SPEED OVER A 1,000 KM CLOSED CIRCUIT WITH 5,000 KG PAYLOAD (USA)
William A. Libby 1,105.68 kmh
Hans H. Miller 687.04 mph
Brian A. Tom
Jeremy S. Agte
Rockwell B-1B
4 GE F101, 30,780 lbs
Edwards AFB, CA
10/25/03

SPEED OVER A 1,000 KM CLOSED CIRCUIT WITH 10,000 KG PAYLOAD (USA)
William A. Libby 1,105.68 kmh
Hans H. Miller 687.04 mph
Brian A. Tom
Jeremy S. Agte
Rockwell B-1B
4 GE F101, 30,780 lbs
Edwards AFB, CA
10/25/03

A B-1B Lancer releases its inert payload after successfully completing a speed record run at Edwards AFB in October.

Photo: Steve Zapka, courtesy of U.S. Air Force

SPEED OVER A 1,000 KM CLOSED CIRCUIT WITH 15,000 KG PAYLOAD (USA)
William A. Libby 1,105.68 kmh
Hans H. Miller 687.04 mph
Brian A. Tom
Jeremy S. Agte
Rockwell B-1B
4 GE F101, 30,780 lbs
Edwards AFB, CA
10/25/03

SPEED OVER A 1,000 KM CLOSED CIRCUIT WITH 20,000 KG PAYLOAD (USA)
William A. Libby 1,105.68 kmh
Hans H. Miller 687.04 mph
Brian A. Tom
Jeremy S. Agte
Rockwell B-1B
4 GE F101, 30,780 lbs
Edwards AFB, CA
10/25/03

SPEED OVER A 1,000 KM CLOSED CIRCUIT WITH 25,000 KG PAYLOAD (USA)
William A. Libby 1,105.68 kmh
Hans H. Miller 687.04 mph
Brian A. Tom
Jeremy S. Agte
Rockwell B-1B
4 GE F101, 30,780 lbs
Edwards AFB, CA
10/25/03

SPEED OVER A 2,000 KM CLOSED CIRCUIT WITH 5,000 KG PAYLOAD (USA)
Robert Chamberlain, Pilot 1,078.20 kmh
Michael Waters, Copilot 669.93 mph
Richard Fisher, Crew Member
Nathan Gray, Crew Member
B-1B, Heavy Bomber, S/N 58
4 GE F101-GE-102, 30,780 lbs
Palmdale, CA
7/4/87

SPEED OVER A 2,000 KM CLOSED CIRCUIT WITH 10,000 KG PAYLOAD (USA)
Robert Chamberlain, Pilot 1,078.20 kmh
Michael Waters, Copilot 669.93 mph
Richard Fisher, Crew Member
Nathan Gray, Crew Member
B-1B, Heavy Bomber, S/N 58
4 GE F101-GE-102, 30,780 lbs
Palmdale, CA
7/4/87

SPEED OVER A 2,000 KM CLOSED CIRCUIT WITH 15,000 KG PAYLOAD (USA)
Robert Chamberlain, Pilot 1,078.20 kmh
Michael Waters, Copilot 669.93 mph
Richard Fisher, Crew Member
Nathan Gray, Crew Member
B-1B, Heavy Bomber, S/N 58
4 GE F101-GE-102, 30,780 lbs
Palmdale, CA
7/4/87

SPEED OVER A 2,000 KM CLOSED CIRCUIT WITH 20,000 KG PAYLOAD (USA)
Robert Chamberlain, Pilot 1,078.20 kmh
Michael Waters, Copilot 669.93 mph
Richard Fisher, Crew Member
Nathan Gray, Crew Member
B-1B, Heavy Bomber, S/N 58
4 GE F101-GE-102, 30,780 lbs
Palmdale, CA
7/4/87

SPEED OVER A 2,000 KM CLOSED CIRCUIT WITH 25,000 KG PAYLOAD (USA)
Robert Chamberlain, Pilot 1,078.20 kmh
Michael Waters, Copilot 669.93 mph
Richard Fisher, Crew Member
Nathan Gray, Crew Member
B-1B, Heavy Bomber, S/N 58
4 GE F101-GE-102, 30,780 lbs
Palmdale, CA
7/4/87

SPEED OVER A 2,000 KM CLOSED CIRCUIT WITH 30,000 KG PAYLOAD (USA)
Robert Chamberlain, Pilot 1,078.20 kmh
Michael Waters, Copilot 669.93 mph
Richard Fisher, Crew Member
Nathan Gray, Crew Member
B-1B, Heavy Bomber, S/N 58
4 GE F101-GE-102, 30,780 lbs
Palmdale, CA
7/4/87

SPEED OVER A 5,000 KM CLOSED CIRCUIT WITHOUT PAYLOAD (USA)
MAJ H. Hedgpeth, Pilot 1,054.21 kmh
LTC Robt Chamberlain, Copilot 655.09 mph
CPT Alexander Ivanchishin, Crew Member
CPT Daniel Novick, Crew Member
B-1B, Heavy Bomber, S/N 70
4 GE F101-GE-102, 30,780 lbs
Palmdale, CA
9/17/87

SPEED OVER A 5,000 KM CLOSED CIRCUIT WITH 1,000 KG PAYLOAD (USA)
MAJ H. Hedgpeth, Pilot 1,054.21 kmh
LTC Robt Chamberlain, Copilot 655.09 mph
CPT Alexander Ivanchishin, Crew Member
CPT Daniel Novick, Crew Member
B-1B, Heavy Bomber, S/N 70
4 GE F101-GE-102, 30,780 lbs
Palmdale, CA
9/17/87

SPEED OVER A 5,000 KM CLOSED CIRCUIT WITH 2,000 KG PAYLOAD (USA)
MAJ H. Hedgpeth, Pilot 1,054.21 kmh
LTC Robt Chamberlain, Copilot 655.09 mph
CPT Alexander Ivanchishin, Crew Member
CPT Daniel Novick, Crew Member
B-1B, Heavy Bomber, S/N 70
4 GE F101-GE-102, 30,780 lbs
Palmdale, CA
9/17/87

SPEED OVER A 5,000 KM CLOSED CIRCUIT WITH 5,000 KG PAYLOAD (USA)
MAJ H. Hedgpeth, Pilot 1,054.21 kmh
LTC Robt Chamberlain, Copilot 655.09 mph
CPT Alexander Ivanchishin, Crew Member
CPT Daniel Novick, Crew Member
B-1B, Heavy Bomber, S/N 70
4 GE F101-GE-102, 30,780 lbs
Palmdale, CA
9/17/87

SPEED OVER A 5,000 KM CLOSED CIRCUIT WITH 10,000 KG PAYLOAD (USA)
MAJ H. Hedgpeth, Pilot 1,054.21 kmh
LTC Robt Chamberlain, Copilot 655.09 mph
CPT Alexander Ivanchishin, Crew Member
CPT Daniel Novick, Crew Member
B-1B, Heavy Bomber, S/N 70
4 GE F101-GE-102, 30,780 lbs
Palmdale, CA
9/17/87

SPEED OVER A 5,000 KM CLOSED CIRCUIT WITH 15,000 KG PAYLOAD (USA)
MAJ H. Hedgpeth, Pilot 1,054.21 kmh
LTC Robt Chamberlain, Copilot 655.09 mph
CPT Alexander Ivanchishin, Crew Member
CPT Daniel Novick, Crew Member
B-1B, Heavy Bomber, S/N 70
4 GE F101-GE-102, 30,780 lbs
Palmdale, CA
9/17/87

SPEED OVER A 5,000 KM CLOSED CIRCUIT WITH 20,000 KG PAYLOAD (USA)
MAJ H. Hedgpeth, Pilot 1,054.21 kmh
LTC Robt Chamberlain, Copilot 655.09 mph
CPT Alexander Ivanchishin, Crew Member
CPT Daniel Novick, Crew Member
B-1B, Heavy Bomber, S/N 70
4 GE F101-GE-102, 30,780 lbs
Palmdale, CA
9/17/87

SPEED OVER A 5,000 KM CLOSED CIRCUIT WITH 25,000 KG PAYLOAD (USA)
MAJ H. Hedgpeth, Pilot 1,054.21 kmh
LTC Robt Chamberlain, Copilot 655.09 mph
CPT Alexander Ivanchishin, Crew Member
CPT Daniel Novick, Crew Member
B-1B, Heavy Bomber, S/N 70
4 GE F101-GE-102, 30,780 lbs
Palmdale, CA
9/17/87

SPEED OVER A 5,000 KM CLOSED CIRCUIT WITH 30,000 KG PAYLOAD (USA)
MAJ H. Hedgpeth, Pilot 1,054.21 kmh
LTC Robt Chamberlain, Copilot 655.09 mph
CPT Alexander Ivanchishin, Crew Member
CPT Daniel Novick, Crew Member
B-1B, Heavy Bomber, S/N 70
4 GE F101-GE-102, 30,780 lbs
Palmdale, CA
9/17/87

SPEED OVER A 10,000 KM CLOSED CIRCUIT WITHOUT PAYLOAD (USA) ★
Capt. R. F. Lewandowski, Commander 956.93 kmh
Capt. Kevin T. Kalen, Pilot 594.61 mph
Lt. Col. Timothy W. Van Splunder, OSO
Capt. Jerrold A. Wangberg, DSO
Rockwell International B-1B
4 GE F101, 12,000 lbs
Grand Forks, ND, Monroeville, AL, Mullan, ID
4/7-8/94

SPEED AROUND THE WORLD, EASTBOUND (WITH REFUELING IN FLIGHT) (USA) ★
Col. Douglas L. Raaberg 1,015.76 kmh
Capt. Ricky W. Carver 631.16 mph
Capt. Gerald V. Goodfellow
Capt. Kevin D. Clotfelter
Rockwell International B-1B
4 GE F101-GE-102, 30,780 lbs
Elapsed Time: 36 hrs 13 min 36 sec
Dyess AFB, Abilene, TX
6/3/95

TIME TO CLIMB

TIME TO 3,000 METERS (USA)
Capt Jeffry F. Smith 1 min
Capt Tracy A. Sharp 59.97 sec
Capt Bryan S. Ferguson
Rockwell International B-1B
4 GE F101-GE-102, 30,780 lbs
Grand Forks AFB, ND
2/29/92

TIME TO 3,000 METERS WITH 5,000 KG PAYLOAD (USA)
Maj. Scott L. Grunwald 3 min
Richard M. Cooper 55 sec
McDonnell Douglas C-17A
4 P&W F117-PW-100, 41,700 lbs
8/26/93

TIME TO 6,000 METERS (USA)
Capt Jeffry F. Smith 2 min
Capt Tracy A. Sharp 39.27 sec
Capt Bryan S. Ferguson
Rockwell International B-1B
4 GE F101-GE-102, 30,780 lbs
Grand Forks AFB, ND
2/29/92

TIME TO 6,000 METERS WITH 5,000 KG PAYLOAD (USA)
Maj. Scott L. Grunwald 7 min
Richard M. Cooper 31 sec
McDonnell Douglas C-17A
4 P&W F117-PW-100, 41,700 lbs
8/26/93

TIME TO 9,000 METERS (USA)
Capt Jeffry F. Smith 3 min
Capt Tracy A. Sharp 47.75 sec
Capt Bryan S. Ferguson
Rockwell International B-1B
4 GE F101-GE-102, 30,780 lbs
Grand Forks AFB, ND
2/29/92

TIME TO 9,000 METERS WITH 5,000 KG PAYLOAD (USA)
Maj. Scott L. Grunwald 14 min
Richard M. Cooper 32 sec
McDonnell Douglas C-17A
4 P&W F117-PW-100, 41,700 lbs
8/26/93

TIME TO 12,000 METERS (USA)
Capt Tracy A. Sharp 9 min
Maj John E. Alexander 42.85 sec
Capt Paul S. Ellia
Rockwell International B-1B
4 GE F101-GE-102, 30,780 lbs
Grand Forks AFB, ND
3/18/92

CLASS C-1.R (200,000-250,000 KG / 440,924-551,155 LBS) GROUP III (JET ENGINE)

DISTANCE

DISTANCE WITHOUT LANDING (AUSTRALIA)
Bruce Simpson 16,901.26 km
David Collier 10,501.95 mi
Bruce Van Eyle
James Peach
Airbus A330-200
2 GE CF6-80, 68,500 lbs
Toulouse, France to Melbourne, Australia
12/25/02

ALTITUDE

ALTITUDE (USSR)
L. V. Kozlov, Pilot 13,894 m
M. I. Pozdnyakov, Copilot 45,584 ft
Tupolev Tu-160
10/31/90

ALTITUDE IN HORIZONTAL FLIGHT (USSR)
L. V. Kozlov, Pilot 12,150 m
M. I. Pozdnyakov, Copilot 39,862 ft
Tupolev Tu-160
10/31/90

ALTITUDE WITH 1,000 KG PAYLOAD (USSR)
L. V. Kozlov, Pilot 13,894 m
M. I. Pozdnyakov, Copilot 45,584 ft
Tupolev Tu-160
10/31/90

ALTITUDE WITH 2,000 KG PAYLOAD (USSR)
L. V. Kozlov, Pilot 13,894 m
M. I. Pozdnyakov, Copilot 45,584 ft
Tupolev Tu-160
10/31/90

ALTITUDE WITH 5,000 KG PAYLOAD (USSR)
L. V. Kozlov, Pilot 13,894 m
M. I. Pozdnyakov, Copilot 45,584 ft
Tupolev Tu-160
10/31/90

ALTITUDE WITH 10,000 KG PAYLOAD (USSR)
L. V. Kozlov, Pilot 13,894 m
M. I. Pozdnyakov, Copilot 45,584 ft
Tupolev Tu-160
10/31/90

ALTITUDE WITH 15,000 KG PAYLOAD (USSR)
L. V. Kozlov, Pilot 13,894 m
M. I. Pozdnyakov, Copilot 45,584 ft
Tupolev Tu-160
10/31/90

ALTITUDE WITH 20,000 KG PAYLOAD (USSR)
L. V. Kozlov, Pilot 13,894 m
M. I. Pozdnyakov, Copilot 45,584 ft
Tupolev Tu-160
10/31/90

ALTITUDE WITH 25,000 KG PAYLOAD (USSR)
L. V. Kozlov, Pilot 13,894 m
M. I. Pozdnyakov, Copilot 45,584 ft
Tupolev Tu-160
10/31/90

ALTITUDE WITH 30,000 KG PAYLOAD (USSR)
L. V. Kozlov, Pilot 13,894 m
M. I. Pozdnyakov, Copilot 45,584 ft
Tupolev Tu-160
10/31/90

ALTITUDE WITH 35,000 KG PAYLOAD (USA)
Capt. Pamela A. Melroy 10,733 m
William R. Casey 35,213 ft
McDonnell Douglas C-17
4 P&W F117-PW-100, 41,700 lbs
12/18/92

ALTITUDE WITH 40,000 KG PAYLOAD (USA)
Capt. Pamela A. Melroy 10,733 m
William R. Casey 35,213 ft
McDonnell Douglas C-17
4 P&W F117-PW-100, 41,700 lbs
12/18/92

ALTITUDE WITH 45,000 KG PAYLOAD (USA)
Capt. Pamela A. Melroy 10,733 m
William R. Casey 35,213 ft
McDonnell Douglas C-17
4 P&W F117-PW-100, 41,700 lbs
12/18/92

ALTITUDE WITH 50,000 KG PAYLOAD (USA)
Capt. Pamela A. Melroy 10,733 m
William R. Casey 35,213 ft
McDonnell Douglas C-17
4 P&W F117-PW-100, 41,700 lbs
12/18/92

ALTITUDE WITH 55,000 KG PAYLOAD (USA)
Capt. Pamela A. Melroy 10,733 m
William R. Casey 35,213 ft
McDonnell Douglas C-17
4 P&W F117-PW-100, 41,700 lbs
12/18/92

ALTITUDE WITH 60,000 KG PAYLOAD (USA)
Capt. Pamela A. Melroy 10,733 m
William R. Casey 35,213 ft
McDonnell Douglas C-17
4 P&W F117-PW-100, 41,700 lbs
12/18/92

GREATEST PAYLOAD CARRIED TO 2,000 METERS (USA)
Capt. Pamela A. Melroy 60,519.10 kg
William R. Casey 133,422.00 lbs
McDonnell Douglas C-17
4 P&W F117-PW-100, 41,700 lbs
12/18/92

SPEED

SPEED OVER A 1,000 KM CLOSED CIRCUIT WITHOUT PAYLOAD (USSR)
L. V. Kozlov, Pilot 1,731.40 kmh
M. I. Pozdnyakov, Copilot 1,075.84 mph
Tupolev Tu-160
10/31/90

SPEED OVER A 1,000 KM CLOSED CIRCUIT WITH 1,000 KG PAYLOAD (USSR)
L. V. Kozlov, Pilot 1,731.40 kmh
M. I. Pozdnyakov, Copilot 1,075.84 mph
Tupolev Tu-160
10/31/90

SPEED OVER A 1,000 KM CLOSED CIRCUIT WITH 2,000 KG PAYLOAD (USSR)
L. V. Kozlov, Pilot 1,731.40 kmh
M. I. Pozdnyakov, Copilot 1,075.84 mph
Tupolev Tu-160
10/31/90

SPEED OVER A 1,000 KM CLOSED CIRCUIT WITH 5,000 KG PAYLOAD (USSR)
L. V. Kozlov, Pilot 1,731.40 kmh
M. I. Pozdnyakov, Copilot 1,075.84 mph
Tupolev Tu-160
10/31/90

SPEED OVER A 1,000 KM CLOSED CIRCUIT WITH 10,000 KG PAYLOAD (USSR)
L. V. Kozlov, Pilot 1,731.40 kmh
M. I. Pozdnyakov, Copilot 1,075.84 mph
Tupolev Tu-160
10/31/90

SPEED OVER A 1,000 KM CLOSED CIRCUIT WITH 15,000 KG PAYLOAD (USSR)
L. V. Kozlov, Pilot 1,731.40 kmh
M. I. Pozdnyakov, Copilot 1,075.84 mph
Tupolev Tu-160
10/31/90

SPEED OVER A 1,000 KM CLOSED CIRCUIT WITH 20,000 KG PAYLOAD (USSR)
L. V. Kozlov, Pilot 1,731.40 kmh
M. I. Pozdnyakov, Copilot 1,075.84 mph
Tupolev Tu-160
10/31/90

SPEED OVER A 1,000 KM CLOSED CIRCUIT WITH 25,000 KG PAYLOAD (USSR)
L. V. Kozlov, Pilot 1,731.40 kmh
M. I. Pozdnyakov, Copilot 1,075.84 mph
Tupolev Tu-160
10/31/90

SPEED OVER A 1,000 KM CLOSED CIRCUIT WITH 30,000 KG PAYLOAD (USSR)
L. V. Kozlov, Pilot 1,731.40 kmh
M. I. Pozdnyakov, Copilot 1,075.84 mph
Tupolev Tu-160
10/31/90

SPEED OVER A 2,000 KM CLOSED CIRCUIT WITHOUT PAYLOAD (USSR)
Nail Sattarov, Pilot 1,195.70 kmh
Alexandre Medvedev, Copilot 742.97 mph
Tupolev Tu-160
5/22/90

SPEED OVER A 2,000 KM CLOSED CIRCUIT WITH 1,000 KG PAYLOAD (USSR)
Nail Sattarov, Pilot 1,195.70 kmh
Alexandre Medvedev, Copilot 742.97 mph
Tupolev Tu-160
5/22/90

SPEED OVER A 2,000 KM CLOSED CIRCUIT WITH 2,000 KG PAYLOAD (USSR)
Nail Sattarov, Pilot 1,195.70 kmh
Alexandre Medvedev, Copilot 742.97 mph
Tupolev Tu-160
5/22/90

SPEED OVER A 2,000 KM CLOSED CIRCUIT WITH 5,000 KG PAYLOAD (USSR)
Nail Sattarov, Pilot 1,195.70 kmh
Alexandre Medvedev, Copilot 742.97 mph
Tupolev Tu-160
5/22/90

SPEED OVER A 2,000 KM CLOSED CIRCUIT WITH 10,000 KG PAYLOAD (USSR)
Nail Sattarov, Pilot 1,195.70 kmh
Alexandre Medvedev, Copilot 742.97 mph
Tupolev Tu-160
5/22/90

SPEED OVER A 2,000 KM CLOSED CIRCUIT WITH 15,000 KG PAYLOAD (USSR)
Nail Sattarov, Pilot 1,195.70 kmh
Alexandre Medvedev, Copilot 742.97 mph
Tupolev Tu-160
5/22/90

SPEED OVER A 2,000 KM CLOSED CIRCUIT WITH 20,000 KG PAYLOAD (USSR)
Nail Sattarov, Pilot 1,195.70 kmh
Alexandre Medvedev, Copilot 742.97 mph
Tupolev Tu-160
5/22/90

SPEED OVER A 2,000 KM CLOSED CIRCUIT WITH 25,000 KG PAYLOAD (USSR)
Nail Sattarov, Pilot 1,195.70 kmh
Alexandre Medvedev, Copilot 742.97 mph
Tupolev Tu-160
5/22/90

SPEED OVER A 2,000 KM CLOSED CIRCUIT WITH 30,000 KG PAYLOAD (USSR)
Nail Sattarov, Pilot 1,195.70 kmh
Alexandre Medvedev, Copilot 742.97 mph
Tupolev Tu-160
5/22/90

SPEED OVER A 5,000 KM CLOSED CIRCUIT WITHOUT PAYLOAD (USSR)
Vladimir Pavlov, Pilot 920.95 kmh
Valery Selivanov, Copilot 572.25 mph
Tupolev Tu-160
5/24/90

SPEED OVER A 5,000 KM CLOSED CIRCUIT WITH 1,000 KG PAYLOAD (USSR)
Vladimir Pavlov, Pilot 920.95 kmh
Valery Selivanov, Copilot 572.25 mph
Tupolev Tu-160
5/24/90

SPEED OVER A 5,000 KM CLOSED CIRCUIT WITH 2,000 KG PAYLOAD (USSR)
Vladimir Pavlov, Pilot 920.95 kmh
Valery Selivanov, Copilot 572.25 mph
Tupolev Tu-160
5/24/90

SPEED OVER A 5,000 KM CLOSED CIRCUIT WITH 5,000 KG PAYLOAD (USSR)
Vladimir Pavlov, Pilot 920.95 kmh
Valery Selivanov, Copilot 572.25 mph
Tupolev Tu-160
5/24/90

SPEED OVER A 5,000 KM CLOSED CIRCUIT WITH 10,000 KG PAYLOAD (USSR)
Vladimir Pavlov, Pilot 920.95 kmh
Valery Selivanov, Copilot 572.25 mph
Tupolev Tu-160
5/24/90

SPEED OVER A 5,000 KM CLOSED CIRCUIT WITH 15,000 KG PAYLOAD (USSR)
Vladimir Pavlov, Pilot 920.95 kmh
Valery Selivanov, Copilot 572.25 mph
Tupolev Tu-160
5/24/90

SPEED OVER A 5,000 KM CLOSED CIRCUIT WITH 20,000 KG PAYLOAD (USSR)
Vladimir Pavlov, Pilot 920.95 kmh
Valery Selivanov, Copilot 572.25 mph
Tupolev Tu-160
5/24/90

SPEED OVER A 5,000 KM CLOSED CIRCUIT WITH 25,000 KG PAYLOAD (USSR)
Vladimir Pavlov, Pilot 920.95 kmh
Valery Selivanov, Copilot 572.25 mph
Tupolev Tu-160
5/24/90

SPEED OVER A 5,000 KM CLOSED CIRCUIT WITH 30,000 KG PAYLOAD (USSR)
Vladimir Pavlov, Pilot 920.95 kmh
Valery Selivanov, Copilot 572.25 mph
Tupolev Tu-160
5/24/90

SPEED OVER A 10,000 KM CLOSED CIRCUIT WITHOUT PAYLOAD (USA) ★
Capt. Michael S. Menser, Commander 964.95 kmh
Capt. Brian P. Gallagher, Pilot 599.59 mph
Capt. Robert P. Boman, OSO
Capt. Matthew E. Grant, WSO
Rockwell International B-1B
4 GE F101, 12,000 lbs
Grand Forks, ND, Monroeville, AL, Mullan, ID
4/7-8/94

SPEED OVER A 10,000 KM CLOSED CIRCUIT WITH 1,000 KG PAYLOAD (USA)
Capt. Russell F. Mathers 884.26 kmh
Capt. Daniel G. Manuel, Jr. 549.45 mph
Capt. Henry C. Jenkins, Jr.
1st Lt. Ralph DeLatour
Capt. Allen D. Patton
Boeing B-52H
8 P&W TF-33, 17,100 lbs
Edwards AFB, CA
8/26/95

SPEED OVER A 10,000 KM CLOSED CIRCUIT WITH 2,000 KG PAYLOAD (USA)
Capt. Russell F. Mathers 884.26 kmh
Capt. Daniel G. Manuel, Jr. 549.45 mph
Capt. Henry C. Jenkins, Jr.
1st Lt. Ralph DeLatour
Capt. Allen D. Patton
Boeing B-52H
8 P&W TF-33, 17,100 lbs
Edwards AFB, CA
8/26/95

SPEED OVER A 10,000 KM CLOSED CIRCUIT WITH 5,000 KG PAYLOAD (USA) ✯
Capt. Russell F. Mathers 884.26 kmh
Capt. Daniel G. Manuel, Jr. 549.45 mph
Capt. Henry C. Jenkins, Jr.
1st Lt. Ralph DeLatour
Capt. Allen D. Patton
Boeing B-52H
8 P&W TF-33, 17,100 lbs
Edwards AFB, CA
8/26/95

CLASS C-1.S (250,000-300,000 KG (551,155-661,386 LBS)
GROUP III (JET ENGINE)

DISTANCE

DISTANCE WITHOUT LANDING (USA)
Frank P. Santoni Jr. 20,045.06 km
Richard A. Austin 12,455.40 mi
John E. Cashman, Charles A. Hovland,
Izham Ismail, Joseph M. MacDonald,
James C. McRoberts, Rodney M. Skaar
Boeing 777-200-IGW
2 RR6 Trent 892, 90,000 lbs
Seattle, WA to Kuala Lumpur, Malaysia
4/2/97

ALTITUDE

ALTITUDE (USSR)
B.I. Veremey, Pilot 14,000 m
G.N. Chapoval, Copilot 45,932 ft
Tupolev Tu-160
11/3/89

ALTITUDE IN HORIZONTAL FLIGHT (USSR)
B.I. Veremey, Pilot 11,250 m
G.N. Chapoval, Copilot 36,909 ft
Tupolev Tu-160
11/3/89

ALTITUDE WITH 1,000 KG PAYLOAD (USSR)
B.I. Veremey, Pilot 14,000 m
G.N. Chapoval, Copilot 45,932 ft
Tupolev Tu-160
11/3/89

ALTITUDE WITH 2,000 KG PAYLOAD (USSR)
B.I. Veremey, Pilot 14,000 m
G.N. Chapoval, Copilot 45,932 ft
Tupolev Tu-160
11/3/89

ALTITUDE WITH 5,000 KG PAYLOAD (USSR)
B.I. Veremey, Pilot 14,000 m
G.N. Chapoval, Copilot 45,932 ft
Tupolev Tu-160
11/3/89

ALTITUDE WITH 10,000 KG PAYLOAD (USSR)
B.I. Veremey, Pilot 14,000 m
G.N. Chapoval, Copilot 45,932 ft
Tupolev Tu-160
11/3/89

ALTITUDE WITH 15,000 KG PAYLOAD (USSR)
B.I. Veremey, Pilot 14,000 m
G.N. Chapoval, Copilot 45,932 ft
Tupolev Tu-160
11/3/89

ALTITUDE WITH 20,000 KG PAYLOAD (USSR)
B.I. Veremey, Pilot 14,000 m
G.N. Chapoval, Copilot 45,932 ft
Tupolev Tu-160
11/3/89

ALTITUDE WITH 25,000 KG PAYLOAD (USSR)
B.I. Veremey, Pilot 14,000 m
G.N. Chapoval, Copilot 45,932 ft
Tupolev Tu-160
11/3/89

ALTITUDE WITH 30,000 KG PAYLOAD (USSR)
B.I. Veremey, Pilot 14,000 m
G.N. Chapoval, Copilot 45,932 ft
Tupolev Tu-160
11/3/89

ALTITUDE WITH 35,000 KG PAYLOAD (USA)
Capt. John B. Norton 11,172 m
Charles N. Walls 36,653 ft
McDonnell Douglas C-17
4 P&W F117-PW-100, 41,700 lbs
12/16/92

ALTITUDE WITH 40,000 KG PAYLOAD (USA)
Capt. John B. Norton 11,172 m
Charles N. Walls 36,653 ft
McDonnell Douglas C-17
4 P&W F117-PW-100, 41,700 lbs
12/16/92

ALTITUDE WITH 45,000 KG PAYLOAD (USA)
Capt. John B. Norton 11,172 m
Charles N. Walls 36,653 ft
McDonnell Douglas C-17
4 P&W F117-PW-100, 41,700 lbs
12/16/92

ALTITUDE WITH 50,000 KG PAYLOAD (USA)
Capt. John B. Norton 11,172 m
Charles N. Walls 36,653 ft
McDonnell Douglas C-17
4 P&W F117-PW-100, 41,700 lbs
12/16/92

ALTITUDE WITH 55,000 KG PAYLOAD (USA)
Capt. John B. Norton 11,172 m
Charles N. Walls 36,653 ft
McDonnell Douglas C-17
4 P&W F117-PW-100, 41,700 lbs
12/16/92

ALTITUDE WITH 60,000 KG PAYLOAD (USA) ✯
Capt. John B. Norton 11,172 m
Charles N. Walls 36,653 ft
McDonnell Douglas C-17
4 P&W F117-PW-100, 41,700 lbs
12/16/92

ALTITUDE WITH 70,000 KG PAYLOAD (USA)
Capt. Pamela A. Melroy 9,805 m
Richard M. Cooper 32,169 ft
McDonnell Douglas C-17A
4 P&W F117-PW-100, 41,700 lbs
10/8/93

GREATEST MASS CARRIED TO 2,000 METERS (USA)
Capt. Pamela A. Melroy 73,039 kg
Richard M. Cooper 161,023 lbs
McDonnell Douglas C-17A
4 P&W F117-PW-100, 41,700 lbs
10/8/93

SPEED

SPEED OVER A 2,000 KM CLOSED CIRCUIT WITHOUT PAYLOAD (USSR)
B.I. Veremey, Pilot 1,678.00 kmh
G.N. Chapoval, Copilot 1,042.66 mph
Tupolev Tu-160
11/3/89

SPEED OVER A 2,000 KM CLOSED CIRCUIT WITH 1,000 KG PAYLOAD (USSR)
B.I. Veremey, Pilot 1,678.00 kmh
G.N. Chapoval, Copilot 1,042.66 mph
Tupolev Tu-160
11/3/89

SPEED OVER A 2,000 KM CLOSED CIRCUIT WITH 2,000 KG PAYLOAD (USSR)
B.I. Veremey, Pilot 1,678.00 kmh
G.N. Chapoval, Copilot 1,042.66 mph
Tupolev Tu-160
11/3/89

SPEED OVER A 2,000 KM CLOSED CIRCUIT WITH 5,000 KG PAYLOAD (USSR)
B.I. Veremey, Pilot 1,678.00 kmh
G.N. Chapoval, Copilot 1,042.66 mph
Tupolev Tu-160
11/3/89

SPEED OVER A 2,000 KM CLOSED CIRCUIT WITH 10,000 KG PAYLOAD (USSR)
B.I. Veremey, Pilot 1,678.00 kmh
G.N. Chapoval, Copilot 1,042.66 mph
Tupolev Tu-160
11/3/89

SPEED OVER A 2,000 KM CLOSED CIRCUIT WITH 15,000 KG PAYLOAD (USSR)
B.I. Veremey, Pilot 1,678.00 kmh
G.N. Chapoval, Copilot 1,042.66 mph
Tupolev Tu-160
11/3/89

SPEED OVER A 2,000 KM CLOSED CIRCUIT WITH 20,000 KG PAYLOAD (USSR)
B.I. Veremey, Pilot 1,678.00 kmh
G.N. Chapoval, Copilot 1,042.66 mph
Tupolev Tu-160
11/3/89

SPEED OVER A 2,000 KM CLOSED CIRCUIT WITH 25,000 KG PAYLOAD (USSR)
B.I. Veremey, Pilot 1,678.00 kmh
G.N. Chapoval, Copilot 1,042.66 mph
Tupolev Tu-160
11/3/89

SPEED OVER A 2,000 KM CLOSED CIRCUIT WITH 30,000 KG PAYLOAD (USSR)
B.I. Veremey, Pilot 1,678.00 kmh
G.N. Chapoval, Copilot 1,042.66 mph
Tupolev Tu-160
11/3/89

SPEED AROUND THE WORLD, EASTBOUND (USA)
★
Frank P. Santoni Jr. 889.20 kmh
Richard A. Austin 552.52 mph
John E. Cashman, Charles A. Hovland,
Izham Ismail, Joseph M. MacDonald,
James C. McRoberts, Rodney M. Skaar
Boeing 777-200-IGW
2 RR6 Trent 892, 90,000 lbs
Seattle, WA
4/2/97

CLASS C-1.T (MORE THAN 300,000 KG / 661,386 LBS)
GROUP III (JET ENGINE)

DISTANCE

DISTANCE WITHOUT LANDING (AUSTRALIA)
David Massy-Greene 17,039.00 km
Qantas Airways 10,588.03 mi
Boeing 747-400
8/16-17/89

DISTANCE OVER A CLOSED CIRCUIT WITHOUT LANDING (USSR)
Vladimir Tersky, Pilot 20,150.92 km
Yuri Resnitsky, Copilot 12,521.78 mi
AN-124
4 D-18T, 23,400 kg
Podmoskovnoye, USSR
5/6-7/87

ALTITUDE

ALTITUDE (USSR)
Alexandr V. Galounenko, Capt 12,340 m
Sergei A. Borbik, Copilot 40,486 ft
Sergei F. Netchaev, Navigator
Antonov 225 "Mriya"
3/22/89

ALTITUDE IN HORIZONTAL FLIGHT (USSR)
Alexandr V. Galounenko, Capt 11,600 m
Sergei A. Borbik, Copilot 38,058 ft
Sergei F. Netchaev, Navigator
Antonov 225 "Mriya"
3/22/89

ALTITUDE WITH 1,000 KG PAYLOAD (USSR)
Alexandr V. Galounenko, Capt 12,340 m
Sergei A. Borbik, Copilot 40,486 ft
Sergei F. Netchaev, Navigator
Antonov 225 "Mriya"
3/22/89

ALTITUDE WITH 2,000 KG PAYLOAD (USSR)
Alexandr V. Galounenko, Capt 12,340 m
Sergei A. Borbik, Copilot 40,486 ft
Sergei F. Netchaev, Navigator
Antonov 225 "Mriya"
3/22/89

ALTITUDE WITH 5,000 KG PAYLOAD (USSR)
Alexandr V. Galounenko, Capt 12,340 m
Sergei A. Borbik, Copilot 40,486 ft
Sergei F. Netchaev, Navigator
Antonov 225 "Mriya"
3/22/89

ALTITUDE WITH 10,000 KG PAYLOAD (USSR)
Alexandr V. Galounenko, Capt 12,340 m
Sergei A. Borbik, Copilot 40,486 ft
Sergei F. Netchaev, Navigator
Antonov 225 "Mriya"
3/22/89

ALTITUDE WITH 15,000 KG PAYLOAD (USSR)
Alexandr V. Galounenko, Capt 12,340 m
Sergei A. Borbik, Copilot 40,486 ft
Sergei F. Netchaev, Navigator
Antonov 225 "Mriya"
3/22/89

ALTITUDE WITH 20,000 KG PAYLOAD (USSR)
Alexandr V. Galounenko, Capt 12,340 m
Sergei A. Borbik, Copilot 40,486 ft
Sergei F. Netchaev, Navigator
Antonov 225 "Mriya"
3/22/89

ALTITUDE WITH 25,000 KG PAYLOAD (USSR)
Alexandr V. Galounenko, Capt 12,340 m
Sergei A. Borbik, Copilot 40,486 ft
Sergei F. Netchaev, Navigator
Antonov 225 "Mriya"
3/22/89

ALTITUDE WITH 30,000 KG PAYLOAD (USSR)
Alexandr V. Galounenko, Capt 12,340 m
Sergei A. Borbik, Copilot 40,486 ft
Sergei F. Netchaev, Navigator
Antonov 225 "Mriya"
3/22/89

ALTITUDE WITH 35,000 KG PAYLOAD (USSR)
Alexandr V. Galounenko, Capt 12,340 m
Sergei A. Borbik, Copilot 40,486 ft
Sergei F. Netchaev, Navigator
Antonov 225 "Mriya"
3/22/89

ALTITUDE WITH 40,000 KG PAYLOAD (USSR)
Alexandr V. Galounenko, Capt 12,340 m
Sergei A. Borbik, Copilot 40,486 ft
Sergei F. Netchaev, Navigator
Antonov 225 "Mriya"
3/22/89

ALTITUDE WITH 45,000 KG PAYLOAD (USSR)
Alexandr V. Galounenko, Capt 12,340 m
Sergei A. Borbik, Copilot 40,486 ft
Sergei F. Netchaev, Navigator
Antonov 225 "Mriya"
3/22/89

ALTITUDE WITH 50,000 KG PAYLOAD (USSR)
Alexandr V. Galounenko, Capt 12,340 m
Sergei A. Borbik, Copilot 40,486 ft
Sergei F. Netchaev, Navigator
Antonov 225 "Mriya"
3/22/89

ALTITUDE WITH 55,000 KG PAYLOAD (USSR)
Alexandr V. Galounenko, Capt 12,340m
Sergei A. Borbik, Copilot 40,486 ft
Sergei F. Netchaev, Navigator
Antonov 225 "Mriya"
3/22/89

ALTITUDE WITH 60,000 KG PAYLOAD (USSR)
Alexandr V. Galounenko, Capt 12,340 m
Sergei A. Borbik, Copilot 40,486 ft
Sergei F. Netchaev, Navigator
Antonov 225 "Mriya"
3/22/89

ALTITUDE WITH 65,000 KG PAYLOAD (USSR)
Alexandr V. Galounenko, Capt 12,340 m
Sergei A. Borbik, Copilot 40,486 ft
Sergei F. Netchaev, Navigator
Antonov 225 "Mriya"
3/22/89

ALTITUDE WITH 70,000 KG PAYLOAD (USSR)
Alexandr V. Galounenko, Capt 12,340 m
Sergei A. Borbik, Copilot 40,486 ft
Sergei F. Netchaev, Navigator
Antonov 225 "Mriya"
3/22/89

ALTITUDE WITH 75,000 KG PAYLOAD (USSR)
Alexandr V. Galounenko, Capt 12,340 m
Sergei A. Borbik, Copilot 40,486 ft
Sergei F. Netchaev, Navigator
Antonov 225 "Mriya"
3/22/89

ALTITUDE WITH 80,000 KG PAYLOAD (USSR)
Alexandr V. Galounenko, Capt 12,340 m
Sergei A. Borbik, Copilot 40,486 ft
Sergei F. Netchaev, Navigator
Antonov 225 "Mriya"
3/22/89

ALTITUDE WITH 85,000 KG PAYLOAD (USSR)
Alexandr V. Galounenko, Capt 12,340 m
Sergei A. Borbik, Copilot 40,486 ft
Sergei F. Netchaev, Navigator
Antonov 225 "Mriya"
3/22/89

ALTITUDE WITH 90,000 KG PAYLOAD (USSR)
Alexandr V. Galounenko, Capt 12,340 m
Sergei A. Borbik, Copilot 40,486 ft
Sergei F. Netchaev, Navigator
Antonov 225 "Mriya"
3/22/89

ALTITUDE WITH 95,000 KG PAYLOAD (USSR)
Alexandr V. Galounenko, Capt 12,340 m
Sergei A. Borbik, Copilot 40,486 ft
Sergei F. Netchaev, Navigator
Antonov 225 "Mriya"
3/22/89

ALTITUDE WITH 100,000 KG PAYLOAD (USSR)
Alexandr V. Galounenko, Capt 12,340 m
Sergei A. Borbik, Copilot 40,486 ft
Sergei F. Netchaev, Navigator
Antonov 225 "Mriya"
3/22/89

ALTITUDE WITH 105,000 KG PAYLOAD (USSR)
Alexandr V. Galounenko, Capt 12,340 m
Sergei A. Borbik, Copilot 40,486 ft
Sergei F. Netchaev, Navigator
Antonov 225 "Mriya"
3/22/89

ALTITUDE WITH 110,000 KG PAYLOAD (USSR)
Alexandr V. Galounenko, Capt 12,340 m
Sergei A. Borbik, Copilot 40,486 ft
Sergei F. Netchaev, Navigator
Antonov 225 "Mriya"
3/22/89

ALTITUDE WITH 115,000 KG PAYLOAD (USSR)
Alexandr V. Galounenko, Capt 12,340 m
Sergei A. Borbik, Copilot 40,486 ft
Sergei F. Netchaev, Navigator
Antonov 225 "Mriya"
3/22/89

ALTITUDE WITH 120,000 KG PAYLOAD (USSR)
Alexandr V. Galounenko, Capt 12,340 m
Sergei A. Borbik, Copilot 40,486 ft
Sergei F. Netchaev, Navigator
Antonov 225 "Mriya"
3/22/89

ALTITUDE WITH 125,000 KG PAYLOAD (USSR)
Alexandr V. Galounenko, Capt 12,340 m
Sergei A. Borbik, Copilot 40,486 ft
Sergei F. Netchaev, Navigator
Antonov 225 "Mriya"
3/22/89

ALTITUDE WITH 130,000 KG PAYLOAD (USSR)
Alexandr V. Galounenko, Capt 12,340 m
Sergei A. Borbik, Copilot 40,486 ft
Sergei F. Netchaev, Navigator
Antonov 225 "Mriya"
3/22/89

ALTITUDE WITH 135,000 KG PAYLOAD (USSR)
Alexandr V. Galounenko, Capt 12,340 m
Sergei A. Borbik, Copilot 40,486 ft
Sergei F. Netchaev, Navigator
Antonov 225 "Mriya"
3/22/89

ALTITUDE WITH 140,000 KG PAYLOAD (USSR)
Alexandr V. Galounenko, Capt 12,340 m
Sergei A. Borbik, Copilot 40,486 ft
Sergei F. Netchaev, Navigator
Antonov 225 "Mriya"
3/22/89

ALTITUDE WITH 145,000 KG PAYLOAD (USSR)
Alexandr V. Galounenko, Capt 12,340 m
Sergei A. Borbik, Copilot 40,486 ft
Sergei F. Netchaev, Navigator
Antonov 225 "Mriya"
3/22/89

ALTITUDE WITH 150,000 KG PAYLOAD (USSR)
Alexandr V. Galounenko, Capt 12,340 m
Sergei A. Borbik, Copilot 40,486 ft
Sergei F. Netchaev, Navigator
Antonov 225 "Mriya"
3/22/89

ALTITUDE WITH 155,000 KG PAYLOAD (USSR)
Alexandr V. Galounenko, Capt 12,340 m
Sergei A. Borbik, Copilot 40,486 ft
Sergei F. Netchaev, Navigator
Antonov 225 "Mriya"
3/22/89

ALTITUDE WITH 160,000 KG PAYLOAD (UKRAINE)
Olexander Halunenko 10,750 m
Anatolii Moisseiev 35,269 ft
Antonov An-225 Mriya
6 Progress D-18T, 51,590 lbs
Kyiv, Ukraine
9/11/01

ALTITUDE WITH 165,000 KG PAYLOAD (UKRAINE)
Olexander Halunenko 10,750 m
Anatolii Moisseiev 35,269 ft
Antonov An-225 Mriya
6 Progress D-18T, 51,590 lbs
Kyiv, Ukraine
9/11/01

ALTITUDE WITH 170,000 KG PAYLOAD (UKRAINE)
Olexander Halunenko 10,750 m
Anatolii Moisseiev 35,269 ft
Antonov An-225 Mriya
6 Progress D-18T, 51,590 lbs
Kyiv, Ukraine
9/11/01

ALTITUDE WITH 175,000 KG PAYLOAD (UKRAINE)
Olexander Halunenko 10,750 m
Anatolii Moisseiev 35,269 ft
Antonov An-225 Mriya
6 Progress D-18T, 51,590 lbs
Kyiv, Ukraine
9/11/01

ALTITUDE WITH 180,000 KG PAYLOAD (UKRAINE)
Olexander Halunenko 10,750 m
Anatolii Moisseiev 35,269 ft
Antonov An-225 Mriya
6 Progress D-18T, 51,590 lbs
Kyiv, Ukraine
9/11/01

ALTITUDE WITH 185,000 KG PAYLOAD (UKRAINE)
Olexander Halunenko 10,750 m
Anatolii Moisseiev 35,269 ft
Antonov An-225 Mriya
6 Progress D-18T, 51,590 lbs
Kyiv, Ukraine
9/11/01

ALTITUDE WITH 190,000 KG PAYLOAD (UKRAINE)
Olexander Halunenko 10,750 m
Anatolii Moisseiev 35,269 ft
Antonov An-225 Mriya
6 Progress D-18T, 51,590 lbs
Kyiv, Ukraine
9/11/01

ALTITUDE WITH 195,000 KG PAYLOAD (UKRAINE)
Olexander Halunenko 10,750 m
Anatolii Moisseiev 35,269 ft
Antonov An-225 Mriya
6 Progress D-18T, 51,590 lbs
Kyiv, Ukraine
9/11/01

ALTITUDE WITH 200,000 KG PAYLOAD (UKRAINE)
Olexander Halunenko 10,750 m
Anatolii Moisseiev 35,269 ft
Antonov An-225 Mriya
6 Progress D-18T, 51,590 lbs
Kyiv, Ukraine
9/11/01

ALTITUDE WITH 205,000 KG PAYLOAD (UKRAINE)
Olexander Halunenko 10,750 m
Anatolii Moisseiev 35,269 ft
Antonov An-225 Mriya
6 Progress D-18T, 51,590 lbs
Kyiv, Ukraine
9/11/01

ALTITUDE WITH 210,000 KG PAYLOAD (UKRAINE)
Olexander Halunenko 10,750 m
Anatolii Moisseiev 35,269 ft
Antonov An-225 Mriya
6 Progress D-18T, 51,590 lbs
Kyiv, Ukraine
9/11/01

ALTITUDE WITH 215,000 KG PAYLOAD (UKRAINE)
Olexander Halunenko 10,750 m
Anatolii Moisseiev 35,269 ft
Antonov An-225 Mriya
6 Progress D-18T, 51,590 lbs
Kyiv, Ukraine
9/11/01

ALTITUDE WITH 220,000 KG PAYLOAD (UKRAINE)
Olexander Halunenko 10,750 m
Anatolii Moisseiev 35,269 ft
Antonov An-225 Mriya
6 Progress D-18T, 51,590 lbs
Kyiv, Ukraine
9/11/01

ALTITUDE WITH 225,000 KG PAYLOAD (UKRAINE)
Olexander Halunenko 10,750 m
Anatolii Moisseiev 35,269 ft
Antonov An-225 Mriya
6 Progress D-18T, 51,590 lbs
Kyiv, Ukraine
9/11/01

ALTITUDE WITH 230,000 KG PAYLOAD (UKRAINE)
Olexander Halunenko 10,750 m
Anatolii Moisseiev 35,269 ft
Antonov An-225 Mriya
6 Progress D-18T, 51,590 lbs
Kyiv, Ukraine
9/11/01

ALTITUDE WITH 235,000 KG PAYLOAD (UKRAINE)
Olexander Halunenko 10,750 m
Anatolii Moisseiev 35,269 ft
Antonov An-225 Mriya
6 Progress D-18T, 51,590 lbs
Kyiv, Ukraine
9/11/01

ALTITUDE WITH 240,000 KG PAYLOAD (UKRAINE)
Olexander Halunenko 10,750 m
Anatolii Moisseiev 35,269 ft
Antonov An-225 Mriya
6 Progress D-18T, 51,590 lbs
Kyiv, Ukraine
9/11/01

ALTITUDE WITH 245,000 KG PAYLOAD (UKRAINE)
Olexander Halunenko 10,750 m
Anatolii Moisseiev 35,269 ft
Antonov An-225 Mriya
6 Progress D-18T, 51,590 lbs
Kyiv, Ukraine
9/11/01

ALTITUDE WITH 250,000 KG PAYLOAD (UKRAINE)
Olexander Halunenko 10,750 m
Anatolii Moisseiev 35,269 ft
Antonov An-225 Mriya
6 Progress D-18T, 51,590 lbs
Kyiv, Ukraine
9/11/01

GREATEST PAYLOAD CARRIED TO 2,000 METERS (UKRAINE)
Olexander Halunenko 253,820 kg
Anatolii Moisseiev 559,576 lbs
Antonov An-225 Mriya
6 Progress D-18T, 51,590 lbs
Kyiv, Ukraine
9/11/01

SPEED

SPEED OVER A 1,000 KM CLOSED CIRCUIT WITHOUT PAYLOAD (UKRAINE)
Olexander Halunenko 763.20 kmh
Anatolii Moisseiev 474.23 mph
Antonov An-225 Mriya
6 Progress D-18T, 51,590 lbs
Kyiv, Ukraine
9/11/01

SPEED OVER A 1,000 KM CLOSED CIRCUIT WITH 1,000 KG PAYLOAD (UKRAINE)
Olexander Halunenko 763.20 kmh
Anatolii Moisseiev 474.23 mph
Antonov An-225 Mriya
6 Progress D-18T, 51,590 lbs
Kyiv, Ukraine
9/11/01

SPEED OVER A 1,000 KM CLOSED CIRCUIT WITH 2,000 KG PAYLOAD (UKRAINE)
Olexander Halunenko 763.20 kmh
Anatolii Moisseiev 474.23 mph
Antonov An-225 Mriya
6 Progress D-18T, 51,590 lbs
Kyiv, Ukraine
9/11/01

SPEED OVER A 1,000 KM CLOSED CIRCUIT WITH 5,000 KG PAYLOAD (UKRAINE)
Olexander Halunenko 763.20 kmh
Anatolii Moisseiev 474.23 mph
Antonov An-225 Mriya
6 Progress D-18T, 51,590 lbs
Kyiv, Ukraine
9/11/01

SPEED OVER A 1,000 KM CLOSED CIRCUIT WITH 10,000 KG PAYLOAD (UKRAINE)
Olexander Halunenko 763.20 kmh
Anatolii Moisseiev 474.23 mph
Antonov An-225 Mriya
6 Progress D-18T, 51,590 lbs
Kyiv, Ukraine
9/11/01

SPEED OVER A 1,000 KM CLOSED CIRCUIT WITH 15,000 KG PAYLOAD (UKRAINE)
Olexander Halunenko 763.20 kmh
Anatolii Moisseiev 474.23 mph
Antonov An-225 Mriya
6 Progress D-18T, 51,590 lbs
Kyiv, Ukraine
9/11/01

SPEED OVER A 1,000 KM CLOSED CIRCUIT WITH 20,000 KG PAYLOAD (UKRAINE)
Olexander Halunenko 763.20 kmh
Anatolii Moisseiev 474.23 mph
Antonov An-225 Mriya
6 Progress D-18T, 51,590 lbs
Kyiv, Ukraine
9/11/01

SPEED OVER A 1,000 KM CLOSED CIRCUIT WITH 25,000 KG PAYLOAD (UKRAINE)
Olexander Halunenko 763.20 kmh
Anatolii Moisseiev 474.23 mph
Antonov An-225 Mriya
6 Progress D-18T, 51,590 lbs
Kyiv, Ukraine
9/11/01

SPEED OVER A 1,000 KM CLOSED CIRCUIT WITH 30,000 KG PAYLOAD (UKRAINE)
Olexander Halunenko 763.20 kmh
Anatolii Moisseiev 474.23 mph
Antonov An-225 Mriya
6 Progress D-18T, 51,590 lbs
Kyiv, Ukraine
9/11/01

SPEED OVER A 1,000 KM CLOSED CIRCUIT WITH 35,000 KG PAYLOAD (UKRAINE)
Olexander Halunenko 763.20 kmh
Anatolii Moisseiev 474.23 mph
Antonov An-225 Mriya
6 Progress D-18T, 51,590 lbs
Kyiv, Ukraine
9/11/01

SPEED OVER A 1,000 KM CLOSED CIRCUIT WITH 40,000 KG PAYLOAD (UKRAINE)
Olexander Halunenko 763.20 kmh
Anatolii Moisseiev 474.23 mph
Antonov An-225 Mriya
6 Progress D-18T, 51,590 lbs
Kyiv, Ukraine
9/11/01

SPEED OVER A 1,000 KM CLOSED CIRCUIT WITH 45,000 KG PAYLOAD (UKRAINE)
Olexander Halunenko 763.20 kmh
Anatolii Moisseiev 474.23 mph
Antonov An-225 Mriya
6 Progress D-18T, 51,590 lbs
Kyiv, Ukraine
9/11/01

SPEED OVER A 1,000 KM CLOSED CIRCUIT WITH 50,000 KG PAYLOAD (UKRAINE)
Olexander Halunenko 763.20 kmh
Anatolii Moisseiev 474.23 mph
Antonov An-225 Mriya
6 Progress D-18T, 51,590 lbs
Kyiv, Ukraine
9/11/01

SPEED OVER A 1,000 KM CLOSED CIRCUIT WITH 55,000 KG PAYLOAD (UKRAINE)
Olexander Halunenko 763.20 kmh
Anatolii Moisseiev 474.23 mph
Antonov An-225 Mriya
6 Progress D-18T, 51,590 lbs
Kyiv, Ukraine
9/11/01

SPEED OVER A 1,000 KM CLOSED CIRCUIT WITH 60,000 KG PAYLOAD (UKRAINE)
Olexander Halunenko 763.20 kmh
Anatolii Moisseiev 474.23 mph
Antonov An-225 Mriya
6 Progress D-18T, 51,590 lbs
Kyiv, Ukraine
9/11/01

SPEED OVER A 1,000 KM CLOSED CIRCUIT WITH 65,000 KG PAYLOAD (UKRAINE)
Olexander Halunenko 763.20 kmh
Anatolii Moisseiev 474.23 mph
Antonov An-225 Mriya
6 Progress D-18T, 51,590 lbs
Kyiv, Ukraine
9/11/01

SPEED OVER A 1,000 KM CLOSED CIRCUIT WITH 70,000 KG PAYLOAD (UKRAINE)
Olexander Halunenko 763.20 kmh
Anatolii Moisseiev 474.23 mph
Antonov An-225 Mriya
6 Progress D-18T, 51,590 lbs
Kyiv, Ukraine
9/11/01

SPEED OVER A 1,000 KM CLOSED CIRCUIT WITH 75,000 KG PAYLOAD (UKRAINE)
Olexander Halunenko 763.20 kmh
Anatolii Moisseiev 474.23 mph
Antonov An-225 Mriya
6 Progress D-18T, 51,590 lbs
Kyiv, Ukraine
9/11/01

SPEED OVER A 1,000 KM CLOSED CIRCUIT WITH 80,000 KG PAYLOAD (UKRAINE)
Olexander Halunenko 763.20 kmh
Anatolii Moisseiev 474.23 mph
Antonov An-225 Mriya
6 Progress D-18T, 51,590 lbs
Kyiv, Ukraine
9/11/01

SPEED OVER A 1,000 KM CLOSED CIRCUIT WITH 85,000 KG PAYLOAD (UKRAINE)
Olexander Halunenko 763.20 kmh
Anatolii Moisseiev 474.23 mph
Antonov An-225 Mriya
6 Progress D-18T, 51,590 lbs
Kyiv, Ukraine
9/11/01

SPEED OVER A 1,000 KM CLOSED CIRCUIT WITH 90,000 KG PAYLOAD (UKRAINE)
Olexander Halunenko 763.20 kmh
Anatolii Moisseiev 474.23 mph
Antonov An-225 Mriya
6 Progress D-18T, 51,590 lbs
Kyiv, Ukraine
9/11/01

SPEED OVER A 1,000 KM CLOSED CIRCUIT WITH 95,000 KG PAYLOAD (UKRAINE)
Olexander Halunenko 763.20 kmh
Anatolii Moisseiev 474.23 mph
Antonov An-225 Mriya
6 Progress D-18T, 51,590 lbs
Kyiv, Ukraine
9/11/01

SPEED OVER A 1,000 KM CLOSED CIRCUIT WITH 100,000 KG PAYLOAD (UKRAINE)
Olexander Halunenko 763.20 kmh
Anatolii Moisseiev 474.23 mph
Antonov An-225 Mriya
6 Progress D-18T, 51,590 lbs
Kyiv, Ukraine
9/11/01

SPEED OVER A 1,000 KM CLOSED CIRCUIT WITH 105,000 KG PAYLOAD (UKRAINE)
Olexander Halunenko 763.20 kmh
Anatolii Moisseiev 474.23 mph
Antonov An-225 Mriya
6 Progress D-18T, 51,590 lbs
Kyiv, Ukraine
9/11/01

SPEED OVER A 1,000 KM CLOSED CIRCUIT WITH 110,000 KG PAYLOAD (UKRAINE)
Olexander Halunenko 763.20 kmh
Anatolii Moisseiev 474.23 mph
Antonov An-225 Mriya
6 Progress D-18T, 51,590 lbs
Kyiv, Ukraine
9/11/01

SPEED OVER A 1,000 KM CLOSED CIRCUIT WITH 115,000 KG PAYLOAD (UKRAINE)
Olexander Halunenko 763.20 kmh
Anatolii Moisseiev 474.23 mph
Antonov An-225 Mriya
6 Progress D-18T, 51,590 lbs
Kyiv, Ukraine
9/11/01

SPEED OVER A 1,000 KM CLOSED CIRCUIT WITH 120,000 KG PAYLOAD (UKRAINE)
Olexander Halunenko 763.20 kmh
Anatolii Moisseiev 474.23 mph
Antonov An-225 Mriya
6 Progress D-18T, 51,590 lbs
Kyiv, Ukraine
9/11/01

SPEED OVER A 1,000 KM CLOSED CIRCUIT WITH 125,000 KG PAYLOAD (UKRAINE)
Olexander Halunenko 763.20 kmh
Anatolii Moisseiev 474.23 mph
Antonov An-225 Mriya
6 Progress D-18T, 51,590 lbs
Kyiv, Ukraine
9/11/01

SPEED OVER A 1,000 KM CLOSED CIRCUIT WITH 130,000 KG PAYLOAD (UKRAINE)
Olexander Halunenko 763.20 kmh
Anatolii Moisseiev 474.23 mph
Antonov An-225 Mriya
6 Progress D-18T, 51,590 lbs
Kyiv, Ukraine
9/11/01

SPEED OVER A 1,000 KM CLOSED CIRCUIT WITH 135,000 KG PAYLOAD (UKRAINE)
Olexander Halunenko 763.20 kmh
Anatolii Moisseiev 474.23 mph
Antonov An-225 Mriya
6 Progress D-18T, 51,590 lbs
Kyiv, Ukraine
9/11/01

SPEED OVER A 1,000 KM CLOSED CIRCUIT WITH 140,000 KG PAYLOAD (UKRAINE)
Olexander Halunenko 763.20 kmh
Anatolii Moisseiev 474.23 mph
Antonov An-225 Mriya
6 Progress D-18T, 51,590 lbs
Kyiv, Ukraine
9/11/01

SPEED OVER A 1,000 KM CLOSED CIRCUIT WITH
145,000 KG PAYLOAD (UKRAINE)
Olexander Halunenko 763.20 kmh
Anatolii Moisseiev 474.23 mph
Antonov An-225 Mriya
6 Progress D-18T, 51,590 lbs
Kyiv, Ukraine
9/11/01

SPEED OVER A 1,000 KM CLOSED CIRCUIT WITH
150,000 KG PAYLOAD (UKRAINE)
Olexander Halunenko 763.20 kmh
Anatolii Moisseiev 474.23 mph
Antonov An-225 Mriya
6 Progress D-18T, 51,590 lbs
Kyiv, Ukraine
9/11/01

SPEED OVER A 1,000 KM CLOSED CIRCUIT WITH
155,000 KG PAYLOAD (UKRAINE)
Olexander Halunenko 763.20 kmh
Anatolii Moisseiev 474.23 mph
Antonov An-225 Mriya
6 Progress D-18T, 51,590 lbs
Kyiv, Ukraine
9/11/01

SPEED OVER A 1,000 KM CLOSED CIRCUIT WITH
160,000 KG PAYLOAD (UKRAINE)
Olexander Halunenko 763.20 kmh
Anatolii Moisseiev 474.23 mph
Antonov An-225 Mriya
6 Progress D-18T, 51,590 lbs
Kyiv, Ukraine
9/11/01

SPEED OVER A 1,000 KM CLOSED CIRCUIT WITH
165,000 KG PAYLOAD (UKRAINE)
Olexander Halunenko 763.20 kmh
Anatolii Moisseiev 474.23 mph
Antonov An-225 Mriya
6 Progress D-18T, 51,590 lbs
Kyiv, Ukraine
9/11/01

SPEED OVER A 1,000 KM CLOSED CIRCUIT WITH
170,000 KG PAYLOAD (UKRAINE)
Olexander Halunenko 763.20 kmh
Anatolii Moisseiev 474.23 mph
Antonov An-225 Mriya
6 Progress D-18T, 51,590 lbs
Kyiv, Ukraine
9/11/01

SPEED OVER A 1,000 KM CLOSED CIRCUIT WITH
175,000 KG PAYLOAD (UKRAINE)
Olexander Halunenko 763.20 kmh
Anatolii Moisseiev 474.23 mph
Antonov An-225 Mriya
6 Progress D-18T, 51,590 lbs
Kyiv, Ukraine
9/11/01

SPEED OVER A 1,000 KM CLOSED CIRCUIT WITH
180,000 KG PAYLOAD (UKRAINE)
Olexander Halunenko 763.20 kmh
Anatolii Moisseiev 474.23 mph
Antonov An-225 Mriya
6 Progress D-18T, 51,590 lbs
Kyiv, Ukraine
9/11/01

SPEED OVER A 1,000 KM CLOSED CIRCUIT WITH
185,000 KG PAYLOAD (UKRAINE)
Olexander Halunenko 763.20 kmh
Anatolii Moisseiev 474.23 mph
Antonov An-225 Mriya
6 Progress D-18T, 51,590 lbs
Kyiv, Ukraine
9/11/01

SPEED OVER A 1,000 KM CLOSED CIRCUIT WITH
190,000 KG PAYLOAD (UKRAINE)
Olexander Halunenko 763.20 kmh
Anatolii Moisseiev 474.23 mph
Antonov An-225 Mriya
6 Progress D-18T, 51,590 lbs
Kyiv, Ukraine
9/11/01

SPEED OVER A 1,000 KM CLOSED CIRCUIT WITH
195,000 KG PAYLOAD (UKRAINE)
Olexander Halunenko 763.20 kmh
Anatolii Moisseiev 474.23 mph
Antonov An-225 Mriya
6 Progress D-18T, 51,590 lbs
Kyiv, Ukraine
9/11/01

SPEED OVER A 1,000 KM CLOSED CIRCUIT WITH
200,000 KG PAYLOAD (UKRAINE)
Olexander Halunenko 763.20 kmh
Anatolii Moisseiev 474.23 mph
Antonov An-225 Mriya
6 Progress D-18T, 51,590 lbs
Kyiv, Ukraine
9/11/01

SPEED OVER A 1,000 KM CLOSED CIRCUIT WITH
205,000 KG PAYLOAD (UKRAINE)
Olexander Halunenko 763.20 kmh
Anatolii Moisseiev 474.23 mph
Antonov An-225 Mriya
6 Progress D-18T, 51,590 lbs
Kyiv, Ukraine
9/11/01

SPEED OVER A 1,000 KM CLOSED CIRCUIT WITH
210,000 KG PAYLOAD (UKRAINE)
Olexander Halunenko 763.20 kmh
Anatolii Moisseiev 474.23 mph
Antonov An-225 Mriya
6 Progress D-18T, 51,590 lbs
Kyiv, Ukraine
9/11/01

SPEED OVER A 1,000 KM CLOSED CIRCUIT WITH
215,000 KG PAYLOAD (UKRAINE)
Olexander Halunenko 763.20 kmh
Anatolii Moisseiev 474.23 mph
Antonov An-225 Mriya
6 Progress D-18T, 51,590 lbs
Kyiv, Ukraine
9/11/01

SPEED OVER A 1,000 KM CLOSED CIRCUIT WITH
220,000 KG PAYLOAD (UKRAINE)
Olexander Halunenko 763.20 kmh
Anatolii Moisseiev 474.23 mph
Antonov An-225 Mriya
6 Progress D-18T, 51,590 lbs
Kyiv, Ukraine
9/11/01

SPEED OVER A 1,000 KM CLOSED CIRCUIT WITH
225,000 KG PAYLOAD (UKRAINE)
Olexander Halunenko 763.20 kmh
Anatolii Moisseiev 474.23 mph
Antonov An-225 Mriya
6 Progress D-18T, 51,590 lbs
Kyiv, Ukraine
9/11/01

SPEED OVER A 1,000 KM CLOSED CIRCUIT WITH
230,000 KG PAYLOAD (UKRAINE)
Olexander Halunenko 763.20 kmh
Anatolii Moisseiev 474.23 mph
Antonov An-225 Mriya
6 Progress D-18T, 51,590 lbs
Kyiv, Ukraine
9/11/01

SPEED OVER A 1,000 KM CLOSED CIRCUIT WITH 235,000 KG PAYLOAD (UKRAINE)
Olexander Halunenko 763.20 kmh
Anatolii Moisseiev 474.23 mph
Antonov An-225 Mriya
6 Progress D-18T, 51,590 lbs
Kyiv, Ukraine
9/11/01

SPEED OVER A 1,000 KM CLOSED CIRCUIT WITH 240,000 KG PAYLOAD (UKRAINE)
Olexander Halunenko 763.20 kmh
Anatolii Moisseiev 474.23 mph
Antonov An-225 Mriya
6 Progress D-18T, 51,590 lbs
Kyiv, Ukraine
9/11/01

SPEED OVER A 1,000 KM CLOSED CIRCUIT WITH 245,000 KG PAYLOAD (UKRAINE)
Olexander Halunenko 763.20 kmh
Anatolii Moisseiev 474.23 mph
Antonov An-225 Mriya
6 Progress D-18T, 51,590 lbs
Kyiv, Ukraine
9/11/01

SPEED OVER A 1,000 KM CLOSED CIRCUIT WITH 250,000 KG PAYLOAD (UKRAINE)
Olexander Halunenko 763.20 kmh
Anatolii Moisseiev 474.23 mph
Antonov An-225 Mriya
6 Progress D-18T, 51,590 lbs
Kyiv, Ukraine
9/11/01

SPEED OVER A 2,000 KM CLOSED CIRCUIT WITHOUT PAYLOAD (USSR)
Alexandr V. Galounenko, Capt 813.09 kmh
Sergei A. Borbik, Copilot 505.25 mph
Sergei F. Netchaev, Navigator
Antonov 225 "Mriya"
3/22/89

SPEED OVER A 2,000 KM CLOSED CIRCUIT WITH 1,000 KG PAYLOAD (USSR)
Alexandr V. Galounenko, Capt 813.09 kmh
Sergei A. Borbik, Copilot 505.25 mph
Sergei F. Netchaev, Navigator
Antonov 225 "Mriya"
3/22/89

SPEED OVER A 2,000 KM CLOSED CIRCUIT WITH 2,000 KG PAYLOAD (USSR)
Alexandr V. Galounenko, Capt 813.09 kmh
Sergei A. Borbik, Copilot 505.25 mph
Sergei F. Netchaev, Navigator
Antonov 225 "Mriya"
3/22/89

SPEED OVER A 2,000 KM CLOSED CIRCUIT WITH 5,000 KG PAYLOAD (USSR)
Alexandr V. Galounenko, Capt 813.09 kmh
Sergei A. Borbik, Copilot 505.25 mph
Sergei F. Netchaev, Navigator
Antonov 225 "Mriya"
3/22/89

SPEED OVER A 2,000 KM CLOSED CIRCUIT WITH 10,000 KG PAYLOAD (USSR)
Alexandr V. Galounenko, Capt 813.09 kmh
Sergei A. Borbik, Copilot 505.25 mph
Sergei F. Netchaev, Navigator
Antonov 225 "Mriya"
3/22/89

SPEED OVER A 2,000 KM CLOSED CIRCUIT WITH 15,000 KG PAYLOAD (USSR)
Alexandr V. Galounenko, Capt 813.09 kmh
Sergei A. Borbik, Copilot 505.25 mph
Sergei F. Netchaev, Navigator
Antonov 225 "Mriya"
3/22/89

SPEED OVER A 2,000 KM CLOSED CIRCUIT WITH 20,000 KG PAYLOAD (USSR)
Alexandr V. Galounenko, Capt 813.09 kmh
Sergei A. Borbik, Copilot 505.25 mph
Sergei F. Netchaev, Navigator
Antonov 225 "Mriya"
3/22/89

SPEED OVER A 2,000 KM CLOSED CIRCUIT WITH 25,000 KG PAYLOAD (USSR)
Alexandr V. Galounenko, Capt 813.09 kmh
Sergei A. Borbik, Copilot 505.25 mph
Sergei F. Netchaev, Navigator
Antonov 225 "Mriya"
3/22/89

SPEED OVER A 2,000 KM CLOSED CIRCUIT WITH 30,000 KG PAYLOAD (USSR)
Alexandr V. Galounenko, Capt 813.09 kmh
Sergei A. Borbik, Copilot 505.25 mph
Sergei F. Netchaev, Navigator
Antonov 225 "Mriya"
3/22/89

SPEED OVER A 2,000 KM CLOSED CIRCUIT WITH 35,000 KG PAYLOAD (USSR)
Alexandr V. Galounenko, Capt 813.09 kmh
Sergei A. Borbik, Copilot 505.25 mph
Sergei F. Netchaev, Navigator
Antonov 225 "Mriya"
3/22/89

SPEED OVER A 2,000 KM CLOSED CIRCUIT WITH 40,000 KG PAYLOAD (USSR)
Alexandr V. Galounenko, Capt 813.09 kmh
Sergei A. Borbik, Copilot 505.25 mph
Sergei F. Netchaev, Navigator
Antonov 225 "Mriya"
3/22/89

SPEED OVER A 2,000 KM CLOSED CIRCUIT WITH 45,000 KG PAYLOAD (USSR)
Alexandr V. Galounenko, Capt 813.09 kmh
Sergei A. Borbik, Copilot 505.25 mph
Sergei F. Netchaev, Navigator
Antonov 225 "Mriya"
3/22/89

SPEED OVER A 2,000 KM CLOSED CIRCUIT WITH 50,000 KG PAYLOAD (USSR)
Alexandr V. Galounenko, Capt 813.09 kmh
Sergei A. Borbik, Copilot 505.25 mph
Sergei F. Netchaev, Navigator
Antonov 225 "Mriya"
3/22/89

SPEED OVER A 2,000 KM CLOSED CIRCUIT WITH 55,000 KG PAYLOAD (USSR)
Alexandr V. Galounenko, Capt 813.09 kmh
Sergei A. Borbik, Copilot 505.25 mph
Sergei F. Netchaev, Navigator
Antonov 225 "Mriya"
3/22/89

SPEED OVER A 2,000 KM CLOSED CIRCUIT WITH 60,000 KG PAYLOAD (USSR)
Alexandr V. Galounenko, Capt 813.09 kmh
Sergei A. Borbik, Copilot 505.25 mph
Sergei F. Netchaev, Navigator
Antonov 225 "Mriya"
3/22/89

SPEED OVER A 2,000 KM CLOSED CIRCUIT WITH 65,000 KG PAYLOAD (USSR)
Alexandr V. Galounenko, Capt 813.09 kmh
Sergei A. Borbik, Copilot 505.25 mph
Sergei F. Netchaev, Navigator
Antonov 225 "Mriya"
3/22/89

SPEED OVER A 2,000 KM CLOSED CIRCUIT WITH 70,000 KG PAYLOAD (USSR)
Alexandr V. Galounenko, Capt 813.09 kmh
Sergei A. Borbik, Copilot 505.25 mph
Sergei F. Netchaev, Navigator
Antonov 225 "Mriya"
3/22/89

SPEED OVER A 2,000 KM CLOSED CIRCUIT WITH 75,000 KG PAYLOAD (USSR)
Alexandr V. Galounenko, Capt 813.09 kmh
Sergei A. Borbik, Copilot 505.25 mph
Sergei F. Netchaev, Navigator
Antonov 225 "Mriya"
3/22/89

SPEED OVER A 2,000 KM CLOSED CIRCUIT WITH 80,000 KG PAYLOAD (USSR)
Alexandr V. Galounenko, Capt 813.09 kmh
Sergei A. Borbik, Copilot 505.25 mph
Sergei F. Netchaev, Navigator
Antonov 225 "Mriya"
3/22/89

SPEED OVER A 2,000 KM CLOSED CIRCUIT WITH 85,000 KG PAYLOAD (USSR)
Alexandr V. Galounenko, Capt 813.09 kmh
Sergei A. Borbik, Copilot 505.25 mph
Sergei F. Netchaev, Navigator
Antonov 225 "Mriya"
3/22/89

SPEED OVER A 2,000 KM CLOSED CIRCUIT WITH 90,000 KG PAYLOAD (USSR)
Alexandr V. Galounenko, Capt 813.09 kmh
Sergei A. Borbik, Copilot 505.25 mph
Sergei F. Netchaev, Navigator
Antonov 225 "Mriya"
3/22/89

SPEED OVER A 2,000 KM CLOSED CIRCUIT WITH 95,000 KG PAYLOAD (USSR)
Alexandr V. Galounenko, Capt 813.09 kmh
Sergei A. Borbik, Copilot 505.25 mph
Sergei F. Netchaev, Navigator
Antonov 225 "Mriya"
3/22/89

SPEED OVER A 2,000 KM CLOSED CIRCUIT WITH 100,000 KG PAYLOAD (USSR)
Alexandr V. Galounenko, Capt 813.09 kmh
Sergei A. Borbik, Copilot 505.25 mph
Sergei F. Netchaev, Navigator
Antonov 225 "Mriya"
3/22/89

SPEED OVER A 2,000 KM CLOSED CIRCUIT WITH 105,000 KG PAYLOAD (USSR)
Alexandr V. Galounenko, Capt 813.09 kmh
Sergei A. Borbik, Copilot 505.25 mph
Sergei F. Netchaev, Navigator
Antonov 225 "Mriya"
3/22/89

SPEED OVER A 2,000 KM CLOSED CIRCUIT WITH 110,000 KG PAYLOAD (USSR)
Alexandr V. Galounenko, Capt 813.09 kmh
Sergei A. Borbik, Copilot 505.25 mph
Sergei F. Netchaev, Navigator
Antonov 225 "Mriya"
3/22/89

SPEED OVER A 2,000 KM CLOSED CIRCUIT WITH 115,000 KG PAYLOAD (USSR)
Alexandr V. Galounenko, Capt 813.09 kmh
Sergei A. Borbik, Copilot 505.25 mph
Sergei F. Netchaev, Navigator
Antonov 225 "Mriya"
3/22/89

SPEED OVER A 2,000 KM CLOSED CIRCUIT WITH 120,000 KG PAYLOAD (USSR)
Alexandr V. Galounenko, Capt 813.09 kmh
Sergei A. Borbik, Copilot 505.25 mph
Sergei F. Netchaev, Navigator
Antonov 225 "Mriya"
3/22/89

SPEED OVER A 2,000 KM CLOSED CIRCUIT WITH 125,000 KG PAYLOAD (USSR)
Alexandr V. Galounenko, Capt 813.09 kmh
Sergei A. Borbik, Copilot 505.25 mph
Sergei F. Netchaev, Navigator
Antonov 225 "Mriya"
3/22/89

SPEED OVER A 2,000 KM CLOSED CIRCUIT WITH 130,000 KG PAYLOAD (USSR)
Alexandr V. Galounenko, Capt 813.09 kmh
Sergei A. Borbik, Copilot 505.25 mph
Sergei F. Netchaev, Navigator
Antonov 225 "Mriya"
3/22/89

SPEED OVER A 2,000 KM CLOSED CIRCUIT WITH 135,000 KG PAYLOAD (USSR)
Alexandr V. Galounenko, Capt 813.09 kmh
Sergei A. Borbik, Copilot 505.25 mph
Sergei F. Netchaev, Navigator
Antonov 225 "Mriya"
3/22/89

SPEED OVER A 2,000 KM CLOSED CIRCUIT WITH 140,000 KG PAYLOAD (USSR)
Alexandr V. Galounenko, Capt 813.09 kmh
Sergei A. Borbik, Copilot 505.25 mph
Sergei F. Netchaev, Navigator
Antonov 225 "Mriya"
3/22/89

SPEED OVER A 2,000 KM CLOSED CIRCUIT WITH 145,000 KG PAYLOAD (USSR)
Alexandr V. Galounenko, Capt 813.09 kmh
Sergei A. Borbik, Copilot 505.25 mph
Sergei F. Netchaev, Navigator
Antonov 225 "Mriya"
3/22/89

SPEED OVER A 2,000 KM CLOSED CIRCUIT WITH 150,000 KG PAYLOAD (USSR)
Alexandr V. Galounenko, Capt 813.09 kmh
Sergei A. Borbik, Copilot 505.25 mph
Sergei F. Netchaev, Navigator
Antonov 225 "Mriya"
3/22/89

SPEED OVER A 2,000 KM CLOSED CIRCUIT WITH 155,000 KG PAYLOAD (USSR)
Alexandr V. Galounenko, Capt 813.09 kmh
Sergei A. Borbik, Copilot 505.25 mph
Sergei F. Netchaev, Navigator
Antonov 225 "Mriya"
3/22/89

SPEED AROUND THE WORLD, EASTBOUND (USA)
Clay Lacy, Captain 1,003.53 kmh
Verne Jobst, Co-Captain 623.59 mph
Boeing 747 SP
4 P&W JT9D-7A, 46,250 lbs
Seattle, WA to Seattle, WA
01/29-30/88

SPEED AROUND THE WORLD OVER BOTH THE EARTH'S POLES (USSR)
Lev V. Kozlov, Captain 689.10 kmh
Yuri P. Resnitsky, Copilot 428.18 mph
Oleg I. Pripuskov, Copilot
Anatoly V. Andronov, Copilot
Antonov-124
Melbourne, South Pole, Rio de Janeiro, Casablanca, North Pole, Ussurijsk, Melbourne
12/1-4/90

Rocket Engine Aircraft

The rocket-powered airplane has followed a rather strange path of development, starting out as a military interceptor of dubious value and then being transformed into a research machine of enormous potential.

When the first Messerschmitt Me-163 Comet rocket planes blasted into World War II skies during the summer of 1944, their speed and radical appearance struck fear into the hearts of Allied airmen. They were over 100 mph faster than any opposing airplanes and there seemed to be no way to counter them.

But the new aircraft had serious shortcomings: they had a duration of only a few minutes, and their extremely hazardous fuel killed more German pilots than did Allied guns. As a combat airplane, the rocket was limited. But the possibility for great speed with a small engine was so obvious that, while no further plans were made for military rocket fighters, they soon began to attack the frontiers of scientific progress.

In 1947, Chuck Yeager became the first human to fly faster than the speed of sound when he piloted the bright orange Bell XS-1 (later called the X-1) to a speed of 670 mph. There was never any thought of turning this winged bullet into a combat airplane; its function was to investigate the conditions near, at, and beyond the speed of sound.

Developments of the X-1 eventually led to the North American X-15, which has flown as fast as 4,500 mph (Mach 6.7) and more than 60 miles high. These rocket airplanes contributed greatly to the knowledge of high-speed flight and to solutions of the problems of traveling in this previously unknown environment.

Rocket-powered research flights were conducted in the most practical way possible, by launching the rocket-plane from a flying carrier already traveling several hundred miles per hour, thousands of feet above the ground. This enabled the rocket plane to conserve its precious fuel for even higher speeds and altitudes.

CLASS C-1, GROUP IV, ROCKET ENGINE

LAUNCHED FROM AN AIRCRAFT

ALTITUDE (USA)
MAJ Robert M. White, USAF — 95,936 m
North American X-15-1 — 314,750 ft
1 Reaction Motors XLR-99, 50,000 lbs
Edwards AFB, CA
7/17/62

LAUNCHED FROM THE GROUND

ALTITUDE (FRANCE)
R. Carpentier — 24,217 m
SO "Trident" — 79,452 ft
2 Turbomeca, 5,500 kg
lus 2 SPER rockets, 1,500 kg
5/2/58

DISTANCE WITHOUT LANDING* ✯
Terence T. Cunningham — 122 ft
Pterodactyl Pfledge
Watsonville, CA
4 Emperor Norton Rockets
10/24/90

CLASS C-1.A/O, GROUP IV, ROCKET ENGINE
Less than 300 kg (661 lbs)

DISTANCE WITHOUT LANDING* ✯
Terence T. Cunningham — 122 ft
Pterodactyl Pfledge
4 Emperor Norton Rockets
Watsonville, CA
10/24/90

ALTITUDE*
Terence T. Cunningham — 165 ft
Pterodactyl Pfledge
4 Emperor Norton Rockets
Watsonville, CA
10/24/90

Speed Over a Recognized Course

The primary reason for most flying is traveling from one place to another, whether for sport, business, or pleasure. Recognizing this, the FAI has established a major set of record categories for those who have made the fastest flights between pairs of cities, non-stop or with one or more intermediate landings.

Within the United States, there is considerable latitude in choosing pairs of cities. The pilot may use any two city pairs (including geographical features) if they are approved in advance by the NAA. If two cities within the same country are used, they must be at least 400 kilometers (249 statute miles) apart to qualify for a world record.

Some of the existing records do not conform to the latest rules, but since they were established when there was greater freedom, they will remain in effect. In many cases, these records were set in series, as part of a much longer flight which passed over the relevant cities, and so multiple records were claimed.

The Gulfstream G550.

CLASS C-1, LANDPLANES

WEST TO EAST, TRANSCONTINENTAL
***Group II* * ★**
Joseph J. Ritchie 546.81 mph
Steve Fossett
Piaggio P.180 Avanti
2 P&WC PT6A, 850 shp
2/6/03

WEST TO EAST, TRANSCONTINENTAL *Group III* * ★
Lt Col Ed Yeilding, Pilot 2,124.51 mph
Lt Col J. T. Vida, RSO
Lockheed SR-71
2 P&W J-58, 34,000 lbs
3/6/90

* * *

ACAPULCO/BANGKOK (FRANCE)
Claude Delorme, Captain 1,211.43 kmh
Jean Boye, Co-Captain 752.75 mph
BAe/Aérospatiale Concorde
10/13/92

ACAPULCO/GUAM (FRANCE)
Claude Delorme, Captain 1,411.33 kmh
Jean Boye, Co-Captain 876.96 mph
BAe/Aérospatiale Concorde
10/13/92

ACAPULCO/HONOLULU (FRANCE)
Claude Delorme, Captain 1,775.76 kmh
Jean Boye, Co-Captain 1,103.40 mph
BAe/Aérospatiale Concorde
10/12/92

ACAPULCO/JESUS CARRANZA (USA)
Marie E. McMillan 230.55 kmh
Beech Bonanza F33A 143.26 mph
1 Continental IO-520B, 285 bhp
2/3/84

ACAPULCO/LAS VEGAS (UK)
Capt John Hutchinson 1,046.95 kmh
British Airways 650.54 mph
BAe/Aérospatiale Concorde
3/16/90

ADANA/IZMIR (CANADA)
Dawn Bartsch 342.00 kmh
Gordon Bartsch 212.51 mph
Cessna 421B
9/21/97

ADDISON, TX/PHOENIX, AZ (USA)
JOINT RECORD
William H. Wisner, Pilot 266.04 kmh
Janice Sullivan, Copilot 165.32 mph
Bonanza S35, N6226
and
Daniel Webb, Pilot
Richard Meyerhoff, Copilot
Bonanza A36, N4504S
5/28/89

ADELAIDE/DARWIN (USA)
Danford A. Bookout 268.56 kmh
Piper Lance PA 32R-300 166.88 mph
1 Lycoming IO-540-K, 300 bhp
6/19-20/86

AGANA/BAHRAIN (FRANCE)
Claude Delorme, Captain 1,119.60 kmh
Jean Boye, Co-Captain 695.69 mph
BAe/Aérospatiale Concorde
10/13/92

AGANA/BANGKOK (FRANCE)
Claude Delorme, Captain 1,592.94 kmh
Jean Boye, Co-Captain 989.80 mph
BAe/Aérospatiale Concorde
10/13/92

AGANA/HILO (USA)
Harold Curtis 900.04 kmh
Gulfstream III 559.26 mph
2 RR Spey 511-8
7 hrs 4 min 12 sec
1/9/82

AGANA/HONOLULU (USA)
Frank P. Santoni Jr. 1,009.30 kmh
Richard A. Austin 627.15 mph
John E. Cashman, Charles A. Hovland,
Izham Ismail, Joseph M. MacDonald,
James C. McRoberts, Rodney M. Skaar
Boeing 777-200-IGW
2 RR6 Trent 892, 90,000 lbs
Seattle, WA
4/2/97

AGANA/LISBON (FRANCE)
Claude Delorme, Captain 972.00 kmh
Jean Boye, Co-Captain 603.97 mph
BAe/Aérospatiale Concorde
10/13/92

ALBUQUERQUE/LAS VEGAS (USA)
Danford A. Bookout 258.84 kmh
Piper Lance, N8352C 160.84 mph
1 Lycoming IO-540K, 300 bhp
10/2/87

AMMAN/TRABZON (CANADA)
Dawn Bartsch 288.82 kmh
Gordon Bartsch 179.46 mph
Cessna 421B
9/19/97

AMSTERDAM/BRUSSELS (USA)
Robert Fero 850.50 kmh
N. American T-39 Sabreliner 528.48 mph
2 P&W J-60P-3
9 min 45.1 sec
6/25/61

ANADYR/FAIRBANKS (USA)
Allyn Caruso, Pilot 402.83 kmh
David Norgart, Copilot 250.30 mph
John Gallichon, Navigator
John Miller, Navigator
Dana Lovell, Mechanic
Fairchild Metroliner III
2 Garrett TPE331-11, 1,100 shp
6/5/91

ANCHORAGE/CHICAGO (USA)
LTC G. A. Andrews, USAF 843.49 kmh
Convair B-58A Hustler 524.12 mph
4 GE-J-79
5 hrs 26 min 33.9 sec
10/16/63

ANCHORAGE/LONDON (USA)
MAJ S. J. Kubesch, USAF 1,330.80 kmh
Convair B-58A Hustler 826.91 mph
4 GE J-79
5 hrs 24 min 54 sec
10/16/63

ANCHORAGE/SAPPORO (USA)
Stephen M. Johnson — 767.95 kmh
Kevin A. Bixler — 477.18 mph
IAI 1126 Galaxy
2 P&WC PW306A, 5,700 lbs
2/17/02

ANKARA/LANGKAWI (USA)
Raymond A. Wellington — 870.97 kmh
William J. Watters — 541.19 mph
Alberto L. Moros
Gulfstream G550
2 BMW R-R BR710, 15,385 lbs
10/1/03

AQABA/VENICE (UK)
Geoffrey Mussett — 1,172.05 kmh
BAe/Aérospatiale Concorde — 728.28 mph
11/11/99

ASCENSION/BRIZE NORTON (UK)
W/Cdr K. D. Filbey — 795.78 kmh
Lockheed Tristar — 494.49 mph
3 RB 211, 50,000 lbs
5/15/85

ASCENSION/LONDON (UK)
Sqn/Ldr John Knapp — 827.55 kmh
VC 10 C — 514.24 mph
4 RR Conway Mk 301, 22,500 lbs
12/21-22/87

ASCENSION/MOUNT PLEASANT (UK)
W/Cdr K. D. Filbey — 758.51 kmh
Lockheed Tristar — 471.33 mph
3 RB 211, 50,000 lbs
5/12/85

ASCENSION/PORT STANLEY (UK)
Ft/Lt John Halstead — 821.64 kmh
VC 10 C — 510.57 mph
4 RR Conway Mk 301, 22,500 lbs
12/20/87

ASHEVILLE/DAYTONA BEACH (USA)
Reginald E. Moody, Jr. — 211.25 kmh
Piper PA-28-180 — 131.26 mph
1 Lycoming O-360, 180 bhp
8/31/95

ASHEVILLE/KITTY HAWK (USA)
Lawrence G. Manofsky — 318.31 kmh
Lake LA-250 Renegade — 197.79 mph
1 Lycoming IO-540, 250 bhp
12/11/95

ASPEN/WASHINGTON (USA)
Peter T. Reynolds, Pilot — 812.33 kmh
James R. McClellan, Copilot — 504.75 mph
Kirk A. Vining, Copilot
Learjet 31A
2 Garrett TFE731-2, 3,500 lbs
8/25/94

ATHENS/CAIRO (USA)
Harold Curtis — 792.00 kmh
Gulfstream III — 492.15 mph
2 RR Spey 511-8
1 hr 21 min 57 sec
1/8/82

ATHENS/NICE (USA)
JOINT RECORD
William H. Wisner, Pilot — 235.80 kmh
Janice Sullivan, Copilot — 146.53 mph
Bonanza S35, N6226
and
Daniel Webb, Pilot
Richard Meyerhoff, Copilot
Bonanza A36, N4504S
6/22/89

ATHENS/TAIPEI (USA)
Clay Lacy, Captain — 984.31 kmh
Verne Jobst, CoCaptain — 611.65 mph
Boeing 747, N149UA
4 P&W JT9D-7A, 46,250 lbs
1/29-30/88

ATLANTA/KITTY HAWK (USA)
Mack Secord, Pilot — 229.32 kmh
Samuel Friedman, Copilot — 142.50 mph
Cessna 182P, N6835M
1 Continental O-470S
12/16/88

ATLANTA/PARIS (USA)
Vern F. Peterson — 553.15 kmh
Lockheed C-130E Hercules — 343.70 mph
4 Allison T56-A-7A
12 hrs 43 min 48.5 sec
5/30-31/63

ATLANTA/SWINDON (USA)
Arlen D. Rens, Pilot — 666.25 kmh
Jeffrey Bullen, Copilot — 413.99 mph
Edward J. Delehant, III, Copilot
Timothy L. Gomez, Aircraft Systems Specialist
Lockheed Martin C-130J-30
4 RR Allison AE 2100, 4,591 shp
12/8/99

AUCKLAND/LONDON (UK)
J. W. Burton — 962.58 kmh
BAe/Aérospatiale Concorde — 598.15 mph
4 Olympus 593/610, 38,000 lbs
4/7-8/86

BAHRAIN/CAIRO (USA)
Allen E. Paulson, Captain — 788.00 kmh
K. C. Edgecomb, Colin B. Allen — 489.67 mph
John Salamankas, Jefferson Bailey, crew
Gulfstream IV, N440GA
2 RR TAY MK610-8
6/14/87

BAHRAIN/COLOMBO (USA)
Harold Curtis — 864.00 kmh
Gulfstream III — 536.89 mph
2 RR Spey 511-8
4 hr 16 min 0 sec
1/9/82

BAHRAIN/LUXOR (USA)
JOINT RECORD
William H. Wisner, Pilot — 254.52 kmh
Janice Sullivan, Copilot — 158.16 mph
Bonanza S35, N6226
and
Daniel Webb, Pilot
Richard Meyerhoff, Copilot
Bonanza A36, N4504S
6/18/89

BAHRAIN/PARIS (USA)
Allen E. Paulson, Captain — 813.64 kmh
K. C. Edgecomb, Colin B. Allen — 505.60 mph
John Salamankas, Jefferson Bailey, crew
Gulfstream IV, N440GA
2 RR TAY MK610-8
6/14/87

BAHRAIN/LISBON (FRANCE)
Claude Delorme, Captain — 1,530.73 kmh
Jean Boye, Co-Captain — 951.15 mph
BAe/Aérospatiale Concorde
10/13/92

BALI/COLOMBO (UK)
Capt Geoffrey Mussett 1,600.68 kmh
British Airways 994.61 mph
BAe/Aérospatiale Concorde
3/31/90

BALI/MOMBASA (UK)
Capt Geoffrey Mussett 1,300.63 kmh
British Airways 808.17 mph
BAe/Aérospatiale Concorde
3/31/90

BALTIMORE/MOSCOW (USA)
COL James B. Swindal, USAF 906.64 kmh
Boeing VC-137 563.36 mph
4 P&W JT3D-3
8 hrs 33 min 45.4 sec
5/19/63

BALTIMORE/OSLO (USA)
COL James B. Swindal, USAF 951.31 kmh
Boeing VC-137 591.12 mph
4 P&W JT3D-3
6 hrs 58 min 27.1 sec
5/19/63

BALTIMORE/STOCKHOLM (USA)
COL James B. Swindal, USAF 944.30 kmh
Boeing VC-137 (707) 586.76 mph
4 P&W JT3D-3
6 hrs 58 min 27.1 sec
5/19/63

BANDAR SERI BEGAWAN/SEOUL (USA)
Alberto L. Moros 788.95 kmh
William J. Watters 490.23 mph
Raymond A. Wellington
Gulfstream G550
2 BMW R-R BR710, 15,385 lbs
10/3/03

BANGKOK/BAHRAIN (FRANCE)
Claude Delorme, Captain 1,380.68 kmh
Jean Boye, Co-Captain 857.91 mph
BAe/Aérospatiale Concorde
10/13/92

BANGKOK/LISBON (FRANCE)
Claude Delorme, Captain 1,175.58 kmh
Jean Boye, Co-Captain 730.47 mph
BAe/Aérospatiale Concorde
10/13/92

BANGKOK/MADRAS (UK)
Max Robinson 1,408.38 kmh
BAe/Aérospatiale Concorde 875.12 mph
11/2/99

BANGKOK/SEATTLE (USA)
Andrew C. Messer 879.13 kmh
John H. Armstrong 546.26 mph
Susan P. Darcy
Edward T. Hoit
Naruj Komalarajun
Boeing 777
2 P&W PW4084, 84,000 lbs
4/30/95

BANGKOK/TAIPEI (USA)
Clay Lacy, Captain 916.85 kmh
Verne Jobst, CoCaptain 569.74 mph
Boeing 747, N149UA
4 P&W JT9D-7A, 46,250 lbs
1/30/88

BANGOR/LUXOR (USA)
Tim Mellon, Pilot 413.28 kmh
Tom Glista, Copilot 256.81 mph
Piper Aerostar 602P, N602PC
2 Lycoming IO-540
4/27-28/87

BANGOR/MANCHESTER (USA)
Tom Glista, Pilot 469.08 kmh
Tim Mellon, Copilot 291.49 mph
Piper Aerostar 602P, N602PC
2 Lycoming IO-540
4/27-28/87

BANGOR/WICHITA (USA)
Richard Trissell 914.07 kmh
William S. Dirks 567.97 mph
Cessna Model 750, Citation X
2 Allison GMA 3007C, 6,000 lbs
6/15/95

BARCELONA/NICE (USA)
James P. Hanson 280.23 kmh
Cessna 208 Caravan 174.12 mph
1 P&W PT-6A, 600 shp
6/6/96

BASSETERRE/CHARLOTTE AMALIE (USA)
Marie E. McMillan 329.75 kmh
Beech Bonanza F33A 204.90 mph
1 Continental IO-520B, 285 bhp
2/20/84

BASSETERRE/HARRIS (USA)
Marie E. McMillan 195.64 kmh
Beech Bonanza F33A 121.57 mph
1 Continental IO-520B, 285 bhp
2/21/84

BASSETERRE/NEW CASTLE (USA)
Marie E. McMillan 299.18 kmh
Beech Bonanza F33A 185.90 mph
1 Continental IO-520B, 285 bhp
2/21/84

BASSETERRE/POINTE A PITRE (USA)
Marie E. McMillan 303.50 kmh
Beech Bonanza F33A 188.59 mph
1 Continental IO-520B, 285 bhp
2/21/84

BASSETERRE/ST. JOHNS (USA)
Marie E. McMillan 359.36 kmh
Beech Bonanza F33A 223.40 mph
1 Continental IO-520B, 285 bhp
2/21/84

BEDFORD, MA/ADDISON, TX (USA)
JOINT RECORD
William H. Wisner, Pilot 282.60 kmh
Janice Sullivan, Copilot 175.61 mph
Bonanza S35, N6226
and
Daniel Webb, Pilot
Richard Meyerhoff, Copilot
Bonanza A36, N4504S
6/26/89

BEDFORD, MA/GOOSE BAY (USA)
Anne B. Baddour 386.28 kmh
Margrit Budert-Waltz 240.03 mph
Beech Baron, N3707N
2 Continental, 325 bhp
12/20/88

BEDFORD, MA/MOSCOW (USA)
Robert S. McKenney 860.64 kmh
Ahmed M. Ragheb 534.77 mph
Gulfstream IV-SP
2 R-R Tay 611, 13,850 lbs
8/13/01

BEDFORD, MA/NARSSARSSUAQ (USA)
Anne B. Baddour 315.72 kmh
Margrit Budert-Waltz 196.19 mph
Beech Baron, N3707N
2 Continental, 325 bhp
12/20-21/88

BEIJING/HONOLULU (USA)
Brooke Knapp 942.23 kmh
Gulfstream III 585.50 mph
2 RR Spey MK511-8, 11,400 lbs
2/14-15/84

BEIJING/HOUSTON (USA)
Larry S. Mueller 869.32 kmh
Robert S. McKenney 540.17 mph
Rick B. Gowthrop
Gulfstream V
2 BMW-RR BR710-48, 14,750 lbs
9/30/99

BEIJING/LOS ANGELES (USA)
Brooke Knapp 931.79 kmh
Gulfstream III 579.01 mph
2 RR Spey MK511-8, 11,400 lbs
2/14-15/84

BEIJING/NEW YORK (USA)
Robert S. McKenney 795.06 kmh
William J. Watters 494.03 mph
Raymond A. Wellington
Gulfstream G550
2 BMW R-R BR710, 15,385 lbs
12/14/03

BEIJING/TOKYO (USA)
Brooke Knapp 887.92 kmh
Gulfstream III 551.75 mph
2 RR Spey MK511-8, 11,400 lbs
2/14/84

BEIJING/WASHINGTON (USA)
Brooke Knapp 928.66 kmh
Gulfstream III 577.07 mph
2 RR Spey MK511-8, 11,400 lbs
2/14-15/84

BEIRUT/KARACHI (PAKISTAN)
A. Baig 1,020.31 kmh
Boeing 720B 633.9 mph
4 P&W JT3D-1
1/2/62

BELFAST/GANDER (UK)
R. P. Beamont 774.25 kmh
Electric Canberra B. 481.09 mph
Mark 2, WD940
2 RR Avon RA-3
4 hrs 18 min 24.4 sec
8/31/51

BELFAST/PARIS (USA)
Rick E. Rowe 824.16 kmh
William A. Landrum 512.11 mph
Learjet 31A
2 AlliedSignal TFE731, 3,500 lbs
6/14/01

BERLIN/HANOI (GERMANY)
A. Henke 243.01 kmh
Focke-Wulf FW 200 Condor 151.00 mph
4 BMW-132-L
34 hrs 17 min 27 sec
11/28/38

BERLIN/MOSCOW (USA)
Allyn Caruso, Pilot 409.48 kmh
David Norgart, Copilot 254.44 mph
John Gallichon, Navigator
John Miller, Navigator
Dana Lovell, Mechanic
Fairchild Metroliner III
2 Garrett TPE331-11, 1,100 shp
6/1/91

BERLIN/NEW YORK (GERMANY)
A. Henke 255.49 kmh
Focke-Wulf FW 200 Condor 158.75 mph
4 BMW-132-L
24 hrs 56 min 12 sec
8/10/38

BERLIN/PARIS (USA)
Warren Breeding 707.23 kmh
Fred J. Coffin 439.45 mph
Canadair Challenger 604
2 GE CF34, 8,729 lbs
6/14/01

BERLIN/TOKYO (GERMANY)
A. Henke 192.30 kmh
Focke-Wulf FW 200 Condor 119.49 mph
4 BMW-132-L
46 hrs 18 min 19 sec
12/28/38

BERMUDA/LONDON (ENGLAND)
CPT M. Gudmundsson 920.76 kmh
Boeing 707-465 572.13 mph
4 RR, Conway 508
6 hrs 2 min 16 sec
5/6/62

BERMUDA/NEW YORK (UK)
Capt. Peter Horton 1,228.73 kmh
BAe/Aérospatiale Concorde 763.49 mph
1/2/93

BIRMINGHAM/NEW YORK (UK)
Stuart Robertson 1,100.05 kmh
BAe/Aérospatiale Concorde 683.54 mph
9/27/93

BIRMINGHAM/SHANNON (UK)
CPT J. W. Burton 671.27 kmh
BAe/Aérospatiale Concorde 417.13 mph
4 RR Olympus, 38,000 lbs
9/16/85

BOMBAY/BANGKOK (USA)
Clay Lacy, Captain 960.54 kmh
Verne Jobst, CoCaptain 596.88 mph
Boeing 747, N149UA
4 P&W JT9D-7A, 46,250 lbs
1/29-30/88

BOSTON/BONN (USA)
Jacqueline Cochran 562.56 kmh
Lockheed Jetstar 349.56 mph
4 P&W JT12A-6
10 hrs 15 min 56 sec
4/22/62

BOSTON/GANDER (USA)
Harold Curtis, Pilot 936.00 kmh
William Mack, Copilot 581.63 mph
Robert Dann Hardt, Copilot
Gulfstream III
2 RR Spey 511-8
1 hr 33 min 50 sec
1/8/82

BOSTON/LONDON (USA)
Jacqueline Cochran 558.50 kmh
Lockheed Jetstar 347.04 mph
4 P&W JT12A-6
9 hrs 25 min 54.3 sec
4/22/62

BOSTON/MOSCOW (USA)
Col James B. Swindal, USAF 905.42 kmh
Boeing VC-137 562.60 mph
4 P&W JT3D-3
7 hrs 58 min 15.7 sec
5/19/63

BOSTON/OSLO (USA)
Col James B. Swindal USAF 952.63 kmh
Boeing VC-137 591.94 mph
4 P&W JT3D-3
5 hrs 54 min 14.7 sec
5/19/63

BOSTON/PARIS (USA)
James P. Dwyer 911.74 kmh
Robert W. Agostino 566.53 mph
Rodney R. Lundy
Bombardier Challenger 300
2 Honeywell AS907, 6,700 lbs
6/12/03

BOSTON/SHANNON (USA)
Jacqueline Cochran 565.45 kmh
Lockheed Jetstar 351.36 mph
4 P&W JT12A-6
8 hrs 13 min 38 sec
4/22/62

BOSTON/STOCKHOLM (USA)
COL James B. Swindal, USAF 944.87 kmh
Boeing VC-137 587.12 mph
4 P&W JT3D-3
6 hrs 22 min 54.1 sec
5/19/63

BRANDON/BUTTRESS (CANADA)
Lyle Johnson 76.06 kmh
Challenger II Special 47.26 mph
8/29/98

BRANDON/MOOSE JAW (CANADA)
Lyle Johnson 77.41 kmh
Challenger II Special 48.10 mph
8/29/98

BRIDGETON/GRENVILLE (USA)
Marie E. McMillan 289.63 kmh
Beech Bonanza F33A 179.98 mph
1 Continental IO-520B, 520 bhp
2/24/84

BRISBANE/LOS ANGELES (USA)
Wilson A. Miles, Jr. 907.61 kmh
Alberto Moros 563.96 mph
Richard D. Robbins
Gulfstream V
2 BMW R-R BR710, 14,750 lbs
2/17/01

BRIZE NORTON/ASCENSION (UK)
W/Cdr K. D. Filbey 808.45 kmh
Lockheed Tristar 502.36 mph
3 RB 211, 50,000 lbs
5/11-12/85

BRIZE NORTON/MOUNT PLEASANT (UK)
W/Cdr K. D. Filbey 687.97 kmh
Lockheed Tristar 427.50 mph
3 RB 211, 50,000 lbs
5/11-12/85

BROMMA/ROVANIEMI (SWEDEN)
Hans U. von der Esch 126.31 kmh
Piper PA18-150 78.48 mph
6/25/91

BRUSSELS/AMSTERDAM (USA)
Robert Fero 864.41 kmh
North American T-39 Sabreliner 537.12 mph
2 P&W J-60P-3
9 min 33 sec
6/28/61

BRUSSELS/PARIS (BELGIUM)
Bernard Neefs 1,576.43 kmh
Lockheed F-104G Super Starfighter 979.54 mph
1 GE J-79-11A
6/6/63

BUCHAREST/CHISINAU (ROMANIA)
Constantin Manolache 210.35 kmh
Gheorgue Militaru 130.71 mph
Moravan Cehia ZLIN 142
1 Moravan Otrokovice M 337 AK, 180 bhp
11/8/96

BUENOS AIRES/CHRISTCHURCH (USA)
Jack L. Martin, Fred L. Austin 692.85 kmh
Harrison Finch, Robert N. Buck 430.52 mph
James R. Gannett
Boeing 707-320C
4 P&W JT3
14 hrs 17 min 17.1 sec
11/17/65

BUENOS AIRES/FT LAUDERDALE (USA)
Gene E. Allen 693.53 kmh
Frank G. Leskaukas 430.94 mph
David M. Singer
Dassault Falcon 900EX
3 AlliedSignal TFE731, 4,750 lbs
1/4/98

BUENOS AIRES/WASHINGTON (USA)
GEN C. E. LeMay, USAF 758.72 kmh
Boeing KC-135 Stratotanker 471.45 mph
P&W J-57-P43W
11 hrs 3 min 57.38 sec
11/13/57

BUTTRESS/BRANDON (CANADA)
Lyle Johnson 93.31 kmh
Challenger II Special 57.98 mph
8/29/98

CAIRO/BAHRAIN (USA)
Harold Curtis 1,008.00 kmh
Gulfstream III 626.37 mph
2 RR Spey 511-8
2 hrs 29 min 37 sec
1/8-9/82

CAIRO/LISBON (UK)
Richard Owen 1,336.61 kmh
BAe/Aérospatiale Concorde 830.53 mph
4 RR Olympus 593, 38,000 lbs
3/2/00

CAIRO/LONDON (USA)
COL Norman L. Mitchell 729.59 kmh
Grumman Gulfstream II, N700ST 453.34 mph
2 RR Spey 511-8
4 hrs 49 min 16.5 sec
3/15/75

CAIRO/LOS ANGELES (USA)
Donald L. Mullin 775.95 kmh
DC-8-72 482.15 mph
4 CFM 56 24,000 lbs
3/29-30/83

CAIRO/NAIROBI (USA)
Douglas G. Matthews 688.68 kmh
Learjet 36 427.95 mph
2 Garrett, TFE-731-2-2B, 3,500 lbs
7/16/84

CAIRO/PARIS (USA)
Allen E. Paulson, Captain 825.87 kmh
K. C. Edgecomb, John Salamankas, 513.19 mph
Jefferson Bailey; Pilots
Colin B. Allen, Flight Engineer
Gulfstream IV, N440GA
2 RR TAY MK610-8
6/14/87

CAIRO/WASHINGTON (UK)
Richard Owen 1,122.50 kmh
BAe/Aérospatiale Concorde 697.49 mph
4 RR Olympus 593, 38,000 lbs
3/2/00

CALDWELL/YOUNGSTOWN (USA)
Steve Kahn 261.07 kmh
Mooney 201 162.22 mph
1 Lycoming IO-360, 200 bhp
11/9/91

CALGARY/THE PAS (CANADA)
Dawn Bartsch 271.85 kmh
Gordon Bartsch 168.92 mph
Cessna 421B
9/5/97

CANNES/LONDON (USA)
Susan Nealey 290.59 kmh
Faith Hillman 180.56 mph
Cessna 310
2 Continental IO-520, 285 bhp
7/14/92

CANOUAN/KINGSTOWN (USA)
Marie E. McMillan 304.14 kmh
Beech Bonanza F33A 188.98 mph
Continental IO-520B, 285 bhp
2/27/84

CANOUAN/LOVELL'S VILLAGE (USA)
Marie E. McMillan 363.32 kmh
Beech Bonanza F33A 225.75 mph
Continental IO-520B, 285 bhp
2/27/84

CANOUAN/PORT ELIZABETH (USA)
Marie E. McMillan 253.76 kmh
Beech Bonanza F33A 157.68 mph
Continental IO-520B, 285 bhp
2/27/84

CAPETOWN/AUCKLAND (USA)
CPT Walter H. Mullikin 860.09 kmh
CPT William P. Monan, 534.43 mph
Pan American World Airways, Inc.
Boeing 7475P, N533PA
10/29-30/77

CAPETOWN/LONDON (UK)
Brian Walpole 1,192.48 kmh
BAe/Aérospatiale Concorde 740.99 mph
4 RR Olympus, 38,000 lbs
8 hrs 7 min 39 sec
3/29/85

CARACAS/JACKSONVILLE (USA)
William A. Landrum 740.63 kmh
Owen G. Zahnle 460.20 mph
Learjet 45
2 AlliedSignal TFE731-20, 3,500 lbs
6/5/99

CARACAS/NEW ORLEANS (USA)
Scott S. Evans 757.15 kmh
Jerry W. Blessing 470.47 mph
IAI 1125 Astra SPX
2 AlliedSignal TFE731, 4,250 lbs
9/1/01

CARDIFF/OSHKOSH (USA)
Susan Nealey 94.84 kmh
Faith Hillman 58.93 mph
Cessna 310
2 Continental IO-520, 285 bhp
7/19-22/92

CARDIFF/REYKJAVIK (USA)
Susan Nealey 181.64 kmh
Faith Hillman 112.87 mph
Cessna 310
2 Continental IO-520, 285 bhp
7/19/92

CARLSBAD/COLUMBIA (USA)
Douglas G. Matthews 785.52 kmh
Learjet 36 488.12 mph
2 Garrett, TFE-731-2-2B, 3,500 lbs
7/28/84

CASTRIES/BASSETERRE (USA)
Marie E. McMillan 251.66 kmh
Beech Bonanza F33A 156.37 mph
Continental IO-520B, 285 bhp
2/28/84

CASTRIES/BRIDGETOWN (USA)
Marie E. McMillan 248.37 kmh
Beech Bonanza F33A 154.37 mph
Continental IO-520B, 285 bhp
2/28/84

CASTRIES/CODRINGTON (USA)
Marie E. McMillan 273.57 kmh
Beech Bonanza F33A 169.99 mph
Continental IO-520B, 285 bhp
2/28/84

CASTRIES/GRAND BOURG (USA)
Marie E. McMillan 303.53 kmh
Beech Bonanza F33A 188.60 mph
Continental IO-520B, 285 bhp
2/28/84

CASTRIES/GUSTAVIA (USA)
Marie E. McMillan 265.58 kmh
Beech Bonanza F33A 165.02 mph
Continental IO-520B, 285 bhp
2/28/84

CASTRIES/HARRIS (USA)
Marie E. McMillan 279.10 kmh
Beech Bonanza F33A 173.42 mph
Continental IO-520B, 285 bhp
2/28/84

CASTRIES/MARTINIQUE (USA)
Marie E. McMillan 373.30 kmh
Beech Bonanza F33A 231.96 mph
Continental IO-520B, 285 bhp
2/28/84

CASTRIES/NEW CASTLE (USA)
Marie E. McMillan 247.50 kmh
Beech Bonanza F33A 153.71 mph
Continental IO-520B, 285 bhp
2/28/84

CASTRIES/PHILLIPSBURG (USA)
Marie E. McMillan 261.63 kmh
Beech Bonanza F33A 162.57 mph
Continental IO-520B, 285 bhp
2/28/84

CASTRIES/POINTE A PITRE (USA)
Marie E. McMillan 254.84 kmh
Beech Bonanza F33A 158.35 mph
Continental IO-520B, 285 bhp
2/28/84

CASTRIES/ST. JOHNS (USA)
Marie E. McMillan 263.66 kmh
Beech Bonanza F33A 163.83 mph
Continental IO-520B, 285 bhp
2/28/84

CHARLOTTE AMALIE/BASSETERRE (USA)
Marie E. McMillan 187.69 kmh
Beech Bonanza F33A 116.69 mph
Continental IO-520B, 285 bhp
2/20/84

CHARLOTTE AMALIE/GRAND CASE (USA)
Marie E. McMillan 219.19 kmh
Beech Bonanza F33A 136.19 mph
Continental IO-520B, 285 bhp
2/20/84

CHARLOTTE AMALIE/GUSTAVIA (USA)
Marie E. McMillan 141.04 kmh
Beech Bonanza 87.64 mph
Continental IO-520B, 285 bhp
2/20/84

CHARLOTTE AMALIE/KINGSTON (USA)
Marie E. McMillan 209.44 kmh
Beech Bonanza F33A 130.14 mph
Continental IO-520B, 285 bhp
3/1/84

CHARLOTTE AMALIE/ORANJESTAD (USA)
Marie E. McMillan 186.02 kmh
Beech Bonanza F33A 115.59 mph
Continental IO-520B, 285 bhp
2/20/84

CHARLOTTE AMALIE/PHILLIPSBURG (USA)
Marie E. McMillan 197.87 kmh
Beech Bonanza F33A 122.95 mph
Continental IO-520B, 285 bhp
2/20/84

CHARLOTTE AMALIE/ROADTOWN (USA)
Marie E. McMillan 317.01 kmh
Beech Bonanza F33A 197.98 mph
Continental IO-520B, 285 bhp
2/20/84

CHARLOTTE AMALIE/THE BOTTOMS (USA)
Marie E. McMillan 180.53 kmh
Beech Bonanza F33A 112.17 mph
Continental IO-520B, 285 bhp
2/20/84

CHARLOTTE AMALIE/THE VALLEY (USA)
Marie E. McMillan 241.86 kmh
Beech Bonanza F33A 150.28 mph
Continental IO-520B, 285 bhp
2/20/84

CHEYENNE/TEXARKANA (USA)
Danford A. Bookout 369.36 kmh
Piper Lance, N8352C 229.52 mph
1 Lycoming IO-540K, 300 bhp
10/5/87

CHICAGO/BARCELONA (USA)
Ehab Hanna 841.84 kmh
Lynn Upham 523.10 mph
Gulfstream IV
2 R-R Tay 611, 13,850 lbs
6/23/03

CHICAGO/FT. PIERCE (USA)
Robert E. Humphrey 277.92 kmh
Charles F. Riedl 172.70 mph
Piper Comanche, N9262P
1 IO 540 D4A5
2/4/89

CHICAGO/NEW YORK (USA)
Brooke Knapp 905.40 kmh
Gulfstream III, 562.62 mph
2 RR Sperry MK-511, 11,400 lbs
7/16-17/84

CHICAGO/TETERBORO (USA)
Ehab Hanna 796.95 kmh
Lynn Upham 495.20 mph
Gulfstream IV
2 R-R Tay 611, 13,850 lbs
6/19/03

CHICAGO/WASHINGTON (USA)
James M. Frankard 460.80 kmh
Glasair III 286.33 mph
1 Lycoming IO-540, 300 bhp
4/24/96

CHISINAU/BUCHAREST (ROMANIA)
Constantin Manolache 220.87 kmh
Gheorgue Militaru 137.24 mph
Moravan Cehia ZLIN 142
1 Moravan Otrokovice M 337 AK, 180 bhp
11/8/96

CHRISTCHURCH/HONOLULU (USA)
Jack L. Martin, 866.88 kmh
Fred L. Austin, 538.65 mph
Harrison Finch, Robert N. Buck
& James R. Gannett
Boeing 707-320C, N332FT
4 P&W JT3
9 hrs 1 min 24.2 sec
11/17/65

CHRISTCHURCH/MCMURDO STATION (USA)
Brooke Knapp 796.68 kmh
Gulfstream III 495.06 mph
2 RR Spey MK511-8, 11,400 lbs
11/15-16/83

CLEVELAND/TEXARKANA (USA)
Danford A. Bookout 250.56 kmh
Piper Lance, N8352C 155.70 mph
1 Lycoming IO-540K, 300 bhp
11/10/87

CLIFTON/KINGSTOWN (USA)
Marie E. McMillan 323.60 kmh
Beech Bonanza F33A 201.08 mph
Continental IO-520B, 285 bhp
2/27/84

CLIFTON/PORT ELIZABETH (USA)
Marie E. McMillan 303.85 kmh
Beech Bonanza F33A 188.80 mph
Continental IO-520B, 285 bhp
2/27/84

CODRINGTON/GUSTAVIA (USA)
Marie E. McMillan 290.28 kmh
Beech Bonanza F33A 180.37 mph
Continental IO-520B, 285 bhp
2/28/84

CODRINGTON/PHILIPSBURG (USA)
Marie E. McMillan 348.04 kmh
Beech Bonanza F33A 216.26 mph
Continental IO-520B, 285 bhp
2/28/84

COLOMBO/JAKARTA (USA)
Arnold Palmer 698.81 kmh
Learjet 36 434.22 mph
2 Garrett TFE-731-2-2B, 3,500 lbs
5/18/76

COLOMBO/SINGAPORE (USA)
Harold Curtis 756.00 kmh
Gulfstream III 469.78 mph
2 RR Spey 511-8
3 hrs 34 min 6 sec
1/9/82

COLORADO SPRINGS/WINSTON-SALEM (USA)
Richard S. Wimbish 557.83 kmh
Piper Aerostar 346.62 mph
2 Lycoming IO-540, 340 bhp
06/02/90

COLUMBUS/HARRISBURG (USA)
Robert Olszewski, Pilot 1,140.90 kmh
Robert Gastall, Copilot 708.95 mph
DC-9-31
2 P&W JT8D-7B, 14,000 lbs
12/27/85

COPENHAGEN/LONDON (UK)
Jan Zurakowski 805.75 kmh
Gloster Meteor MK F8 500.67 mph
2 RR Derwent V
1 hr 11 min 17 sec
4/4/50

COZUMEL/GRAND CAYMAN (USA)
Marie E. McMillan 277.20 kmh
Beech Bonanza F33A 172.25 mph
1 Continental IO-520 BA, 285 bhp
2/5/84

DALLAS/ATLANTA (USA)
Steve Fossett 1,180.28 kmh
Douglas A. Travis 733.39 mph
Cessna 750 Citation X
2 R-R AE3007C, 6,442 lbs
2/5/03

DALLAS/FRANKFURT (USA)
Roger E. Ruch 888.00 kmh
DC-10-30 552.32 mph
3 GE CF6-6, 40,000 lbs
5/23-24/86

DALLAS/JACKSONVILLE (USA)
William H. Cox, Pilot 479.32 kmh
Martha A. Morrell, Copilot 297.83 mph
Mooney TLS
1 Lycoming TIO-540, 270 bhp
3/25/94

DALLAS/LONDON (USA)
Rick B. Gowthrop 978.02 kmh
H. Randy Gaston 607.71 mph
William J. Watters
Gulfstream V
2 BMW R-R BR710, 14,750 lbs
7/21/00

DALLAS/TOKYO (USA)
Andrew S. Milewski 751.16 kmh
Yves Tessier 466.75 mph
Durwood J. Heinrich
Bombardier BD-700 Global Express
2 BMW R-R BR710, 14,750 lbs
4/22/02

DAYTON/JOHNSTOWN (USA)
Robert Olszewski 1,139.86 kmh
DC-9-31, N942VJ 708.27 mph
2 P&W JT8D07B, 14,000 lbs
1/5/88

DAYTON/KITTY HAWK (USA)
Maurice Eldridge, Pilot 1,196.26 kmh
Jeffrey Thomas, Copilot 743.36 mph
Convair F-106
1 P&W J-75, 16,100 lbs
12/17/83

DAYTON/WASHINGTON (USA)
Carl A. La Rue 618.68 kmh
John Slais 384.43 mph
Lancair IV P
1 Continental TSIO-550B, 350 bhp
10/21/97

DEER LAKE/VICTORIA (CANADA)
Michael H. Leslie 20.63 kmh
Joseph Leslie 12.82 mph
Daniel J. Leslie
Cessna 172
1 Lycoming O-320, 160 bhp
8/15/00

DENVER/BOSTON (USA)
Arnold Palmer 856.61 kmh
Learjet 36 532.27 mph
2 Garrett TFE-731-2-2B, 3,500 lbs
5/17/76

DENVER/HARTFORD (USA)
Kenneth P. Wolf, Pilot 387.72 kmh
Matthew Wolf, Copilot 240.93 mph
Aerostar, N8217J
2 Lycoming IO-540-S1A5
11/27/88

DES MOINES/WASHINGTON (USA)
Dan Vollum, Pilot 790.00 kmh
Don Vollum, Copilot 491.03 mph
Jet Commander
2 GE CJ610-5, 2,950 lbs
12/4/86

DETROIT/SEATTLE (USA)
William S. Demray 239.43 kmh
Jeffrey Seit 148.78 mph
Piper PA-30C
2 Lycoming IO-320, 160 bhp
4/22/95

DHAHRAN/BANGKOK (USA)
Clay Lacy, Captain 1,005.80 kmh
Verne Jobst, CoCaptain 625.01 mph
Boeing 747, N149UA
4 P&W JT9D-7A, 46,250 lbs
1/29/88

DHAHRAN/BOMBAY (USA)
Clay Lacy, Captain 1,070.20 kmh
Verne Jobst, CoCaptain 665.03 mph
Boeing 747, N149UA
4 P&W JT9D-7A, 46,250 lbs
1/29-30/88

DUBAI/BAHRAIN (USA)
Allen E. Paulson, Captain 724.40 kmh
K. C. Edgecomb, Colin B. Allen 450.14 mph
John Salamankas, Jefferson Bailey; crew
Gulfstream IV, N440GA
2 RR TAY MK610-8
6/14/87

DUBAI/CAIRO (USA)
Allen E. Paulson, Captain 774.95 kmh
K. C. Edgecomb, Colin B. Allen 503.91 mph
John Salamankas, Jefferson Bailey; crew
Gulfstream IV, N440GA
2 RR TAY MK610-8
6/14/87

DUBAI/DHABRAN (USA)
Allen E. Paulson, Captain 738.60 kmh
K. C. Edgecomb, Colin B. Allen 458.96 mph
John Salamankas, Jefferson Bailey; crew
Gulfstream IV, N440GA
2 RR TAY MK610-8
6/14/87

DUBAI/HONG KONG (USA)
Raymond A. Wellington 970.96 kmh
Robert S. McKenney 603.33 mph
William J. Watters
Gulfstream G550
2 BMW R-R BR710, 15,385 lbs
12/11/03

DUBAI/MAUI (USA)
Allen E. Paulson, Captain 944.28 kmh
Robert K. Smyth, CoCaptain 586.78 mph
John Salamankas, CoCaptain
Jeff Bailey, CoCaptain
Gulfstream IV, N400GA
2 RR TAY MK611-8
2/27/88

DUBAI/PARIS (USA)
Allen E. Paulson, Captain 802.21 kmh
K.C. Edgecomb, Colin B. Allen 498.49 mph
John Salamankas, Jefferson Bailey; crew
Gulfstream IV, N440GA
2 RR TAY MK610-8
6/14/87

DUBAI/TAIPEI (USA)
Allen E. Paulson, Captain 992.24 kmh
Robert K. Smyth, CoCaptain 616.75 mph
John Salamankas, CoCaptain
Jeff Bailey, CoCaptain
Gulfstream IV, N400GA
2 RR TAY MK611-8
2/27/88

DUBLIN/SAN ANTONIO (USA)
Wilson A. Miles, Jr. 964.64 kmh
William J. Watters 599.40 mph
Gulfstream V
2 BMW R-R BR710, 14,750 lbs
7/2/01

EDINBURGH/LONDON (UK)
1st Lt. Russ Peart 1,208.95 kmh
Jaguar Gr. 1 750.89 mph
2 RR Adour, 7,040 lbs
26 min 25 sec
9/9/77

EDMONTON/MIDWAY (USA)
Allen E. Paulson, Captain 840.21 kmh
K. C. Edgecomb, Colin B. Allen 522.11 mph
John Salamankas, Jefferson Bailey; crew
Gulfstream IV, N440GA
2 RR TAY MK610-8
6/12-13/87

EDMONTON/MINNEAPOLIS (USA)
Allyn Caruso, Pilot 413.79 kmh
David Norgart, Copilot 257.12 mph
John Gallichon, Navigator
John Miller, Navigator
Dana Lovell, Mechanic
Fairchild Metroliner III
2 Garrett TPE331-11, 1,100 shp
6/6/91

EDMONTON/NOME (USA)
Allen E. Paulson, Captain 853.62 kmh
K. C. Edgecomb, Colin B. Allen 530.41 mph
John Salamankas, Jefferson Bailey; crew
Gulfstream IV, N440GA
2 RR TAY MK610-8
6/12-13/87

FAIRBANKS/EDMONTON (USA)
Allyn Caruso, Pilot 424.56 kmh
David Norgart, Copilot 263.81 mph
John Gallichon, Navigator
John Miller, Navigator
Dana Lovell, Mechanic
Fairchild Metroliner III
2 Garrett TPE331-11, 1,100 shp
6/5-6/91

FAIRBANKS/LOS ANGELES (USA)
Brooke Knapp 939.96 kmh
Gulfstream III 584.09 mph
2 RR Spey MK511-8, 11,400 lbs
11/18/83

FAYETTEVILLE/CAMBRIDGE (USA)
Arlen D. Rens, Pilot 661.43 kmh
Edward J. Delehant, III, Copilot 410.99 mph
Richard L. White, Aircraft Systems Specialist
Lockheed Martin C-130J
4 R-R Allision AE2100, 4,591 shp
2/13/00

FLORENCE/CINCINNATI (USA)
James P. Mayer 810.66 kmh
Charles B. Smith 503.72 mph
Bombardier BD-700 Global Express
2 BMW R-R BR710, 14,750 lbs
9/29/02

FT. LAUDERDALE/MYRTLE BEACH (USA)
Mike Hackney 305.65 kmh
Mooney M20R Ovation 189.92 mph
1 Continental IO-550, 280 bhp
7/15/99

FORT WORTH/ANCHORAGE (USA)
Scott S. Evans 774.04 kmh
Bradley S. Robinson 480.96 mph
IAI 1126 Galaxy
2 P&WC PW306A, 5,700 lbs
2/16/02

FORT WORTH/ATLANTA (USA)
Joseph J. Ritchie 927.40 kmh
Steve Fossett 576.26 mph
Piaggio P.180 Avanti
2 P&WC PT6A, 850 shp
2/6/03

FORT WORTH/MADRID (USA)
MAJ Clyde P. Evely, USAF 929.30 kmh
Boeing B-52H, No. 60040 577.44 mph
8 P&W TF-33P-3
8 hrs 35 min 24.43 sec
1/11/62

FORT WORTH/ST. LOUIS (USA)
Edwin Payne 400.05 kmh
Republic P-47 Thunderbolt 248.58 mph
1 P&W R-2800
2 hrs 17 min 37.9 sec
6/21/63

FORT WORTH/WASHINGTON (USA)
MAJ Clyde P. Evely, USAF 972.75 kmh
Boeing B-52H, No. 60040 604.44 mph
8 P&W TF-33P-3
2 hrs 26.66 sec
1/11/62

FRANKFURT/SEATTLE (USA)
Susan P. Darcy 896.38 kmh
Andrew C. Messer 556.98 mph
William F. Royce
Charles L. Gebhardt, III
Boeing 777-200
2 P&W PW4084, 84,000 lbs
9/13/95

FREDERICK/GOOSE BAY (USA)
Susan Nealey
Faith Hillman
Cessna 310
2 Continental IO-520, 285 bhp
7/8/92
368.76 kmh
229.13 mph

FRESNO/FREDERICK (USA)
Susan Nealey
Faith Hillman
Cessna 310
2 Continental IO-520, 285 bhp
7/5/92
355.83 kmh
221.10 mph

GANDER/BELFAST (UK)
Roland P. Beamont
English Electric Canberra
VX 185
2 RR Avon
8/26/52
974.50 kmh
605.52 mph

GANDER/BERLIN (USA)
Timothy Cwik, Pilot
Eleanor Carlisle, Copilot
William Ellsworth, flt Engineer
Aerospatiale ART-42
2 P&WC PW120, 2,000 shp
4/21-22/89
518.40 kmh
322.13 mph

GANDER/BONN (USA)
Jacqueline Cochran
Lockheed Jetstar, N172L
4 P&W JT12A-6
5 hrs 54 min 17.6 sec
4/22/62
728.26 kmh
452.52 mph

GANDER/GENEVA (USA)
Robert W. Agostino
Rick E. Rowe
Learjet 60
2 P&WC PW305, 4,600 lbs
6/7/95
774.67 kmh
481.36 mph

GANDER/HALIFAX (USA)
JOINT RECORD
William H. Wisner, Pilot
Janice Sullivan, Copilot
Bonanza S35, N6226
and
Daniel Webb, Pilot
Richard Meyerhoff, Copilot
Bonanza A36, N4504S
6/25/89
311.40 kmh
193.50 mph

GANDER/KEFLAVIK (USA)
Rick E. Rowe
William A. Landrum
Learjet 31A
2 AlliedSignal TFE731, 3,500 lbs
6/12/01
755.99 kmh
469.75 mph

GANDER/LONDON (USA)
Harold Curtis, Pilot
William Mack, Copilot
Robert Dann Hardt, Copilot
Gulfstream III
2 RR Spey 511-8
4 hrs 24 min 42 sec
1/8/82
828.00 kmh
514.52 mph

GANDER/NEWPORT NEWS (USA)
Douglas Matthews, Pilot
Doyle Sanderlin, Copilot
Learjet 35A
2 Garrett TFE 731-2-2B, 3,500 lbs
6/5-6/85
716.40 kmh
445.17 mph

GANDER/NICE (SAUDI ARABIA)
Aziz Ojjeh
Canadair Challenger 601
2 GE CF-34, 8,650 lbs
7/23-24/84
855.72 kmh
531.74 mph

GANDER/PARIS (UK)
Alain F. George
Dassault Mystère-Falcon 200
2 Garrett ATF 3-6A-4C, 5,200 lbs each
2/9/03
918.16 kmh
570.52 mph

GANDER/SHANNON (USA)
Jacqueline Cochran
Lockheed Jetstar, N172L
3 hrs 52 min 4 sec
4/22/62
822.69 kmh
511.20 mph

GASTONIA, NC/KITTY HAWK (USA)
H. S. Kendall Willis, Jr., Pilot
Julius S. Palermo, Copilot
Cessna 182 RG
1 Lycoming O-540, 235 bhp
12/16/90
296.93 kmh
184.50 mph

GENEVA/HELSINKI*
Susan Nealey
Faith Hillman
Cessna 310
2 Continental IO-520, 285 bhp
6/20/92
199.88 mph

GENEVA/ROME (USA)
Harold Curtis
Gulfstream III
2 RR Spey 511-8
50 min 28 sec
1/8/82
792.04 kmh
492.15 mph

GEORGETOWN/COZUMEL (USA)
Marie E. McMillan
Beech Bonanza F33A
1 Continental IO-520B, 285 bhp
3/3/84
304.52 kmh
189.22 mph

GIBRALTAR/LONDON (UK)
CPT A. C. P. Carver
DeHavilland Hornet F.MarkIII
2 RR Merlin 130
2 hrs 30 min 21 sec
9/19/49
701.49 kmh
435.88 mph

GODTHAB/CANNES (USA)
Susan Nealey
Faith Hillman
Cessna 310
2 Continental IO-520, 285 bhp
7/12/92
364.32 kmh
226.37 mph

GODTHAB/SHANNON (USA)
Susan Nealey
Faith Hillman
Cessna 310
2 Continental IO-520, 285 bhp
7/12/92
377.77 kmh
234.73 mph

GOOSE BAY/BERLIN (USA)
Steve Fossett
Douglas A. Travis
Cessna 750 Citation X
2 R-R AE 3007C, 6,442 lbs
10/8/03
1,045.97 kmh
649.93 mph

GOOSE BAY/GODTHAB (USA)
Susan Nealey 328.21 kmh
Faith Hillman 203.94 mph
Cessna 310
2 Continental IO-520, 285 bhp
7/9/92

GOOSE BAY/KRAKOW (USA)
Steve Fossett 1,037.43 kmh
Douglas A. Travis 644.63 mph
Cessna 750 Citation X
2 R-R AE 3007C, 6,442 lbs
10/8/03

GOOSE BAY/LOS ANGELES (USA)
Ehab Hanna 743.00 kmh
Lynn Upham 461.68 mph
Gulfstream IV
2 R-R Tay 611, 13,850 lbs
7/8/03

GOOSE BAY/NARSSARSSUAQ (USA)
Anne B. Baddour 346.60 kmh
Margrit Budert-Waltz 241.76 mph
Beech Baron, N3707N
2 Continental, 325 bhp
12/20-21/88

GOOSE BAY/PARIS (USA)
Clay Lacy 870.92 kmh
Brian Kirkdoffer 541.16 mph
Gary Meermans
Hal Fishman
Gulfstream GIISP
2 RR Spey Mk 511-8, 11,400 lbs
6/9/95

GOOSE BAY/REYKJAVIK (USA)
Susan Nealey 365.47 kmh
Faith Hillman 227.09 mph
Cessna 310
2 Continental IO-520, 285 bhp
6/16/92

GOOSE BAY/SHANNON (USA)
Susan Nealey 277.42 kmh
Faith Hillman 172.38 mph
Cessna 310
2 Continental IO-520, 285 bhp
6/16/92

GRAND-BOURG/BASSETERRE (USA)
Marie E. McMillan 236.24 kmh
Beech Bonanza F33A 146.79 mph
1 Continental IO-520B, 285 bhp
2/28/84

GRAND-BOURG/BRIDGETOWN (USA)
Marie E. McMillan 211.47 kmh
Beech Bonanza F33A 131.40 mph
1 Continental IO-520B, 285 bhp
2/23/84

GRAND-BOURG/CASTRIES (USA)
Marie E. McMillan 220.66 kmh
Beech Bonanza F33A 137.11 mph
1 Continental IO-520B, 285 bhp
2/23/84

GRAND-BOURG/CODRINGTON (USA)
Marie E. McMillan 250.68 kmh
Beech Bonanza F33A 155.77 mph
1 Continental IO-520B, 285 bhp
2/28/84

GRAND-BOURG/GUSTAVIA (USA)
Marie E. McMillan 255.55 kmh
Beech Bonanza F33A 158.79 mph
1 Continental IO520B, 285 bhp
2/28/84

GRAND-BOURG/HARRIS (USA)
Marie E. McMillan 278.61 kmh
Beech Bonanza F33A 173.12 mph
1 Continental IO-520B, 285 bhp
2/28/84

GRAND-BOURG/MARTINIQUE (USA)
Marie E. McMillan 252.19 kmh
Beech Bonanza F33A 156.70 mph
1 Continental IO-520B, 285 bhp
2/23/84

GRAND-BOURG/NEW CASTLE (USA)
Marie E. McMillan 228.52 kmh
Beech Bonanza F33A 141.99 mph
1 Continental IO-520B, 285 bhp
2/28/84

GRAND-BOURG/PHILIPSBURG (USA)
Marie E. McMillan 253.14 kmh
Beech Bonanza F33A 157.29 mph
1 Continental IO-520B, 285 bhp
2/28/84

GRAND-BOURG/PORTSMOUTH (USA)
Marie E. McMillan 214.91 kmh
Beech Bonanza F33A 133.54 mph
1 Continental IO-520B, 285 bhp
2/23/84

GRAND-BOURG/ST. JOHNS (USA)
Marie E. McMillan 227.06 kmh
Beech Bonanza F33A 141.09 mph
1 Continental IO-520B, 285 bhp
2/28/84

GRAND-CASE/BASSETERRE (USA)
Marie E. McMillan 206.29 kmh
Beech Bonanza F33A 128.18 mph
1 Continental IO-520B, 285 bhp
2/20/84

GRAND-CASE/CHARLOTTE AMALIE (USA)
Marie E. McMillan 162.71 kmh
Beech Bonanza F33A 101.10 mph
1 Continental IO-520B, 285 bhp
2/20/84

GRAND-CASE/CHRISTIANSTED (USA)
Marie E. McMillan 193.19 kmh
Beech Bonanza F33A 120.04 mph
1 Continental IO-520B, 285 bhp
2/20/84

GRAND-CASE/ORANJESTAD (USA)
Marie E. McMillan 240.31 kmh
Beech Bonanza F33A 149.32 mph
1 Continental IO-520B, 285 bhp
2/20/84

GRAND-CASE/PHILIPSBURG (USA)
Marie E. McMillan 294.72 kmh
Beech Bonanza F33A 183.13 mph
1 Continental IO-520B, 285 bhp
2/20/84

GRAND-CASE/THE BOTTOMS (USA)
Marie E. McMillan 359.87 kmh
Beech Bonanza F33A 223.61 mph
1 Continental IO-520B, 285 bhp
2/20/84

GRAND CAYMAN/MONTEGO BAY (USA)
Marie E. McMillan 254.31 kmh
Beech Bonanza F33A 158.03 mph
1 Continental IO-520 BA, 285 bhp
2/9/84

GRAND FORKS/MONROEVILLE (USA)
Capt. Michael S. Menser, Commander 1,008.37 kmh
Capt. Brian P. Gallagher, Pilot 626.57 mph
Capt. Robert P. Boman, OSO
Capt. Matthew E. Grant, WSO
Rockwell International B-1B
4 GE F101, 12,000 lbs
4/7-8/94

GRAND FORKS/PORTLAND, ME (USA)
Richard W. Taylor, Pilot 459.99 kmh
Clayton L. Scott, Copilot 285.82 mph
Piper Aerostar
2 Lycoming IO-540, 350 bhp
5/24/92

GRAND RAPIDS/MAHE (USA)
Robert Kay, CoCaptain 850.63 kmh
Roger K. Profit, CoCaptain 528.59 mph
Colin Williamson, CoCaptain
Boeing 767ER
2 GE CF6-80
7/25-26/89

GRENVILLE/CANOUAN (USA)
Marie E. McMillan 298.75 kmh
Beech Bonanza F33A 186.64 mph
1 Continental IO-520B, 285 bhp
2/27/84

GRENVILLE/CASTRIES (USA)
Marie E. McMillan 300.94 kmh
Beech Bonanza F33A 187.00 mph
1 Continental IO-520B, 285 bhp
2/27/84

GRENVILLE/CLIFTON (USA)
Marie E. McMillan 263.75 kmh
Beech Bonanza F33A 163.89 mph
1 Continental IO-520B, 285 bhp
2/27/84

GRENVILLE/KINGSTOWN (USA)
Marie E. McMillan 291.64 kmh
Beech Bonanza F33A 181.22 mph
1 Continental IO-520B, 285 bhp
2/27/84

GRENVILLE/LOVELL'S VILLAGE (USA)
Marie E. McMillan 306.29 kmh
Beech Bonanza F33A 190.32 mph
1 Continental IO-520B, 285 bhp
2/27/84

GRENVILLE/PETITE MARTINIQUE (USA)
Marie E. McMillan 300.12 kmh
Beech Bonanza F33A 186.48 mph
1 Continental IO-520B, 285 bhp
2/27/84

GRENVILLE/PORT ELIZABETH (USA)
Marie E. McMillan 306.20 kmh
Beech Bonanza F33A 190.27 mph
1 Continental IO-520B, 285 bhp
2/27/84

GRITVYKEN/PORT STANLEY (UK)
A. Holman 417.97 kmh
Lockheed Hercules 259.71 mph
4 T56-A7A
12/1/84

GUAYAQUIL/SANTIAGO (USA)
Paulo S. Jancitsky 870.65 kmh
Kevin A. Bixler 540.99 mph
IAI 1126 Galaxy
2 P&WC PW306A, 5,700 lbs
3/31/02

GULFPORT/TEXARKANA (USA)
Danford A. Bookout 209.16 kmh
Piper Lance, N8352C 129.97 mph
1 Lycoming IO-540K, 300 bhp
1/8/88

GUSTAVIA/CHARLOTTE AMALIE (USA)
Marie E. McMillan 391.82 kmh
Beech Bonanza F33A 243.46 mph
1 Continental IO-520B, 285 bhp
2/20/84

GUSTAVIA/PHILIPSBURG (USA)
Marie E. McMillan 274.59 kmh
Beech Bonanza F33A 170.62 mph
1 Continental IO-520B, 285 bhp
2/28/84

HALIFAX/BEDFORD, MA (USA)
JOINT RECORD
William H. Wisner, Pilot 316.44 kmh
Janice Sullivan, Copilot 196.64 mph
Bonanza S35, N6226
and
Daniel Webb, Pilot
Richard Meyerhoff, Copilot
Bonanza A36, N4504S
6/26/89

HALIFAX/PORT LOUIS (USA)
Edmund Hepner 863.41 kmh
James Dunham 530.31 mph
Gerry Delich
Ranjit Appa
Boeing 767-200ER, 3B-NAL
4/17-18/88

HALIFAX/TOLUCA (MEXICO)
José Alfredo Gastelum Arce 671.28 kmh
Jose Manuel Muradas Rodriguez 415.62 mph
Dassault Falcon 2000
2 GE/Honeywell CFE738, 5,888 lbs
11/14/02

HAMILTON/CAPE TOWN (USA)
Bernard R. Ozbolt 834.56 kmh
Henry K. Gibson 518.57 mph
Christian M. Kennedy
Gulfstream V
2 BMW-RR BR710-48, 14,750 lbs
10/6/99

HAMILTON, ON/WASHINGTON (CANADA)
George Thompson 297.36 kmh
Piper Aerostar, C-GSMO 184.78 mph
2 Lycoming IO-540, 350 bhp
5/18/88

HARRIS/BASSETERRE (USA)
Marie E. McMillan 193.37 kmh
Beech Bonanza F33A 120.15 mph
1 Continental IO-520B, 285 bhp
2/28/84

HARRIS/CODRINGTON (USA)
Marie E. McMillan 316.32 kmh
Beech Bonanza F33A 196.55 mph
1 Continental IO-520B, 285 bhp
2/28/84

HARRIS/GUSTAVIA (USA)
Marie E. McMillan 242.22 kmh
Beech Bonanza F33A 150.51 mph
1 Continental IO-520B, 285 bhp
2/28/84

HARRIS/NEW CASTLE (USA)
Marie E. McMillan 167.73 kmh
Beech Bonanza F33A 104.22 mph
1 Continental IO-520B, 285 bhp
2/28/84

HARRIS/PHILIPSBURG (USA)
Marie E. McMillan — 238.55 kmh / 148.23 mph
Beech Bonanza F33A
1 Continental IO-520B, 285 bhp
2/28/84

HARRIS/ST. JOHNS (USA)
Marie E. McMillan — 332.83 kmh / 206.81 mph
Beech Bonanza F33A
1 Continental IO-520B, 285 bhp
2/28/84

HAVANA/WASHINGTON (USA)
W. W. Edmondson — 563.80 kmh / 350.32 mph
North American P-51D, NX-4E
Packard RR Merlin, V-1650
11/27/47

HELSINKI/MOSCOW*
Susan Nealey — 148.87 mph
Faith Hillman
Cessna 310
2 Continental IO-520, 285 bhp
6/21/92

HELSINKI/NOME (USA)
Susan Nealey — 35.38 kmh / 21.98 mph
Faith Hillman
Cessna 310
2 Continental IO-520, 285 bhp
6/21-28/92

HILLSBORO/CANOUAN (USA)
Marie E. McMillan — 204.70 kmh / 127.19 mph
Beech Bonanza F33A
1 Continental IO-520B, 285 bhp
2/27/84

HILLSBORO/KINGSTOWN (USA)
Marie E. McMillan — 233.16 kmh / 144.88 mph
Beech Bonanza F33A
1 Continental IO-520B, 285 bhp
2/27/84

HILLSBORO/LOVELL'S VILLAGE (USA)
Marie E. McMillan — 230.14 kmh / 143.00 mph
Beech Bonanza F33A
1 Continental IO-520B, 285 bhp
2/27/84

HILLSBORO/PORT ELIZABETH (USA)
Marie E. McMillan — 210.26 kmh / 130.65 mph
Beech Bonanza F33A
1 Continental IO-520B, 285 bhp
2/27/84

HILO/CHICAGO (USA)
Harold Curtis — 792.00 kmh / 492.13 mph
Gulfstream III
2 RR Spey 511-8
8 hrs 16 min 48 sec
1/10/82

HILTON HEAD/MAUI (USA)
Borden F. Schofield — 807.15 kmh / 501.54 mph
E. Bruce Robinson
Geoffrey M. Foster
Andrew Milewski
Bombardier BD 700
2 BMW RR BR-710, 14,690 lbs
5/5/99

HILTON HEAD/MILAN (USA)
Robert S. McKenney — 892.62 kmh / 554.65 mph
Larry S. Mueller
Gulfstream V
2 BMW R-R BR710, 14,750 lbs
1/31/01

HOLLYWOOD/LAKELAND*
Victor P. Rameau — 110.79 mph
American General AG-5B Tiger
1 Lycoming O-360, 180 bhp
4/16/96

HONG KONG/BEIJING (USA)
James J. Keller — 848.28 kmh / 527.10 mph
Dwight V. Whitaker, III
Gulfstream V
2 BMW RR BR710-48, 14,750 lbs
4/12/97

HONG KONG/GUAM (USA)
Douglas G. Matthews — 718.92 kmh / 446.74 mph
Learjet 36
2 Garrett, TFE-731-2-2B, 3,500 lbs
7/22/84

HONG KONG/HONOLULU (USA)
Allen E. Paulson, Captain — 1,029.60 kmh / 639.79 mph
Robert K. Smyth, CoCaptain
John Salamankas, CoCaptain
Jeff Bailey, CoCaptain
Gulfstream IV, N400GA
2 RR TAY MK611-8
2/27/88

HONG KONG/PHUKET, THAILAND (USA)
JOINT RECORD
William H. Wisner, Pilot — 205.92 kmh / 127.96 mph
Janice Sullivan, Copilot
Bonanza S35, N6226
and
Daniel Webb, Pilot
Richard Meyerhoff, Copilot
Bonanza A36, N4504S
6/11/89

HONOLULU/BAHRAIN (FRANCE)
Claude Delorme, Captain — 1,020.67 kmh / 634.21 mph
Jean Boye, Co-Captain
BAe/Aérospatiale Concorde
10/13/92

HONOLULU/BANGKOK (FRANCE)
Claude Delorme, Captain — 1,348.36 kmh / 837.83 mph
Jean Boye, Co-Captain
BAe/Aérospatiale Concorde
10/13/92

HONOLULU/BUENOS AIRES (USA)
George R. Jansen — 851.50 kmh / 529.10 mph
McDonnell Douglas DC-10
3 JT9D-25W
14 hrs 18 min
10/12/72

HONOLULU/CARLSBAD (USA)
Douglas G. Matthews — 798.12 kmh / 495.95 mph
Learjet 36
2 Garrett TFE-731-2-2B, 3,500 lbs
7/28/84

HONOLULU/DENVER (USA)
Arnold Palmer — 781.05 kmh / 485.32 mph
Learjet 36
2 Garrett TFE-731-2-2B, 3500 lbs
6 hrs 54 min 28 sec
5/19/76

HONOLULU/GUAM (FRANCE)
Claude Delorme, Captain — 1,813.42 kmh / 1,126.81 mph
Jean Boye, Co-Captain
BAe/Aérospatiale Concorde
10/13/92

HONOLULU/KUALA LUMPUR (MALAYSIA)
Khairi Mohamad — 881.79 kmh
Roland A. Thomas — 547.92 mph
McDonnell Douglas DC-10-30
3 GE CF6-50C, 40,000 lbs
12 hrs 26 min 14.5 sec
8/13-14/76

HONOLULU/LONDON (USA)
Jack L Martin — 837.58 kmh
Fred L. Austin — 520.45 mph
Robert N. Buck
James R. Gannett
Boeing 707-320C, N332FT
4 P&W JT3
13 hrs 53 min 49 sec
11/15/65

HONOLULU/MAJURO (USA)
JOINT RECORD
William H. Wisner, Pilot — 313.66 kmh
Janice Sullivan, Copilot — 194.85 mph
Bonanza S35, N6226
and
Daniel Webb, Pilot
Richard Meyerhoff, Copilot
Bonanza A36, N4504S
5/31-6/1/89

HONOLULU/MONTREAL (CANADA)
Capt. C. H. Simpson — 937.08 kmh
Capt. E. E. Doyle — 582.30 mph
Boeing 747-238B, VH-ECA
1/12/88

HONOLULU/OAKLAND (USA)
Mark Huffstutler
Cessna Citation 500 — 582.44 kmh
2 P&W JT15D, 2,200 lbs — 361.91 mph
12/31/94

HONOLULU/PAGO PAGO (USA)
Brooke Knapp — 815.98 kmh
Gulfstream III — 507.05 mph
2 RR Spey, MK511-8, 11,400 lbs
11/15/83

HONOLULU/ST LOUIS (USA)
Ray E. Slaughter — 858.60 kmh
DC-9-41 — 533.53 mph
2 P&W JT-8D, 15,000 lbs
2/7-8/83

HONOLULU/SAN FRANCISCO (USA)
Brooke Knapp, Pilot — 988.89 kmh
James Topalian, Copilot — 614.50 mph
Paul Broyles, Copilot
James Magill, crew chief
Gates Learjet 35A
2 Garrett TFE 731-2-2B, 3,500 lbs
2/18/83

HONOLULU/SOUTH POLE (USA)
Brooke Knapp
Gulfstream III — 669.53 kmh
2 RR Spey — 416.05 mph
MK511-8 11,400 lbs
11/15-16/83

HONOLULU/WASHINGTON (USA)
Brooke Knapp — 950.56 kmh
Gulfstream III — 590.68 mph
2 RR Spey MK511-8, 11,400 lbs
2/15/83

HOUSTON/CAIRO (USA)
Larry S. Mueller — 926.35 kmh
John D. Mackay — 575.60 mph
Gulfstream V
2 BMW R-R BR710, 14,750 lbs
3/18/01

HOUSTON/CLEVELAND (USA)
Danford A. Bookout — 305.64 kmh
Piper Lance, N8352C — 189.92 mph
1 Lycoming IO-540K, 300 bhp
11/9/87

HOUSTON/DUBAI (USA)
Allen E. Paulson, Captain — 876.96 kmh
Robert K. Smyth, CoCaptain — 539.35 mph
John Salamankas, CoCaptain
Jeff Bailey, CoCaptain
Gulfstream IV, N400GA
2 RR TAY MK611-8
2/26-27/88

HOUSTON/LONDON (UK)
CPT David Deadman
John Hill-Paul — 983.11 kmh
Peter Moore — 610.87 mph
Boeing 707-320AC
4 P&W JT-3D-3B, 8,617 kg
10/29-30/77

HOUSTON/SHANNON (USA)
Allen E. Paulson, Captain — 950.04 kmh
Robert K. Smyth, CoCaptain — 590.35 mph
John Salamankas, CoCaptain
Jeff Bailey, CoCaptain
Gulfstream IV, N400GA
2 RR TAY MK611-8
2/26/88

HOUSTON/TEXARKANA (USA)
Danford A. Bookout — 306.36 kmh
Piper Lance, N8352C — 190.37 mph
1 Lycoming IO-540K, 300 bhp
2/2/88

HOWELL, MI/OSHKOSH (USA)
Kenneth P. Rittenhouse, Jr., Pilot — 833.77 kmh
Kirk J. Appell, Crewmember — 518.08 mph
William Lee Walker, Jr., Crewmember
Learjet 35A
2 Garrett TFE731, 3,500 lbs
07/31/90

INDIANAPOLIS/PHILADELPHIA (USA)
Robert Olszewski, Pilot — 975.26 kmh
Steven Wilson First Officer — 606.03 mph
USAir DC 9-31, N935VJ
2 P&W JT8D9B, 14,500 lbs
12/20-21/87

INDIANAPOLIS/TEXARKANA (USA)
Danford A. Bookout — 273.24 kmh
Piper Lance, N8352C — 169.79 mph
1 Lycoming IO-540K, 300 bhp
8/12/87

IQALUIT/KUUJJUAQ (USA)
James P. Hanson — 248.11 kmh
Cessna 208 Caravan — 154.16 mph
1 P&W PT-6A, 600 shp
5/16/96

IRKUTSK/KHABAROVSK (USA)
Allyn Caruso, Pilot — 377.00 kmh
David Norgart, Copilot — 234.25 mph
John Gallichon, Navigator
John Miller, Navigator
Dana Lovell, Mechanic
Fairchild Metroliner III
2 Garrett TPE331-11, 1,100 shp
6/4/91

IRKUTSK/NOME (USA)
Susan Nealey — 82.97 kmh
Faith Hillman — 51.56 mph
Cessna 310
2 Continental IO-520, 285 bhp
6/26-28/92

IRKUTSK/YAKUTSK*
Susan Nealey
Faith Hillman
Cessna 310
2 Continental IO-520, 285 bhp
6/26/92
197.07 mph

JACKSONVILLE/LOS ANGELES (USA)
Rodney R. Lundy
David S. Ryan
Learjet 45
2 AlliedSignal TFE731-20, 3,500 lbs
6/6/99
754.39 kmh
468.75 mph

JACKSONVILLE/MIAMI (USA)
Mac McConnell
Cougar, N791GA
2 Lycoming 0-320, 160 bhp
12/2/88
364.68 kmh
226.61 mph

JACKSONVILLE/SAN DIEGO (USA)
Steve Fossett, Pilot
Darrin L. Adkins, Copilot
Cessna 750 Citation X
2 R-R AE 3007C, 6,442 lbs
9/17/00
952.67 kmh
591.96 mph

JAKARTA/MANILA (USA)
Arnold Palmer
Learjet 36
2 Garrett TFE-731-2-2B, 3,500 lbs
5/19/76
723.11 kmh
449.32 mph

JAKARTA/MELBOURNE (USA)
Paulo S. Jancitsky
Lanny C. Harwell
IAI 1126 Galaxy
2 P&WC PW306A, 5,700 lbs
3/6/02
898.79 kmh
558.48 mph

JAKARTA/SAN FRANCISCO (USA)
Frank P. Santoni Jr.
Richard A. Austin
John E. Cashman, Charles A. Hovland,
Izham Ismail, Joseph M. MacDonald,
James C. McRoberts, Rodney M. Skaar
Boeing 777-200-IGW
2 RR6 Trent 892, 90,000 lbs
Seattle, WA
4/2/97
888.95 kmh
552.37 mph

JOHANNESBURG/DUBLIN (USA)
William J. Watters
Wilson A. Miles, Jr.
Gulfstream V
2 BMW R-R BR710, 14,750 lbs
6/28/01
820.26 kmh
509.68 mph

JOHANNESBURG/MANILA (USA)
Rick B. Gowthrop
Wilson A. Miles, Jr.
Gulfstream V
2 BMW RR BR710-48, 14,750 lbs
11/10/98
805.96 kmh
500.80 mph

JUNEAU/ORLANDO (USA)
James T. Brown
Donald M. Majors
IAI 1125 Astra SPX
2 Allied Signal TFE731, 4,250 lbs
11/13/96
810.31 kmh
503.50 mph

KAILUA/NADI (UK)
Max Robinson
BAe/Aérospatiale Concorde
10/27/99
1,725.02 kmh
1,071.87 mph

KAILUA/SYDNEY (UK)
Max Robinson
BAe/Aérospatiale Concorde
10/27/99
1,237.75 kmh
769.10 mph

KANSAS CITY/RENO (USA)
R. A. "Bob" Hoover, Pilot
Robert Morgenthaler, Copilot
Sabreliner
2 Garrett TFE-731, 3,700 lbs
9/10/86
706.32 kmh
438.90 mph

KANSAS CITY/WASHINGTON (USA)
Lt Col Ed Yeilding, Pilot
Lt Col J. T. Vida, RSO
Lockheed SR-71
2 P&W J-58, 34,000 lbs
03/06/90
3,502.08 kmh
2,176.08 mph

KEFLAVIK/BERLIN (USA)
Allyn Caruso, Pilot
David Norgart, Copilot
John Gallichon, Navigator
John Miller, Navigator
Dana Lovell, Mechanic
Fairchild Metroliner III
2 Garrett TPE331-11, 1,100 shp
5/31/91
439.45 kmh
273.06 mph

KEFLAVIK/PARIS (USA)
Rodney R. Lundy
J. Shawn Christian
Learjet 45
2 AlliedSignal TFE731, 3,500 lbs
6/14/01
835.29 kmh
519.02 mph

KHABAROVSK/MAGADAN (USA)
Allyn Caruso, Pilot
David Norgart, Copilot
John Gallichon, Navigator
John Miller, Navigator
Dana Lovell, Mechanic
Fairchild Metroliner III
2 Garrett TPE331-11, 1,100 shp
6/4-5/91
402.05 kmh
249.82 mph

KILLEEN/BATON ROUGE*
Marlon D. Brose, Pilot
Douglas M. Whitson II, Copilot
Cessna 172
1 Lycoming O-320, 160 bhp
4/26/93
100.72 mph

KINGSTON/GEORGETOWN (USA)
Marie E. McMillan
Beech Bonanza F33A
1 Continental IO-520B, 285 bhp
3/2/84
330.09 kmh
205.11 mph

KIRUNA/SEATTLE (USA)
Michael E. Hewett
Richard K. Nelson
Boeing 777
2 P&W PW4084, 77,000 lbs
11/12/94
795.28 kmh
494.16 mph

KOTA KINABALU/DUBAI (USA)
Allen E. Paulson, Captain
K. C. Edgecomb, Colin B. Allen
John Salamankas, Jefferson Bailey; crew
Gulfstream IV, N440GA
2 RR TAY MK610-8
6/13-14/87
897.97 kmh
558.00 mph

KOTA KINABALU/MADRAS (USA)
Allen E. Paulson, Captain
K. C. Edgecomb, Colin B. Allen
John Salamankas, Jefferson Bailey; crew
Gulfstream IV, N440GA
2 RR TAY MK610-8
6/13/87
899.21 kmh
558.74 mph

KUALA LUMPUR/HONIARA (USA)
David J. Dwyer 730.49 kmh
David A. Montgomery 453.90 mph
Dassault Falcon 900EX
3 AlliedSignal TFE731-60, 5,000 lbs
7/12/98

KULUSUK/SONDRE STROMFJORD (USA)
James P. Hanson 278.38 kmh
Cessna 208 Caravan 172.98 mph
1 P&W PT-6A, 600 shp
5/15/96

KUUJJUAQ/CHICAGO (USA)
Susan Nealey 281.92 kmh
Faith Hillman 175.17 mph
Cessna 310
2 Continental IO-520, 285 bhp
7/22/92

KUUJJUAQ/IQALUIT (CANADA)
Terrance L. Jantzi 259.00 kmh
Lauren V. Jantzi 160.93 mph
Van's RV-6
6/4/99

KUUJJUAQ/OSHKOSH (USA)
Susan Nealey 287.84 kmh
Faith Hillman 178.85 mph
Cessna 310
2 Continental IO-520, 285 bhp
7/22/92

LAJES/LISBON (USA)
George P. Eremea 805.38 kmh
N American 282-40 Sabreliner 500.43 mph
2 P&W JT12A-6A
1 hr 55 min 45.1 sec
10/26/63

LAS VEGAS/CLEVELAND (USA)
Raymon B. Fogg 426.24 kmh
Piper Aerostar 601P 264.87 mph
5/29/89

LAS VEGAS/HERMOSILLO (USA)
Marie E. McMillan 262.40 kmh
Beech Bonanza F33A 163.05 mph
1 Continental IO-520 BA, 285 bhp
1/31/84

LAS VEGAS/HONOLULU (UK)
Capt John Hutchinson 1,258.55 kmh
British Airways 782.02 mph
BAe/Aérospatiale Concorde
3/16/90

LAS VEGAS/RENO (USA)
Danford A. Bookout 280.44 kmh
Piper Lance, N8352C 174.26 mph
1 Lycoming IO-540K, 300 bhp
10/2/87

LAS VEGAS/VENICE (USA)
William J. Watters 923.18 kmh
Richard D. Robbins 573.64 mph
Gulfstream V
2 BMW R-R BR710, 14,750 lbs
7/28/01

LATROBE/KITTY HAWK (USA)
Carl A. La Rue 549.06 kmh
John Slais 341.17 mph
Lancair IV P
1 Continental TSIO-550B, 350 bhp
10/21/97

LEXINGTON/WASHINGTON (USA)
Philip B. Boyer 469.46 kmh
Cessna 425 291.70 mph
2 P&W PT-6, 450 shp
2/24/91

LIHUE/SINGAPORE (USA)
Bernard R. Ozbolt 740.58 kmh
Charles M. Barr II 460.17 mph
Gary M. Freeman
Gulfstream V
2 BMW Rolls-Royce BR710-48, 14,750 lbs
2/21/98

LISBON/ACAPULCO (FRANCE)
Claude Delorme, Captain 1,227.40 kmh
Jean Boye, Co-Captain 762.67 mph
BAe/Aérospatiale Concorde
10/12/92

LISBON/BUENOS AIRES (USA)
Jack L Martin, Fred L Austin, 807.39 kmh
Harrison Finch, Robert N. Buck, 501.69 mph
James R. Gannett
Boeing 707-320C, N332FT
4 P&W JT-3
11 hrs 56 min 29 sec
11/16/65

LISBON/FRANKFURT (USA)
George P. Eremea 785.11 kmh
N American 282-40 Sabreliner 487.83 mph
2 P&W JT12A-6A
2 hrs 24 min 46.1 sec
10/27/63

LISBON/HONOLULU (FRANCE)
Claude Delorme, Captain 1,058.43 kmh
Jean Boye, Co-Captain 657.67 mph
BAe/Aérospatiale Concorde
10/13/92

LISBON/NEW YORK (UK)
Geoffrey Mussett 1,668.37 kmh
BAe/Aérospatiale Concorde 1,036.67 mph
11/14/99

LISBON/PARIS (FRANCE)
Claude Delorme, Captain 1,008.41 kmh
Jean Boye, Co-Captain 626.59 mph
BAe/Aérospatiale Concorde
10/14/92

LISBON/SANTO DOMINGO (FRANCE)
Claude Delorme, Captain 1,761.61 kmh
Jean Boye, Co-Captain 1,094.61 mph
BAe/Aérospatiale Concorde
10/12/92

LONDON/ADEN (UK)
John R. Ward 948.50 kmh
Avro Vulcan 589.36 mph
4 Bristol Siddeley Olympus
6 hrs 13 min 59.9 sec
3/30/62

LONDON/AMSTERDAM (UK)
LT J. R. S. Overbury 919.76 kmh
Hawker Sea Hawk P.B.3 571.51 mph
1 RR Nene Mark-101
23 min 30.9 sec
7/29/54

LONDON/ASCENSION (UK)
Ft/Lt John Halstead 824.35 kmh
VC 10 C 512.25 mph
4 RR Conway Mk 301, 22,500 lbs
12/19/87

LONDON/ATHENS (CANADA)
MAJ D. B. O'Connor
Dassault Falcon 20
2 GE CF 700-C
2/27/69
600.69 kmh
373.25 mph

LONDON/AUCKLAND (UK)
J. W. Burton
BAe/Aérospatiale Concorde
4 Olympus 593/610, 38,000 lbs
4/5-6/86
983.22 kmh
610.97 mph

LONDON/BAGDAD (UK)
J. Finch
Vickers Valiant B.I WP209
2 RR Avon RA-14
4 hrs 51 min 28.8 sec
7/31/55
842.46 kmh
523.48 mph

LONDON/BASRA, IRAQ (UK)
R. L. E. Burton
English Electric Canberra
P. R. MK III
2 RR Avon RA-3
5 hrs 11 min 5.6 sec
10/8/53
876.01 kmh
544.33 mph

LONDON/BEIJING (USA)
Brooke Knapp
Gulfstream III
2 RR Spey MK511-8, 11,400 lbs
2/13-14/84
688.85 kmh
689.47 mph

LONDON/BEIRUT (PAKISTAN)
A. Baig
Boeing 720B
4 P&W JT3D-1
1/1/62
964.50 kmh
599.30 mph

LONDON/BERMUDA (UK)
CPT G. N. Henderson
Boeing 707-465
4 RR Conway 508
6 hrs 53 min 01 sec
5/6/72
802.97 kmh
498.94 mph

LONDON/BONN (USA)
Jacqueline Cochran
Lockheed Jetstar, N172L
4 P&W JT12A-6
4/22/62
620.49 kmh
385.56 mph

LONDON/BOSTON (UK)
Mike Bannister
BAe/Aérospatiale Concorde
4 RR Olympus, 38,000 lbs
10/8/03
1,689.47 kmh
1,049.79 mph

LONDON/BRUSSELS (UK)
D.W. Morgan
Vickers Armstrong Supermarine Swift
1 RR Avon RA7
18 min 3.3 sec
7/10/52
1,071.65 kmh
665.89 mph

LONDON/BUENOS AIRES (USA)
Jack L. Martin,
Fred L. Austin,
Harrison Finch,
Robert N. Buck & James R. Gannett
Boeing 707-320C, N332FT
4 P&W JT3
16 hrs 40 min 44.6 sec
11/16/65
668.40 kmh
415.33 mph

LONDON/CAIRO (UK)
J. Cunningham
H.S. Trident IE
3 RR Spey
3 hrs 37 min 44 sec
4/13/65
968.99 kmh
602.10 mph

LONDON/CALCUTTA (CANADA)
Maj D. B. O'Connor
2 GE CF 700
Dassault Falcon 20
13 hrs 29 min 51 sec
12/27/69
590.61 kmh
366.99 mph

LONDON/CAPETOWN (UK)
G. G. Petty
English Electric
Canberra B MK II WH699
2 RR Avon RA3
12 hrs 21 min 3.9 sec
12/17/53
783.08 kmh
486.58 mph

LONDON/CHRISTCHURCH (UK)
R. L. E. Burton
English Electric
Canberra P.R. MK III, WE139
2 RR Avon RA-3
23 hrs 50 min 42 sec
10/8/53
795.89 kmh
493.54 mph

LONDON/COLOMBO, CEYLON (UK)
L. M. Hodges
English Electric
Canberra P.R. MK VII
2 RR Avon RA-7
10 hrs 25 min 21.5 sec
10/8/53
836.00 kmh
519.47 mph

LONDON/COPENHAGEN (UK)
Jan Zurakowski
Gloster Meteor MK. F8 V2468
2 RR Derwent V
1 hr 5 min 5 sec
4/4/50
871.34 kmh
541.42 mph

LONDON/CORFU (USA)
Brooke Knapp, Pilot
James Topalian, Copilot
Paul Broyles, Copilot
James Magill, crew chief
Gates Learjet 35A
2 Garrett TFE 731-2-2B, 3,500 lbs
2/17/83
750.70 kmh
466.48 mph

LONDON/DARWIN (UK)
L. M. Whittington
English Elec
Canberra VX 181
2 RR Avon
22 hrs 21.8 sec
1/27/53
629.55 kmh
391.18 mph

LONDON/DUBAI (USA)
Bernard R. Ozbolt
Henry K. Gibson
Robert S. McKenney
Gulfstream V
2 BMW RR BR 710-48, 14,750 lbs
11/13/97
838.06 kmh
520.75 mph

LONDON/EDINBURGH (UK)
JOINT RECORD
W/Cdr C. Spink
Flt. Lt. I. T. Gale
2 Phantom F4J
2 P&W J79/10B, 17,820 lbs
7/1/87
1,191.76 kmh
740.56 mph

LONDON/HALIFAX (MEXICO)
José Alfredo Gastelum Arce
Jose Manuel Muradas Rodriguez
Dassault Falcon 2000
2 GE/Honeywell CFE738, 5,888 lbs
11/13/02
668.87 kmh
415.62 mph

LONDON/HONG KONG (USA)
James J. Keller
Dwight V. Whitaker, III
Gulfstream V
2 BMW RR BR710-48, 14,750 lbs
4/10/97
794.63 kmh
493.76 mph

LONDON/HOUSTON (UK)
R. Cottrell, Capt.
A. Moore, Flight Officer
E. Goodyear, Engineering Officer
D. Kemp, Navigating Officer
Boeing 707-320C
4 P&W JT-3D-3B, 8,617 kg
10/24/77
760.55 kmh
472.58 mph

LONDON/JOHANNESBURG (UK)
J. Cunningham
de Havilland Comet
Series 3 G-ANLO
4 RR Avon RA 29
12 hrs 59 min 7.3 sec
10/23/57
698.31 kmh
433.91 mph

LONDON/KARACHI (PAKISTAN)
A. Baig
Boeing 720B
6 hrs 43 min 51 sec
1/2/62
938.79 kmh
583.36 mph

LONDON/KHARTOUM (UK)
J. Cunningham
de Havilland Comet
Series 3 G-ANLO
4 RR Avon RA 29
5 hrs 51 min 14.8 sec
10/16/57
842.35 kmh
523.41 mph

LONDON/KUWAIT (UK)
CPT A. W. Hebborn
Hawker Siddeley Comet 4C
4 RR Avon MK 525
6 hrs 25 sec
2/2/64
773.87 kmh
480.85 mph

LONDON/LA VALETTE, MALTA (UK)
CDR Harry Bennett, A.F.C.
Hawker Hunter, F.G.A.9
1 RR Avon MK 207
2 hrs 3 min 8 sec
6/17/58
1,019.24 kmh
633.32 mph

LONDON/LISBON (USA)
Jack L Martin, Fred L Austin,
Harrison Finch
Boeing 707/320C N332FT
4 P&W JT3
2 hrs 18 min 51.2 sec
11/16/65
677.14 kmh
420.76 mph

LONDON/LOS ANGELES (USA)
CPT Harold B. Adams, USAF
Lockheed SR-71
2 P&W JT11D-20
3 hrs 47 min 39 sec
9/13/74
2,310.35 kmh
1,435.59 mph

LONDON/LUXOR (USA)
Brooke Knapp, Pilot
James Topalian, Copilot
Paul Broyles, Copilot
James Magill, crew chief
Gates Learjet 35A
2 Garrett TFE 731-2-2B, 3,500 lbs
2/17/83
775.73 kmh
482.04 mph

LONDON/MELBOURNE (UK)
W. Baillie
BEA Vickers Viscount
700 G-AMAV
4 RR Dart 503 FDA 3
35 hrs 46 min 47.6 sec
10/8/53
472.52 kmh
293.61 mph

LONDON/MOSCOW (USA)
Brooke Knapp
Gulfstream III
RR Spey MK511-8, 11,400 lbs
2/14/84
733.50 kmh
455.80 mph

LONDON/NAIROBI (UK)
H. P. Connolly
English Electric Canberra
2 RR Avon
9 hrs 55 min 16.7 sec
9/28/52
687.61 kmh
427.26 mph

LONDON/NEW YORK (UK)
Capt. Peter Horton
British Airways
BAe/Aérospatiale Concorde
3/14/90
1,644.02 kmh
1,021.54 mph

LONDON/NOVOSIBIRSK (USA)
Brooke Knapp
Gulfstream III
2 RR Spey MK511-8, 11,400 lbs
2/14/84
678.89 kmh
421.86 mph

LONDON/PARIS (UK)
M. J. Lithgow
Vickers Armstrong
Supermarine Swift MKIV
1 RR Avon
7/5/53
1,077.42 kmh
669.40 mph

LONDON/PERTH (UK)
W/CDR J. L. Upritchard
VC 10K3
4 Conway 550B, 22,500 lbs
4/8/87
914.07 kmh
568.00 mph

LONDON/PORT STANLEY (UK)
Group Capt Chris Lumb
VC 10 C
4 RR Conway Mk 301, 22,500 lbs
12/19-20/87
805.78 kmh
500.71 mph

LONDON/RAWALPINDI (PAKISTAN)
CPT Shaukat
Boeing 747-282B
4 P&W 46,150 lbs
7 hrs 57 sec
4/26/76
862.38 kmh
535.85 mph

LONDON/REYKJAVIK (UK)
Lord D. G. Trefgarne
C. B. G. Masefield
de Havilland D. H. 90 Dragonfly
2 Gipsy Major 10 MKI
32 hrs 36 min 36 sec
7/31-8/1/65
57.97 kmh
36.02 mph

LONDON/ROME (UK)
A. W. Bedford — 911.01 kmh / 566.07 mph
Hawker Hunter T. 7-P 1101
1 RR Avon MK J21A
1 hr 34 min 28.5 sec
10/20/56

LONDON/SHANNON (UK)
Capt. Peter Horton — 846.98 kmh / 526.29 mph
BAe/Aérospatiale Concorde
12/31/92

LONDON/SINGAPORE (CHINA)
Yoon Choon Choy — 757.34 kmh / 532.75 mph
Boeing 747/312
4 JT9D-7R4G2, 53,580 lbs
10/29-30/84

LONDON/SYDNEY (UK)
CPT Hector McMullen — 998.59 kmh / 620.50 mph
British Airways
BAe/Aérospatiale Concorde
4 RR Snecma Olympus, 38,000 lbs
17 hrs 3 min 45 sec
2/13-14/85

LONDON/TOKYO (USA)
Brooke Knapp — 731.09 kmh / 454.30 mph
Gulfstream III
2 RR Spey MK511-8, 11,400 lbs
2/14/84

LONDON/TOLUCA (MEXICO)
José Alfredo Gastelum Arce — 620.00 kmh / 385.25 mph
Jose Manuel Muradas Rodriguez
Dassault Falcon 2000
2 GE/Honeywell CFE738, 5,888 lbs
11/14/02

LONDON/TRIPOLI (UK)
L. C. E. DeVinge — 866.02 kmh / 538.12 mph
English Elect. Canberra BMK
2 RR Avon MKI
2 hrs 41 min 49.5 sec
2/15/52

LONDON/WASHINGTON (UK)
Christopher Joseph — 1,647.94 kmh / 1,023.98 mph
Capell Morley
BAe/Aérospatiale Concorde
4 RR Olympus 593, 38,000 lbs
3 hrs 34 min 48 sec
5/29/76

LONDON/WELLINGTON (UK)
N. H. d'Aeth — 313.27 kmh / 194.66 mph
Modified Avro Lancaster
4 RR Merlin XXLV/LL
59 hrs 50 min 00 sec
8/21/46

LONG BEACH/WASHINGTON (USA)
Richard W. Taylor, Pilot — 410.07 kmh / 254.80 mph
Starr M. Tavenner, Copilot
Piper Aerostar
2 Lycoming IO-540, 350 bhp
11/6/92

LOS ANGELES/ALBUQUERQUE (USA)
William H. Cox, Pilot — 544.47 kmh / 338.32 mph
Martha A. Morrell, Copilot
Mooney TLS
1 Lycoming TIO-540, 270 bhp
3/25/94

LOS ANGELES/ANNAPOLIS (USA)
Charles Hunter — 728.28 kmh / 452.55 mph
Citation 2SP, N1UH
2 P&W JT 15 D-4, 2,500 lbs
12/21/87

LOS ANGELES/BELGRADE (YUGOSLAVIA)
Slobodan Tanaskovic — 914.96 kmh / 568.19 mph
McDonnell Douglas DC-10-30
3 GE CF6
11 hrs 19 min 2 sec
12/12/78

LOS ANGELES/CHICAGO (USA)
Ehab Hanna — 786.71 kmh / 488.84 mph
Lynn Upham
Gulfstream IV
2 R-R Tay 611, 13,850 lbs
6/18/03

LOS ANGELES/CHRISTCHURCH (USA)
Brooke Knapp — 696.60 kmh / 432.87 mph
Gulfstream III
2 RR Spey MK511-8, 11,400 lbs
11/15/83

LOS ANGELES/DALLAS *Group I* (USA)
William H. Cox, Pilot — 524.24 kmh / 325.75 mph
Martha A. Morrell, Copilot
Mooney TLS
1 Lycoming TIO-540, 270 bhp
3/25/94

LOS ANGELES/DALLAS *Group III* (USA)
Clay Lacy — 993.95 kmh / 617.61 mph
Alex A. Kvassay
Gulfstream II SP
2 RR Spey MK 511-8, 11,400 lbs
9/22/97

LOS ANGELES/DAYTONA BEACH (USA)
Robert Hoover — 659.52 kmh / 409.83 mph
P-51 Mustang
1 RR -1650, 1,450 bhp
3/28/85

LOS ANGELES/GERONA (SPAIN)
Carlos Moreno — 860.51 kmh / 534.70 mph
McDonnell Douglas DC-10-30
3 GE CF6-50
10 hrs 47 min 30 sec
3/28/73

LOS ANGELES/GOOSE BAY (USA)
Clay Lacy — 1,126.57 kmh / 700.02 mph
Brian Kirkdoffer
Gary Meermans
Hal Fishman
Gulfstream GIISP
2 RR Spey Mk 511-8, 11,400 lbs
6/8/95

LOS ANGELES/HONG KONG (USA)
George R. Jansen — 792.00 kmh / 492.13 mph
McDonnell Douglas DC-10-40
3 JT9D-25W
14 hrs 45 min
10/9/72

LOS ANGELES/HONOLULU (USA)
Steve Fossett, Pilot — 981.44 kmh / 609.84 mph
Darrin L. Adkins, Copilot
Cessna 750 Citation X
2 R-R AE 3007C, 6,442 lbs
3/23/00

LOS ANGELES/JACKSONVILLE (USA)
William H. Cox, Pilot — 483.12 kmh / 300.20 mph
Martha A. Morrell, Copilot
Mooney TLS
1 Lycoming TIO-540, 270 bhp
3/25/94

LOS ANGELES/KITTY HAWK (USA)
Clay Lacy 1,019.22 kmh
Alex A. Kvassay 633.31 mph
Gulfstream IISP
2 R-R Spey Mk511-8, 11,400 lbs
12/15/03

LOS ANGELES/LONDON (UK)
John D. Ryder 1,002.38 kmh
McDonnell Douglas DC-10-30 622.85 mph
3 GE CF6
8 hrs 44 min 36 sec
4/22/77

LOS ANGELES/MCMURDO STATION (USA)
Brooke Knapp 627.48 kmh
Gulfstream III 389.91 mph
2 RR Spey MK511-8; 11,400 lbs
11/15-16/83

LOS ANGELES/NEW YORK *Group I* (USA)
Calvin B. Early 452.10 kmh
Beechcraft Bonanza M35 280.92 mph
1 Continental IO-550, 300 bhp
3/3/97

LOS ANGELES/NEW YORK *Group III* (USA)
CPT Robert G. Sowers, USAF 1,954.79 kmh
Convair B-58 1,214.65 mph
Hustler, No. 59-2458
4 GE J-79 5B
2 hrs 58.71 sec
3/5/62

LOS ANGELES/PAGO PAGO (USA)
Brooke Knapp 724.68 kmh
Gulfstream III 450.32 mph
2 RR Spey MK511-8, 11,400 lbs
11/15/83

LOS ANGELES/PARIS *Group I* (FRANCE)
C. Billett 444.68 kmh
Douglas DC-6 276.31 mph
4 P&W R-2800
20 hrs 26 min
5/28/53

LOS ANGELES/PARIS *Group III* (USA)
Clay Lacy 943.97 kmh
Brian Kirkdoffer 586.55 mph
Gary Meermans
Hal Fishman
Gulfstream GIISP
2 RR Spey Mk 511-8, 11,400 lbs
6/9/95

LOS ANGELES/PRESTWICK (UK)
Godfrey Bowles 913.85 kmh
McDonnell Douglas DC-10-30 567.84 mph
3 GE CF
6 hrs 1 min 49 sec
3/13/77

LOS ANGELES/ROME (ITALY)
Armando Tarroni 884.68 kmh
McDonnell Douglas DC-10-30 549.71 mph
3 GE CF6-50
11 hrs 31 min 45 sec
3/23/73

LOS ANGELES/STOCKHOLM (USA)
J. J. Armstrong 410.14 kmh
Douglas DC-7C, LN-MOE 254.85 mph
4 Wright
21 hrs 39 min 11.88 sec
11/15/56

LOS ANGELES/SOUTH POLE (USA)
Brooke Knapp 574.92 kmh
Gulfstream III 357.25 mph
2 RR Spey MK511-8, 11,400 lbs
11/15-16/83

LOS ANGELES/TETERBORO (USA)
Ehab Hanna 812.20 kmh
John D. Gibson 504.68 mph
Gulfstream IV
2 R-R Tay 611, 13,850 lbs
7/29/03

LOS ANGELES/TOKYO (USA)
A. G. Heimerdinger 767.04 kmh
Douglas DC-8-61, N80704 476.62 mph
4 P&W JT3D-3B
11 hrs 32 min 19 sec
8/15-16/66

LOS ANGELES/WASHINGTON (USA)
Lt Col Ed Yeilding, Pilot 3,451.78 kmh
Lt Col J. T. Vida, RSO 2,144.83 mph
Lockheed SR-71
2 P&W J-58, 34,000 lbs
03/06/90

LOS ANGELES/WICHITA (USA)
Clay Lacy 987.69 kmh
Alex A. Kvassay 613.72 mph
Gulfstream IISP
2 R-R Spey Mk511-8, 11,400 lbs
12/15/03

LUBBOCK/TEXARKANA (USA)
Danford A. Bookout 346.68 kmh
Piper Lance, N8352C 215.43 mph
1 Lycoming IO-540K, 300 bhp
3/14/88

LUXOR/ATHENS (USA)
JOINT RECORD
William H. Wisner, Pilot 235.44 kmh
Janice Sullivan, Copilot 146.30 mph
Bonanza S35, N6226
and
Daniel Webb, Pilot
Richard Meyerhoff, Copilot
Bonanza A36, N4504S
6/20/89

MADRAS/DUBAI (USA)
Allen E. Paulson, Captain 896.21 kmh
K. C. Edgecomb, Colin B. Allen 556.90 mph
John Salamankas, Jefferson Bailey; crew
Gulfstream IV, N440GA
2 RR TAY MK610-8
6/13-14/87

MADRAS/NAIROBI (UK)
Max Robinson 1,481.35 kmh
BAe/Aérospatiale Concorde 920.47 mph
11/4/99

MADRID/LONDON (MEXICO)
José Alfredo Gastelum Arce 780.00 kmh
Jose Manuel Muradas Rodriguez 484.67 mph
Dassault Falcon 2000
2 GE/Honeywell CFE738, 5,888 lbs
11/11/02

MADRID/NEW YORK (USA)
Robert Fero 598.48 kmh
North American 371.88 mph
T-39 Sabreliner, No. 60-3478
2 P&W J-60P-3
9 hrs 38 min 46 sec
7/17/61

MADRID/OAKLAND (SPAIN)
CPT R. Bay, Jr. — 812.76 kmh, 505.02 mph
McDonnell Douglas DC-10-30
3 GE CF6-50C12, 3,496 kg
6/5/79

MAGADAN/ANADYR (USA)
Allyn Caruso, Pilot — 381.14 kmh, 236.82 mph
David Norgart, Copilot
John Gallichon, Navigator
John Miller, Navigator
Dana Lovell, Mechanic
Fairchild Metroliner III
2 Garrett TPE331-11, 1,100 shp
6/5/91

MAJURO/SAIPAN (USA)
JOINT RECORD
William H. Wisner, Pilot — 304.92 kmh, 189.48 mph
Janice Sullivan, Copilot
Bonanza S35, N6226
and
Daniel Webb, Pilot
Richard Meyerhoff, Copilot
Bonanza A36, N4504S
6/2-3/89

MANCHESTER/LUXOR (USA)
Tim Mellon, Pilot — 399.24 kmh, 248.09 mph
Tom Glista, Copilot
Piper Aerostar 602P, N602PC
2 Lycoming IO-540
4/28/87

MANILA/GUAM (USA)
Harold Curtis — 792.00 kmh, 492.14 mph
Gulfstream III
2 RR Spey 511-8
7 hrs 4 min 12 sec
1/9/82

MANILA/HONG KONG (USA)
JOINT RECORD
William H. Wisner, Pilot — 286.56 kmh, 178.07 mph
Janice Sullivan, Copilot
Bonanza S35, N6226
and
Daniel Webb, Pilot
Richard Meyerhoff, Copilot
Bonanza A36, N4504S
6/8-9/89

MANILA/KOTA KINABALU (USA)
Allen E. Paulson, Captain — 808.06 kmh, 502.10 mph
K. C. Edgecomb, Colin B. Allen
John Salamankas, Jefferson Bailey; crew
Gulfstream IV, N440GA
2 RR TAY MK610-8
6/13/87

MANILA/PARIS (USA)
Allen E. Paulson, Captain — 821.30 kmh, 510.33 mph
K. C. Edgecomb, Colin B. Allen
John Salamankas, Jefferson Bailey; crew
Gulfstream IV, N440GA
2 RR TAY MK610-8
6/13-14/87

MANILA/SINGAPORE (USA)
Stephen M. Johnson — 605.37 kmh, 376.16 mph
Kevin A. Bixler
IAI 1126 Galaxy
2 P&WC PW306A, 5,700 lbs
2/24/02

MANILA/WAKE ISLAND (USA)
Arnold Palmer — 783.00 kmh, 486.56 mph
Learjet 36
2 Garrett TFE-731-2-2B
6 hrs 13 min 30 sec
5/19/76

MARTINIQUE/BASSETERE (USA)
Marie E. McMillan — 243.10 kmh, 151.05 mph
Beech Bonanza F33A
1 Continental IO-520B, 285 bhp
2/28/84

MARTINIQUE/BRIDGETOWN (USA)
Marie E. McMillan — 204.27 kmh, 126.93 mph
Beech Bonanza F33A
1 Continental IO-520B, 285 bhp
2/23/84

MARTINIQUE/CODRINGTON (USA)
Marie E. McMillan — 262.55 kmh, 163.14 mph
Beech Bonanza F33A
1 Continental IO-520B, 285 bhp
2/28/84

MARTINIQUE/GRAND-BOURG (USA)
Marie E. McMillan — 283.57 kmh, 176.20 mph
Beech Bonanza F33A
1 Continental IO-520B, 285 bhp
2/28/84

MARTINIQUE/GUSTAVIA (USA)
Marie E. McMillan — 258.55 kmh, 160.65 mph
Beech Bonanza F33A
1 Continental IO-520B, 285 bhp
2/28/84

MARTINIQUE/HARRIS (USA)
Marie E. McMillan — 269.32 kmh, 167.35 mph
Beech Bonanza F33A
1 Continental IO-520B, 285 bhp
2/28/84

MARTINIQUE/NEW CASTLE (USA)
Marie E. McMillan — 237.81 kmh, 147.77 mph
Beech Bonanza F33A
1 Continental IO-520B, 285 bhp
2/28/84

MARTINIQUE/PHILIPSBURG (USA)
Marie E. McMillan — 255.38 kmh, 158.68 mph
Beech Bonanza F33A
1 Continental IO-520B, 285 bhp
2/28/84

MARTINIQUE/POINTE A PITRE (USA)
Marie E. McMillan — 233.69 kmh, 145.21 mph
Beech Bonanza F33A
1 Continental IO-520B, 285 bhp
2/28/84

MARTINIQUE/ST. JOHNS (USA)
Marie E. McMillan — 247.21 kmh, 153.61 mph
Beech Bonanza F33A
1 Continental IO-520B, 285 bhp
2/28/84

MAYOTTE/MOMBASA (FRANCE)
Claude LePrince, Pilot — 187.74 kmh, 114.63 mph
Henri Corderoy du Tiers
Cessna 310R
7/8/88

MAUI/HOUSTON (USA)
Allen E. Paulson, Captain — 1,077.12 kmh, 669.32 mph
Robert K. Smyth, CoCaptain
John Salamankas, CoCaptain
Jeff Bailey, CoCaptain
Gulfstream IV, N400GA
2 RR TAY MK611-8
2/27-28/88

MAUI/TOKYO (USA)
Geoffrey M. Foster 794.83 kmh
E. Bruce Robinson 493.88 mph
Andrew Milewski
Borden F. Schofield
Bombardier BD 700
2 BMW RR BR-710, 14,690 lbs
5/7/99

MCCOMB, MS/WASHINGTON (USA)
J. S. Hancock, Pilot 435.60 kmh
George Hancock, Copilot 270.68 mph
Piper Aerostar, N713BC
2 Lycoming IO-540
5/16-17/88

MCMURDO STATION/PUNTA ARENAS (USA)
Brooke Knapp 668.88 kmh
Gulfstream III 415.64 mph
2 RR Spey MK511-8, 11,400 lbs
11/16/83

MCMURDO STATION/RECIFE (USA)
Brooke Knapp 682.27 kmh
Gulfstream III 423.96 mph
2 RR Spey MK511-8, 11,400 lbs
11/16/83

MEADOWCREEK/KITTY HAWK (USA)
Dean A. Del Bene, Pilot 329.04 kmh
James A. Leonard, Copilot 204.47 mph
Mooney Super 21, N3364X
1 Lycoming IO-360 A1A
12/16/88

MEXICO CITY/MADRID (USA)
Peter T. Reynolds 938.72 kmh
E. Bruce Robinson 583.29 mph
Christopher D. Hall
Borden F. Schofield
Bombardier BD 700
2 BMW RR BR710, 14,690 lbs
5/2/99

MEXICO CITY/PARIS (MEXICO)
Alfonso Ponce Alcocer 925.49 kmh
Jaime Soto Werschitz 575.07 mph
Jorge O. Castillero Uribe
Boeing 767-200ER
9/24/98

MIAMI/CARACAS (USA)
Jerry W. Blessing 765.84 kmh
Scott S. Evans 475.87 mph
IAI 1125 Astra SPX
2 AlliedSignal TFE731, 4,250 lbs
9/1/01

MIAMI/LONDON (UK)
Brian Walpole 1,179.44 kmh
BAe/Aérospatiale Concorde 732.90 mph
4 RR Olympus, 38,000 lbs
9/17/86

MIAMI/NEW ORLEANS (USA)
Jerry W. Blessing 832.12 kmh
Scott S. Evans 517.05 mph
IAI 1125 Astra SPX
2 AlliedSignal TFE731, 4,250 lbs
8/31/01

MIAMI/SEATTLE (USA)
Robert W. Agostino 752.92 kmh
James P. Dwyer 467.84 mph
Eugene A. Cernan
Bombardier Challenger 300
2 Honeywell AS907, 6,700 lbs
6/10/03

MIDWAY/KOTA KINABALU (USA)
Allen E. Paulson, Captain 807.15 kmh
K. C. Edgecomb, 501.54 mph
John Salamankas, Jefferson Bailey; Pilots
Colin B. Allen, Flight Engineer
Gulfstream IV, N440GA
2 RR TAY MK610-8
6/13/87

MIDWAY/MANILA (USA)
Allen E. Paulson, Captain 807.00 kmh
K.C. Edgecomb, 501.47 mph
John Salamankas, Jefferson Bailey; Pilots
Colin B. Allen, Flight Engineer
Gulfstream IV, N440GA
2 RR TAY MK610-8
6/13/87

MILLINOCKET, MA/GOOSE BAY (USA)
Anne B. Baddour 389.16 kmh
Margrit Budert-Waltz 241.82 mph
Beech Baron, N3707N
2 Continental, 325 bhp
12/20/88

MILLINOCKET, MA/NARSSARSSUAQ (USA)
Anne B. Baddour 270.00 kmh
Margrit Budert-Waltz 167.78 mph
Beech Baron, N3707N
2 Continental, 325 bhp
12/21/88

MINITATLIN/COZUMEL (USA)
Marie E. McMillan 177.95 kmh
Beech Bonanza F33A 110.58 mph
1 Continental IO-520 BA, 285 bhp
2/4/84

MINNEAPOLIS/SAN ANTONIO (USA)
Allyn Caruso, Pilot 413.44 kmh
David Norgart, Copilot 256.90 mph
John Gallichon, Navigator
John Miller, Navigator
Dana Lovell, Mechanic
Fairchild Metroliner III
2 Garrett TPE331-11, 1,100 shp
6/6/91

MOJAVE/GANDER (SAUDI ARABIA)
Aziz Ojjeh 816.48 kmh
Canadair Challenger 601 507.36 mph
2 GE CF-34, 8,650 lbs
7/23-24/84

MOJAVE/RENO (USA)
Lyle T. Shelton 618.85 kmh
Grumman F8F Bearcat 384.54 mph
1 Wright R3350, 3,250 bhp
9/8/91

MONROEVILLE/GRAND FORKS (USA)
Capt. R. F. Lewandowski, Commander 956.42 kmh
Capt. Kevin T. Kalen, Pilot 594.29 mph
Lt. Col. Timothy W. Van Splunder, OSO
Capt. Jerrold A. Wangberg, DSO
Rockwell International B-1B
4 GE F101, 12,000 lbs
4/8/94

MONROEVILLE/MULLAN (USA)
Capt. Michael S. Menser, Commander 920.72 kmh
Capt. Brian P. Gallagher, Pilot 572.11 mph
Capt. Robert P. Boman, OSO
Capt. Matthew E. Grant, WSO
Rockwell International B-1B
4 GE F101, 12,000 lbs
4/7/94

MONTREAL/KEFLAVIK (USA)
Rodney R. Lundy 765.70 kmh
J. Shawn Christian 475.78 mph
Learjet 45
2 AlliedSignal TFE731, 3,500 lbs
6/14/01

MONTREAL/PARIS (USA)
Rodney R. Lundy 692.91 kmh
J. Shawn Christian 430.55 mph
Learjet 45
2 AlliedSignal TFE731, 3,500 lbs
6/14/01

MULLAN/MONROEVILLE (USA)
Capt. Michael S. Menser, Commander 993.93 kmh
Capt. Brian P. Gallagher, Pilot 617.59 mph
Capt. Robert P. Boman, OSO
Capt. Matthew E. Grant, WSO
Rockwell International B-1B
4 GE F101, 12,000 lbs
4/7/94

MONTEGO BAY/PORT AU PRINCE (USA)
Marie E. McMillan 208.19 kmh
Beech Bonanza F33A 129.37 mph
1 Continental IO-520 BA, 285 bhp
2/14/84

MONTEGO BAY/SANTO DOMINGO (USA)
Marie E. McMillan 228.24 kmh
Beech Bonanza F33A 141.83 mph
1 Continental IO-520 BA, 285 bhp
2/14/84

MONTERREY/TUCSON (USA)
Marie E. McMillan 185.85 kmh
Beech Bonanza F33A 115.54 mph
1 Continental IO-520B, 285 bhp
3/7/84

MONT JOLI, CANADA/GOOSE BAY (USA)
Anne B. Baddour 469.44 kmh
Margrit Budert-Waltz 291.71 mph
Beech Baron, N3707N
2 Continental, 325 bhp
12/20/88

MOSCOW/BALTIMORE (USA)
COL James B. Swindal, USAF 792.28 kmh
Boeing VC-137 No. 26000 492.30 mph
4 P&W JT3D-3
9 hrs 47 min 53.2 sec
5/20-21/63

MOSCOW/BEIJING (USA)
Andrew Milewski 857.62 kmh
E. Bruce Robinson 532.90 mph
Bombardier BD-700 Global Express
2 BMW-RR BR710, 14,750 lbs
9/22/99

MOSCOW/BOSTON (USA)
COL James B. Swindal, USAF 800.21 kmh
Boeing VC-137 No. 26000 497.21 mph
4 P&W JT3D-3
9 hrs 1 min 7.8 sec
5/20-21/63

MOSCOW/HONOLULU (USA)
Brooke Knapp 831.17 kmh
Gulfstream III 516.49 mph
2 RR Spey MK511-8, 11,400 lbs
2/14-15/84

MOSCOW/IRKUTSK *Group I* (USA)
Susan Nealey 343.39 kmh
Faith Hillman 213.37 mph
Cessna 310
2 Continental IO-520, 285 bhp
6/23/92

MOSCOW/IRKUTSK *Group II* (USA)
Allyn Caruso, Pilot 338.35 kmh
David Norgart, Copilot 210.24 mph
John Gallichon, Navigator
John Miller, Navigator
Dana Lovell, Mechanic
Fairchild Metroliner III
2 Garrett TPE331-11, 1,100 shp
6/3/91

MOSCOW/LOS ANGELES (USA)
Larry S. Mueller 812.77 kmh
Henry K. Gibson 505.03 mph
Robert S. McKenney
Gulfstream V
2 BMW RR BR710-48, 14,750 lbs
12/11/97

MOSCOW/NEW YORK (USA)
COL James B. Swindal, USAF 797.14 kmh
Boeing VC-137 No. 26000 495.32 mph
4 P7W JT3D-3
9 hrs 24 min 48 sec
5/20-21/63

MOSCOW/NOME (USA)
Susan Nealey 51.17 kmh
Faith Hillman 31.79 mph
Cessna 310
2 Continental IO-520, 285 bhp
6/23-28/92

MOSCOW/NOVOSIBIRSK (USA)
Brooke Knapp 769.05 kmh
Gulfstream III 477.89 mph
2 RR Spey MK511-8, 11,400 lbs
2/14/84

MOSCOW/OMSK*
Susan Nealey 236.34 mph
Faith Hillman
Cessna 310
2 Continental IO-520, 285 bhp
6/23/92

MOSCOW/PHILADELPHIA (USA)
COL James B. Swindal, USAF 795.22 kmh
Boeing VC-137 No. 26000 494.13 mph
4 P&W JT3D-3
9 hrs 35 min 54.9 sec
5/20-21/63

MOSCOW/SVERDLOVSK (USA)
Allyn Caruso, Pilot 414.52 kmh
David Norgart, Copilot 257.57 mph
John Gallichon, Navigator
John Miller, Navigator
Dana Lovell, Mechanic
Fairchild Metroliner III
2 Garrett TPE331-11, 1,100 shp
6/3/91

MOSCOW/TOKYO (USA)
Brooke Knapp 766.48 kmh
Gulfstream III 476.30 mph
2 RR Spey MK511-8, 11,400 lbs
2/14/84

MOSCOW/WASHINGTON (USA)
COL James B. Swindal, USAF 788.67 kmh
Boeing VC-137 No. 26000 490.06 mph
4 P&W JT3D-3
9 hrs 54 min 48.5 sec
5/20-21/63

MOUNT MITCHELL/KITTY HAWK (USA)
Lawrence G. Manofsky 312.38 kmh
Lake Renegade 250 194.10 mph
1 Lycoming IO-540, 250 bhp
12/20/96

MOUNT PLEASANT/ASCENSION (UK)
W/Cdr K. D. Filbey 863.14 kmh
Lockheed Tristar 536.34 mph
3 RB 211, 50,000 lbs
5/14/85

MOUNT PLEASANT/BRIZE NORTON (UK)
W/Cdr K. D. Filbey 732.16 kmh
Lockheed Tristar 454.96 mph
3 RB 211, 50,000 lbs
5/14-15/85

MUSCAT/BOMBAY (SAUDI ARABIA)
Aziz Ojjeh 739.00 kmh
Canadair Challenger 601 459.22 mph
2 GE CF-34, 8,650 lbs
7/22/84

MUSCAT/RANGOON (USA)
Douglas G. Matthews 633.24 kmh
Learjet 36 393.49 mph
2 Garrett TFE-731-2-2B, 3,500 lbs
7/19/84

MUSCAT/SINGAPORE (SAUDI ARABIA)
Aziz Ojjeh 677.88 kmh
Canadair Challenger 601 421.24 mph
2 GE CF-34, 8,650 lbs
7/22/84

NADI/SYDNEY (UK)
Max Robinson 1,649.09 kmh
BAe/Aérospatiale Concorde 1,024.69 mph
10/27/99

NAGOYA/HONG KONG (USA)
Stephen M. Johnson 591.05 kmh
Kevin A. Bixler 367.26 mph
IAI 1126 Galaxy
2 P&WC PW306A, 5,700 lbs
2/20/02

NAIROBI/AQABA (UK)
Geoffrey Mussett 1,030.58 kmh
BAe/Aérospatiale Concorde 640.37 mph
11/9/99

NAIROBI/BOMBAY (USA)
William Arnott, Clay Lacy 861.01 kmh
United Airlines 535.01 mph
McDonnell Douglas DC-8
4 P&W
5 hrs 16 min 55.1 sec
6/30/73

NAIROBI/MUSCAT (USA)
Douglas G. Matthews 703.08 kmh
Learjet 36 436.89 mph
2 Garrett TFE-731-2-2B, 3,500 lbs
7/19/84

NANDI/TAHITI (USA)
Douglas G. Matthews 847.80 kmh
Learjet 36 526.82 mph
2 Garrett TFE-731-2-2B, 3,500 lbs
7/24/84

NARSSARSSUAQ/KEFLAVIK (USA)
Anne B. Baddour 338.40 kmh
Margrit Budert-Waltz 210.28 mph
Beech Baron, N3707N
2 Continental, 325 bhp
12/21/88

NARSSARSSUAQ/REYKJAVIK (USA)
Anne B. Baddour 335.52 kmh
Margrit Budert-Waltz 208.49 mph
Beech Baron, N3707N
2 Continental, 325 bhp
12/21/88

NASHVILLE/TOKYO (USA)
Gregory S. Sheldon 778.69 kmh
Ahmed M. Ragheb 483.85 mph
John A. Mullican
Gulfstream V
2 BMW R-R BR710, 14,750 lbs
11/6/02

NEW CASTLE/HARRIS (USA)
Marie E. McMillan 182.93 kmh
Beech Bonanza F33A 113.67 mph
1 Continental IO-520B, 285 bhp
2/21/84

NEW CASTLE/PHILIPSBURG (USA)
Marie E. McMillan 328.34 kmh
Beech Bonanza F33A 204.02 mph
1 Continental IO-520B, 285 bhp
2/28/84

NEW CASTLE/POINTE A PITRE (USA)
Marie E. McMillan 312.83 kmh
Beech Bonanza F33A 194.38 mph
1 Continental IO-520B, 285 bhp
2/21/84

NEW DELHI/BAHRAIN (USA)
JOINT RECORD
William H. Wisner, Pilot 231.84 kmh
Janice Sullivan, Copilot 144.07 mph
Bonanza S35, N6226
and
Daniel Webb, Pilot
Richard Meyerhoff, Copilot
Bonanza A36, N4504S
6/16/89

NEW ORLEANS/BONN (USA)
Jacqueline Cochran 603.69 kmh
Lockheed Jetstar 375.12 mph
4 P&W JT12A-6
13 hrs 10 min 31 sec
4/22/62

NEW ORLEANS/BOSTON (USA)
Jacqueline Cochran 752.01 kmh
Lockheed Jetstar 467.28 mph
4 P&W JT12A-6
3 hrs 10 min 31 sec
4/22/62

NEW ORLEANS/GANDER (USA)
Jacqueline Cochran 776.34 kmh
Lockheed Jetstar 482.40 mph
4 P&W JT12A-6
4 hrs 42 min 52.9 sec
4/22/62

NEW ORLEANS/LONDON (USA)
Jacqueline Cochran 603.69 kmh
Lockheed Jetstar 375.12 mph
4 P&W JT12A-6
12 hrs 20 min 14.9 sec
4/22/62

NEW ORLEANS/MIAMI (USA)
Jerry W. Blessing 835.74 kmh
Scott S. Evans 519.30 mph
IAI 1125 Astra SPX
2 AlliedSignal TFE731, 4,250 lbs
8/31/01

NEW ORLEANS/NEW YORK (USA)
Jacqueline Cochran 748.53 kmh
Lockheed Jetstar 465.12 mph
4 P&W JT21A-6
2 hrs 31 min 8.5 sec
4/22/62

NEW ORLEANS/PARIS (USA)
Jacqueline Cochran — 607.75 kmh
Lockheed Jetstar — 377.64 mph
4 P&W JT12A-6
12 hrs 42 min 3.9 sec
4/22/62

NEW ORLEANS/SEATTLE (USA)
Jerry W. Blessing — 794.16 kmh
Scott S. Evans — 493.47 mph
IAI 1125 Astra SPX
2 AlliedSignal TFE731, 4,250 lbs
9/2/01

NEW ORLEANS/SHANNON (USA)
Jacqueline Cochran — 614.12 kmh
Lockheed Jetstar — 381.60 mph
4 P&W JT12A-6
12 hrs 8 min 8.7 sec
4/22/62

NEW ORLEANS/TEXARKANA (USA)
Danford A. Bookout — 272.52 kmh
Piper Lance, N8352C — 169.34 mph
1 Lycoming IO-540K, 300 bhp
12/18/87

NEW ORLEANS/WASHINGTON (USA)
Jacqueline Cochran — 746.22 kmh
Lockheed Jetstar — 463.68 mph
4 P&W J512A-6
2 hrs 5 min 2.4 sec
4/22/62

NEW YORK/BERLIN (GERMANY)
A. Henke — 320.91 kmh
Fock-Wulf FW-200 C — 199.40 mph
4 BMW-132 L
19 hrs 55 min 1 sec
8/13/38

NEW YORK/BIRMINGHAM (UK)
CPT J.W. Burton — 1,697.15 kmh
BAe/Aérospatiale Concorde — 1,054.61 mph
4 RR Olympus, 38,000 lbs
9/9/85

NEW YORK/BONN (USA)
Jacqueline Cochran — 570.09 kmh
Lockheed Jetstar — 354.24 mph
4 P&W JT12A-6
10 hrs 39 min 12.5 sec
4/22/62

NEW YORK/BOSTON (USA)
Harold Curtis — 828.00 kmh
Gulfstream III — 514.52 mph
2 RR Spey 511-8
1/8/82

NEW YORK/CARACAS (USA)
Owen G. Zahnle — 793.08 kmh
William A. Landrum — 492.79 mph
Learjet 45
2 AlliedSignal TFE731-20, 3,500 lbs
6/5/99

NEW YORK/DUBAI (USA)
Timothy P. Freise — 853.36 kmh
Christian M. Kennedy — 530.25 mph
Henry K. Gibson
Gary M. Freeman
Gulfstream V
2 BMW-RR BR710-48, 14,750 lbs
11/13/99

NEW YORK/EDINBURGH (UK)
John Cook — 1,729.38 kmh
BAe/Aérospatiale Concorde — 1,074.64 mph
4 Olympus 593, 38,000 lbs
7/13/87

NEW YORK/GANDER (USA)
George P. Eremea — 840.01 kmh
N American Sabreliner — 521.96 mph
2 P&W JT12A-61
2 hrs 6 min 49.7 sec
10/25/63

NEW YORK/GOOSE BAY (USA)
Allyn Caruso, Pilot — 405.37 kmh
David Norgart, Copilot — 251.88 mph
John Gallichon, Navigator
John Miller, Navigator
Dana Lovell, Mechanic
Fairchild Metroliner III
2 Garrett TPE331-11, 1,100 shp
5/30/91

NEW YORK/HONG KONG (UK)
Paul Horsting — 830.96 kmh
Boeing 747-400 — 516.33 mph
7/5/98

NEW YORK/KARACHI (PAKISTAN)
CPT Mian Abdul Aziz — 895.38 kmh
McDonnell Douglas DC-10-30, — 556.37 mph
AP-AX
3 GE CF6-50C
10/20-21/74

NEW YORK/LONDON (USA)
MAJ James V Sullivan, USAF — 2,908.03 kmh
Lockheed SR-71 Blackbird — 1,806.96 mph
2 P&W JT11D-20
9/1/74

NEW YORK/LOS ANGELES (USA)
CPT Robert G. Sowers, USAF — 1,741.00 kmh
Convair B-58 Hustler — 1,081.80 mph
4 GE J70-5B
2 hrs 15 min 50.08 sec
3/5/62

NEW YORK/MOSCOW (USA)
COL James B. Swindal, USAF — 907.86 kmh
Boeing VC-137 — 564.12 mph
4 P&W JT3D-3
8 hrs 15 min 54.1 sec
5/19/63

NEW YORK/OSLO (USA)
COL James B. Swindal, USAF — 954.56 kmh
Boeing VC-137 — 593.13 mph
4 P&W JT3D-3
6 hrs 11 min 58.8 sec
5/19/63

NEW YORK/PARIS (USA)
MAJ W. R. Payne — 1,753.06 kmh
Convair B-58 Hustler — 1,089.36 mph
4 GE J79-5
3 hrs 19 min 44.53 sec
5/26/61

NEW YORK/SHANNON (USA)
Jacqueline Cochran — 575.30 kmh
Lockheed Jetstar — 357.47 mph
4 P&W JT12A-6
8 hrs 36 min 57.5 sec
4/22/62

NEW YORK/STOCKHOLM (USA)
COL James B. Swindal, USAF — 946.79 kmh
Boeing VC-137 — 588.31 mph
4 P&W JT3D-3
6 hrs 40 min 36 sec
5/19/63

NEW YORK/VANCOUVER (UK)
Roger Mills — 1,004.67 kmh
BAe/Aérospatiale Concorde — 624.27 mph
10/23/99

NEW YORK/ZURICH (USA)
Brooke Knapp 888.12 kmh
Gulfstream III 551.88 mph
2 RR Spey MK-511, 11,400 lbs
7/2/84

NICE/ABU DHABI (SAUDI ARABIA)
Aziz Ojjeh 618.56 kmh
Canadair Challenger 601 446.51 mph
2 GE CF-34, 8,650 lbs
7/22/84

NICE/DHARAN (SAUDI ARABIA)
Aziz Ojjeh 714.24 kmh
Canadair Challenger 601 443.83 mph
2 GE CF-34, 8,650 lbs
7/22/84

NICE/MUSCAT (SAUDI ARABIA)
Aziz Ojjeh 710.28 kmh
Canadair Challenger 601 441.37 mph
2 GE CF-34, 8,650 lbs
7/22/84

NICE/SANTA MARIA (USA)
JOINT RECORD
William H. Wisner, Pilot 255.60 kmh
Janice Sullivan, Copilot 158.83 mph
Bonanza S35, N6226
and
Daniel Webb, Pilot
Richard Meyerhoff, Copilot
Bonanza A36, N4504S
6/24/89

NOME/MIDWAY (USA)
Allen E. Paulson, Captain 830.00 kmh
K. C. Edgecomb, Colin B. Allen 515.73 mph
John Salamankas, Jefferson Bailey; crew
Gulfstream IV, N440GA
2 RR TAY MK610-8
6/13/87

NOME/VICTORIA (USA)
Susan Nealey 318.12 kmh
Faith Hillman 197.67 mph
Cessna 310
2 Continental IO-520, 285 bhp
6/30/92

NORTH POLE/HAMBURG (SWITZERLAND)
Felix Estermann 241.11 kmh
Mooney 201, N5789B 149.83 mph
1 Lycoming IO-360, 200 bhp
8/29-30/87

NORTH POLE/LOS ANGELES (USA)
Brooke Knapp 723.24 kmh
Gulfstream III 449.42 mph
2 RR Spey MK511-8, 11,400 lbs
11/18/83

NORTH POLE/OSLO (USA)
Howard Mayes, Donald Preston 879.49 kmh
United Airlines 546.49 mph
McDonnell Douglas DC-8
4 P&W
3 hrs 48 min 27.7 sec
6/25/73

NORTH POLE/SOUTH POLE *Group I* (USA)
Elgen M. Long 59.71 kmh
Piper Navajo, N9097Y 37.10 mph
11/8-22/71

NORTH POLE/SOUTH POLE *Group III* (USA)
Capt. W. Mullikin, 732.77 kmh
Capt. A. A. Frink, 458.68 mph
S. Beckett, F. Cassaniti, E. Shields
Pan American World Airways
Boeing 747 SP, N533PA
4 P&W JT9D-7A
10/29-30/77

NOUMEA/MELBOURNE (AUSTRALIA)
Ian Haig 677.08 kmh
Australian Airlines 420.72 mph
Boeing 737-400
7/24/92

NOVOSIBIRSK/BEIJING (USA)
Brooke Knapp 844.98 kmh
Gulfstream III 525.07 mph
2 RR Spey MK511-8, 11,400 lbs
2/14/84

NOVOSIBIRSK/HONOLULU (USA)
Brooke Knapp 880.24 kmh
Gulfstream III 546.98 mph
2 RR Spey MK511-8, 11,400 lbs
2/14-15/84

NOVOSIBIRSK/IRKUTSK (USA)
Allyn Caruso, Pilot 370.82 kmh
David Norgart, Copilot 230.41 mph
John Gallichon, Navigator
John Miller, Navigator
Dana Lovell, Mechanic
Fairchild Metroliner III
2 Garrett TPE331-11, 1,100 shp
6/3/91

NOVOSIBIRSK/TOKYO (USA)
Brooke Knapp 815.11 kmh
Gulfstream III 506.50 mph
2 RR Spey MK511, 11,500 lbs
2/14/84

OKLAHOMA CITY/WASHINGTON (USA)
Donald Wyvell 466.20 kmh
Mooney 252, N252KM 289.70 mph
1 Continental SE TS 360MB, 210 bhp
1/14/88

OMSK/IRKUTSK*
Susan Nealey 216.67 mph
Faith Hillman
Cessna 310
2 Continental IO-520, 285 bhp
6/23/92

ORANJESTAD/CHARLOTTE AMALIE (USA)
Marie E. McMillan 238.39 kmh
Beech Bonanza F33A 148.13 mph
1 Continental IO-520B, 285 bhp
2/20/84

ORANJESTAD/CHRISTIANSTED (USA)
Marie E. McMillan 277.88 kmh
Beech Bonanza F33A 172.67 mph
1 Continental IO-520B, 285 bhp
2/20/84

ORANJESTAD/GUSTAVIA (USA)
Marie E. McMillan 136.69 kmh
Beech Bonanza F33A 84.93 mph
1 Continental IO-520B, 285 bhp
2/20/84

ORANJESTAD/HARRIS (USA)
Marie E. McMillan 181.46 kmh
Beech Bonanza F33A 112.76 mph
1 Continental IO-520B, 285 bhp
2/21/84

ORANJESTAD/POINTE A PITRE (USA)
Marie E. McMillan — 262.63 kmh
Beech Bonanza F33A — 163.19 mph
1 Continental IO-520B, 285 bhp
2/21/84

ORANJESTAD/ST. JOHNS (USA)
Marie E. McMillan — 267.49 kmh
Beech Bonanza F33A — 166.21 mph
1 Continental IO-520B, 285 bhp
2/21/84

OSHKOSH/TEXARKANA (USA)
Danford A. Bookout — 335.52 kmh
Piper Lance, N8352C — 208.49 mph
1 Lycoming IO-540K, 300 bhp
8/3/87

OSHKOSH/PONTIAC (USA)
Kenneth P. Rittenhouse, Jr. — 988.50 kmh
David L. Arnold — 614.22 mph
William Lee Walker, Jr.
Dassault DA-10
2 Allied Signal TFE-731, 3,230 lbs
8/8/96

OSLO/BALTIMORE (USA)
COL James B. Swindal, USAF — 807.71 kmh
Boeing VC-137 — 501.88 mph
4 P&W JT3D-3
7 hrs 39 min 20.9 sec
5/20-21/63

OSLO/BOSTON (USA)
COL James B. Swindal, USAF — 818.01 kmh
Boeing VC-137 — 508.28 mph
4 P&W JT3D-3
6 hrs 52 min 34.9 sec
5/20-21/63

OSLO/NEW YORK (USA)
COL James B. Swindal, USAF — 813.71 kmh
Boeing VC-137 — 505.62 mph
4 P&W JT3D-3
7 hrs 16 min 21 sec
5/20-21/63

OSLO/PHILADELPHIA (USA)
COL James B. Swindal, USAF — 811.40 kmh
Boeing VC-137 — 504.18 mph
4 P&W JT3D-3
7 hrs 27 min 19 sec
5/20-21/63

OSLO/WASHINGTON (USA)
COL James B. Swindal, USAF — 802.75 kmh
Boeing VC-137 — 498.81 mph
4 P&W JT3D-3
7 hrs 46 min 18.7 sec
5/20-21/63

OTTAWA/ABBOTSFORD (CANADA)
Michael H. Leslie — 73.17 kmh
Joseph Leslie — 45.47 mph
Daniel J. Leslie
Cessna 172
1 Lycoming O-320, 160 bhp
8/10/00

OTTAWA/IQALUIT (CANADA)
Terrance L. Jantzi — 190.13 kmh
Lauren V. Jantzi — 118.14 mph
Van's RV-6
6/4/99

OTTAWA/KUUJJUAQ (CANADA)
Terrance L. Jantzi — 198.58 kmh
Lauren V. Jantzi — 123.39 mph
Van's RV-6
6/4/99

OTTAWA/LONDON (UK)
I. G. Broom — 799.56 kmh
English Electric — 496.82 mph
Canberra "Aeries IV"
2 RR Avon Mark I
6 hrs 42 min 12 sec
6/27/55

OTTAWA/VICTORIA (CANADA)
Graham A. Palmer — 43.67 kmh
Cessna 140 — 27.13 mph
1 Continental C-85, 85 bhp
9/21/95

OTTAWA/WABUSH (CANADA)
Terrance L. Jantzi — 253.89 kmh
Lauren V. Jantzi — 157.76 mph
Van's RV-6
6/4/99

OWEN SOUND/QUEBEC (CANADA)
C. Nelson Gain — 146.39 kmh
Ultravia Aero Pelican GS — 90.96 mph
1 Rotax 912, 80 bhp
6/26/96

PAGO PAGO/CHRISTCHURCH (USA)
Brooke Knapp — 749.88 kmh
Gulfstream III — 465.97 mph
2 RR Spey MK511-8, 11,400 lbs
1/15/83

PALERMO/TUNIS (USA)
James P. Hanson — 213.71 kmh
Cessna 208 Caravan — 132.79 mph
1 P&W PT-6A, 600 shp
5/4/96

PALMA DE MALLORCA/GOOSE BAY (USA)
Ehab Hanna — 746.09 kmh
Lynn Upham — 463.60 mph
Gulfstream IV
2 R-R Tay 611, 13,850 lbs
7/7/03

PALMA DE MALLORCA/LOS ANGELES (USA)
Ehab Hanna — 715.30 kmh
Lynn Upham — 444.47 mph
Gulfstream IV
2 R-R Tay 611, 13,850 lbs
7/8/03

PALM BEACH/TEL AVIV (USA)
Christian M. Kennedy — 949.79 kmh
James R. Cannon — 590.17 mph
Henry K. Gibson
Bruce A. Yates
Gulfstream V
2 BMW Rolls-Royce BR 710-48, 14,750 lbs
1/11/98

PALM ISLAND/CANOUAN (USA)
Marie E. McMillan — 368.02 kmh
Beech Bonanza F33A — 228.68 mph
1 Continental IO-520B, 285 bhp
2/27/84

PALM ISLAND/LOVELL'S VILLAGE (USA)
Marie E. McMillan — 335.09 kmh
Beech Bonanza F33A — 208.21 mph
1 Continental IO-520B, 285 bhp
2/27/84

PALM ISLAND/PORT ELIZABETH (USA)
Marie E. McMillan — 269.75 kmh
Beech Bonanza F33A — 167.61 mph
1 Continental IO-520B, 285 bhp
2/27/84

PARIS/ANTANANARIVO (FRANCE)
Genin
Caudron Simoun
1 Renault, 180 bhp
57 hrs 35 min 21 sec
12/18/35
151.90 kmh
94.39 mph

PARIS/BONN (USA)
Jacqueline Cochran
Lockheed Jetstar
4 P&W JT12A-6 Turbo
26 min 5.5 sec
4/22/62
926.98 kmh
576.0 mph

PARIS/DUBAI (USA)
Allen E. Paulson, Captain
K. C. Edgecomb, Colin B. Allen
John Salamankas, Jefferson Bailey; Pilots
Colin B. Allen, Flight Engineer
Gulfstream IV, N440GA
2 RR TAY MK610-8
6/12-14/87
794.12 kmh
493.44 mph

PARIS/EDMONTON (USA)
Allen E. Paulson, Captain
K. C. Edgecomb,
John Salamankas, Jefferson Bailey; Pilots
Colin B. Allen, Flight Engineer
Gulfstream IV, N440GA
2 RR TAY MK610-8
6/12/87
836.00 kmh
519.46 mph

PARIS/FRANKFURT (USA)
George P. Eremea
NA 282-40 Sabreliner
2 P&W JT12A-6A
10/27/63
804.61 kmh
499.96 mph

PARIS/GENEVA (USA)
Harold Curtis
Gulfstream III
2 RR Spey 511-8
38 min 28 sec
1/8/82
612.00 kmh
380.30 mph

PARIS/GRENOBLE (FRANCE)
Jean-Yves Dupont
Cessna Cardinal
1 Lycoming IO-360 AIB6, 203 bhp
10/7/78
215.67 kmh
134.01 mph

PARIS/HANOI (FRANCE)
A. Japy
Caudron Simoun
Renault 60-01N6-71
11/15/36
180.20 kmh
111.97 mph

PARIS/LISBON (FRANCE)
Claude Delorme, Captain
Jean Boye, Co-Captain
BAe/Aérospatiale Concorde
10/11/92
1,172.29 kmh
728.43 mph

PARIS/LONDON (UK)
Michael J. Lithgow
Vickers Armstrong
Supermarine Swift MK IV, WK 198
1 RR Avon RA-7
7/5/63
1,069.29 kmh
664.42 mph

PARIS/MANILA (USA)
Allen E. Paulson, Captain
K. C. Edgecomb, Colin B. Allen
John Salamankas, Jefferson Bailey; Pilots
Colin B. Allen, Flight Engineer
Gulfstream IV, N440GA
2 RR TAY MK610-8
6/12-13/87
804.14 kmh
499.66 mph

PARIS/MIDWAY (USA)
Allen E. Paulson, Captain
K. C. Edgecomb,
John Salamankas, Jefferson Bailey; Pilots
Colin B. Allen, Flight Engineer
Gulfstream IV, N440GA
2 RR TAY MK610-8
6/12-13/87
819.68 kmh
509.35 mph

PARIS/NICE (FRANCE)
B. Muselli
Mystere IV N "Avon"
41 min 55.8 sec
6/18/55
982.43 kmh
610.45 mph

PARIS/NOME (USA)
Allen E. Paulson, Captain
K. C. Edgecomb,
John Salamankas, Jefferson Bailey; Pilots
Colin B. Allen, Flight Engineer
Gulfstream IV, N440GA
2 RR TAY MK610-8
6/12-13/87
820.41 kmh
509.77 mph

PARIS/REYKJAVIK (USA) *Group I*
Charles C. Mack
Beech Bonanza
1 Continental IO-520, 285 bhp
7/3/89
264.25 kmh
164.20 mph

PARIS/REYKJAVIK (USA) *Group III*
Douglas Matthews, Pilot
Doyle Sanderlin, Copilot
Learjet 35A
2 Garrett TFE 731-2-2B, 3,500 lbs
6/4/85
749.52 kmh
465.75 mph

PARIS/SAIGON (FRANCE)
Maryse Hilsz
Caudron Simoun C-635
Renault
92 hrs 36 min
12/23/37
109.30 kmh
67.92 mph

PARIS/SAVANNAH (USA)
Dwight V. Whitaker, III
Edward D. Mendenhall
Gulfstream V
2 BMW RR BR710-48, 14,750 lbs
6/22/97
818.97 kmh
508.88 mph

PARIS/TEHERAN (USA)
Arnold Palmer
Gates Learjet 36
Garrett TFE-731-2-2B
5 hrs 38 min 15 sec
5/18/76
745.27 kmh
465.22 mph

PERRYTON, TX/COFFEYVILLE, KS (USA)
William C. Wright, Pilot
Michael LaFrance, Copilot
1948 Luscombe Model 11A, NC1666B
7/2/89
230.40 kmh
143.17 mph

PERTH/BRISBANE (USA)
Steve Fossett, Pilot
Alexander M. Tai, Copilot
Cessna 750 Citation X
2 R-R AE 3007C, 6,442 lbs
7/28/01
1,134.69 kmh
705.06 mph

PERTH/HOBART (USA)
Steve Fossett, Pilot
Alexander M. Tai, Copilot
Cessna 750 Citation X
2 R-R AE 3007C, 6,442 lbs
7/30/01
1,194.17 kmh
742.02 mph

PETITE MARTINIQUE/CASTRIES (USA)
Marie E. McMillan 305.25 kmh
Beech Bonanza F33A 189.67 mph
1 Continental IO-520B, 285 bhp
2/27/84

PETITE MARTINIQUE/CLIFTON (USA)
Marie E. McMillan 321.60 kmh
Beech Bonanza F33A 199.83 mph
1 Continental IO-520B, 285 bhp
2/27/84

PETITE MARTINIQUE/KINGSTOWN (USA)
Marie E. McMillan 291.22 kmh
Beech Bonanza F33A 180.95 mph
1 Continental IO-520B, 285 bhp
2/27/84

PETITE MARTINIQUE/LOVELL'S VILLAGE (USA)
Marie E. McMillan 326.44 kmh
Beech Bonanza F33A 202.84 mph
1 Continental IO-520B, 285 bhp
2/27/84

PETITE MARTINIQUE/PORT ELIZABETH (USA)
Marie E. McMillan 271.72 kmh
Beech Bonanza F33A 168.84 mph
1 Continental IO-520B, 285 bhp
2/27/84

PHILADLEPHIA/BERLIN (USA)
Timothy Cwik, Pilot 442.80 kmh
Eleanor Carlisle, Copilot 275.15 mph
William Ellsworth, Flt Engineer
Aerospatiale ATR-42
2 P&W 120, 2,000 shp
4/21-22/89

PHILADLEPHIA/GANDER (USA)
Timothy Cwik, Pilot 532.44 kmh
Eleanor Carlisle, Copilot 330.85 mph
William Ellsworth, Flt Engineer
Aerospatiale ATR-42
2 P&W 120, 2,000 shp
4/21-22/89

PHILADELPHIA/MOSCOW (USA)
Col James B. Swindal, USAF 907.62 kmh
Boeing VC-137 563.97 mph
4 P&W JT3D-3
8 hrs 24 min 36.2 sec
5/19/63

PHILADELPHIA/OSLO (USA)
Col James B. Swindal, USAF 953.80 kmh
Boeing VC-137 592.66 mph
4 P&W JT3D-3
6 hrs 20 min 31 sec
5/19/63

PHILADELPHIA/STOCKHOLM (USA)
Col James B. Swindal, USAF 946.10 kmh
Boeing VC-137 587.88 mph
4 P&W JT3D-3
6 hrs 49 min 11.6 sec
5/19/63

PHILIPSBURG/BRIDGETOWN (USA)
Marie E. McMillan 232.75 kmh
Beech Bonanza F33A 144.63 mph
1 Continental IO-520B, 285 bhp
2/23/84

PHILIPSBURG/CHARLOTTE AMALIE (USA)
Marie E. McMillan 392.44 kmh
Beech Bonanza F33A 243.85 mph
1 Continental IO-520B, 285 bhp
3/1/84

PHILIPSBURG/GUSTAVIA (USA)
Marie E. McMillan 55.70 kmh
Beech Bonanza F33A 34.61 mph
1 Continental IO-520B, 285 bhp
2/20/84

PHILIPSBURG/HARRIS (USA)
Marie E. McMillan 190.91 kmh
Beech Bonanza F33A 118.63 mph
1 Continental IO-520B, 285 bhp
2/21/84

PHILIPSBURG/KINGSTON (USA)
Marie E. McMillan 223.22 kmh
Beech Bonanza F33A 138.70 mph
1 Continental IO-520B, 285 bhp
3/1/84

PHILIPSBURG/POINTE A PITRE (USA)
Marie E. McMillan 248.56 kmh
Beech Bonanza F33A 154.44 mph
1 Continental IO-520B, 285 bhp
2/21/84

PHILIPSBURG/ST. JOHNS (USA)
Marie E. McMillan 228.56 kmh
Beech Bonanza F33A 142.02 mph
1 Continental IO-520B, 285 bhp
2/21/84

PHOENIX/SANTA BARBARA (USA)
JOINT RECORD
William H. Wisner, Pilot 348.12 kmh
Janice Sullivan, Copilot 216.32 mph
Bonanza S35, N6226
and
Daniel Webb, Pilot
Richard Meyerhoff, Copilot
Bonanza A36, N4504S
5/28-29/89

PHUKET/NEW DELHI (USA)
JOINT RECORD
William H. Wisner, Pilot 282.60 kmh
Janice Sullivan, Copilot 175.61 mph
Bonanza S35, N6226
and
Daniel Webb, Pilot
Richard Meyerhoff, Copilot
Bonanza A36, N4504S
6/13/89

PITTSBURGH/GANDER (USA)
Robert W. Agostino 852.11 kmh
Damon K. Griebel 529.47 mph
Learjet 60
2 P&W PW 305A, 4,600 lbs
6/12/97

PITTSBURGH/PARIS (USA)
Robert W. Agostino 782.48 kmh
Damon K. Griebel 486.21 mph
Learjet 60
2 P&W PW 305A, 4,600 lbs
6/12/97

POINT BARROW/HAMBURG (SWITZERLAND)
Felix Estermann 235.22 kmh
Mooney 201, N5789B 146.17 mph
1 Lycoming 0-360, 180 bhp
8/29-30/87

POINT BARROW/HELSINKI (USA)
Charles C. Mack 277.56 kmh
Beech Bonanza, N18279 172.47 mph
1 Continental IO-520, 285 bhp
6/21-22/89

POINT BARROW/NORTH POLE (SWITZERLAND)
Felix Estermann 224.57 kmh
Mooney 201, N5789B 139.54 mph
1 Lycoming 0-360, 180 bhp
8/29-30/87

POINTE A PITRE/BASSETERRE (USA)
Marie E. McMillan 277.74 kmh
Beech Bonanza F33A 172.58 mph
1 Continental IO-520B, 285 bhp
2/28/84

POINTE A PITRE/BRIDGETOWN (USA)
Marie E. McMillan 232.75 kmh
Beech Bonanza F33A 144.63 mph
1 Continental IO-520B, 285 bhp
2/23/84

POINTE A PITRE/CASTRIES (USA)
Marie E. McMillan 254.49 kmh
Beech Bonanza F33A 158.13 mph
1 Continental IO-520B, 285 bhp
2/23/84

POINTE A PITRE/CODRINGTON (USA)
Marie E. McMillan 312.78 kmh
Beech Bonanza F33A 194.35 mph
1 Continental IO-520B, 285 bhp
2/28/84

POINTE A PITRE/GUSTAVIA (USA)
Marie E. McMillan 294.50 kmh
Beech Bonanza F33A 182.99 mph
1 Continental IO-520B, 285 bhp
2/28/84

POINTE A PITRE/MARTINIQUE (USA)
Marie E. McMillan 303.70 kmh
Beech Bonanza F33A 188.71 mph
1 Continental IO-520B, 285 bhp
2/23/84

POINTE A PITRE/NEW CASTLE (USA)
Marie E. McMillan 268.20 kmh
Beech Bonanza F33A 166.65 mph
1 Continental IO-520B, 285 bhp
2/28/84

POINTE A PITRE/PHILIPSBURG (USA)
Marie E. McMillan 286.22 kmh
Beech Bonanza F33A 177.85 mph
1 Continental IO-520B, 285 bhp
2/28/84

POINTE A PITRE/PORTSMOUTH (USA)
Marie E. McMillan 343.10 kmh
Beech Bonanza F33A 213.19 mph
1 Continental IO-520B, 285 bhp
2/23/84

POINTE A PITRE/ST. JOHNS (USA)
Marie E. McMillan 292.35 kmh
Beech Bonanza F33A 181.66 mph
1 Continental IO-520B, 285 bhp
2/28/84

PORT ANGELES/HILO (AUSTRALIA)
Ian Haig 686.35 kmh
Australian Airlines 426.48 mph
Boeing 737-400
7/21/92

PORTLAND, OR/DES MOINES (USA)
Dan Vollum, Pilot 795.60 kmh
Don Vollum, Copilot 494.39 mph
Jet Commander
2 GE CJ610-5, 2,950 lbs
12/3/86

PORTLAND, OR/GRAND FORKS (USA)
Richard W. Taylor, Pilot 455.47 kmh
Clayton L. Scott, Copilot 283.01 mph
Piper Aerostar
2 Lycoming IO-540, 350 bhp
5/23/92

PORTLAND, OR/PORTLAND, ME (USA)
Richard W. Taylor, Pilot 440.82 kmh
Clayton L. Scott, Copilot 273.91 mph
Piper Aerostar
2 Lycoming IO-540, 350 bhp
5/24/92

PORTLAND, OR/WASHINGTON (USA)
Dan Vollum, Pilot 702.36 kmh
Don Vollum, Copilot 436.45 mph
Jet Commander
2 GE CJ610-5, 2,950 lbs
12/3-4/86

PORTSMOUTH/BASSETERRE (USA)
Marie E. McMillan 219.10 kmh
Beech Bonanza F33A 136.10 mph
1 Continental IO-520B, 285 bhp
2/28/84

PORTSMOUTH/CODRINGTON (USA)
Marie E. McMillan 238.07 kmh
Beech Bonanza F33A 147.93 mph
1 Continental IO-520B, 285 bhp
2/28/84

PORTSMOUTH/GRAND-BOURG (USA)
Marie E. McMillan 210.73 kmh
Beech Bonanza F33A 130.94 mph
1 Continental IO-520B, 285 bhp
2/28/84

PORTSMOUTH/GUSTAVIA (USA)
Marie E. McMillan 239.98 kmh
Beech Bonanza F33A 149.12 mph
1 Continental IO-520B, 285 bhp
2/28/84

PORTSMOUTH/HARRIS (USA)
Marie E. McMillan 239.26 kmh
Beech Bonanza F33A 148.67 mph
1 Continental IO-520B, 285 bhp
2/28/84

PORTSMOUTH/NEW CASTLE (USA)
Marie E. McMillan 210.67 kmh
Beech Bonanza F33A 130.90 mph
1 Continental IO-520B, 285 bhp
2/28/84

PORTSMOUTH/PHILIPSBURG (USA)
Marie E. McMillan 238.45 kmh
Beech Bonanza F33A 148.17 mph
1 Continental IO-520B, 285 bhp
2/28/84

PORTSMOUTH/POINTE A PITRE (USA)
Marie E. McMillan 167.26 kmh
Beech Bonanza F33A 103.93 mph
1 Continental IO-520B, 285 bhp
2/28/84

PORTSMOUTH/ST. JOHNS (USA)
Marie E. McMillan 216.29 kmh
Beech Bonanza F33A 134.39 mph
1 Continental IO-520B, 285 bhp
2/28/84

PORT STANLEY/ASCENSION (UK)
Ft/Lt John Halstead 912.03 kmh
VC 10 C 566.74 mph
4 RR Conway Mk 301, 22,500 lbs
12/21/87

PORT STANLEY/GRITVYKEN (UK)
A. Holman 571.14 kmh
Lockheed Hercules 354.89 mph
4 T56-A7A
12/1/84

PRESQUE ISLE, ME/GOOSE BAY (USA)
Anne B. Baddour 397.08 kmh
Margrit Budert-Waltz 246.75 mph
Beech Baron, N3707N
2 Continental, 325 bhp
12/20/88

PRESQUE ISLE, ME/NARSSARSSUAQ (USA)
Anne B. Baddour 266.76 kmh
Margrit Budert-Waltz 165.76 mph
Beech Baron, N3707N
2 Continental, 325 bhp
12/20-21/88

PUERTO VALLARTA/ACAPULCO (USA)
Marie E. McMillan 282.96 kmh
Beech Bonanza F33A 175.83 mph
1 Continental IO-520 BA, 285 bhp
2/3/84

PUNTA ARENAS/RECIFE (USA)
Brooke Knapp 851.18 kmh
Gulfstream III 528.92 mph
2 RR Spey MK511-8, 11,400 lbs
1/16/83

QUEBEC CITY/OWEN SOUND (CANADA)
C. Nelson Gain 151.02 kmh
Ultravia Aero Pelican GS 93.84 mph
1 Rotax 912, 80 bhp
6/26/96

QUEBEC CITY/TORONTO (CANADA)
Terrance L. Jantzi 281.78 kmh
RV-6 175.09 mph
8/22/98

RECIFE/TENERIFE (USA)
Brooke Knapp 857.67 kmh
Gulfstream III 532.96 mph
2 RR Spey MK511-8, 11,400 lbs
11/17/83

RECIFE/TRONDHEIM (USA)
Brooke Knapp 728.28 kmh
Gulfstream III 452.55 mph
2 RR Spey MK511-8, 11,400 lbs
11/17-11/18/83

REDMOND/OSHKOSH (USA)
David Schroder 502.02 kmh
Lancair IV-P 311.94 mph
1 Continental TSIO-550, 350 bhp
7/20/01

RENO/CHEYENNE (USA)
Danford A. Bookout 352.44 kmh
Piper Lance, N8352C 219.01 mph
1 Lycoming IO-540K, 300 bhp
10/4/87

RENO/LAS VEGAS (USA)
Anthony Ferrante 444.24 kmh
Barbara Ferrante 276.05 mph
Piper Aerostar, N442AF
2 Lycoming 540
5/22/89

REYKJAVIK/BANGOR (USA)
Charles C. Mack 224.29 kmh
Beech Bonanza, N18279 139.37 mph
1 Continental IO-520, 285 bhp
7/17-18/89

REYKJAVIK/GANDER (USA)
Douglas Matthews, Pilot 678.60 kmh
Doyle Sanderlin, Copilot 421.68 mph
Learjet 35A
2 Garrett TFE 731-2-2B, 3,500 lbs
6/5/85

REYKJAVIK/KULUSUK (USA)
James P. Hanson 281.27 kmh
Cessna 208 Caravan 174.77 mph
1 P&W PT-6A, 600 shp
5/14/96

REYKJAVIK/KUUJJUAQ (USA)
Susan Nealey 312.05 kmh
Faith Hillman 193.90 mph
Cessna 310
2 Continental IO-520, 285 bhp
7/20/92

REYKJAVIK/NEW YORK (UK)
Lord D.G. Trefgarne 54.43 kmh
Charles B. G. Masefield 33.82 mph
de Havilland D.H. 90 Dragonfly
2 Gipsy Major 10 MK I
77 hrs 13 min 30 sec
8/2-5/64

REYKJAVIK/OSHKOSH (USA)
Susan Nealey 157.93 kmh
Faith Hillman 98.13 mph
Cessna 310
2 Continental IO-520, 285 bhp
7/20-22/92

REYKJAVIK/SHANNON (USA)
Susan Nealey 274.82 kmh
Faith Hillman 170.76 mph
Cessna 310
2 Continental IO-520, 285 bhp
6/16/92

REYKJAVIK/STRASBOURG (CANADA)
Dawn Bartsch 363.78 kmh
Gordon Bartsch 226.04 mph
Cessna 421B
9/11/97

RIO DE JANEIRO/LONDON (UK)
W. Brunn, Captain 895.56 kmh
J. M. A. Sharp, First Officer 556.47 mph
L. N. Griffiths, Flight Engineer
McDonnell Douglas DC-10-30
3 GE CF6-50C, 51,000 lbs
10 hrs 2 min 11 sec
1/23/78

RIO DE JANEIRO/LOS ANGELES (USA)
George R. Jansen 854.44 kmh
McDonnell Douglas DC-10 530.93 mph
3 P&W JT9D-25W
11 hrs 51 min 25 sec
10/14/72

ROADTOWN/CHARLOTTE AMALIE (USA)
Marie E. McMillan 309.95 kmh
Beech Bonanza F33A 192.59 mph
1 Continental IO-520B, 285 bhp
2/20/84

ROME/ADDIS ABABA (ITALY)
M. Lauldi 390.97 kmh
Fiat Br. 20L 242.94 mph
2 Fiat Asso. 80
11 hrs 20 min
10/25/56

ROME/ATHENS *Group II* (USA)
James P. Hanson 267.80 kmh
Cessna 208 Caravan 166.40 mph
1 P&W PT-6A, 600 shp
4/28/96

ROME/ATHENS *Group III* (USA)
Harold Curtis 900.04 kmh
Gulfstream III 559.26 mph
2 RR Spey 511-8
1 hr 10 min 48 sec
1/8/82

ROME/LONDON (UK)
A. W. Bedford 859.29 kmh
Hawker Hunter T.7 533.93 mph
1 RR Avon MK 121A
1 hr 40 min 7 sec
10/25/56

ROME/RIO DE JANEIRO (ITALY)
A. Biseo 221.96 kmh
Savoia-79 I-BISE 137.92 mph
3 Alfa Romeo 126 RC 34, 750 bhp
41 hrs 32 min
1/24/38

ROME/TEL AVIV (CANADA)
Dawn Bartsch 384.54 kmh
Gordon Bartsch 238.94 mph
Cessna 421B
9/16/97

ROMEOVILLE/KITTY HAWK (USA)
Robert E. Humphrey, Pilot 329.76 kmh
Charles F. Riedl, Copilot 204.91 mph
Commanche, N9262P
1 Lycoming IO-540-D4A5
12/16/88

ROVANIEMI/MURMANSK (SWEDEN)
Hans U. von der Esch 150.59 kmh
Piper PA18-150 93.57 mph
6/25/91

ST. JOHN'S, ANTIGUA/BASSETERRE (USA)
Marie E. McMillan 366.35 kmh
Beech Bonanza F33A 227.64 mph
1 Continental IO-520B, 285 bhp
2/28/84

ST. JOHN'S, ANTIGUA/DAKAR (USA)
COL Norman L. Mitchell 929.89 kmh
Gulfstream II, N700ST 577.80 mph
2 RR 511-8
2/27/75

ST. JOHN'S, ANTIGUA/HARRIS (USA)
Marie E. McMillan 355.50 kmh
Beech Bonanza F33A 220.89 mph
1 Continental IO-520B, 285 bhp
2/21/84

ST. JOHN'S, ANTIGUA/PHILIPSBURG (USA)
Marie E. McMillan 307.94 kmh
Beech Bonanza F33A 191.35 mph
1 Continental IO-520B, 285 bhp
2/28/84

ST. JOHN'S, ANTIGUA/POINTE A PITRE (USA)
Marie E. McMillan 352.01 kmh
Beech Bonanza F33A 218.73 mph
1 Continental IO-520B, 285 bhp
2/21/84

ST. JOHN'S, NF/LAJES, AZORES (USA)
George P. Eremea 806.23 kmh
A 282-40 Sabreliner 500.96 mph
2 P&W JT12A-6A
2 hrs 50 min 2.4 sec
10/26/63

ST. JOHN'S, NF/LISBON (USA)
George P. Eremea 759.24 kmh
A 282-40 Sabreliner 471.76 mph
2 P&W JT12A-6A
4 hrs 45 min 59.4 sec
10/26/63

ST. JOHN'S, NF/VICTORIA (CANADA)
Michael H. Leslie 16.13 kmh
Joseph Leslie 10.02 mph
Daniel J. Leslie
Cessna 172
1 Lycoming O-320, 160 bhp
8/15/00

ST. LOUIS/AUGUSTA (USA)
James R. Lyle 323.97 kmh
Piper PA-32R-301 201.30 mph
1 Lycoming IO-540, 300 bhp
11/7/93

ST. LOUIS/CALGARY *
Edward L. Taylor 490.58 mph
Michael W. Wyss
McDonnell Douglas DC-9-41
2 P&W JT8D, 16,000 lbs
5/17/85

ST. LOUIS/CINCINNATI (USA)
Lt Col Ed Yeilding, Pilot 3,524.38 kmh
Lt Col J. T. Vida, RSO 2,189.94 mph
Lockheed SR-71
2 P&W J-58, 34,000 lbs
03/06/90

ST. LOUIS/PARIS (USA)
Donald L. Mullin 910.50 kmh
McDonnell Douglas DC-10 565.76 mph
3 GE CF-6-6
7 hrs 45 min 27 sec
6/1-2/71

ST. LOUIS/READING (USA)
R. A. "Bob" Hoover, Pilot 921.24 kmh
Robert Morgenthaler, Copilot 572.46 mph
Sabreliner
2 Garrett TFE-731, 3,700 lbs
8/22/86

ST. LOUIS/WASHINGTON (USA)
Mark Patiky 615.96 kmh
Mooney 252 TSE, N252YD 382.75 mph
1 Continental TSIO-36-MBI, 210 bhp
1/7/87

SAIPAN/MANILA (USA)
JOINT RECORD
William H. Wisner, Pilot 308.52 kmh
Janice Sullivan, Copilot 191.71 mph
Bonanza S35, N6226
and
Daniel Webb, Pilot
Richard Meyerhoff, Copilot
Bonanza A36, N4504S
6/4-5/89

SAN ANTONIO/NEW YORK (USA)
Allyn Caruso, Pilot 395.18 kmh
David Norgart, Copilot 245.55 mph
John Gallichon, Navigator
John Miller, Navigator
Dana Lovell, Mechanic
Fairchild Metroliner III
2 Garrett TPE331-11, 1,100 shp
6/7/91

SAN CRISTOBAL/KINGSTON (USA)
Marie E. McMillan 304.12 kmh
Beech Bonanza F33A 188.97 mph
1 Continental IO-520B, 285 bhp
3/1/84

SAN CRISTOBAL/PORT AU PRINCE (USA)
Marie E. McMillan 311.29 kmh
Beech Bonanza F33A 193.43 mph
1 Continental IO-520B, 285 bhp
3/1/84

SAN DIEGO/CHARLESTON (USA) *Group II*
Joseph J. Ritchie 880.01 kmh
Steve Fossett 546.81 mph
Piaggio P.180 Avanti
2 P&WC PT6A, 850 shp
2/6/03

SAN DIEGO/CHARLESTON (USA) *Group III*
Steve Fossett 1,169.71 kmh
Douglas A. Travis 726.83 mph
Cessna 750 Citation X
2 R-R AE3007C, 6,442 lbs
2/5/03

SAN DIEGO/JACKSONVILLE (USA)
Spencer Lane, Pilot 491.00 kmh
Marc Mosier, Copilot 305.09 mph
Socata TBM-700
1 P&W PT6A, 700 shp
9/26/92

SAN DIEGO/NEW YORK (USA)
William L. Mack 912.37 kmh
Harold Curtis 566.92 mph
Gulfstream III
2 RR Spey 511-8
4 hrs 17 min
1/22/82

SAN DIEGO/WASHINGTON (USA)
Robert W. Agostino 849.91 kmh
David S. Ryan 528.11 mph
Learjet 45
2 AlliedSignal TFE731-20, 3,500 lbs
11/23/98

SAN FRANCISCO/CHICAGO (USA)
Steve Fossett, Pilot 1,104.66 kmh
Darrin L. Adkins, Copilot 686.40 mph
Cessna 750 Citation X
2 R-R AE 3007C, 6,442 lbs
4/6/00

SAN FRANCISCO/GANDER (USA)
Brooke Knapp, Pilot 814.44 kmh
James Topalian, Copilot 506.09 mph
Paul Broyles, Copilot
James Magill, Crew Chief
Gates Learjet 35A
2 Garrett TFE 731-2-2B, 3,500 lbs
2/16/83

SAN FRANCISCO/LAS VEGAS (USA)
Dwight V. Whitaker, III, Pilot 764.67 kmh
Harry G. Butler, III, Copilot 475.14 mph
Gulfstream IV
2 RR Tay Mk 611-8, 13,850 lbs
3/5/92

SAN FRANCISCO/LONDON (USA)
CPT Walter H. Mullikin 898.26 kmh
Pan American World Airways 558.15 mph
(CPT Albert A. Frink,
Chief, Relief Airmen)
Boeing 747 SP, N533PA
4 P&W JT9D-7A
11 hrs 16 min
10/28-29/77

SAN FRANCISCO/NEW YORK (USA)
Steve Fossett, Pilot 1,115.51 kmh
Darrin L. Adkins, Copilot 693.14 mph
Cessna 750 Citation X
2 R-R AE 3007C, 6,442 lbs
4/6/00

SAN FRANCISCO/SYRACUSE (USA)
Alan M. Marcum 347.71 kmh
Mooney 252TSE 216.05 mph
1 Continental TSIO-360, 210 bhp
6/10/99

SAN FRANCISCO/WASHINGTON (USA)
Alan W. Gerharter 486.20 kmh
Mooney M 20K 302.11 mph
1 Continental, 210 bhp
8 hrs 4 min 25 sec
1/7/80

SAN JOSE/NEW YORK (USA)
Charles D. Kissner, Pilot 376.85 kmh
Paul Bertorelli, Copilot 234.16 mph
Piper PA46-310P, N4319M
1 Continental TSIO-520, 310 bhp
5/4/92

SAN JOSE/WASHINGTON (USA)
Timothy Mellon 418.27 kmh
Piper Aerostar 602P, N602PC 259.91 mph
2 Lycoming IO-540
5/18/88

SAN JUAN/KINGSTON (USA)
Marie E. McMillan 203.72 kmh
Beech Bonanza F33A 126.58 mph
1 Continental IO-520B, 285 bhp
3/1/84

SAN JUAN/PORT AU PRINCE (USA)
Marie E. McMillan 162.98 kmh
Beech Bonanza F33A 101.27 mph
1 Continental IO-520B, 285 bhp
3/1/84

SANTA ANA/WASHINGTON (USA)
Charles D. Kissner, Pilot 417.64 kmh
Steven A. Silver, Copilot 259.51 mph
Piper PA46-310P, N4319M
1 Continental TSIO-520, 310 bhp
1/27/91

SANTA BARBARA/HILO (USA)
JOINT RECORD
William H. Wisner, Pilot 277.56 kmh
Janice Sullivan, Copilot 172.48 mph
Bonanza S35, N6226
and
Daniel Webb, Pilot
Richard Meyerhoff, Copilot
Bonanza A36, N4504S
5/29-30/89

SANTA BARBARA/HONOLULU (USA)
JOINT RECORD
William H. Wisner, Pilot 262.08 kmh
Janice Sullivan, Copilot 162.86 mph
Bonanza S35, N6226
and
Daniel Webb, Pilot
Richard Meyerhoff, Copilot
Bonanza A36, N4504S
5/29-30/89

SANTA BARBARA/MAUI (USA)
JOINT RECORD
William H. Wisner, Pilot 266.76 kmh
Janice Sullivan, Copilot 165.76 mph
Bonanza S35, N6226
and
Daniel Webb, Pilot
Richard Meyerhoff, Copilot
Bonanza A36, N4504S
5/29-30/89

SANTA MARIA/GANDER (USA)
JOINT RECORD
William H. Wisner, Pilot 282.60 kmh
Janice Sullivan, Copilot 175.61 mph
Bonanza S35, N6226
and
Daniel Webb, Pilot
Richard Meyerhoff, Copilot
Bonanza A36, N4504S
5/29-30/89

SANTA MARIA/ST. JOHNS (USA)
William H. Wisner, Pilot 247.32 kmh
Janice Sullivan, Copilot 153.68 mph
Beech Bonanza, N6826Q
1 Continental, 285 bhp
7/4/86

SANTO DOMINGO/ACAPULCO (FRANCE)
Claude Delorme, Captain 1,404.43 kmh
Jean Boye, Co-Captain 872.67 mph
BAe/Aérospatiale Concorde
10/12/92

SANTO DOMINGO/GUAM (FRANCE)
Claude Delorme, Captain 1,194.65 kmh
Jean Boye, Co-Captain 742.32 mph
BAe/Aérospatiale Concorde
10/13/92

SANTO DOMINGO/HONOLULU (FRANCE)
Claude Delorme, Captain 1,259.07 kmh
Jean Boye, Co-Captain 782.35 mph
BAe/Aérospatiale Concorde
10/13/92

SANTORINI/PALERMO (USA)
James P. Hanson 240.80 kmh
Cessna 208 Caravan 149.62 mph
1 P&W PT-6A, 600 shp
5/3/96

SAO PAULO/TETERBORO (USA)
Richard J. Iudice, Pilot 694.31 kmh
Robert D. Phillips, Copilot 431.42 mph
Dassault Falcon 900EX
3 Allied Signal TFE731-60, 5,000 lbs
6/3/98

SAULT STE. MARIE/FORT FRANCIS (CANADA)
Terrance L. Jantzi 235.59 kmh
Joël Jantzi 146.39 mph
Van's RV-6
11/14/98

SAULT STE. MARIE/SEPT-ILES (USA)
James P. Hanson 250.72 kmh
Cessna 208 Caravan 155.79 mph
1 P&W PT-6A, 600 shp
4/8/96

SAULT STE. MARIE/WINNIPEG (CANADA)
Terrance L. Jantzi 213.76 kmh
Joël Jantzi 132.82 mph
Van's RV-6
11/14/98

SAVANNAH/ANKARA (USA)
William J. Watters 901.40 kmh
Raymond A. Wellington 560.10 mph
Alberto L. Moros
Gulfstream G550
2 BMW R-R BR710, 15,385 lbs
9/29/03

SAVANNAH/DUBAI (USA)
William J. Watters 883.84 kmh
Raymond A. Wellington 549.19 mph
Ahmed M. Ragheb
Gulfstream G550
2 BMW R-R BR710, 15,385 lbs
12/6/03

SAVANNAH/GUAYAQUIL (USA)
Paulo S. Jancitsky 829.78 kmh
Kevin A. Bixler 515.60 mph
IAI 1126 Galaxy
2 P&WC PW306A, 5,700 lbs
3/31/02

SAVANNAH/PARIS (USA)
Robert K. Smyth, Captain 838.80 kmh
Edward D. Mendenhall, CoCaptain 521.23 mph
William Parker, Flight Engineer
Gulfstream IV, N440GA
2 RR TAY MK610-8
6/11/87

SEATTLE/ATHENS (USA)
Clay Lacy, Captain 1,083.75 kmh
Verne Jobst, CoCaptain 673.44 mph
Boeing 747, N149UA
4 P&W JT9D-7A, 46,250 lbs
1/29/88

SEATTLE/CHARLOTTE (USA)
Gene F. Sharp, Captain 794.65 kmh
Robert S. Hall, Captain 493.77 mph
USAir Boeing 767-200ER
2 GE CF6-80C2B2, 52,000 lbs
05/22/90

SEATTLE/CHICAGO (USA)
Clay Lacy, Captain 1,059.56 kmh
Verne Jobst, CoCaptain 658.41 mph
Boeing 747, N149UA
4 P&W JT9D-7A, 46,250 lbs
1/29/88

SEATTLE/COLORADO SPRINGS (USA)
Richard W. Taylor, Pilot 453.41 kmh
Clayton L. Scott, Copilot 281.73 mph
Piper Aerostar
2 Lycoming IO-540, 305 bhp
05/25/90

SEATTLE/DAYTON (USA)
Clay Lacy, Captain 1,058.76 kmh
Verne Jobst, CoCaptain 657.91 mph
Boeing 747, N149UA
4 P&W JT9D-7A, 46,250 lbs
1/29/88

SEATTLE/FORT WORTH (USA)
MAJ Clyde P. Evely, USAF 889.16 kmh
Boeing B-52H Stratofortress 552.60 mph
8 P&W TF-33P-3
3 hrs 24.62 sec
1/11/62

SEATTLE/FRANKFURT (USA)
Michael Fox, Captain 851.06 kmh
John Lloyd, Co-Captain 528.82 mph
Orme Dockins, Co-Captain
Khalidkhan Asmakhan, 1st Officer
Royal Brunei Airlines Boeing 767-200ER
2 P&W 4050
07/09/90

SEATTLE/GENEVA (USA)
Charles L. Gebhardt, III 876.16 kmh
William F. Royce 544.42 kmh
Andrew C. Messer
Susan P. Darcy
Boeing 777-200
2 P&W PW4084, 84,000 lbs
9/7/95

SEATTLE/INDIANAPOLIS (USA)
Richard W. Taylor, Pilot 429.33 kmh
Stephen R. Taylor, Copilot 266.77 mph
Piper Aerostar
2 Lycoming IO-540, 305 bhp
3/23/95

SEATTLE/LONG BEACH (USA)
Richard W. Taylor, Pilot 429.33 kmh
Stephen R. Taylor, Copilot 266.77 mph
Piper Aerostar 602P
2 Lycoming IO-540, 305 bhp
3/23/95

SEATTLE/LAS VEGAS (USA)
Richard W. Taylor, Pilot 431.28 kmh
Clayton L. Scott, Copilot 268.00 mph
Piper Aerostar 602P, N6903E
2 Lycoming IO-540-AA1A5M
5/21/89

SEATTLE/LISBON (USA)
Clay Lacy, Captain 1,109.15 kmh
Verne Jobst, CoCaptain 689.22 mph
Boeing 747, N149UA
4 P&W JT9D-7A, 46,250 lbs
1/29/88

SEATTLE/LONDON (UK)
G. A. C. Gray 844.96 kmh
Boeing 757-236 525.06 mph
2 RR RB211-535C, 37,400 lbs
5/15-16/86

SEATTLE/LONG BEACH*
Richard W. Taylor, Pilot 235.53 mph
Clayton L. Scott, Copilot
Piper Aerostar
2 Lycoming IO-540, 305 bhp
9/25/90

SEATTLE/LOS ANGELES (USA)
Stephen Lutz, Pilot 1,028.12 kmh
Charles King, Copilot 638.84 mph
Western Airlines
Boeing 737-200
2 P&W JT80-91, 14,000 lbs
12/13/84

SEATTLE/MADRID (USA)
Clay Lacy, Captain 1,106.24 kmh
Verne Jobst, CoCaptain 687.42 mph
Boeing 747, N149UA
4 P&W JT9D-7A, 46,250 lbs
1/29/88

SEATTLE/NAIROBI (USA)
Michael Fox, Captain 835.29 kmh
John Lloyd, Co-Captain 519.02 mph
Orme Dockins, Co-Captain
Khalidkhan Asmakhan, 1st Officer
Royal Brunei Airlines Boeing 767-200ER
2 P&W 4050
07/09/90

SEATTLE/NEW ORLEANS (USA)
Scott S. Evans 878.26 kmh
Jerry W. Blessing 545.72 mph
IAI 1125 Astra SPX
2 AlliedSignal TFE731, 4,250 lbs
9/3/01

SEATTLE/NORTH POLE (USA)
Howard Mayes, Clay Lacy 899.25 kmh
United Airlines 558.77 mph
McDonnell Douglas DC-8
4 P&W
5 hrs 14 min 51.6 sec
6/25/73

SEATTLE/OSHKOSH (USA)
Richard W. Taylor, Pilot 388.20 kmh
Clayton L. Scott, Copilot 241.22 mph
Piper Aerostar
2 Lycoming IO-540, 305 bhp
07/28/90

SEATTLE/PARIS (USA)
John E. Cashman 890.80 kmh
John K. Higgins 553.52 kmh
Charles A. Hovland
Boeing 777-200
2 P&W PW4084, 84,000 lbs
6/11/95

SEATTLE/PORTLAND, ME (USA)
Richard W. Taylor, Pilot 408.02 kmh
Clayton L. Scott, Copilot 253.53 mph
Piper Aerostar
2 Lycoming IO-540, 350 bhp
5/24/92

SEATTLE/SINGAPORE (USA)
Robert N. Sweeney 792.50 kmh
Charles P. Little 492.43 mph
Abdul Aziz Bin Buang, Eric Khaw, Freddie Koh,
Victor Oh Kim Song
Boeing 777-200
2 RR Trent 884, 84,000 lbs
5/7/97

SEATTLE/WASHINGTON *Group I* (USA)
Jack Lewis, Pilot 405.00 kmh
James Hodges, Copilot 251.67 mph
Piper Aerostar 700PC, N700JL
2 Lycoming IO-540
5/18/88

SEATTLE/WASHINGTON *Group III* (USA)
Richard R. Severson 1,033.22 kmh
Ronald L. Weight 642.01 mph
LeRoy Herrman
Cessna 750
2 Allison AE 3007C, 6,400 lbs
5/5/97

SEOUL/ORLANDO (USA)
Wilson A. Miles, Jr. 836.87 kmh
Anthony J. Briotta 520.01 mph
Thomas C. Horne
William J. Watters
Gulfstream G550
2 BMW R-R BR710, 15,385 lbs
10/3/03

SEPT-ILES/GOOSE BAY *Group I* (USA)
Anne B. Baddour 470.52 kmh
Margrit Budert-Waltz 292.38 mph
Beech Baron, N3707N
2 Continental, 325 bhp
12/20/88

SEPT-ILES/GOOSE BAY *Group II* (USA)
James P. Hanson 291.82 kmh
Cessna 208 Caravan 181.33 mph
1 P&W PT-6A, 600 shp
4/8/96

SEVILLA/ROME (CANADA)
Dawn Bartsch 406.68 kmh
Gordon Bartsch 252.70 mph
Cessna 421B
9/15/97

SHANNON/BANGOR (USA)
Richard Trissell — 838.65 kmh
William S. Dirks — 521.11 mph
Cessna Model 750, Citation X
2 Allison GMA 3007C, 6,000 lbs
6/15/95

SHANNON/BERMUDA (UK)
Capt. Peter Horton — 1,757.42 kmh
BAe/Aérospatiale Concorde — 1,092.01 mph
1/1/93

SHANNON/BONN (USA)
Jacqueline Cochran — 685.96 kmh
Lockheed Jetstar — 426.24 mph
4 P&W JT12A-6
1 hr 38 min 15.8 sec
4/22/62

SHANNON/CANNES (USA)
Susan Nealey — 384.69 kmh
Faith Hillman — 239.03 mph
Cessna 310
2 Continental IO-520, 285 bhp
7/12/92

SHANNON/DUBAI (USA)
Allen E. Paulson, Captain — 836.28 kmh
Robert K. Smyth, CoCaptain — 519.66 mph
John Salamankas, CoCaptain
Jeff Bailey, CoCaptain
Gulfstream IV, N400GA
2 RR TAY MK611-8
2/26-27/88

SHANNON/LONDON (USA)
Jacqueline Cochran — 767.07 kmh
Lockheed Jetstar — 476.64 mph
4 P&W JT12A-6
47 min 56.5 sec
4/22/62

SHANNON/NEW YORK (UK)
CPT J. W. Burton — 1,643.67 kmh
BAe/Aérospatiale Concorde — 1,021.38 mph
4 RR Olympus, 38,000 lbs
9/16/85

SHANNON/PARIS (USA)
Jacqueline Cochran — 765.91 kmh
Lockheed Jetstar — 475.92 mph
4 P&W JT12A-6
1 hr 10 min 10.8 sec
4/22/62

SINGAPORE/DARWIN (UK)
J. Finch — 834.21 kmh
Vickers Armstrong — 518.35 mph
Valiant B1 WP209
2 RR Avon RA 14
4 hrs 50.1 sec
7/13/55

SINGAPORE/GUAM (SAUDI ARABIA)
Aziz Ojjeh — 759.60 kmh
Canadair Challenger — 472.02 mph
GE CF-34 8,650 lbs
7/22-7/23/84

SINGAPORE/MANILA (USA)
Harold Curtis — 828.04 kmh
Gulfstream III — 514.52 mph
2 RR Spey 511-8
2 hrs 46 min 54 sec
1/9/82

SIOUX FALLS/PIERRE*
John C. Penney — 401.22 mph
Grumman F8F Bearcat
1 Wright R3350, 3,250 bhp
8/3/94

SONDRE STROMFJORD/IQALUIT (USA)
James P. Hanson — 268.30 kmh
Cessna 208 Caravan — 166.71 mph
1 P&W PT-6A, 600 shp
5/16/96

SOUTH POLE/NORTH POLE (USSR)
Lev V. Kozlov, Captain — 676.88 kmh
Yuri P. Resnitsky, Copilot — 420.59 mph
Oleg I. Pripuskov, Copilot
Anatoly V. Andronov, Copilot
Antonov-124
12/1-4/90

STEPHENVILLE/MADRID (MEXICO)
Jose Manuel Muradas Rodriguez — 816.43 kmh
José Alfredo Gastelum Arce — 507.31 mph
Dassault Falcon 2000
2 GE/Honeywell CFE738, 5,888 lbs
11/8/02

STOCKHOLM/BALTIMORE (USA)
COL James B. Swindal, USAF — 803.86 kmh
Boeing VC-137 — 499.50 mph
4 P&W JT3D-3
8 hrs 11 min 33.3 sec
5/20-21/63

STOCKHOLM/BOSTON (USA)
COL James B. Swindal, USAF — 813.42 kmh
Boeing VC-137 — 506.44 mph
4 P&W JT3D-3
7 hrs 24 min 45.6 sec
5/20-21/63

STOCKHOLM/NEW YORK (USA)
COL James B. Swindal USAF — 809.53 kmh
Boeing VC-137 — 503.02 mph
4 P&W JT3D-3
7 hrs 48 min 31.1 sec
5/20-21/63

STOCKHOLM/PHILADELPHIA (USA)
COL James B. Swindal, USAF — 807.34 kmh
Boeing VC-137 — 501.66 mph
4 P&W JT3D-3
7 hrs 59 min 31 sec
5/20-21/63

STOCKHOLM/WASHINGTON (USA)
COL James B. Swindal, USAF — 799.29 kmh
Boeing VC-137 — 496.65 mph
4 P&W JT3D-3
8 hrs 18 min 30.8 sec
5/20-21/63

STORNOWAY/REYKJAVIK (USA)
James P. Hanson — 291.16 kmh
Cessna 208 Caravan — 180.92 mph
1 P&W PT-6A, 600 shp
5/13/96

STORNOWAY/VAGAR (USA)
James P. Hanson — 304.83 kmh
Cessna 208 Caravan — 189.41 mph
1 P&W PT-6A, 600 shp
5/13/96

STRASBOURG/SEVILLA (CANADA)
Dawn Bartsch — 318.49 kmh
Gordon Bartsch — 197.90 mph
Cessna 421B
9/13/97

SYDNEY/AGANA (UK)
Max Robinson — 1,512.09 kmh
BAe/Aérospatiale Concorde — 939.57 mph
10/30/99

SYDNEY/BALI (UK)
Capt. G. Mussett 1,331.01 kmh
British Airways 827.05 mph
BAe/Aérospatiale Concorde
3/19/90

SYDNEY/BANGKOK (UK)
Max Robinson 849.83 kmh
BAe/Aérospatiale Concorde 528.06 mph
10/30/99

SYDNEY/LONDON (UK)
CPT Hector McMullen 837.67 kmh
British Airways 520.50 mph
BAe/Aérospatiale Concorde
4 RR Snecma Olympus, 38,000 lbs
20 hrs 20 min 25 sec
2/15-2/16/85

SYDNEY/PARIS (UK)
CPT Hector McMullen 940.39 kmh
British Airways 584.35 mph
BAe/Aérospatiale Concorde
4 RR Snecma Olympus, 38,000 lbs
18 hrs 2 min 15 sec
2/15/85

TAIPEI/HOUSTON (USA)
Allen E. Paulson, Captain 929.88 kmh
Robert K. Smyth, CoCaptain 577.83 mph
John Salamankas, CoCaptain
Jeff Bailey, CoCaptain
Gulfstream IV, N400GA
2 RR TAY MK611-8
2/27-28/88

TAIPEI/MAUI (USA)
Allen E. Paulson, Captain 1,079.29 kmh
Robert K. Smyth, CoCaptain 670.67 mph
John Salamankas, CoCaptain
Jeff Bailey, CoCaptain
Gulfstream IV, N400GA
2 RR TAY MK611-8
2/27/88

TAIPEI/SEATTLE (USA)
Clay Lacy, Captain 1,094.85 kmh
Verne Jobst, CoCaptain 680.34 mph
Boeing 747, N149UA
4 P&W JT9D-7A, 46,250 lbs
1/30/88

TAHITI/HONOLULU (USA)
Douglas G. Matthews 714.96 kmh
Learjet 36 444.28 mph
2 Garrett TFE-731-2-2B, 3,500 lbs
7/26/84

TEL AVIV/BOSTON (USA)
Christian M. Kennedy 770.11 kmh
James R. Cannon 478.52 mph
Henry K. Gibson
Bruce A. Yates
Gulfstream V
2 BMW Rolls-Royce BR 710-48, 14,750 lbs
1/11/98

TENERIFE/NORTH POLE (USA)
Brooke Knapp 661.18 kmh
Gulfstream III 410.85 mph
2 RR Spey MK511-8, 11,400 lbs
11/17-11/18/83

TENERIFE/TRONDHEIM (USA)
Brooke Knapp 763.20 kmh
Gulfstream III 474.25 mph
2 RR Spey MK511-8, 11,400 lbs
11/17-18/83

TETERBORO/CHICAGO (USA)
Ehab Hanna 762.79 kmh
Lynn Upham 473.98 mph
Gulfstream IV
2 R-R Tay 611, 13,850 lbs
6/21/03

TETERBORO/LIHUE (USA)
Bernard R. Ozbolt 825.65 kmh
Charles M. Barr II 513.03 mph
Gary M. Freeman
Gulfstream V
2 BMW Rolls-Royce BR710-48, 14,750 lbs
2/20/98

TETERBORO/MOSCOW (USA)
Andrew Milewski 922.92 kmh
E. Bruce Robinson 573.47 mph
Bombardier BD-700 Global Express
2 BMW-RR BR710, 14,750 lbs
9/19/99

TETERBORO/NEW ORLEANS (USA)
Scott S. Evans 806.99 kmh
Jerry W. Blessing 501.44 mph
IAI 1125 Astra SPX
2 AlliedSignal TFE731, 4,250 lbs
8/31/01

TETERBORO/TOKYO (USA)
E. Bruce Robinson 826.69 kmh
Andrew Milewski 513.68 mph
Roger Noble
Borden F. Schofield
Bombardier BD 700
2 BMW RR BR-710, 14,690 lbs
5/13/99

TETERBORO/VAN NUYS (USA)
Thomas W. Owens 861.98 kmh
Christian M. Kennedy 535.61 mph
Gulfstream V
2 BMW R-R BR710, 14,750 lbs
7/3/00

TEXARKANA/ALBUQUERQUE (USA)
Danford A. Bookout 301.32 kmh
Piper Lance, N8352C 187.24 mph
1 Lycoming IO-540K, 300 bhp
10/2/87

TEXARKANA/GULFPORT (USA)
Danford A. Bookout 329.04 kmh
Piper Lance, N8352C 204.46 mph
1 Lycoming IO-540K, 300 bhp
1/8/88

TEXARKANA/HOUSTON (USA)
Danford A. Bookout 274.68 kmh
Piper Lance, N8352C 170.69 mph
1 Lycoming IO-540K, 300 bhp
2/2/88

TEXARKANA/INDIANAPOLIS (USA)
Danford A. Bookout 333.36 kmh
Piper Lance, N8352C 207.15 mph
1 Lycoming IO-540K, 300 bhp
8/11/87

TEXARKANA/KITTY HAWK (USA)
Thomas Chames 323.31 kmh
Piper Lance, N8352C 200.91 mph
1 Lycoming IO-50 KIAS, 300 bhp
12/16/87

TEXARKANA/LUBBOCK (USA)
Danford A. Bookout 278.64 kmh
Piper Lance, N8352C 173.15 mph
1 Lycoming IO-540K, 300 bhp
3/14/88

TEXARKANA/NEW ORLEANS (USA)
Danford A. Bookout — 293.40 kmh
Piper Lance, N8352C — 182.32 mph
1 Lycoming IO-540K, 300 bhp
12/18/87

TEXARKANA/ORLANDO (USA)
Danford A. Bookout — 290.60 kmh
Piper Lance, N8352C — 180.30 mph
1 Lycoming IO-540K, 300 bhp
1/18/88

TEXARKANA/OSHKOSH (USA)
Danford A. Bookout — 302.76 kmh
Piper Lance, N8352C — 188.14 mph
1 Lycoming IO-540K, 300 bhp
7/29/87

THE BOTTOMS/CHARLOTTE AMALIE (USA)
Marie E. McMillan — 169.28 kmh
Beech Bonanza F33A — 105.19 mph
1 Continental IO-520B, 285 bhp
2/20/84

THE BOTTOMS/CHRISTIANSTED (USA)
Marie E. McMillan — 182.85 kmh
Beech Bonanza F33A — 113.61 mph
1 Continental IO-520B, 285 bhp
2/20/84

THE BOTTOMS/GUSTAVIA (USA)
Marie E. McMillan — 89.29 kmh
Beech Bonanza F33A — 55.48 mph
1 Continental IO-520B, 285 bhp
2/20/84

THE BOTTOMS/HARRIS (USA)
Marie E. McMillan — 183.34 kmh
Beech Bonanza F33A — 113.92 mph
1 Continental IO-520B, 285 bhp
2/21/84

THE BOTTOMS/NEW CASTLE (USA)
Marie E. McMillan — 188.34 kmh
Beech Bonanza F33A — 117.03 mph
1 Continental IO-520B, 285 bhp
2/21/84

THE BOTTOMS/POINTE A PITRE (USA)
Marie E. McMillan — 250.57 kmh
Beech Bonanza F33A — 155.70 mph
1 Continental IO-520B, 285 bhp
2/21/84

THE PAS/IQALUIT (CANADA)
Dawn Bartsch — 356.95 kmh
Gordon Bartsch — 221.80 mph
Cessna 421B
9/5/97

THE VALLEY/CHARLOTTE AMALIE (USA)
Marie E. McMillan — 148.72 kmh
Beech Bonanza F33A — 92.41 mph
1 Continental IO-520B, 285 bhp
2/20/84

THE VALLEY/CHRISTIANSTED (USA)
Marie E. McMillan — 176.72 kmh
Beech Bonanza F33A — 109.81 mph
1 Continental IO-520B, 285 bhp
2/20/84

THE VALLEY/GRAND-CASE (USA)
Marie E. McMillan — 178.90 kmh
Beech Bonanza F33A — 111.16 mph
1 Continental IO-520B, 285 bhp
2/20/84

THE VALLEY/ORANJESTAD (USA)
Marie E. McMillan — 207.61 kmh
Beech Bonanza F33A — 129.00 mph
1 Continental IO-520B, 285 bhp
2/20/84

THE VALLEY/THE BOTTOMS (USA)
Marie E. McMillan — 260.13 kmh
Beech Bonanza F33A — 161.64 mph
1 Continental IO-520b, 285 bhp
2/20/84

TOKYO/ALBUQUERQUE (USA)
Wayne Altman, Pilot — 933.22 kmh
Robert Kirksey, Copilot — 579.87 mph
Gulfstream IV
2 RR Tay Mk 611-8, 13,850 lbs
3/3/93

TOKYO/ANCHORAGE (USA)
MAJ S. J. Kubesch, USAF — 1,759.73 kmh
Convair B-58A Hustler — 1,093.44 mph
4 GE J-79
10/16/63

TOKYO/BEIJING (USA)
LTC Royce Grones, USAF — 512.64 kmh
MAJ Robyn S. Read, USAF — 318.55 mph
C-135
4 TF-33 P5, 16,000 lbs
10/3-4/85

TOKYO/CHICAGO (USA)
LTC G. A. Andrews, USAF — 1,173.61 kmh
Convair B-58A Hustler — 729.25 mph
4 GE J-79 Engines
8 hrs 38 min 42 sec
10/16/63

TOKYO/DALLAS (USA)
Andrew S. Milewski — 915.19 kmh
Yves Tessier — 568.67 mph
Durwood J. Heinrich
Bombardier BD-700 Global Express
2 BMW R-R BR710, 14,750 lbs
4/26/02

TOKYO/FORT WORTH (USA)
MAJ Clyde P. Evely, USAF — 885.26 kmh
Boeing B-52H Stratofortress — 550.08 mph
8 P&W TF-33P-3
11 hrs 41 min 24.69 sec
1/10-11/62

TOKYO/HONOLULU (USA)
Brooke Knapp — 1,085.60 kmh
Gulfstream III — 674.59 mph
2 RR Spey MK511-8, 11,400 lbs
2/14-15/84

TOKYO/LAS VEGAS (USA)
Dwight V. Whitaker, III, Pilot — 880.01 kmh
Harry G. Butler, III, Copilot — 546.81 mph
Gulfstream IV
2 RR Tay Mk 611-8, 13,850 lbs
3/5/92

TOKYO/LONDON (USA)
MAJ S. J. Kubesch, USAF — 1,114.81 kmh
Convair B-58A Hustler — 692.70 mph
4 GE J-79
8 hrs 35 min 20.4 sec
10/16/63

TOKYO/LOS ANGELES (USA) ✯
Gregory S. Sheldon — 1,036.19 kmh
Ahmed M. Ragheb — 643.86 mph
John A. Mullican
Gulfstream V
2 BMW R-R BR710, 14,750 lbs
11/7/02

TOKYO/MADRID (USA)
MAJ Clyde P. Evely, USAF 529.12 kmh
Boeing B-52H Stratofortress 328.78 mph
8 P&W TF-33P-3
20 hrs 22 min 12 sec
1/10-11/62

TOKYO/MONTREAL (USA)
E. Bruce Robinson 911.29 kmh
Andrew Milewski 566.24 mph
Roger Noble
Borden F. Schofield
Bombardier BD 700
2 BMW RR BR-710, 14,690 lbs
5/14/99

TOKYO/MOSCOW (USA)
Larry S. Mueller 835.50 kmh
Christian M. Kennedy 519.15 mph
Gulfstream V
2 BMW RR BR710-48, 14,750 lbs
8/22/97

TOKYO/PALM BEACH (USA)
Gregory S. Sheldon 952.43 kmh
John D. Mackay 591.81 mph
Neil E. Vernon
Gulfstream G550
2 BMW R-R BR710, 15,385 lbs
11/10/03

TOKYO/PARIS (USA)
Richard D. Robbins 781.18 kmh
William J. Watters 485.40 mph
Christian M. Kennedy
Gulfstream V
2 BMW RR BR710-48, 14,750 lbs
6/10/99

TOKYO/SAN FRANCISCO (USA)
Dwight V. Whitaker, III, Pilot 895.35 kmh
Harry G. Butler, III, Copilot 556.34 mph
Gulfstream IV
2 RR Tay Mk 611-8, 13,850 lbs
3/5/92

TOKYO/SEATTLE (USA)
MAJ Clyde P. Evely, USAF 884.11 kmh
Boeing B-52H Stratofortress 549.36 mph
8 P&W TF-33P-3
8 hrs 43 min 40.83 sec
1/10-11/62

TOKYO/TETERBORO (USA)
Borden F. Schofield 909.75 kmh
E. Bruce Robinson 565.29 mph
Geoffrey M. Foster
Andrew Milewski
Bombardier BD 700
2 BMW RR BR-710, 14,690 lbs
5/9/99

TOKYO/WASHINGTON (USA)
Gregory S. Sheldon 913.38 kmh
Robert S. McKenney 567.55 mph
Gulfstream V
2 BMW R-R BR710, 14,750 lbs
3/5/02

TOKYO/WINNIPEG (USA)
A.G. Heimerdinger 820.95 kmh
Douglas DC-8-61 510.12 mph
P&W JT3D-3B
10 hrs 57 min 51 sec
8/18/66

TOLUCA/MADRID (MEXICO)
Jose Manuel Muradas Rodriguez 759.02 kmh
José Alfredo Gastelum Arce 471.63 mph
Dassault Falcon 2000
2 GE/Honeywell CFE738, 5,888 lbs
11/8/02

TOLUCA/STEPHENVILLE (MEXICO)
Jose Manuel Muradas Rodriguez 833.08 kmh
José Alfredo Gastelum Arce 517.65 mph
Dassault Falcon 2000
2 GE/Honeywell CFE738, 5,888 lbs
11/8/02

TORONTO/FORT FRANCIS (CANADA)
Terrance L. Jantzi 180.29 kmh
Joël Jantzi 112.02 mph
Van's RV-6
11/14/98

TORONTO/NASSAU (CANADA)
Louis Sytsma 619.63 kmh
Kevin Spackman 385.02 mph
DHC-8 Dash 8 Q400
2 P&W PW150, 5100 shp
10/10/00

TORONTO/QUEBEC CITY (CANADA)
Terrance L. Jantzi 309.71 kmh
Van's RV-6 192.44 mph
8/22/98

TORONTO/SAULT STE. MARIE (CANADA)
Terrance L. Jantzi 204.76 kmh
Joël Jantzi 127.23 mph
Van's RV-6
11/14/98

TORONTO/WINNIPEG (CANADA)
Terrance L. Jantzi 179.65 kmh
Joël Jantzi 111.63 mph
Van's RV-6
11/14/98

TOULOUSE/SANTIAGO (FRANCE)
Patrick Baudry 787.62 kmh
Airbus A330-222 489.40 mph
3/21/98

TRABZON/ADANA (CANADA)
Dawn Bartsch 309.38 kmh
Gordon Bartsch 192.24 mph
Cessna 421B
9/20/97

TRENTON/THULE*
Capt. James C. Fleming, USAF 486.31 mph
Lockheed C-141A, #70003
4 P&W TF-33-P-7
12/15/78

TRONDHEIM/FAIRBANKS (USA)
Brooke Knapp 743.04 kmh
Gulfstream III 461.72 mph
2 RR Spey MK511-8, 11,400 lbs
11/18/83

TRONDHEIM/LOS ANGELES (USA)
Brooke Knapp 658.80 kmh
Gulfstream III 409.38 mph
2 RR Spey MK511-8 11,400 lbs
11/18/83

TRUK/NANDI(USA)
Douglas G. Matthews 714.24 kmh
Learjet 36 443.83 mph
2 Garrett TFE-731-2-2B, 3,500 lbs
7/24/84

TULSA/POINT BARROW (USA)
Mary G. Kelly, CoCaptain 66.61 kmh
Joe Cunningham, CoCaptain 41.39 mph
Piper Cherokee, N6544YR
1 Lycoming 0-360
8/10-15/88

TUNIS/PALMA DE MAJORCA (USA)
James P. Hanson 254.08 kmh
Cessna 208 Caravan 157.88 mph
1 P&W PT-6A, 600 shp
5/5/96

ULAANBAATAR/CHICAGO (USA)
Larry S. Mueller 766.13 kmh
William J. Watters 476.05 mph
Kent R. Crenshaw
Gulfstream V
2 BMW R-R BR710, 14,750 lbs
6/27/00

VAGAR/REYKJAVIK (USA)
James P. Hanson 347.09 kmh
Cessna 208 Caravan 215.67 mph
1 P&W PT-6A, 600 shp
5/13/96

VALDOSTA, GA/WASHINGTON (USA)
Paul Neuda 434.16 kmh
Piper Aerostar, N110JC 269.79 mph
2 Lycoming IO-540
5/17/88

VANCOUVER/KAILUA (UK)
Max Robinson 1,595.75 kmh
BAe/Aérospatiale Concorde 991.55 mph
10/24/99

VAN NUYS/FT. LAUDERDALE (USA)
Mark Huffstutler 607.67 kmh
Citation Eagle XP 377.58 mph
2 P&W JT15D, 2,200 lbs
2/29/92

VAN NUYS/WHITE PLAINS (USA)
Benjamin M. Budzowski 958.22 kmh
Mark O. Schlegel 595.40 mph
Cessna Model 750, Citation X
2 Allison GMA 3007C, 6,000 lbs
5/11/95

VENICE/LISBON (UK)
Geoffrey Mussett 965.43 kmh
BAe/Aérospatiale Concorde 599.89 mph
11/14/99

VENICE/NEW YORK (UK)
Geoffrey Mussett 909.16 kmh
BAe/Aérospatiale Concorde 564.92 mph
11/14/99

VICTORIA/DEER LAKE (CANADA)
Daniel J. Leslie 93.19 kmh
Joseph Leslie 57.91 mph
Michael H. Leslie
Cessna 172
1 Lycoming O-320, 160 bhp
7/31/00

VICTORIA/FRESNO (USA)
Susan Nealey 350.14 kmh
Faith Hillman 217.56 mph
Cessna 310
2 Continental IO-520, 285 bhp
7/1/92

VICTORIA/OTTAWA (CANADA)
Graham A. Palmer 20.92 kmh
Cessna 140 12.99 mph
1 Continental C-85, 85 bhp
9/21/95

VICTORIA/PRINCE RUPERT (CANADA)
Daniel T. Bartie 232.50 kmh
Thomas L. Bartie 144.47 mph
Cessna 172Q
1 Avco-Lycoming, 180 bhp
6/3/00

VICTORIA/ST. JOHN'S, NF (CANADA)
Daniel J. Leslie 53.51 kmh
Michael H. Leslie 33.25 mph
Joseph Leslie
Cessna 172N
1 Lycoming O-320, 160 bhp
8/2/00

WABUSH/IQALUIT (CANADA)
Terrance L. Jantzi 209.09 kmh
Lauren V. Jantzi 129.92 mph
Van's RV-6
6/4/99

WABUSH/KUUJJUAQ (CANADA)
Terrance L. Jantzi 253.74 kmh
Lauren V. Jantzi 157.67 mph
Van's RV-6
6/4/99

WASHINGTON/ADDIS ABABA (USA)
CPT Zeleke Demissie 870.79 kmh
Boeing 767-200 ER 541.11 mph
2 P&W JT 9D 7R 4E 50,000 lbs
6/1/84

WASHINGTON/BONN (USA)
Jacqueline Cochran 577.62 kmh
Lockheed Jetstar 358.92 mph
4 P&W JT12A-6
11 hrs 5 min 12.1 sec
4/22/62

WASHINGTON/BOSTON (USA)
Jacqueline Cochran 767.07 kmh
Lockheed Jetstar 476.64 mph
4 P&W JT12A-6
49 min 29.7 sec
4/22/62

WASHINGTON/CARACAS (UK)
J. W. Hackett 791.69 kmh
English Electric Canberra T4 491.93 mph
RR Avon RA-3
4 hrs 10 min 59.75 sec
2/22/58

WASHINGTON/CHICAGO*
James M. Frankard 204.19 mph
Glasair III
1 Lycoming IO-540, 300 bhp
4/27/96

WASHINGTON/DUBAI (USA)
Bernard R. Ozbolt 895.59 kmh
Henry K. Gibson 556.49 mph
Robert S. McKenney
Gulfstream V
2 BMW RR BR 710-48, 14,750 lbs
11/13/97

WASHINGTON/GANDER (USA)
Jacqueline Cochran 800.68 kmh
Lockheed Jetstar 497.52 mph
4 P&W JT12A-6
2 hrs 37 min 48.4 sec
4/22/62

WASHINGTON/GOOSE BAY (USA)
Susan Nealey 290.71 kmh
Faith Hillman 180.64 mph
Cessna 310
2 Continental IO-520, 285 bhp
6/15/92

WASHINGTON/HAVANA (USA)
W. W. Edmondson — 506.09 kmh
North American P-51D, NX-4E — 314.07 mph
1 Packard RR Merlin V-1650
3 hrs 37 min 28.6 sec
11/25/47

WASHINGTON/JUNEAU (USA)
James T. Brown — 693.31 kmh
Donald M. Majors — 430.80 mph
IAI 1125 Astra SPX
2 Allied Signal TFE731, 4,250 lbs
11/12/96

WASHINGTON/LONDON (UK)
Brian Walpole — 1,712.49 kmh
BAe/Aérospatiale Concorde — 1,064.14 mph
4 RR Olympus, 38,000 lbs
9/17/86

WASHINGTON/MADRID (USA)
MAJ Clyde P. Evely, USAF — 922.35 kmh
Boeing B-52H Stratofortress — 573.12 mph
8 P&W TF-33P-3
6 hrs 36 min 38.98 sec
1/10/62

WASHINGTON/MOSCOW (USA)
COL James B. Swindal, USAF — 948.13 kmh
Boeing VC-137 — 589.14 mph
4 P&W JT3D, 18,000 lbs
5/19/63

WASHINGTON/NEW YORK (USA)
Jacqueline Cochran — 757.80 kmh
Lockheed Jetstar — 470.88 mph
4 P&W JT12A-6
26 min 9.6 sec
4/22/62

WASHINGTON/NOVOSIBIRSK (USA)
Brooke Knapp — 749.51 kmh
Gulfstream III — 465.75 mph
2 RR Spey MK511-8, 11,400 lbs
2/13-14/84

WASHINGTON/OSLO (USA)
COL James B. Swindal, USAF — 948.13 kmh
Boeing VC-137 — 589.14 mph
4 P&W JT3D
6 hrs 34 min 49.9 sec
5/19/63

WASHINGTON/PARIS (USA)
MAJ W.R. Payne, USAF — 1,687.69 kmh
Convair B-58 Hustler — 1,048.68 mph
4 GE J-79 5B
3 hrs 39 min 49 sec
5/26/61

WASHINGTON/SHANNON (USA)
Jacqueline Cochran — 583.99 kmh
Lockheed Jetstar — 362.88 mph
4 P&W JT12A-6
9 hrs 2 min 53.6 sec
4/22/62

WASHINGTON/STOCKHOLM (USA)
COL James B. Swindal, USAF — 940.77 kmh
Boeing VC-137 — 584.56 mph
4 P&W JT3D-3
7 hrs 3 min 33.4 sec
5/19/63

WELLINGTON, NZ/LONDON (UK)
A. F. Clouston — 134.30 kmh
de Havilland 88 Comet — 83.45 mph
2 D. H. Gipsy VI
140 hrs 12 min
3/20/38

WICHITA/BANGOR (USA)
Richard Trissell — 944.41 kmh
Benjamin M. Budzowski — 586.82 mph
Russell W. Meyer
Cessna Model 750, Citation X
2 Allison GMA 3007C, 6,000 lbs
6/6/95

WICHITA/GENEVA (USA)
Robert W. Agostino — 770.59 kmh
Rick E. Rowe — 478.82 mph
Learjet 60
2 P&WC PW305, 4,600 lbs
6/7/95

WICHITA/KITTY HAWK (USA)
Clay Lacy — 1,054.01 kmh
Alex A. Kvassay — 654.93 mph
Gulfstream IISP
2 R-R Spey Mk511-8, 11,400 lbs
12/15/03

WICHITA/PARIS (USA)
Peter T. Reynolds — 895.28 kmh
Alain Lacharite — 556.30 mph
Geoff M. Foster
Jeff Kirdeikis
Bombardier BD 700
2 BMW RR BR710, 14,970 lbs
6/12/97

WINSTON SALEM, NC/LAS VEGAS (USA)
Richard S. Wimbish, Pilot — 341.64 kmh
Dorothy T. Wimbish, Copilot — 212.30 mph
Piper Aerostar, N6899X
2 Lycoming IO-540 AA1A5
5/19/89

YAKUTSK/ANADYR (USA)
Susan Nealey — 317.14 kmh
Faith Hillman — 197.06 mph
Cessna 310
2 Continental IO-520, 285 bhp
6/28/92

YAKUTSK/NOME (USA)
Susan Nealey — 236.97 kmh
Faith Hillman — 147.24 mph
Cessna 310
2 Continental IO-520, 285 bhp
6/28/92

YAKUTSK/VICTORIA (USA)
Susan Nealey — 118.04 kmh
Faith Hillman — 73.35 mph
Cessna 310
2 Continental IO-520, 285 bhp
6/27-30/92

YOKOTA/BEIJING (USA)
LTC Royce G. Grones, USAF — 512.75 kmh
MAJ Robyn S. Read, USAF — 318.63 mph
Boeing C-135C 61-2669
4 P&W TF33-P5
4 hrs
10/3/85

YOUNGSTOWN/FARMINGDALE (USA)
Steve Kahn — 254.80 kmh
Mooney 201 — 158.33 mph
1 Lycoming IO-360, 200 bhp
11/9/91

ROUND TRIP

AUSTIN/FT. STOCKTON/AUSTIN (USA)
Charles I. Fitzsimmons — 405.00 kmh
Mooney 252TSE, N252YM — 251.67 mph
1 Continental TSIO-360, 210 bhp
3/7/87

BELFAST/GANDER/BELFAST (UK)
R. P. Beamont 663.04 kmh
English Elec Canberra, VX185 411.99 mph
2 RR Avon
10 hrs 3 min 29.28 sec
8/26/52

BUTTRESS/BRANDON/BUTTRESS (CANADA)
Lyle Johnson 70.95 kmh
Challenger II Special 44.09 mph
8/29/98

LONDON/AUCKLAND/LONDON (UK)
J. W. Burton 537.79 kmh
BAe/Aérospatiale Concorde 334.18 mph
4 Olympus 593/610, 38,000 lbs
4/5-8/86

LONDON/CAPETOWN/LONDON (UK)
Capt Brian Walpole 625.55 kmh
BAe/Aérospatiale Concorde 388.71 mph
4 RR Olympus, 38,000 lbs
30 hours 59 min 12 sec
3/28-29/85

LONDON/COPENHAGEN/LONDON (UK)
Jan Zurakowski 770.20 kmh
Gloster Meteor MK F8 478.58 mph
2 RR Derwent
2 hrs 29 min 8 sec
4/4/50

LONDON/NEW YORK/LONDON (UK)
J. W. Hackett 774.94 kmh
English Elec Canberra PR MK7 481.52 mph
2 RR Avon
14 hrs 21 min 45.5 sec
8/23/55

LONDON/SYDNEY/LONDON (UK)
CPT Hector McMullen 551.03 kmh
British Airways 342.40 mph
BAe/Aérospatiale Concorde
4 RR Snecma Olympus, 38,000 lbs
61 hrs 50 min 30 sec
2/13-2/16/85

LOS ANGELES/NEW YORK/LOS ANGELES (USA)
CPT Robert G. Sowers, USAF 1,681.71 kmh
Convair B-58 Hustler 1,044.46 mph
4 GE J-79-5B
4 hrs 41 min 14.98 sec
3/5/62

OWEN SOUND/QUEBEC/OWEN SOUND (CANADA)
C. Nelson Gain 137.49 kmh
Ultravia Aero Pelican GS 85.43 mph
1 Rotax 912, 80 bhp
6/26/96

RENO/GALVESTON/RENO (USA)
William S. Dirks 876.66 kmh
Jeffrey C. Brollier 544.73 mph
Cessna Citation X
2 Allison AE 3007C, 6,400 lbs
Reno, NV - Galveston, TX
9/13/97

TORONTO/QUEBEC CITY/TORONTO (CANADA)
Terrance L. Jantzi 193.42 kmh
RV-6 120.19 mph
8/22/98

VICTORIA/OTTAWA/VICTORIA (CANADA)
Graham A. Palmer 20.92 kmh
Cessna 140 12.99 mph
1 Continental C-85, 85 bhp
9/21/95

• • •

CLASS C-1.A/O (UNDER 300 kg / 661 lbs)
GROUP I (PISTON ENGINE)

ATLANTA/CEDAR KEY (USA)
Peter H. Burgher 10.66 kmh
Quicksilver MX 6.62 mph
1 Cuyuna, 30 bhp
7/28-30/82

ATLANTA/ST. PETERSBURG (USA)
Peter H. Burgher 9.30 kmh
Quicksilver MX 5.78 mph
1 Cuyuna, 30 bhp
7/28-31/82

BRANDON/BUTTRESS (CANADA)
Lyle Johnson 76.06 kmh
Challenger II Special 47.26 mph
8/29/98

BRANDON/MOOSE JAW (CANADA)
Lyle Johnson 77.41 kmh
Challenger II Special 48.10 mph
8/29/98

BUTTRESS/BRANDON (CANADA)
Lyle Johnson 93.31 kmh
Challenger II Special 57.98 mph
8/29/98

CHATTANOOGA/CEDAR KEY (USA)
Peter H. Burgher 10.69 kmh
Quicksilver MX 6.64 mph
1 Cuyuna, 30 bhp
7/27-30/82

CHATTANOOGA/ST PETERSBURG (USA)
Peter H. Burgher 9.54 kmh
Quicksilver MX 5.93 mph
1 Cuyuna, 30 bhp
7/27-31/82

CINCINNATI/ATLANTA (USA)
Peter H. Burgher 12.90 kmh
Quicksilver MX 8.02 mph
1 Cuyuna, 30 bhp
7/26-28/82

CINCINNATI/CEDAR KEY (USA)
Peter H. Burgher 11.52 kmh
Quicksilver MX 7.16 mph
1 Cuyuna, 30 bhp
7/26-30/82

CINCINNATI/CHATTANOOGA (USA)
Peter H. Burgher 14.20 kmh
Quicksilver MX 8.82 mph
1 Cuyuna, 30 bhp
7/26-27/82

CINCINNATI/KNOXVILLE (USA)
Peter H. Burgher 14.40 kmh
Quicksilver MX 8.95 mph
1 Cuyuna, 30 bhp
7/26-27/82

CINCINNATI/ST. PETERSBURG (USA)
Peter H. Burgher 10.51 kmh
Quicksilver MX 6.53 mph
1 Cuyuna, 30 bhp
7/26-31/82

DAYTON/ATLANTA (USA)
Peter H. Burgher 14.00 kmh
Quicksilver MX 8.70 mph
1 Cuyuna, 30 bhp
7/26-28/82

DAYTON/CEDAR KEY (USA)
Peter H. Burgher 12.04 kmh
Quicksilver MX 7.48 mph
1 Cuyuna, 30 bhp
7/26-30/82

DAYTON/CHATTANOOGA (USA)
Peter H. Burgher 15.70 kmh
Quicksilver MX 9.76 mph
1 Cuyuna, 30 bhp
7/26-27/82

DAYTON/KNOXVILLE (USA)
Peter H. Burgher 16.10 kmh
Quicksilver MX 10.00 mph
1 Cuyuna, 30 bhp
7/26-27/82

DAYTON/ST. PETERSBURG (USA)
Peter H. Burgher 10.93 kmh
Quicksilver MX 6.79 mph
1 Cuyuna, 30 bhp
7/26-31/82

DETROIT/ATLANTA (USA)
Peter H. Burgher 12.80 kmh
Quicksilver MX 7.95 mph
1 Cuyuna, 30 bhp
7/25-28/82

DETROIT/CEDAR KEY (USA)
Peter H. Burgher 11.70 kmh
Quicksilver MX 7.27 mph
1 Cuyuna, 30 bhp
7/25-30/82

DETROIT/CHATTANOOGA (USA)
Peter H. Burgher 13.60 kmh
Quicksilver MX 8.45 mph
1 Cuyuna, 30 bhp
7/25-27/82

DETROIT/CINCINNATI (USA)
Peter H. Burgher 13.30 kmh
Quicksilver MX 8.26 mph
1 Cuyuna, 30 bhp
7/25-26/82

DETROIT/DAYTON (USA)
Peter H. Burgher 25.44 kmh
Quicksilver MX 15.80 mph
1 Cuyuna, 30 bhp
7/25/82

DETROIT/KNOXVILLE (USA)
Peter H. Burgher 13.30 kmh
Quicksilver MX 8.26 mph
1 Cuyuna, 30 bhp
7/25-27/82

DETROIT/ST. PETERSBURG (USA)
Peter H. Burgher 10.80 kmh
Quicksilver MX 6.71 mph
1 Cuyuna, 30 bhp
7/25-31/82

KNOXVILLE/ATLANTA *
Peter H. Burgher 7.16 mph
Quicksilver MX
1 Cuyuna, 30 bhp
7/27-28/82

KNOXVILLE/CEDAR KEY (USA)
Peter H. Burgher 10.55 kmh
Quicksilver MX 6.56 mph
1 Cuyuna, 30 bhp
7/27-30/82

KNOXVILLE/ST PETERSBURG (USA)
Peter H. Burgher 9.52 kmh
Quicksilver MX 5.92 mph
1 Cuyuna, 30 bhp
7/27-31/82

LONDON/PARIS (FRANCE)
A. Flotard 138.98 kmh
Marcos J5 86.35 mph
9/1/91

PARIS/LONDON (FRANCE)
A. Flotard 175.35 kmh
Marcos J5 108.95 mph
9/1/91

PRAGUE/KRAKOW (FRANCE)
Alain Flotard 148.23 kmh
Marcos 92.10 mph
7/17/90

SAN DIEGO/KITTY HAWK (USA)
Joanne Anderson 4.15 kmh
Quicksilver MXL 2.58 mph
Rotax-Bombardier, 33.5 bhp
3/29-5/6/83

SAN FRANCISO/NEW YORK (USA)
Joe Tong 10.60 kmh
CGS Hawk 6.59 mph
1 Cuyuna, 30 bhp
6/22-7/8/83

STRASBOURG/PRAGUE (FRANCE)
Alain Flotard 169.14 kmh
Marcos 105.09 mph
7/16/90

WARSAW/PRAGUE (FRANCE)
Alain Flotard 149.38 kmh
Marcos 92.82 mph
7/20/90

ROUND TRIP

BUTTRESS/BRANDON/BUTTRESS (CANADA)
Lyle Johnson 70.95 kmh
Challenger II Special 44.09 mph
8/29/98

CLASS C-1.A (300-500 kg / 661-1,102 lbs)
GROUP I (PISTON ENGINE)

ADELAIDE/BRISBANE (AUSTRALIA)
Michael Coates 215.13 kmh
TL-2000 Sting Carbon 133.67 mph
5/2/02

AJACCIO/LONDON (UK)
Peter John Calvert 155.37 kmh
Rutan Varieze 96.54 mph
1 Continental R.R. 0-200-A, 100 bhp
6/26/80

AMSTERDAM/LONDON (UK)
Victor Davies 246.11 kmh
Taylor JT2 152.92 mph
9/15/90

BERLIN/LONDON (UK)
Victor Davies 131.05 kmh
Taylor JT2 81.43 mph
7/26/92

BERNE/MALTA (UK)
Victor Davies 59.36 kmh
Taylor JT2 36.88 mph
6/12-13/93

BRUSSELS/LONDON (UK)
Victor Davies
Taylor JT2
9/5/90
216.89 kmh
134.76 mph

BRISBANE/MELBOURNE (AUSTRALIA)
Lennard S. Dyson
Viking Dragonfly
9/16-17/85
55.12 kmh
34.25 mph

BRISBANE/TOWNSVILLE (GERMANY)
Wilhelm Ewig
WD D4 BK Fascination
1 Rotax 912 ULS, 100 bhp
10/9/02
202.78 kmh
126.00 mph

BURKETOWN/DARWIN (GERMANY)
Wilhelm Ewig
WD D4 BK Fascination
1 Rotax 912 ULS, 100 bhp
10/10/02
193.55 kmh
120.27 mph

COFF'S HARBOUR/LORD HOWE ISLAND (NEW ZEALAND)
John Bolton-Riley
Sky Arrow 480T
8/5/97
151.64 kmh
94.22 mph

COPENHAGEN/HELSINKI (UK)
Victor Davies
Taylor JT2
9/15/91
34.34 kmh
21.33 mph

COPENHAGEN/LONDON (UK)
Victor Davies
Taylor JT2
9/18/91
22.50 kmh
13.98 mph

COPENHAGEN/OSLO (UK)
Victor Davies
Taylor JT2
9/15/91
9.75 kmh
6.05 mph

COPENHAGEN/STOCKHOLM (UK)
Victor Davies
Taylor JT2
9/15/91
234.99 kmh
146.01 mph

DARWIN/BROOME (GERMANY)
Wilhelm Ewig
WD D4 BK Fascination
1 Rotax 912 ULS, 100 bhp
10/11/02
218.77 kmh
135.94 mph

DAYTON/OSHKOSH (USA)
Jack Halbeisen
Pioneer Flightstar
1 Kawasaki 440A, 37 bhp
7/24-25/85
15.07 kmh
9.36 mph

DAYTON/ST LOUIS (USA)
Jack Halbeisen
Pioneer Flightstar
1 Kawasaki 440A, 37 bhp
7/31-8/1/84
16.73 kmh
10.40 mph

DUBLIN/LONDON (UK)
Victor Davies
Taylor JT2
10/4/92
203.35 kmh
126.35 mph

EDINBURGH/LONDON (UK)
Victor Davies
Taylor JT2
6/3/93
176.93 kmh
109.93 mph

ELLINGTON/DAYTON (USA)
Jack Halbeisen
Pioneer Flightstar
1 Kawasaki 440A, 37 bhp
7/25-29/84
10.19 kmh
6.33 mph

GLASGOW/LEICESTER (UK)
Graeme Park
Denney Aerocraft Kitfox III
1 Rotax 912, 80 bhp
6/20/02
119.56 kmh
74.29 mph

GLASGOW/LONDON (UK)
Graeme Park
Denney Aerocraft Kitfox III
1 Rotax 912, 80 bhp
6/20/02
118.31 kmh
73.51 mph

GUERNSEY/LONDON (UK)
Victor Davies
Taylor JT2
10/8/92
192.72 kmh
119.75 mph

HELSINKI/COPENHAGEN (UK)
Victor Davies
Taylor JT2
9/16/91
20.08 kmh
12.47 mph

HELSINKI/LONDON (UK)
Victor Davies
Taylor JT2
9/16/91
20.39 kmh
12.66 mph

HELSINKI/OSLO (UK)
Victor Davies
Taylor JT2
9/16/91
35.28 kmh
21.92 mph

HELSINKI/STOCKHOLM (UK)
Victor Davies
Taylor JT2
9/16/91
199.89 kmh
124.20 mph

INVERCARGILL/KAITAIA (NEW ZEALAND)
Ian R. Todd
Titan Tornado II 912
4/12-13/98
128.42 kmh
79.80 mph

INVERCARGILL/RANGIORA (NEW ZEALAND)
Julie Cook
Avid Mark III
3/12/98
111.44 kmh
69.25 mph

KAITAIA/NORFOLK ISLAND (NEW ZEALAND)
John Bolton-Riley
Sky Arrow 480T
7/25/97
168.30 kmh
104.58 mph

KITTY HAWK/DAYTON (USA)
Jack Halbeisen
Pioneer Flightstar
1 Kawasaki 440A, 37 bhp
9/23-28/84
6.90 kmh
4.29 mph

LINCOLN/SIOUX CITY (USA)
Robert L. Carlisle
Bowers FB-1A, N4688T
1 Continental A-65, 65 bhp
5/20/73
160.30 kmh
99.61 mph

LISBON/LONDON (UK)
Victor Davies
Taylor JT2
9/16/92
28.51 kmh
17.71 mph

LISBON/MADRID (UK)
Victor Davies
Taylor JT2
9/16/92
94.20 kmh
58.53 mph

LONDON/AMSTERDAM (UK)
Victor Davies
Taylor JT2
9/15/90
211.92 kmh
131.68 mph

LONDON/BERLIN (UK)
Victor Davies 159.21 kmh
Taylor JT2 98.92 mph
7/25/92

LONDON/BERNE (UK)
Victor Davies 119.26 kmh
Taylor JT2 74.10 mph
6/9/93

LONDON/BRUSSELS (UK)
Victor Davies 234.46 kmh
Taylor JT2 145.68 mph
9/5/90

LONDON/COPENHAGEN (UK)
Victor Davies 177.09 kmh
Taylor JT2 110.03 mph
9/14/91

LONDON/DUBLIN (UK)
Victor Davies 210.35 kmh
Taylor JT2 130.70 mph
4/10/92

LONDON/EDINBURGH (UK)
Victor Davies 21.03 kmh
Taylor JT2 13.06 mph
5/31/93

LONDON/GIBRALTAR (UK)
Paul Wright 140.75 kmh
Quickie 2 87.45 mph
9/13/90

LONDON/GUERNSEY (UK)
Victor Davies 203.08 kmh
Taylor JT2 126.19 mph
10/8/92

LONDON/HELSINKI (UK)
Victor Davies 39.00 kmh
Taylor JT2 24.23 mph
9/14/91

LONDON/LISBON (UK)
Victor Davies 30.63 kmh
Taylor JT2 19.03 mph
9/9/92

LONDON/LUXEMBOURG (UK)
Victor Davies 184.15 kmh
Taylor JT2 114.43 mph
6/9/93

LONDON/MADRID (UK)
Victor Davies 49.75 kmh
Taylor JT2 30.91 mph
9/9/92

LONDON/MALTA (UK)
Victor Davies 21.26 kmh
Taylor JT2 13.21 mph
6/9-13/93

LONDON/OSLO (UK)
Victor Davies 16.79 kmh
Taylor JT2 10.43 mph
9/14/91

LONDON/PARIS (UK)
Victor Davies 202.64 kmh
Taylor JT2 125.91 mph
7/14/90

LONDON/STOCKHOLM (UK)
Victor Davies 62.68 kmh
Taylor JT2 38.94 mph
9/14/91

LORD HOWE ISLAND/COFF'S HARBOUR (NEW ZEALAND)
John Bolton-Riley 141.39 kmh
Sky Arrow 480T 87.86 mph
7/27/97

LORD HOWE ISLAND/NORFOLK ISLAND (NEW ZEALAND)
John Bolton-Riley 149.06 kmh
Sky Arrow 480T 92.62 mph
8/6/97

LUXEMBOURG/BERNE (UK)
Victor Davies 199.48 kmh
Taylor JT2 123.95 mph
6/9/93

LUXEMBOURG/MALTA (UK)
Victor Davies 17.93 kmh
Taylor JT2 11.14 mph
6/9-13/93

MADRID/A CORUÑA (SPAIN)
Miguel Angel Gordillo 191.44 kmh
Dyn'Aero MCR01 118.95 mph
1 Rotax 912S, 100 bhp
4/19/01

MADRID/LISBON (UK)
Victor Davies 129.04 kmh
Taylor JT2 80.18 mph
9/11/92

MADRID/LONDON (UK)
Victor Davies 39.24 kmh
Taylor JT2 24.38 mph
9/17/92

MALTA/BERNE (UK)
Victor Davies 27.28 kmh
Taylor JT2 16.95 mph
6/16-18/93

MALTA/LONDON (UK)
Peter John Calvert 145.05 kmh
Rutan Varieze 90.13 mph
1 Continental R.R. 0-200-A, 100 bhp
6/26/80

MALTA/ROME (UK)
Victor Davies 123.95 kmh
Taylor JT2 77.02 mph
6/16/93

MANILLA/BRISBANE (GERMANY)
Wilhelm Ewig 213.06 kmh
WD D4 BK Fascination 132.39 mph
1 Rotax 912 ULS, 100 bhp
10/8/02

MEMPHIS/KITTY HAWK (USA)
Jack Halbeisen 4.68 kmh
Pioneer Flightstar 2.91 mph
1 Kawasaki 440A, 37 bhp
9/11-22/84

NORFOLK ISLAND/AUCKLAND (NEW ZEALAND)
John Bolton-Riley 150.56 kmh
Sky Arrow 480T 93.55 mph
8/7/97

NORFOLK ISLAND/LORD HOWE ISLAND (NEW ZEALAND)
John Bolton-Riley 157.04 kmh
Sky Arrow 480T 97.58 mph
7/26/97

OAKLAND/LOS ANGELES (USA)
Lowell E. Richards — 76.32 kmh
Pioneer Flightstar 494 — 47.42 mph
1 Kawasaki TA44DA, 35 shp
4/6/84

OSHKOSH/SANTA PAULA (USA)
C. Michael Bowers — 50.04 kmh
Piper Cub — 31.09 mph
1 Continental, 85 bhp
7/29-31/91

OSLO/COPENHAGEN (UK)
Victor Davies — 22.60 kmh
Taylor JT2 — 14.04 mph
9/17/91

OSLO/LONDON (UK)
Victor Davies — 17.81 kmh
Taylor JT2 — 11.06 mph
9/17/91

OWEN SOUND/QUEBEC (CANADA)
C. Nelson Gain — 146.39 kmh
Ultravia Aero Pelican GS — 90.96 mph
1 Rotax 912, 80 bhp
6/26/96

OXFORD/PALMA DE MAJORCA (USA)
Simon P. Johnson — 46.41 kmh
Fournier RF-4B — 28.84 mph
5/31/98

PARIS/LONDON (UK)
Capt. Victor Davies — 221.41 kmh
Taylor JT2 — 137.57 mph
7/14/90

PERTH/MUNDRABILLA (GERMANY)
Wilhelm Ewig — 213.51 kmh
WD D4 BK Fascination — 132.67 mph
1 Rotax 912 ULS, 100 bhp
10/13/02

PONTOISE/ROVANIEMI (FRANCE)
Denis Guillotel — 238.69 kmh
DG1 — 148.31 mph
1 Jabiru 3300, 120 bhp
8/15/02

QUEBEC/OWEN SOUND (CANADA)
C. Nelson Gain — 151.02 kmh
Ultravia Aero Pelican GS — 93.84 mph
1 Rotax 912, 80 bhp
6/26/96

RANGIORA/INVERCARGILL (NEW ZEALAND)
Julie Cook — 112.46 kmh
Avid Mark III — 69.88 mph
3/13/98

ROME/BERNE (UK)
Victor Davies — 15.31 kmh
Taylor JT2 — 9.51 mph
6/16-18/93

ROSWELL/SAN DIEGO (USA)
Jack Halbeisen — 11.16 kmh
Pioneer Flightstar — 6.93 mph
1 Kawasaki 440A, 37 bhp
8/12-16/84

ROVANIEMI/PONTOISE (FRANCE)
Denis Guillotel — 227.24 kmh
DG1 — 141.20 mph
1 Jabiru 3300, 120 bhp
8/19/02

SACRAMENTO/LOS ANGELES (USA)
Lowell E. Richards — 67.00 kmh
Pioneer Flightstar 494 — 41.63 mph
1 Kawasaki TA44DA, 35 shp
4/6/84

ST LOUIS/ROSWELL (USA)
Jack Halbeisen — 10.07 kmh
Pioneer Flightstar — 6.25 mph
1 Kawasaki 440A, 37 bhp
8/2-8/84

SAN DIEGO/MEMPHIS (USA)
Jack Halbeisen — 4.35 kmh
Pioneer Flightstar — 2.70 mph
1 Kawasaki 440A, 37 bhp
8/17-9/10/84

SAN FRANCISCO/CHICAGO (USA)
Lowell E. Richards — 29.16 kmh
Pioneer Flightstar — 18.12 mph
1 Kawasaki
TS44 OA, 35 bhp
6/6-10/84

SAN FRANCISCO/LOS ANGELES (USA)
C. Michael Bowers — 104.33 kmh
Piper J-3 Cub, N38003 — 64.83 mph
1 Lycoming C-85, 85 bhp
5/25/87

SAN FRANCISCO/NEW YORK (USA)
Lowell E. Richards — 24.44 kmh
Pioneer Flightstar — 15.18 mph
1 Kawasaki TS44 OA, 35 bhp
6/6-12/84

SAN FRANCISCO/SALT LAKE CITY (USA)
Lowell E. Richards — 19.07 kmh
Pioneer Flightstar — 11.85 mph
1 Kawasaki TS44 OA, 35 bhp
6/6-8/84

SHEPPARTON/SYDNEY (GERMANY)
Wilhelm Ewig — 248.65 kmh
WD D4 BK Fascination — 154.50 mph
1 Rotax 912 ULS, 100 bhp
10/14/02

SIOUX FALLS/OSHKOSH (USA)
Robert G. Ray, Jr. — 156.98 kmh
Taylorcraft BD-12D — 97.54 mph
1 Continental A-65, 65 bhp
7/29/94

STOCKHOLM/COPENHAGEN (UK)
Victor Davies — 216.32 kmh
Taylor JT2 — 134.41 mph
9/18/91

STOCKHOLM/HELSINKI (UK)
Victor Davies — 223.52 kmh
Taylor JT2 — 138.88 mph
9/16/91

STOCKHOLM/LONDON (UK)
Victor Davies — 30.38 kmh
Taylor JT2 — 18.87 mph
9/18/91

TOWNSVILLE/BURKETOWN (GERMANY)
Wilhelm Ewig — 207.92 kmh
WD D4 BK Fascination — 129.19 mph
1 Rotax 912 ULS, 100 bhp
10/9/02

ROUND TRIP

CHINO/HEMET/CHINO*
Margaret T. Wilson
Welsh Rabbit, N3599G
1 Continental A65, 65 bhp
4/19/80
75.95 mph

COPENHAGEN/HELSINKI/COPENHAGEN (UK)
Victor Davies
Taylor JT2
9/15/91
24.78 kmh
15.39 mph

COPENHAGEN/OSLO/COPENHAGEN (UK)
Victor Davies
Taylor JT2
9/15/91
13.24 kmh
8.22 mph

COPENHAGEN/STOCKHOLM/COPENHAGEN (UK)
Victor Davies
Taylor JT2
9/15/91
14.65 kmh
9.10 mph

FARGO/BISMARK/FARGO (USA)
Steven J. Adams
Boredom Fighter, N317MA
1 Continental, 65 bhp
8/19/89
144.00 kmh
89.48 mph

INVERCARGILL/RANGIORA/INVERCARGILL (NEW ZEALAND)
Julie Cook
Avid Mark III
3/13/98
111.44 kmh
69.25 mph

LONDON/BERLIN/LONDON (UK)
Victor Davies
Taylor JT2
7/26/92
57.05 kmh
35.44 mph

LONDON/COPENHAGEN/LONDON (UK)
Victor Davies
Taylor JT2
9/14/91
13.00 kmh
8.07 mph

LONDON/DUBLIN/LONDON (UK)
Victor Davies
Taylor JT2
10/4/92
136.95 kmh
85.09 mph

LONDON/EDINBURGH/LONDON (UK)
Victor Davies
Taylor JT2
5/31-6/3/93
13.87 kmh
8.61 mph

LONDON/GUERNSEY/LONDON (UK)
Victor Davies
Taylor JT2
10/8/92
139.90 kmh
86.93 mph

LONDON/HELSINKI/LONDON (UK)
Victor Davies
Taylor JT2
9/14/91
26.47 kmh
16.44 mph

LONDON/LISBON/LONDON (UK)
Victor Davies
Taylor JT2
9/11/92
14.09 kmh
8.75 mph

LONDON/MADRID/LONDON (UK)
Victor Davies
Taylor JT2
9/9/92
11.23 kmh
6.97 mph

LONDON/OSLO/LONDON (UK)
Victor Davies
Taylor JT2
9/14/91
17.03 kmh
10.58 mph

Mark Stolzberg (l) and Fred Coon (r), at the conclusion of their record-setting transcontinental flight in a Grumman Cheetah.

LONDON/PARIS/LONDON (UK)
Victor Davies
Taylor JT2
7/14/90
132.05 kmh
82.05 mph

LONDON/STOCKHOLM/LONDON (UK)
Victor Davies
Taylor JT2
9/14/91
20.76 kmh
12.89 mph

MADRID/LISBON/MADRID (UK)
Victor Davies
Taylor JT2
9/11/92
8.15 kmh
5.06 mph

OWEN SOUND/QUEBEC/OWEN SOUND (CANADA)
C. Nelson Gain
Ultravia Aero Pelican GS
1 Rotax 912, 80 bhp
6/26/96
137.49 kmh
85.43 mph

STOCKHOLM/HELSINKI/STOCKHOLM (UK)
Victor Davies
Taylor JT2
9/16/91
148.11 kmh
92.03 mph

CLASS C-1.B (500-1,000 kg / 1,102-2,205 lbs)
GROUP I (PISTON ENGINE)

EAST TO WEST, TRANSCONTINENTAL *
D. Wayne Woollard
Gary Tuovinen
Alon A-2 Aircoupe
1 Continental C90, 90 bhp
9/30/03
45.42 mph

WEST TO EAST, TRANSCONTINENTAL * ★
Fred M. Coon
Mark Stolzberg
Grumman AA-5A Cheetah
1 Lycoming O-320, 150 bhp
10/16/03
159.78 mph

Superseded
*WEST TO EAST, TRANSCONTINENTAL **
Assaf Stoler
D. Laksen Sirimanne
Diamond DA 40-180 Diamond Star
1 Lycoming IO-360, 180 bhp
7/18/03
136.79 mph

***

ABBEVILLE/MAIDENHEAD (UK)
Michael Ashfield
Piper PA-28-140
1 Lycoming O-320, 140 bhp
7/15/94
163.72 kmh
101.73 mph

ADELAIDE/AUCKLAND (AUSTRALIA)
Jon Johanson
Van's RV-4
1 Lycoming O-320, 160 bhp
2/6/96
302.17 kmh
187.76 mph

ADELAIDE/HOBART (AUSTRALIA)
Jon Johanson
Van's RV-4
1 Lycoming O-320, 160 bhp
12/20/94
311.41 kmh
193.50 mph

ADELAIDE/MELBOURNE (AUSTRALIA)
Jon Johanson
Van's RV-4
1 Lycoming O-320, 160 bhp
10/28/97
278.90 kmh
173.30 mph

ADELAIDE/OSHKOSH (AUSTRALIA)
Jon Johanson
Van's RV-4
1 Lycoming O-320, 160 bhp
7/27/95
21.32 kmh
13.24 mph

ADELAIDE/STAUNING (AUSTRALIA)
Jon Johanson
Van's RV-4
1 Lycoming, 160 bhp
6/9/00
26.44 kmh
16.43 mph

AGADIR/DAKAR (SWITZERLAND)
Hans G. Schmid
Long Ez
11/8/98
227.51 kmh
141.37 mph

ALBANY/ATLANTA*
William C. Brodbeck
Lark Commander, N1970L
1 Lycoming 0-360-A2F
6/9/70
133.21 mph

ALBANY/CARIBOU (USA)
Monie W. Pease
Grumman Tiger
1 Lycoming O-360, 180 bhp
9/28/90
259.90 kmh
161.49 mph

ALBANY/NEW YORK CITY*
Scott Olszewski, Pilot
Robert Olszewski, Copilot
Piper Arrow, N185X
1 Lycoming, 200 bhp
12/17/87
196.64 mph

ALBANY/READING*
William C. Brodbeck
Lark Commander, N1970L
1 Lycoming 0-360-A2F
6/9/70
123.68 mph

ALBUQUERQUE/LAS VEGAS (USA)
James Cline
Glasair SH-2
1 Lycoming O-320 D1D, 160 bhp
5/23/85
316.44 kmh
196.64 mph

ALDERNEY/CHERBOURG (UK)
Martin Evans
Jodel Mascaret D-150
1 Continental O-200, 100 bhp
6/16/85
43.51 kmh
27.04 mph

ALDERNEY/REDHILL (UK)
Martin Evans
Jodel Mascaret D-150
1 Continental O-200, 100 bhp
6/16/85
34.93 kmh
21.71 mph

ALLENTOWN/DAYTON (USA)
Edward S. Figuli, Jr.
Cessna 150
1 Lycoming O-320, 150 bhp
9/5/03
161.63 kmh
100.43 mph

ALLENTOWN/FT. SMITH (USA)
Edward S. Figuli, Jr.
Cessna 150
1 Lycoming O-320, 150 bhp
7/13/02
132.13 kmh
82.10 mph

AMSTERDAM/LONDON (UK)
T. W. Hayhow
Auster Aiglet
1 de Havilland Gipsy Major IG
1 hr 32 min 9.2 sec
4/14/52
234.69 kmh
145.83 mph

ARLINGTON, TN/ATHENS, TN (USA) ★
Dal Scott Anderson — 279.09 kmh
Swift GC-1B — 173.42 mph
1 Continental IO-360, 210 bhp
5/24/91

ARLINGTON, WA/LAKELAND, FL (USA) ★
Lance G. Turk — 339.89 kmh
Glasair I FT — 211.19 mph
1 Lycoming O-320, 160 bhp
04/07/90

ARLINGTON, WA/OSHKOSH (USA)
Lance G. Turk, Pilot — 295.51 kmh
Adam T. Turk, Navigator — 183.62 mph
Glasair I FT
1 Lycoming O-320, 160 bhp
07/27/90

ASCENSION ISLAND/RECIFE (AUSTRALIA)
Jon Johanson — 231.39 kmh
Van's RV-4 — 143.78 mph
1 Lycoming O-320, 160 bhp
7/16/96

ASHEBORO/KITTY HAWK (USA)
Kenneth Clark, Pilot — 227.52 kmh
N. B. Cheek, Copilot — 141.38 mph
Piper Cherokee, N83291
1 Lycoming 0360, 180 bhp
12/16/87

ASHEVILLE/DAYTONA BEACH (USA)
Reginald E. Moody, Jr. — 211.25 kmh
Piper PA-28-180 — 131.26 mph
1 Lycoming O-360, 180 bhp
8/31/95

ATHENS/MADRID (FRANCE)
Christophe de Brichambaut — 203.86 kmh
Long EZ — 126.68 mph
1 Lycoming, 115 CV
9/9/86

ATLANTA/KITTY HAWK (USA)
Linda Dickerson, Pilot — 211.02 kmh
John Eslinger, Copilot — 131.13 mph
Piper Warrior, N8466M
1 Lycoming 0320-D3G, 160 bhp
12/16/86

ATLANTA/READING (USA)
William C. Brodbeck — 202.00 kmh
Lark Commander, N1970L — 125.52 mph
1 Lycoming 0-360-A2F
5 hrs 11 min 52.2 sec
6/9/70

ATLANTA/TALLAHASSEE (USA)
Bryan W. Robinson — 264.14 kmh
Pitts Special S-1 — 164.13 mph
1 Lycoming IO-360, 180 bhp
1 hr 21 min 45.6 sec
10/8/72

AUSTIN/BIMINI (USA)
Donald P. Taylor — 250.92 kmh
Thorp T-18, N455DT — 155.91 mph
1 Lycoming 0-360-A2A, 180 bhp
7 hrs 29 min
5/24-25/80

BARBADOS/OSHKOSH (AUSTRALIA)
Jon Johanson — 17.51 kmh
Van's RV-4 — 10.88 mph
1 Lycoming O-320, 160 bhp
7/30/96

BERNE/LONDON (UK)
T. W. Hayhow — 178.67 kmh
Auster Aiglet — 111.02 mph
1 de Havilland Gipsy Major IG
4 hrs 11 min 7.6 sec
6/7/52

BILLUND/REYKJAVIK (BRAZIL)
Andre Joseph Deberdt — 230.85 kmh
Long-EZ — 143.44 mph
1 Lycoming O-235, 118 bhp
7/25/95

BRIDGEPORT/OSHKOSH (USA)
David H. Faile Jr., Pilot — 108.61 kmh
Jim O'Keefe, Copilot — 67.48 mph
Christen Eagle II
1 Lycoming IO-360, 200 bhp
7/28/93

BRISBANE/PERTH (AUSTRALIA)
Magna Liset — 263.33 kmh
Rutan Long EZ — 163.62 mph
3/3/90

BRISBANE/TOWNSVILLE (AUSTRALIA)
Jon Johanson — 314.12 kmh
Van's RV-4 — 195.19 mph
1 Lycoming O-320, 160 bhp
10/28/97

BROWNSVILLE/CORPUS CHRISTI (USA)
Rodolfo Faccini — 176.09 kmh
Pitts Special S-1C HK848-Z — 109.42 mph
1 Lycoming IO-360, 180 bhp
1 hr 11 min 58.9 sec
9/19/74

BRUSSELS/LONDON (UK)
T. W. Hayhow — 223.32 kmh
Auster Aiglet — 133.76 mph
1 de Havilland Gipsy Major IG
1 hr 26 min 38.6 sec
4/12/52

BUCHAREST/KIEV (ROMANIA)
Monolache, Onciu & Otoiu — 177.94 kmh
I. A. R. 813 — 100.57 mph
1 Walter Minor 4 111
4 hrs 11 min 43 sec
7/4/58

BURLINGTON/ALBANY*
Scott Olszewski, Pilot — 207.82 mph
Robert Olszewski, Copilot
Piper Arrow, N185X
1 Lycoming, 200 bhp
12/17/87

BURLINGTON/NEW YORK (USA)
R. Scott Olszewski, Pilot — 320.04 kmh
Robert F. Olszewski, Copilot — 198.87 mph
Arrow, N1585X
1 Lycoming PA 28R-200, 200 bhp
12/17/87

CALDWELL/PORTLAND, ME (USA)
Thomas A. Schror — 194.79 kmh
Cessna 172N — 121.04 mph
1 Lycoming O-320, 160 bhp
10/19/95

CAPETOWN/BARBADOS (AUSTRALIA)
Jon Johanson — 64.78 kmh
Van's RV-4 — 40.25 mph
1 Lycoming O-320, 160 bhp
7/18/96

CAPETOWN/LONDON (UK)
Alex Henshaw 243.74 kmh
Percival Mew Gull G-AEXF 151.46 mph
1 de Havilland Gipsy VI-2
39 hrs 36 min
2/7/39

CAPETOWN/OSHKOSH (AUSTRALIA)
Jon Johanson 31.52 kmh
Van's RV-4 19.59 mph
1 Lycoming O-320, 160 bhp
7/30/96

CAPETOWN/RECIFE (AUSTRALIA)
Jon Johanson 59.80 kmh
Van's RV-4 37.16 mph
1 Lycoming O-320, 160 bhp
7/16/96

CARDIFF/DUBLIN (UK)
William S. Allen 239.52 kmh
Long-Eze 148.84 mph
1 Lycoming 0235-C2A, 115 bhp
8/15/87

CARNARVON/PERTH (AUSTRALIA)
Jon Johanson 228.03 kmh
Van's RV-4 141.69 mph
1 Lycoming O-320, 160 bhp
10/29/97

CHARLESTON/LAKELAND (USA)
J. Allen Miles 195.63 kmh
Gulfstream American AA5A 121.56 mph
1 Lycoming O-320, 150 bhp
4/4/92

CHARLESTON/NEW ORLEANS*
J. Allen Miles 114.92 mph
Gulfstream American AA5A
1 Lycoming O-320, 150 bhp
10/22/91

CHERBOURG/REDHILL (UK)
Martin Evans 88.43 kmh
Jodel Mascaret D-150 54.95 mph
1 Continental O-200, 100 bhp
6/16/85

CHICAGO/BOSTON (USA) ✯
Henry M. Bouley Jr. 646.63 kmh
Questair Venture 401.79 mph
1 PMA 550-TTV, 350 bhp
1/26/93

CHRISTIANSTED/ROADTOWN (U.S. VIRGIN ISLANDS)
John Stuart-Jervis 183.69 kmh
Cessna 172 114.14 mph
1 Lycoming O-320, 150 bhp
2/9/95

CINCINNATI/LAKELAND (USA)
Rudolph Siegel 332.39 kmh
Long EZ 206.54 mph
1 RR O-240, 130 bhp
4/17/93

CLINTON/WARRENTON (USA)
John Roy "J. D." Doggett 157.24 kmh
Cessna 150M, N6482K 97.72 mph
1 Continental O-200, 100 bhp
5/9/87

COLLEGE PARK/KITTY HAWK*
John W. Strong, Pilot 221.02 kmh
Leonard J. Stone, Copilot 137.33 mph
Stinson 108
1 Franklin 6A4-150, 150 bhp
12/17/95

COLOGNE/MUNICH (W. GERMANY)
Berthold Sessler 97.50 kmh
Putzer-Elster B 60.59 mph
1 Continental C90, 90 bhp
4/1/82

COLUMBUS, MS/ATHENS, TN (USA) ✯
Charles E. Nelson 298.56 kmh
Swift GC-1B 185.51 mph
1 Continental IO-360, 210 bhp
5/24/91

COPENHAGEN/LONDON (UK)
T. W. Hayhow 219.42 kmh
Auster Aiglet 136.34 mph
1 de Havilland Gipsy Major IG
4 hrs 21 min 45 sec
10/11/52

CORPUS CHRISTI/HOUSTON (USA)
Rodolfo Faccini 179.92 kmh
Pitts Special S-1C, HK848-Z 111.80 mph
1 Lycoming IO-360, 180 bhp
1 hr 38 min 22.4 sec
9/19/74

COVINGTON, VA/KITTY HAWK (USA)
Ed Covington 354.60 kmh
Ashly Covington 220.34 mph
Glasair, N3EC
1 Lycoming IO-320
12/16/88

DAKAR/RECIFE (SWITZERLAND)
Hans Georg Schmid 230.83 kmh
Long-EZ 143.43 mph
11/10/98

DARWIN/CARNARVON (AUSTRALIA)
Jon Johanson 226.61 kmh
Van's RV-4 140.81 mph
1 Lycoming O-320, 160 bhp
10/29/97

DARWIN/SOUTHAMPTON (UK)
Anthony J. E. Smith 25.85 kmh
CASA 1 131 E 16.06 kmh
4/28 - 5/22/89

DARWIN/SYDNEY (AUSTRALIA)
John Fisher 20.83 kmh
de Havilland DH-82A, Tiger Moth 12.94 mph
1 de Havilland Gipsy Major 1C, 135 bhp
10/12/96

DAVID/PUNTA ARENAS (USA)
Rodolfo Faccini 198.30 kmh
Pitts Special S-1C, HK848-Z 123.22 mph
1 Lycoming IO-360, 180 bhp
1 hr 34 min 53.1 sec
9/16/74

DAYTON/KITTY HAWK (USA)
Robert F. Olszewski 346.29 kmh
Piper PA 28R-200; 215.18 mph
1 Lycoming IO-360C, 200 bhp
12/17/83

DAYTONA BEACH/OSHKOSH (USA)
JOINT RECORD
Charles E. Cheeseman — 62.53 kmh / 38.85 mph
Aeromot AMT-200 Super Ximango
I Rotax 912A, 80 bhp
and
Keith E. Phillips
Aeromot AMT-200 Super Ximango
I Rotax 912A, 80 bhp
and
Heinz G. Peier
Aeromot AMT-200 Super Ximango
I Rotax 912A, 80 bhp
7/25-26/95

DECATUR/KITTY HAWK (USA)
William G. Castlen, Pilot — 337.02 kmh / 209.41 mph
Jennifer L. Castlen, Copilot
Glasair SH-2R
I Lycoming IO-360, 180 bhp
12/17/93

DENVER/OSHKOSH (USA)
Rex A. Victor, Pilot — 173.27 kmh / 107.67 mph
Susan J. Victor, Copilot
Varga 2150A
I Lycoming O-320, 150 bhp
7/25/94

DES MOINES/LAFAYETTE (USA)
Patricia M. Dennehy — 248.40 kmh / 154.36 mph
Cessna 170 B
I Continental C-145, 145 bhp
9/22/83

DES MOINES/OSHKOSH (USA)
Jerome A. Karrels — 235.30 kmh / 146.20 mph
Grumman American AA5
I Lycoming O-320, 150 bhp
8/1/91

DINARD/ST. PETER (UK)
David O'Byrne — 184.52 kmh / 114.66 mph
Cessna 152
I Lycoming 0-235-L2C, 110 bhp
11/30/85

DOTHAN, AL/ATHENS, TN (USA) ☆
William E. Jennings — 249.88 kmh / 155.27 mph
Swift GC-1B
I Continental IO-360, 210 bhp
5/25/91

DUBLIN/LONDON (UK)
T. W. Hayhow — 201.59 kmh / 125.26 mph
Auster Aiglet
I de Havilland Gipsy Major IG
2 hrs 17 min 20.9 sec
4/13/52

DULUTH/OSHKOSH (USA)
Stanley H. Mick — 171.79 kmh / 106.74 mph
Cessna 152
I Lycoming IO-235, 110 bhp
8/7/93

DUNEDIN/PARAPARAUMU (NEW ZEALAND)
Robyn N. Harris — 186.13 kmh / 115.66 mph
Piper PA-28-140 Cherokee
6/6/98

DUNGENESS/LE TOUQUET (UK)
John S. Hemmings — 154.25 kmh / 95.85 mph
Sipa 903
I Continental C90 14F, 91 bhp
11/28/85

DUNEDIN/NELSON (NEW ZEALAND)
Maree Mazure — 164.42 kmh / 102.16 mph
Vera Perry
Pam Peters
Cessna 172P
I Lycoming O-320, 160 bhp
9/4/95

DURBAN/BARBADOS (AUSTRALIA)
Jon Johanson — 33.10 kmh / 20.57 mph
Van's RV-4
I Lycoming O-320, 160 bhp
7/18/96

DURBAN/OSHKOSH (AUSTRALIA)
Jon Johanson — 23.70 kmh / 14.73 mph
Van's RV-4
I Lycoming O-320, 160 bhp
7/30/96

EDINBURGH/LONDON (UK)
Michael P. Hallam — 167.40 kmh / 104.02 mph
Jodel DR 1051
5/7/88

ELIZABETHTON, TN/WARRENTON, VA (USA)
John Roy Doggett — 198.36 kmh / 123.26 mph
Cessna 150M, N6482K
I Continental O-200, 100 bhp
9/8/88

EUREKA/LONGYEARBYEN (AUSTRALIA)
Jon Johanson — 206.27 kmh / 128.17 mph
Van's RV-4
I Lycoming, 160 bhp
6/7/00

FAIRBANKS/OSHKOSH (USA)
Stanley H. Mick — 44.79 kmh / 27.83 mph
Cessna 152
I Lycoming IO-235, 110 bhp
8/7/93

FARMINGDALE/BUFFALO (USA)
Eric T. Saper — 197.60 kmh / 122.78 mph
Cessna 172N
I Lycoming O-320, 160 bhp
9/30/95

FLAGSTAFF, AZ/PALM SPRINGS, CA (USA)
L. B. Jones — 198.65 kmh / 123.43 mph
Piper Warrior
I Lycoming O-320, 160 bhp
10/26/90

FLORENCE/KITTY HAWK (USA)
Roger Bowman — 384.48 kmh / 238.92 mph
Glasair II, N360RB
I Lycoming IO-360-B1E, 180 bhp
12/16/89

FREDERICK/ATHENS (USA)
William L. Gruber — 102.28 kmh / 63.55 mph
Aeronca Champion 7AC
I Continental A-65, 65 bhp
4/14/92

FREDERICK/KITTY HAWK (USA)
John W. Strong, Pilot — 218.83 kmh / 135.97 mph
Leonard J. Stone, Copilot
Stinson 108
I Franklin 6A4-150, 150 bhp
12/17/95

FROBISHER/RESOLUTE (USA)
Donald P. Taylor — 15.48 kmh / 9.61 mph
Thorp T-18, N455DT
I Lycoming 0-360, A1A, 180 bhp
7/21-26/83

GENEVA/OSHKOSH (SWITZERLAND)
Felix J. Estermann 48.23 kmh
Glastar 29.97 mph
1 Continental IO-240, 125 bhp
7/22-28/96

GREENVILLE, IL/KITTY HAWK (USA)
Bill Weder, Pilot 155.53 kmh
Edward Faiss, Copilot 96.64 mph
Piper PA-22-108 Colt
1 Lycoming O-235, 108 bhp
12/16/95

GREENVILLE, SC/KITTY HAWK (USA)
Robert Olszewski 342.72 kmh
Piper PA28R-200 212.97 mph
1 Lycoming IO-360-C1C, 200 bhp
12/16/85

GREENWOOD, MS/DALTON, GA (USA)
William E. Jennings, Pilot 261.36 kmh
Geraldine Jennings, Copilot 162.41 mph
Globe Swift G1CB, N90373
1 Continental IO360, 210 bhp
5/30/87

GUAM/ANCHORAGE (USA)
Donald P. Taylor 18.05 kmh
Thorp T-18, N455DT 11.22 mph
1 Lycoming O-360-A2A, 180 bhp
548 hrs 27 min
8/31-9/23/76

HICKORY/KITTY HAWK (USA)
Norman F. Carden III 223.05 kmh
American Aviation AA-1A 138.59 mph
1 Lycoming O-235, 108 bhp
1/18/98

HILO/NORFOLK ISLAND (AUSTRALIA)
Jon Johanson 81.88 kmh
Van's RV-4 50.88 mph
1 Lycoming O-320, 160 bhp
9/20/96

HOBART/MELBOURNE (AUSTRALIA)
Jon Johanson 225.05 kmh
Van's RV-4 139.83 mph
1 Lycoming O-320, 160 bhp
12/20/94

HOUSTON/ATLANTA (USA)
R. Stuart Hagedorn 177.86 kmh
Michelle D. Trobaugh 110.52 mph
Grumman AA-5A Cheetah
1 Lycoming IO-320, 160 bhp
12/20/03

HOUSTON/DALLAS (USA)
Gerry Griffin, Pilot 202.95 kmh
Larry Griffin, Copilot 126.11 mph
Piper SuperCub
1 Lycoming O-360, 180 bhp
5/17/91

HOUSTON/DENVER (USA)
Larry Griffin, Pilot 182.88 kmh
Gerry Griffin, Copilot 113.64 mph
Piper SuperCub, N1937G
1 Lycoming O-360, 180 bhp
5/30/87

HOUSTON/TYLER (USA)
Rodolfo Faccini 206.97 kmh
Pitts Special S-1C, HK848-Z 128.61 mph
1 Lycoming IO-360
1 hr 23 min 52.3 sec
9/23/74

INDIANAPOLIS/ASHEVILLE (USA)
Kristan R. Maynard 248.44 kmh
Aviat A-1 Husky 154.37 mph
1 Lycoming O-360, 180 bhp
4/7/99

INVERCARGILL/MELBOURNE (AUSTRALIA)
Jon Johanson 258.44 kmh
Van's RV-4 160.59 mph
1 Lycoming O-320, 160 bhp
2/15/96

IXTEPEC/VERACRUZ (USA)
Rodolfo Faccini 140.90 kmh
Pitts Special S-1C HK848-Z 87.55 mph
1 Lycoming IO-360, 180 bhp
2 hrs 16 min 00.4 sec
9/18/74

JACKSONVILLE/FT. LAUDERDALE (USA)
Robert E. Alley 234.72 kmh
Piper Archer, N8256X 145.85 mph
1 Lycoming 0-360-A4M, 180 bhp
12/2/88

JOHNSON CITY/KITTY HAWK (USA)
William J. Anderson 193.50 kmh
Cessna 172M 120.23 mph
1 Lycoming O-320-E2D, 150 bhp
12/17/97

JOHNSON CITY/WARRENTON (USA)
John Doggett 178.10 kmh
Cessna 150M 110.67 mph
1 Continental O-200, 100 bhp
7/1/88

KAHULUI/SACREMENTO (USA)
JOINT RECORD
John Koch 228.96 kmh
Longeze, N388TT 142.28 mph
and
Ed Roman
Longeze, N38AJ
1 Lycoming 2-235, 115 bhp
6/18-19/87

KANSAS CITY/FREDERICK (USA)
Kent S. Jackson 481.92 kmh
Lancair Legacy 299.45 mph
1 Continetnal IO-550, 310 bhp
11/16/03

KARACHI/BAHRAIN (AUSTRALIA)
Clive R. Canning 66.53 kmh
Thorp T-18 43.34 mph
1 Lycoming 0-320-E2A
25 hrs 20 min 40 sec
6/24-25/76

KEFLAVIK/ASTURIAS (SPAIN)
Miguel Angel Gordillo Urquia 215.13 kmh
Dyn'Aero MCR01 133.67 mph
1 Rotax 912S, 100 bhp
8/1/01

KEFLAVIK/GLASGOW (USA)
Donald P. Taylor 262.16 kmh
Thorp T-18, N455DT 163.30 mph
1 Lycoming O-360-A2A, 180 bhp
5 hrs 10 min
8/7/76

KENOSHA/INUVIK (USA)
James K. Jackson, Sr., Pilot 51.63 kmh
Thomas Zentz, Copilot 32.08 mph
Lancair 360
1 Lycoming O-360, 180 bhp
7/5-8/94

KEY WEST/CANCUN (USA)
August Belmont 161.56 kmh
American Champion 7ECA Citabria 100.39 mph
1 Lycoming O-235, 118 bhp
5/20/01

KIEV/BUCHAREST (ROMANIA)
Calota, Manolache & Onci 172.35 kmh
I.A.R. 813 107.10 mph
1 Walter Minor 4 III
4 hrs 19 min 52 sec
7/26/58

KIEV/MOSCOW (ROMANIA)
Calota, Onciu & Otoiu 177.34 kmh
I. A. R.813 110.19 mph
1 Walter Minor 4 I I I
4 hrs 5 min 34 sec
7/5/58

KITTY HAWK/HOLLYWOOD, FL (USA)
Haakon E. Weise 243.97 kmh
Long-EZ 151.60 mph
1 Lycoming O-320, 160 bhp
12/15/91

KITTY HAWK/SALINAS (USA)
D. Wayne Woollard 73.09 kmh
Gary Tuovinen 45.42 mph
Alon A-2 Aircoupe
1 Continental C90, 90 bhp
9/30/03

KUALA LUMPUR/CAIRO (MALAYSIA)
S. B. Reynolds 31.08 kmh
KZIII 9.96 mph
1 Blackburn Cirrus Minor
256 hrs 15 min 11 sec
10/31-11/10/64

KUALA LUMPUR/CALCUTTA (MALAYSIA)
S. B. Reynolds 48.05 kmh
ZIII 9.85 mph
Blackburn Cirrus Minor
53 hrs 56 min 42 sec
10/31-11/2/64

KUALA LUMPUR/KARACHI (MALAYSIA)
S. B. Reynolds 41.17 kmh
KZIII 25.58 mph
1 Blackburn Cirrus Minor
107 hrs 39 min 40 sec
10/31-11/4/64

KUALA LUMPUR/LONDON (MALAYSIA)
S. B. Reynolds 15.93 kmh
KZIII 9.90 mph
1 Blackburn Cirrus Minor
662 hrs 49 min 13 sec
10/31/64

KUUJJUAQ/IQALUIT (CANADA)
Terrance L. Jantzi 259.00 kmh
Lauren V. Jantzi 160.93 mph
Van's RV-6
6/4/99

LAFAYETTE/MORRISTOWN (USA)
Patricia M. Dennehy 229.03 kmh
Cessna 170 B 142.32 mph
1 Continental C-145, 145 bhp
9/22/83

LAFAYETTE/WHEELING (USA)
Patricia M. Dennehy 243.11 kmh
Cessna 170 B 151.07 mph
1 Continental C-145, 145 bhp
9/22/83

LANSING/ATHENS, TN (USA) ★
Drew G. Seguin 167.54 kmh
Swift GC-1B 104.10 mph
1 Continental O-300, 145 bhp
5/23/91

LANSING/EVANSVILLE (USA)
Kevin C. O'Malley 162.11 kmh
Cessna 172 Skyhawk 100.73 mph
1 Lycoming O-320, 160 bhp
9/11/99

LAS VEGAS/ST. LOUIS (USA)
Monie W. Pease 208.43 kmh
Grumman Tiger 129.51 mph
1 Lycoming O-360, 180 bhp
9/26/90

LATROBE/KITTY HAWK (USA)
James E. Polen 222.29 kmh
Piper PA-28-181 138.12 mph
1 Lycoming O-360, 180 bhp
6/21/97

LEXINGTON/LAKELAND*
Jim Sprowl 188.86 mph
Long-EZ
1 Lycoming O-235, 125 bhp
4/7/94

LINCOLN/SIOUX FALLS (USA)
Robert L. Carlisle, Pilot 242.18 kmh
Jon C. Howser, Copilot 150.49 mph
Cessna 152, N48862 "Jason"
1 Lycoming, 110 bhp
1 hr 15 min 10 sec
12/17/78

LONDON/AMSTERDAM (UK)
T. W. Hayhow 199.22 kmh
Auster Aiglet 123.78 mph
1 de Havilland Gipsy Major IG
1 hr 48 min 34 sec
4/14/52

LONDON/BERNE (UK)
T. W. Hayhow 227.09 kmh
Auster Aiglet 114.11 mph
1 de Havilland Gipsy Major
3 hrs 17 min 34.5 sec
4/14/52

LONDON/BRUSSELS (UK)
T. W. Hayhow 176.96 kmh
Auster Aiglet 109.96 mph
1 de Havilland Gipsy Major
1 hr 49 min 20.5 sec
4/12/52

LONDON/CAPETOWN (UK)
Alex Henshaw 244.88 kmh
Percival Mew Gull G-AEXF 152.16 mph
1 de Havilland Gipsy VI-2
2/6/39

LONDON/COPENHAGEN (UK)
T. W. Hayhow 190.09 kmh
Auster Aiglet 118.11 mph
1 de Havilland Gipsy Major
5 hrs 2 min 8.5 sec
5/30/52

LONDON/DARWIN (AUSTRALIA)
John Fisher 17.65 kmh
de Havilland DH-82A Tiger Moth 10.97 mph
1 de Havilland Gipsy Major 1C, 135 bhp
10/4/96

LONDON/DUBAI (AUSTRALIA)
John Fisher — 14.46 kmh / 8.99 mph
de Havilland DH-82A, Tiger Moth
1 de Havilland Gipsy Major 1C, 135 bhp
9/17/96

LONDON/DUBLIN (UK)
T. W. Hayhow — 203.27 kmh / 126.31 mph
Auster Aiglet
1 de Havilland Gipsy Major IG
2 hrs 16 min 12.9 sec
4/13/52

LONDON/IRAKLION (AUSTRALIA)
John Fisher — 14.26 kmh / 8.86 mph
de Havilland DH-82A, Tiger Moth
1 de Havilland Gipsy Major 1C, 135 bhp
9/9/96

LONDON/LUXEMBOURG (UK)
T. W. Hayhow — 197.52 kmh / 122.73 mph
Auster Aiglet
1 de Havilland Gipsy Major IG
2 hrs 29 min 2.8 sec
4/12/52

LONDON/MELBOURNE (AUSTRALIA)
Jon Johanson — 61.02 kmh / 37.91 mph
Van's RV-4
1 Lycoming O-320, 160 bhp
8/31/95

LONDON/MOSCOW (UK)
Jonathan Elwes — 14.82 kmh / 9.20 mph
Tiger Moth
4/18-27/89

LONDON/PARIS (USA)
Donald P. Taylor — 248.24 kmh / 154.25 mph
Thorp T-18,
1 Lycoming IO-360
1 hr 22 min 54.5 sec
9/12/73

LONDON/RANGOON (AUSTRALIA)
John Fisher — 15.03 kmh / 9.34 mph
de Havilland DH-82A Tiger Moth
1 de Havilland Gipsy Major 1C, 135 bhp
9/26/96

LONDON/SINGAPORE (AUSTRALIA)
John Fisher — 16.40 kmh / 10.19 mph
de Havilland DH-82A, Tiger Moth
1 de Havilland Gipsy Major 1C, 135 bhp
9/29/96

LONDON/STOCKHOLM (UK)
T. W. Hayhow — 216.38 kmh / 134.45 mph
Auster Aiglet
1 de Havilland Gipsy Major IG
8/9/52

LONDON/SYDNEY (AUSTRALIA)
John Fisher — 17.43 kmh / 10.83 mph
de Havilland DH-82A Tiger Moth
1 de Havilland Gipsy Major 1C, 135 bhp
10/12/96

LONDON/THE HAGUE (UK)
T. W. Hayhow — 227.38 kmh / 141.29 mph
Auster Aiglet
1 de Havilland Gipsy Major IG
1 hr 21 min 40.2 sec
4/11/52

LONG BEACH/LAS VEGAS (USA)
Wilfred K. Taylor — 325.71 kmh / 202.39 mph
Thorp T-18
1 Lycoming 0-360A-2A
5/17/72

LONG BEACH/SACRAMENTO (USA)
Donald P. Taylor — 262.04 kmh / 162.83 mph
Thorp T-18 N455DT
1 Lycoming 0-360A-2A
2 hrs 18 min 00 sec
4/5/72

LORD HOWE ISLAND/SYDNEY (AUSTRALIA)
Jon Johanson — 276.39 kmh / 171.74 mph
Van's RV-4
1 Lycoming O-320, 160 bhp
9/24/96

LOS ANGELES/GUADALAJARA (USA)
David L. Kolstad — 298.08 kmh / 185.23 mph
VariEZE, N506D
1 Lycoming O-235L2C 115 bhp
7 hrs 9 min 18 sec
11/22/86

LOS ANGELES/LAS VEGAS (USA)
Donald P. Taylor — 296.19 kmh / 184.04 mph
Thorp T-18 N455DT
1 Lycoming 0-360-A2A
1 hr 6 min 14 sec
4/28/79

LOS ANGELES/PORTLAND, ME (USA)
Richard Theriault — 29.75 kmh / 18.49 mph
Steen Skybolt 001
1 Lycoming 0-360, 180 bhp
142 hrs 13 min
9/10-16/78

LOS ANGELES/WASHINGTON (USA)
LTC Thomas M. Lee, Pilot — 15.34 kmh / 9.53 mph
Douglas M. Scott, Copilot
Piper L-4H
1 Continental, 65 bhp
7/31-8/10/95

LUXEMBOURG/AJACCIO (LUXEMBOURG)
Patrick Louis — 217.07 kmh / 134.88 mph
Robin DR 400/180
1 Lycoming, 180 bhp
9/28/02

LUXEMBOURG/LONDON (UK)
T. W. Hayhow — 221.12 kmh / 137.40 mph
Auster Aiglet
1 de Havilland Gipsy Major IG
2 hrs 13 min 8.2 sec
4/12/52

LYON/AJACCIO (FRANCE)
Bernard Bonneval — 217.92 kmh / 135.41 mph
Morane Saulnier MS 893 E
1 Lycoming 0-360-A3A, 180 bhp
4/28/84

MADRID/LONDON (UK)
T. W. Hayhow — 184.95 kmh / 114.92 mph
Auster Aiglet
1 de Havilland Gipsy Major IG
6 hrs 49 min 1.9 sec
8/7/52

MADRID/PARIS (FRANCE)
Christophe de Brichambaut — 235.19 kmh / 146.15 mph
Long EZ
1 Lycoming, 115 CV
9/10/86

MAIDENHEAD/ABBEVILLE (UK)
Michael Ashfield — 167.09 kmh / 103.82 mph
Piper PA-28-140
1 Lycoming O-320, 140 bhp
7/15/94

MANAGUA/TEGUCIGALPA (USA)
Rodolfo Faccini
Pitts Special S-1C, HK848-Z
1 Lycoming IO-360
1 hr 45 min 21 sec
9/17/74
136.39 kmh
84.75 mph

MANASSAS/CHARLESTON, SC (USA)
James H. Heffernan
Cessna 172
1 Lycoming O-320, 160 bhp
11/8/92
202.74 kmh
125.98 mph

MANASSAS/GLENS FALLS (USA)
James A. Way
Piper PA-28-180
1 Lycoming O-360, 180 bhp
9/11/98
188.70 kmh
117.25 mph

MANASSAS/KITTY HAWK (USA)
John W. Rathgeber
Cessna 182, N5928B
1 Continental O-470
12/16/88
248.78 kmh
154.59 mph

MANILA/CHEJU (SPAIN)
Miguel Angel Gordillo Urquia
Dyn'Aero MCR01
1 Rotax 912S, 100 bhp
7/9/01
211.31 kmh
131.30 mph

MANSFIELD/TERRA HAUTE (USA)
Robert A. Brown
Piper PA-28-180 Cherokee
1 Lycoming O-360, 180 bhp
6/16/99
237.77 kmh
147.74 mph

MARSHALL ISLANDS/HONOLULU (USA)
JOINT RECORD
John Koch
Longeze, N388TT
Ed Roman
Longeze, N38AJ
1 Lycoming 2-235, 115 bhp
6/15-16/87
223.56 kmh
138.92 mph

MARSEILLE-SHOREHAM/LONDON (AUSTRALIA)
Clive R. Canning
Thorp T-18
1 Lycoming 0-320-E2A
4 hrs 35 min 59 sec
7/1/76
200.18 kmh
124.39 mph

MAUI/OAKLAND (USA)
Donald P. Taylor
Thorp T-18, N4550T
1 Lycoming 0-336-A2A, 180 bhp
16 hrs 32 min 10 sec
10/12/80
226.79 kmh
140.92 mph

MAURITIUS/DURBAN (AUSTRALIA)
Jon Johanson
Van's RV-4
1 Lycoming O-320, 160 bhp
7/1/96
240.70 kmh
149.56 mph

MEDELLIN/TURBO, COLUMBIA (USA)
Rodolfo Faccini
Pitts Special S-1C, HK 848-Z
1 Lycoming IO-360
1 hr 42 min 30.5 sec
9/15/74
141.05 kmh
87.64 mph

MELBOURNE/ADELAIDE (AUSTRALIA)
Jon Johanson
Van's RV-4
1 Lycoming O-320, 160 bhp
12/20/94
277.25 kmh
172.27 mph

MELBOURNE/BRISBANE (AUSTRALIA)
Clive R. Canning
Thorp T-18
1 Lycoming 0-320-E2A
6 hrs 25 min 58 sec
6/12/76
219.18 kmh
136.19 mph

MELBOURNE/CANBERRA (AUSTRALIA)
Raymond James Pearson
Glasair II-FT
1 Lycoming O-320, 160 hp
8/26/03
326.06 kmh
202.60 mph

MELBOURNE/DURBAN (AUSTRALIA)
Jon Johanson
Van's RV-4
1 Lycoming O-320, 160 bhp
7/1/96
37.77 kmh
23.47 mph

MELBOURNE/MAURITIUS (AUSTRALIA)
Jon Johanson
Van's RV-4
1 Lycoming O-320, 160 bhp
6/30/96
37.33 kmh
23.20 mph

MELBOURNE/RIVERSIDE (USA)
Donald P. Taylor
Thorp T-18, N455DT
1 Lycoming 0-366-A2A, 180 bhp
737 hrs 55 min
9/13-10/13/80
17.44 kmh
10.84 mph

MELBOURNE/SYDNEY (AUSTRALIA)
Raymond James Pearson
Glasair II-FT
1 Lycoming O-320, 160 hp
8/26/03
335.04 kmh
208.18 mph

MELBOURNE-SHOREHAM/LONDON (AUSTRALIA)
Clive R. Canning
Thorp T-18
1 Lycoming 0-320-E2A
444 hrs 54 min 3 sec
6/12-7/1/76
38.17 kmh
23.72 mph

MEMPHIS/NEW ORLEANS (USA)
James S. Jenkins, Pilot
Les Seago, Copilot
Cessna C-172
1 Lycoming 320, 160 bhp
12/17-18/82
210.02 kmh
130.51 mph

MIDWAY/ADAK (USA)
Donald P. Taylor
Thorp T-18, N455DT
1 Lycoming 0-360-A2A, 180 bhp
12 hrs 58 min
9/20-21/76
203.38 kmh
126.37 mph

MINOT/DULUTH (USA)
Stanley H. Mick
Cessna 152
1 Lycoming IO-235, 110 bhp
8/7/93
150.59 kmh
93.57 mph

MONTEREY/HILO (AUSTRALIA)
Jon Johanson
Van's RV-4
1 Lycoming O-320, 160 bhp
9/15/96
247.77 kmh
153.96 mph

MONTEREY/SYDNEY (AUSTRALIA)
Jon Johanson
Van's RV-4
1 Lycoming O-320, 160 bhp
9/24/96
54.52 kmh
33.88 mph

MONTREAL/BURLINGTON (USA)
R. Scott Olszewski, Pilot
Robert F. Olszewski, Copilot
Piper Arrow, N1585X
1 Lycoming IO-360, 200 bhp
9/13/86
273.85 kmh
170.17 mph

MOSCOW/KIEV (ROMANIA)
Calota, Manolache and Otoiu
I. A. R. 813
1 Walter Minor 4 III
4 hrs 41 min 58 sec
7/25/58
160.80 kmh
99.91 mph

MOSCOW/LONDON (UK)
Jonathan Elwes
Tiger Moth
4/18-27/89
11.69 kmh
7.26 mph

MUNICH/HANOVER (W. GERMANY)
Hermann Greiner
Putzer-Elster B
1 Continental C90-12F, 95 bhp
4/2/82
98.60 kmh
61.27 mph

NAPLES/CANNES (AUSTRALIA)
Barry H. A. Markham
de Havilland D.H.82A Tiger Moth
1 de Havilland Gipsy Major, 130 bhp
6/19/98
128.48 kmh
79.83 mph

NATAL/DAKAR (BRAZIL)
André J. Deberdt
Long-EZ
1 Lycoming O-235, 118 bhp
4/22/95
249.50 kmh
155.03 mph

NATAL/TENERIFE (BRAZIL)
Andre Joseph Deberdt
Long-EZ
1 Lycoming O-235, 118 bhp
7/5/95
241.10 kmh
149.81 mph

NELSON/DUNEDIN (NEW ZEALAND)
Vera Perry
Maree Mazure
Pam Peters
Cessna 172P
1 Lycoming O-320, 160 bhp
9/3/95
202.46 kmh
125.80 mph

NEW ORLEANS/NASHVILLE (USA)
James S. Jenkins, Pilot
Les Seago, Copilot
Cessna C-172
1 Lycoming 320, 160 bhp
12/18-19/83
227.11 kmh
141.12 mph

NEW YORK/MARTHA'S VINEYARD*
Phil Scott
Cessna 152
1 Lycoming, 110 bhp
6/15/95
93.12 mph

NORFOLK ISLAND/LORD HOWE ISLAND (AUSTRALIA)
Jon Johanson
Van's RV-4
1 Lycoming O-320, 160 bhp
9/23/96
217.18 kmh
134.95 mph

NORTH POLE/OSHKOSH (USA)
Donald P. Taylor
Thorp T-18, N455DT
1 Lycoming 0-360-A1A, 180 bhp
7/31-8/6/83
35.25 kmh
21.90 mph

OAKLAND/MAUI (USA)
Donald P. Taylor
Thorp T-18, N455DT
1 Lycoming 0-360-A2A, 180 bhp
14 hrs 36 min
8/15-17/80
259.00 kmh
160.94 mph

OKLAHOMA CITY/TOPEKA (USA)
Daniel R. Stroud, Pilot
Kristian R. Kennedy, Copilot
Piper PA-38-112
1 Lycoming O-235, 112 bhp
1/20/96
190.85 kmh
118.59 mph

OSHKOSH/BERLIN (SWITZERLAND)
Felix J. Estermann
Glastar
1 Continental IO-240, 125 bhp
5/5-11/96
50.05 kmh
31.10 mph

OSHKOSH/MELBOURNE (USA)
Donald P. Taylor
Thorp T-18, N455DT
1 Lycoming 0-360-A2A, 180 bhp
708 hrs 52 min
8/9-9/8/80
21.92 kmh
13.61 mph

OSHKOSH/NORTH POLE (USA)
Donald P. Taylor
Thorp T-18, N455DT
1 Lycoming 0-360-A1A, 180 bhp
7/16-31/83
14.26 kmh
8.86 mph

OSHKOSH/SPRINGFIELD (USA)
Raymond J. Johnson
Varga 2150A
1 Lycoming O-320, 150 bhp
8/6/92
177.24 kmh
110.13 mph

OSHKOSH/SYDNEY (AUSTRALIA)
Jon Johanson
Van's RV-4
1 Lycoming O-320, 160 bhp
9/24/96
14.59 kmh
9.07 mph

OSHKOSH/VAN NUYS (USA)
Jennifer Riley (aka Jenny Parks)
Cessna 152
1 Lycoming O-235, 110 bhp
8/1/91
97.28 kmh
60.45 mph

OTTAWA/IQALUIT (CANADA)
Terrance L. Jantzi
Lauren V. Jantzi
Van's RV-6
6/4/99
190.13 kmh
118.14 mph

OTTAWA/KUUJJUAQ (CANADA)
Terrance L. Jantzi
Lauren V. Jantzi
Van's RV-6
6/4/99
198.58 kmh
123.39 mph

OTTAWA/VICTORIA (CANADA)
Graham A. Palmer
Cessna 140
1 Continental C-85, 85 bhp
9/21/95
43.67 kmh
27.13 mph

OTTAWA/WABUSH (CANADA)
Terrance L. Jantzi
Lauren V. Jantzi
Van's RV-6
6/4/99
253.89 kmh
157.76 mph

PALM SPRINGS/FLAGSTAFF (USA)
L. B. Jones
Piper Warrior
1 Lycoming O-320, 160 bhp
10/26/90
195.90 kmh
121.72 mph

PALM SPRINGS/PHOENIX (CANADA)
John N. Leggatt, Jr. 159.21 kmh
Citabria 98.93 mph
1 Lycoming IO-320
2 hrs 34 min 18 sec
5/27/74

PALM SPRINGS/SCOTTSDALE (USA)
Michael W. Campbell 373.28 kmh
Lancair 320 231.94 mph
1 Lycoming O-320, 160 bhp
3/3/94

PANAMA/DAVID (USA)
Rodolfo Faccini 190.61 kmh
Pitts Special S-1C, HK848-Z 118.44 mph
1 Lycoming IO-360
1 hr 43 min 24.8 sec
9/16/74

PARAPARAUMU/DUNEDIN (NEW ZEALAND)
Robyn N.Harris 193.39 kmh
Piper PA-28-140 Cherokee 120.17 mph
6/5/98

PARIS/ATHENS (FRANCE)
Christophe de Brichambaut 237.13 kmh
Long EZ 147.35 mph
1 Lycoming, 115 CV
9/7/86

PARIS/CANNES (FRANCE)
Thibaut Chantegret 315.39 kmh
Xavier Beck 195.97 mph
Stelio Frati Falco F8L
11/7/99

PARIS/LA BAULE (FRANCE)
Joseph Blond 239.19 kmh
Mooney 21 148.62 mph
1 Lycoming IO-360
8/13/70

PARIS/LONDON (UK)
T. W. Hayhow 188.02 kmh
Auster Aiglet 116.83 mph
1 de Haviland Gipsy Major IG
4/11/52

PARIS/ST. LOUIS, SENEGAL (FRANCE)
Denis Guillotel 140.44 kmh
DG1 87.26 mph
1 Rotax 912, 80 bhp
10/2/00

PARIS/TUNIS (FRANCE)
Christophe de Brichambaut 235.93 kmh
Rutan Long Eze 146.61 mph
1 Lycoming 0-235-N2A, 116 bhp
7/13/85

PARIS/VENICE (FRANCE)
A. Rebellon 205.62 kmh
Supercab G.Y. 30 127.77 mph
1 Continental C90-12 F
4 hrs 6 min 34 sec
7/18/55

PERPIGNAN/ST. LOUIS, SENEGAL (FRANCE)
Denis Guillotel 217.00 kmh
DG1 134.84 mph
1 Rotax 912, 80 bhp
10/2/00

PERTH/ADELAIDE (AUSTRALIA)
Jon Johanson 263.51 kmh
Van's RV-4 163.74 mph
1 Lycoming O-320, 160 bhp
10/30/97

PERTH/BARBADOS (AUSTRALIA)
Jon Johanson 31.37 kmh
Van's RV-4 19.49 mph
1 Lycoming O-320, 160 bhp
7/18/96

PERTH/BRISBANE (AUSTRALIA)
Magna Liset 259.17 kmh
Rutan Long EZ 161.04 mph
3/8/90

PERTH/CHITTAGONG (AUSTRALIA)
Barry H. A. Markham 12.36 kmh
de Havilland D.H.82A Tiger Moth 7.68 mph
1 de Havilland Gipsy Major, 130 bhp
5/18/98

PERTH/COCOS ISLAND (AUSTRALIA)
Jon Johanson 37.30 kmh
Van's RV-4 23.18 mph
1 Lycoming O-320, 160 bhp
6/28/96

PERTH/DUBAI (AUSTRALIA)
Barry H. A. Markham 11.33 kmh
de Havilland D.H.82A Tiger Moth 7.04 mph
1 de Havilland Gipsy Major, 130 bhp
5/29/98

PERTH/DURBAN (AUSTRALIA)
Jon Johanson 48.64 kmh
Van's RV-4 30.22 mph
1 Lycoming O-320, 160 bhp
7/1/96

PERTH/IRAKLION (AUSTRALIA)
Barry H. A. Markham 10.38 kmh
de Havilland D.H.82A Tiger Moth 6.45 mph
1 de Havilland Gipsy Major, 130 bhp
6/13/98

PERTH/JAKARTA (AUSTRALIA)
Barry H. A. Markham 13.73 kmh
de Havilland D.H.82A Tiger Moth 8.53 mph
1 de Havilland Gipsy Major, 130 bhp
5/5/98

PERTH/KERKIRA (AUSTRALIA)
Barry H. A. Markham 10.68 kmh
de Havilland D.H.82A Tiger Moth 6.64 mph
1 de Havilland Gipsy Major, 130 bhp
6/14/98

PERTH/LONDON (AUSTRALIA)
Barry H. A. Markham 10.16 kmh
de Havilland D.H.82A Tiger Moth 6.31 mph
1 de Havilland Gipsy Major, 130 bhp
6/24/98

PERTH/MAURITIUS (AUSTRALIA)
Jon Johanson 46.43 kmh
Van's RV-4 28.85 mph
1 Lycoming O-320, 160 bhp
6/30/96

PERTH/OSHKOSH (AUSTRALIA)
Jon Johanson 26.27 kmh
Van's RV-4 16.32 mph
1 Lycoming O-320, 160 bhp
7/30/96

PERTH/RANGOON (AUSTRALIA)
Barry H. A. Markham 12.01 kmh
de Havilland D.H.82A Tiger Moth 7.46 mph
1 de Havilland Gipsy Major, 130 bhp
5/16/98

PERTH/SINGAPORE (AUSTRALIA)
Barry H. A. Markham 11.66 kmh
de Havilland D.H.82A Tiger Moth 7.24 mph
1 de Havilland Gipsy Major, 130 bhp
5/10/98

PFLUGERVILLE. TX/FLOYDADA, TX (USA)
Charles I. Fitzsimmons 204.48 kmh
Tomahawk, N25417 127.06 mph
1 Lycoming, 112 bhp
10/15/88

PHOENIX/NORTH POLE (USA)
Donald P. Taylor 15.33 kmh
Thorp T-18, N455DT 9.53 mph
1 Lycoming 0-360-A1A, 180 bhp
7/14-31/83

PHOENIX/PALM SPRINGS (USA)
Alton K. Marsh, Pilot 171.19 kmh
David W. Weigelt, Copilot 106.37 mph
Piper Tri-Pacer
1 Lycoming O-320, 160 bhp
10/15/98

PHOENIX/WICHITA (USA)
Donald P. Taylor 236.16 kmh
Thorp T-18; N455DT 146.75 mph
1 Lycoming 0-360-A1A, 180 bhp
7/14/83

PICAYUNE, MS/ARLINGTON, WA (USA)
Robert J. Steil 256.32 kmh
Glasair RG, N86BS 159.28 mph
1 Lycoming 0-320, 160 bhp
7/15-16/88

PORTO ALEGRE/DAYTONA BEACH (USA)
JOINT RECORD
Charles E. Cheeseman 32.43 kmh
Aeromot AMT-200 Super Ximango 20.15 mph
1 Rotax 912A, 80 bhp
and
Keith E. Phillips
Aeromot AMT-200 Super Ximango
1 Rotax 912A, 80 bhp
and
Heinz G. Peier
Aeromot AMT-200 Super Ximango
1 Rotax 912A, 80 bhp
7/11-20/95

PORTO ALEGRE/OSHKOSH (USA)
JOINT RECORD
Charles E. Cheeseman 24.92 kmh
Aeromot AMT-200 Super Ximango 15.48 mph
1 Rotax 912A, 80 bhp
and
Keith E. Phillips
Aeromot AMT-200 Super Ximango
1 Rotax 912A, 80 bhp
and
Heinz G. Peier
Aeromot AMT-200 Super Ximango
1 Rotax 912A, 80 bhp
7/11-26/95

PUNTA ARENAS/MANAGUA (USA)
Rodolfo Faccini 188.56 kmh
Pitts Special S-1C, HK848-Z 117.17 mph
1 Lycoming IO-360 Engine
1 hr 31 min 18.3 sec
9/16/74

QUEBEC CITY/TORONTO (CANADA)
Terrance L. Jantzi 281.78 kmh
RV-6 175.09 mph
8/22/98

RANGOON/CHITTAGONG (AUSTRALIA)
Barry H. A. Markham 157.23 kmh
de Havilland D.H.82A Tiger Moth 97.70 mph
1 de Havilland Gipsy Major, 130 bhp
5/18/98

RANGOON/SINGAPORE (AUSTRALIA)
John Fisher 37.53 kmh
de Havilland DH-82A, Tiger Moth 23.32 mph
1 de Havilland Gipsy Major 1C, 135 bhp
9/29/96

RECIFE/OSHKOSH (AUSTRALIA)
Jon Johanson 26.17 kmh
Van's RV-4 16.26 mph
1 Lycoming O-320, 160 bhp
7/30/96

RECIFE/RIO DE JANEIRO (SWITZERLAND)
Hans Georg Schmid 223.13 kmh
Long-EZ 138.65 mph
11/12/98

RECIFE/SAL (SWITZERLAND)
Hans Georg Schmid 226.76 kmh
Long-EZ 140.90 mph
12/17/98

RECIFE/SAO PAULO (BRAZIL)
André J. Deberdt 248.16 kmh
Long-EZ 154.19 mph
1 Lycoming O-235, 118 bhp
4/30/95

REYKJAVIK/GANDER (BRAZIL)
Andre Joseph Deberdt 223.99 kmh
Long-EZ 139.18 mph
1 Lycoming O-235, 118 bhp
7/26/95

REYKJAVIK/STAVENGER (GERMANY)
Heiner Neumann 251.58 kmh
Speed Canard 156.32 mph
8/14/90

RICHMOND/CLEMSON (USA)
Hughey A. Woodle Jr. 152.57 kmh
Cessna 120 94.80 mph
1 Continental, 100 bhp
11/18/97

RICHMOND/KITTY HAWK (USA)
Ed Covington, Pilot 320.04 kmh
Hank Worth, Copilot 198.87 mph
Glasair, N3EC
1 Lycoming R680D-5, 265 bhp
12/17/87

RIVERSIDE/AUSTIN (USA)
Donald P. Taylor 263.22 kmh
Thorp T-18, N455DT 163.56 mph
1 Lycoming 0-366-A2A, 180 bhp
7 hrs 4 min
5/24-25/80

RIVERSIDE/BIMINI (USA)
Donald P. Taylor 256.58 kmh
Thorp T-18, N455DT 159.43 mph
1 Lycoming 0-366-A2A, 180 bhp
14 hrs 33 min
5/24-25/80

RIVERSIDE, CA/MELBOURNE (USA)
Donald P. Taylor 22.83 kmh
Thorp T-18, N455DT 14.19 mph
1 Lycoming 0-366-A2A, 180 bhp
563 hrs 39 min
8/10-9/8/80

RIVERSIDE/NORTH POLE (USA)
Donald P. Taylor 14.40 kmh
Thorp T-18, N455DT 8.95 mph
1 Lycoming 0-360-A1A, 180 bhp
7/13-31/83

RIVERSIDE/OSHKOSH (USA)
Donald P. Taylor 211.51 kmh
Thorp T-18, N455DT 131.43 mph
Lycoming 0-360
12 hrs 49 min 41.3 sec
8/1/73

ROADTOWN/CHRISTIANSTED (U.S. VIRGIN ISLANDS)
John Stuart-Jervis 163.47 kmh
Cessna 172 101.58 mph
1 Lycoming O-320, 150 bhp
2/9/95

ROANOKE/KITTY HAWK (USA)
Lloyd L. Otey 334.44 kmh
Swift GC1B, N2459B 207.82 mph
1 Continental IO-360D, 210 bhp
12/19/88

ROME/ATHENS (USA)
Donald P. Taylor 224.68 kmh
Thorp T-18, N455DT 139.61 mph
1 Lycoming 0-360
4 hrs 41 min 11.7 sec
9/15/73

ST. ANNE/DINARD (UK)
David O'Byrne 155.78 kmh
Cessna F152 96.80 mph
1 Lycoming 0-235-L2C, 110 bhp
11/30/85

ST. HELIER/DINARD (UK)
David O'Byrne 158.21 kmh
Cessna F152 98.31 mph
1 Lycoming 0-235-L2C, 110 bhp
11/30/85

ST. JACOB, IL/KITTY HAWK (USA)
Henry Huelsmann, Pilot 212.05 kmh
Bill Weder, Copilot 131.76 mph
Piper Tomahawk
1 Lycoming, 112 bhp
12/16/90

ST. JEAN DE TERRE NEUVE/BIARRITZ (FRANCE)
André Georges Lafitte 125.00 kmh
Ultravia Pelican 77.67 mph
6/23-25/91

ST. JEAN DE TERRE NEUVE/SANTA MARIA (FRANCE)
André Georges Lafitte 186.06 kmh
Ultravia Pelican 115.61 mph
6/23-25/91

ST. LOUIS/ALBANY (USA)
Monie W. Pease 210.98 kmh
Grumman Tiger 131.10 mph
1 Lycoming O-360, 180 bhp
9/27/90

ST. LOUIS/SOUTH BEND (USA)
Stanley H. Mick 200.21 kmh
Cessna 152 124.40 mph
1 Lycoming O-235, 110 bhp
7/14/92

SAL/FERNANDO DE NORONHA (BRAZIL)
André J. Deberdt 257.86 kmh
Long-EZ 160.22 mph
1 Lycoming O-235, 118 bhp
4/28/95

SALINAS/EASTPORT (USA)
D. Wayne Woollard, Pilot 84.43 kmh
Gary Tuovinen, Copilot 52.46 mph
Alon A-2 Aircoupe
1 Continental C90, 90 bhp
6/30/00

SALINAS/LOS ANGELES (USA)
George G. Butts 161.84 kmh
Champion 7ECA Citabria 100.56 mph
1 Lycoming O-235, 115 bhp
6/11/00

SAN CARLOS/SANTA BARBARA (USA)
Karen T. Morss 202.90 kmh
Vance E. Cochrane 126.07 mph
Diamond Katana DA20
1 Rotax 912, 79 bhp
7/1/97

SAN DIEGO/LAS VEGAS (USA)
Monie W. Pease 203.97 kmh
Grumman Tiger 126.74 mph
1 Lycoming O-360, 180 bhp
9/25/90

SAN FRANCISCO/LOS ANGELES (USA)
Randal L. Kruger 397.80 kmh
Kevin A. Munson 247.18 mph
Lancair 360
1 Lycoming IO-360, 200 bhp
1/25/03

SAN SALVADOR/TAPACHULA (USA)
Rodolfo Faccini 183.43 kmh
Pitts Special S-1C, HK848-Z 113.98 mph
1 Lycoming IO-360
1 hr 57 min 13.9 sec
9/17/74

SANTA ANA, CA/BANGOR, ME (USA)
J. Nelson 147.64 kmh
Cessna C-150M 91.73 mph
1 Continental O-200, 100 bhp
29 hrs 32 min
11/1-2/76

SANTA ANA/KITTY HAWK (USA)
Fred M. Coon 257.15 kmh
Mark Stolzberg 159.78 mph
Grumman AA-5A Cheetah
1 Lycoming O-320, 150 bhp
10/16/03

Superseded
SANTA ANA/KITTY HAWK (USA)
Assaf Stoler 220.14 kmh
D. Laksen Sirimanne 136.79 mph
Diamond DA 40-180 Diamond Star
1 Lycoming IO-360, 180 bhp
7/18/03

SANTA MARIA/BIARRITZ (FRANCE)
André Georges Lafitte 184.72 kmh
Ultravia Pelican 114.78 mph
6/23-25/91

SANTA ROSA/LAS VEGAS (USA)
Carl F. Friberg 149.83 kmh
Cessna 172L 93.10 mph
1 Lycoming O-320, 150 bhp
10/11/92

SAO PAULO/NATAL (BRAZIL)
André J. Deberdt 234.07 kmh
Long-EZ 145.44 mph
1 Lycoming O-235, 118 bhp
4/21/95

SAULT STE. MARIE/FORT FRANCIS (CANADA)
Terrance L. Jantzi 235.59 kmh
Joël Jantzi 146.39 mph
Van's RV-6
11/14/98

SAULT STE. MARIE/WINNIPEG (CANADA)
Terrance L. Jantzi 213.76 kmh
Joël Jantzi 132.82 mph
Van's RV-6
11/14/98

SHOREHAM/LONDON-MELBOURNE (AUSTRALIA)
Clive R. Canning 32.60 kmh
Thorp T-18 20.21 mph
1 Lycoming 0-320-E2A
520 hrs 58 min 26 sec
10/1-23/76

SHUTTLE LANDING STRIP/KITTY HAWK (USA)
Donald R. Conover 189.36 kmh
Glasair, N136DC 117.67 mph
1 Lycoming 0-320
12/16/88

SINGAPORE/DARWIN (AUSTRALIA)
John Fisher 32.54 kmh
de Havilland DH-82A, Tiger Moth 20.22 mph
1 de Havilland Gipsy Major 1C, 135 bhp
10/4/96

SINGAPORE/HAT YAI, THAILAND (AUSTRALIA)
Clive R. Canning 201.31 kmh
Thorp T-18 125.09 mph
1 Lycoming 0-320-E2A
3 hrs 42 min 33 sec
6/20/76

SINGAPORE/SYDNEY (AUSTRALIA)
John Fisher 21.50 kmh
de Havilland DH-82A, Tiger Moth 13.36 mph
1 de Havilland Gipsy Major 1C, 135 bhp
10/12/96

SLIDELL, LA/LAKELAND, FL (USA)
King "Bud" Herron 289.80 kmh
Glasair I TD 180.07 mph
1 Lycoming O-320, 150 bhp
04/07/90

SONOMA VALLEY/ORLANDO (USA)
Peter A. Karpaty 114.56 kmh
Cessna Skyhawk 71.18 mph
1 Lycoming O-320, 160 bhp
10/3/93

SOUTHAMPTON/CHERBOURG (UK)
John C. L. Fell 125.92 kmh
Piper PA-22-108 Colt 78.24 mph
9/4/97

SPARTANBURG/KITTY HAWK (USA)
Robert Olszewski 342.72 kmh
Piper PA28R-200 212.97 mph
1 Lycoming IO-360-C1C, 200 bhp
12/16/85

STATESBORO, GA/ATHENS, TN (USA) ★
Asa V. Brown 321.85 kmh
Swift GC-1B 199.99 mph
1 Lycoming O-360, 180 bhp
5/24/91

STAVANGER, NORWAY/HAMBURG (GERMANY)
Michael Schulz 172.28 kmh
Taifun 17-E 107.05 mph
Limbach L 2000 EB 1
9/28/81

STOCKHOLM/LONDON (UK)
T. W. Hayhow 164.47 kmh
Auster Aiglet 102.20 mph
1 de Havilland Gypsy Major IG
8 hrs 43 min 18.5 sec
8/9/52

STRASBOURG/BIARRITZ (FRANCE)
J. P. Weiss 195.22 kmh
Jodel DR 1050 121.30 mph
1 Continental, 101 bhp
4 hrs 34 min 46 sec
9/18/71

STUTTGART/HAMBURG (W. GERMANY)
Karl U. Volkel 259.20 kmh
MBB 209 Monsun 161.07 mph
1 Lycoming IO-320-D1B, 160 bhp
6/1/83

SYDNEY/AUCKLAND (USA)
Donald P. Taylor 260.28 kmh
Thorp T-18, N455DT 161.73 mph
1 Lycoming 0-366-A2A, 180 bhp
8 hrs 18 min
9/23-24/80

SYDNEY/BRISBANE (AUSTRALIA)
Jon Johanson 281.52 kmh
Van's RV-4 174.93 mph
1 Lycoming O-320, 160 bhp
10/28/97

SYDNEY/MELBOURNE (AUSTRALIA)
Sean Peter Tanner 276.33 kmh
Van's RV-8 171.70 mph
1 Lycoming, 180 bhp
8/2/02

TALLADEGA, AL/LAKELAND, FL (USA)
Bill McClintock, Pilot 368.02 kmh
Deborah McClintock, Copilot 228.67 mph
Glasair I TD SH-2
Lycoming O-320, 160 bhp
04/04/90

TAPACHULA/IXTEPEC (USA)
Rodolfo Faccini 220.29 kmh
Pitts Special S-1C, HK848-Z 136.88 mph
1 Lycoming IO-360
1 hr 34 min 52.5 sec
9/18/74

TEGUCIGALPA/SAN SALVADOR (USA)
Rodolfo Faccini 161.12 kmh
Pitts Special S-1C, HK848-Z 100.12 mph
1 Lycoming IO-360
1 hr 21 min 30.6 sec
9/17/74

TENERIFE/LE CASTELLET (BRAZIL)
Andre Joseph Deberdt 244.66 kmh
Long-EZ 152.02 mph
1 Lycoming O-235, 118 bhp
7/7/95

TETERBORO/WILLIAMSPORT *
Joshua C. Ramo 126.26 mph
Piper PA-28R-201 Arrow III
1 Lycoming IO-360, 200 bhp
11/23/00

THE HAGUE/LONDON (UK)
T. W. Hayhow 188.53 kmh
Auster Aiglet 117.15 mph
1 de Havilland Gipsy Major IG
1 hr 38 min 30.1 sec
4/11/52

TOPEKA/ST. LOUIS (USA)
Stanley H. Mick 207.41 kmh
Cessna 152 128.88 mph
1 Lycoming O-235, 110 bhp
7/14/92

TORONTO/BUFFALO (USA)
Robert F. Olszewski 337.18 kmh
Mooney M201 209.52 mph
Lycoming IO-360, 200 bhp
5/5/83

TORONTO/FORT FRANCIS (CANADA)
Terrance L. Jantzi 180.29 kmh
Joël Jantzi 112.02 mph
Van's RV-6
11/14/98

TORONTO/MONTREAL (CANADA)
Bernard Runstedler 176.27 kmh
Cessna 152 109.52 mph
1 Lycoming O-235, 110 bhp
6/6/95

TORONTO/QUEBEC CITY (CANADA)
Terrance L. Jantzi 309.71 kmh
RV-6 192.44 mph
8/22/98

TORONTO/SAULT STE. MARIE (CANADA)
Terrance L. Jantzi 204.76 kmh
Joël Jantzi 127.23 mph
Van's RV-6
11/14/98

TORONTO/WINNIPEG (CANADA)
Terrance L. Jantzi 179.65 kmh
Joël Jantzi 111.63 mph
Van's RV-6
11/14/98

TOWNSVILLE/DARWIN (AUSTRALIA)
Jon Johanson 315.33 kmh
Van's RV-4 195.94 mph
1 Lycoming O-320, 160 bhp
10/29/97

TUNIS/PARIS (FRANCE)
Christophe de Brichambaut 236.56 kmh
Rutan Long Eze 147.00 mph
1 Lycoming 0-235-N2A, 116 bhp
7/13/85

TURBO/PANAMA (USA)
Rodolfo Faccini 198.19 kmh
Pitts Special S-1C, HK848-Z 123.15 mph
1 Lycoming IO-360
9/16/74

TYLER, TX/OSHKOSH (USA)
Harry Bergman 146.57 kmh
Piper Super Cub 91.07 mph
1 Lycoming O-320, 150 bhp
07/25/90

TYLER/SHERMAN, TX (USA)
Rodolfo Faccini 190.53 kmh
Pitts Special S-1C, HK848-Z 118.39 mph
1 Lycoming IO-360
59 min 11.7 sec
9/23/74

VERACRUZ/TAMPICO (USA)
Rodolfo Faccini 215.31 kmh
Pitts Special S-1C, HK848-Z 133.79 mph
1 Lycoming IO-360
1 hr 43 min 11.9 sec
9/18/74

VERO BEACH/SAVANNAH (USA)
David P. Shook 168.59 kmh
Piper Cadet PA-28-161 104.76 mph
1 Lycoming O-320, 160 bhp
3/15/93

VICTORIA/LA PAZ, MEXICO (USA)
Earl J. Sharritt 238.52 kmh
Mooney MK 20 148.21 mph
1 Lycoming IO-320
12 hrs 18 min 12.5 sec
5/28/75

VICTORIA/OTTAWA (CANADA)
Graham A. Palmer 46.80 kmh
Cessna 140 29.08 mph
1 Continental C-85, 85 bhp
9/10/95

VICTORIA/TIJUANA (USA)
Earl J. Sharritt 240.98 kmh
Mooney MK 20 149.73 mph
Lycoming IO-320
7 hrs 46 min
9/2/73

VIDALIA, GA/ATHENS, TN (USA) ★
James M. Jones 315.13 kmh
Swift GC-1B 195.81 mph
1 Lycoming O-360, 180 bhp
5/24/91

VIRGINIA HIGHLANDS/WARRENTON, VA (USA)
John Roy Doggett 160.59 kmh
Cessna 150M, N6428K 99.78 mph
1 Continental 2-200, 100 bhp
6/27/88

WABUSH/IQALUIT (CANADA)
Terrance L. Jantzi 209.09 kmh
Lauren V. Jantzi 129.92 mph
Van's RV-6
6/4/99

WABUSH/KUUJJUAQ (CANADA)
Terrance L. Jantzi 253.74 kmh
Lauren V. Jantzi 157.67 mph
Van's RV-6
6/4/99

WALVIS BAY/ASCENSION ISLAND (AUSTRALIA)
Jon Johanson 257.25 kmh
Van's RV-4 159.85 mph
1 Lycoming O-320, 160 bhp
7/15/96

WARRENTON, VA/ATHENS, TN (USA)
John Roy Doggett 155.88 kmh
Cessna 150M, N6482K 96.86 mph
1 Continental O-200, 100 bhp
10/14/88

WARRENTON, VA/GREENVILLE, TN (USA)
John Roy Doggett 174.24 kmh
Cessna 150M, N6482K 108.27 mph
1 Continental, 100 bhp
6/30/89

WARRENTON, VA/MORRISTOWN, TN (USA)
John Roy Doggett 198.88 kmh
Cessna 150M, N6482K 119.23 mph
1 Continental O-200, 100 bhp
1/23/89

WARRENTON, VA/ROGERSVILLE, TN*
John Roy Doggett 105.15 mph
Cessna 150M
1 Continental O-200, 100 bhp
04/19/90

WASHINGTON/OTTAWA (USA)
Donald P. Taylor — 245.97 kmh
Thorp T-18, N455DT — 152.84 mph
1 Lycoming IO-360
8/25/73

WELLINGTON/MELBOURNE (AUSTRALIA)
Jon Johanson — 215.68 kmh
Van's RV-4 — 134.01 mph
1 Lycoming O-320, 160 bhp
2/5/95

WHEELING/MORRISTOWN (USA)
Patricia M. Dennehy — 247.98 kmh
Cessna 170 B — 154.10 mph
1 Continental C-145, 145 bhp
9/22/83

WICHITA/DES MOINES (USA)
Patricia M. Dennehy — 205.47 kmh
Cessna 170 B — 127.68 mph
1 Continental C-145, 145 bhp
9/22/83

WICHITA/KANSAS CITY, MO (USA)
Gene Soucy — 334.67 kmh
Bede BD-4 — 207.96 mph
1 Lycoming IO-360-A4A
48 min
3/16/72

WILKSBORO/WARRENTON (USA)
John Roy Doggett — 157.68 kmh
Cessna 150M — 97.98 mph
1 Continental O-200, 100 bhp
4/26/86

WILLIAMSPORT/NEW YORK *
Joshua C. Ramo — 154.08 mph
Piper PA-28R-201 Arrow III
1 Lycoming IO-360, 200 bhp
11/23/00

WILMINGTON/KITTY HAWK (USA)
Robert J. Swanson — 175.66 kmh
Ercoupe 415-C — 109.15 mph
1 Continental C85-12F, 85 bhp
12/17/93

WINSTON SALEM/KITTY HAWK (USA)
MayCay Beeler, Pilot — 577.44 kmh
Richard Gritter, Copilot — 358.82 mph
Questair Venture, N62V
1 Continental IO-550-G1B, 280 bhp
12/16/89

ROUND TRIP

BRIDGEPORT/AUGUSTA/BRIDGEPORT (USA)
David H. Faile, Jr. — 203.81 kmh
Christen Eagle II — 126.65 mph
1 Lycoming IO-360, 200 bhp
8/31/86

BUFFALO/TORONTO/BUFFALO (USA)
Robert F. Olszewski — 312.63 kmh
Mooney M201 — 194.26 mph
Lycoming IO-360, 200 bhp
5/5/83

CHRISTIANSTED/ROADTOWN/CHRISTIANSTED (U.S. VIRGIN ISLANDS)
John Stuart-Jervis — 139.12 kmh
Cessna 172 — 86.45 mph
1 Lycoming O-320, 150 bhp
2/9/95

FARGO/BISMARK/FARGO (USA)
Eugene N. Rausch — 289.08 kmh
Rausch Special RV4, N319ER — 179.63 mph
9/10/89

LONDON/AMSTERDAM/LONDON (UK)
T. W. Hayhow — 214.76 kmh
Auster Aiglet — 133.45 mph
1 de Havilland Gipsy Major IG
3 hrs 21 min 24.5 sec
4/14/52

LONDON/BERNE/LONDON (UK)
T. W. Hayhow — 188.32 kmh
Auster Aiglet — 117.02 mph
1 de Havilland Gipsy Major IG
7 hrs 56 min 30.5 sec
6/7/52

LONDON/BRUSSELS/LONDON (UK)
T. W. Hayhow — 196.68 kmh
Auster Aiglet — 122.21 mph
1 de Havilland Gipsy Major IG
3 hrs 16 min 45 sec
4/12/52

LONDON/COPENHAGEN/LONDON (UK)
T. W. Hayhow — 166.30 kmh
Auster Aiglet — 103.33 mph
1 de Havilland Gipsy Major IG
11 hrs 30 min 43.9 sec
5/30/52

LONDON/DUBLIN/LONDON (UK)
T. W. Hayhow — 201.60 kmh
Auster Aiglet — 125.26 mph
1 de Havilland Gipsy Major IG
4 hrs 34 min 42.9 sec
4/13/52

LONDON/LUXEMBOURG/LONDON (UK)
T. W. Hayhow — 208.44 kmh
Auster Aiglet — 129.52 mph
1 de Havilland Gipsy Major IG
4 hrs 42 min 28.2 sec
4/12/52

LONDON/PARIS/LONDON (UK)
T. W. Hayhow — 182.38 kmh
Auster Aiglet — 113.32 mph
1 de Havilland Gipsy Major IG
3 hrs 44 min 59.9 sec
4/11/52

LONDON/STOCKHOLM/LONDON (UK)
T. W. Hayhow — 182.13 kmh
Auster Aiglet — 113.17 mph
1 de Havilland Gipsy Major IG
15 hrs 45 min 8.5 sec
8/9/52

LONDON/THE HAGUE/LONDON (UK)
T. W. Hayhow — 205.00 kmh
Auster Aiglet — 127.38 mph
1 de Havilland Gipsy Major IG
3 hrs 1 min 10.5 sec
4/11/52

LOS ANGELES/LAS VEGAS/LOS ANGELES (USA)
Donald P. Taylor — 278.30 kmh
Thorp T-18, N455DT — 172.92 mph
1 Lycoming 0-366-A2A, 180 bhp
2 hrs 39 min 7 sec
4/28/79

LOS ANGELES/SACRAMENTO/LOS ANGELES (USA)
Donald P. Taylor — 256.25 kmh
Thorp T-18, N455DT — 159.54 mph
1 Lycoming 0-360-A2A
5/6/78

MAIDENHEAD/ABBEVILLE/MAIDENHEAD (UK)
Michael Ashfield — 80.17 kmh
Piper PA-28-140 — 49.81 mph
1 Lycoming O-320, 140 bhp
7/15/94

MELBOURNE/SYDNEY/MELBOURNE (AUSTRALIA)
Sean Peter Tanner — 175.28 kmh
Van's RV-8 — 108.91 mph
1 Lycoming, 180 bhp
8/2/02

PALM SPRINGS/FLAGSTAFF/PALM SPRINGS (USA)
L. B. Jones — 187.06 kmh
Piper Warrior — 116.23 mph
1 Lycoming O-320, 160 bhp
10/26/90

PARIS/TUNIS/PARIS (FRANCE)
Christophe de Brichambaut — 236.24 kmh
Rutan Long Eze — 146.79 mph
1 Lycoming 0-235-N2A, 116 bhp
7/13/85

PARAPARAUMU/DUNEDIN/PARAPARAUMU (NEW ZEALAND)
Robyn N. Harris — 162.30 kmh
Piper PA-28-140 Cherokee — 100.85 mph
6/6/98

RIVERSIDE/ALBUQUERQUE/RIVERSIDE (USA)
Donald P. Taylor — 214.76 kmh
Thorp T-18, N455DT — 133.45 mph
1 Lycoming 0-360-A2A, 180 bhp
9 hrs 45 sec
6/30/73

RIVERSIDE/MELBOURNE/RIVERSIDE (USA)
Donald P. Taylor — 18.15 kmh
Thorp T-18, N455DT — 11.28 mph
1 Lycoming 0-366-A2A, 180 bhp
1,417 hrs 57 min
8/9-10/13/80

TORONTO/QUEBEC CITY/TORONTO (CANADA)
Terrance L. Jantzi — 193.42 kmh
RV-6 — 120.19 mph
8/22/98

TRENTON/LEBANON/TRENTON (USA)
Stephan R. Smith, Pilot — 163.74 kmh
Edward R. Wenglicki, Copilot — 101.74 mph
Piper PA-28-140
1 Lycoming O-320, 150 bhp
12/2/92

VICTORIA/OTTAWA/VICTORIA (CANADA)
Graham A. Palmer — 20.92 kmh
Cessna 140 — 12.99 mph
1 Continental C-85, 85 bhp
9/21/95

YELLOWKNIFE/EUREKA (AUSTRALIA)
Jon Johanson — 232.98 kmh
Van's RV-4 — 144.76 mph
1 Lycoming, 160 bhp
6/6/00

CLASS C-1.C (1,000-1,750 kg / 2,205-3,858 lbs) GROUP I (PISTON ENGINE)

WEST TO EAST, TRANSCONTINENTAL *
Lloyd C. Harmon — 222.18 mph
Peter M. Cohen
Mooney M20R Ovation
1 Continental IO-550, 280 bhp
3/29/02

***

ABBOTSFORD/BILLINGS (CANADA)
Frank J. Quigg — 160.26 kmh
Waco Classic F-5 — 99.58 mph
1 Jacobs R-755, 275 bhp
9/4/93

ACAPULCO/GUADALAJARA (USA)
Marie E. McMillan, Pilot — 293.04 kmh
Gloria May, Navigator — 182.09 mph
Beech Bonanza F33A
1 Continental IO-520
1 hr 55 min
1/30/81

ACAPULCO/JESUS CARRANZA (USA)
Marie E. McMillan — 230.55 kmh
Beech Bonanza F33A — 143.26 mph
1 Continental IO-520B, 285 bhp
2/3/84

ADANA/IZMIR (CANADA)
Mathias Stinnes — 291.84 kmh
Gérard Guillaumaud — 181.34 mph
Glasair II-S FT
9/21/97

ADELAIDE/ALICE SPRINGS (USA)
William H. Wisner, Pilot — 302.04 kmh
Janice Sullivan, Copilot — 187.68 mph
Beech Bonanza, N6826Q
1 Continental, 285 bhp
6/18-19/86

ADELAIDE/CANBERRA (AUSTRALIA)
John W. Chesbrough — 320.90 kmh
Marguerite Chesbrough — 199.39 mph
Mooney M20J
11/21/92

ADELAIDE/INVERCARGILL (AUSTRALIA)
Jon Johanson — 315.06 kmh
Van's RV-4 — 195.77 mph
1 Lycoming IO-360, 180 bhp
12/6/03

ADELAIDE/MELBOURNE (AUSTRALIA)
Eric H. Wheatley — 312.36 kmh
Mooney M20K — 194.09 mph
9/14/89

ADELAIDE/MCMURDO STATION (AUSTRALIA)
Jon Johanson — 97.05 kmh
Van's RV-4 — 60.30 mph
1 Lycoming IO-360, 180 bhp
12/7/03

ADELAIDE/PERTH (AUSTRALIA)
John W. Chesbrough — 246.23 kmh
Mooney M20J-201 — 153.00 mph
1 Lycoming I0-360, 200 bhp
5/5/96

ADELAIDE/SOUTH POLE (AUSTRALIA)
Jon Johanson — 110.56 kmh
Van's RV-4 — 68.70 mph
1 Lycoming IO-360, 180 bhp
12/8/03

ADELAIDE/SYDNEY (AUSTRALIA)
Leslie Robert Thompson — 238.84 kmh
James Bernie Yeo — 148.41 mph
Cessna 210N
11/30/89

AGANA, GUAM/BATON ROUGE (USA)
Hypolite T. Landry, Jr. — 72.30 kmh
Beechcraft Bonanza — 44.93 mph
1 Lycoming IO-520-B
170 hrs 7 min
5/18-25/69

AGANA, GUAM/SACRAMENTO (USA)
Alvin Marks — 125.46 kmh
Cessna 210, N942SM — 77.96 mph
1 Continental TSIO-520-C
75 hrs 51 min 43 sec
4/13-16/69

AGRA/HO CHI MINH CITY (USA)
Kenneth P. Johnson, Pilot
Larry Cioppi, Copilot
Glasair III
1 Lycoming IO-540, 300 bhp
5/11/94
490.47 kmh
304.76 mph

AITUTAKI/RAROTONGA (BRAZIL)
Gerard Moss
EMBRAER EMB-721
12/21/91
210.54 kmh
130.82 mph

ALBANY/BERMUDA (USA)
Millard Harmon
Beech Bonanza 36
1 Continental 10-520BA, 285 bhp
10/29/85
345.55 kmh
214.72 mph

ALBANY/GOOSE BAY (USA)
Millard Harmon
Beech Bonanza 36
1 Continental IO-520-BA, 285 bhp
6/29/84
318.96 kmh
198.20 mph

ALBANY/KITTY HAWK (USA)
Millard Harmon, Pilot
286.50 kmh

Ruth Harmon, Copilot
Beech Bonanza 36
1 Continental IO-520 BA, 285 bhp
12/16/84
178.02 mph

ALBANY/MONT JOLI (USA)
Millard Harmon
Beech Bonanza 36
1 Continental IO-520-BA, 285 bhp
6/13/85
288.53 kmh
179.28 mph

ALBANY/OSHKOSH (USA)
Millard Harmon
Beech Bonanza 36
1 Continental IO-520-BA, 285 bhp
8/2/84
267.48 kmh
166.21 mph

ALBUQUERQUE/KERRVILLE (USA)
Glenn A. Watts
Mooney M20F
1 Lycoming IO-360, 200 bhp
10/6/92
248.71 kmh
154.54 mph

Pilots Mike Dolabi (l) and John Eslinger (r) congratulate each other on their record flights from Amarillo to Wichita and back on December 17th.

ALGER/TOUSSUS LE NOBLE (FRANCE)
Michel Gelin 292.40 kmh
Piper PA-32 181.70 mph
1 Lycoming IO-540, 300 CV
5/20/86

ALICE SPRINGS/DARWIN (USA)
William H. Wisner, Pilot 307.08 kmh
Janice Sullivan, Copilot 190.82 mph
Beech Bonanza, N6826Q
1 Continental, 285 bhp
6/20/86

ALICE SPRINGS/MELBOURNE (AUSTRALIA)
Graeme J. Wakeling, Pilot 316.60 kmh
Graeme D. Lowe 196.73 mph
Beechcraft Bonanza, VH-CHX
1 Continental, 285 bhp
5 hrs 58 min 58 sec
8/22-23/71

ALICE SPRINGS/SYDNEY (AUSTRALIA)
Peter H. Norvill 264.70 kmh
Mooney M20J 164.47 mph
6/21/92

ALTA FLORESTA/BOA VISTA (BRAZIL)
Marcelo Bellodi 450.51 kmh
Noberto Bellodi 279.93 mph
Lancair IV-P
1 Continental, 350 bhp
7/22/00

AMARILLO/LITTLE ROCK (USA)
W. K. McGehee, Jr. 233.06 kmh
Cessna 172XP 144.83 mph
Continental IO-360, 195 bhp
5/9/85

AMARILLO/WICHITA (USA)
John C. Eslinger 276.65 kmh
Mike Dolabi 171.90 mph
Beechcraft Bonanza A36
1 Continental IO-550, 300 bhp
12/17/03

AMMAN/ROME (FRANCE)
Jean-Michel Masson, Pilot 193.83 kmh
Jean Le Ber, Copilot 120.44 mph
TB 20 No 513
1 Lycoming IO-540-C4D5D, 250 bhp
3/25/87

AMMAN/TRABZON (CANADA)
Mathias Stinnes 265.86 kmh
Gérard Guillaumaud 165.20 mph
Glasair II-S FT
9/19/97

ANADYR/NOME (CANADA)
Mathias Stinnes 479.51 kmh
Heinz Bittermann 297.95 mph
Glasair III
6/28/92

ANCHORAGE/CALGARY (USA)
Kenneth P. Johnson, Pilot 350.95 kmh
Larry Cioppi, Copilot 218.07 mph
Glasair III
1 Lycoming IO-540, 300 bhp
5/22/94

ANNAPOLIS/KITTY HAWK*
David F. Rogers, Pilot 229.72 mph
Michael J. Bangert, Copilot
Beechcraft Bonanza E33A
1 Continental IO-520, 285 bhp
12/12/93

ANTALYA/ISTANBUL (ROMANIA)
JOINT RECORD
Constantin Manolache 200.80 kmh
Alexandru Popovici 124.77 mph
Morovan Cehia ZLIN 142
Moravan Otrokovice M 337 AK, 180 bhp
and
Traian Gheorghiu
Gheorghe Militaru
Morovan Cehia ZLIN 142
Moravan Otrokovice M 337 AK, 180 bhp
and
Filip Stelian
Morovan Cehia ZLIN 142
Moravan Otrokovice M 337 AK, 180 bhp
9/17/97

APPLETON/SIOUX CITY (USA)
Lowell G. Redler 220.76 kmh
Cessna 182 137.17 mph
1 Continental O-470, 230 bhp
8/3/92

ASHEVILLE/KITTY HAWK (USA)
Lawrence G. Manofsky, Pilot 318.31 kmh
Linda Morgan, Copilot 197.79 mph
Lake LA-250 Renegade
1 Lycoming IO-540, 250 bhp
12/11/95

ASHEVILLE/OSHKOSH (USA)
Ted Rogers 270.59 kmh
Rutan Defiant 168.14 mph
1 Lycoming IO-360, 180 bhp
07/24/90

ASHLAND/KERRVILLE (USA)
David L. Harrington 247.07 kmh
Mooney M-20E 153.52 mph
1 Lycoming IO-360, 200 bhp
10/6/92

ATHENS/AUCKLAND (NEW ZEALAND)
Clifford V. Tait 191.55 kmh
Beech Bonanza A36 119.03 mph
Continental IO-520-BB, 285 bhp
4/6-9/82

ATHENS/BAHRAIN (NEW ZEALAND)
Clifford V. Tait 237.90 kmh
Beech Bonanza A36 147.82 mph
Continental IO-520-BB, 285 bhp
4/6/82

ATHENS/BRISBANE (NEW ZEALAND)
Clifford V. Tait 187.96 kmh
Beech Bonanza A36 116.80 mph
Continental IO-520-BB, 285 bhp
4/6-9/82

ATHENS/CALCUTTA (NEW ZEALAND)
Clifford V. Tait 230.26 kmh
Beech Bonanza A36 143.08 mph
Continental IO-520-BB, 285 bhp
4/6-7/82

ATHENS/DARWIN (NEW ZEALAND)
Clifford V. Tait 202.24 kmh
Beech Bonanza A35 125.67 mph
Continental IO-520-BB, 285 bhp
4/6-9/82

ATHENS/LUXOR (CANADA)
Frank J. Quigg 160.06 kmh
Waco Classic F-5 99.46 mph
1 Jacobs R-755, 275 bhp
9/17/93

ATLANTA/DAYTON*
Charles A. Kohler
Lancair IV-P
1 Continental TSIO-550, 350 bhp
7/19/95
309.94 mph

ATLANTA/DAYTONA BEACH (USA)
Michael J. Friese, Pilot
Ashley Hendrix, Copilot
Piper PA-32R-301T
1 Lycoming TIO-540, 300 bhp
6/7/95
316.82 kmh
196.86 mph

ATLANTA/KITTY HAWK (USA)
Nancy Toon, Pilot
Jim Ross, Copilot
Beech Duchess, N1804S
2 Lycoming IO-360, 180 bhp
12/16/86
295.32 kmh
183.88 mph

ATLANTA/PANAMA CITY (USA)
John W. Doane
George B. Doane, IV
Cessna 182B
1 Continental O-470, 230 bhp
4/4/02
232.14 kmh
144.24 mph

ATLANTA/SPRINGFIELD, MO (USA)
Nancy Toon, Pilot
Daniel Emin, Copilot
Beech Duchess
2 Lycoming 0-360, 180 bhp
9/11/86
261.36 kmh
162.41 mph

ATLANTIC CITY/OAKLAND (USA)
S. Dan Brodie
Navion Model A, N707YU
1 Continental IO-470-H
18 hrs 43 min 51 sec
6/14/67
219.61 kmh
134.46 mph

AUBURN/OSHKOSH (USA)
Jeff Bosonetto, Pilot
Chris Komlodi, Copilot
Comanche, N7315Y
2 Lycoming IO 320-BIA
7/28/89
226.73 kmh
140.89 mph

AUCKLAND/CHRISTCHURCH (NEW ZEALAND)
Christopher R. Toms
Lancair IV
4/4/96
397.60 kmh
247.06 mph

AUCKLAND/NELSON (NEW ZEALAND)
Peter Locke
Christopher Croucher
Piper PA-28-161
1 Lycoming O-360, 180 bhp
9/4/95
251.57 kmh
156.31 mph

AUCKLAND/NORFOLK ISLAND (NEW ZEALAND)
Sue Campbell
Dawson Bowles
Richard Neave
Mooney M20
1 Lycoming, 200 bhp
6/16/00
270.58 kmh
168.13 mph

AUCKLAND/SYDNEY (AUSTRALIA)
Peter William Hodgens
Kevin Charles Haydon
Lancair IV
1 Continental IO-550, 280 bhp
1/18/02
363.45 kmh
225.84 mph

AUCKLAND/WELLINGTON (NEW ZEALAND)
Christopher R. Toms
Lancair IV
4/5/96
420.04 kmh
261.00 mph

AUGSBURG/KAHULUI (GERMANY)
Dietrich Schmitt
Beechcraft Bonanza F33A
1 Continental IO-550, 300 bhp
6/1/95
86.48 kmh
53.73 mph

AUGSBURG/LITTLE ROCK (GERMANY)
Dietrich Schmitt
Beechcraft Bonanza F33A
1 Continental IO-550, 300 bhp
5/29/95
92.46 kmh
57.45 mph

AUGSBURG/LONG BEACH (GERMANY)
Dietrich Schmitt
Beechcraft Bonanza F33A
1 Continental IO-550, 300 bhp
5/30/95
87.36 kmh
54.28 mph

AUGSBURG/NORFOLK (GERMANY)
Dietrich Schmitt
Beechcraft Bonanza F33A
1 Continental IO-550, 300 bhp
5/28/95
159.04 kmh
98.82 mph

AUGSBURG/SANTA MARIA (GERMANY)
Dietrich Schmitt
Beechcraft Bonanza F33A
1 Continental IO-550, 300 bhp
5/26/95
248.29 kmh
154.28 mph

AUSTIN/ADELAIDE (USA)
William H. Wisner, Pilot
Janice Sullivan, Copilot
Beech Bonanza, N6826Q
1 Continental, 285 bhp
6/11-19/86
88.56 kmh
55.03 mph

AUSTIN/BATON ROUGE (USA)
Stuart C. Goldberg, Pilot
Jonathan Sabin, Copilot
Beechcraft Bonanza V35B
1 Continental IO-550, 300 bhp
5/8/93
388.49 kmh
241.39 mph

AUSTIN/BROWNSVILLE (USA)
Tressie Groten, Pilot
Tom O. Hutchison, Copilot
Beechcraft Sundowner 180
1 Lycoming O-360, 180 bhp
3/4/99
129.18 kmh
80.27 mph

AUSTIN/EL PASO (USA)
G. R. Wood
Beech Bonanza V-35-TC
1 Continental TSIO-520-D, 285 bhp
2 hrs 59 min 18 sec
6/17/76
283.89 kmh
176.40 mph

AUSTIN/PHOENIX (USA)
G. R. Wood
Beechcraft Bonanza V-35-TC
1 Continental TSIO-520-D, 285 bhp
5 hrs 13 sec
6/17/76
279.25 kmh
173.51 mph

AUSTIN/SAN JOSE, CA (USA)
William H. Wisner, Pilot
Janice Sullivan, Copilot
Beech Bonanza, N6826Q
1 Continental, 285 bhp
6/11-12/86
236.52 kmh
146.97 mph

BAHRAIN/AUCKLAND (NEW ZEALAND)
Clifford V. Tait
Beech Bonanza A36
1 Continental IO-520-BB, 285 bhp
4/6-9/82
189.20 kmh
117.57 mph

BAHRAIN/BRISBANE (NEW ZEALAND)
Clifford V. Tait
Beech Bonanza A36
1 Continental IO-520-BB, 285 bhp
4/6-9/82
185.42 kmh
115.22 mph

BAHRAIN/CALCUTTA (NEW ZEALAND)
Clifford V. Tait
Beech Bonanza A36
Continental IO-520-BB, 285 bhp
4/6-7/82
278.25 kmh
172.89 mph

BAHRAIN/DARWIN (NEW ZEALAND)
Clifford V. Tait
Beech Bonanza A36
Continental IO-520-BB, 285 bhp
4/6-9/82
203.88 kmh
125.70 mph

BAHRAIN/MALTA (GERMANY)
Dietrich Schmitt
Beech A36 Bonanza
7/15/97
249.74 kmh
155.18 mph

BAHRAIN/MANNHEIM (GERMANY)
Dietrich Schmitt
Beech A36 Bonanza
7/16/97
109.21 kmh
67.86 mph

BALI/KUALA LUMPUR (USA)
William H. Wisner, Pilot
Janice Sullivan, Copilot
Beech Bonanza, N6826Q
Continental, 285 bhp
2/23-24/86
264.60 kmh
164.42 mph

BALIKPAPAN/MEDAN (GERMANY)
Dietrich Schmitt
Beech A36 Bonanza
7/7/97
300.84 kmh
186.93 mph

BANGKOK/BATON ROUGE (USA)
Hypolite T. Landry, Jr.
Beechcraft Bonanza, N5842K
1 Continental IO-520-B
214 hrs 2 min
5/16-25/69
69.74 kmh
43.34 mph

BANGKOK/SACRAMENTO (USA)
Alvin Marks
Cessna 210, N942SM
1 Continental TSIO-520-C
214 hrs 2 min
4/10-16/69
89.22 kmh
55.44 mph

BANGOR/CANNES (FRANCE)
Henri Chorosz
Glasair II-S FT
1 Lycoming IO-360, 180 bhp
6/24/01
308.39 kmh
191.62 mph

BANGOR/DALLAS (USA)
William H. Wisner, Pilot
Cheryl M. Davis, Copilot
Bonanza, N6826A
1 Continental, 285 bhp
5/14-15/87
264.96 kmh
164.65 mph

BANGOR/GANDER (AUSTRALIA)
Gary Thomas Burns
Alexander Bernard Schenk
Lancair IV
8/5/98
506.49 kmh
314.72 mph

BANGOR/HONOLULU (GERMANY)
Dietrich Schmitt
Beech A36 Bonanza
6/21-27/97
64.35 kmh
39.99 mph

BANGOR/LONG BEACH (GERMANY)
Dietrich Schmitt
Beech A36 Bonanza
6/22/97
134.47 kmh
83.56 mph

BANGOR/SHANNON (AUSTRALIA)
Gary Thomas Burns
Alexander Bernard Schenk
Lancair IV
8/5/98
552.53 kmh
343.33 mph

BANGOR/WICHITA (GERMANY)
Dietrich Schmitt
Beech A36 Bonanza
6/21/97
227.33 kmh
141.26 mph

BARCELONA/COLOGNE-BONN (W. GERMANY)
Richard Flohr
Mooney M20K
1 Continental TSIO-360-GB1, 210 bhp
2/27/83
233.05 kmh
144.81 mph

BAR HARBOR/ST. JOHN'S (USA)
Clifford A. Pulis
Beechcraft Bonanza E-33
1 Continental IO-470, 225 bhp
10/20/93
251.11 kmh
156.03 mph

BAR HARBOR/SYDNEY (USA)
Clifford A. Pulis
Beechcraft Bonanza E-33
1 Continental IO-470, 225 bhp
10/20/93
296.95 kmh
184.51 mph

BASSETERRE/CHARLOTTE AMALIE (USA)
Marie E. McMillan
Beech Bonanza F33A
1 Continental IO-520B, 285 bhp
2/20/84
329.75 kmh
204.90 mph

BASSETERRE/HARRIS (USA)
Marie E. McMillan
Beech Bonanza F33A
1 Continental IO-520B, 285 bhp
2/21/84
195.64 kmh
121.57 mph

BASSETERRE/NEW CASTLE (USA)
Marie E. McMillan
Beech Bonanza F33A
1 Continental IO-520B, 285 bhp
2/21/84
299.18 kmh
185.90 mph

BASSETERRE/POINTE A PITRE (USA)
Marie E. McMillan
Beech Bonanza F33A
1 Continental IO-520B, 285 bhp
2/21/84
303.50 kmh
188.59 mph

BASSETERRE/ST. JOHNS (USA)
Marie E. McMillan
Beech Bonanza F33A
1 Continental IO-520B, 285 bhp
2/21/84
359.36 kmh
223.40 mph

BATON ROUGE/ATHENS (USA)
Hypolite T. Landry, Jr.
Beechcraft Bonanza, N5842K
1 Continental IO-520-B
130 hrs 6 min 50 sec
5/2-7/69
59.60 kmh
37.04 mph

BATON ROUGE/BANGKOK (USA)
Hypolite T. Landry, Jr.
Beechcraft Bonanza, N5842K
1 Continental IO-520-B
314 hrs 21 min 50 sec
5/2-15/69
46.97 kmh
29.19 mph

BATON ROUGE/HAMILTON (USA)
Hypolite T. Landry, Jr., M.D
Beechcraft Bonanza, N5842K
1 Continental IO-520-B
9 hrs 30 min 50 sec
5/2/69
260.10 kmh
161.61 mph

BATON ROUGE/KARACHI (USA)
Hypolite T. Landry, Jr.
Beechcraft Bonanza, N5842K
1 Continental IO-520-B
240 hrs 16 min 50 sec
5/2-12/69
55.95 kmh
34.77 mph

BATON ROUGE/MADRID (USA)
Hypolite T. Landry, Jr.
Beechcraft Bonanza, N5842K
1 Continental IO-520-B
104 hrs 18 min 40 sec
5/2-6/69
73.69 kmh
45.79 mph

BATON ROUGE/NEW DELHI (USA)
Hypolite T. Landry, Jr.
Beechcraft Bonanza, N5842K
1 Continental IO-520-B
126 hrs 9 min 50 sec
5/2-12/69
53.21 kmh
33.06 mph

BATON ROUGE/SAN JOSE DEL CABO (USA)
Jerry L. Payne
James S. Little
Beechcraft Bonanza V35B
1 Continental IO-520, 285 bhp
7/17/97
238.02 kmh
147.90 mph

BATON ROUGE/SANTA MARIA, AZORES (USA)
Hypolite T. Landry, Jr.
Beechcraft Bonanza, N5842K
1 Continental IO-520-B
84 hrs 14 min 9 sec
5/2-5/69
69.94 kmh
43.46 mph

BATON ROUGE/TEHRAN (USA)
Hypolite T. Landry, Jr.
Beechcraft Bonanza, N5842K
1 Continental IO-520-B
197 hrs 40 min 50 sec
5/2-10/69
59.11 kmh
36.73 mph

BEAUMONT/MOBILE (USA)
J. Steven Rayburn
James J. Giordano
Ryan Navion L-17B
1 Continental IO-470, 260 bhp
12/20/02
260.79 kmh
162.05 mph

BEAUMONT/PENSACOLA (USA)
James J. Giordano
J. Steven Rayburn
Ryan Navion L-17B
1 Continental IO-470, 260 bhp
12/18/03
284.09 kmh
176.53 mph

BEAUMONT/SAN ANTONIO (USA)
J. Steven Rayburn
James J. Giordano
Ryan Navion L-17B
1 Continental IO-470, 260 bhp
1/1/00
249.68 kmh
155.14 mph

BEDFORD/CHARLOTTETOWN (USA)
Anne B. Baddour, Pilot
Patricia M. Thrasher, Copilot
Beechcraft Sierra B24R
1 Lycoming, 200 bhp
8/13/91
184.25 kmh
114.49 mph

BEDFORD/ST. JOHN (USA)
Anne B. Baddour, Pilot
Patricia M. Thrasher, Copilot
Beechcraft Sierra B24R
1 Lycoming, 200 bhp
8/13/91
217.01 kmh
134.84 mph

35 years ago - Dr. Hypolite Landry with his V-tail Beechcraft Bonanza, which he flew around the world in 1969, setting several speed records in the process.

BEDFORD/SYDNEY (USA)
Anne B. Baddour, Pilot — 193.03 kmh
Patricia M. Thrasher, Copilot — 119.94 mph
Beechcraft Sierra B24R
1 Lycoming, 200 bhp
8/13/91

BELFAST/LONDON (UK)
Sheila Scott — 337.96 kmh
Piper Comanche 400 — 210.01 mph
1 Lycoming IO-720
1 hr 3 min 44.1 sec
5/20/65

BERLIN/TOUSSUS LE NOBLE (FRANCE)
Georges Simon — 236.37 kmh
Piper PA30 — 146.87 mph
6/20/91

BERMUDA/ST. THOMAS (USA)
Millard Harmon — 267.12 kmh
Beech Bonanza 36 — 165.99 mph
1 Continental 10-520BA, 285 bhp
10/31/85

BIARRITZ/CANNES (FRANCE)
Daniel Muraro — 324.00 kmh
Piper Commanche PA 24-400 — 201.34 mph
1 Lycoming, 260 bhp
7/31/87

BIARRITZ/PARIS (FRANCE)
Christophe Toscas — 395.16 kmh
Michel Rasquin — 245.54 mph
LCT30
1 Continental, 280 bhp
5/27/00

BIARRITZ/PORTO (NETHERLANDS)
Robert Hermans — 187.66 kmh
Piper Comanche PA30B — 116.61 mph
2 Lycoming IO-320, 160 bhp
8/6/85

BIARRITZ/STRASBOURG (FRANCE)
Daniel Muraro — 345.24 kmh
Piper Commanche PA 24-400 — 214.53 mph
1 Lycoming, 260 bhp
7/29/87

BILLINGS/LEWISTOWN (USA)
Amos Babb — 238.68 kmh
Piper Commanche, N110LF — 148.32 mph
1 Lycoming, 0-540-A1A5, 250 bhp
9/1/87

BILLINGS/MINNEAPOLIS (CANADA)
Frank J. Quigg — 202.08 kmh
Waco Classic F-5 — 125.57 mph
1 Jacobs R-755, 275 bhp
9/5/93

BIRMINGHAM/REYKJAVIK (USA)
Millard Harmon — 292.66 kmh
Beech Bonanza 36 — 181.86 mph
1 Continental IO-520-BA, 285 bhp
6/23/85

BLENHEIM/HAMILTON (NEW ZEALAND)
Bernard R. Scherer — 249.10 kmh
Piper PA-28-180 — 154.78 mph
1 Lycoming O-360, 180 bhp
8/27/95

BLENHEIM/INVERCARGILL (NEW ZEALAND)
John Nickolas Darragh Matheson — 144.21 kmh
Wayne John Matheson — 89.61 mph
Cessna 150M
1 Continental O-200, 100 bhp
2/17/02

BLENHEIM/WHANGAREI (NEW ZEALAND)
Glenn P. Armstrong — 160.92 kmh
Wayne J. Matheson — 99.99 mph
Piper PA-28-181
1 Lycoming O-360-A4A, 180 bhp
2/27/97

BLOCK ISLAND/BERMUDA (USA)
Millard Harmon — 352.57 kmh
Beech Bonanza 36 — 219.09 mph
1 Continental 10-520BA, 285 bhp
10/29/85

BLUFF/BLENHEIM (NEW ZEALAND)
Wayne J. Matheson — 173.37 kmh
Glenn P. Armstrong — 107.73 mph
Piper PA-28-181
1 Lycoming O-360-A4A, 180 bhp
2/26/97

BOA VISTA/ORANJESTAD (BRAZIL)
Marcelo Bellodi — 456.70 kmh
Noberto Bellodi — 283.78 mph
Lancair IV-P
1 Continental, 350 bhp
7/22/00

BOMBAY/DUBAI (USA)
William H. Wisner, Pilot — 281.52 kmh
Janice Sullivan, Copilot — 174.94 mph
Beech Bonanza, N6826Q
1 Continental, 285 bhp
6/29/86

BONN-KÖLN/PARIS (FRANCE)
Guy-Eric Oumier — 325.19 kmh
Marc Pontet — 202.06 mph
Mooney M20R Ovation
1 Continental IO-550, 280 bhp
5/14/95

BORINQUEN/FT. LAUDERDALE (USA)
Millard Harmon — 294.13 kmh
Beech Bonanza 36 — 182.77 mph
1 Continental 10-520BA, 285 bhp
11/1/85

BOSTON/CHARLESTON, WV (USA)
William H. Wisner, Pilot — 267.84 kmh
Janice Sullivan, Copilot — 166.44 mph
Beech Bonanza, N6826Q
1 Continental, 285 bhp
7/5/86

BOSTON/GOOSE BAY (USA)
Anne Bridge Baddour, Pilot — 355.32 kmh
Margrit Orlowski, Copilot — 220.80 mph
Mooney 231
1 Continental TSIO-360-LB1B, 210 bhp
10/2/85

BOSTON/REYKJAVIK (USA)
Anne Bridge Baddour, Pilot — 154.08 kmh
Margrit Orlowski, Copilot — 95.75 mph
Mooney 231
1 Continental TSIO-360-LB1B, 210 bhp
10/2-3/85

BOURNEMOUTH/GUERNSEY (GUERNSEY)
James Buchanan — 243.09 kmh
Cessna 182 — 151.05 mph
1 Continental O-470 U, 230 bhp
2/11/85

BOZEMAN/SPOKANE (USA)
Kenneth J. Morton 246.10 kmh
Beechcraft C-35 152.91 mph
1 Continental E-185-11, 205 bhp
7/17/95

BREST/BIARRITZ (FRANCE)
Daniel Muraro 268.63 kmh
Piper Commanche PA 24-400 166.92 mph
1 Lycoming, 260 bhp
7/30/87

BRIDGETON/GRENVILLE (USA)
Marie E. McMillan 289.63 kmh
Beech Bonanza F33A 179.98 mph
1 Continental IO-520B, 520 bhp
2/24/84

BRISBANE/AUCKLAND (NEW ZEALAND)
Clifford V. Tait 237.81 kmh
Beech Bonanza A36 147.77 mph
1 Continental IO-520-BB, 285 bhp
4/9/82

BRISBANE/DARWIN (AUSTRIALIA)
Peter H. Norvill, Pilot 191.94 kmh
Timothy J. Holland, Copilot 119.27 mph
Cessna XP Hawk
1 Continental 10-360-K, 195 bhp
4/25-26/84

BRISBANE/LORD HOWE ISLAND (NEW ZEALAND)
Christopher R. Toms 438.07 kmh
Allan Roberts 272.20 mph
Lancair IV
1 Lycoming TIO-540, 350 bhp
4/1/97

BRISBANE/MELBOURNE (NEW ZEALAND)
Christopher R. Toms 414.05 kmh
Allan Roberts 257.28 mph
Lancair IV
1 Lycoming TIO-540, 350 bhp
3/28/97

BRISBANE/PAGO PAGO (AUSTRALIA)
Gary Thomas Burns 520.17 kmh
Alexander Bernard Schenk 323.22 mph
Lancair IV
7/15/98

BRISBANE/PERTH (AUSTRALIA)
William Wesley Finlen 300.51 kmh
Beechcraft Bonanza V35B 186.72 mph
1 Continental IO-520, 285 bhp
7/19/01

BRISBANE/ROCKHAMPTON (NEW ZEALAND)
Christopher R. Toms 408.93 kmh
Allan Roberts 254.10 mph
Lancair IV
1 Lycoming TIO-540, 350 bhp
3/22/97

BRISBANE/SYDNEY (AUSTRALIA)
Gary T. Burns 514.15 kmh
Peter Lindsay 319.48 mph
Lancair IV
1/15/99

BRISBANE/TOWNSVILLE (AUSTRALIA)
Tevor C. J. Potter 295.10 kmh
Cessna 210N 183.38 mph
1 Continental IO-520 L, 300 bhp
2/28/86

BROWNSVILLE/KEY WEST (USA)
Clyde C. McDonald 267.12 kmh
Cessna 182 165.99 mph
1 Continental O470R, 230 bhp
Elapsed Time: 5 hrs 55 min 18 sec
3/1/86

BRUSSELS/DUBLIN (IRELAND)
Gerald W. Connolly 332.80 kmh
Piper Twin Comanche PA39 206.79 mph
2 Lycoming IO-360
2 hrs 22 min
3/8/75

BRUSSELS/LONDON (UK)
Sheila Scott 329.45 kmh
Piper Comanche 400 204.71 mph
1 Lycoming IO-720
58 min 47.7 sec
5/19/65

BUCHAREST/CHISINAU (ROMANIA)
Constantin Manolache 210.35 kmh
Gheorgue Militaru 130.71 mph
Moravan Cehia ZLIN 142
1 Moravan Otrokovice M 337 AK, 180 bhp
11/8/96

BUCHAREST/ISTANBUL (ROMANIA)
JOINT RECORD
Constantin Manolache 203.71 kmh
Traian Gheorghiu 126.58 mph
Morovan Cehia ZLIN 142
Moravan Otrokovice M 337 AK, 180 bhp
and
Gheorghe Militaru
Morovan Cehia ZLIN 142
Moravan Otrokovice M 337 AK, 180 bhp
and
Alexandru Popovici
Filip Stelian
Morovan Cehia ZLIN 142
Moravan Otrokovice M 337 AK, 180 bhp
9/13/97

BUDAPEST/COLOGNE-BONN (W. GERMANY)
Richard Flohr 261.80 kmh
Mooney M2OK 162.67 mph
1 Continental, 210 bhp
11/29/81

BUENOS AIRES/LOS ANGELES (USA)
Jeremy N. White 39.30 kmh
Beechcraft Bonanza F33A 24.42 mph
1 Continental IO-520, 285 bhp
7/18-29/91

CAIRNS/MUNDA (BRAZIL)
Gerard Moss 247.05 kmh
EMBRAER EMB-721 153.50 mph
10/4/91

CAIRO/PALMA DE MAJORCA (USA)
William H. Wisner, Pilot 261.00 kmh
Janice Sullivan, Copilot 162.18 mph
Beech Bonanza, N6826Q
1 Continental, 285 bhp
7/1/86

CALAIS/DOVER (UK)
Michael J. Baker 256.39 kmh
Piper PA-28R-201 159.31 mph
1 Lycoming IO-360, 200 bhp
10/9/95

CALAIS/LONDON (UK)
Michael J. Baker 233.08 kmh
Piper PA-28R-201 144.82 mph
1 Lycoming IO-360, 200 bhp
10/9/95

CALCUTTA/AUCKLAND (NEW ZEALAND)
Clifford V. Tait 205.26 kmh
Beech Bonanza A36 127.55 mph
1 Continental IO-520-BB, 285 bhp
4/7-9/82

CALCUTTA/BRISBANE (NEW ZEALAND)
Clifford V. Tait 201.40 kmh
Beech Bonanza A36 125.15 mph
1 Continental IO-520-BB, 285 bhp
4/7-9/82

CALCUTTA/DARWIN (NEW ZEALAND)
Clifford V. Tait 247.84 kmh
Beech Bonanza A36 154.01 mph
1 Continental IO-520-BB, 285 bhp
4/7-8/82

CALCUTTA/PENANG (NEW ZEALAND)
Clifford V. Tait 272.17 kmh
Beech Bonanza A36 169.12 mph
1 Continental IO-520-BB, 285 bhp
4/7/82

CALGARY/MONTREAL (USA)
Kenneth P. Johnson, Pilot 415.96 kmh
Larry Cioppi, Copilot 258.46 mph
Glasair III
1 Lycoming IO-540, 300 bhp
5/24/94

CAMBERLEY/DE KOOY (UK)
Geoffrey G. Boot 216.63 kmh
Piper PA-32R-301 134.61 mph
11/10/98

CAMBERLEY/PARIS (UK)
John J. Evendon 214.43 kmh
Cessna C182 133.25 mph
1 Continental O-470U, 230 bhp
11/14/85

CAMBRIDGE/KITTY HAWK (USA)
Carl La Rue 476.31 kmh
John Slais 295.97 mph
Lancair IV-P
1 Continental TSIO-550, 350 bhp
9/13/03

CANBERRA/BRISBANE (AUSTRALIA)
John W. Chesbrough 227.11 kmh
Marguerite Chesbrough 141.11 mph
Mooney M20J
11/25/92

CANBERRA/MELBOURNE (AUSTRALIA)
Leslie R. Thompson, Pilot 265.06 kmh
James Bernie Yeo, Copilot 164.71 mph
Cessna 210N
11/30/89

CANCUN/MATAMOROS (USA)
William H. Wisner, Pilot 290.52 kmh
Fred Perez, Copilot 180.53 mph
Beechcraft E185, N299Z
2 P&W R985, 450 bhp
1/7/87

CANNES/BIARRITZ (FRANCE)
Christophe Toscas 344.97 kmh
Michel Rasquin 214.35 mph
LCT30
1 Continental, 280 bhp
5/27/00

CANNES/BREST (FRANCE)
Daniel Muraro 306.48 kmh
Piper Commanche PA 24-400 190.44 mph
1 Lycoming, 260 bhp
7/30/87

CANNES/DEAUVILLE (FRANCE)
Daniel Robert-Bancharelle 235.44 kmh
Scintex ML 250 146.29 mph
1 Lycoming
3 hrs 30 min 59 sec
7/17/70

CANNES/PARIS (FRANCE)
Daniel Robert-Bancharelle 233.41 kmh
Scintex ML 250 145.03 mph
1 Lycoming
2 hrs 55 min 34 sec
7/17/70

CANOUAN/KINGSTOWN (USA)
Marie E. McMillan 304.14 kmh
Beech Bonanza F33A 188.98 mph
Continental IO-520B, 285 bhp
2/27/84

CANOUAN/LOVELL'S VILLAGE (USA)
Marie E. McMillan 363.32 kmh
Beech Bonanza F33A 225.75 mph
Continental IO-520B, 285 bhp
2/27/84

CANOUAN/PORT ELIZABETH (USA)
Marie E. McMillan 253.76 kmh
Beech Bonanza F33A 157.68 mph
Continental IO-520B, 285 bhp
2/27/84

CAPE CANAVERAL/KITTY HAWK (USA)
Ronald White 248.26 kmh
Cessna 182 RG 154.26 mph
1 Continental O-470, 230 bhp
12/16/84

CAPE TOWN/HARARE (UK)
Bryan Eccles 313.77 kmh
Mooney M20K 194.98 mph
9/9/88

CAPETOWN/JOHANNESBURG (UK)
Bryan Eccles 410.45 kmh
Mooney 20K 255.05 mph
8/9-13/88

CAPETOWN/LONDON (UK)
Sheila Scott 143.02 kmh
Piper Comanche 400 88.87 mph
1 Lycoming IO-720
7/29/67

CARBONDALE, IL/OSHKOSH (USA)
Patrick J. Harris 251.26 kmh
Mooney M20C 156.12 mph
1 Lycoming O-360-A1A
07/27/90

CARDIFF/DUBLIN (UK)
M. D. Evans 168.39 kmh
Piper PA-28 104.64 mph
1 Lycoming, 180 bhp
12/27/85

CARNARVON/BRISBANE (AUSTRALIA)
Peter H. Norvill 179.77 kmh
Cessna XP Hawk 111.71 mph
1 Continental IO-360-KB, 195 bhp
9/30-10/1/83

CARNARVON/PERTH (AUSTRALIA)
Eric H. Wheatley 310.75 kmh
Mooney M20K 193.09 mph
9/15/89

CASABLANCA/BARCELONA (W. GERMANY)
Richard Flohr 320.18 kmh
Mooney M20K 198.95 mph
1 Continental TSIO-360-GB1, 210 bhp
2/26/83

CASABLANCA/COLOGNE-BONN (W. GERMANY)
Richard Flohr 92.47 kmh
Mooney M20K 57.46 mph
1 Continental TSIO-360-GB1, 210 bhp
2/27/83

CASTRIES/BASSETERRE (USA)
Marie E. McMillan 251.66 kmh
Beech Bonanza F33A 156.37 mph
Continental IO-520B, 285 bhp
2/28/84

CASTRIES/BRIDGETOWN (USA)
Marie E. McMillan 248.37 kmh
Beech Bonanza F33A 154.37 mph
Continental IO-520B, 285 bhp
2/28/84

CASTRIES/CODRINGTON (USA)
Marie E. McMillan 273.57 kmh
Beech Bonanza F33A 169.99 mph
Continental IO-520B, 285 bhp
2/28/84

CASTRIES/GRAND BOURG (USA)
Marie E. McMillan 303.53 kmh
Beech Bonanza F33A 188.60 mph
Continental IO-520B, 285 bhp
2/28/84

CASTRIES/GUSTAVIA (USA)
Marie E. McMillan 265.58 kmh
Beech Bonanza F33A 165.02 mph
Continental IO-520B, 285 bhp
2/28/84

CASTRIES/HARRIS (USA)
Marie E. McMillan 279.10 kmh
Beech Bonanza F33A 173.42 mph
Continental IO-520B, 285 bhp
2/28/84

CASTRIES/MARTINIQUE (USA)
Marie E. McMillan 373.30 kmh
Beech Bonanza F33A 231.96 mph
Continental IO-520B, 285 bhp
2/28/84

CASTRIES/NEW CASTLE (USA)
Marie E. McMillan 247.50 kmh
Beech Bonanza F33A 153.71 mph
Continental IO-520B, 285 bhp
2/28/84

CASTRIES/PHILLIPSBURG (USA)
Marie E. McMillan 261.63 kmh
Beech Bonanza F33A 162.57 mph
Continental IO-520B, 285 bhp
2/28/84

CASTRIES/POINTE A PITRE (USA)
Marie E. McMillan 254.84 kmh
Beech Bonanza F33A 158.35 mph
Continental IO-520B, 285 bhp
2/28/84

CASTRIES/ST. JOHNS (USA)
Marie E. McMillan 263.66 kmh
Beech Bonanza F33A 163.83 mph
Continental IO-520B, 285 bhp
2/28/84

CHARLESTON WV/AUSTIN (USA)
William H. Wisner, Pilot 196.56 kmh
Janice Sullivan, Copilot 122.14 mph
Beech Bonanza, N6826Q
1 Continental, 285 bhp
7/6/86

CHARLOTTE AMALIE/BASSETERRE (USA)
Marie E. McMillan 187.79 kmh
Beech Bonanza F33A 116.69 mph
Continental IO-520B, 285 bhp
2/20/84

CHARLOTTE AMALIE/GRAND CASE (USA)
Marie E. McMillan 219.19 kmh
Beech Bonanza F33A 136.19 mph
Continental IO-520B, 285 bhp
2/20/84

CHARLOTTE AMALIE/GUSTAVIA (USA)
Marie E. McMillan 141.04 kmh
Beech Bonanza 87.64 mph
Continental IO-520B, 285 bhp
2/20/84

CHARLOTTE AMALIE/KINGSTON (USA)
Marie E. McMillan 209.44 kmh
Beech Bonanza F33A 130.14 mph
Continental IO-520B, 285 bhp
3/1/84

CHARLOTTE AMALIE/ORANJESTAD (USA)
Marie E. McMillan 186.02 kmh
Beech Bonanza F33A 115.59 mph
Continental IO-520B, 285 bhp
2/20/84

CHARLOTTE AMALIE/PHILLIPSBURG (USA)
Marie E. McMillan 197.87 kmh
Beech Bonanza F33A 122.95 mph
Continental IO-520B, 285 bhp
2/20/84

CHARLOTTE AMALIE/ROADTOWN (USA)
Marie E. McMillan 317.01 kmh
Beech Bonanza F33A 197.98 mph
Continental IO-520B, 285 bhp
2/20/84

CHARLOTTE AMALIE/THE BOTTOMS (USA)
Marie E. McMillan 180.53 kmh
Beech Bonanza F33A 112.17 mph
Continental IO-520B, 285 bhp
2/20/84

CHARLOTTE AMALIE/THE VALLEY (USA)
Marie E. McMillan 241.86 kmh
Beech Bonanza F33A 150.28 mph
Continental IO-520B, 285 bhp
2/20/84

CHARLOTTETOWN/KENNEBUNK (USA)
Anne B. Baddour, Pilot 165.15 kmh
Patricia M. Thrasher, Copilot 102.62 mph
Beechcraft Sierra B24R
1 Lycoming, 200 bhp
8/13/91

CHARLOTTETOWN/PORTSMOUTH (USA)
Anne B. Baddour, Pilot 166.18 kmh
Patricia M. Thrasher, Copilot 103.26 mph
Beechcraft Sierra B24R
1 Lycoming, 200 bhp
8/13/91

CHICAGO/COLUMBUS (USA)
Marie Christensen 436.68 kmh
Beech Bonanza, N36TU 271.35 mph
1 Continental 520, 300 bhp
4/19/89

CHICAGO/KITTY HAWK (USA)
James M. Frankard 423.86 kmh
Glasair III 263.37 mph
1 Lycoming IO-540, 300 bhp
4/24/96

CHICAGO/NEW YORK (USA)
Morris Wortman 541.71 kmh
Mooney TLS 336.60 mph
1 Lycoming TIO-540, 270 bhp
10/22/93

CHICAGO/PITTSBURGH (USA)
Marie Christensen 430.56 kmh
Beech Bonanza, N36TU 267.55 mph
1 Continental 520, 300 bhp
4/19/89

CHICAGO/SHREVEPORT, LA (USA)
Marie Christensen 360.36 kmh
Beech Bonanza B36, N36TU 223.93 mph
1 Continental TSIO 520, 300 bhp
4/11/88

CHICAGO/WASHINGTON (USA)
James M. Frankard 460.80 kmh
Glasair III 286.33 mph
1 Lycoming IO-540, 300 bhp
4/24/96

CHISINAU/BUCHAREST (ROMANIA)
Constantin Manolache 220.87 kmh
Gheorgue Militaru 137.24 mph
Moravan Cehia ZLIN 142
1 Moravan Otrokovice M 337 AK, 180 bhp
11/8/96

CHRISTCHURCH/HAMILTON (NEW ZEALAND)
Bernard R. Scherer 214.80 kmh
Piper PA-28-180 133.47 mph
1 Lycoming O-360, 180 bhp
8/27/95

CINCINNATI/LAKELAND (USA)
Paul F. Siegel Jr. 451.90 kmh
Glasair III 280.79 mph
1 Lycoming IO-540, 300 bhp
4/17/93

CLEARFIELD/KITTY HAWK (USA)
Rodney H. Bowers 228.40 kmh
Cessna 177B 141.92 mph
1 Lycoming O-360, 180 bhp
12/17/93

CLIFTON/KINGSTOWN (USA)
Marie E. McMillan 323.60 kmh
Beech Bonanza F33A 201.08 mph
Continental IO-520B, 285 bhp
2/27/84

CLIFTON/PORT ELIZABETH (USA)
Marie E. McMillan 303.85 kmh
Beech Bonanza F33A 188.80 mph
Continental IO-520B, 285 bhp
2/27/84

CODRINGTON/GUSTAVIA (USA)
Marie E. McMillan 290.28 kmh
Beech Bonanza F33A 180.37 mph
Continental IO-520B, 285 bhp
2/28/84

CODRINGTON/PHILIPSBURG (USA)
Marie E. McMillan 348.04 kmh
Beech Bonanza F33A 216.26 mph
Continental IO-520B, 285 bhp
2/28/84

COFF'S HARBOUR/LORD HOWE ISLAND (NEW ZEALAND)
Sue Campbell 294.60 kmh
Dawson Bowles 183.06 mph
Richard Neave
Mooney 201
1 Lycoming, 200 bhp
6/30/00

COLLEGE PARK/KITTY HAWK*
Jeremy R. Stapley 92.57 mph
Cessna Skyhawk
1 Lycoming O-320, 160 bhp
7/16/98

COLOGNE-BONN/BUDAPEST (W. GERMANY)
Richard Flohr 340.15 kmh
Mooney M20K 211.36 mph
1 Continental, 210 bhp
11/27/81

COLOGNE-BONN/GENOA (W. GERMANY)
Richard Flohr 261.33 kmh
Mooney M20K 162.39 mph
1 Continental TSIO-360-GB1, 210 bhp
5/31/84

COLOGNE-BONN/PARIS (W. GERMANY)
Richard Flohr 316.29 kmh
Mooney M20K 196.54 mph
1 Continental TSIO-360-GB1, 210 bhp
3/14-15/84

COLOGNE-BONN/PRAGUE (W. GERMANY)
Richard Flohr 297.21 kmh
Mooney M20K 184.69 mph
1 Continental TSIO 360-GB1, 210 bhp
6/10/83

COLOGNE-BONN/WARSAW (W. GERMANY)
Richard Flohr 219.23 kmh
Mooney M20K 136.23 mph
1 Continental TSIO-360-GB1, 210 bhp
4/27/84

COLOMBO/BAHRAIN (GERMANY)
Dietrich Schmitt 259.20 kmh
Beech A36 Bonanza 161.06 mph
7/13/97

COLOMBO/MALTA (GERMANY)
Dietrich Schmitt 119.30 kmh
Beech A36 Bonanza 74.13 mph
7/13-15/97

COLOMBO/MANNHEIM (GERMANY)
Dietrich Schmitt 92.01 kmh
Beech A36 Bonanza 57.17 mph
7/13-16/97

COLORADO SPRINGS/NEW YORK (USA)
Calvin B. Early 462.54 kmh
Beechcraft Bonanza M35 287.40 mph
1 Continental IO-550, 300 bhp
3/3/97

COLUMBIA/PORTLAND (USA)
Stephen H. Humphrey, Pilot 291.70 kmh
E. Leland Humphrey, Copilot 181.25 mph
Piper PA-32RT-300 Lance II
1 Lycoming IO-540, 300 bhp
7/14/00

COLUMBUS/DAYTONA BEACH*
James A. Terry, Jr. 116.67 mph
Mark W. Houser
Piper PA-28-161 Cadet
1 Lycoming O-320, 160 bhp
5/23/94

COLUMBUS/KITTY HAWK (USA)
Carl A. La Rue 637.98 kmh
Lancair IV-P 396.42 mph
1 Continental TSIO-550, 350 bhp
2/2/02

COLUMBUS/PITTSBURGH (USA)
Marie Christensen 418.32 kmh
Beech Bonanza, N36TU 259.94 mph
1 Continental 520, 300 bhp
4/19/89

COLUMBUS/WASHINGTON, DC*
Lincoln Cummings 180.55 mph
Cessna 182P, N58650
1 Continental, 230 bhp
9/3/79

COPENHAGEN/LONDON (UK)
Sheila Scott 261.09 kmh
Piper Comanche 260, G-ATOY 162.23 mph
1 Lycoming TO-540
5/25/69

COZUMEL/GRAND CAYMAN (USA)
Marie E. McMillan 277.20 kmh
Beech Bonanza F33A 172.25 mph
1 Continental IO-520 BA, 285 bhp
2/5/84

DAKAR/LAKELAND (FRANCE)
Henri Chorosz 227.34 kmh
Stoddard-Hamilton GL-20 Glasair II FT-S 141.26 mph
4/9/1999

DAKAR/TOULOUSE (FRANCE)
Jean-Pierre Vignel 65.80 kmh
Cessna 172 40.89 mph
1 Lycoming, 150 bhp
4/24/00

DALLAS-FORT WORTH/GANDER (USA)
Samuel W. Marshall, Jr. 100.77 kmh
Cessna 210, N3631Y 62.61 mph
1 Continental IO-470
38 hrs 46 min 47 sec
5/3-5/69

DALLAS/JACKSONVILLE (USA)
William H. Cox, Pilot 479.32 kmh
Martha A. Morrell, Copilot 297.83 mph
Mooney TLS
1 Lycoming TIO-540, 270 bhp
3/25/94

DALLAS/KEY WEST (USA)
William H. Wisner, Pilot 307.27 kmh
Janice Sullivan, Copilot 190.93 mph
Beech Bonanza, N6826Q
1 Continental, 285 bhp
6/8/88

DALLAS/ORLANDO (USA)
Earl R. Epperson 273.44 kmh
Cessna T210L 169.91 mph
1 Continental TSIO-520, 285 bhp
2/16/99

DALLAS/QUEBEC*
William H. Wisner, Pilot 199.68 mph
Lanita Clay, Copilot
Beechcraft Bonanza, N6826Q
06/01/90

DARWIN/ALICE SPRINGS (AUSTRALIA)
Roy Travis 270.40 kmh
Beechcraft Bonanza A36 168.02 mph
4/20/99

DARWIN/AUCKLAND (NEW ZEALAND)
Clifford V. Tait 245.52 kmh
Beech Bonanza A36 152.57 mph
1 Continental IO-520-BB, 285 bhp
4/8-9/82

DARWIN/BALI (USA)
William H. Wisner, Pilot 270.00 kmh
Janice Sullivan, Copilot 167.78 mph
Beech Bonanza, N6826Q
1 Continental, 285 bhp
6/21-22/86

DARWIN/BRISBANE (AUSTRALIA)
Gary Thomas Burns 582.27 kmh
Alexander Bernard Schenk 361.81 mph
Lancair IV
8/25/98

DARWIN/CARNARVON (AUSTRIALIA)
Eric H. Wheatley 305.23 kmh
Mooney M20K 189.66 mph
9/15/89

DARWIN/MELBOURNE (AUSTRALIA)
Roy Travis 113.34 kmh
Beechcraft Bonanza A36 70.43 mph
4/15/99

DARWIN/PERTH (AUSTRALIA)
L.K. Smith 277.61 kmh
Piper Comanche 260-B, N8945 172.50 mph
1 Lycoming IO-540-D
11/22/72

DARWIN/SINGAPORE (AUSTRALIA)
David McDonald 184.41 kmh
Peter R. Smith 114.58 mph
Beechcraft Bonanza A36
1 Continental IO-520, 285 bhp
2/13/01

DAYTON/ALBANY (USA)
Millard Harmon 345.24 kmh
Beech Bonanza 36 214.52 mph
1 Continental IO-520 BA, 285 bhp
2/6-7/85

DAYTON/KITTY HAWK (USA)
Carl A. La Rue, Pilot 642.00 kmh
Daniel T. Johnston, Copilot 398.92 mph
Lancair IV-P
1 Continental TSIO-550, 350 bhp
1/6/01

DAYTON/NORFOLK (USA)
Tom Harnish, Pilot 268.56 kmh
Judith Watson, Copilot 166.88 mph
Beech Bonanza BE-35, N4533D
1 Continental E-225, 225 bhp
3/31/88

DAYTON/WASHINGTON (USA)
Carl A. La Rue 618.68 kmh
John Slais 384.43 mph
Lancair IV P
1 Continental TSIO-550B, 350 bhp
10/21/97

DAYTONA BEACH/ALBANY (USA)
Millard Harmon 305.28 kmh
Beech Bonanza 36 189.70 mph
1 Continental 10-520BA, 285 bhp
11/1-2/85

DAYTONA BEACH/CRESTVIEW*
Keith Plumb, Pilot 103.38 mph
John Walsh, Copilot
Cessna 172
1 Lycoming O-320, 160 bhp
4/24/92

DAYTONA BEACH/MONROE*
Keith Plumb, Pilot — 82.68 mph
John Walsh, Copilot
Cessna 172
1 Lycoming O-320, 160 bhp
4/24/92

DAYTONA BEACH/OSHKOSH (USA)
Nancy E. Sliwa, Pilot — 211.37 kmh, 131.34 mph
David V. Lewis, Copilot
Mooney M20J
1 Lycoming IO-360, 260 bhp
7/31/92

DAYTONA BEACH/PENSACOLA*
John F. Wassong, Pilot — 104.50 mph
Eric Quinn, Copilot
Mark W. Ruane, Copilot
Cessna 172P
1 Lycoming O-320, 160 bhp
4/16/93

DEAUVILLE/CANNES (FRANCE)
Daniel Robert-Bancharelle — 275.01 kmh, 170.89 mph
Scintex ML 250
1 Lycoming
3 hrs 00 min 38 sec
7/17/70

DECATUR/KITTY HAWK (USA)
Michael D. Scroggins — 299.66 kmh, 186.20 mph
Piper PA-24-260
1 Lycoming IO-540, 260 bhp
12/17/93

DEER LAKE/VICTORIA (CANADA)
Michael H. Leslie — 20.63 kmh, 12.82 mph
Joseph Leslie
Daniel J. Leslie
Cessna 172
1 Lycoming O-320, 160 bhp
8/15/00

DE KOOY/EPINAL (UK)
Geoffrey G. Boot — 249.26 kmh, 154.88 mph
Piper PA-32R-301
11/11/98

DENVER/ALBANY (USA)
Millard Harmon — 319.32 kmh, 198.43 mph
Beech Bonanza 36
1 Continental IO-520-BA, 285 bhp
6/24/84

DENVER/ALBUQUERQUE*
Stan L. VanderWerf — 161.01 mph
Beechcraft Bonanza F33A
1 Continental IO-520, 285 bhp
4/20/92

DENVER/ATHENS (USA)
L. P. Whistle — 237.60 kmh, 147.64 mph
Eddie Whistle
Beech Bonanza V-35, N27FH
1 Continental TSIO-520-B, 285 bhp
10/20-22/81

DENVER/CHICAGO (USA)
Calvin B. Early — 418.84 kmh, 260.25 mph
Beechcraft Bonanza M35
1 Continental IO-550, 300 bhp
4/17/92

DENVER/KARACHI (USA)
L. P. Whistle — 215.28 kmh, 133.77 mph
Eddie Whistle
Beech Bonanza V-35, N27FH
1 Continental TSIO-520-B, 285 bhp
11/5-9/80

DENVER/LAS VEGAS (USA)
Ian Bentley — 247.23 kmh, 153.62 mph
Piper Dakota
1 Lycoming O-540, 235 bhp
10/11/92

DENVER/LUXOR (USA)
L. P. Whistle — 224.64 kmh, 139.58 mph
Eddie Whistle
Beech Bonanza V-35, N27FH
1 Continental TSIO-520-B, 285 bhp
11/5-8/80

DENVER/MUNCIE (USA)
J. Lance Zellers — 272.87 kmh, 169.55 mph
Piper PA32R-300
1 Lycoming IO-540, 300 bhp
9/12/93

DENVER/NEW YORK (USA)
Ron Roldan, Pilot — 498.35 kmh, 309.66 mph
Bryan Touhey, Copilot
Mooney M20K
1 Continental TSIO-360, 210 bhp
12/19/90

DENVER/OSHKOSH (USA)
Charles Woolacott — 421.53 kmh, 261.93 mph
Mooney M20M
1 Lycoming TIO-540, 270 bhp
7/25/94

DENVER/PALERMO (USA)
L. P. Whistle — 263.35 kmh, 163.44 mph
Eddie Whistle
Beech Bonanza V-35, N27FH
1 Continental TSIO-520-B
38 hrs 16 min 15 sec
11/5-7/80

DENVER/SANTA MARIA (USA)
L. P. Whistle — 263.16 kmh, 163.52 mph
Eddie Whistle
Beech Bonanza V-35, N27FH
1 Continental TSIO-520-B
11/5-6/80

DETROIT/KITTY HAWK (USA)
William S. Demray — 253.33 kmh, 157.41 mph
James R. Townsley
Piper PA-30 Twin Comanche
2 Lycoming IO-320, 160 bhp
12/17/03

DETROIT/SEATTLE (USA)
William S. Demray — 239.43 kmh, 148.78 mph
Jeffrey Seit
Piper PA-30C
2 Lycoming IO-320, 160 bhp
4/22/95

DOUGLAS/CARDIFF (UK)
M. D. Evans — 202.92 kmh, 126.09 mph
Piper PA-28
1 Lycoming, 180 bhp
12/27/85

DOVER/CALAIS (UK)
Michael J. Baker — 259.96 kmh, 161.53 mph
Piper PA-28R-201
1 Lycoming IO-360, 200 bhp
10/9/95

DUBAI/AGRA (USA)
Kenneth P. Johnson, Pilot — 364.80 kmh, 226.67 mph
Larry Cioppi, Copilot
Glasair III
1 Lycoming IO-540, 300 bhp
5/9/94

DUBAI/CAIRO (USA)
William H. Wisner, Pilot — 252.72 kmh
Janice Sullivan, Copilot — 157.04 mph
Beech Bonanza, N6826Q
1 Continental, 285 bhp
6/30/86

DUBLIN/GENEVA (IRELAND)
Michael Slazenger — 261.38 kmh
Piper Twin Comanche — 162.41 mph
2 Lycoming, 160 bhp
4 hrs 33 min 1 sec
3/10/76

DUBLIN/LONDON (UK)
Sheila Scott — 338.33 kmh
Piper Comanche 400 — 210.23 mph
1 Lycoming IO-720
1 hr 21 min 55.7 sec
5/20/65

DUBLIN/PORT ERIN (UK)
M. D. Evans — 158.27 kmh
Piper PA-28 — 98.35 mph
1 Lycoming, 180 bhp
12/27/85

DUBLIN/RONALDSWAY (IRELAND)
Gerald W. Connolly — 316.31 kmh
Piper Comanche PA39 — 196.54 mph
2 Lycoming IO-360
24 min 25 sec
2/15/75

DULUTH/SAULT STE. MARIE (USA)
Marilyn J. Moody — 352.61 kmh
Janet L. Liberty — 219.10 mph
Beechcraft Bonanza P35
1 Continental IO-470, 260 bhp
6/25/97

DURANGO/PUEBLO*
Billy Morency — 108.72 mph
Cessna 172L, N9898Q
6/12/88

DUSSELDORF/HELSINKI (W. GERMANY)
Wilhelm Heller — 405.40 kmh
Cessna P210 N Centurion — 251.90 mph
Continental TSIO-520-P, 310 bhp
1/6/83

DUSSELDORF/KORFU (W. GERMANY)
Wilhelm Heller — 336.15 kmh
Cessna P210 N Centurion — 208.87 mph
Continental TSIO-520-P, 310 bhp
5/29/82

DUSSELDORF/LISBON (USA)
William H. Wisner, Pilot — 251.28 kmh
Cheryl M. Davis, Copilot — 156.15 mph
Bonanza, N6826A
1 Continental, 285 bhp
5/11/87

DUSSELDORF/REYKJAVIK (W. GERMANY)
Wilhelm Heller — 369.16 kmh
Cessna P210N — 229.40 mph
Continental TSIO-520-P, 310 bhp
11/6/82

EGLESBACH/TUNIS (W. GERMANY)
Horst Regar — 177.80 kmh
Beechcraft Bonanza V-35 — 110.50 mph
1 Continental IO-520 BA
8 hrs 13 min 44 sec
10/28/76

EL PASO/GUADALAJARA (USA)
Marie E. McMillan — 263.70 kmh
Beech Bonanaza F 33 A — 163.86 mph
Continental IO-520BA, 285 bhp
5/4/83

EL PASO/LAS VEGAS (USA)
Marie E. McMillan — 285.85 kmh
Beech Bonanaza F 33 A — 177.62 mph
Continental IO-520BA, 285 bhp
5/9/83

EL PASO/PHOENIX (USA)
G. R. Wood — 272.30 kmh
Beechcraft Bonanza V-35-TC — 169.19 mph
1 Continental TSIO-52-D, 285 bhp
2 hrs 59 min 18 sec
6/17/76

EPINAL/PALMA DE MAJORCA (UK)
Geoffrey G. Boot — 204.02 kmh
Piper PA-32R-301 — 126.77 mph
11/13/98

EUREKA/NORTH POLE*
Millard Harmon — 159.05 mph
Beech Bonanza 36, N7710R
1 Continental IO-520-BA, 285 bhp
8/8/86

EUREKA/RESOLUTE*
Millard Harmon — 162.41 mph
Beech Bonanza 36, N7710R
1 Continental IO-520-BA, 285 bhp
8/8/86

EVERETT/LOUISVILLE*
Thomas F. Smith — 171.46 mph
Piper Twin Comanche
2 Lycoming IO-320, 160 bhp
8/24/92

FARMINGTON/PRESCOTT*
Anganette Morency — 106.93 mph
Cessna 172L, N9898Q
6/19/88

FISHER'S ISLAND/PORTLAND, ME*
Frederick G. Herbert — 154.82 mph
Cessna 182R
1 Avco Lycoming O-540, 235 bhp
9/27/97

FLORENCE/ALBANY (USA)
Millard Harmon — 299.89 kmh
Beech Bonanza 36 — 186.35 mph
1 Continental 10-520BA, 285 bhp
11/1-2/85

FT. LAUDERDALE/ALBANY (USA)
Millard Harmon — 285.89 kmh
Beech Bonanza 36 — 177.65 mph
1 Continental 10-520BA, 285 bhp
11/1/85

FT. LAUDERDALE/MYRTLE BEACH (USA)
Mike Hackney — 305.65 kmh
Mooney M20R Ovation — 189.92 mph
1 Continental IO-550, 280 bhp
7/15/99

FT. MYERS/NEW ORLEANS (USA)
Peter H. Burgher — 297.13 kmh
Piper PA-30 — 184.63 mph
2 Lycoming O-320, 160 bhp
10/23/91

FORT WALTON BEACH/KITTY HAWK (USA)
William G. Castlen 344.83 kmh
Michael D. Scroggins 214.27 mph
Cirrus SR22
1 Continental IO-550, 310 bhp
12/17/03

FORT WORTH/SAN DIEGO (USA)
Martin V. Case Jr., Pilot 272.17 kmh
Mark A. Case, Copilot 169.12 mph
Piper Comanche
1 Lycoming O-540, 250 bhp
8/25/93

FRANKFORT/KITTY HAWK (USA)
Kevin Janiak, Pilot 330.92 kmh
Dean A. Del Bene, Copilot 205.62 mph
Piper PA-30 Comanche
2 Lycoming IO-320, 160 bhp
12/16/95

FRANKFURT/LISBON (USA)
Millard Harmon 202.68 kmh
Beech Bonanza 36 125.95 mph
1 Continental IO-520-BA, 285 bhp
7/11/84

FREDERICK/GOOSE BAY (CANADA)
Mathias Stinnes 399.27 kmh
Heinz Bittermann 248.09 mph
Glasair III
7/5/92

FRESNO/FREDERICK (CANADA)
Mathias Stinnes 453.86 kmh
Heinz Bittermann 282.02 mph
Glasair III
7/8/92

FRESNO/LAS VEGAS (USA)
Harold Kindsvater, Pilot 363.24 kmh
Wayne Easley, Copilot 225.72 mph
Cessna P210, N734VM
1 Continental TSIO-520-P, 300 bhp
4/30/88

FROBISHER BAY/ALBANY*
Millard Harmon 144.29 mph
Beech Bonanza 36, N7710R
1 Continental IO-520-BA, 285 bhp
8/10/86

FROBISHER BAY/RESOLUTE*
Millard Harmon 141.16 mph
Beech Bonanza 36, N7710R
1 Continental IO-520-BA, 285 bhp
8/6/86

FULLERTON, CA/PALM SPRINGS*
Henry Ilves, Pilot 143.95 mph
Joyce Ilves, Copilot
Piper Comanche
1 Lycoming IO-540, 260 bhp
10/24/90

GAITHERSBURG/KITTY HAWK (USA)
William H. Ottley, Pilot 219.96 kmh
John Meyers, Copilot 136.68 mph
Beech Bonanza F33A
1 Continental IO 520, 285 bhp
12/16/85

GANDER/BRUSSELS (W. GERMANY)
Margrit Orlowski 338.90 kmh
Mooney 20 K 210.59 mph
1 Continental TSIO-360-GB-1, 210 bhp
1/30-31/82

GANDER/COLOGNE-BONN (W. GERMANY)
Margrit Orlowski 334.50 kmh
Mooney 201 207.85 mph
1 Lycoming 0-360 A4M, 200 bhp
2/8-9/82

GANDER/COPENHAGEN (W. GERMANY)
Wilhelm Heller 273.74 kmh
Rockwell Commander 114A 170.09 mph
1 Lycoming, 260 bhp
11/9/79

GANDER/DALLAS-FORT WORTH (USA)
Samuel W. Marshall, Jr. 90.47 kmh
Cessna 210, N3631Y 56.22 mph
1 Continental IO-470
43 hrs 12 min
5/6-8/69

GANDER/MUNICH (W. GERMANY)
Dietrich Schmitt 242.29 kmh
Beechcraft Bonanza 150.55 mph
Continental TSIO-520-BB, 300 bhp
6/23-24/77

GANDER/PARIS (W. GERMANY)
Margrit Orlowski 343.60 kmh
Mooney 20K 213.50 mph
Continental TSIO-360-GB4, 210 bhp
4/23-24/82

GANDER/SHANNON (AUSTRALIA)
Gary Thomas Burns 573.16 kmh
Alexander Bernard Schenk 356.14 mph
Lancair IV
8/5/98

GANDER/STOCKHOLM (SWEDEN)
Olof Calert 267.90 kmh
Piper Commanche PA-24-180 166.56 mph
1 Continental, 180 bhp
17 hrs 12 min
7/18-19/71

GANDER/ZURICH (W. GERMANY)
Margrit Orlowski 362.40 kmh
Mooney 20K 225.20 mph
Continental TSIO-360-6B, 210 bhp
4/11-12/83

GASTONIA/KITTY HAWK (USA)
H. S. Kendal Willis, Pilot 276.45 kmh
Robert R. Beitel, Copilot 171.78 mph
Cessna 182RF, N2706C
1 Lycoming 0-540-J-3C5D
12/17/88

GENEVA/DUBLIN (IRELAND)
Michael Slazenger 257.91 kmh
Piper Twin Commanche 160.25 mph
2 Lycoming 160
4 hrs 36 min 8 sec
10/3/76

GENEVA/HELSINKI (CANADA)
Mathias Stinnes 370.08 kmh
Heinz Bittermann 229.95 mph
Glasair III
6/20/92

GEORGETOWN/COZUMEL (USA)
Marie E. McMillan 304.52 kmh
Beech Bonanza F33A 189.22 mph
1 Continental IO-520B, 285 bhp
3/3/84

GILMER, TX/OSHKOSH (USA)
Stephen E. Dean 196.82 kmh
Beechcraft Bonanza 35-C33A 122.30 mph
1 Continental IO-520, 285 bhp
07/26/90

GOOSE BAY/ALBANY (USA)
Millard Harmon 265.00 kmh
Beech Bonanza 36 164.67 mph
1 Continental IO-520-BA, 285 bhp
6/25/85

GOOSE BAY/DUSSELDORF (W. GERMANY)
Margrit Orlowski 321.35 kmh
Cessna T-210 199.69 mph
1 Continental TSIO-520-R, 310 bhp
7/12/81

GOOSE BAY/KEFLAVIK*
Millard Harmon 168.90 mph
Beech Bonanza 36, N7710R
1 Continental IO-520-BA, 285 bhp
6/30/84

GOOSE BAY/NARSSARSSUAQ (USA)
William H. Wisner, Pilot 303.84 kmh
Cheryl M. Davis, Copilot 188.81 mph
Bonanza, N6826A
1 Continental, 285 bhp
5/5/87

GOOSE BAY/NUUK (CANADA)
Mathias Stinnes 358.01 kmh
Heinz Bittermann 222.46 mph
Glasair III
7/8/92

GOOSE BAY/REYKJAVIK (USA)
Anne Bridge Baddour, Pilot 331.20 kmh
Margrit Orlowski, Copilot 205.81 mph
Mooney 231
1 Continental TSIO-360-LB1B, 210 bhp
10/3/85

GRAND-BOURG/BASSETERRE (USA)
Marie E. McMillan 236.24 kmh
Beech Bonanza F33A 146.79 mph
1 Continental IO-520B, 285 bhp
2/28/84

GRAND-BOURG/BRIDGETOWN (USA)
Marie E. McMillan 211.47 kmh
Beech Bonanza F33A 131.40 mph
1 Continental IO-520B, 285 bhp
2/23/84

GRAND-BOURG/CASTRIES (USA)
Marie E. McMillan 220.66 kmh
Beech Bonanza F33A 137.11 mph
1 Continental IO-520B, 285 bhp
2/23/84

GRAND-BOURG/CODRINGTON (USA)
Marie E. McMillan 250.68 kmh
Beech Bonanza F33A 155.77 mph
1 Continental IO-520B, 285 bhp
2/28/84

GRAND-BOURG/GUSTAVIA (USA)
Marie E. McMillan 255.55 kmh
Beech Bonanza F33A 158.79 mph
1 Continental IO520B, 285 bhp
2/28/84

GRAND-BOURG/HARRIS (USA)
Marie E. McMillan 278.61 kmh
Beech Bonanza F33A 173.12 mph
1 Continental IO-520B, 285 bhp
2/28/84

GRAND-BOURG/MARTINIQUE (USA)
Marie E. McMillan 252.19 kmh
Beech Bonanza F33A 156.70 mph
1 Continental IO-520B, 285 bhp
2/23/84

GRAND-BOURG/NEW CASTLE (USA)
Marie E. McMillan 228.52 kmh
Beech Bonanza F33A 141.99 mph
1 Continental IO-520B, 285 bhp
2/28/84

GRAND-BOURG/PHILIPSBURG (USA)
Marie E. McMillan 253.14 kmh
Beech Bonanza F33A 157.29 mph
1 Continental IO-520B, 285 bhp
2/28/84

GRAND-BOURG/PORTSMOUTH (USA)
Marie E. McMillan 214.91 kmh
Beech Bonanza F33A 133.54 mph
1 Continental IO-520B, 285 bhp
2/23/84

GRAND-BOURG/ST. JOHNS (USA)
Marie E. McMillan 227.06 kmh
Beech Bonanza F33A 141.09 mph
1 Continental IO-520B, 285 bhp
2/28/84

GRAND-CASE/BASSETERRE (USA)
Marie E. McMillan 206.29 kmh
Beech Bonanza F33A 128.18 mph
1 Continental IO-520B, 285 bhp
2/20/84

GRAND-CASE/CHARLOTTE AMALIE (USA)
Marie E. McMillan 162.71 kmh
Beech Bonanza F33A 101.10 mph
1 Continental IO-520B, 285 bhp
2/20/84

GRAND-CASE/CHRISTIANSTED (USA)
Marie E. McMillan 193.19 kmh
Beech Bonanza F33A 120.04 mph
1 Continental IO-520B, 285 bhp
2/20/84

GRAND-CASE/ORANJESTAD (USA)
Marie E. McMillan 240.31 kmh
Beech Bonanza F33A 149.32 mph
1 Continental IO-520B, 285 bhp
2/20/84

GRAND-CASE/PHILIPSBURG (USA)
Marie E. McMillan 294.72 kmh
Beech Bonanza F33A 183.13 mph
1 Continental IO-520B, 285 bhp
2/20/84

GRAND-CASE/THE BOTTOMS (USA)
Marie E. McMillan 359.87 kmh
Beech Bonanza F33A 223.61 mph
1 Continental IO-520B, 285 bhp
2/20/84

GRAND CAYMAN/DALLAS (USA)
William H. Wisner, Pilot 302.14 kmh
Janice Sullivan, Copilot 187.75 mph
Beech Bonanza, N6826Q
1 Continental, 285 bhp
6/11/88

GRAND CAYMAN/MONTEGO BAY (USA)
Marie E. McMillan 254.31 kmh
Beech Bonanza F33A 158.03 mph
1 Continental IO-520 BA, 285 bhp
2/9/84

GRAND TURK/FT. LAUDERDALE (USA)
Millard Harmon 291.01 kmh
Beech Bonanza 36 180.83 mph
1 Continental IO-520BA, 285 bhp
11/1/85

GREENVILLE/LAKELAND (USA)
Daniel L. Allen, Jr. 260.61 kmh
Cessna 182Q 161.93 mph
1 Continental O-470, 230 bhp
4/4/92

GRENVILLE/CANOUAN (USA)
Marie E. McMillan 298.75 kmh
Beech Bonanza F33A 186.64 mph
1 Continental IO-520B, 285 bhp
2/27/84

GRENVILLE/CASTRIES (USA)
Marie E. McMillan 300.94 kmh
Beech Bonanza F33A 187.00 mph
1 Continental IO-520B, 285 bhp
2/27/84

GRENVILLE/CLIFTON (USA)
Marie E. McMillan 263.75 kmh
Beech Bonanza F33A 163.89 mph
1 Continental IO-520B, 285 bhp
2/27/84

GRENVILLE/KINGSTOWN (USA)
Marie E. McMillan 291.64 kmh
Beech Bonanza F33A 181.22 mph
1 Continental IO-520B, 285 bhp
2/27/84

GRENVILLE/LOVELL'S VILLAGE (USA)
Marie E. McMillan 306.29 kmh
Beech Bonanza F33A 190.32 mph
1 Continental IO-520B, 285 bhp
2/27/84

GRENVILLE/PETITE MARTINIQUE (USA)
Marie E. McMillan 300.12 kmh
Beech Bonanza F33A 186.48 mph
1 Continental IO-520B, 285 bhp
2/27/84

GRENVILLE/PORT ELIZABETH (USA)
Marie E. McMillan 306.20 kmh
Beech Bonanza F33A 190.27 mph
1 Continental IO-520B, 285 bhp
2/27/84

GUADALCANAL/BRISBANE (USA)
William H. Wisner, Pilot 258.12 kmh
Janice Sullivan, Copilot 160.40 mph
Beech Bonanza, N6826Q
1 Continental, 285 bhp
6/16-17/86

GUADALAJARA/EL PASO (USA)
Marie E. McMillan 273.80 kmh
Beech Bonanza F33A 170.10 mph
Continental IO-520-BA, 285 bhp
5/8/83

GUADALAJARA/MEXICO CITY (USA)
Marie E. McMillan, Pilot 287.64 kmh
Gloria May, Navigator 178.73 mph
Beech Bonanza F33A
1 Continental IO-520
1 hr 36 min
1/28/81

GUADALAJARA/PUERTO VALLARTA (USA)
Marie E. McMillan, Pilot 289.44 kmh
Gloria May, Navigator 179.85 mph
Beech Bonanza F33A
1 Continental IO-520
42 min 17 sec
1/31/81

GUADALCANAL/RABAUL, NB (USA)
Geraldine L. Mock 234.91 kmh
Cessna P206, N155JM 145.97 mph
1 Continental IO-520
4 hrs 26 min
10/29-30/69

GUAM/BALIKPAPAN (GERMANY)
Dietrich Schmitt 295.24 kmh
Beech A36 Bonanza 183.45 mph
7/5/97

GUAM/MEDAN (GERMANY)
Dietrich Schmitt 89.47 kmh
Beech A36 Bonanza 55.59 mph
7/4-7/97

GUERNSEY/BOURNEMOUTH (GUERNSEY)
James Buchanan 190.97 kmh
Cessna 182 118.67 mph
1 Continental O-470 U, 230 bhp
2/11/85

GUERNSEY/COLOGNE-BONN (W. GERMANY)
Richard Flohr 311.54 kmh
Mooney M20K 193.59 mph
1 Continental TSIO 360-GB1, 210 bhp
6/15/83

GUERNSEY/SHANNON (GUERNSEY)
Philip C. Blows 226.74 kmh
Cessna F182Q 140.90 mph
1 Continental O-470-U, 230 bhp
9/26/85

GUERNSEY/SOUTHAMPTON (GUERNSEY)
David S. Innes 281.93 kmh
Rockwell Commander 114 175.19 mph
1 Lycoming IO-540, 260 bhp
2/8/85

GULFPORT/BEAUMONT (USA)
James J. Giordano 230.99 kmh
J. Steven Rayburn 143.53 mph
Ryan Navion
1 Continental E-225, 225 bhp
12/20/01

GULFPORT/KITTY HAWK (USA)
Charles S. Horton, Jr. 416.81 kmh
Piper Comanche 400 258.99 mph
1 Lycoming IO-720, 400 bhp
12/8/96

GULFPORT/LAKELAND (USA)
Charles S. Horton, Jr. 435.68 kmh
Piper PA-24 Comanche 400 270.72 mph
1 Lycoming IO-720, 400 bhp
4/13/99

GUSTAVIA/CHARLOTTE AMALIE (USA)
Marie E. McMillan 391.82 kmh
Beech Bonanza F33A 243.46 mph
1 Continental IO-520B, 285 bhp
2/20/84

GUSTAVIA/PHILIPSBURG (USA)
Marie E. McMillan 274.59 kmh
Beech Bonanza F33A 170.62 mph
1 Continental IO-520B, 285 bhp
2/28/84

HAMILTON/BLENHEIM (NEW ZEALAND)
Bernard R. Scherer 179.30 kmh
Piper PA-28-180 111.41 mph
1 Lycoming O-360, 180 bhp
8/25/95

HAMILTON/CHRISTCHURCH (NEW ZEALAND)
Bernard R. Scherer 173.10 kmh
Piper PA-28-180 107.55 mph
1 Lycoming O-360, 180 bhp
8/25/95

HANOVER/BROCKVILLE (CANADA)
David M. Gullacher 191.15 kmh
Stinson 108 118.77 mph
1 Lycoming O-435, 190 bhp
2/22/01

HANOVER/CORNWALL (CANADA)
David M. Gullacher 211.75 kmh
Stinson 108 131.57 mph
1 Lycoming O-435, 190 bhp
2/7/02

HARBOUR GRACE/LONDONDERRY (USA)
Mary G. Kelly, Pilot 28.44 kmh
Joe Cunningham, Copilot 17.67 mph
Cessna 172RG, N6449V
1 Lycoming 0360, 180 bhp
5/22-27/87

HARRIS/BASSETERRE (USA)
Marie E. McMillan 193.37 kmh
Beech Bonanza F33A 120.15 mph
1 Continental IO-520B, 285 bhp
2/28/84

HARRIS/CODRINGTON (USA)
Marie E. McMillan 316.32 kmh
Beech Bonanza F33A 196.55 mph
1 Continental IO-520B, 285 bhp
2/28/84

HARRIS/GUSTAVIA (USA)
Marie E. McMillan 242.22 kmh
Beech Bonanza F33A 150.51 mph
1 Continental IO-520B, 285 bhp
2/28/84

HARRIS/NEW CASTLE (USA)
Marie E. McMillan 167.73 kmh
Beech Bonanza F33A 104.22 mph
1 Continental IO-520B, 285 bhp
2/28/84

HARRIS/PHILIPSBURG (USA)
Marie E. McMillan 238.55 kmh
Beech Bonanza F33A 148.23 mph
1 Continental IO-520B, 285 bhp
2/28/84

HARRIS/ST. JOHNS (USA)
Marie E. McMillan 332.83 kmh
Beech Bonanza F33A 206.81 mph
1 Continental IO-520B, 285 bhp
2/28/84

HARTFORD/BERMUDA (USA)
Millard Harmon 350.94 kmh
Beech Bonanza 36 218.08 mph
1 Continental 10-520BA, 285 bhp
10/29/85

HAVANA/MIAMI (USA)
Danford A. Bookout, Pilot 209.52 kmh
Joe N. Oliver, Copilot 130.20 mph
Piper Lance, N8352C
1 Lycoming IO-540K, 300 bhp
3/24/87

HELSINKI/ALBANY (USA)
Millard Harmon 210.96 kmh
Beech Bonanza 36, N7710R 131.09 mph
1 Continental IO-520BA, 285 bhp
6/26-27/87

HELSINKI/BIRMINGHAM (USA)
Millard Harmon 284.86 kmh
Beech Bonanza 36 177.01 mph
1 Continental IO-520-BA, 285 bhp
6/22/85

HELSINKI/DUBLIN (FINLAND)
Aki Soukas, Pilot 36.58 kmh
Heimo Nieminen, Copilot 22.73 mph
Valmet L-70
1 Lycoming AEIO-360-A1B6, 200 bhp
8/5-7/84

HELSINKI/MOSCOW (CANADA)
Mathias Stinnes 304.14 kmh
Heinz Bittermann 188.98 mph
Glasair III
6/21/92

HELSINKI/ODENSE (FINLAND)
Aki Soukas, Pilot 35.64 kmh
Heimo Nieminen, Copilot 22.15 mph
Valmet L-70
1 Lycoming AEIO-360-A1B6, 200 bhp
8/5-7/84

HELSINKI/STOCKHOLM (FINLAND)
Aki Soukas, Pilot 165.12 kmh
Heimo Nieminen, Copilot 102.61 mph
Valmet L-70
1 Lycoming AEIO-360-A1B6, 200 bhp
8/5-7/84

HERMOSILLO/MAZATLAN (USA)
Marie E. McMillan, Pilot 317.88 kmh
Gloria May, Navigator 197.52 mph
Beech Bonanza F33A
1 Continental IO-520
2 hrs 33 min 50 sec
1/27/81

HERMOSILLO/TUCSON (USA)
Marie E. McMillan, Pilot 311.04 mph
Gloria May, Navigator 193.27 mph
Beech Bonanza F33A
1 Continental IO-520
1 hr 4 min 50 sec
2/2/81

HIBBING, MN/OSHKOSH (USA)
William M. Forsythe 344.51 kmh
Mooney M20K 214.07 mph
1 Continental IO-360, 210 bhp
07/29/90

HILLSBORO/CANOUAN (USA)
Marie E. McMillan 204.70 kmh
Beech Bonanza F33A 127.19 mph
1 Continental IO-520B, 285 bhp
2/27/84

HILLSBORO/KINGSTOWN (USA)
Marie E. McMillan 233.16 kmh
Beech Bonanza F33A 144.88 mph
1 Continental IO-520B, 285 bhp
2/27/84

HILLSBORO/LOVELL'S VILLAGE (USA)
Marie E. McMillan 230.14 kmh
Beech Bonanza F33A 143.00 mph
1 Continental IO-520B, 285 bhp
2/27/84

HILLSBORO/PORT ELIZABETH (USA)
Marie E. McMillan 210.26 kmh
Beech Bonanza F33A 130.65 mph
1 Continental IO-520B, 285 bhp
2/27/84

HILO/LAKELAND (FRANCE)
Henri Chorosz 288.56 kmh
Glasair II-S FT 179.30 mph
1 Lycoming IO-360, 180 bhp
3/30/00

HILO/OSHKOSH (FRANCE)
Henri Chorosz 292.45 kmh
Glasair II-S FT 181.72 mph
1 Lycoming IO-360, 180 bhp
7/22/99

HILO/SAN FRANCISCO (AUSTRALIA)
Gary Thomas Burns 476.51 kmh
Alexander Bernard Schenk 296.09 mph
Lancair IV
7/17/98

HOBART/CANBERRA (AUSTRALIA)
Matthew Green 122.75 kmh
Cessna 172 RG 76.27 mph
1 Lycoming O-360, 180 bhp
5/17/96

HOBART/CHRISTCHURCH (AUSTRALIA)
Peter H. Norvill 200.31 kmh
Cessna XP Hawk 124.47 mph
1 Continental 10-360K, 195 bhp
1/21-22/86

HO CHI MINH CITY/OKINAWA (USA)
Kenneth P. Johnson, Pilot 382.08 kmh
Larry Cioppi, Copilot 237.41 mph
Glasair III
1 Lycoming IO-540, 300 bhp
5/14/94

HONIARA/SANTO (BRAZIL)
Gerard Moss 168.61 kmh
EMBRAER EMB-721 104.76 mph
10/25/91

HONOLULU/BALIKPAPAN (GERMANY)
Dietrich Schmitt 104.32 kmh
Beech A36 Bonanza 64.82 mph
7/1-5/97

HONOLULU/BATON ROUGE, LA (USA)
Hypolite T. Landry, Jr. 132.83 kmh
Beechcraft Bonanza, N5842K 82.54 mph
1 Continental IO-520-B
50 hrs 6 min 13 sec
5/23-25/69

HONOLULU/COLOMBO (GERMANY)
Dietrich Schmitt 61.01 kmh
Beech A36 Bonanza 37.91 mph
7/1-10/97

HONOLULU/DENVER (USA)
L. P. Whistle 305.64 kmh
Eddie Whistle 189.92 mph
Beechcraft Bonanza, N27FH
1 Continental TSIO-520-B
17 hrs 35 min 20 sec
10/30-31/81

HONOLULU/GUAM (GERMANY)
Dietrich Schmitt 144.34 kmh
Beech A36 Bonanza 89.69 mph
7/1-3/97

HONOLULU/MAJURO (USA)
William H. Wisner, Pilot 314.64 kmh
Janice Sullivan, Copilot 195.52 mph
Beech Bonanza, N6826Q
1 Continental, 285 bhp
6/14-15/86

HONOLULU/MEDAN (GERMANY)
Dietrich Schmitt 81.64 kmh
Beech A36 Bonanza 50.73 mph
7/1-7/97

HONOLULU/OSHKOSH (USA)
Michael Ferguson 217.44 kmh
Beech Bonanza D-35 135.12 mph
1 Continental E-225-8, 225 bhp
7/26-28/85

HONOLULU/SACRAMENTO (USA)
Alvin Marks 301.57 kmh
Cessna 210, N942SM 187.39 mph
1 Continental TSIO-520-C
13 hrs 16 min 43 sec
4/16/69

HONOLULU/SALT LAKE CITY*
Michael Ferguson 136.91 mph
Beech Bonanza D-35
1 Continental E-225-8, 225 bhp
7/26-27/85

HONOLULU/SAN FRANCISCO (USA)
Mike Hance 228.24 kmh
Cessna C-177B 141.83 mph
1 Lycoming 0360, 180 bhp
2/18-19/86

HUNTSVILLE/KITTY HAWK (USA)
Ralph E. Hood, Pilot 260.49 kmh
Clyde McDonald, Copilot 161.87 mph
Chris McGee, Navigator
Cessna 182, N42849
1 Continental O-470-R, 230 bhp
12/16/86

INDIANAPOLIS/KITTY HAWK (USA)
Paul E. Lowe, Pilot 452.74 kmh
Robert F. Olszewski, Copilot 281.33 mph
Mooney M20K
1 Continental TSIO-360, 210 bhp
12/17/83

INVERCARGILL/MCMURDO STATION (AUSTRALIA)
Jon Johanson 208.47 kmh
Van's RV-4 129.54 mph
1 Lycoming IO-360, 180 bhp
12/7/03

INVERCARGILL/SOUTH POLE (AUSTRALIA)
Jon Johanson 222.50 kmh
Van's RV-4 138.26 mph
1 Lycoming IO-360, 180 bhp
12/8/03

IRAKLION/KHARTOUM (UK)
Bryan Eccles 311.90 kmh
Mooney 20K 193.82 mph
8/6/88

IRKUTSK/YAKUTSK (CANADA)
Mathias Stinnes 339.96 kmh
Heinz Bittermann 211.24 mph
Glasair III
6/26/92

ISLIP/CANNES (FRANCE)
Henri Chorosz 293.83 kmh
Glasair II 182.58 mph
1 Lycoming IO-360, 180 bhp
5/22/97

ISLIP/PARIS (FRANCE)
Henri Chorosz 317.17 kmh
Glasair II 197.08 mph
1 Lycoming IO-360, 180 bhp
7/20/95

ISTANBUL/BUCHAREST (ROMANIA)
JOINT RECORD
Constantin Manolache 195.04 kmh
Filip Stelian 121.19 mph
Morovan Cehia ZLIN 142
Moravan Otrokovice M 337 AK, 180 bhp
and
Traian Gheorghiu
Alexandru Popovici
Morovan Cehia ZLIN 142
Moravan Otrokovice M 337 AK, 180 bhp
and
Gheorghe Militaru
Morovan Cehia ZLIN 142
Moravan Otrokovice M 337 AK, 180 bhp
9/20/97

ISTANBUL/DUBAI (USA)
Kenneth P. Johnson, Pilot 348.47 kmh
Larry Cioppi, Copilot 216.53 mph
Glasair III
1 Lycoming IO-540, 300 bhp
5/7/94

JACKSON, MS/ATLANTA (USA)
Nancy Toon 329.40 kmh
Beech Duchess 204.69 mph
2 Lycoming 0-360, 180 bhp
11/9/86

JACKSONVILLE/ALBANY (USA)
Millard Harmon 305.30 kmh
Beech Bonanza 36 189.72 mph
1 Continental 10-520BA, 285 bhp
11/1-2/85

JERSEY/PRAGUE (UK)
John P. Frewer 279.36 kmh
Commander 115TC 173.59 mph
1 Lycoming TIO-540, 270 bhp
5/18/02

JOHANNESBURG/CAPE TOWN (UK)
Bryan Eccles 302.66 kmh
Mooney 20K 188.07 mph
8/6/88

JOHANNESBURG/LONDON (UK)
Sheila Scott 160.95 kmh
Piper Comanche 260, G-ATOY 100.00 mph
1 Lycoming IO-720
55 hrs 45 min 51 sec
10/2-4/69

JOHANNNESBURG/NAIROBI (UK)
Sheila Scott 272.31 kmh
Piper Comanche 260, G-ATOY 169.20 mph
1 Lycoming IO-720
11 hrs 8 min 48 sec
10/2/69

JOSHUA TREE/WILLITS (USA)
Park W. Richardson 246.73 kmh
Rockwell Commander 112TC
1 Lycoming, 210 bhp
6/25/92

KAILUA/LIHUE (USA)
Bruce J. Mayes, Pilot 258.58 kmh
Jerry V. Proctor, Copilot 160.67 mph
Grumman Aerospace AA-5B
1 Lycoming O-360, 180 bhp
5/30/93

KAITAIA/BRISBANE (NEW ZEALAND)
Christopher R. Toms 227.64 kmh
Allan Roberts 141.45 mph
Lancair IV
1 Lycoming TIO-540, 350 bhp
3/18/97

KAITAIA/NORFOLK ISLAND (NEW ZEALAND)
Christopher R. Toms 418.07 kmh
Allan Roberts 259.78 mph
Lancair IV
1 Lycoming TIO-540, 350 bhp
3/18/97

KALGOORLIE/NORTHAM (AUSTRALIA)
Robyn Stewart 309.74 kmh
Mooney M20 192.46 mph
4/13/99

KANSAS CITY/KITTY HAWK (USA)
Robert F. Olszewski 471.31 kmh
Clasair III, N540RG 292.87 mph
1 Lycoming 0-540, 300 bhp
12/16/88

KANSAS CITY/WASHINGTON (USA)
Mark Patiky 602.54 kmh
Mooney 252 TSE, N252TD 374.48 mph
1 Continental TSIO-36-MBI, 210 bhp
1/7/87

KARACHI/CALCUTTA (UK)
K. M. C. Brooksbank 302.66 kmh
Sokol Super Aero 45 188.03 mph
7 hrs 51 min 27 min
8/9-15/70

KARACHI/LONDON (UK)
K. M. C. Brooksbank 41.25 kmh
Sokol Super Aero 45 25.63 mph
153 hrs 9 min 20 sec
8/9-15/70

KARACHI/SACRAMENTO (USA)
Alvin Marks 70.91 kmh
Cessna 210, N942SM 44.63 mph
1 Continental TSIO-520-C
179 hrs 34 min 43 sec
4/9-16/69

KENNEDY SPACE CENTER/ALBANY (USA)
Millard Harmon 294.48 kmh
Beech Bonanza 36 182.99 mph
1 Continental 10-520BA, 285 bhp
11/1-2/85

KERIKERI/PARAPARAUMU (NEW ZEALAND)
Robyn N. Harris 185.49 kmh
Penelope A. Haines 115.26 mph
Cessna 172N
12/30/97

KEY WEST/DETROIT (USA)
William S. Demray 236.91 kmh
James R. Townsley 147.21 mph
Piper PA-30 Twin Comanche
2 Lycoming IO-320, 160 bhp
12/20/03

KEY WEST/GEORGE TOWN *
Robert M. Byrom 119.53 mph
Cessna R172K
1 Continental IO-360, 195 bhp
6/13/01

KHARTOUM/NAIROBI (UK)
Bryan Eccles 302.68 kmh
Mooney 20K 188.09 mph
8/6/88

KINGSTON/GEORGETOWN (USA)
Marie E. McMillan 330.09 kmh
Beech Bonanza F33A 205.11 mph
1 Continental IO-520B, 285 bhp
3/2/84

KITTY HAWK/ALBANY (USA)
Millard Harmon, Pilot — 270.00 kmh
Ruth Harmon, Copilot — 167.77 mph
Beech Bonanza 36
1 Continental IO-520 BA, 285 bhp
12/17/84

KITTY HAWK/ATLANTA (USA)
Michelle D. Trobaugh — 115.87 kmh
R. Stuart Hagedorn — 72.00 mph
Grumman AA-5A Cheetah
1 Lycoming IO-320, 160 bhp
12/23/03

KITTY HAWK/CHICAGO*
James M. Frankard — 198.74 mph
Glasair III
1 Lycoming IO-540, 300 bhp
4/27/96

KITTY HAWK/COLLEGE PARK*
Sean A. Salazar — 99.60 mph
Cessna Skyhawk
1 Lycoming O-320, 160 bhp
7/16/98

KITTY HAWK/KEY WEST (USA)
William S. Demray — 199.27 kmh
James R. Townsley — 123.82 mph
Piper PA-30 Twin Comanche
2 Lycoming IO-320, 160 bhp
12/17/03

KITTY HAWK/NEW YORK (USA)
Alan Feierstein — 334.86 kmh
Piper PA-24-260 — 208.07 mph
1 Lycoming IO-540, 260 bhp
12/16/96

KITTY HAWK/TEXARKANA (USA)
Wayne Hooker — 261.13 kmh
Piper Lance, N8352C — 162.27 mph
1 Lycoming IO-540 KIAS, 300 bhp
12/17/87

KITTY HAWK/WASHINGTON *
John Meyers — 160.84 mph
Beech Bonanza F33A
1 Continental IO-520, 285 bhp
12/17/85

KITTY HAWK/WINONA*
James M. Frankard — 203.79 mph
Glasair III
1 Lycoming IO-540, 300 bhp
4/27/96

KUALA LUMPUR/KARACHI (UK)
K. M. C. Brooksbank — 52.00 kmh
Sokol Super Aero 45 — 32.21 mph
85 hrs 7 min 2 sec
8/3-15/70

KUALA LUMPUR/LONDON (UK)
K. M. C. Brooksbank — 37.44 kmh
Sokol Super Aero 45 — 22.26 mph
282 hrs 1 min 24 sec
8/3-7/70

KUALA LUMPUR/SRI LANKA (USA)
William H. Wisner, Pilot — 250.56 kmh
Janice Sullivan, Copilot — 155.70 mph
Beech Bonanza, N6826Q
1 Continental, 285 bhp
6/25-26/86

KUSHIRO/SEATTLE (USA)
Robert S. Mucklestone — 174.42 kmh
Cessna T210 — 108.38 mph
1 Continental IO-520
39 hrs 20 min 7.4 sec
9/2-4/75

LA BAULE/PARIS (FRANCE)
Joseph Blond — 300.55 kmh
Mooney MK-21 — 186.75 mph
1 Lycoming IO-360
1 hr 18 min 21 sec
8/13/70

LAKE TAHOE/CHICAGO (USA)
Mark W. Bronson — 113.22 kmh
Daniel S. Dawson — 70.35 mph
Cessna R172K
1 Continental IO-360, 210 bhp
7/23/01

LANGENTHAL/OSHKOSH (SWITZERLAND)
Felix J. Estermann — 128.05 kmh
Däetwyler MD-3 — 79.56 mph
1 Lycoming O-320, 160 bhp
7/8-10/94

LA PAZ/LOS ANGELES (USA)
Jeremy N. White — 49.98 kmh
Beechcraft Bonanza F33A — 31.05 mph
1 Continental IO-520, 285 bhp
7/22-29/91

LARNACA/MUSCAT (AUSTRALIA)
Gary Thomas Burns — 440.26 kmh
Alexander Bernard Schenk — 273.56 mph
Lancair IV
8/19/98

LAS PALMAS/BARCELONA (W. GERMANY)
Richard Flohr — 265.41 kmh
Mooney M20K — 164.92 mph
1 Lycoming TSIO-360-GB1, 210 bhp
2/26/83

LAS PALMAS/CASABLANCA (W. GERMANY)
Richard Flohr — 293.15 kmh
Mooney M20K — 182.15 mph
1 Lycoming TSIO-360-GB1, 210 bhp
2/26/83

LAS PALMAS/COLOGNE-BONN (W. GERMANY)
Richard Flohr — 109.71 kmh
Mooney M20K — 68.17 mph
1 Lycoming TSIO-360-GB1, 210 bhp
2/26-27/83

LAS VEGAS/EL PASO (USA)
Marie E. McMillan — 309.70 kmh
Beech Bonanza F33A, — 192.45 mph
Continental 10-520BA, 285 bhp
5/3/83

LAS VEGAS/FRESNO (USA)
Wayne Easley — 326.88 kmh
Harold Kindsvater, Copilot — 230.12 mph
Cessna P210, N734VM
1 Continental TSIO-520-P, 300 bhp
7/19/88

LAS VEGAS/HERMOSILLO (USA)
Marie E. McMillan — 262.40 kmh
Beech Bonanza F33A — 163.05 mph
1 Continental IO-520 BA, 285 bhp
1/31/84

LAS VEGAS/LOS ANGELES (USA)
Hal Fishman — 344.51 kmh
SIAI Marchetti SF.260, N730W — 214.07 mph
1 hr 4 min 1 sec
10/17/68

LAS VEGAS/SALT LAKE CITY (USA)
Vincent F. Latona — 315.54 kmh / 196.06 mph
Beechcraft Bonanza B36TC
1 Continental GTSIO-550, 300 bhp
8/4/93

LAS VEGAS/SAN DIEGO (USA)
Charles Freer — 330.53 kmh / 205.38 mph
Cessna TU206-G
1 Continental, 285 bhp
1 hr 15 min
5/16/80

LAS VEGAS/TUCSON (USA)
Marie E. McMillan, Pilot — 327.24 kmh / 203.34 mph
Gloria May, Navigator
Beech Bonanza F33A
1 hr 47 min 37 sec
1/26/81

LATROBE/KITTY HAWK (USA)
Carl A. La Rue — 549.06 kmh / 341.17 mph
John Slais
Lancair IV P
1 Continental TSIO-550B, 350 bhp
10/21/97

LEADVILLE/DEATH VALLEY (USA)
Michael W. Headrick — 237.44 kmh / 147.53 mph
Piper PA-24
1 Lycoming O-540, 250 bhp
6/27/95

LEICESTER/LYON (UK)
Richard M. Clarke — 229.32 kmh / 142.50 mph
Cessna 177
1 Lycoming 10-360-AIB6, 200 bhp
6/20/84

LE TOUQUET/OSTEND (UK)
Colin W. Yarnton — 203.88 kmh / 126.69 mph
Mooney M20F
1 Lycoming, 200 bhp
11/19-21/85

LIEGE/COLOGNE-BONN (W. GERMANY)
Richard Flohr — 498.67 kmh / 309.87 mph
Mooney M20K
1 Continental TSIO 360-GB1, 210 bhp
3/22/83

LIHUE/KAILUA (USA)
Bruce J. Mayes, Pilot — 233.78 kmh / 145.26 mph
Jerry V. Proctor, Copilot
Grumman Aerospace AA-5B
1 Lycoming O-360, 180 bhp
5/30/93

LILLE/AJACCIO (FRANCE)
Paul Laurent Zaccarie — 265.16 kmh / 164.77 mph
Cessna 172 RG
1 Lycoming 0360 SEA6, 183 bhp
4/28/84

LINCOLN/NEW ORLEANS (USA)
William A. Hamilton, Pilot — 289.55 kmh / 179.92 mph
Penny Rafferty Hamilton, Copilot
Cessna T210L
1 Continental TSIO-520, 285 bhp
10/22/91

LISBON/CANNES (UK)
Geoffrey G. Boot — 195.12 kmh / 121.24 mph
Rushmeyer R90-230RG
1 IO-540, 230 bhp
11/29/94

LISBON/LONDON (UK)
Sheila Scott — 244.00 kmh / 151.61 mph
Piper Comanche PA-24
1 Lycoming IO-720
6 hrs 25 min 17 sec
6/20/66

LISBON/LOS ANGELES (USA)
Jeremy N. White — 86.49 kmh / 53.74 mph
Beechcraft Bonanza F33A
1 Continental 520, 285 bhp
06/15-06/19/90

LISBON/ST. JOHNS (AUSTRALIA)
Peter H. Norvill — 59.82 kmh / 37.17 mph
Cessna 172K
1 Continental IO-360K, 195 bhp
5/7-6/11/89

LISBON/SANTA MARIA (USA)
William H. Wisner, Pilot — 288.00 kmh / 178.96 mph
Cheryl M. Davis, Copilot
Bonanza, N6826A
1 Continental, 285 bhp
5/12/87

LISBON/VALENCIA (UK)
Geoffrey G. Boot — 288.58 kmh / 179.32 mph
Rushmeyer R90-230RG
1 IO-540, 230 bhp
11/29/94

LITTLE ROCK/AMARILLO (USA)
W. K. McGehee, Jr. — 229.44 kmh / 142.57 mph
Cessna 172XP
Continental IO-360, 195 bhp
5/9/85

LITTLE ROCK/KAHULUI (GERMANY)
Dietrich Schmitt — 159.64 kmh / 99.19 mph
Beechcraft Bonanza F33A
1 Continental IO-550, 300 bhp
6/1/95

LITTLE ROCK/LONG BEACH (GERMANY)
Dietrich Schmitt — 281.25 kmh / 174.76 mph
Beechcraft Bonanza F33A
1 Continental IO-550, 300 bhp
5/30/95

LITTLE ROCK/NASHVILLE (USA)
Danford A. Bookout — 299.98 kmh / 186.41 mph
Piper Lance PA32 R-300
1 Lycoming IO-540K, 300 bhp
10/22/85

LITTLE ROCK/WASHINGTON (USA)
Danford A. Bookout — 286.28 kmh / 177.89 mph
Piper Lance PA32 R-300
1 Lycoming IO-540K, 300 bhp
10/22/85

LODZ/STRASBOURG (FRANCE)
Philippe Beloc, Pilot — 149.29 kmh / 92.77 mph
Jacques Genza, Copilot
DR 400:160 Robin
1 Lycoming 0-320, 160 cv
8/17/86

LONDON/AJACCIO (UK)
Frank Spriggs — 304.67 kmh / 189.31 mph
Cessna Centurion C210M
1 Continental, 310 bhp
2/15/80

LONDON/ATHENS (NEW ZEALAND)
Clifford V. Tait — 222.04 kmh / 137.97 mph
Beech Bonanza A36
Continental IO-520-BB, 285 bhp
4/5-6/82

LONDON/AUCKLAND (NEW ZEALAND)
Clifford V. Tait 197.62 kmh
Beech Bonanza A36 122.80 mph
Continental IO-520-BB, 285 bhp
4/5-9/82

LONDON/BAHRAIN (NEW ZEALAND)
Clifford V. Tait 210.57 kmh
Beech Bonanza A36 130.85 kmh
1 Continental IO-520-BB, 285 bhp
4/5-6/82

LONDON/BELFAST (UK)
Sheila Scott 320.29 kmh
Piper Comanche 400 198.02 mph
1 Lycoming IO-720
1 hr 36 min 47.7 sec
5/20/65

LONDON/BENGHAZI (UK)
Sheila Scott 255.36 kmh
Piper Comanche 260-B, G-ATOY 158.67 mph
1 Lycoming IO-720
10 hrs 43 min 18 sec
6/29/67

LONDON/BIARRITZ (UK)
Roger H. Theaker 248.95 kmh
Piper Comanche PA-30-B, G-BAWU 154.70 mph
2 Lycoming IO-320-C1A, 160 bhp
8/6/85

LONDON/BRISBANE (NEW ZEALAND)
Clifford V. Tait 178.03 kmh
Beech Bonanza A36 110.63 mph
1 Continental IO-520-BB, 285 bhp
4/5-9/82

LONDON/BRUSSELS (UK)
Sheila Scott 333.58 kmh
Piper Comanche 400 207.26 mph
Lycoming IO-720
58 min 4.1 sec
5/19/65

LONDON/CAIRO (UK)
Graham A. Ponsford 111.71 kmh
Geoffrey E. Tyler 69.41 mph
Clive E. Ponsford
Piper PA-32 Cherokee SIX 300
1 Lycoming IO-540, 300 bhp
10/18/02

LONDON/CALAIS (UK)
Michael J. Baker 245.66 kmh
Piper PA-28R-201 152.64 mph
1 Lycoming IO-360, 200 bhp
10/9/95

LONDON/CALCUTTA (NEW ZEALAND)
Clifford V. Tait 201.55 kmh
Beech Bonanza A36 125.24 mph
1 Continental IO-520-BB, 285 bhp
4/5-7/82

LONDON/CAPETOWN (UK)
Sheila Scott 130.43 kmh
Piper Comanche 260-B 81.04 mph
1 Lycoming IO-720
72 hrs 18 min 26 sec
7/9/67

LONDON/COLOGNE-BONN (W. GERMANY)
Richard Flohr 319.93 kmh
Mooney M20K 198.80 mph
1 Continental TSIO-360-GB1, 210 bhp
10/12/84

LONDON/DARWIN (AUSTRALIA)
R. P. Bennett 220.50 kmh
Piper Comanche 260-B, VH-MDD 137.01 mph
1 Lycoming IO-540
62 hrs 53 min 53 sec
12/18-21/69

LONDON/DUBLIN (UK)
Sheila Scott 300.71 kmh
Piper Comanche 400 186.85 mph
1 Lycoming IO-720
1 hr 32 min 10.7 sec
5/20/65

LONDON/ERIE (UK)
Igor Best-Devereux 39.85 kmh
SIAI-Marchetti SF.260 24.76 mph
7/5-11/92

LONDON/FIJI ISLANDS (UK)
Sheila Scott 34.60 kmh
Piper Comanche 400 21.49 mph
1 Lycoming IO-720
5/18-6/7/66

LONDON/IRAKLION (UK)
Bryan Eccles 286.67 kmh
Mooney 20K 178.14 mph
8/6/88

LONDON/ISTANBUL (USA)
Robert S. Mucklestone 247.66 kmh
Cessna T-210 153.89 mph
1 Continental IO-520
8/26/75

LONDON/LE TOUQUET (UK)
Colin W. Yarnton 226.51 kmh
Mooney M20F 140.75 mph
1 Lycoming, 200 bhp
11/19-21/85

LONDON/MALTA (UK)
Frank Spriggs 286.21 kmh
Cessna Centurion C210M 177.84 mph
1 Continental, 310 bhp
2/15/80

LONDON, ON/MONCTON (CANADA)
Frank J. Quigg 188.89 kmh
Waco Classic F-5 117.37 mph
1 Jacobs R-755, 275 bhp
9/7/93

LONDON/NAIROBI (UK)
Bryan Eccles 204.83 kmh
Mooney 20K 127.28 mph
8/6/88

LONDON/NEW YORK (UK)
Sheila Scott 215.91 kmh
Piper Comanche 260, G-ATOY 134.15 mph
1 Lycoming IO-540
5/4-5/69

LONDON/ORAN (UK)
Bryan Eccles 322.08 kmh
Mooney M20K 200.14 mph
8/12/89

LONDON/PARIS (UK)
Sheila Scott 324.94 kmh
Piper Comanche 400 201.90 mph
1 Lycoming IO-720
1 hr 3 min 12.5 sec
5/19/65

LONDON/PARIS *
Robert J. Moriarty 273.37 mph
Beech Bonanza V35TC
1 Continental TSIO-520, 285 bhp
6/27/85

LONDON/ROME (UK)
Sheila Scott 258.13 kmh
Piper Comanche 260-B 160.39 mph
1 Lycoming IO-720
5 hrs 35 min 57.4 sec
5/18/66

LONDON/ROSKILDE (UK)
Geoffrey G. Boot 364.08 kmh
Mooney M20R Ovation 226.22 mph
1 Continental IO-550, 280 bhp
6/26/95

LONDON/SINGAPORE (USA)
Robert S. Mucklestone 96.08 kmh
Cessna T-210 59.70 mph
1 Continental IO-520
113 hrs 8 min 33.2 sec
8/26-30/75

LONDON/THE HAGUE (UK)
Sheila Scott 330.01 kmh
Piper Comanche 400 205.11 mph
1 Lycoming IO-720
7/19/65

LONDON/TRIPOLI (UK)
Sheila Scott 251.81 kmh
Piper Comanche 260-B 156.46 mph
1 Lycoming IO-720
7/6/67

LONG BEACH/BALIKPAPAN (GERMANY)
Dietrich Schmitt 63.22 kmh
Beech A36 Bonanza 39.28 mph
6/26-7/5/97

LONG BEACH/GUAM (GERMANY)
Dietrich Schmitt 61.19 kmh
Beech A36 Bonanza 38.02 mph
6/26-7/3/97

LONG BEACH/HONOLULU (GERMANY)
Dietrich Schmitt 292.34 kmh
Beech A36 Bonanza 181.65 mph
6/27/97

LONG BEACH/KAHULUI (GERMANY)
Dietrich Schmitt 315.72 kmh
Beechcraft Bonanza F33A 196.17 mph
1 Continental IO-550, 300 bhp
6/1/95

LONG BEACH/LAS VEGAS (USA)
Leonard Perry 318.44 kmh
Beechcraft Bonanza F-33 197.87 mph
1 Continental O-470
1 hr 10 min 30 sec
5/17/72

LONG BEACH/MAJURO ATOLL (GERMANY)
Dietrich Schmitt 59.88 kmh
Beech A36 Bonanza 37.21 mph
6/26-7/2/97

LONG BEACH/MEDAN (GERMANY)
Dietrich Schmitt 56.07 kmh
Beech A36 Bonanza 34.84 mph
6/26-7/7/97

LONG BEACH/SALINAS (USA)
Barbara A. Hartman, Pilot 259.12 kmh
Richard L. Double, Copilot 161.01 mph
Cessna 182
1 Continental O-470, 230 bhp
10/17/92

LONG BEACH/YOLO, CA (USA)
Daniel K. Dowling 310.32 kmh
Cessna 182RG, N756EG 192.83 mph
1 Lycoming TIO-540-L3C5D
5/6/89

LORD HOWE ISLAND/AUCKLAND (NEW ZEALAND)
Christopher R. Toms 425.11 kmh
Allan Roberts 264.15 mph
Lancair IV
1 Lycoming TIO-540, 350 bhp
4/2/97

LORD HOWE ISLAND/BRISBANE (NEW ZEALAND)
Christopher R. Toms 414.22 kmh
Allan Roberts 257.38 mph
Lancair IV
1 Lycoming TIO-540, 350 bhp
3/18/97

LORD HOWE ISLAND/COFF'S HARBOUR (NEW ZEALAND)
Richard Neave 261.38 kmh
Dawson Bowles 162.41 mph
Sue Campbell
Mooney M20
1 Lycoming, 200 bhp
6/18/00

LORD HOWE ISLAND/NORFOLK ISLAND (NEW ZEALAND)
Richard Neave 276.57 kmh
Dawson Bowles 171.85 mph
Sue Campbell
Mooney 201
1 Lycoming, 200 bhp
7/3/00

LOS ANGELES/ALBUQUERQUE (USA)
William H. Cox, Pilot 544.47 kmh
Martha A. Morrell, Copilot 338.32 mph
Mooney TLS
1 Lycoming TIO-540, 270 bhp
3/25/94

LOS ANGELES/BOSTON (USA)
Stuart C. Goldberg 368.60 kmh
Beechcraft Bonanza V35B 229.04 mph
1 Continental IO-550, 300 bhp
9/6/93

LOS ANGELES/COLORADO SPRINGS (USA)
Calvin B. Early 434.69 kmh
Beechcraft Bonanza M35 270.10 mph
1 Continental IO-550, 300 bhp
3/2/97

LOS ANGELES/DALLAS (USA)
William H. Cox, Pilot 524.24 kmh
Martha A. Morrell, Copilot 325.75 mph
Mooney TLS
1 Lycoming TIO-540, 270 bhp
3/25/94

LOS ANGELES/JACKSONVILLE (USA)
William H. Cox, Pilot 483.12 kmh
Martha A. Morrell, Copilot 300.20 mph
Mooney TLS
1 Lycoming TIO-540, 270 bhp
3/25/94

LOS ANGELES/LAKELAND (USA)
Dave Morss 584.92 kmh
Lancair IV 363.45 mph
1 Continental TSIO-550, 350 bhp
4/18/93

LOS ANGELES/LAS VEGAS (USA)
Barry J. Schiff 306.48 kmh
Aero Commander 200, N2914T 190.44 mph
1 Continental IO-520-A
1 hr 11 min 57.8 sec
4/30/67

LOS ANGELES/MIAMI (USA)
G. R. Wood 387.00 kmh
Beechcraft Bonanza V35TC 240.60 mph
1 Continental TSIO-520D
9 hrs 43 min 12 sec
2/11-12/79

LOS ANGELES/NEW YORK (USA)
Todd Duhnke 477.54 kmh
Mooney 252 296.74 mph
1 Continental TSIO-360-MBI
5/1/86

LOS ANGELES/PHOENIX (USA)
David G. Riggs 502.52 kmh
Nathan H. East 312.25 mph
Lancair IV-P
1 Continental TSIO-550, 350 bhp
12/17/03

LUSAKA, CONGO/LONDON (UK)
D. J. Irvin 122.30 kmh
Beechcraft Bonanza V-35 75.99 mph
1 Continental IO-520-B
64 hrs 57 min 25 sec
12/6-8/69

LUXOR/RIYADH (CANADA)
Frank J. Quigg 163.58 kmh
Waco Classic F-5 101.64 mph
1 Jacobs R-755, 275 bhp
9/18/93

MADRID/COLOGNE-BONN (W. GERMANY)
Richard Flohr 250.76 kmh
Mooney M20K 155.81 mph
1 Continental TSIO-360-GB1, 210 bhp
8/1/83

MADRID/LISBON*
JOINT RECORD
William H. Wisner, Pilot 185.60 mph
Lanita Clay, Copilot
Beechcraft Bonanza, N6826Q
and
Frank Haile, Pilot
Wallace Hedgren, Copilot
Beechcraft Bonanza, N4FH
and
Jeremy N. White
Beechcraft Bonanza, N222JW
06/13/90

MADRID/LONDON (UK)
Sheila Scott 244.25 kmh
Piper Comanche 260-B 151.77 mph
1 Lycoming IO-720
5 hrs 6 min 30 sec
11/20/67

MAGENTA/NADI (NEW ZEALAND)
Mat Wakelin 243.64 kmh
Cessna 177RG 151.39 mph
1 Lycoming IO-360, 200 bhp
9/13/95

MAJURO ATOLL/BALIKPAPAN (GERMANY)
Dietrich Schmitt 103.77 kmh
Beech A36 Bonanza 64.48 mph
7/2-5/97

MAJURO ATOLL/COLOMBO (GERMANY)
Dietrich Schmitt 56.02 kmh
Beech A36 Bonanza 34.81 mph
7/2-10/97

MAJURO ATOLL/GUADALCANAL (USA)
William H. Wisner, Pilot 305.64 kmh
Janice Sullivan, Copilot 189.92 mph
Beech Bonanza, N6826Q
1 Continental, 285 bhp
6/15-16/86

MAJURO ATOLL/GUAM (GERMANY)
Dietrich Schmitt 313.91 kmh
Beech A36 Bonanza 195.05 mph
7/3/97

MAJURO ATOLL/MEDAN (GERMANY)
Dietrich Schmitt 76.64 kmh
Beech A36 Bonanza 47.62 mph
7/2-7/97

MALTA/LONDON (UK)
Sheila Scott 241.31 kmh
Piper Comanche 260-B 149.94 mph
1 Lycoming IO-720
8 hrs 43 min 15 sec
8/1/67

MALTA/MANNHEIM (GERMANY)
Dietrich Schmitt 213.40 kmh
Beech A36 Bonanza 132.60 mph
7/16/97

MANCHESTER/DUBLIN (IRELAND)
Gerald W. Connolly 340.12 kmh
Piper Twin Comanche PA-39 211.34 mph
2 Lycoming IO-360
46 min 35 sec
2/15/75

MANILA/BATON ROUGE, LA (USA)
Hypolite T. Landry, Jr. 73.33 kmh
Beechcraft Bonanza N5842K 45.57 mph
1 Continental IO-520-B
190 hrs 15 min
5/17-25/69

MANILA/SACRAMENTO (USA)
Alvin Marks 115.34 kmh
Cessna 210, N9425M 71.73 mph
1 Continental TSIO-520-C
7 hrs 23 min 43 sec
4/12-16/69

MANNHEIM/BANGOR (GERMANY)
Dietrich Schmitt 94.52 kmh
Beech A36 Bonanza 58.73 mph
6/18-20/97

MANNHEIM/LONG BEACH (GERMANY)
Dietrich Schmitt 89.02 kmh
Beech A36 Bonanza 55.31 mph
6/18-22/97

MANNHEIM/SANTA MARIA (GERMANY)
Dietrich Schmitt 246.06 kmh
Beech A36 Bonanza 152.89 mph
6/18/97

MANNHEIM/WICHITA (GERMANY)
Dietrich Schmitt 94.29 kmh
Beech A36 Bonanza 58.59 mph
6/18-21/97

MANSTON/BERLIN (UK)
Philip Wadsworth 179.55 kmh
Grumman AA-5 111.57 mph
1 Lycoming O-360, 180 bhp
8/26/02

MANSTON/REIMS (UK)
David Wooldridge 194.44 kmh
Scottish Aviation Bulldog T1 120.83 mph
1 Lycoming 0-360, 200 bhp
6/20/84

MARRAKECH/ISTANBUL (USA)
Kenneth P. Johnson, Pilot 370.64 kmh
Larry Cioppi, Copilot 230.30 mph
Glasair III
1 Lycoming IO-540, 300 bhp
5/5/94

MARTHA'S VINEYARD/BERMUDA (USA)
Millard Harmon 371.92 kmh
Beech Bonanza 36 231.11 mph
1 Continental 10-520BA, 285 bhp
10/29/85

MARTINIQUE/BASSETERE (USA)
Marie E. McMillan 243.10 kmh
Beech Bonanza F33A 151.05 mph
1 Continental IO-520B, 285 bhp
2/28/84

MARTINIQUE/BRIDGETOWN (USA)
Marie E. McMillan 204.27 kmh
Beech Bonanza F33A 126.93 mph
1 Continental IO-520B, 285 bhp
2/23/84

MARTINIQUE/CODRINGTON (USA)
Marie E. McMillan 262.55 kmh
Beech Bonanza F33A 163.14 mph
1 Continental IO-520B, 285 bhp
2/28/84

MARTINIQUE/GRAND-BOURG (USA)
Marie E. McMillan 283.57 kmh
Beech Bonanza F33A 176.20 mph
1 Continental IO-520B, 285 bhp
2/28/84

MARTINIQUE/GUSTAVIA (USA)
Marie E. McMillan 258.55 kmh
Beech Bonanza F33A 160.65 mph
1 Continental IO-520B, 285 bhp
2/28/84

MARTINIQUE/HARRIS (USA)
Marie E. McMillan 269.32 kmh
Beech Bonanza F33A 167.35 mph
1 Continental IO-520B, 285 bhp
2/28/84

MARTINIQUE/NEW CASTLE (USA)
Marie E. McMillan 237.81 kmh
Beech Bonanza F33A 147.77 mph
1 Continental IO-520B, 285 bhp
2/28/84

MARTINIQUE/PHILIPSBURG (USA)
Marie E. McMillan 255.38 kmh
Beech Bonanza F33A 158.68 mph
1 Continental IO-520B, 285 bhp
2/28/84

MARTINIQUE/POINTE A PITRE (USA)
Marie E. McMillan 233.69 kmh
Beech Bonanza F33A 145.21 mph
1 Continental IO-520B, 285 bhp
2/28/84

MARTINIQUE/ST. JOHNS (USA)
Marie E. McMillan 247.21 kmh
Beech Bonanza F33A 153.61 mph
1 Continental IO-520B, 285 bhp
2/28/84

MATAMOROS/VERA CRUZ (USA)
William H. Wisner, Pilot 321.48 kmh
Fred Perez, Copilot 199.77 mph
Beechcraft E185, N299Z
2 P&W R985, 450 bhp
2/28/87

MAZATLAN/HERMOSILLO (USA)
Marie E. McMillan, Pilot 332.28 kmh
Gloria May, Navigator 206.47 mph
Beech Bonanza F33A
1 Continental IO-520
2 hrs 27 min 10 sec
2/2/81

MAZATLAN/PUERTO VALLARTA (USA)
Marie E. McMillan, Pilot 305.28 kmh
Gloria May, Navigator 189.69 mph
Beech Bonanza F33A
1 Continental IO-520
57 min 59 sec
1/27/81

MCMURDO STATION/INVERCARGILL (AUSTRALIA)
Jon Johanson 284.19 kmh
Van's RV-4 176.59 mph
1 Lycoming IO-360, 180 bhp
12/15/03

MEDAN/BAHRAIN (GERMANY)
Dietrich Schmitt 64.98 kmh
Beech A36 Bonanza 40.38 mph
7/9-13/97

MEDAN/COLOMBO (GERMANY)
Dietrich Schmitt 235.62 kmh
Beech A36 Bonanza 146.41 mph
7/10/97

MEDAN/MALTA (GERMANY)
Dietrich Schmitt 68.70 kmh
Beech A36 Bonanza 42.69 mph
7/9-15/97

MEDAN/MANNHEIM (GERMANY)
Dietrich Schmitt 60.43 kmh
Beech A36 Bonanza 37.55 mph
7/9-16/97

MELBOURNE/ADELAIDE (AUSTRALIA)
Leslie R. Thompson, Pilot 257.82 kmh
James Bernie Yeo, Copilot 160.21 mph
Cessna 210N
11/30/89

MELBOURNE/ALICE SPRINGS (AUSTRALIA)
Roy Travis 291.56 kmh
Beechcraft Bonanza A36 181.17 mph
4/15/99

MELBOURNE/BRISBANE (AUSTRALIA)
Eric H. Wheatley 316.31 kmh
Mooney M20K 196.54 mph
9/14/89

MELBOURNE/DARWIN (AUSTRALIA)
Roy Travis 294.87 kmh
Beechcraft Bonanza A36 183.20 mph
4/15/99

MELBOURNE/PERTH (AUSTRALIA)
Roy Travis 236.39 kmh
Beechcraft Bonanza A36 146.89 mph
1 Continental IO-520, 285 bhp
6/15/96

MELBOURNE/SYDNEY (NEW ZEALAND)
Christopher R. Toms 486.75 kmh
Allan Roberts 302.45 mph
Lancair IV
1 Lycoming TIO-540, 350 bhp
3/31/97

MELBOURNE/TOWNSVILLE (AUSTRALIA)
Eric H. Wheatley 246.38 kmh
Mooney M20K 153.09 mph
9/14/89

MEMPHIS/WASHINGTON (USA)
Danford A. Bookout 291.98 kmh
Piper Lance PA32 R-300 181.44 mph
1 Lycoming IO-540K, 300 bhp
10/22/85

MEXICO CITY/ACAPULCO (USA)
Marie E. McMillan, Pilot 294.12 kmh
Gloria May, Navigator 182.76 mph
Beech Bonanza F33A
1 Continental IO-520
1 hr 2 min 45 sec
1/30/81

MIAMI/HAVANA (USA)
Danford A. Bookout, Pilot 207.00 kmh
Joe N. Oliver, Copilot 128.63 mph
Piper Lance, N8352C
1 Lycoming IO-540K, 300 bhp
3/22/87

MINITATLIN/COZUMEL (USA)
Marie E. McMillan 177.95 kmh
Beech Bonanza F33A 110.58 mph
1 Continental IO-520 BA, 285 bhp
2/4/84

MINNEAPOLIS/LONDON, ON (CANADA)
Frank J. Quigg 212.89 kmh
Waco Classic F-5 132.28 mph
1 Jacobs R-755, 275 bhp
9/6/93

MINOT, ND/OSHKOSH (USA)
Stephen P. Conners 245.50 kmh
Ceecy C. Nucker 152.54 mph
Cessna 177
1 Lycoming O-360, 180 bhp
07/29/90

MOBILE/AUSTIN (USA)
G. R. Wood 160.48 kmh
Beechcraft Bonanza V-35-TC 99.71 mph
1 Continental TSIO-520-D, 285 bhp
5 hrs 47 min 41 sec
6/16/76

MOBILE/EL PASO (USA)
G. R. Wood 202.20 kmh
Beechcraft Bonanza V-35-TC 125.63 mph
1 Continental TSIO-520-D, 285 bhp
8 hrs 41 min 19 sec
6/16-17/76

MOBILE/PHOENIX (USA)
G. R. Wood 215.52 kmh
Beechcraft Bonanza V-35-TC 133.91 mph
Continental TSIO-520-D, 285 bhp
10 hrs 34 min 35 sec
6/6-17/76

MONCTON/BEDFORD (USA)
Anne B. Baddour, Pilot 169.61 kmh
Patricia M. Thrasher, Copilot 105.39 mph
Beechcraft Sierra B24R
1 Lycoming, 200 bhp
8/13/91

MONCTON/GANDER (USA)
Richard Gilson 377.28 kmh
Mooney M20K, N252YM 234.44 mph
1 Continental 360 MBI, 210 bhp
5/13/87

MONCTON/GOOSE BAY (CANADA)
Frank J. Quigg 193.57 kmh
Waco Classic F-5 120.28 mph
1 Jacobs R-755, 275 bhp
9/8/93

MONTEGO BAY/PORT AU PRINCE (USA)
Marie E. McMillan 208.19 kmh
Beech Bonanza F33A 129.37 mph
1 Continental IO-520 BA, 285 bhp
2/14/84

MONTEGO BAY/SANTO DOMINGO (USA)
Marie E. McMillan 228.24 kmh
Beech Bonanza F33A 141.83 mph
1 Continental IO-520 BA, 285 bhp
2/14/84

MONTERREY/TUCSON (USA)
Marie E. McMillan 185.85 kmh
Beech Bonanza F33A 115.54 mph
1 Continental IO-520B, 285 bhp
3/7/84

MONTGOMERY/DAYTONA BEACH (USA)
Scott Tarves, Pilot 216.92 kmh
Thomas Jordan, Copilot 134.78 mph
Michael Wiggins, E-RAU Training Manager
Cessna 172
1 Lycoming O-320, 160 bhp
04/29/90

MONT JOLI/GOOSE BAY (USA)
Millard Harmon 281.90 kmh
Beech Bonanza 36 175.18 mph
1 Continental IO-520-BA, 285 bhp
6/13/85

MONTPELLIER/ATHENS (CANADA)
Frank J. Quigg 186.94 kmh
Waco Classic F-5 116.16 mph
1 Jacobs R-755, 275 bhp
9/16/93

MONTREAL/ST. JOHN'S (USA)
Kenneth P. Johnson, Pilot 452.70 kmh
Larry Cioppi, Copilot 281.29 mph
Glasair III
1 Lycoming IO-540, 300 bhp
5/1/94

MOSCOW/ALBANY (USA)
Millard Harmon 137.52 kmh
Beech Bonanza 36, N7710R 85.45 mph
1 Continental IO-520BA, 285 bhp
6/25-27/87

MOSCOW/OMSK (CANADA)
Mathias Stinnes 459.05 kmh
Heinz Bittermann 285.24 mph
Glasair III
6/23/92

MOUNT MITCHELL/KITTY HAWK (USA)
Lawrence G. Manofsky 312.38 kmh
Lake Renegade 250 194.10 mph
1 Lycoming IO-540, 250 bhp
12/20/96

MUSCAT/BOMBAY (CANADA)
Frank J. Quigg 158.50 kmh
Waco Classic F-5 98.49 mph
1 Jacobs R-755, 275 bhp
9/21/93

MUSCAT/PHUKET (AUSTRALIA)
Gary Thomas Burns 433.31 kmh
Alexander Bernard Schenk 269.25 mph
Lancair IV
8/21/98

NADI/NOUMEA (NEW ZEALAND)
Dee Bond Wakelin 209.34 kmh
Mat Wakelin 130.08 mph
Cessna 177 Cardinal RG
11/4/99

NADI/SYDNEY (AUSTRALIA)
Peter H. Norvill 43.31 kmh
Cessna 172K 26.91 mph
1 Continental IO-360K, 195 bhp
5/7-6/11/89

NAIROBI/CAPETOWN (UK)
Sheila Scott 221.79 kmh
Piper Comanche 260, G-ATOY 137.81 mph
1 Lycoming IO-540
18 hrs 30 min 57 sec
9/18-19/69

NAIROBI/JOHANNESBURG (UK)
Bryan Eccles 355.29 kmh
Mooney 20K 220.77 mph
8/9-13/88

NAIROBI/LONDON (UK)
Sheila Scott 196.17 kmh
Piper Comanche 260, G-ATOY 121.89 mph
1 Lycoming IO-54
10/3-4/69

NANTUCKET/BERMUDA (USA)
Millard Harmon 379.41 kmh
Beech Bonanza 36 235.76 mph
1 Continental 10-520BA, 285 bhp
10/29/85

NARSSARSSUAQ/GOOSE BAY (USA)
Millard Harmon 272.20 kmh
Beech Bonanza 36 169.15 mph
1 Continental IO-520-BA, 285 bhp
6/24/85

NARSSARSSUAQ/REYKJAVIK (USA)
William H. Wisner, Pilot 290.16 kmh
Cheryl M. Davis, Copilot 180.31 mph
Bonanza, N6826A
1 Continental, 285 bhp
5/5/87

NASHVILLE/WASHINGTON (USA)
Danford A. Bookout 279.36 kmh
Piper Lance PA32 R-300 173.59 mph
1 Lycoming IO-540K, 300 bhp
10/22/85

NASHVILLE/MACON (USA)
James S. Jenkins, Pilot 387.00 kmh
James Andrews, Copilot 234.89 mph
Twin Comanche, N8226Y
2 Lycoming, 160 bhp
12/17/87

NASSAU/MONCTON (SWITZERLAND)
Hansjakob Beldi 267.20 kmh
Beechcraft A36 166.03 mph
1 Continental IO-520, 285 bhp
4/29/91

NATAL/DAKAR (UK)
Sheila Scott 272.99 kmh
Piper Comanche 260, G-ATOY 169.63 mph
1 Lycoming IO-540
11 hrs 4 min
11/18/67

NELSON/AUCKLAND (NEW ZEALAND)
Christopher Croucher 212.70 kmh
Peter Locke 132.16 mph
Piper PA-28-161
1 Lycoming O-360, 180 bhp
9/3/95

NELSON/WANAKA (NEW ZEALAND)
Graham J. Cochrane 228.94 kmh
Vincent J. Wills 142.25 mph
Cessna 172RG
1 Lycoming O-360, 180 bhp
9/3/95

NEW BEDFORD/KERRVILLE (USA)
Jack Rosen 267.80 kmh
Mooney M20F 166.40 mph
1 Lycoming IO-360, 200 bhp
10/7/92

NEW BEDFORD/SALZBURG (USA)
Jack Rosen 240.12 kmh
Mooney M20F 149.20 mph
1 Lycoming IO-360, 200 bhp
5/22/92

NEW CASTLE/HARRIS (USA)
Marie E. McMillan 182.93 kmh
Beech Bonanza F33A 113.67 mph
1 Continental IO-520B, 285 bhp
2/21/84

NEW CASTLE/PHILIPSBURG (USA)
Marie E. McMillan 328.34 kmh
Beech Bonanza F33A 204.02 mph
1 Continental IO-520B, 285 bhp
2/28/84

NEW CASTLE/POINTE A PITRE (USA)
Marie E. McMillan 312.83 kmh
Beech Bonanza F33A 194.38 mph
1 Continental IO-520B, 285 bhp
2/21/84

NEW DELHI/SACRAMENTO (USA)
Alvin Marks 72.56 kmh
Cessna 210, N942SM 45.93 mph
1 Continental TSIO-520-C
166 hrs 23 min 43 sec
4/10-16/69

NEW ORLEANS/FT. MYERS (USA)
Peter H. Burgher 255.30 kmh
Piper PA-30 158.63 mph
2 Lycoming O-320, 160 bhp
10/27/91

NEW ORLEANS/KITTY HAWK (USA)
Roger C. Doerr 245.28 kmh
Piper PA-32-301 152.41 mph
1 Lycoming IO-540, 300 bhp
12/16/93

NEW ORLEANS/OSHKOSH (USA)
Charles S. Horton, Jr. 311.30 kmh
Piper Comanche 400 193.43 mph
1 Lycoming IO-720, 400 bhp
07/26/90

NEW ORLEANS/ST. LOUIS (USA)
Charles S. Horton, Jr. 347.76 kmh
Piper Comanche 400 216.09 mph
1 Lycoming IO-720, 400 bhp
8/25/92

NEW YORK/COPENHAGEN (UK)
Sheila Scott 186.58 kmh
Piper Comanche 260, G-ATOY 115.93 mph
1 Lycoming IO-540
45 hrs 1 min 46 sec
5/22-24/69

NEW YORK/GOOSE BAY (UK)
Sheila Scott 289.80 kmh
Piper Comanche 260, G-ATOY 180.07 mph
1 Lycoming IO-540
6 hrs 56 sec
5/22/69

NEW YORK/KITTY HAWK (USA)
Millard Harmon, Pilot 284.01 kmh
Ruth Harmon, Copilot 176.47 mph
Beech Bonanza 36
1 Continental IO-520 BA, 285 bhp
12/16/84

NEW YORK/LOS ANGELES (USA)
Judy Wagner 309.65 kmh
Beechcraft Bonanza E33C 192.41 mph
1 Continental IO-520-B
6/28/70

NEW YORK/LOUISVILLE (USA)
Alan Feierstein 278.46 kmh
Piper PA-24-260 173.02 mph
1 Lycoming IO-540, 260 bhp
8/23/92

NEW YORK/PARIS (USA)
Robert Moriarty 322.85 kmh
Beech Bonanza C-35 200.62 mph
1 Continental IO-520 285 bhp
3/10-11/84

NOME/SITKA (CANADA)
Mathias Stinnes 306.92 kmh
Heinz Bittermann 190.71 mph
Glasair III
6/29/92

NORFOLK/KAHULUI (GERMANY)
Dietrich Schmitt 131.01 kmh
Beechcraft Bonanza F33A 81.40 mph
1 Continental IO-550, 300 bhp
6/1/95

NORFOLK/LAKELAND (USA)
James J. Hanrahan 219.57 kmh
Luellyn S. Leonard 136.43 mph
Beech D45
1 Continental O-470, 225 bhp
4/12/96

NORFOLK/LITTLE ROCK (GERMANY)
Dietrich Schmitt 269.33 kmh
Beechcraft Bonanza F33A 167.35 mph
1 Continental IO-550, 300 bhp
5/29/95

NORFOLK/LONG BEACH (GERMANY)
Dietrich Schmitt 135.94 kmh
Beechcraft Bonanza F33A 84.46 mph
1 Continental IO-550, 300 bhp
5/30/95

NORFOLK ISLAND/AUCKLAND (NEW ZEALAND)
Dee Bond Wakelin 247.73 kmh
Mat Wakelin 153.93 mph
Cessna 177 Cardinal RG
11/12/99

NORFOLK ISLAND/LORD HOWE ISLAND (NEW ZEALAND)
Christopher R. Toms 400.90 kmh
Allan Roberts 248.60 mph
Lancair IV
1 Lycoming TIO-540, 350 bhp
3/18/97

NORFOLK ISLAND/MAGENTA (NEW ZEALAND)
Mat Wakelin 236.17 kmh
Cessna 177RG 146.74 mph
1 Lycoming IO-360, 200 bhp
9/11/95

NORTHAM/KALGOORLIE (AUSTRALIA)
Robyn Stewart 239.66 kmh
Mooney M20 148.92 mph
4/13/99

NORTH POLE/ALBANY*
Millard Harmon 123.71 mph
Beech Bonanza 36, N7710R
1 Continental IO-520-BA, 285 bhp
8/10/86

NORTH POLE/EUREKA*
Millard Harmon 149.88 mph
Beech Bonanza 36, N7710R
1 Continental IO-520-BA, 285 bhp
8/8/86

NORTH POLE/SVALBARD (USA)
William H. Wisner, Pilot 294.12 kmh
Cheryl M. Davis, Copilot 182.77 mph
Bonanza, N6826A
1 Continental, 285 bhp
5/9/87

NOUMEA/BRISBANE (AUSTRALIA)
John W. Chesbrough 272.61 kmh
Mooney M20J-201 169.39 mph
1 Lycoming I0-360, 200 bhp
4/28/96

NOUMEA/NADI (BRAZIL)
Gerard Moss 262.03 kmh
EMBRAER EMB-721 162.81 mph
11/12/91

NOUMEA/NORFOLK ISLAND (NEW ZEALAND)
Dee Bond Wakelin 254.09 kmh
Mat Wakelin 157.88 mph
Cessna 177 Cardinal RG
11/11/99

NUUK/CANNES (CANADA)
Mathias Stinnes 397.32 kmh
Heinz Bittermann 246.88 mph
Glasair III
7/12/92

OKINAWA/SENDAI (USA)
Kenneth P. Johnson, Pilot 421.44 kmh
Larry Cioppi, Copilot 261.87 mph
Glasair III
1 Lycoming IO-540, 300 bhp
5/17/94

OKLAHOMA CITY/ATCHISON (USA)
Mary Gayle Kelly 224.33 kmh
William J. Cunningham 139.39 mph
Piper PA-28-180
1 Lycoming O-360, 180 bhp
7/24/97

OMSK/IRKUTSK (CANADA)
Mathias Stinnes 377.45 kmh
Heinz Bittermann 234.54 mph
Glasair III
6/24/92

ONTARIO/PHOENIX (USA)
Lee White, Pilot 383.35 kmh
Larry Randlett, Copilot 238.20 mph
Mooney M20K
1 Continental TSIO-520, 310 bhp
5/29/00

ORAN/LONDON (UK)
Bryan Eccles 329.04 kmh
Mooney M20K 204.47 mph
8/12/89

ORANJESTAD/CHARLOTTE AMALIE (USA)
Marie E. McMillan 238.39 kmh
Beech Bonanza F33A 148.13 mph
1 Continental IO-520B, 285 bhp
2/20/84

ORANJESTAD/CHRISTIANSTED (USA)
Marie E. McMillan 277.88 kmh
Beech Bonanza F33A 172.67 mph
1 Continental IO-520B, 285 bhp
2/20/84

ORANJESTAD/GUSTAVIA (USA)
Marie E. McMillan 136.69 kmh
Beech Bonanza F33A 84.93 mph
1 Continental IO-520B, 285 bhp
2/20/84

ORANJESTAD/HARRIS (USA)
Marie E. McMillan 181.46 kmh
Beech Bonanza F33A 112.76 mph
1 Continental IO-520B, 285 bhp
2/21/84

ORANJESTAD/MIAMI (BRAZIL)
Marcelo Bellodi 401.79 kmh
Noberto Bellodi 249.66 mph
Lancair IV-P
1 Continental, 350 bhp
7/22/00

ORANJESTAD/POINTE A PITRE (USA)
Marie E. McMillan 262.63 kmh
Beech Bonanza F33A 163.19 mph
1 Continental IO-520B, 285 bhp
2/21/84

ORANJESTAD/ST. JOHNS (USA)
Marie E. McMillan 267.49 kmh
Beech Bonanza F33A 166.21 mph
1 Continental IO-520B, 285 bhp
2/21/84

OSHKOSH/HELENA (USA)
Michael Ferguson, Pilot 186.52 kmh
Brenda Spivey, Copilot 115.90 mph
Beech Bonanza D-25
1 Continental E-225-8, 225 bhp
8/2-3/85

OSHKOSH/LANCASTER (USA)
Charles Zerphey 188.99 kmh
Cessna 172 117.43 mph
1 Lycoming O-320, 160 bhp
7/31/90

OSHKOSH/LANGENTHAL (SWITZERLAND)
Felix J. Estermann 86.71 kmh
Däetwyler MD-3 53.88 mph
1 Lycoming O-320, 160 bhp
8/12-16/92

OSHKOSH/NEW YORK (USA)
Donald Wyvell 386.40 kmh
Mooney 252TS, N252KM 240.11 mph
1 Continental SE TS 260 MB, 210 bhp
8/2/87

OSHKOSH/TOPEKA (USA)
Ilse de Vries 281.64 kmh
Beechcraft Bonanza A36 175.00 mph
1 Continental IO-550, 300 bhp
8/5/92

OSHKOSH/TULSA (USA)
Dick Gossen 411.27 kmh
Glasair III 255.55 mph
1 Lycoming IO-540, 300 bhp
7/30/91

OSHKOSH/WARREN (USA)
Dennis H. Bohn 272.18 kmh
Cessna T210M 169.13 mph
1 Continental TSIO-520, 285 bhp
7/30/94

OSHKOSH/WASHINGTON, DC*
Thatcher A. Stone, Pilot 188.35 mph
Jay P. B. Falatko, Copilot
Beechcraft Bonanza
1 Continental, 250 bhp
7/27/91

OSLO/ALBANY (USA)
Millard Harmon 220.68 kmh
Beech Bonanza 36, N7710R 137.13 mph
1 Continental IO-520BA, 285 bhp
6/26-27/87

OSLO/FRANKFURT (USA)
Millard Harmon 266.76 kmh
Beech Bonanza 36 165.77 mph
1 Continental IO-520-BA, 285 bhp
7/9/84

OSLO/HELSINKI (USA)
Millard Harmon 284.67 kmh
Beech Bonanza 36 176.89 mph
1 Continental IO-520-BA, 285 bhp
6/16/85

OSTEND/PARIS (UK)
Colin W. Yarnton 170.14 kmh
Mooney M20F 105.72 mph
1 Lycoming, 200 bhp
11/19-21/85

OTTAWA/ABBOTSFORD (CANADA)
Michael H. Leslie 73.17 kmh
Joseph Leslie 45.47 mph
Daniel J. Leslie
Cessna 172
1 Lycoming O-320, 160 bhp
8/10/00

OTTAWA/PALM SPRINGS, CA (CANADA)
Rene Goldberger 70.26 kmh
Beechcraft Duchess 76 43.66 mph
2 Lycoming O-360, 180 bhp
10/20/90

OXNARD, CA/OSHKOSH (USA)
Margaret Bird, Pilot 333.93 kmh
Thomas Bird, Copilot 207.49 mph
Beechcraft Bonanza B36TC
1 Continental IO-520, 310 bhp
7/30/90

PAGO PAGO/AITUTAKI (BRAZIL)
Gerard Moss 234.15 kmh
EMBRAER EMB-721 145.49 mph
12/19/91

PAGO PAGO/HILO (AUSTRALIA)
Gary Thomas Burns 433.28 kmh
Alexander Bernard Schenk 269.23 mph
Lancair IV
7/16/98

PALM BEACH/ALBANY (USA)
Millard Harmon 288.41 kmh
Beech Bonanza 36 179.22 mph
1 Continental 10-520BA, 285 bhp
11/1-2/85

PALM BEACH/KITTY HAWK (USA)
Richard L. Kane 255.95 kmh
Cirrus SR22 159.04 mph
1 Continental IO-550, 310 bhp
12/15/03

PALM BEACH/WASHINGTON (USA)
Millard Harmon 279.36 kmh
Beech Bonanza 36 173.59 mph
1 Continental 10-520BA, 285 bhp
11/1/85

PALM ISLAND/CANOUAN (USA)
Marie E. McMillan 368.02 kmh
Beech Bonanza F33A 228.68 mph
1 Continental IO-520B, 285 bhp
2/27/84

PALM ISLAND/LOVELL'S VILLAGE (USA)
Marie E. McMillan 335.09 kmh
Beech Bonanza F33A 208.21 mph
1 Continental IO-520B, 285 bhp
2/27/84

PALM ISLAND/PORT ELIZABETH (USA)
Marie E. McMillan 269.75 kmh
Beech Bonanza F33A 167.61 mph
1 Continental IO-520B, 285 bhp
2/27/84

PALM SPRINGS/CHICAGO (USA)
Charles Woolacott 414.80 kmh
Mooney M20M 257.74 mph
1 Lycoming TIO-540, 270 bhp
7/25/92

PALMA DE MAJORCA/LYDD (UK)
Geoffrey G. Boot 230.74 kmh
Piper PA-32R-301 143.38 mph
11/14/98

PALMA DE MAJORCA/SANTA MARIA (USA)
William H. Wisner, Pilot 294.48 kmh
Janice Sullivan, Copilot 182.99 mph
Beech Bonanza, N6826Q
1 Continental, 285 bhp
7/3/86

PARIS/AJACCIO (FRANCE)
Rene le Roux 315.55 kmh
Beechcraft F33A 196.08 mph
1 Continental IO-520-BB, 285 bhp
4/28/84

PARIS/BEIJING (BELGIUM)
Patrice Saillez, Pilot 25.14 kmh
Jean de Broqueville, Copilot 15.62 mph
Beech Bonanza A36
1 Continental IO-550B, 300 bhp
2/28-3/27/87

PARIS/BIARRITZ (FRANCE)
Jean Yves Dupont 206.37 kmh
Claude Martin, passenger 128.23 mph
Gardan GY 80
1 Lycoming 0-360-AIA
3 hrs 12 min 36 sec
5/21/76

PARIS/BONN-KÖLN (FRANCE)
Guy-Eric Oumier 339.36 kmh
Marc Pontet 210.86 mph
Mooney M20R Ovation
1 Continental IO-550, 280 bhp
5/14/95

PARIS/BROMMA (SWEDEN)
Hans Ulrisk von der Esch, Pilot 134.27 kmh
Jan Olof Friksman, Copilot 83.43 mph
Piper PA-32 RT
6/9/90

PARIS/CAMBERLEY (UK)
John J. Evendon 232.72 kmh
Cessna C182 144.61 mph
1 Continental O-470U, 230 bhp
11/14/85

PARIS/CANNES (FRANCE)
Christophe Toscas 368.22 kmh
Michel Rasquin 228.80 mph
LCT30
1 Continental, 280 bhp
5/27/00

PARIS/COLOGNE-BONN (W. GERMANY)
Richard Flohr 235.20 kmh
Mooney M20K 146.15 mph
1 Continental TSIO-360-GB1, 210 bhp
3/14-15/84

PARIS/LONDON (FRANCE)
Richard Fenwick 365.90 kmh
Beech Bonanza V35 227.37 mph
1 Continental, 285 bhp
3/25/84

PARIS/VIENNA (AUSTRIA)
Erwin Pettirsch 272.85 kmh
Cessna 210M 169.55 mph
1 Lycoming TSIO 260, 300 bhp
6/21/87

PASCUAS/ROBINSON ISLAND (BRAZIL)
Gerard Moss 272.96 kmh
EMBRAER EMB-721 169.60 mph
1/2/92

PENANG/DARWIN (NEW ZEALAND)
Clifford V. Tait 276.53 kmh
Beech Bonanza A36 171.83 mph
Continental IO-520-BB, 285 bhp
4/7-8/82

PEORIA/KITTY HAWK*
Linda K. Schumm, Pilot 132.78 mph
Rosemary A. Emhoff, Copilot
Cessna 177B
1 Lycoming O-360, 180 bhp
12/16/93

PERPIGNAN/GUERNSEY (UK)
John P. Frewer 247.66 kmh
Commander 115TC 153.89 mph
1 Lycoming TIO-540, 270 bhp
5/23/02

PERTH/ADELAIDE (AUSTRALIA)
Eric H. Wheatley 365.15 kmh
Mooney M20K 226.89 mph
9/14/89

PERTH/BRISBANE (AUSTRALIA)
William Wesley Finlen 304.04 kmh
Beechcraft Bonanza V35B 188.92 mph
1 Continental IO-520, 285 bhp
7/22/01

PERTH/CANBERRA (AUSTRALIA)
John W. Chesbrough 275.13 kmh
Marguerite Chesbrough 170.95 mph
Mooney M20J
11/21/92

PERTH/MELBOURNE (AUSTRALIA)
Eric H. Wheatley 344.05 kmh
Mooney M20K 213.78 mph
9/14/89

PERTH/SYDNEY (AUSTRALIA)
Eric H. Wheatley 380.83 kmh
Mooney M20K 236.63 mph
7/24/89

PETITE MARTINIQUE/CASTRIES (USA)
Marie E. McMillan 305.25 kmh
Beech Bonanza F33A 189.67 mph
1 Continental IO-520B, 285 bhp
2/27/84

PETITE MARTINIQUE/CLIFTON (USA)
Marie E. McMillan 321.60 kmh
Beech Bonanza F33A 199.83 mph
1 Continental IO-520B, 285 bhp
2/27/84

PETITE MARTINIQUE/KINGSTOWN (USA)
Marie E. McMillan 291.22 kmh
Beech Bonanza F33A 180.95 mph
1 Continental IO-520B, 285 bhp
2/27/84

PETITE MARTINIQUE/LOVELL'S VILLAGE (USA)
Marie E. McMillan 326.44 kmh
Beech Bonanza F33A 202.84 mph
1 Continental IO-520B, 285 bhp
2/27/84

PETITE MARTINIQUE/PORT ELIZABETH (USA)
Marie E. McMillan 271.72 kmh
Beech Bonanza F33A 168.84 mph
1 Continental IO-520B, 285 bhp
2/27/84

PETROPAVLOVSK/ANCHORAGE (USA)
Kenneth P. Johnson, Pilot 363.81 kmh
Larry Cioppi, Copilot 226.06 mph
Glasair III
1 Lycoming IO-540, 300 bhp
5/21/94

PHILIPSBURG/BRIDGETOWN (USA)
Marie E. McMillan 232.75 kmh
Beech Bonanza F33A 144.63 mph
1 Continental IO-520B, 285 bhp
2/23/84

PHILIPSBURG/CHARLOTTE AMALIE (USA)
Marie E. McMillan 392.44 kmh
Beech Bonanza F33A 243.85 mph
1 Continental IO-520B, 285 bhp
3/1/84

PHILIPSBURG/GUSTAVIA (USA)
Marie E. McMillan 55.70 kmh
Beech Bonanza F33A 34.61 mph
1 Continental IO-520B, 285 bhp
2/20/84

PHILIPSBURG/HARRIS (USA)
Marie E. McMillan 190.91 kmh
Beech Bonanza F33A 118.63 mph
1 Continental IO-520B, 285 bhp
2/21/84

PHILIPSBURG/KINGSTON (USA)
Marie E. McMillan 223.22 kmh
Beech Bonanza F33A 138.70 mph
1 Continental IO-520B, 285 bhp
3/1/84

PHILIPSBURG/POINTE A PITRE (USA)
Marie E. McMillan 248.56 kmh
Beech Bonanza F33A 154.44 mph
1 Continental IO-520B, 285 bhp
2/21/84

PHILIPSBURG/ST. JOHNS (USA)
Marie E. McMillan 228.56 kmh
Beech Bonanza F33A 142.02 mph
1 Continental IO-520B, 285 bhp
2/21/84

PHOENIX/ALBANY (USA)
Millard Harmon 272.16 kmh
Beech Bonanza 36 169.12 mph
1 Continental IO-520-BA, 285 bhp
8/7-8/84

PHOENIX/CHICAGO (USA)
David E. Wright 187.49 kmh
Beechcraft Bonanza A35 116.50 mph
1 Continental E-225, 225 bhp
4/19/99

PHOENIX/LOUISVILLE (USA)
Larry K. Clark 270.16 kmh
Piper Twin Comanche 167.87 mph
2 Lycoming IO-320, 160 bhp
8/22/92

PHOENIX/ONTARIO (USA)
Larry Randlett, Pilot 340.56 kmh
Lee White, Copilot 211.61 mph
Mooney M20K
1 Continental TSIO-520, 310 bhp
5/29/00

PHOENIX/OSHKOSH (USA)
James M. Frankard 396.55 kmh
Glasair III 246.40 mph
1 Lycoming IO-540, 300 bhp
3/17/95

PHOENIX/SAN DIEGO (USA)
Nathan H. East 447.07 kmh
David G. Riggs 277.80 mph
Lancair IV-P
1 Continental TSIO-550, 350 bhp
12/18/03

PHUKET/DARWIN (AUSTRALIA)
Gary Thomas Burns 470.03 kmh
Alexander Bernard Schenk 292.06 mph
Lancair IV
8/23/98

PITTSBURGH/COLUMBUS, OHIO (USA)
Robert L. Wick 249.64 kmh
Beechcraft Bonanza B35 155.12 mph
1 Continental E225-81
1 hr 2 min 42.1 sec
7/14/71

POINTE A PITRE/BASSETERRE (USA)
Marie E. McMillan 277.74 kmh
Beech Bonanza F33A 172.58 mph
1 Continental IO-520B, 285 bhp
2/28/84

POINTE A PITRE/BRIDGETOWN (USA)
Marie E. McMillan 232.75 kmh
Beech Bonanza F33A 144.63 mph
1 Continental IO-520B, 285 bhp
2/23/84

POINTE A PITRE/CASTRIES (USA)
Marie E. McMillan — 254.49 kmh
Beech Bonanza F33A — 158.13 mph
1 Continental IO-520B, 285 bhp
2/23/84

POINTE A PITRE/CODRINGTON (USA)
Marie E. McMillan — 312.78 kmh
Beech Bonanza F33A — 194.35 mph
1 Continental IO-520B, 285 bhp
2/28/84

POINTE A PITRE/GUSTAVIA (USA)
Marie E. McMillan — 294.50 kmh
Beech Bonanza F33A — 182.99 mph
1 Continental IO-520B, 285 bhp
2/28/84

POINTE A PITRE/MARTINIQUE (USA)
Marie E. McMillan — 303.70 kmh
Beech Bonanza F33A — 188.71 mph
1 Continental IO-520B, 285 bhp
2/23/84

POINTE A PITRE/NEW CASTLE (USA)
Marie E. McMillan — 268.20 kmh
Beech Bonanza F33A — 166.65 mph
1 Continental IO-520B, 285 bhp
2/28/84

POINTE A PITRE/PHILIPSBURG (USA)
Marie E. McMillan — 286.22 kmh
Beech Bonanza F33A — 177.85 mph
1 Continental IO-520B, 285 bhp
2/28/84

POINTE A PITRE/PORTSMOUTH (USA)
Marie McMillan — 343.10 kmh
Beech Bonanza F33A — 213.19 mph
1 Continental IO-520B, 285 bhp
2/23/84

POINTE A PITRE/ST. JOHNS (USA)
Marie E. McMillan — 292.35 kmh
Beech Bonanza F33A — 181.66 mph
1 Continental IO-520B, 285 bhp
2/28/84

PORTLAND, ME/GOOSE BAY (USA)
Anne Bridge Baddour, Pilot — 366.48 kmh
Margrit Orlowski, Copilot — 227.73 mph
Mooney 231
1 Continental TSIO-360-LB1B, 210 bhp
10/2/85

PORTLAND, ME/REYKJAVIK (USA)
Anne Bridge Baddour, Pilot — 151.56 kmh
Margrit Orlowski, Copilot — 94.18 mph
Mooney 231
1 Continental TSIO-360-LB1B, 210 bhp
10/2-3/85

PORTLAND, OR/PORTLAND, ME (USA)
Lloyd C. Harmon — 359.65 kmh
Peter M. Cohen — 223.48 mph
Mooney M20R Ovation
1 Continental IO-550, 280 bhp
3/29/02

PORTO/LONDON (NETHERLANDS)
Robert Hermans — 184.11 kmh
Piper Comanche PA30B — 114.40 mph
2 Lycoming IO-320, 160 bhp
8/6/85

PORTSMOUTH/BASSETERRE (USA)
Marie E. McMillan — 219.10 kmh
Beech Bonanza F33A — 136.10 mph
1 Continental IO-520B, 285 bhp
2/28/84

PORTSMOUTH/CODRINGTON (USA)
Marie E. McMillan — 238.07 kmh
Beech Bonanza F33A — 147.93 mph
1 Continental IO-520B, 285 bhp
2/28/84

PORTSMOUTH/GRAND-BOURG (USA)
Marie E. McMillan — 210.73 kmh
Beech Bonanza F33A — 130.94 mph
1 Continental IO-520B, 285 bhp
2/28/84

PORTSMOUTH/GUSTAVIA (USA)
Marie E. McMillan — 239.98 kmh
Beech Bonanza F33A — 149.12 mph
1 Continental IO-520B, 285 bhp
2/28/84

PORTSMOUTH/HARRIS (USA)
Marie E. McMillan — 239.26 kmh
Beech Bonanza F33A — 148.67 mph
1 Continental IO-520B, 285 bhp
2/28/84

PORTSMOUTH/NEW CASTLE (USA)
Marie E. McMillan — 210.67 kmh
Beech Bonanza F33A — 130.90 mph
1 Continental IO-520B, 285 bhp
2/28/84

PORTSMOUTH/PHILIPSBURG (USA)
Marie E. McMillan — 238.45 kmh
Beech Bonanza F33A — 148.17 mph
1 Continental IO-520B, 285 bhp
2/28/84

PORTSMOUTH/POINTE A PITRE (USA)
Marie E. McMillan — 167.26 kmh
Beech Bonanza F33A — 103.93 mph
1 Continental IO-520B, 285 bhp
2/28/84

PORTSMOUTH/ST. JOHNS (USA)
Marie E. McMillan — 216.29 kmh
Beech Bonanza F33A — 134.39 mph
1 Continental IO-520B, 285 bhp
2/28/84

PORT VILA/NOUMEA (BRAZIL)
Gerard Moss — 344.17 kmh
EMBRAER EMB-721 — 213.85 mph
11/9/91

PRAGUE/COLOGNE-BONN (W. GERMANY)
Richard Flohr — 278.73 kmh
Mooney M20K — 173.19 mph
1 Continental TSIO-360-GB1, 210 bhp
6/12/83

PRAGUE/PERPIGNAN (UK)
John P. Frewer — 246.30 kmh
Commander 115TC — 153.04 mph
1 Lycoming TIO-540, 270 bhp
5/21/02

PRESCOTT/KERRVILLE (USA)
Albert R. Kondall, Pilot — 391.54 kmh
Albert R. Kondall II, Copilot — 243.29 mph
Mooney M20K
1 Continental TSIO-360, 210 bhp
10/4/92

PRESTWICK/COPENHAGEN (W. GERMANY)
Wilhelm Heller — 393.64 kmh
Cessna Centurion P210N — 244.56 mph
1 Continental TSIO-520-P, 310 bhp
9/30/80

PRESTWICK/MONTPELLIER (CANADA)
Frank J. Quigg 165.87 kmh
Waco Classic F-5 103.07 mph
1 Jacobs R-755, 275 bhp
9/15/93

PUERTO VALLARTA/ACAPULCO (USA)
Marie E. McMillan 282.96 kmh
Beech Bonanza F33A 175.83 mph
1 Continental IO-520 BA, 285 bhp
2/3/84

PUERTO VALLARTA/GUADALAJARA (USA)
Marie E. McMillan, Pilot 274.32 kmh
Gloria May, Navigator 170.45 mph
Beech Bonanza F33A
1 Continental IO-520
44 min 35 sec
1/28/81

PUERTO VALLARTA/MAZATLAN (USA)
Marie E. McMillan, Pilot 311.76 kmh
Gloria May, Navigator 193.72 mph
Beech Bonanza F33A
1 Continental IO-520
56 min 45 sec
1/31/81

PUERTO VALLARTA/MEXICO CITY (USA)
Marie E. McMillan, Pilot 173.88 kmh
Gloria May, Navigator 108.04 mph
Beech Bonanza F33A
1 Continental IO-520
3 hrs 48 min 35 sec
1/28/81

QUEBEC/FROBISHER BAY*
Millard Harmon 165.99 mph
Beech Bonanza 36, N7710R
1 Continental IO-520-BA, 285 bhp
8/5/86

QUEBEC/TORONTO (CANADA)
Steven Anderson 217.76 kmh
Derrick A. Hobson 135.31 mph
Cessna 172 RG
1 Lycoming O-360, 180 bhp
11/16/96

QUONSET/BERMUDA (USA)
Millard Harmon 364.12 kmh
Beech Bonanza 36 226.27 mph
1 Continental 10-520BA, 285 bhp
10/29/85

RALEIGH/ALBANY (USA)
Millard Harmon 295.44 kmh
Beech Bonanza 36 183.59 mph
1 Continental 10-520BA, 285 bhp
11/1-2/85

RALEIGH/LAKELAND*
Charles A. Kohler 273.84 mph
Doug Beary
Lancair IV-P
1 Continental TSIO-550, 350 bhp
4/8/95

RAPID CITY/JACKSON (USA)
Ronald E. Brodowicz, Pilot 325.80 kmh
Vera M. Virchow, Crewmember 202.44 mph
Cessna P210
1 Continental TSIO-520-P, 310 bhp
8/15/92

RAPID CITY, SD/OSHKOSH (USA)
N. Colin Lind 283.99 kmh
Beechcraft BE-36 176.46 mph
1 Continental IO-520, 285 bhp
07/21/90

RAROTONGA/PAPEETE (BRAZIL)
Gerard Moss 222.51 kmh
EMBRAER EMB-721 138.26 mph
12/27/91

REDDING/ALLENTOWN (USA)
Kenneth P. Johnson, Pilot 403.38 kmh
Larry Cioppi, Copilot 250.65 mph
Glasair III
1 Lycoming IO-540, 300 bhp
4/3/94

REDLANDS, CA/OSHKOSH (USA)
Fredrick E. Foster 292.35 kmh
Beechcraft Bonanza V35TC 181.66 mph
1 Continental IO-520, 285 bhp
07/30/90

RENO/OSHKOSH (USA)
Thomas J. Gribbin 341.60 kmh
Cessna T-210 212.26 mph
1 Continental TSIO-520, 310 bhp
7/28/93

REDMOND/OSHKOSH (USA)
David Schroder 502.02 kmh
Lancair IV-P 311.94 mph
1 Continental TSIO-550, 350 bhp
7/20/01

RESOLUTE/EUREKA*
Millard Harmon 142.35 mph
Beech Bonanza 36, N7710R
1 Continental IO-520-BA, 285 bhp
8/7/86

RESOLUTE/FROBISHER BAY*
Millard Harmon 180.75 mph
Beech Bonanza 36, N7710R
1 Continental IO-520-BA, 285 bhp
8/9/86

REYKJAVIK/ALBANY (USA)
Millard Harmon 232.56 kmh
Beech Bonanza 36, N7710R 144.15 mph
1 Continental IO-520BA, 285 bhp
6/26-27/87

REYKJAVIK/EDINBURGH (USA)
William R. Grider 303.50 kmh
William H. Cox 188.58 mph
Piper PA-28RT-201T
1 Continental TSIO-360, 200 bhp
6/25/02

REYKJAVIK/HELSINKI (USA)
Millard Harmon 272.52 kmh
Beech Bonanza 36, N7710R 169.34 mph
1 Continental IO-520BA, 285 bhp
6/17-18/87

REYKJAVIK/NARSSARSSUAQ (USA)
Millard Harmon 276.18 kmh
Beech Bonanza 36 171.62 mph
1 Continental IO-520-BA, 285 bhp
6/24/85

REYKJAVIK/OSLO (USA)
Millard Harmon 273.60 kmh
Beech Bonanza 36 170.02 mph
1 Continental IO-520-BA, 285 bhp
7/1/84

REYKJAVIK/PRESTWICK, SCOTLAND (W. GERMANY)
Wilhelm Heller 303.22 kmh
Cessna P210N 188.42 mph
1 Continental TSIO-520-P, 310 bhp
4/18/81

REYKJAVIK/STRASBOURG (CANADA)
Mathias Stinnes 337.34 kmh
Gérard Guillaumaud 209.61 mph
Glasair II-S FT
9/11/97

RIBEIRÃO PRETO/ALTA FLORESTA (BRAZIL)
Marcelo Bellodi 411.03 kmh
Noberto Bellodi 255.40 mph
Lancair IV-P
1 Continental, 350 bhp
7/22/00

RIBEIRÃO PRETO/MIAMI (BRAZIL)
Marcelo Bellodi 450.51 kmh
Noberto Bellodi 219.53 mph
Lancair IV-P
1 Continental, 350 bhp
7/22/00

RICHMOND/ALBANY (USA)
Millard Harmon 303.12 kmh
Beech Bonanza 36 188.36 mph
1 Continental 10-520BA, 285 bhp
11/1-2/85

RICHMOND/KITTY HAWK (USA)
Pete Womack, Pilot 325.08 kmh
George Preston, Copilot 202.00 mph
Bellanca N505A
1 Lycoming 0-545A
12/16/88

RIO DE JANEIRO/RECIFE (BRAZIL)
Marcelo Bellodi 400.95 kmh
Lancair IV-P 249.14 mph
1 Continental, 350 bhp
6/3/00

RIYADH/MUSCAT (CANADA)
Frank J. Quigg 143.06 kmh
Waco Classic F-5 88.89 mph
1 Jacobs R-755, 275 bhp
9/19/93

ROADTOWN/CHARLOTTE AMALIE (USA)
Marie E. McMillan 309.95 kmh
Beech Bonanza F33A 192.59 mph
1 Continental IO-520B, 285 bhp
2/20/84

ROBINSON ISLAND/SANTIAGO (BRAZIL)
Gerard Moss 257.46 kmh
EMBRAER EMB-721 159.97 mph
1/24/92

ROCKLAND/ELMIRA (USA)
Edward B. Sleeper 211.35 kmh
James J. Nolan 131.33 mph
Cessna 172K
1 Lycoming O-360, 180 bhp
7/19/03

ROME/LARNACA (AUSTRALIA)
Gary Thomas Burns 497.81 kmh
Alexander Bernard Schenk 309.32 mph
Lancair IV
8/17/98

ROME/PARIS (FRANCE)
Jean Le Ber, Pilot 227.07 kmh
Jean-Michel Masson, Copilot 141.10 mph
TB 20 No 513
1 Lycoming IO-540-C4D5D, 250 bhp
3/27/87

ROME/TEL AVIV (CANADA)
Mathias Stinnes 339.76 kmh
Gérard Guillaumaud 211.12 mph
Glasair II-S FT
9/16/97

RONALDSWAY/DUBLIN (IRELAND)
Gerald W. Connolly 325.10 kmh
Piper Twin Comanche PA-39 202.00 mph
2 Lycoming IO-360
23 min 46 sec
2/15/75

ROSKILDE/LONDON (UK)
Geoffrey G. Boot 341.74 kmh
Mooney M20R Ovation 212.34 mph
1 Continental IO-550, 280 bhp
6/27/95

ROVANIEMI/MOURMANSK (FRANCE)
Sylvie Simon 218.00 kmh
Piper PA30 135.45 mph
6/26/91

SACRAMENTO/ATHENS (USA)
Alvin Marks 111.54 kmh
Cessna 210, N942SM 69.31 mph
1 Continental TSIO-520-C
96 hrs 50 min
4/3-7/69

SACRAMENTO/HAMILTON, BERMUDA (USA)
Alvin Marks 157.81 kmh
Cessna 210, N942SM 98.06 mph
1 Continental TSIO-520-C
32 hrs 27 min
4/3-4/69

SACRAMENTO/KARACHI (USA)
Alvin Marks 92.69 kmh
Cessna 210, N942SM 57.60 mph
1 Continental TSIO-520-C
139 hrs 9 min
4/3-9/69

SACRAMENTO/LOS ANGELES (USA)
Dan Richland 451.08 kmh
Beech Bonanza V35B 280.30 mph
1 Continental
IO-520- BA3, 285 bhp
4/20/84

SACRAMENTO/MADRID (USA)
Alvin Marks 127.45 kmh
Cessna 210, N942SM 79.20 mph
1 Continental TSIO-520-C
72 hrs 15 min
4/3-6/69

SACRAMENTO/SANTA MARIA, AZORES (USA)
Alvin Marks 140.15 kmh
Cessna 210, N942SM 87.89 mph
1 Continental TSIO-520-C
56 hrs 47 min 52 sec
4/3-5/69

SACRAMENTO/TEHERAN (USA)
Alvin Marks 99.15 kmh
Cessna 210, N942SM 61.61 mph
1 Continental TSIO-520-C
118 hrs 15 min
4/3-8/69

SACRAMENTO/WICHITA (USA)
G. R. Wood, Jr., Pilot 373.77 kmh
Louis Wood, Copilot 232.25 mph
Beech Bonanza
1 Continental TSIO-520-D, 285 bhp
5 hrs 42 min
6/8/80

ST. JACOB, IL/KITTY HAWK (USA)
Ken Bohannon, Pilot 329.04 kmh
Bill Weder, Copilot 204.47 mph
Larry Mills, Navigator
Piper Arrow, N2DW
12/16/89

ST. JOHN, NB/BEDFORD (USA)
Anne B. Baddour, Pilot 161.18 kmh
Patricia M. Thrasher, Copilot 100.15 mph
Beechcraft Sierra B24R
1 Lycoming, 200 bhp
8/13/91

ST. JOHN, NB/KENNEBUNK (USA)
Anne B. Baddour, Pilot 153.96 kmh
Patricia M. Thrasher, Copilot 95.67 mph
Beechcraft Sierra B24R
1 Lycoming, 200 bhp
8/13/91

ST. JOHN'S, ANTIGUA/BASSETERRE (USA)
Marie E. McMillan 366.35 kmh
Beech Bonanza F33A 227.64 mph
1 Continental IO-520B, 285 bhp
2/28/84

ST. JOHN'S, ANTIGUA/HARRIS (USA)
Marie E. McMillan 355.50 kmh
Beech Bonanza F33A 220.89 mph
1 Continental IO-520B, 285 bhp
2/21/84

ST. JOHN'S, ANTIGUA/PHILIPSBURG (USA)
Marie E. McMillan 307.94 kmh
Beech Bonanza F33A 191.35 mph
1 Continental IO-520B, 285 bhp
2/28/84

ST. JOHN'S, ANTIGUA/POINTE A PITRE (USA)
Marie E. McMillan 352.01 kmh
Beech Bonanza F33A 218.73 mph
1 Continental IO-520B, 285 bhp
2/21/84

ST. JOHN'S, NF/ALBANY (USA)
Millard Harmon 241.56 kmh
Beech Bonanza 36 150.11 mph
1 Continental IO-520-BA, 285 bhp
7/15/84

ST. JOHN'S, NF/BOSTON (USA)
William H. Wisner, Pilot 250.56 kmh
Janice Sullivan, Copilot 155.70 mph
Beech Bonanza, N6826Q
1 Continental, 285 bhp
7/5/86

ST. JOHN'S, NF/SANTA MARIA (USA)
Kenneth P. Johnson, Pilot 449.63 kmh
Larry Cioppi, Copilot 279.38 mph
Glasair III
1 Lycoming IO-540, 300 bhp
5/2/94

ST. JOHN'S, NF/VICTORIA (CANADA)
Michael H. Leslie 16.13 kmh
Joseph Leslie 10.02 mph
Daniel J. Leslie
Cessna 172
1 Lycoming O-320, 160 bhp
8/15/00

ST. JOHN'S, NF/WAUKEGAN (USA)
Charles Classen, Pilot 200.97 kmh
Phillip Greth, Copilot 124.88 mph
Bonanza G-35, N4493D
1 Continental E-225, 225 bhp
6/16/88

ST. JOHN'S/NADI (AUSTRALIA)
Peter H. Norvill 63.58 kmh
Cessna 172K 39.51 mph
1 Continental IO-360K, 195 bhp
5/7-6/11/89

ST. JOSEPH/ST. LOUIS (USA)
James H. Wilmes 291.01 kmh
Beechcraft Bonanza K35 180.82 mph
1 Continental IO-470, 250 bhp
7/7/99

ST. LOUIS/AUGUSTA (USA)
James R. Lyle 323.97 kmh
Piper PA-32R-301 201.30 mph
1 Lycoming IO-540, 300 bhp
11/7/93

ST. LOUIS/DAYTON (USA)
Coy G. Jacob 488.88 kmh
Mooney M20K/252, N261MS 303.79 mph
1 Continental TSOI-36-MBI
10/4/88

ST. LOUIS/OSHKOSH*
Verne Jobst 78.95 mph
"Spirit of St. Louis" replica, NX211
1 Continental W670, 220 bhp
10/4/92

ST. LOUIS/TAMPA (USA)
Coy G. Jacob 419.76 kmh
Mooney M20K/252, N261MS 260.84 mph
1 Continental TSOI-36-MBI
12/11/88

ST. LOUIS/WASHINGTON (USA)
Mark Patiky 615.96 kmh
Mooney 252 TSE, N272TD 382.75 mph
1 Continental TSIO-36-MBI, 210 bhp
1/7/87

ST. THOMAS/ALBERT TOWN (USA)
Millard Harmon 288.36 kmh
Beech Bonanza 36 179.19 mph
1 Continental 10-520BA, 285 bhp
11/1/85

ST. THOMAS/BIMINI (USA)
Millard Harmon 286.72 kmh
Beech Bonanza 36 178.17 mph
1 Continental 10-520BA, 285 bhp
11/1/85

ST. THOMAS/FT. LAUDERDALE (USA)
Millard Harmon 287.52 kmh
Beech Bonanza 36 178.66 mph
1 Continental 10-520BA, 285 bhp
11/1/85

ST. THOMAS/GRAND TURK (USA)
Millard Harmon 283.06 kmh
Beech Bonanza 36 175.90 mph
1 Continental 10-520BA, 285 bhp
11/1/85

ST. THOMAS/NASSAU (USA)
Millard Harmon 288.90 kmh
Beech Bonanza 36 179.52 mph
1 Continental 10-520BA, 285 bhp
11/1/85

SAIPAN/TAIPEI (USA)
Jerry Tsai 178.20 kmh
Cessna 182 RG 110.73 mph
1 Lycoming 0-540-J3C5D, 235 bhp
7/18-19/84

SAL/FUNCHAL (SWEDEN)
H. U. von der Esch 224.06 kmh
Piper 32 RT 139.23 mph
1 Lycoming IO-540, 300 bhp
1/20/85

SALEM/TEXARKANA (USA)
Danford A. Bookout 252.48 kmh
Piper Lance PA32 R-300 156.89 mph
1 Lycoming IO-540K, 300 bhp
10/25/85

SAN ANTONIO/EL PASO (USA)
William Gunn, Pilot 296.64 kmh
John Eslinger, Copilot 184.33 mph
Cessna C-210K, N850BJ
1 Continental TSIO 520H
7/16/87

SAN CARLOS/SAN DIEGO (USA)
Frank J. Shelton 277.92 kmh
Piper Comanche 172.69 mph
1 Lycoming O-540, 250 bhp
8/23/93

SAN CRISTOBOL/KINGSTON (USA)
Marie E. McMillan 304.12 kmh
Beech Bonanza F33A 188.97 mph
1 Continental IO-520B, 285 bhp
3/1/84

SAN CRISTOBOL/PORT AU PRINCE (USA)
Marie E. McMillan 311.29 kmh
Beech Bonanza F33A 193.43 mph
1 Continental IO-520B, 285 bhp
3/1/84

SAN DIEGO/ALBUQUERQUE (USA)
David L. Smith, Pilot 568.04 kmh
Stuart W. Ott, Copilot 352.96 mph
Beechcraft Bonanza S35
1 Lycoming TIO-540, 350 bhp
3/9/90

SAN DIEGO/PHOENIX (USA)
David L. Smith, Pilot 587.64 kmh
Stuart W. Ott, Copilot 365.14 mph
Beechcraft Bonanza S35
1 Lycoming TIO-540, 350 bhp
3/9/90

SAN DIEGO/PRESCOTT (USA)
David L. Smith, Pilot 600.64 kmh
Stuart W. Ott, Copilot 373.22 mph
Beechcraft Bonanza S35
1 Lycoming TIO-540, 350 bhp
4/11/91

SAN DIEGO/SAN FRANCISCO (USA)
Harold Kindsvater, Pilot 304.20 kmh
Wayne Easley, Copilot 189.03 mph
1 Continental TSIO-520-P, 300 bhp
5/2/88

SAN FRANCISCO/BATON ROUGE (USA)
Hypolite T. Landry, Jr. 245.45 kmh
Beechcraft Bonanza, N5842K 152.52 mph
1 Continental IO-520-B
12 hrs 6 min
6/25/69

SAN FRANCISCO/CHICAGO (USA)
Calvin B. Early 411.69 kmh
Beechcraft Bonanza M35 255.81 mph
1 Continental IO-550, 300 bhp
4/17/92

SAN FRANCISCO/DENVER (USA)
Dave Morss 579.86 kmh
Lancair IV 360.31 mph
1 Continental TSIO-550, 350 bhp
2/20/91

SAN FRANCISCO/LAS VEGAS (USA)
Robert H. Hornauer 252.06 kmh
Aérospatiale TB 20 156.62 mph
1 Lycoming IO-540, 260 bhp
10/10/92

SAN FRANCISCO/LOS ANGELES (USA)
Dave Morss 603.74 kmh
Lancair IV 375.14 mph
1 Continental TSIO-550, 350 bhp
12/30/92

SAN FRANCISCO/NEW YORK (USA) ✯
Calvin B. Early 419.49 kmh
Beechcraft Bonanza M35 260.66 mph
1 Continental IO-550, 300 bhp
4/17/92

SAN FRANCISCO/SAN DIEGO (USA)
Wayne Easley, Pilot 437.80 kmh
Harold Kindsvater, Copilot 272.17 mph
Cessna P210
1 Continental TSIO-520-P, 300 bhp
5/1/88

SAN FRANCISCO/SYRACUSE (USA)
Alan M. Marcum 347.71 kmh
Mooney 252TSE 216.05 mph
1 Continental TSIO-360, 210 bhp
6/10/99

SAN FRANCISCO/TAIPEI (USA)
Jerry Tsai 18.14 kmh
Cessna 182 RG 11.27 mph
1 Lycoming 0-540-J3C5D, 235 bhp
6/25-7/19/84

SAN FRANCISCO/WASHINGTON (USA)
Alan W. Gerharter 486.20 kmh
Mooney M 20K 302.11 mph
1 Continental, 210 bhp
8 hrs 4 min 25 sec
1/7/80

SAN JOSE/DENVER (USA)
Chris Verbil 378.06 kmh
Socata Trinidad TB-21 234.91 mph
1 Lycoming TIO-540, 250 bhp
5/6/96

SAN JOSE/HONOLULU (USA)
William H. Wisner, Pilot 257.76 kmh
Janice Sullivan, Copilot 160.17 mph
Beech Bonanza, N6826Q
1 Continental, 285 bhp
6/12-13/86

SAN JOSE/LOS ANGELES (USA)
Dorian A. De Maio 336.63 kmh
Beechcraft Bonanza E33A 209.17 mph
1 Continental IO-520, 285 bhp
3/27/99

SAN JOSE DEL CABO/BATON ROUGE (USA)
James S. Little 226.78 kmh
Jerry L. Payne 140.91 mph
Beechcraft Bonanza V35B
1 Continental IO-520, 285 bhp
7/21/97

SAN JUAN/BIMINI (USA)
Millard Harmon 290.05 kmh
Beech Bonanza 36 180.24 mph
1 Continental 10-520BA, 285 bhp
11/1/85

SAN JUAN/FT. LAUDERDALE (USA)
Millard Harmon 291.14 kmh
Beech Bonanza 36 180.91 mph
1 Continental 10-520BA, 285 bhp
11/1/85

SAN JUAN/KINGSTON (USA)
Marie E. McMillan 203.72 kmh
Beech Bonanza F33A 126.58 mph
1 Continental IO-520B, 285 bhp
3/1/84

SAN JUAN/PORT AU PRINCE (USA)
Marie E. McMillan 162.98 kmh
Beech Bonanza F33A 101.27 mph
1 Continental IO-520B, 285 bhp
3/1/84

SANFORD, FL/NEW ORLEANS (USA)
Thomas J. Connolly, Pilot 236.36 kmh
Henry R. Lehrer, Copilot 146.86 mph
Cessna Skylane 182P
1 Continental O-470, 230 bhp
4/3/91

SANTA ANA/GRAND CANYON (USA)
Glenn E. Davis 268.63 kmh
Beechcraft BE-76 166.92 mph
2 Lycoming O-360, 180 bhp
5/25/91

SANTA ANA/SACRAMENTO (USA)
Robert R. Whiton 294.06 kmh
Cessna T210 182.72 mph
1 Continental TSIO-510-R, 310 bhp
10/8/92

SANTA FE/ATLANTA (USA)
Nancy Toon, Pilot 269.23 kmh
Daniel Emin, Copilot 167.33 mph
Beech Duchess
2 Lycoming 0-360, 180 bhp
9/14-15/86

SANTA FE/TULSA (USA)
Nancy Toon, Pilot 314.64 kmh
Daniel Emin, Copilot 195.52 mph
Beech Duchess
2 Lycoming 0-360, 180 bhp
9/14/86

SANTA MARIA/BANGOR (GERMANY)
Dietrich Schmitt 262.98 kmh
Beech A36 Bonanza 163.41 mph
6/20/97

SANTA MARIA/COLOGNE-BONN (W. GERMANY)
Richard Flohr 55.47 kmh
Mooney M20K 34.47 mph
1 Continental TSIO-360-GB1, 210 bhp
1/6-8/83

SANTA MARIA/HONOLULU (GERMANY)
Dietrich Schmitt 71.86 kmh
Beech A36 Bonanza 44.65 mph
6/20-27/97

SANTA MARIA/KAHULUI (GERMANY)
Dietrich Schmitt 102.13 kmh
Beechcraft Bonanza F33A 63.46 mph
1 Continental IO-550, 300 bhp
6/1/95

SANTA MARIA/LISBON (W. GERMANY)
Michael Schultz 282.49 kmh
Porsche-Mooney M20K 175.54 mph
1 Porsche PFM 3200
1/8/86

SANTA MARIA/LITTLE ROCK (GERMANY)
Dietrich Schmitt 96.23 kmh
Beechcraft Bonanza F33A 59.79 mph
1 Continental IO-550, 300 bhp
5/29/95

SANTA MARIA/LONG BEACH (GERMANY)
Dietrich Schmitt 134.41 kmh
Beech A36 Bonanza 83.52 mph
6/20-22/97

SANTA MARIA/MADRID (W. GERMANY)
Richard Flohr 328,69 kmh
Mooney M20K 204.24 mph
1 Continental TSIO-360-GB1, 210 bhp
1/6/83

SANTA MARIA/MARRAKECH (USA)
Kenneth P. Johnson, Pilot 398.40 kmh
Larry Cioppi, Copilot 247.55 mph
Glasair III
1 Lycoming IO-540, 300 bhp
5/3/94

SANTA MARIA/NORFOLK (GERMANY)
Dietrich Schmitt 256.24 kmh
Beechcraft Bonanza F33A 159.22 mph
1 Continental IO-550, 300 bhp
5/28/95

SANTA MARIA/ST. JOHNS (USA)
William H. Wisner, Pilot 398.83 kmh
Cheryl M. Davis, Copilot 247.86 mph
Bonanza, N6826A
1 Continental, 285 bhp
5/13/87

SANTA MARIA/WICHITA (GERMANY)
Dietrich Schmitt 160.64 kmh
Beech A36 Bonanza 99.82 mph
6/21/97

SARASOTA/FREEPORT (USA)
William Heckel, Pilot 318.60 kmh
Eugene Gauch, Copilot 197.98 mph
Cessna 182RG, N78XP
1 Lycoming 0-540, 235 bhp
11/7/88

SARASOTA/NASSAU (USA)
Eugene Gauch, Pilot 290.16 kmh
William Heckel, Copilot 180.31 mph
Cessna 182RG, N78XP
1 Lycoming 0-540, 235 bhp
11/29/88

SAVANNAH/ALBANY (USA)
Millard Harmon 303.94 kmh
Beech Bonanza 36 188.87 mph
1 Continental 10-520BA, 285 bhp
11/1-2/85

SEATTLE/DENVER (USA)
Christopher L. Wooldridge 367.22 kmh
Mooney M20K 228.18 mph
1 Continental TSIO-360, 210 bhp
6/9/92

SEATTLE/OSHKOSH (USA)
Robert I. Dempster 160.12 kmh
Diane W. Dempster 99.49 mph
Piper PA-18-150 Super Cub
1 Lycoming O-320, 150 bhp
8/23/00

SEATTLE/SAN DIEGO (USA)
Alan Jackson 284.52 kmh
Piper Twin Comanche 176.79 mph
2 Lycoming IO-320, 160 bhp
8/21/93

SEATTLE/WASHINGTON (USA)
Robert S. Mucklestone 268.10 kmh
Cessna T-210, N6922R 166.59 mph
1 Continental IO-520
13 hrs 57 min 45 sec
4/27/75

SENDAI/PETROPAVLOVSK (USA)
Kenneth P. Johnson, Pilot 365.44 kmh
Larry Cioppi, Copilot 227.07 mph
Glasair III
1 Lycoming IO-540, 300 bhp
5/19/94

SEVILLA/ROME (CANADA)
Mathias Stinnes 336.37 kmh
Gérard Guillaumaud 209.01 mph
Glasair II-S FT
9/15/97

SHAFER/KITTY HAWK (USA)
Ken Bohannon 329.04 kmh
Piper Arrow PA28R-201 204.47 mph
12/16/89

SHANNON/COLOGNE (W. GERMANY)
Wilhelm Heller 449.82 kmh
H. Greiner, Crew 279.52 mph
Beech B 36TC
2/23/89

SHANNON/DUSSELDORF (W. GERMANY)
Wilhelm Heller 360.39 kmh
Beech Bonanza V35B 223.95 mph
1 Continental IO-520-BA, 285 bhp
7/20/85

SHANNON/GANDER (UK)
Sheila Scott 247.52 kmh
Piper Comanche 260, G-ATOY 153.79 mph
1 Lycoming IO-540
12 hrs 50 min 39.6 sec
5/4-5/69

SHANNON/GUERNSEY (GUERNSEY)
Christopher W. Shaw 212.66 kmh
Cessna F182Q 132.15 mph
1 Continental O-470-U, 230 bhp
10/1/85

SHANNON/OTTAWA (UK)
Sheila Scott 116.81 kmh
Piper Comanche 260, G-ATOY 72.58 mph
1 Lycoming IO-720
16 hrs 47 min 5 sec
10/13-14/67

SHANNON/PARIS (USA)
Richard Gilson 427.68 kmh
Mooney M20K, N252YM 265.76 mph
1 Continental 360 MBI, 210 bhp
5/17/87

SHANNON/TORONTO (UK)
Sheila Scott 121.24 kmh
Piper Comanche 260, G-ATOY 75.34 mph
1 Lycoming IO-720
18 hrs 16 min 36 sec
10/3-14/67

SHEYMA/HAKODATE (USA)
Robert Mucklestone, Pilot 243.00 kmh
Megan Kruse, Copilot 151.00 mph
Cessna T-210, N6922R
1 Continental TSIO-520
9/16-17/87

SINGAPORE/CALCUTTA (AUSTRALIA)
David McDonald 258.60 kmh
Peter R. Smith 160.68 mph
Beechcraft Bonanza A36
1 Continental IO-520, 285 bhp
2/15/01

SINGAPORE/LISBON (AUSTRALIA)
Peter H. Norvill 74.58 kmh
Cessna 172K 46.34 mph
1 Continental IO-360K, 195 bhp
5/7-6/11/89

SINGAPORE/SACRAMENTO (USA)
Alvin Marks 113.07 kmh
Cessna 210, N942SM 70.26 mph
1 Continental TSIO-520-C
4/11-16/69

SIOUX FALLS/OSHKOSH (USA)
Richard O. Middlen 411.64 kmh
Michael D. Blumer 255.78 mph
Glasair III
1 Lycoming IO-540, 325 bhp
7/25/91

SITKA/VICTORIA (CANADA)
Mathias Stinnes 364.76 kmh
Heinz Bittermann 226.65 mph
Glasair III
6/29/92

SOUTHEND/ROTTERDAM (UK)
Stephen Hunt 177.37 kmh
Cardinal C177 110.22 mph
1 Lycoming DAR-1FS-EU, 200 bhp
8/17/85

SOUTH POLE/MCMURDO STATION (AUSTRALIA)
Jon Johanson 308.87 kmh
Van's RV-4 191.92 mph
1 Lycoming IO-360, 180 bhp
12/8/03

SPOKANE/TAMPA (USA)
Darwin Conrad, Pilot 361.13 kmh
Deral Green, Copilot 224.39 mph
Mooney M20J
1 Lycoming TIO-360, 200 bhp
02/23/90

SRI LANKA/BOMBAY (USA)
William H. Wisner, Pilot 258.20 kmh
Janice Sullivan, Copilot 160.44 mph
Beech Bonanza, N6826Q
1 Continental, 285 bhp
6/27/86

STOCKTON/CHARLOTTE (USA)
Robert W. Prichard 327.87 kmh
Donald W. Blackley 203.72 mph
Beechcraft Bonanza B36TC
1 Continental TSIO-520, 300 bhp
4/28/96

STRASBOURG/CANNES (FRANCE)
Daniel Muraro 265.90 kmh
Piper Commanche PA 24-400 165.23 mph
1 Lycoming, 260 bhp
7/30/87

STRASBOURG/SEVILLA (CANADA)
Mathias Stinnes 315.12 kmh
Gérard Guillaumaud 195.81 mph
Glasair II-S FT
9/13/97

SVALBARD/NORTH POLE (USA)
William H. Wisner, Pilot 299.52 kmh
Cheryl M. Davis, Copilot 186.12 mph
Bonanza, N6826A
1 Continental, 285 bhp
5/9/87

SYDNEY/ADELAIDE (AUSTRALIA)
John W. Chesbrough — 303.30 kmh
Mooney M20J-201 — 188.46 mph
1 Lycoming IO-360, 200 bhp
5/4/96

SYDNEY/ALICE SPRINGS (AUSTRALIA)
Peter H. Norvill — 280.21 kmh
Mooney M20J — 174.11 mph
6/20/92

SYDNEY/AUCKLAND (AUSTRALIA)
Kevin Charles Haydon — 414.54 kmh
Peter William Hodgens — 257.58 mph
Lancair IV
1 Continental IO-550, 280 bhp
1/16/02

SYDNEY/BRISBANE (AUSTRALIA)
Gary T. Burns — 588.58 kmh
Peter Lindsay — 365.73 mph
Lancair IV
1/15/99

SYDNEY/DARWIN (AUSTRALIA)
David McDonald — 281.23 kmh
Peter R. Smith — 174.74 mph
Beechcraft Bonanza A36
1 Continental IO-520, 285 bhp
2/11/01

SYDNEY/LUSAKA, CONGO (UK)
O. J. Irwin — 65.88 kmh
Beechcraft Bonanza — 40.93 mph
1 Continental IO-520
180 hrs 24 min 47.5 sec
1/8-15/70

SYDNEY/MELBOURNE (AUSTRALIA)
Leslie R. Thompson, Pilot — 253.59 kmh
James Bernie Yeo, Copilot — 157.58 mph
Cessna 210N
11/30/89

SYDNEY/PERTH (AUSTRALIA)
Peter William Hodgens — 323.52 kmh
Kevin Charles Haydon — 201.02 mph
Lancair IV
1 Continental IO-550, 280 bhp
4/26/01

SYDNEY/TOWNSVILLE (AUSTRALIA)
Eric H. Wheatley — 262.49 kmh
Mooney M20K — 163.10 mph
9/14/89

TAIPEI/SEATTLE (USA)
Robert S. Mucklestone — 144.51 kmh
1 Continental IO-520 — 89.80 mph
67 hrs 24 min 47.5 sec
9/1-4/75

TARAWA/GUADALCANAL (USA)
Geraldine Mock — 252.84 kmh
Cessna P 206, N155JM — 157.11 mph
1 Continental IO-520
7 hrs 22 min 29 sec
10/28-29/69

TAYLORVILLE/OSHKOSH (USA)
Eric A. Overby, Pilot — 273.53 kmh
Robert E. Overby, Copilot — 169.96 mph
Beechcraft Bonanza F33A
1 Continental IO-520, 285 bhp
8/3/92

TEXARKANA/KITTY HAWK (USA)
Danford A. Bookout, Pilot — 304.92 kmh
Don Shilling, Copilot — 189.48 mph
Piper Lance, N8352C
1 Lycoming IO-540K, 300 bhp
12/16/86

TEXARKANA/NASHVILLE (USA)
Danford A. Bookout — 284.56 kmh
Piper Lance PA32 R-300 — 176.82 mph
1 Lycoming IO-540K, 300 bhp
10/22/85

TEXARKANA/WASHINGTON (USA)
Danford A. Bookout — 286.92 kmh
Piper Lance PA32 R-300 — 178.29 mph
1 Lycoming IO-540K, 300 bhp
10/22/85

THE BOTTOMS/CHARLOTTE AMALIE (USA)
Marie E. McMillan — 169.28 kmh
Beech Bonanza F33A — 105.19 mph
1 Continental IO-520B, 285 bhp
2/20/84

THE BOTTOMS/CHRISTIANSTED (USA)
Marie E. McMillan — 182.85 kmh
Beech Bonanza F33A — 113.61 mph
1 Continental IO-520B, 285 bhp
2/20/84

THE BOTTOMS/GUSTAVIA (USA)
Marie E. McMillan — 89.29 kmh
Beech Bonanza F33A — 55.48 mph
1 Continental IO-520B, 285 bhp
2/20/84

THE BOTTOMS/HARRIS (USA)
Marie E. McMillan — 183.34 kmh
Beech Bonanza F33A — 113.92 mph
1 Continental IO-520B, 285 bhp
2/21/84

THE BOTTOMS/NEW CASTLE (USA)
Marie E. McMillan — 188.34 kmh
Beech Bonanza F33A — 117.03 mph
1 Continental IO-520B, 285 bhp
2/21/84

THE BOTTOMS/POINTE A PITRE (USA)
Marie E. McMillan — 250.57 kmh
Beech Bonanza F33A — 155.70 mph
1 Continental IO-520B, 285 bhp
2/21/84

THE HAGUE/LONDON (UK)
Sheila Scott — 338.41 kmh
Piper Comanche 400 — 210.27 mph
1 Lycoming IO-720
5/19/65

THE VALLEY/CHARLOTTE AMALIE (USA)
Marie E. McMillan — 148.72 kmh
Beech Bonanza F33A — 92.41 mph
1 Continental IO-520B, 285 bhp
2/20/84

THE VALLEY/CHRISTIANSTED (USA)
Marie E. McMillan — 176.72 kmh
Beech Bonanza F33A — 109.81 mph
1 Continental IO-520B, 285 bhp
2/20/84

THE VALLEY/GRAND-CASE (USA)
Marie E. McMillan — 178.90 kmh
Beech Bonanza F33A — 111.16 mph
1 Continental IO-520B, 285 bhp
2/20/84

THE VALLEY/ORANJESTAD (USA)
Marie E. McMillan 207.61 kmh
Beech Bonanza F33A 129.00 mph
1 Continental IO-520B, 285 bhp
2/20/84

THE VALLEY/THE BOTTOMS (USA)
Marie E. McMillan 260.13 kmh
Beech Bonanza F33A 161.64 mph
1 Continental IO-520B, 285 bhp
2/20/84

TOLUCA/PUERTO ESCONDIDO (MEXICO)
Jorge Cornish 337.69 kmh
Mooney 231 M20K 209.84 mph
1 Continental TSIO-GBI, 210 bhp
11/28/86

TORONTO/QUEBEC CITY (CANADA)
Steven Anderson 233.15 kmh
Derrick A. Hobson 144.87 mph
Cessna 172 RG
1 Lycoming O-360, 180 bhp
11/16/96

TOTEGEGIE/PASCUAS (BRAZIL)
Gerard Moss 225.17 kmh
EMBRAER EMB-721 139.91 mph
1/16/92

TOULOUSE/CORFU (GREECE)
Efstratios Haralambakis 287.5 kmh
Georges Benecos 178.64 mph
Socata TB 20 Trinidad
1 Lycoming IO-540, 250 bhp
1/29/02

TOUSSUS-LE-NOBLE/ALGER (FRANCE)
Michel Lemonnier 260.33 kmh
Piper PA-32 161.77 mph
1 Lycoming IO-540, 300 cv
5/16/86

TOWNSVILLE/BRISBANE (AUSTRALIA)
Tevor C. J. Potter 253.88 kmh
Cessna 210N 157.76 mph
1 Continental IO-520 L, 300 bhp
3/1/86

TOWNSVILLE/DARWIN (AUSTRIALIA)
Eric H. Wheatley 295.65 kmh
Mooney M20K 183.71 mph
9/15/89

TRABZON/ADANA (CANADA)
Mathias Stinnes 259.70 kmh
Gérard Guillaumaud 161.37 mph
Glasair II-S FT
9/20/97

TROMSO/DUSSELDORF (USA)
William H. Wisner, Pilot 231.84 kmh
Cheryl M. Davis, Copilot 144.07 mph
Bonanza, N6826A
1 Continental, 285 bhp
5/10-11/87

TUCSON/GUAYMAS (USA)
Chanda S. Budhabhatti, Pilot 257.58 kmh
Barbara L. Harper, Copilot 160.05 mph
Piper PA-28-235
1 Lycoming O-540, 235 bhp
7/14/98

TUCSON/JUAREZ (USA)
Charles T. Adair 191.55 kmh
Cherokee D, N919CA 119.03 mph
1 Lycoming 0-360A4A, 180 bhp
11/2/88

TUCSON/HERMOSILLO (USA)
Marie E. McMillan, Pilot 315.00 kmh
Gloria May, Navigator 195.73 mph
Beech Bonanza F33A
1 Continental IO-520-BA
1 hr 4 min
1/26/81

TUCSON/LAS VEGAS (USA)
Marie E. McMillan, Pilot 319.68 kmh
Gloria May, Navigator 198.64 mph
Beech Bonanza F33A
1 Continental IO-520 BA
1 hr 50 min 13 sec
2/3/81

TULSA/ATLANTA (USA)
Nancy Toon, Pilot 297.36 kmh
Daniel Emin, Copilot 184.78 mph
Beech Duchess
2 Lycoming 0-360, 180 bhp
9/14-15/86

TUNIS/EGELSBACH (W. GERMANY)
Horst Regar 290.16 kmh
Beechcraft Bonanza V25 180.30 mph
1 Continental IO-520-BA
5 hrs 2 min
10/31/76

TWENTYNINE PALMS/ST. CHARLES (USA)
Marvin A. Robinson, Pilot 258.33 kmh
Kenneth Rapier, Copilot 160.52 mph
Mooney M20J
1 Lycoming IO-360, 200 bhp
7/17/93

UNST/ST. MARY'S (UK)
Robin Clark 38.67 kmh
Piper Warrior II 24.03 mph
6/26/99

USHUAIA/ROTHERA (UK)
Polly M. A. Vacher 164.72 kmh
Piper PA-28-236 102.35 mph
1 Lycoming O-540, 235hp
11/29/03

VAGAR/OSLO (USA)
Millard Harmon 282.22 kmh
Beech Bonanza 36 175.37 mph
1 Continental IO-520-BA, 285 bhp
6/15/85

VALENCIA/CANNES (UK)
Geoffrey G. Boot 286.88 kmh
Rushmeyer R90-230RG 178.26 mph
1 IO-540, 230 bhp
11/29/94

VANCOUVER/HALIFAX (CANADA)
Robin Morris 195.10 kmh
Zenair Tri-Z CH 300 121.22 mph
1 Lycoming 0-360-2F, 180 bhp
7/1-2/78

VANCOUVER/NORTH BAY (CANADA)
Robin Morris 192.30 kmh
Zenair Tri-Z CH 300 119.48 mph
1 Lycoming 0-360-2F, 180 bhp
7/1-2/78

VANCOUVER/WINNIPEG (CANADA)
Robin Morris 203.68 kmh
Zenair Tri-Z CH 300 126.55 mph
1 Lycoming 0-360-2F, 180 bhp
7/1-2/78

VERA CRUZ/MERIDA (USA)
William H. Wisner, Pilot 338.76 kmh
Fred Perez, Copilot 210.51 mph
Beechcraft E185, N299Z
2 P&W R985, 450 bhp
3/1/87

VERO BEACH/ALBANY (USA)
Millard Harmon 292.19 kmh
Beech Bonanza 36 181.57 mph
1 Continental 10-520BA, 285 bhp
11/1-2/85

VICTORIA/DEER LAKE (CANADA)
Daniel J. Leslie 93.19 kmh
Joseph Leslie 57.91 mph
Michael H. Leslie
Cessna 172
1 Lycoming O-320, 160 bhp
7/31/00

VICTORIA/FRESNO (CANADA)
Mathias Stinnes 417.17 kmh
Heinz Bittermann 259.21 mph
Glasair III
7/1/92

VICTORIA/PRINCE RUPERT (CANADA)
Daniel T. Bartie 232.50 kmh
Thomas L. Bartie 144.47 mph
Cessna 172Q
1 Avco-Lycoming, 180 bhp
6/300

VICTORIA/ST. JOHN'S, NF (CANADA)
Daniel J. Leslie 53.51 kmh
Michael H. Leslie 33.25 mph
Joseph Leslie
Cessna 172N
1 Lycoming O-320, 160 bhp
8/2/00

VIENNA/BUDAPEST (AUSTRIA)
Erwin Pettirsch 259.28 kmh
Cessna 210 M 161.12 mph
1 Continental TSIO-520, 305 bhp
6/27/86

WAKE ISLAND/BATON ROUGE (USA)
Hypolite T. Landry, Jr. 67.76 kmh
Beechcraft Bonanza, N5842K 42.11 mph
148 hrs 14 min
5/19-25/69

WAKE ISLAND/SACRAMENTO (USA)
Alvin Marks 130.66 kmh
Cessna 210, N9425M 81.19 mph
1 Continental TSIO-520-C
54 hrs 41 min 23.2 sec
4/14-16/69

WANAKA/NELSON (NEW ZEALAND)
Graham J. Cochrane 195.73 kmh
Vincent J. Wills 121.62 mph
Cessna 172RG
1 Lycoming O-360, 180 bhp
9/4/95

WASHINGTON/ALBANY (USA)
Millard Harmon 328.32 kmh
Beech Bonanza 36 204.02 mph
1 Continental 10-520BA, 285 bhp
11/1-2/85

WASHINGTON/CHICAGO*
James M. Frankard 204.19 mph
Glasair III
1 Lycoming IO-540, 300 bhp
4/27/96

WASHINGTON/COLOGNE-BONN (W. GERMANY)
Richard Flohr 365.59 kmh
Mooney M20K 227.16 mph
1 Continental TSIO-360-GB1, 210 bhp
10/11-12/80

WASHINGTON/GOOSE BAY (USA)
Millard Harmon 263.16 kmh
Beech Bonanza 36, N7710R 163.53 mph
1 Continental IO-520BA, 285 bhp
6/16/87

WASHINGTON/HAVANA (USA)
Millard Harmon 277.05 kmh
Beechcraft Bonanza BE-36 172.15 mph
1 Continental IO-520BA, 285 bhp
3/30/97

WASHINGTON/HELSINKI (USA)
Millard Harmon 194.40 kmh
Beech Bonanza 36, N7710R 120.80 mph
1 Continental IO-520BA, 285 bhp
6/16-18/87

WASHINGTON/MIAMI (USA)
Ronald White 223.56 kmh
Cessna 182 RG 138.56 mph
1 Continental O-470, 230 bhp
12/14/84

WASHINGTON/MOSCOW (USA)
Millard Harmon 187.92 kmh
Beech Bonanza 36, N7710R 116.77 mph
1 Continental IO-520BA, 285 bhp
6/16-18/87

WASHINGTON/NARSSARSSUAQ (USA)
Millard Harmon 200.52 kmh
Beech Bonanza 36, N7710R 124.60 mph
1 Continental IO-520BA, 285 bhp
6/16-17/87

WASHINGTON/PARIS (USA)
Robert Moriarty 315.79 kmh
Beech Bonanza C-35 196.22 mph
1 Continental 10-520, 285 bhp
3/10-11/84

WASHINGTON/REYKJAVIK (USA)
Millard Harmon 181.80 kmh
Beech Bonanza 36, N7710R 112.97 mph
1 Continental IO-520BA, 285 bhp
6/16-17/87

WASHINGTON/ROME, NY (USA)
Jon Coile 230.00 kmh
Russell C. Coile 142.91 mph
Piper PA-32-260
1 Lycoming O-540, 260 bhp
4/28/96

WASHINGTON/TEXARKANA (USA)
Danford A. Bookout 244.54 kmh
Piper Lance PA32 R-300 151.96 mph
1 Lycoming IO-540K, 300 bhp
10/25/85

WAUKEGAN/SAN JOSE (USA)
Charles Classen, Pilot 204.33 kmh
Phillip Greth, Copilot 126.97 mph
Bonanza G-35, N4493D
1 Continental E-225, 225 bhp
5/27/88

WEVELGEM/NEW BEDFORD (USA)
Jack Rosen 253.73 kmh
Mooney M20F 157.66 mph
1 Lycoming IO-360, 200 bhp
5/26/92

WHITEHORSE/NORTHWAY (USA)
Terry Winkler 221.99 kmh
Cessna 177 Cardinal RG 137.93 mph
1 Lycoming IO-360, 200 bhp
6/17/99

WICHITA/AMARILLO (USA)
Mike Dolabi 288.00 kmh
John C. Eslinger 178.95 mph
Beechcraft Bonanza A36
1 Continental IO-550, 300 bhp
12/17/03

WICHITA/LONG BEACH (GERMANY)
Dietrich Schmitt 258.17 kmh
Beech A36 Bonanza 160.42 mph
6/22/97

WICHITA/OSHKOSH (USA)
Michael C. Elsenrath 335.30 kmh
David J. Copeland 208.35 mph
Beech Bonanza A36
1 Continental IO-550, 300 bhp
7/27/03

WICHITA/WASHINGTON (USA)
Timothy T. Coons 320.68 kmh
Mooney M20K 199.26 mph
1 Continental TSIO-360, 210 bhp
8/27/97

WILMINGTON/KITTY HAWK (USA)
Merrill Mirman, Pilot 272.02 kmh
Ken Smith, Copilot 169.03 mph
Al Rappa, Navigator
Cessna 177RG, N2079Q
12/16/89

WINONA/SEATTLE (USA)
James M. Frankard 351.83 kmh
Glasair III 218.62 mph
1 Lycoming IO-540, 300 bhp
10/7/95

WINSTON-SALEM/KITTY HAWK (USA)
William P. Heitman 330.12 kmh
Beechcraft Bonanza 36 205.12 mph
1 Continental IO-520, 285 bhp
8/19/92

YAKUTSK/ANADYR (CANADA)
Mathias Stinnes 347.47 kmh
Heinz Bittermann 215.91 mph
Glasair III
6/27/92

YORK/KITTY HAWK (USA)
Mihail G. Baxevanis, Pilot 214.73 kmh
Soteris Baxevanis, Copilot 150.21 mph
Cessna 182RG, N3665C
1 Lycoming IO-540, 235 bhp
12/16/86

ZURICH/AJACCIO (SWITZERLAND)
Willi Schwarz 296.52 kmh
Mooney, M20J 184.26 mph
1 Lycoming, IO-36O, 220 bhp
1/28/84

ROUND TRIP

ALICE SPRINGS/DARWIN/ALICE SPRINGS (AUSTRALIA)
Roy Travis 21.29 kmh
Beechcraft Bonanza A36 13.23 mph
4/21/99

AUSTIN/BATON ROUGE/AUSTIN (USA)
Stuart C. Goldberg, Pilot 360.79 kmh
Jonathan Sabin, Copilot 224.18 mph
Beechcraft Bonanza V35B
1 Continental IO-550, 300 bhp
5/8/93

AUSTIN/CORPUS CHRISTI/AUSTIN (USA)
Charles I. Fitzsimmons 383.40 kmh
Mooney 252TSE 238.24 mph
1 Continental TSIO-36-BI, 210 bhp
12/6/86

BEDFORD/SYDNEY/BEDFORD (USA)
Anne B. Baddour, Pilot 170.23 kmh
Patricia M. Thrasher, Copilot 105.78 mph
Beechcraft Sierra B24R
1 Lycoming, 200 bhp
8/13/91

BRISBANE/SYDNEY/BRISBANE (AUSTRALIA)
Gary T. Burns 531.03 kmh
Peter Lindsay 329.97 mph
Lancair IV
1/15/99

BRUNSWICK/SYDNEY/BRUNSWICK (USA)
Anne B. Baddour, Pilot 167.17 kmh
Patricia M. Thrasher, Copilot 103.87 mph
Beechcraft Sierra B24R
1 Lycoming, 200 bhp
8/13/91

BUFFALO/NEW YORK/BUFFALO (USA)
Morris Wortman 342.37 kmh
Mooney TLS 212.74 mph
1 Lycoming TIO-540, 270 bhp
5/16/93

CAMBERLEY/PARIS/CAMBERLEY (UK)
John J. Evendon 199.26 kmh
Cessna C182 123.82 mph
1 Continental O-470U, 230 bhp
11/14/85

CANBERRA/MELBOURNE-ADELAIDE/CANBERRA (AUSTRALIA)
Leslie R. Thompson, Pilot 220.02 kmh
James Bernie Yeo, Copilot 136.72 mph
Cessna 210N
11/30/89

CASPER/CHEYENNE/CASPER (USA)
Norm Harvey, Pilot 248.11 kmh
Ellen Noonan, Copilot 154.17 mph
Cessna 195A
1 Jacobs R-755B2, 275 bhp
7/20/95

COLUMBUS/PITTSBURGH/COLUMBUS (USA)
Robert L. Wick 288.20 kmh
Beech Bonanza B35 179.08 mph
1 Continental E255-81
1 hr 48 min 37.6 sec
7/14/71

DUBLIN/GENEVA/DUBLIN (IRELAND)
Michael Slazenger 259.63 kmh
Piper Twin Comanche PA 39 161.33 mph
2 Lycoming 0-320, 160 bhp
9 hrs 10 min
3/10/76

JOHANNESBURG/CAPE TOWN/JOHANNESBURG (UK)
Bryan Eccles 344.45 kmh
Mooney 20K 214.05 mph
8/9-13/88

KAILUA/LIHUE/KAILUA (USA)
Bruce J. Mayes, Pilot 214.90 kmh
Jerry V. Proctor, Copilot 133.53 mph
Grumman Aerospace AA-5B
1 Lycoming O-360, 180 bhp
5/30/93

LONDON/BELFAST/LONDON (UK)
Sheila Scott 328.89 kmh
Piper Comanche 400 204.34 mph
1 Lycoming IO-720
3 hrs 8 min 31.8 sec
5/20/65

LONDON/BRUSSELS/LONDON (UK)
Sheila Scott 329.98 kmh
Piper Comanche 400 205.03 mph
1 Lycoming IO-720
1 hr 57 min 24.1 sec
5/19/65

LONDON/CALAIS/LONDON (UK)
Michael J. Baker 217.08 kmh
Piper PA-28R-201 134.88 mph
1 Lycoming IO-360, 200 bhp
10/9/95

LONDON/DUBLIN/LONDON (UK)
Sheila Scott 318.42 kmh
Piper Comanche 400 197.85 mph
1 Lycoming IO-720
2 hrs 54 min 6.4 sec
5/20/65

LONDON/PARIS/LONDON (UK)
Sheila Scott 324.52 kmh
Piper Comanche 400 201.65 mph
1 Lycoming IO-720
2 hrs 6 min 34.8 sec
5/19/65

LONDON/THE HAGUE/LONDON (UK)
Sheila Scott 332.54 kmh
Piper Comanche 400 206.62 mph
1 Lycoming IO-720
1 hr 51 min 48.8 sec
5/19/65

LONGYEARBYEN/NORTH POLE/LONGYEARBYEN (UK)
Peter Riggs 236.74 kmh
Piper PA-24 147.10 mph
8/2/91

MELBOURNE/ALICE SPRINGS/MELBOURNE (AUSTRALIA)
Roy Travis 24.55 kmh
Beechcraft Bonanza A36 15.25 mph
4/21/99

MELBOURNE/DARWIN/MELBOURNE (AUSTRALIA)
Roy Travis 41.28 kmh
Beechcraft Bonanza A36 25.65 mph
4/21/99

NEW YORK/LOS ANGELES/NEW YORK (USA) ★
Stuart C. Goldberg 319.09 kmh
Beechcraft Bonanza V35B 198.27 mph
1 Continental IO-550, 300 bhp
9/5-6/93

NORTHAM/KALGOORLIE/NORTHAM (AUSTRALIA)
Robyn Stewart 245.89 kmh
Mooney M20 152.79 mph
4/13/99

PORTSMOUTH/SYDNEY/PORTSMOUTH (USA)
Anne B. Baddour, Pilot 170.13 kmh
Patricia M. Thrasher, Copilot 105.71 mph
Beechcraft Sierra B24R
1 Lycoming, 200 bhp
8/13/91

ROCHESTER/NEW YORK/ROCHESTER (USA)
Morris Wortman 366.02 kmh
Mooney TLS 227.43 mph
1 Lycoming TIO-540, 270 bhp
5/16/93

RONALDSWAY/DUBLIN/RONALDSWAY (IRELAND)
Gerald W. Connolly 320.65 kmh
Piper Twin Comanche PA 39 199.24 mph
2 Lycoming IO-360
48 min 11 sec
2/15/75

SAN DIEGO/TUCSON/SAN DIEGO (USA)
Jeffrey R. Acord 313.09 kmh
James P. Blasingame 194.55 mph
Beechcraft Turbo Bonanza B36TC
1 Continental TSIO-520, 300 bhp
10/7/03

SEATTLE/VANCOUVER/SEATTLE (USA)
Robert F. Olszewski 417.24 kmh
Glasair III, N540RG 259.27 mph
1 Lycoming IO-540, 300 bhp
6/21/88

SVALBARD/NORTH POLE/SVALBARD (USA)
William H. Wisner, Pilot 295.20 kmh
Cheryl M. Davis, Copilot 183.44 mph
Bonanza, N6826A
1 Continental, 285 bhp
5/9/87

SYDNEY/MELBOURNE-ADELAIDE/SYDNEY (AUSTRALIA)
Leslie R. Thompson, Pilot 220.93 kmh
James Bernie Yeo, Copilot 137.29 mph
Cessna 210N
11/30/89

TORONTO/QUEBEC CITY/TORONTO (CANADA)
Steven Anderson 176.28 kmh
Derrick A. Hobson 109.54 mph
Cessna 172 RG
1 Lycoming O-360, 180 bhp
11/16/96

CLASS C-1.D (1,750-3,000 kg / 3,858-6,614 lbs)
GROUP I (PISTON ENGINE)

ADAK/SHEMYA (USA)
Jeremie Kaelin, Pilot 328.32 kmh
David Kaelin, Copilot 204.02 mph
Beech Baron, N22JJ
2 Continental IO-520
5/8-9/89

ADELAIDE/DARWIN (AUSTRALIA)
Trevor K. Brougham 290.56 kmh
Beechcraft Baron B55 180.54 mph
2 Continental IO-470-L
9 hrs 4 min 35 sec
8/1/71

AGANA/MAJURO (USA)
William Kaelin, Pilot 270.00 kmh
Jeremie J. Kaelin, Copilot 167.78 mph
Joseph J. Kaelin, Navigator
Beech Baron, N22JJ
2 Continental IO-520
5/15-16/89

AGANA/PONAPE (QATAR)
Hamad Al-Thani — 262.44 kmh
Piper Seneca II — 163.08 mph
2 Continental TSIO-360-E, 200 bhp
1/19-20/86

ALEXANDRIA/MIAMI (USA)
Danford A. Bookout — 279.03 kmh
Piper Lance PA 32R-300 — 173.38 mph
1 Lycoming IO-540-K, 300 bhp
6/20/85

ANCHORAGE/CALGARY (SWEDEN)
Erik O. Banck — 330.31 kmh
Merĉe M. Inglada — 205.25 mph
Cessna P210N
5/22/94

ANCHORAGE/MUNICH (W. GERMANY)
Dietrich Schmitt — 238.73 kmh
Beech Bonanza V35B — 148.33 mph
1 Continental IO-520-BA, 285 bhp
8/18-19/78

ANDOYA/NORD (UK)
Sheila Scott — 213.61 kmh
Piper Aztec D — 132.74 mph
2 Lycoming 10-540 C4B5, 250 bhp
7 hrs 33 min 24 sec
6/25/71

ANTWERP/BONN (UK)
E. K. Conventry — 309.88 kmh
Piper Navajo — 192.56 mph
2 Lycoming TIO-540, 315 bhp
5/8/85

ANTWERP/COLOGNE (UK)
E. K. Conventry — 325.84 kmh
Piper Navajo — 202.47 mph
2 Lycoming TIO-540, 315 bhp
5/8/85

ANTWERP/LONDON (UK)
E. K. Conventry — 326.69 kmh
Piper Navajo — 203.00 mph
2 Lycoming TIO-540, 315 bhp
5/8/85

ARMAN/KUWAIT (QATAR)
Hamad Al-Thani — 273.01 kmh
Piper Seneca II — 169.64 mph
2 Continental TSIO-360-E, 200 bhp
12/28/85

ATHENS/BOMBAY (AUSTRALIA)
Trevor K. Brougham — 313.70 kmh
Beechcraft Baron B55 — 194.91 mph
2 Continental IO-470-L
16 hrs 20 min 1 sec
8/8-9/71

ATHENS/DARWIN (AUSTRALIA)
Trevor K. Brougham — 289.75 kmh
Beechcraft Baron B55 — 180.04 mph
2 Continental IO-470-L
42 hrs 42 min 57 sec
8/8-10/71

ATHENS/SINGAPORE (AUSTRALIA)
Trevor K. Brougham — 311.98 kmh
Beechcraft Baron B55 — 193.85 mph
2 Continental IO-470-L
29 hrs 7 min 1 sec
8/8-9/71

ATLANTA/APPLETON, WI (USA)
Zack T. Pate — 358.81 kmh
Beechcraft Baron BE58 — 222.95 mph
2 Continental IO-550, 300 bhp
07/26/90

ATLANTA/OSHKOSH (USA)
David R. Britt, Pilot — 328.67 kmh
Stanley J. Anderson, Copilot — 204.22 mph
Beechcraft Baron BE58
2 Continental IO-550, 300 bhp
07/26/91

AUCKLAND/BRISBANE (AUSTRALIA)
Claude Meunier — 269.38 kmh
Robyn Stewart — 167.38 mph
Denis Beresford
Piper PA-60-602P Aerostar
2 Lycoming IO-540, 350 bhp
4/2/02

AUCKLAND/HONOLULU (UK)
Judith M. Chisholm — 87.25 kmh
Cessna T-210 Centurion — 54.21 mph
1 Continental TSIO-520-H
11/28-29/80

AUCKLAND/LONDON (UK)
Judith M. Chisholm — 104.35 kmh
Cessna T-210 Centurion — 64.84 mph
1 Continental TSIO-520-H
11/28-3/12/80

AUCKLAND/NORFOLK ISLAND (AUSTRALIA)
Claude Meunier — 339.34 kmh
Robyn Stewart — 210.86 mph
Denis Beresford
Piper PA-60-602P Aerostar
2 Lycoming IO-540, 350 bhp
4/1/02

AUCKLAND/NOUMEA (AUSTRALIA)
Claude Meunier — 408.17 kmh
Paul Stewart — 253.62 mph
Piper PA-60-602P
2 Lycoming IO-540, 350 bhp
8/17/95

AUCKLAND/SAN FRANCISCO (UK)
Judith M. Chisholm — 84.94 kmh
Cessna T-210 Centurion — 52.80 mph
1 Continental TSIO-520-H
11/28-1/14/80

AUSTIN/ATLANTA (USA)
Moton H. Crockett, Jr. — 348.23 kmh
Piper Aerostar 601P — 216.38 mph
2 Lycoming IO-540, 340 bhp
5/19/91

AUSTIN/SAN JOSE (USA)
Danford A. Bookout — 252.72 kmh
Piper Lance PA 32R-300 — 157.04 mph
1 Lycoming IO-540-K, 300 bhp
6/11-12/86

AZTEC/NASHVILLE (USA)
James S. Jenkins, Pilot — 312.84 kmh
Jim Fulbright, Copilot — 194.39 mph
Piper Aztec
2 Lycoming IO-540, 250 bhp
12/17-18/80

AZTEC/TULSA (USA)
James S. Jenkins, Pilot — 318.60 kmh
Jim Fulbright, Copilot — 197.97 mph
Piper Aztec
2 Lycoming IO-540, 250 bhp
12/17-18/80

BALI/KUALA LUMPUR (USA)
Danford A. Bookout — 232.20 kmh
Piper Lance PA 32R-300 — 144.29 mph
1 Lycoming IO-540-K, 300 bhp
6/23-24/86

BANGOR/AUGSBURG (GERMANY)
Dietrich Schmitt 34.92 kmh
Beechcraft Bonanza F33A 21.70 mph
11/7-14/93

BANGOR/BLACKBUSHE (GERMANY)
Dietrich Schmitt 91.71 kmh
Beechcraft Baron 58 56.98 mph
2 Continental IO-550
2/2/94

BANGOR/GLASGOW (GERMANY)
Dietrich Schmitt 139.03 kmh
Beechcraft Baron 58 86.38 mph
2 Continental IO-550
2/2/94

BANGOR/GOOSE BAY (GERMANY)
Dietrich Schmitt 403.81 kmh
Beechcraft Baron 58 250.91 mph
2 Continental IO-550
2/2/94

BANGOR/NARSSARSSUAQ (GERMANY)
Dietrich Schmitt 304.49 kmh
Beechcraft Baron 58 189.20 mph
2 Continental IO-550
2/2/94

BANGOR/REYKJAVIK (GERMANY)
Dietrich Schmitt 299.37 kmh
Beechcraft Baron 58 186.02 mph
2 Continental IO-550
2/2/94

BATON ROUGE/MIAMI (USA)
Danford A. Bookout 280.00 kmh
Piper Lance PA 32R-300 173.99 mph
1 Lycoming IO-540-K, 300 bhp
6/15/85

BEDFORD/GOOSE BAY (USA)
Michael A. Alper 308.75 kmh
Joseph F. Gatt 191.85 mph
Piper PA-46-310P
Continental TSIO-520, 310 bhp
8/24/92

BELFAST/CARDIFF (UK)
Spencer Flack 220.28 kmh
Piper Seneca PA34-200 136.88 mph
2 Lycoming LIO-360-61E6 & IO-360-CIE6
10/13/83

BELFAST/LONDON (UK)
Spencer Flack 160.12 kmh
Piper Seneca PA34-200 99.50 mph
2 Lycoming LIO-360-C1E6 & IO-360-CIE6
10/13/83

BIRMINGHAM/LAKELAND, FL (USA)
Anthony Ferrante, Pilot 419.85 kmh
Barbara Ferrante, Copilot 260.88 mph
Piper Aerostar
2 Lycoming IO-540, 350 bhp
04/07/90

BOMBAY/DARWIN (AUSTRALIA)
Trevor K. Brougham 288.45 kmh
Beechcraft Baron B55 179.23 mph
2 Continental IO-470-L
25 hrs 30 min 44 sec
8/9-10/71

BOMBAY/DUBAI (USA)
Danford A. Bookout 281.52 kmh
Piper Lance PA 32R-300 174.94 mph
1 Lycoming IO-540-K, 300 bhp
6/29/86

BOMBAY/SINGAPORE (USA)
Amos E. Buettell 345.43 kmh
Cessna 310N, N721X 214.64 mph
2 Lycoming TSIO-540
1 hrs 20 min 1.1 sec
8/6/75

BONN/ANTWERP (UK)
E. K. Conventry 323.66 kmh
Piper Navajo 201.12 mph
2 Lycoming TIO-540, 315 bhp
5/8/85

BONN/BRUGES (UK)
E. K. Conventry 324.23 kmh
Piper Navajo 201.47 mph
2 Lycoming TIO-540, 315 bhp
5/8/85

BONN/LONDON (UK)
E. K. Conventry 324.78 kmh
Piper Navajo 201.81 mph
2 Lycoming TIO-540, 315 bhp
5/8/85

BOSTON/CHARLESTON, WV (USA)
Danford A. Bookout 264.24 kmh
Piper Lance PA 32R-300 164.20 mph
1 Lycoming IO-540-K, 300 bhp
7/4/86

BOSTON/KITTY HAWK (USA)
Dean S. Edmonds 344.88 kmh
Beech Baron, N5522T 214.31 mph
2 Continental TSIO-520L
12/16/88

BRISBANE/ADELAIDE (USA)
Danford A. Bookout 248.40 kmh
Piper Lance PA 32R-300 154.35 mph
1 Lycoming IO-540-K, 300 bhp
6/17-18/86

BRISBANE/NORFOLK ISLAND (AUSTRALIA)
Claude Meunier 392.21 kmh
Denis Beresford 243.71 mph
Robyn Stewart
Piper 602P Aerostar
5/12/99

BRISBANE/SYDNEY (AUSTRALIA)
Peter Wilkins 247.90 kmh
Piper Malibu 154.05 mph
1 Continental TSIO 520-BE, 310 bhp
5/23-24/86

BRUGES/BONN (UK)
E. K. Conventry 328.38 kmh
Piper Navajo 204.05 mph
2 Lycoming TIO-540, 315 bhp
5/8/85

BRUGES/COLOGNE (UK)
E. K. Conventry 343.55 kmh
Piper Navajo 213.48 mph
2 Lycoming TIO-540, 315 bhp
5/8/85

BRUGES/LONDON (UK)
E. K. Conventry 325.47 kmh
Piper Navajo 202.24 mph
2 Lycoming TIO-540, 315 bhp
5/8/85

CAIRO/BAHRAIN (USA)
John F. X. Browne, Pilot 257.62 kmh
Orville Winters, Copilot 160.08 mph
Piper Aztec, PA-23
2 Lycoming IO-540-C4B5, 250 bhp
5/28/84

CAIRO/PALMA DE MAJORCA (USA)
Danford A. Bookout 261.00 kmh
Piper Lance PA 32R-300 162.18 mph
1 Lycoming IO-540-K, 300 bhp
7/1/86

CALGARY/MONTREAL (SWEDEN)
Erik O. Banck 408.53 kmh
Merće M. Inglada 253.85 mph
Cessna P210N
5/24/94

CALGARY/MONTREAL*
Arthur T. Mott, Pilot 237.91 mph
Thomas E. Hatch, Copilot
Beechcraft Baron 58
2 Continental IO-550, 300 bhp
5/24/94

CANNES/LONDON (USA)
Susan Nealey 290.59 kmh
Faith Hillman 180.56 mph
Cessna 310
2 Continental IO-520, 285 bhp
7/14/92

CANTON/KITTY HAWK (USA)
William H. James Jr. 370.79 kmh
Piper PA-34-200T 230.40 mph
2 Continental TSIO-360, 220 bhp
12/18/97

CARACAS/SAN JUAN (USA)
John F. X. Browne 277.92 kmh
Piper Aztec 172.70 mph
2 Lycoming IO-540C455, 250 bhp
3/4/85

CARDIFF/OSHKOSH (USA)
Susan Nealey 94.84 kmh
Faith Hillman 58.93 mph
Cessna 310
2 Continental IO-520, 285 bhp
7/19-22/92

CARDIFF/REYKJAVIK (USA)
Susan Nealey 181.64 kmh
Faith Hillman 112.87 mph
Cessna 310
2 Continental IO-520, 285 bhp
7/19/92

CHARLESTON, WV/LITTLE ROCK (USA)
Danford A. Bookout 271.80 kmh
Piper Lance PA 32R-300 168.90 mph
1 Lycoming IO-540-K, 300 bhp
7/5/86

CHARLESTON, WV/TEXARKANA (USA)
Danford A. Bookout 236.52 kmh
Piper Lance PA 32R-300 146.97 mph
1 Lycoming IO-540-K, 300 bhp
7/5/86

CHEYENNE/NASHVILLE (USA)
Russell H. Hancock 543.55 kmh
Piper Navajo Panther, N800PC 337.74 mph
2 Lycoming TIO-450-12BD, 350 bhp
3 hrs 3 min 11 sec
12/10/78

CHEYENNE/OSHKOSH (USA)
John R. Dewane, Pilot 309.54 kmh
William E. Unternaehrer, Copilot 192.33 mph
Beechcraft Baron B55
2 Continental IO-470, 260 bhp
7/28/94

CHICAGO/ATLANTA (USA)
Francis I. Blair 407.37 kmh
Piper Aerostar 601P 253.12 mph
2 Lycoming IO-540, 325 bhp
5/18/91

CHICAGO/KITTY HAWK (USA)
John P. Henebry 494.92 kmh
Beechcraft Baron P-58, N6056S 307.53 mph
2 Continental, 310 bhp
2 hrs 27 min 32 sec
12/17/78

CHICAGO/ORLANDO (USA)
Earle Boyter, Pilot 386.64 kmh
Douglas Smith, Copilot 240.26 mph
Piper Malibu, N91272
1 Continental TSIO 520-BE, 310 bhp
8/9/87

CHICAGO/TORONTO (USA)
J. Stephen Stout, Pilot 706.68 kmh
Michael Mills, Copilot 439.13 mph
Piper Malibu, N9114B
1 Continental TSIO-520, 310 bhp
1/8/89

CHICAGO/WASHINGTON (USA)
Susan Nealey 352.23 kmh
Faith Hillman 218.87 mph
Cessna 310
2 Continental IO-520, 285 bhp
6/14/92

CHRISTIANSTED/ROADTOWN (U.S. VIRGIN ISLANDS)
David Crisp 297.23 kmh
Piper PA-31-350 184.69 mph
2 Continental TSIO-540, 350 bhp
4/11/97

CHRISTIANSTED/SAN JUAN (U.S. VIRGIN ISLANDS)
Reynaldo Modeste 256.45 kmh
Piper PA-34-200 Seneca II 159.35 mph
2 Continental TSIO-360, 200 bhp
12/1/96

CINCINNATI/LONDON (USA)
Gerald P. Dietrick 302.16 kmh
Windecker Eagle, N4198G 187.75 mph
1 Continental IO-520-C, 285 bhp
21 hrs 4 min 15 sec
4/16-17/80

CINCINNATI/MUNICH (USA)
Gerald P. Dietrick 299.44 kmh
Windecker Eagle, N4198G 186.06 mph
1 Continental IO-520-C, 285 bhp
4/16-17/80

CINCINNATI/PARIS (USA)
Gerald P. Dietrick 304.12 kmh
Windecker Eagle, N4198G 188.97 mph
1 Continental IO-520-C, 285 bhp
21 hrs 53 min 11 sec
6/3-4/79

COLOGNE/ANTWERP (UK)
E. K. Conventry 265.05 kmh
Piper Navajo 164.70 mph
2 Lycoming TIO-540, 315 bhp
5/8/85

COLOGNE/BRUGES (UK)
E. K. Conventry 282.77 kmh
Piper Navajo 175.71 mph
2 Lycoming TIO-540, 315 bhp
5/8/85

COLOGNE/LONDON (UK)
E. K. Conventry — 301.07 kmh
Piper Navajo — 187.08 mph
2 Lycoming TIO-540, 315 bhp
5/8/85

COLOMBO/JAKARTA (UK)
Judith M. Chisholm — 250.14 kmh
Cessna T210 Centurion — 155.43 mph
1 Continental TSIO-520-H
11/20-21/80

COLUMBUS/SAN JUAN (USA)
Geraldine L. Mock — 175.25 kmh
Cessna P-206, N155JM — 108.90 mph
1 Continental IO-520
16 hrs 29 min 17.6 sec
6/24-25/68

CROSS CITY/TEXARKANA (USA)
Danford A. Bookout — 253.27 kmh
Piper Lance PA 32R-300 — 157.37 mph
1 Lycoming IO-540-K, 300 bhp
6/20/85

DAKAR/RIO DE JANERIO (USA)
Richard Norton, Pilot — 303.94 kmh
Calin Rosetti, Pilot — 188.87 mph
Piper Malibu, N26033
1 Continental TSIO 520BE, 310 bhp
1/22-23/87

DALLAS/EL PASO (USA)
Danford A. Bookout — 235.08 kmh
Piper Lance — 146.08 mph
1 Lycoming IO-540K, 300 bhp
5/30-31/85

DARWIN/BALI (USA)
Danford A. Bookout — 270.00 kmh
Piper Lance PA 32R-300 — 167.78 mph
1 Lycoming IO-540-K, 300 bhp
6/21-22/86

DARWIN/HONOLULU (AUSTRALIA)
Trevor K. Brougham — 286.36 kmh
Beechcraft Baron B55 — 177.93 mph
2 Continental IO-470-L
30 hrs 11 min 53 sec
8/4-6/71

DARWIN/LONDON (AUSTRALIA)
Trevor K. Brougham — 189.52 kmh
Beechcraft Baron B55 — 177.93 mph
2 Continental IO-470-L
73 hrs 32 min 11 sec
8/4-7/71

DARWIN/SAN FRANCISCO (AUSTRALIA)
Trevor K. Brougham — 288.52 kmh
Beechcraft Baron B55 — 179.23 mph
2 Continental IO-470-L
42 hrs 46 min 27 sec
8/4-6/71

DARWIN/TORONTO (AUSTRALIA)
Trevor K. Brougham — 277.37 kmh
Beechcraft Baron B55 — 172.34 mph
2 Continental IO-470-L
55 hrs 57 min 46 sec
8/4-7/71

DAYTON/BRIDGEPORT (USA)
John L. Cink — 490.68 kmh
Aerostar 601P — 304.91 mph
2 Lycoming IO-540, 290 bhp
1/16/85

DAYTON/KITTY HAWK (USA)
Craig S. Kern — 377.84 kmh
Beech Baron 95-B55 — 234.78 mph
2 Continental IO-470, 260 bhp
11/7/03

DAYTON/WASHINGTON (USA) ✭
Senator John H. Glenn, Pilot — 369.02 kmh
Phillip S. Woodruff, Copilot — 229.29 mph
Beechcraft Baron 58P
2 Continental TSIO-520, 325 bhp
12/17/96

DAYTONA BEACH/GANDER (QATAR)
Hamad Al-Thani — 299.88 kmh
Piper Seneca II — 186.34 mph
2 Continental TSIO-360-E, 200 bhp
12/20/85

DENPASAR/DARWIN (UK)
David John Hare — 248.55 kmh
Robert Miller — 154.44 mph
Piper PA-23 Aztec
2 Lycoming IO-540, 250 bhp
4/1/01

DENPASAR/KUPANG (UK)
David John Hare — 243.32 kmh
Robert Miller — 151.19 mph
Piper PA-23 Aztec
2 Lycoming IO-540, 250 bhp
3/31/01

DENVER/BOSTON (USA)
Dean S. Edmonds, Pilot — 369.36 kmh
G. Drew, Copilot — 229.52 mph
Beechcraft Baron
2 TSIO-520L, 310 bhp
7/24/84

DENVER/FORT LAUDERDALE (USA)
Paul Salsburg, Pilot — 404.07 kmh
Bill Salsburg, Crewmember — 251.08 mph
Tony Biancarosa, Crewmember
Piper Malibu
1 Continental TSIO-520, 310 bhp
09/24/90

DENVER/OSHKOSH (USA)
Paul F. Siegel Jr., Pilot — 487.69 kmh
Rudolph Siegel, Copilot — 303.04 mph
Piper Aerostar 601B
2 Lycoming IO-540, 350 bhp
7/24/95

DETROIT/NEW YORK (USA)
Theodore C. Patecell — 473.04 kmh
Cessna T337A — 293.95 mph
2 Continental TIO-360-H, 210 bhp
3/23/84

DETROIT/SPITZBERGEN (USA)
John F. X. Browne — 144.00 kmh
Piper Aztec — 89.48 mph
1 Lycoming IO-540-C4B5, 250 bhp
5/25-27/85

DETROIT/WASHINGTON (USA)
J. Stephen Stout, Pilot — 637.20 kmh
Michael Mills, Copilot — 395.96 mph
Piper Malibu, N9114B
1 Continental TSIO-520, 310 bhp
1/4/89

DOHA, QATAR/MADRAS (QATAR)
Hamad Al-Thani — 306.72 kmh
Piper Seneca II — 190.59 mph
2 Continental TSIO-360-E, 200 bhp
1/7/86

DOUGLAS/LYDD (UK)
Geoffrey G. Boot 301.18 kmh
Cessna C340 A 187.14 mph
9/17/97

DUBAI/AGRA (USA)
Arthur T. Mott, Pilot 340.11 kmh
Thomas E. Hatch, Copilot 211.33 mph
Beechcraft Baron 58
2 Continental IO-550, 300 bhp
5/9/94

DUBAI/CAIRO (USA)
Danford A. Bookout 252.72 kmh
Piper Lance PA 32R-300 157.04 mph
1 Lycoming IO-540-K, 300 bhp
6/30/86

EASTER ISLAND/LIMA (USA)
John F. X. Browne 277.63 kmh
Piper Aztec PA 23 172.52 mph
2 Lycoming IO-540-C4B5, 250 bhp
7/4-5/86

EASTER ISLAND/TAHITI (USA)
Richard Norton, Pilot 295.59 kmh
Calin Rosetti, Pilot 183.68 mph
Piper Malibu, N26033
1 Continental TSIO 520BE, 310 bhp
2/7-8/87

EDINBURGH/CARDIFF (UK)
Spencer Flack 139.38 kmh
Piper Seneca PA34-200 86.61 mph
2 Lycoming LIO-360-CIE6 & IO-360-CIE6
10/13/83

EDINBURGH/PARIS*
Stanley H. Mick 165.99 mph
Cessna T303 Crusader
2 Continental TSIO-520, 250 bhp
5/22/94

EL PASO/ALEXANDRIA (USA)
Danford A. Bookout 213.86 kmh
Piper Lance PA 32R-300 132.90 mph
1 Lycoming IO-540-K, 300 bhp
6/19-20/85

EL PASO/BATON ROUGE (USA)
Danford A. Bookout 233.28 kmh
Piper Lance PA 32R-300 144.96 mph
1 Lycoming IO-540-K, 300 bhp
6/19-20/85

EL PASO/DAYTONA BEACH (QATAR)
Hamad Al-Thani 329.91 kmh
Piper Seneca II 204.91 mph
2 Continental TSIO-360-E, 200 bhp
1/29/86

EL PASO/MIAMI (USA)
Danford A. Bookout 239.46 kmh
Piper Lance PA 32R-300 148.79 mph
1 Lycoming IO-540-K, 300 bhp
6/19-20/85

EL PASO/NEW ORLEANS (USA)
Danford A. Bookout 253.02 kmh
Piper Lance PA 32R-300 157.22 mph
1 Lycoming IO-540-K, 300 bhp
6/19-20/85

EL PASO/OGDEN (USA)
Danford A. Bookout 174.96 kmh
Piper L ance 108.72 mph
1 Lycoming IO-540K, 300 bhp
5/31/85

EL PASO/SALT LAKE CITY (USA)
Danford A. Bookout 175.15 kmh
Piper Lance 108.93 mph
1 Lycoming IO-540K, 300 bhp
5/31/85

EL PASO/TUCSON (USA)
Mark E. Ackerman 358.85 kmh
Cessna 310R 222.98 mph
2 Continental IO-520, 285 bhp
11/15/92

EUREKA/SPITZBERGEN (USA)
Jeremie J. Kaelin, Pilot 256.32 kmh
William Kaelin, Copilot 159.28 mph
Joseph J. Kaelin, Copilot
Beech Baron, N22JJ
2 Continental IO-520
6/18/88

EVERETT/OSHKOSH (USA)
Don L. Fitzpatrick, Jr., Pilot 401.56 kmh
Heather L. Fitzpatrick, Copilot 249.51 mph
Piper PA-46-310P Malibu
1 Continental TSIO-520, 310 bhp
7/28/94

FAIRBANKS/SAN FRANCISCO (USA)
Larry Grant, Pilot 227.20 kmh
Susie Grant, Copilot 172.25 mph
Brian Munson, Navigator
Piper Malibu, N4319M
1 Continental TSIO-520BE, 310 bhp
6/20-21/87

FARMINGDALE/FT. LAUDERDALE (USA)
Arthur T. Mott 372.01 kmh
Charlotte M. Mott 231.16 mph
Tim Bastick
Beechcraft Baron BE 58
2 Continental IO-550, 300 bhp
4/7/95

FARMINGDALE/HAMILTON*
Arthur T. Mott 220.97 mph
Thomas E. Hatch
Charlotte M. Mott
Beechcraft Baron BE 58
2 Continental IO-550, 300 bhp
6/14/94

FARMINGDALE/LOS ANGELES (USA)
Arthur T. Mott, Pilot 297.54 kmh
Tim Bastick, Copilot 184.88 mph
Beechcraft Baron BE 58
2 Continental IO-550, 300 bhp
10/27/95

FARMINGDALE/PANAMA CITY (USA)
Arthur T. Mott 300.21 kmh
Tim Bastick 186.54 mph
Robert Mott
Beechcraft Baron BE 58
2 Continental IO-550, 300 bhp
4/30/95

FARO, PORTUGAL/IRAKLION (QATAR)
Hamad Al-Thani 311.76 kmh
Piper Seneca II 193.72 mph
2 Continental TSIO-360-E, 200 bhp
12/26/85

FIJI/PAGO PAGO (USA)
John F. X. Browne 267.05 kmh
Piper Aztec PA 23 164.94 mph
2 Lycoming IO-540-C4B5, 250 bhp
6/26-27/86

FT. WORTH/FARMINGDALE (USA)
Arthur T. Mott 317.75 kmh
Philip Mouyiaris 197.44 mph
Geoffrey T. Mott
Beechcraft Baron BE 58
2 Continental IO-550, 300 bhp
6/16/95

FRANKFURT/VIENNA (UK)
Geoffrey G. Boot 355.38 kmh
Cessna 340 220.82 mph
2 Lycoming TS10-520, 310 bhp
7/11/96

FREDERICK/GOOSE BAY (USA)
Susan Nealey 368.76 kmh
Faith Hillman 229.13 mph
Cessna 310
2 Continental IO-520, 285 bhp
7/8/92

FRESNO/FREDERICK (USA)
Susan Nealey 355.83 kmh
Faith Hillman 221.10 mph
Cessna 310
2 Continental IO-520, 285 bhp
7/5/92

GALLUP/OGDEN (USA)
Danford A. Bookout 257.24 kmh
Piper Lance PA 32R-300 159.84 mph
1 Lycoming IO-540-K, 300 bhp
5/31/85

GALLUP/SALT LAKE CITY (USA)
Danford A. Bookout 263.50 kmh
Piper Lance PA 32R-300 163.74 mph
1 Lycoming IO-540-K, 300 bhp
5/31/85

GALWAY/LYDD (UK)
Geoffrey G. Boot 335.92 kmh
Cessna 340 208.73 mph
6/8/97

GANDER/ATHENS (AUSTRALIA)
Trevor K. Brougham 314.42 kmh
Beechcraft Baron B55 195.37 mph
2 Continental IO-470-L
19 hrs 38 min 33 sec
8/7-8/71

GANDER/BOMBAY (AUSTRALIA)
Trevor K. Brougham 296.95 kmh
Beechcraft Baron B55 183.51 mph
2 Continental IO-470-L
36 hrs 37 min 40 sec
8/7-9/71

GANDER/BOSTON (USA)
Danford A. Bookout 231.84 kmh
Piper Lance PA 32R-300 144.07 mph
1 Lycoming IO-540-K, 300 bhp
7/4/86

GANDER/COPENHAGEN (W. GERMANY)
Dietrich Schmitt 279.92 kmh
Beechcraft Bonanza 173.93 mph
1 Continental IO-520, 285 bhp
6/30/79

GANDER/LISBON (USA)
John F. X. Browne, Pilot 132.93 kmh
Orville Winters, Copilot 82.60 mph
Piper Aztec, PA-23
2 Lycoming IO-540-C4B5, 250 bhp
5/26/84

GANDER/LONDON (AUSTRALIA)
Trevor K. Brougham 368.90 kmh
Beechcraft Baron B55 229.22 mph
2 Continental IO-470-L
10 hrs 26 min 50 sec
8/7/71

GANDER/MANCHESTER (UK)
G. A. Pilling 343.35 kmh
Cessna P210N 213.34 mph
1 Continental TSIO-520-P, 285 bhp
9/29-30/79

GANDER/MUNICH (W. GERMANY)
Herman Krug 366.32 kmh
Piper Malibu 227.63 mph
1 Continental TSIO-520-BE, 310 bhp
5/11-12/84

GANDER/SANTA MARIA (USA)
James F. Nields 351.43 kmh
Beechcraft Baron C55 218.37 mph
2 Continental IO-520-C
7 hrs 43 min 44 sec
6/8/68

GANDER/SINGAPORE (AUSTRALIA)
Trevor K. Brougham 258.08 kmh
Beechcraft Baron B55 177.13 mph
2 Continental IO-470-L
8/7-9/71

GENEVA/HELSINKI (GUATEMALA)
Frederico Bauer 395.79 kmh
Piper PA-60 245.93 mph
6/20/92

GENEVA/HELSINKI*
Susan Nealey 199.88 mph
Faith Hillman
Cessna 310
2 Continental IO-520, 285 bhp
6/20/92

GLASGOW/BLACKBUSHE (GERMANY)
Dietrich Schmitt 261.15 kmh
Beechcraft Baron 58 162.27 mph
2 Continental IO-550
2/4/94

GODTHAB/CANNES (USA)
Susan Nealey 364.32 kmh
Faith Hillman 226.37 mph
Cessna 310
2 Continental IO-520, 285 bhp
7/12/92

GODTHAB/SHANNON (USA)
Susan Nealey 377.77 kmh
Faith Hillman 234.73 mph
Cessna 310
2 Continental IO-520, 285 bhp
7/12/92

GOOSE BAY/AUGSBURG (GERMANY)
Dietrich Schmitt 85.25 kmh
Beechcraft Bonanza F33A 52.97 mph
11/12-14/93

GOOSE BAY/BLACKBUSHE (GERMANY)
Dietrich Schmitt 80.28 kmh
Beechcraft Baron 58 49.88 mph
2 Continental IO-550
2/2/94

GOOSE BAY/GLASGOW (GERMANY)
Dietrich Schmitt 125.06 kmh
Beechcraft Baron 58 77.70 mph
2 Continental IO-550
2/2/94

GOOSE BAY/GODTHAB (USA)
Susan Nealey — 328.21 kmh
Faith Hillman — 203.94 mph
Cessna 310
2 Continental IO-520, 285 bhp
7/9/92

GOOSE BAY/LONDON (UK)
Judith M. Chisholm — 301.88 kmh
Cessna T210 Centurion — 187.58 mph
1 Continental TSIO-520-H
2/3/80

GOOSE BAY/NARSSARSSUAQ (GERMANY)
Dietrich Schmitt — 391.96 kmh
Beechcraft Baron 58 — 243.55 mph
2 Continental IO-550
2/2/94

GOOSE BAY/OLSO (NORWAY)
Poju Stephansen, Pilot — 328.44 kmh
Jan Roang, Copilot — 204.08 mph
Cessna 210N
5/27/94

GOOSE BAY/REYKJAVIK (GERMANY)
Dietrich Schmitt — 384.97 kmh
Beechcraft Bonanza B36TC — 239.21 mph
1 Continental TSIO-520, 300 bhp
2/10/96

GOOSE BAY/SHANNON (USA)
Susan Nealey — 277.42 kmh
Faith Hillman — 172.38 mph
Cessna 310
2 Continental IO-520, 285 bhp
6/16/92

GRAND FORKS/MORRISTOWN (USA)
John J. McNamara, Jr. — 393.48 kmh
Piper Aztec — 244.51 mph
2 Lycoming TIO-540-CIA, 250 bhp
11/21-22/86

GRAND ISLAND/OSHKOSH (USA)
John R. Dewane, Pilot — 362.94 kmh
William E. Unternaehrer, Copilot — 225.52 mph
Beechcraft Baron B55
2 Continental IO-470, 260 bhp
7/31/96

GREENSBORO/OSHKOSH (USA)
Gary S. Jacobs — 272.44 kmh
Beechcraft Baron 58 — 169.29 mph
2 Continental IO-520, 285 bhp
7/29/93

GUADALCANAL/BRISBANE (USA)
Danford A. Bookout — 258.12 kmh
Piper Lance PA 32R-300 — 160.40 mph
1 Lycoming IO-540-K, 300 bhp
6/16-17/86

GUADALCANAL/SYDNEY (USA)
John F. X. Browne — 260.32 kmh
Piper Aztec PA 23 — 161.76 mph
2 Lycoming IO-540-C4B5, 250 bhp
6/20-21/86

HA'IL/BAHRAIN (UK)
Robert Miller — 264.61 kmh
David John Hare — 164.42 mph
Piper PA-23 Aztec
2 Lycoming IO-540, 250 bhp
3/17/01

HAMILTON/FARMINGDALE*
Arthur T. Mott — 216.38 mph
Thomas E. Hatch
Charlotte M. Mott
Beechcraft Baron BE 58
2 Continental IO-550, 300 bhp
6/15/94

HELSINKI/MOSCOW (GUATEMALA)
Frederico Bauer — 312.26 kmh
Piper PA-60 — 194.03 mph
6/21/92

HELSINKI/MOSCOW*
Susan Nealey — 148.87 mph
Faith Hillman
Cessna 310
2 Continental IO-520, 285 bhp
6/21/92

HELSINKI/NOME (USA)
Susan Nealey — 35.38 kmh
Faith Hillman — 21.98 mph
Cessna 310
2 Continental IO-520, 285 bhp
6/21-28/92

HELSINKI/PORTLAND, ME (USA)
Charles C. Mack — 223.64 kmh
Beechcraft Bonanza A36 — 138.96 mph
1 Continental IO-550, 300 bhp
6/30/92

HELSINKI/ZURICH (USA)
Jeremie J. Kaelin, Pilot — 298.42 kmh
William Kaelin, Copilot — 185.45 mph
Joseph J. Kaelin, Copilot
Beech Baron, N22JJ
2 Continental IO-520
6/22/88

HICKORY/KITTY HAWK (USA)
George Johnson — 432.08 kmh
Beech Baron 58 — 268.49 mph
2 Continental IO-550-C, 300 bhp
12/16/85

HO CHI MINH CITY/OKINAWA (USA)
Arthur T. Mott, Pilot — 354.55 kmh
Thomas E. Hatch, Copilot — 220.31 mph
Beechcraft Baron 58
2 Continental IO-550, 300 bhp
5/14/94

HONG KONG/SINGAPORE (FRANCE)
Delio Iglesias, Pilot — 315.13 kmh
Jean Claude De Lassee, Copilot — 195.82 mph
Cessna 310R
2 Continental IO-520M, 285 bhp
3/15/87

HONOLULU/EL PASO (USA)
Danford A. Bookout — 234.31 kmh
Piper Lance PA 32R-300 — 145.59 mph
1 Lycoming IO-540-K, 300 bhp
6/17-18/85

HONOLULU/FAIRBANKS (USA)
Richard Norton, Pilot — 305.50 kmh
Calin Rosetti, Pilot — 189.84 mph
Piper Malibu, N26033
1 Continental TSIO 520BE, 310 bhp
2/10-11/87

HONOLULU/GOOSE BAY (UK)
Judith M. Chisholm — 179.66 kmh
Cessna T210 Centurion — 111.64 mph
1 Continental TSIO-520-H
11/30-12/3/80

HONOLULU/IMPERIAL (USA)
Danford A. Bookout — 231.46 kmh
Piper Lance PA 32R-300 — 143.82 mph
1 Lycoming IO-540-K, 300 bhp
6/17/85

HONOLULU/LONDON (AUSTRALIA)
Trevor K. Brougham — 275.32 kmh
Beechcraft Baron B55 — 171.07 mph
2 Continental IO-470-L
42 hrs 36 min 20 sec
8/6-8/71

HONOLULU/MAJURO (USA)
Danford A. Bookout — 314.64 kmh
Piper Lance PA 32R-300 — 195.52 mph
1 Lycoming IO-540-K, 300 bhp
6/14-15/86

HONOLULU/RENO (USA)
Jeremie J. Kaelin, Pilot — 287.64 kmh
William Kaelin, Copilot — 178.74 mph
Joseph J. Kaelin, Navigator
Beech Baron, N22JJ
2 Continental IO-520
5/20-21/89

HONOLULU/SAN DIEGO (USA)
Danford A. Bookout — 228.32 kmh
Piper Lance PA 32R-300 — 141.87 mph
1 Lycoming IO-540-K, 300 bhp
6/17/85

HONOLULU/SAN FRANCISCO (USA)
Amos E. Buettell — 373.51 kmh
Cessna 310N, N721X — 232.09 mph
2 Lycoming TIO-540
10 hrs 19 min 40.8 sec
9/8/75

HONOLULU/SAN JOSE (QATAR)
Hamad Al-Thani — 327.60 kmh
Piper Seneca II — 203.57 mph
2 Continental TSIO-360-E, 200 bhp
1/25-26/86

HONOLULU/SYDNEY (USA)
Donald T. Holmes, Pilot — 329.40 kmh
Steward Reid, Copilot — 204.68 mph
Piper Malibu PA46-310P
1 Continental TSIO-520BE, 310 bhp
12/17-18/84

HONOLULU/TARAWA (USA)
Geraldine L. Mock — 228.14 kmh
Cessna P-206, N155JM — 141.76 mph
1 Continental IO-520
16 hrs 56 min 38 sec
10/26/69

HONOLULU/TORONTO (AUSTRALIA)
Trevor K. Brougham — 298.92 kmh
Beechcraft Baron B55 — 185.73 mph
2 Continental IO-470-L
25 hrs 6 min 26 sec
8/6-7/71

HONOLULU/TUCSON (USA)
Danford A. Bookout — 221.87 kmh
Piper Lance PA 32R-300 — 137.87 mph
1 Lycoming IO-540-K, 300 bhp
6/17/85

HONOLULU/YUMA (USA)
Danford A. Bookout — 231.00 kmh
Piper Lance PA 32R-300 — 143.54 mph
1 Lycoming IO-540-K, 300 bhp
6/17/85

HONOLULU/WINNIPEG (UK)
Judith M. Chisholm — 227.98 kmh
Cessna T210 Centurion — 141.67 mph
1 Continental TSIO-520-H
11/30-12/2/80

INDIANAPOLIS/BOSTON (USA)
Dean S. Edmonds, Pilot — 437.04 kmh
G. Drew, Copilot — 271.58 mph
Beechcraft Baron
2 TSIO-520L, 310 bhp
7/24/84

IRAKLION/ARMAN (QATAR)
Hamad Al-Thani — 186.48 kmh
Piper Seneca II — 115.88 mph
2 Continental TSIO-360-E, 200 bhp
12/27/86

IRKUTSK/NOME (USA)
Susan Nealey — 82.97 kmh
Faith Hillman — 51.56 mph
Cessna 310
2 Continental IO-520, 285 bhp
6/26-28/92

IRKUTSK/YAKUTSK (GUATEMALA)
Frederico Bauer — 371.49 kmh
Piper PA-60 — 230.83 mph
6/26/92

IRKUTSK/YAKUTSK*
Susan Nealey — 197.07 mph
Faith Hillman
Cessna 310
2 Continental IO-520, 285 bhp
6/26/92

JACKSON/NASHVILLE, TN*
James S. Jenkins — 230.70 mph
Piper Aztec
2 Lycoming IO-540, 250 bhp
12/17/81

JACKSON, TN/OSHKOSH (USA)
Ronald Armand Jenkins — 288.71 kmh
Beechcraft Bonanza B50 — 179.40 mph
2 Lycoming GO-435-C2, 260 bhp
07/26/90

JAKARTA/PORT HEADLAND (UK)
Judith M. Chisholm — 248.51 kmh
Cessna T210 Centurion — 154.42 mph
1 Continental TSIO-520-H
11/21-22/80

JOHNSON ISLAND/HONOLULU (QATAR)
Hamad Al-Thani — 261.00 kmh
Piper Seneca II — 162.18 mph
2 Continental TSIO-360-E, 200 bhp
1/22/86

KANSAS CITY/OSHKOSH (USA)
John R. Dewane, Pilot — 299.39 kmh
William E. Unternaehrer, Copilot — 186.03 mph
Beechcraft Baron B55
2 Continental IO-470, 260 bhp
7/28/95

KIRKSVILLE, MO/COLORADO SPRINGS (USA)
Fred W. Replogle — 411.31 kmh
Piper Aerostar PA-60-601P — 255.57 mph
2 Lycoming IO-540, 290 bhp
05/27/90

KITTY HAWK/LAKELAND (USA)
Theodore C. Patecell — 295.97 kmh
Cessna Skymaster — 183.91 mph
2 Continental TSIO-360, 225 bhp
4/20/98

KITTY HAWK/WASHINGTON (USA)
Philander P. Claxton, III — 443.10 kmh, 275.33 mph
Aerostar 601-P
2 Lycoming IO-540, 290 bhp
50 min 41 sec
8/19/78

KUALA LUMPUR/MADRAS (USA)
Danford A. Bookout — 272.88 kmh, 169.56 mph
Piper Lance PA 32R-300
1 Lycoming IO-540-K, 300 bhp
6/25-26/86

KUPANG/DARWIN (UK)
David John Hare — 256.74 kmh, 159.53 mph
Robert Miller
Piper PA-23 Aztec
2 Lycoming IO-540, 250 bhp
4/1/01

KUSHIRO/ANCHORAGE (USA)
John F. X. Browne, Pilot — 169.56 kmh, 105.36 mph
Orville Winters, Copilot
Piper Aztec, PA-23
2 Lycoming IO-540-C4B5, 250 bhp
6/13-15/84

KUUJJUAQ/CHICAGO (USA)
Susan Nealey — 281.92 kmh, 175.17 mph
Faith Hillman
Cessna 310
2 Continental IO-520, 285 bhp
7/22/92

KUUJJUAQ/OSHKOSH (USA)
Susan Nealey — 287.84 kmh, 178.85 mph
Faith Hillman
Cessna 310
2 Continental IO-520, 285 bhp
7/22/92

KUWAIT/DOHA (QATAR)
Hamad Al-Thani — 317.88 kmh, 197.53 mph
Piper Seneca II
2 Continental TSIO-360-E, 200 bhp
12/31/85

LAKELAND/FARMINGDALE (USA)
Arthur T. Mott — 352.77 kmh, 219.20 mph
Tim Bastick
Charlotte M. Mott
Beechcraft Baron BE 58
2 Continental IO-550, 300 bhp
4/10/95

LA REUNION/PARIS (FRANCE)
Francois Garcon — 278.31 kmh, 172.94 mph
Piper 46
7/8/88

LA VERNE, CA/RENO (USA)
Roger Hyder, Pilot — 362.07 kmh, 224.98 mph
Bobbi Thompson, Copilot
Beechcraft Baron 56TC
2 Lycoming TIO-541, 380 bhp
9/12/91

LIMA/DETROIT (USA)
John F. X. Browne — 220.03 kmh, 136.73 mph
Piper Aztec PA 23
2 Lycoming IO-540-C4B5, 250 bhp
7/5-7/86

LIMA/HOUSTON (USA)
John F. X. Browne — 235.40 kmh, 146.28 mph
Piper Aztec PA 23
2 Lycoming IO-540-C4B5, 250 bhp
7/5-6/86

LONDON/ANTWERP (UK)
E. K. Conventry — 329.93 kmh, 205.01 mph
Piper Navajo
2 Lycoming TIO-540, 315 bhp
5/8/85

LONDON/ATHENS (AUSTRALIA)
Trevor K. Brougham — 272.25 kmh, 169.16 mph
Beechcraft Baron B55
2 Continental IO-470-L
8 hrs 57 min 28 sec
8/8/71

LONDON/AUCKLAND (UK)
Judith M. Chisholm — 101.18 kmh, 62.87 mph
Cessna T210 Centurion
1 Continental TSIO-520-H
11/18-26/80

LONDON/BELFAST (UK)
Spencer Flack — 140.71 kmh, 87.44 mph
Piper Seneca PA34-200
2 Lycoming LIO-360-CIE6 & IO-360-CIE6
10/13/83

LONDON/BOMBAY (AUSTRALIA)
Trevor K. Brougham — 277.83 kmh, 172.63 mph
Beechcraft Baron B55
2 Continental IO-470-L
25 hrs 36 min 48 sec
8/8-9/71

LONDON/BONN (UK)
E. K. Conventry — 321.20 kmh, 199.59 mph
Piper Navajo
2 Lycoming TIO-540, 315 bhp
5/8/85

LONDON/BRUGES (UK)
E. K. Conventry — 313.01 kmh, 194.50 mph
Piper Navajo
2 Lycoming TIO-540, 315 bhp
5/8/85

LONDON/COLOGNE (UK)
E. K. Conventry — 328.40 kmh, 204.06 mph
Piper Navajo
2 Lycoming TIO-540, 315 bhp
5/8/85

LONDON/COLOMBO (UK)
Judith M. Chisholm — 201.11 kmh, 124.96 mph
Cessna T210 Centurion
1 Continental TSIO-520-H
11/18-20/80

LONDON/DARWIN (AUSTRALIA)
Trevor K. Brougham — 266.43 kmh, 165.55 mph
Beechcraft Baron B55
2 Continental IO-470-L
52 hrs 18 min 25 sec
8/8-10/71

LONDON/EDINBURGH (UK)
Spencer Flack — 262.04 kmh, 162.83 mph
Piper Seneca PA34-200
2 Lycoming LIO-360-CIE6 &
IO-360-CIE6, 200 bhp
10/13/83

LONDON/GENEVA (USA)
Stephen A. Oster — 294.33 kmh, 182.90 mph
Piper Aerostar, 601P
2 Lycoming IO-540-S1A5, 290 bhp
3/29/84

LONDON/JAKARTA (UK)
Judith M. Chisholm — 173.27 kmh, 107.66 mph
Cessna T210 Centurion
1 Continental TSIO-520-H
11/18-21/80

LONDON/LARNACA (UK)
Judith M. Chisholm 252.57 kmh
Cessna T210 Centurion 156.94 mph
1 Continental TSIO-520-H
11/18-19/80

LONDON/MILAN (AUSTRALIA)
Adam Boyd Munro 403.75 kmh
Piper Navajo 250.87 mph
2 Lycoming, 350 bhp
4/21/79

LONDON/NEW YORK (UK)
Lord D. G. Trefgarne 44.62 kmh
Charles B. G. Masefield 27.73 mph
de Havilland D.H. 90 Dragonfly
2 Gipsy Major 10 MK-1
7/31-8/5/64

LONDON/PARIS (FRANCE)
Richard Fenwick 360.79 kmh
Beech Bonanza, V-35 224.19 mph
1 Continental IO-550-B, 300 bhp
6/26/85

LONDON/PORT HEADLAND (UK)
Judith M. Chisholm 166.03 kmh
Cessna T210 Centurion 103.17 mph
1 Continental TSIO-520-H
1/18-22/80

LONDON/SINGAPORE (AUSTRALIA)
D. N. Dalton, Pilot 353.67 kmh
K. Carmody, Copilot 219.76 mph
Beechcraft Duke 60
2 Lycoming TIO-540-E
4/24-26/75

LONDON/SPITZBERGEN, NORWAY (UK)
W. E. Cattle, J. Behrman, 189.54 kmh
R. J. Pooley 117.77 mph
Beechcraft Travelair D95
2 Lycoming IO-360-B1B, 180 bhp
7/14/77

LONDON/SYDNEY (UK)
Judith M. Chisholm 123.40 kmh
Cessna T210 Centurion 76.68 mph
1 Continental TSIO-520-H
11/18-24/80

LONG BEACH/YOLO, CA (USA)
Dennis Huntington 318.24 kmh
Beech Baron, N12WB 197.75 mph
2 Lycoming TIO-541-E1B4
5/6/89

LONGYEARBYEN/NORTH POLE (SWITZERLAND)
Robert Plimpton 231.70 kmh
Piper PA-30 143.97 mph
8/2/91

LOS ANGELES/NEW YORK (USA)
Ronald Ancherani 449.29 kmh
Cessna P210N 279.13 mph
1 Lycoming TIO-540-VAD, 350 bhp
5/7/85

LOS ANGELES/PHOENIX (USA)
Roger Hyder 548.44 kmh
Beechcraft Baron 56TC 340.78 mph
2 Lycoming TIO-541, 380 bhp
03/12/90

LOS ANGELES/RICHMOND (USA)
Henry Dainys 392.27 kmh
Aerostar 601 243.74 mph
2 Lycoming IO-540
9 hrs 20 min 25.2 sec
8/2/73

LOS ANGELES/SYDNEY (AUSTRALIA)
Peter Wilkins 272.72 kmh
Piper Malibu, N9221M 169.47 mph
1 Continental TSIO-BE, 310 bhp
3/16-18/87

LOUISVILLE/GANDER (USA)
Jeremie J. Kaelin 195.77 kmh
Piper Twin Comanche PA-30 121.65 mph
2 Lycoming TIO-320
14 hrs 5 min 45 sec
5/30/69

LURAY, VA/KITTY HAWK (USA)
Weldon Britton, Pilot 335.88 kmh
Marthasu Britton, Copilot 208.72 mph
Piper Aztec, N223MB
2 Lycoming IO540-C485, 250 bhp
12/16/87

LUXOR/BAHRAIN (UK)
Robert Miller 266.22 kmh
David John Hare 165.42 mph
Piper PA-23 Aztec
2 Lycoming IO-540, 250 bhp
3/17/01

LUXOR/HA'IL (UK)
Robert Miller 274.51 kmh
David John Hare 170.57 mph
Piper PA-23 Aztec
2 Lycoming IO-540, 250 bhp
3/17/01

LYDD/DIJON (UK)
Geoffrey G. Boot 347.36 kmh
Cessna 340 215.83 mph
2 Lycoming TSIO-520, 310 bhp
3/15/95

LYDD/GALWAY (UK)
Geoffrey G. Boot 316.31 kmh
Cessna 340 196.55 mph
6/6/97

LYDD/LA ROCHELLE (UK)
Geoffrey G. Boot 307.35 kmh
Cessna 340 190.97 mph
2 Lycoming TSIO-520, 310 bhp
3/15/95

LYDD/POITIERS (UK)
Geoffrey G. Boot 258.98 kmh
Cessna 340 160.92 mph
2 Lycoming TSIO-520, 310 bhp
5/24/98

MADRAS/BOMBAY (USA)
Danford A. Bookout 159.84 kmh
Piper Lance PA 32R-300 99.32 mph
1 Lycoming IO-540-K, 300 bhp
6/27/86

MADRAS/SINGAPORE (USA)
John F. X. Browne, Pilot 286.92 kmh
Orville Winters, Copilot 178.29 mph
Piper Aztec PA-23
2 Lycoming IO-540-C4B5, 250 bhp
5/31/84

MAGADAN/ANADYR (GUATEMALA)
Frederico Bauer 413.26 kmh
Piper PA-60 256.79 mph
6/28/92

MAJURO/GUADALCANAL (USA)
Danford A. Bookout 305.54 kmh
Piper Lance PA 32R-300 189.92 mph
1 Lycoming IO-540-K, 300 bhp
6/15-16/86

MAJURO/HONOLULU (USA)
Jeremie J. Kaelin, Pilot 295.92 kmh
William Kaelin, Copilot 183.88 mph
Joseph J. Kaelin, Navigator
Beech Baron, N22JJ
2 Continental IO-520
5/18/89

MAJURO/JOHNSON ISLAND (QATAR)
Hamad Al-Thani 269.28 kmh
Piper Seneca II 167.33 mph
2 Continental TSIO-360-E, 200 bhp
1/21-22/86

MANILA/AGANA (QATAR)
Hamad Al-Thani 264.24 kmh
Piper Seneca II 164.19 mph
2 Continental TSIO-360-E, 200 bhp
1/18-19/86

MARAMBIO/SOUTH POLE (USA)
Richard Norton, Pilot 315.41 kmh
Calin Rosetti, Pilot 196.00 mph
Piper Malibu, N26033
1 Continental TSIO 520BE, 310 bhp
1/31-2/1/87

MARIANNA/TEXARKANA (USA)
Danford A. Bookout 259.53 kmh
Piper Lance PA 32R-300 161.27 mph
1 Lycoming IO-540-K, 300 bhp
6/20/85

MARRAKECH/ISTANBUL (USA)
Arthur T. Mott, Pilot 342.89 kmh
Thomas E. Hatch, Copilot 213.06 mph
Beechcraft Baron 58
2 Continental IO-550, 300 bhp
5/5/94

MCCOOK/NEW YORK (USA)
Larry Grant, Pilot 414.00 kmh
Susie Grant, Copilot 257.26 mph
Piper Malibu, N4319M
1 Continental TSIO-520-BE, 310 bhp
8/15/86

MCCOOK/OSHKOSH (USA)
Brian Munson, Pilot 378.72 kmh
Susie Grant, Copilot 235.34 mph
Kevin Munson, Navigator
Piper Malibu N4319M
1 Continental TSIO-520-BE, 310 bhp
8/1/87

MEMPHIS/BRISTOL, TN*
James S. Jenkins 221.92 mph
Piper Aztec
2 Lycoming IO-540, 250 bhp
12/17/81

MEMPHIS/NASHVILLE, TN (USA)
James S. Jenkins 331.89 kmh
Piper Aztec 206.27 mph
2 Lycoming IO-540, 250 bhp
12/17/81

MEMPHIS/NEW ORLEANS (USA)
James S. Jenkins 381.24 kmh
Piper Aztec PA-23-250 236.90 mph
2 Lycoming IO-540, 250 bhp
12/17-18/83

MIAMI/FARMINGDALE (USA)
Arthur T. Mott 343.53 kmh
Robert Mott 213.46 mph
Tim Bastick
Beechcraft Baron BE 58
2 Continental IO-550, 300 bhp
5/8/95

MIAMI/NEW YORK (USA)
Stephen A. Oster 365.32 kmh
Aerostar 601P 226.99 mph
2 Lycoming, 290 bhp
4 hrs 52 min
11/29/81

MIAMI/SAN JUAN (USA)
John F. X. Browne 278.64 kmh
Piper Aztec 173.15 mph
2 Lycoming IO-540C455, 250 bhp
3/2/85

MIAMI/TEXARKANA (USA)
Danford A. Bookout 242.86 kmh
Piper Lance PA 32R-300 150.91 mph
1 Lycoming IO-540-K, 300 bhp
6/20/85

MOBILE/KITTY HAWK (USA)
Robert B. O'Daniel 319.81 kmh
Beechcraft Baron B55 198.72 mph
2 Continental IO-470, 260 bhp
12/16/93

MOBILE/TEXARKANA (USA)
Danford A. Bookout 369.72 kmh
Piper Lance PA 32R-300 229.74 mph
1 Lycoming IO-540-K, 300 bhp
6/20/85

MONTREAL/GOOSE BAY (NORWAY)
Poju Stephansen, Pilot 416.00 kmh
Jan Roang, Copilot 258.49 mph
Cessna 210N
5/26/94

MONTREAL/OLSO (NORWAY)
Poju Stephansen, Pilot 316.65 kmh
Jan Roang, Copilot 196.75 mph
Cessna 210N
5/26/94

MONTREAL/REYKJAVIK (NORWAY)
Poju Stephansen, Pilot 331.46 kmh
Jan Roang, Copilot 205.95 mph
Cessna 210N
5/26/94

MOSCOW/IRKUTSK (USA)
Susan Nealey 343.39 kmh
Faith Hillman 213.37 mph
Cessna 310
2 Continental IO-520, 285 bhp
6/23/92

MOSCOW/NOME (USA)
Susan Nealey 51.17 kmh
Faith Hillman 31.79 mph
Cessna 310
2 Continental IO-520, 285 bhp
6/23-28/92

MOSCOW/OMSK (GUATEMALA)
Frederico Bauer 471.91 kmh
Piper PA-60 293.23 mph
6/23/92

MOSCOW/OMSK*
Susan Nealey 236.34 mph
Faith Hillman
Cessna 310
2 Continental IO-520, 285 bhp
6/23/92

MOSCOW/PORTLAND*
Charles C. Mack 126.42 mph
Beechcraft Bonanza A36
1 Continental IO-550, 300 bhp
7/4/93

NAPLES/LAKELAND*
Dean S. Edmonds, Jr. 231.11 mph
Beechcraft Baron
2 Continental TSIO-520, 320 bhp
4/19/98

NARSSARSSUAQ/BLACKBUSHE (GERMANY)
Dietrich Schmitt 64.08 kmh
Beechcraft Baron 58 39.81 mph
2 Continental IO-550
2/2/94

NARSSARSSUAQ/GLASGOW (GERMANY)
Dietrich Schmitt 99.24 kmh
Beechcraft Baron 58 61.66 mph
2 Continental IO-550
2/2/94

NARSSARSSUAQ/REYKJAVIK (GERMANY)
Dietrich Schmitt 352.56 kmh
Beechcraft Baron 58 219.07 mph
2 Continental IO-550
2/2/94

NASHVILLE/BRISTOL, TN*
James S. Jenkins 235.45 mph
Piper Aztec
2 Lycoming IO-540, 250 bhp
12/17/81

NASHVILLE/LONDON (USA)
Russell H. Hancock 238.09 kmh
Piper Navajo Panther 147.94 mph
2 Lycoming TIO-540-J2BD, 350 bhp
28 hrs 12 min 36 sec
9/16-17/78

NASHVILLE/PLAINS (USA)
James S. Jenkins, Pilot 270.72 kmh
Bob Archer, Copilot 168.23 mph
Piper Turbo Seneca, N7754C
2 Continental TSIO-260E, 200 bhp
12/17/86

NASHVILLE/SAN DIEGO (USA)
Russell H. Hancock 315.93 kmh
Piper Navajo Panther 198.79 mph
2 Lycoming TIO-540 J2BD, 350 bhp
8 hrs 52 min 12 sec
9/22-23/78

NEW ORLEANS/LAKELAND (USA)
Dean S. Edmonds, Jr. 351.55 kmh
Beechcraft Baron 58TC 218.44 mph
2 Continental TSIO-520, 310 bhp
4/6/91

NEW ORLEANS/NASHVILLE (USA)
James Fulbright 282.23 kmh
James S. Jenkins 175.37 mph
Piper PA 23-250 Aztec
2 Lycoming IO-540, 250 bhp
2 hrs 41 min 31 sec
12/17/79

NEW YORK/GOOSE BAY (USA)
James F. Nields 310.97 kmh
Beechcraft Baron C55, N8282V 193.23 mph
2 Continental IO-520-C
5 hrs 33 min 56.8 sec
8/28/73

NEW YORK/LONDON (USA)
Louise Sacchi 318.92 kmh
Beech Bonanza A-356, N9017V 198.17 mph
1 Continental IO-520
17 hrs 28 min 0.2 sec
6/28/71

NEW YORK/MUNICH (W. GERMANY)
Dieter Schmitt 392.02 kmh
Beech Bonanza A36TC 243.58 mph
1 Continental TSIO-520-UB, 300 bhp
7/12-13/80

NEW YORK/PARIS (FRANCE)
Patrick Fourticq, Pilot 416.173 kmh
Henri Pescarlo, Copilot 258.61 mph
Piper Malibu, N4319M
1 Continental TSIO 520-BE, 310 bhp
2/19/84

NEW YORK/TULSA (USA)
Robert Z. Bliss 293.11 kmh
Piper Aerostar 601P 182.13 mph
2 Lycoming IO-540, 350 bhp
10/15/01

NICEVILLE/CHATTANOOGA (USA)
Andrew Watson 286.74 kmh
Cessna 310R 178.17 mph
8/4/98

NOME/VICTORIA (USA)
Susan Nealey 318.12 kmh
Faith Hillman 197.67 mph
Cessna 310
2 Continental IO-520, 285 bhp
6/30/92

NORD, NORWAY/POINT BARROW (UK)
Sheila Scott 183.73 kmh
Piper Aztec 114.16 mph
2 Lycoming IO-540-C4B5
16 hrs 22 min 11 sec
6/28-29/71

NORFOLK ISLAND/AUCKLAND (AUSTRALIA)
Claude Meunier 424.96 kmh
Denis Beresford 264.06 mph
Robyn Stewart
Piper 602P Aerostar
5/13/99

NORFOLK ISLAND/BRISBANE (AUSTRALIA)
Claude Meunier 321.18 kmh
Robyn Stewart 199.57 mph
Denis Beresford
Piper PA-60-602P Aerostar
2 Lycoming IO-540, 350 bhp
4/2/02

NORFOLK ISLAND/NOUMÉA (AUSTRALIA)
Claude Meunier 368.41 kmh
Denis Beresford 228.92 mph
Robyn Stewart
Piper 602P Aerostar
5/17/99

NORTH POLE/LONGYEARBYEN (USA)
Richard Norton, Pilot 353.52 kmh
Calin Rosetti, Copilot 219.68 mph
Piper Malibu, N26033
1 Continental TSIO-520-B, 310 bhp
6/11-12/87

NOUMEA/BRISBANE (AUSTRALIA)
Claude Meunier 385.12 kmh
Paul Stewart 239.30 mph
Piper PA-60-602P
2 Lycoming IO-540, 350 bhp
8/21/95

OAKLAND/ATLANTA (USA)
Charles D. Kissner, Pilot 362.36 kmh
David M. Alden, Copilot 225.16 mph
Piper PA46-310P, N4319M
1 Continental TSIO-520, 310 bhp
4/16/90

OAKLAND/HONOLULU (USA)
Donald T. Holmes, Pilot
Steward Reid, Copilot
Piper Malibu PA46-310P,
1 Continental TSIO-520BE, 310 bhp
12/5-7/84
322.20 kmh
200.21 mph

OGDEN/OAKLAND (USA)
Danford A. Bookout
Piper Lance PA 32R-300
1 Lycoming IO-540-K, 300 bhp
6/2/85
218.64 kmh
135.85 mph

OGDEN/RENO (USA)
Danford A. Bookout
Piper Lance PA 32R-300
1 Lycoming IO-540-K, 300 bhp
6/2/85
238.68 kmh
148.31 mph

OGDEN/SACRAMENTO (USA)
Dennis Huntington
Beech Baron, N12WB
2 Lycoming TIO-541-E1B4
4/10/89
311.04 kmh
193.28 mph

OKINAWA/SENDAI (USA)
Arthur T. Mott, Pilot
Thomas E. Hatch, Copilot
Beechcraft Baron 58
2 Continental IO-550, 300 bhp
5/17/94
370.01 kmh
229.91 mph

OLYMPIA/RENO (USA)
Bruce J. Mayes
North American AT-6D
1 Pratt & Whitney R-1340, 600 bhp
9/6/03
212.13 kmh
131.81 mph

OMAHA/CINCINNATI (UK)
Douglas R. Cairns
Beech Baron 58
2 Continental IO-520, 285 bhp
9/19/03
369.94 kmh
229.87 mph

OMAHA/MINNEAPOLIS (UK)
Douglas R. Cairns
Beech Baron 58
2 Continental IO-520, 285 bhp
9/15/03
351.93 kmh
218.68 mph

OMSK/IRKUTSK (GUATEMALA)
Frederico Bauer
Piper PA-60
6/23/92
425.03 kmh
264.10 mph

OMSK/IRKUTSK*
Susan Nealey
Faith Hillman
Cessna 310
2 Continental IO-520, 285 bhp
6/23/92
216.67 mph

ORLANDO/BOSTON (USA)
Dean S. Edmonds
Beech Baron
2 Continental TSIO-520 L, 350 bhp
5/20/83
449.50 kmh
297.31 mph

ORLANDO/DALLAS (USA)
Albert H. Wuerz, Jr.
Piper Malibu PA-46-310P
1 TSIO-520-BE, 310 bhp
10/2/83
370.00 kmh
229.92 mph

Douglas Cairns and his record-setting Beech Baron.

OSHKOSH/BANGOR (USA)

Allyn Caruso, Pilot 366.26 kmh
Theodore Littlefield, Crewmember 227.58 mph
John Miller, Crewmember
Cessna 340
2 Continental TSIO-520, 310 bhp
08/01/90

OSHKOSH/TAOS, NM (USA)
Raymond H. Lutz 254.89 kmh
Cessna P210N 158.38 mph
1 TSIO-520, 310 bhp
7/31/91

OSLO/PARIS (USA)
Richard Norton, Pilot 269.64 kmh
Calin Rosetti, Copilot 167.55 mph
Piper Malibu, N26033
1 Continental TSIO-520-B, 310 bhp
6/15/87

PAGO PAGO/TAHITI (USA)
John F. X. Browne 265.86 kmh
Piper Aztec PA 23 165.21 mph
2 Lycoming IO-540-C4B5, 250 bhp
6/27-28/86

PALMA DE MAJORCA/SANTA MARIA (USA)
Danford A. Bookout 253.08 kmh
Piper Lance PA 32R-300 157.26 mph
1 Lycoming IO-540-K, 300 bhp
7/2/86

PALM SPRINGS/CHICAGO (USA)
Leslie R. Travioli 327.37 kmh
Cessna T210 203.42 mph
1 Continental TSIO-520, 310 bhp
7/25/92

PALM SPRINGS/PHOENIX (USA)
David Morss 503.51 kmh
Starkraft SK-700 312.87 mph
2 Continental LTSIO-550, 350 bhp
3/1/96

PARIS/AJACCIO (FRANCE)
Jean-Pierre Burne 422.43 kmh
Cessna 402C 262.50 mph
2 Continental TSIO-520-VB, 330 bhp
4/28/84

PARIS/CINCINNATI (USA)
Gerald P. Dietrick 250.20 kmh
Windecker Eagle I 155.47 mph
1 Continental IO-520C, 285 bhp
6/14-15/81

PARIS/DAKAR (USA)
Richard Norton, Pilot 319.48 kmh
Calin Rosetti, Pilot 198.53 mph
Piper Malibu, N26033
1 Continental TSIO 520BE, 310 bhp
1/21-22/87

PARIS/NEW YORK (USA)
Charles C. Mack 219.60 kmh
Bonanza A36, N18279 136.46 mph
1 Continental IO-520, 285 bhp
8/3-4/88

PARIS/SANTA MARIA (USA)
Richard Norton, Pilot 313.20 kmh
Calin Rosetti, Copilot 194.62 mph
Piper Malibu, N26033
1 Continental TSIO-520-B, 310 bhp
6/22/87

PERTH/ADELAIDE (AUSTRALIA)
Claude Meunier 506.76 kmh
Paul Stewart 314.88 mph
Piper PA-60-602P
2 Lycoming IO-540, 350 bhp
7/19/95

PERTH/BRISBANE (AUSTRALIA)
Claude Meunier 497.61 kmh
Paul Stewart 309.20 mph
Piper PA-60-602P
2 Lycoming IO-540, 350 bhp
7/28/95

PERTH/FORREST (AUSTRALIA)
Claude Meunier 584.37 kmh
Denis Beresford 363.11 mph
Piper PA-60-602P Aerostar
2 Lycoming IO-540, 350 hp each
6/26/03

PERTH/MELBOURNE (AUSTRALIA)
Claude Meunier 418.50 kmh
Denis Beresford 260.04 mph
Piper PA-60-602P
2 Lycoming IO-540, 350 bhp
7/13/95

PERTH/SYDNEY (AUSTRALIA)
Claude Meunier 456.53 kmh
Paul Stewart 283.67 mph
Piper PA-60-602P
2 Lycoming IO-540, 350 bhp
8/13/95

PETROPAVLOVSK/ANCHORAGE (SWEDEN)
Erik O. Banck 324.36 kmh
Merĉe M. Inglada 201.55 mph
Cessna P210N
5/21/94

PETROPAVLOVSK/ANCHORAGE*
Arthur T. Mott, Pilot 205.79 mph
Thomas E. Hatch, Copilot
Beechcraft Baron 58
2 Continental IO-550, 300 bhp
5/21/94

PHILADELPHIA/FT. LAUDERDALE (USA)
Stephen A. Oster 383.73 kmh
Aerostar 601P 238.44 mph
2 Lycoming, 290 bhp
4 hrs 14 min 58 sec
11/24-25/81

PHOENIX/BURBANK (USA)
Russell T. MacFarlane, Pilot 317.41 kmh
Harold A. Katinszky, Copilot 197.23 mph
Cessna-Riley Rocket 310
2 Lycoming TIO-540, 290 bhp
3/7/91

PHOENIX/EL PASO (USA)
David Morss 571.68 kmh
Starkraft SK-700 355.22 mph
2 Continental LTSIO-550, 350 bhp
3/1/96

PHOENIX/OSHKOSH (USA)
John R. Dewane, Pilot 319.08 kmh
William E. Unternaehrer, Copilot 198.26 mph
Beechcraft Baron B55
2 Continental IO-470, 260 bhp
7/30/93

POINT BARROW/NEW YORK (USA)
Francis X. Sommer 244.32 kmh
Beechcraft S-35, N990MD 139.39 mph
1 Continental IO-520
24 hrs 38 min 50 sec
6/23/67

POITIERS/LYDD (UK)
Geoffrey G. Boot 292.65 kmh
Cessna 340 181.84 mph
2 Lycoming TSIO-520, 310 bhp
5/26/98

PONAPE/MAJURO (QATAR)
Hamad Al-Thani 258.12 kmh
Piper Seneca II 160.40 mph
2 Continental TSIO-360-E, 200 bhp
1/20/86

PORT HEADLAND/AUCKLAND (UK)
Judith M. Chisholm 101.47 kmh
Cessna T210 Centurion 63.05 mph
1 Continental TSIO-520-H
11/23-26/80

PORT HEADLAND/HONOLULU (UK)
Judith M. Chisholm 72.79 kmh
Cessna T210 Centurion 45.23 mph
1 Continental TSIO-520-H
11/23-29/80

PORT HEADLAND/SAN FRANCISCO (UK)
Judith M. Chisholm 76.20 kmh
Cessna T210 Centurion 47.35 mph
1 Continental TSIO-520-H
11/23-12/1/80

PORT HEADLAND/SYDNEY (UK)
Judith M. Chisholm 282.06 kmh
Cessna T210 Centurion 175.26 mph
1 Continental TSIO-520-H
11/23-24/80

PUEBLO/DAYTON (USA)
John L. Cink 487.80 kmh
Aerostar 601P 303.11 mph
2 Lycoming IO-540, 290 bhp
1/16/85

RALEIGH/ROCKLAND (USA)
Robert O. Ames 378.35 kmh
Piper Aerostar 602P 235.10 mph
2 Lycoming TIO-540, 350 bhp
5/21/92

RENO/FORT LAUDERDALE (USA)
William Salsburg, Pilot 323.55 kmh
Paul Salsburg, Copilot 201.04 mph
Harry Guberman, Crewmember
Anthony Biancarosa, Crewmember
Piper Malibu
1 Continental TSIO-520, 310 bhp
9/16/91

RENO/LOUISVILLE (USA)
William Kaelin, Pilot 332.28 kmh
Jeremie J. Kaelin, Copilot 206.48 mph
Beech Baron, N22JJ
2 Continental IO-520
5/22/89

RESOLUTE/NORTH POLE (USA)
John F. X. Browne 279.69 kmh
Piper Aztec 173.79 mph
1 Lycoming IO-540-C4B5, 250 bhp
5/27/85

RESOLUTE/SPITZBERGEN (USA)
John F. X. Browne 275.76 kmh
Piper Aztec 171.36 mph
1 Lycoming IO-540-C4B5, 250 bhp
5/27/85

REYKJAVIK/AUGSBURG (GERMANY)
Dietrich Schmitt 328.95 kmh
Beechcraft Bonanza F33A 204.40 mph
11/14/93

REYKJAVIK/BLACKBUSHE (GERMANY)
Dietrich Schmitt 71.57 kmh
Beechcraft Baron 58 44.47 mph
2 Continental IO-550
2/3/94

REYKJAVIK/GLASGOW (GERMANY)
Dietrich Schmitt 248.72 kmh
Beechcraft Baron 58 154.54 mph
2 Continental IO-550
2/3/94

REYKJAVIK/GOTEBORG (USA)
William A. DuFour 342.61 kmh
Beechcraft Baron 58P 212.88 mph
2 Continental IO-520, 300 bhp
6/8/96

REYKJAVIK/KUUJJUAQ (USA)
Susan Nealey 312.05 kmh
Faith Hillman 193.90 mph
Cessna 310
2 Continental IO-520, 285 bhp
7/20/92

REYKJAVIK/NEW YORK (UK)
Lord D. G. Trefgarne 54.426 kmh
C. B. G. Masefield 33.848 mph
de Havilland D.H. 90 Dragonfly
2 Gipsy Major 10 MK-1
8/2-5/65

REYKJAVIK/OSHKOSH (USA)
Susan Nealey 157.93 kmh
Faith Hillman 98.13 mph
Cessna 310
2 Continental IO-520, 285 bhp
7/20-22/92

REYKJAVIK/OSLO (NORWAY)
Poju Stephansen, Pilot 347.15 kmh
Jan Roang, Copilot 215.70 mph
Cessna 210N
5/27/94

REYKJAVIK/SHANNON (USA)
Susan Nealey 274.82 kmh
Faith Hillman 170.76 mph
Cessna 310
2 Continental IO-520, 285 bhp
6/16/92

REYKJAVIK/VAGAR*
Stanley H. Mick 187.88 mph
Cessna T303 Crusader
2 Continental TSIO-520, 250 bhp
5/21/94

RIO GRANDE/MARAMBIO (USA)
Richard Norton, Pilot 337.27 kmh
Calin Rosetti, Pilot 209.58 mph
Piper Malibu, N26033
1 Continental TSIO 520BE, 310 bhp
1/28/87

ROADTOWN/CHRISTIANSTED (U.S. VIRGIN ISLANDS)
David Crisp 332.10 kmh
Piper PA-31-350 206.36 mph
2 Continental TSIO-540, 350 bhp
4/11/97

SACREMENTO/SALT LAKE CITY (USA)
Dennis Huntington 397.08 kmh
Beech Baron, N12WB 246.75 mph
2 Lycoming TIO-541-E1B4
3/31-4/1/89

ST. JOHN'S/DUBLIN (USA)
Robert M. Byrom, Pilot — 288.95 kmh
Robert L. Gray, Copilot — 179.54 mph
Cessna 337G Skymaster
2 Continental IO-360, 210 bhp
7/13/93

ST. JOHN'S/SANTA MARIA (USA)
Arthur T. Mott, Pilot — 385.88 kmh
Thomas E. Hatch, Copilot — 239.77 mph
Beechcraft Baron 58
2 Continental IO-550, 300 bhp
5/2/94

ST. JOHN'S/ZURICH (USA)
Jeremie J. Kaelin — 184.47 kmh
Piper Twin Comanche PA-30 — 114.62 mph
2 Lycoming TIO-320
24 hrs 17 min 14 sec
6/1-2/69

ST. LOUIS/PARIS*
Stanley H. Mick — 30.45 mph
Cessna T303 Crusader
2 Continental TSIO-520, 250 bhp
5/16-22/94

SAN DIEGO/BANGOR (USA)
Stuart C. Goldberg — 379.04 kmh
Beechcraft Bonanza V35B — 235.52 mph
1 Continental IO-550, 300 bhp
10/1/99

SAN DIEGO/EL PASO (USA)
Danford A. Bookout — 261.60 kmh
Piper Lance PA 32R-300 — 162.56 mph
1 Lycoming IO-540-K, 300 bhp
6/17-18/85

SAN DIEGO/SYDNEY, NS (USA)
Stuart C. Goldberg — 357.95 kmh
Beechcraft Bonanza V35B — 222.42 mph
1 Continental IO-550, 300 bhp
4/26/01

SAN DIEGO/TUCSON (USA)
Danford A. Bookout — 189.38 kmh
Piper Lance PA 32R-300 — 117.68 mph
1 Lycoming IO-540-K, 300 bhp
6/17/85

SAN FRANCISCO/ATHENS (AUSTRALIA)
Trevor K. Brougham — 282.21 kmh
Beechcraft Baron B55 — 175.54 mph
Continental IO-470-L
8/6-8/71

SAN FRANCISCO/CHICAGO (USA)
Amos E. Buettell — 417.91 kmh
Cessna 310N, N721X — 259.68 mph
2 Lycoming TIO-540
7 hrs 8 min 47.4 sec
9/11/75

SAN FRANCISCO/FARMINGDALE (USA)
Arthur T. Mott, Pilot — 387.95 kmh
Tim Bastick, Copilot — 241.06 mph
Beechcraft Baron BE 58
2 Continental IO-550, 300 bhp
10/31/95

SAN FRANCISCO/GANDER (AUSTRALIA)
Trevor K. Brougham — 295.07 kmh
Beechcraft Baron B55 — 183.34 mph
2 Continental IO-470-L
18 hrs 31 min 58 sec
8/6-7/71

SAN FRANCISCO/GOOSE BAY (UK)
Judith M. Chisholm — 148.49 kmh
Cessna T210 Centurion — 92.33 mph
1 Continental TSIO-520-H
12/1-2/80

SAN FRANCISCO/HONOLULU (UK)
Sheila Scott — 236.56 kmh
Piper Aztec D — 146.99 mph
2 Lycoming IO-540
15 hrs 2 min 12 sec
7/11-12/71

SAN FRANCISCO/LONDON (AUSTRALIA)
Trevor K. Brougham — 293.94 kmh
Beechcraft Baron B55 — 182.64 mph
2 Continental IO-470-L
8/6-8/71

SAN FRANCISCO/LOS ANGELES (USA)
Dennis Huntington — 554.69 kmh
Beech Baron 56TC, N12WB — 344.68 mph
2 Lycoming TSIO-541-E1, 380 bhp
3/23/87

SAN FRANCISCO/MCCOOK (USA)
Brian Munson, Pilot — 394.56 kmh
Kevin Munson, Copilot — 245.18 mph
Susie Grant, Navigator
Piper Malibu, N4319M
1 Continental TSIO-520-BE, 310 bhp
8/1/87

SAN FRANCISCO/NEW YORK (USA)
Larry Grant, Pilot — 389.88 kmh
Susie Grant, Copilot — 242.27 mph
Piper Malibu, N4319M
1 Continental TSIO-520-BE, 310 bhp
8/15/86

SAN FRANCISCO/OSHKOSH (USA)
Brian Munson, Pilot — 372.96 kmh
Susie Grant, Copilot — 231.76 mph
Kevin Munson, Navigator
Piper Malibu , N4319M
1 Continental TSIO-520-BE, 310 bhp
8/1/87

SAN FRANCISCO/PARIS (FRANCE)
Patrick Fourticq, Pilot — 284.741 kmh
Henri Pescarlo, Copilot — 176.94 mph
Piper Malibu, N4319M
1 Continental TSIO 520-BE, 310 bhp
2/18-19/84

SAN FRANCISCO/SEATTLE (USA)
Jack Ellis — 390.24 kmh
Piper Malibu, N4319M — 242.50 mph
1 Continental TSIO-520BE, 310 bhp
6/13/87

SAN FRANCISCO/TORONTO (AUSTRALIA)
Trevor K. Brougham — 300.78 kmh
Beechcraft Baron B55 — 186.89 mph
2 Continental IO-470-L
12 hrs 11 min 51 sec
8/6-7/71

SAN FRANCISCO/WINNIPEG (UK)
Judith M. Chisholm — 195.35 kmh
Cessna T210 Centurion — 121.38 mph
1 Continental TSIO-520-H
12/1-2/80

SAN JOSE/EL PASO (QATAR)
Hamad Al-Thani — 291.24 kmh
Piper Seneca II — 180.98 mph
2 Continental TSIO-360-E, 200 bhp
1/28/86

SAN JOSE/HONOLULU (USA)
Danford A. Bookout — 257.76 kmh
Piper Lance PA 32R-300 — 160.17 mph
1 Lycoming IO-540-K, 300 bhp
6/12-13/86

SAN JOSE/NEW YORK (USA)
Charles D. Kissner, Pilot — 376.85 kmh
Paul Bertorelli, Copilot — 234.16 mph
Piper PA46-310P, N4319M
1 Continental TSIO-520, 310 bhp
5/4/92

SAN JOSE/OSHKOSH (USA)
Robert W. Hill — 384.16 kmh
Piper Aerostar 601P — 238.71 mph
2 Lycoming IO-540, 290 bhp
7/28/93

SAN JOSE/WASHINGTON (USA)
Kent Munson — 336.6 kmh
Piper Malibu, N4319M — 209.16 mph
Continental TSIO-520-BE
4/27/86

SAN JUAN/CARACAS (USA)
John F. X. Browne — 277.20 kmh
Piper Aztec — 172.25 mph
2 Lycoming IO-540C455, 250 bhp
3/4/85

SAN JUAN/CHRISTIANSTED (U.S. VIRGIN ISLANDS)
David Crisp — 312.00 kmh
Piper PA-31-350 — 193.87 mph
2 Continental TSIO-540, 350 bhp
2/20/97

SAN JUAN/COLUMBUS (USA)
Geraldine L. Mock — 177.55 kmh
Cessna P-206,N155JM — 110.33 mph
1 Continental IO-520
16 hrs 16 min 30 sec
6/24-25/68

SAN JUAN/MIAMI (USA)
John F. X. Browne — 284.40 kmh
Piper Aztec — 176.73 mph
2 Lycoming IO-540C455, 250 bhp
3/5/85

SANTA ANA/WASHINGTON (USA)
Charles D. Kissner, Pilot — 417.64 kmh
Steven A. Silver, Copilot — 259.51 mph
Piper PA46-310P, N4319M
1 Continental TSIO-520, 310 bhp
1/27/91

SANTA BARBARA/HONOLULU (USA)
John F. X. Browne — 290.47 kmh
Piper Aztec PA 23 — 180.50 mph
2 Lycoming IO-540-C4B5, 250 bhp
6/15-16/86

SANTA BARBARA/PUEBLO (USA)
John L. Cink — 359.64 kmh
Aerostar 601P — 223.48 mph
2 Lycoming IO-540, 290 bhp
1/16/85

SANTA MARIA/FARO (QATAR)
Hamad Al-Thani — 312.48 kmh
Piper Seneca II — 194.17 mph
2 Continental TSIO-360-E, 200 bhp
12/24/85

SANTA MARIA/GANDER (USA)
Danford A. Bookout — 205.56 kmh
Piper Lance PA 32R-300 — 127.73 mph
1 Lycoming IO-540-K, 300 bhp
7/3/86

SANTA MARIA/PHILADELPHIA (USA)
Richard Norton, Pilot — 264.24 kmh
Calin Rosetti, Copilot — 164.20 mph
Piper Malibu, N26033
1 Continental TSIO-520-B, 310 bhp
6/23/87

SANTA MARIA/ZURICH (USA)
Jeremie J. Kaelin — 224.56 kmh
Piper Twin Comanche PA-30 — 139.54 mph
2 Lycoming TIO-320
13 hrs 8 min 47 sec
6/1-2/69

SANTIAGO/EASTER ISLAND (USA)
Richard Norton, Pilot — 266.25 kmh
Calin Rosetti, Pilot — 165.45 mph
Piper Malibu, N26033
1 Continental TSIO 520BE, 310 bhp
2/6-7/87

SEATTLE/NEW YORK (USA)
J. Stephen Stout — 417.24 kmh
Malibu PA-46, N9114B — 259.27 mph
1 Continental TSIO-520, 310 bhp
11/23/87

SENDAI/PETROPAVLOVSK-KAMCHATSKIY (USA)
Wilfred T. Tashima, Pilot — 242.25 kmh
Herbert H. Halperin, Copilot — 150.53 mph
Beechcraft Bonanza A36
1 Continental IO-550, 300 bhp
5/19/94

SHANNON/CANNES (USA)
Susan Nealey — 384.69 kmh
Faith Hillman — 239.03 mph
Cessna 310
2 Continental IO-520, 285 bhp
7/12/92

SHEYMA/NIIGATA (USA)
Jeremie J. Kaelin, Pilot — 262.80 kmh
David Kaelin, Copilot — 163.31 mph
Beech Baron, N22JJ
2 Continental IO-520
5/9-10/89

SINGAPORE/DARWIN (AUSTRALIA)
Trevor K. Brougham — 266.46 kmh
Beechcraft Baron B55 — 165.56 mph
2 Continental IO-470-L
12 hrs 38 min 29 sec
8/9-10/71

SINGAPORE/MANILA (QATAR)
Hamad Al-Thani — 239.04 kmh
Piper Seneca II — 148.54 mph
2 Continental TSIO-360-E, 200 bhp
1/16/86

SINGAPORE/PERTH (USA)
Amos E. Buettell — 342.48 kmh
Cessna 310N, N721X — 212.81 mph
2 Lycoming TIO-540
11 hrs 25 min 53.9 sec
8/10/75

SOUTH BEND/AUGSBURG (GERMANY)
Dietrich Schmitt — 41.40 kmh
Beechcraft Bonanza F33A — 25.72 mph
11/7-14/93

SOUTH BEND/BANGOR (GERMANY)
Dietrich Schmitt — 396.41 kmh
Beechcraft Baron 58 — 246.31 mph
2 Continental IO-550
2/2/94

SOUTH BEND/BLACKBUSHE (GERMANY)
Dietrich Schmitt 105.93 kmh
Beechcraft Baron 58 65.82 mph
2 Continental IO-550
2/2/94

SOUTH BEND/GLASGOW (GERMANY)
Dietrich Schmitt 153.00 kmh
Beechcraft Baron 58 95.06 mph
2 Continental IO-550
2/2/94

SOUTH BEND/GOOSE BAY (GERMANY)
Dietrich Schmitt 284.31 kmh
Beechcraft Baron 58 176.66 mph
2 Continental IO-550
2/2/94

SOUTH BEND/NARSSARSSUAQ (GERMANY)
Dietrich Schmitt 264.19 kmh
Beechcraft Baron 58 164.16 mph
2 Continental IO-550
2/2/94

SOUTH BEND/REYKJAVIK (GERMANY)
Dietrich Schmitt 271.43 kmh
Beechcraft Baron 58 168.65 mph
2 Continental IO-550
2/1/94

SOUTH POLE/MARAMBIO (USA)
Richard Norton, Pilot 310.86 kmh
Calin Rosetti, Pilot 193.16 mph
Piper Malibu, N26033
1 Continental TSIO 520BE, 310 bhp
2/1/87

SOUTH POLE/NORTH POLE (USA)
Richard Norton, Pilot 8.64 kmh
Calin Rosetti, Copilot 5.37 mph
Piper Malibu, N26033
1 Continental TSIO-520-B, 310 bhp
2/1-6/11/87

SYDNEY/AUCKLAND (AUSTRALIA)
Claude Meunier 395.16 kmh
Paul Stewart 245.54 mph
Piper PA-60-602P
2 Lycoming IO-540, 350 bhp
8/14/95

SYDNEY/FIJI (USA)
John F. X. Browne 292.32 kmh
Piper Aztec PA 23 181.65 mph
2 Lycoming IO-540-C4B5, 250 bhp
6/25/86

SYDNEY/HONOLULU (UK)
Judith M. Chisholm 90.35 kmh
Cessna T210 Centurion 56.14 mph
1 Continental TSIO-520-H
11/25-29/80

SYDNEY/LOS ANGELES (AUSTRALIA)
Peter Wilkins 309.17 kmh
Piper Malibu, N9221M 192.10 mph
1 Continental TSIO-BE, 310 bhp
3/30-4/1/87

SYDNEY/SAN FRANCISCO (UK)
Judith M. Chisholm 89.91 kmh
Cessna T210 Centurion 55.87 mph
1 Continental TSIO-520-H
11/25-12/1/80

SYDNEY/PERTH (AUSTRALIA)
Peter Wilkins 247.90 kmh
Piper Malibu 154.05 mph
1 Continental TSIO 520-BE, 310 bhp
5/23-24/86

SYDNEY, NS/SAN DIEGO (USA)
Stuart C. Goldberg 308.29 kmh
Beechcraft Bonanza V35B 191.56 mph
1 Continental IO-550, 300 bhp
4/26/01

SYDNEY/WIGRAM (NEW ZEALAND)
C. V. Tait 196.47 kmh
FU 24-950 122.00 mph
1 Lycoming IO-720, 400 bhp
9/9-10/78

TAHITI/EASTER ISLAND (USA)
John F. X. Browne 159.12 kmh
Piper Lance PA 23 98.88 mph
2 Lycoming IO-540-C4B5, 250 bhp
6/30-7/1/86

TAHITI/HONOLULU (USA)
Richard Norton, Pilot 321.18 kmh
Calin Rosetti, Pilot 199.58 mph
Piper Malibu, N26033
1 Continental TSIO 520BE, 310 bhp
2/8-9/87

TALLAHASSE/TEXARKANA (USA)
Danford A. Bookout 265.68 kmh
Piper Lance PA 32R-300 165.09 mph
1 Lycoming IO-540-K, 300 bhp
6/20/85

TAMPA/GANDER (W. GERMANY)
Herman Krug 395.15 kmh
Piper Malibu 245.55 mph
1 Continental TSIO-520-BE, 310 bhp
5/10/84

TAOS, NM/OSHKOSH (USA)
Raymond H. Lutz 287.85 kmh
Cessna P210N 178.86 mph
1 Continental IO-520, 285 bhp
7/25/90

TEXARKANA/ABILENE (USA)
Danford A. Bookout 223.54 kmh
Piper Lance 138.90 mph
1 Lycoming IO-540K, 300 bhp
5/30-31/85

TEXARKANA/AUSTIN (USA)
Danford A. Bookout 304.56 kmh
Piper Lance PA 32R-300 189.25 mph
1 Lycoming IO-540-K, 300 bhp
7/6/86

TEXARKANA/EL PASO (USA)
Danford A. Bookout 223.92 kmh
Piper Lance 139.14 mph
1 Lycoming IO-540K, 300 bhp
5/30-31/85

TEXARKANA/MIDLAND (USA)
Danford A. Bookout 227.28 kmh
Piper Lance 141.22 mph
1 Lycoming IO-540K, 300 bhp
5/30-31/85

TOKYO/POINT BARROW (USA)
Francis X. Sommer 201.86 kmh
Beech Bonanza S35, N990MD 125.43 mph
1 Continental IO-520
26 hrs 34 min
6/19/67

TORONTO/ATHENS (AUSTRALIA)
Trevor K. Brougham 309.71 kmh
Beechcraft Baron B55 192.44 mph
2 Continental IO-470-L
26 hrs 20 min 31 sec
8/7-8/71

TORONTO/BOMBAY (AUSTRALIA)
Trevor K. Brougham — 288.82 kmh / 179.46 mph
Beechcraft Baron B55
2 Continental IO-470-L
43 hrs 19 min 57 sec
8/7-9/71

TORONTO/DARWIN (AUSTRALIA)
Trevor K. Brougham — 222.63 kmh / 138.33 mph
Beechcraft Baron B55
2 Continental IO-470-L
69 hrs 43 min 24 sec
8/7-10/71

TORONTO/LONDON (AUSTRALIA)
Trevor K. Brougham — 338.91 kmh / 210.85 mph
Beechcraft Baron B55
2 Continental IO-470-L
17 hrs 9 min 9 sec
8/7-8/71

TORONTO/SINGAPORE (AUSTRALIA)
Trevor K. Brougham — 268.14 kmh / 167.39 mph
Beechcraft Baron B55
2 Continental IO-470-L
56 hrs 6 min 34 sec
8/7-9/71

TULSA/NASHVILLE (USA)
James S. Jenkins, Pilot — 344.52 kmh / 214.07 mph
Jim Fulbright, Copilot
Piper Aztec
2 Lycoming IO-540, 250 bhp
12/17-18/80

TULSA/NEW YORK (USA)
Robert Z. Bliss — 379.32 kmh / 235.70 mph
Piper Aerostar 601P
2 Lycoming IO-540, 350 bhp
10/17/01

TULSA/WINSTON-SALEM (USA)
Richard S. Wimbish — 468.94 kmh / 291.39 mph
Piper Aerostar 602P
2 Lycoming TIO-540, 350 bhp
10/18/01

VERO BEACH/ATLANTA (USA)
Earl Tindol, Pilot — 440.32 kmh / 273.60 mph
Tom Daugherty, Copilot
Piper Aerostar 700P
2 Lycoming TIO-540, 350 bhp
5/19/91

VICTORIA/FRESNO (USA)
Susan Nealey — 350.14 kmh / 217.56 mph
Faith Hillman
Cessna 310
2 Continental IO-520, 285 bhp
7/1/92

WASHINGTON/BERLIN (FRG)
Dietrich Schmitt — 286.87 kmh / 178.25 mph
Beechcraft F33A
7/7/91

WASHINGTON/GOOSE BAY (USA)
Susan Nealey — 290.71 kmh / 180.64 mph
Faith Hillman
Cessna 310
2 Continental IO-520, 285 bhp
6/15/92

WASHINGTON/HELSINKI (USA)
Charles C. Mack — 317.32 kmh / 197.17 mph
Beechcraft Bonanza A36
1 Continental IO-550, 300 bhp
6/20/92

WASHINGTON/MOSCOW (USA)
Charles C. Mack — 241.91 kmh / 150.31 mph
Beechcraft Bonanza A36
1 Continental IO-550, 300 bhp
6/27/93

WASHINGTON/NANTUCKET*
Dean S. Edmonds, Jr. — 223.56 mph
Beechcraft Baron 58TC
2 Continental TSIO-520, 310 bhp
5/19/91

WICHITA/BANGOR (GERMANY)
Dietrich Schmitt — 338.47 kmh / 210.31 mph
Beechcraft Baron 58
2 Continental IO-550
2/1/94

WICHITA/BLACKBUSHE (GERMANY)
Dietrich Schmitt — 115.50 kmh / 71.76 mph
Beechcraft Baron 58
2 Continental IO-550
2/1/94

WICHITA/GLASGOW (GERMANY)
Dietrich Schmitt — 162.57 kmh / 101.01 mph
Beechcraft Baron 58
2 Continental IO-550
2/1/94

WICHITA/GOOSE BAY (GERMANY)
Dietrich Schmitt — 279.30 kmh / 173.54 mph
Beechcraft Baron 58
2 Continental IO-550
2/1/94

WICHITA/NARSSARSSUAQ (GERMANY)
Dietrich Schmitt — 261.04 kmh / 162.20 mph
Beechcraft Baron 58
2 Continental IO-550
2/1/94

WICHITA/REYKJAVIK (GERMANY)
Dietrich Schmitt — 266.52 kmh / 165.60 mph
Beechcraft Baron 58
2 Continental IO-550
2/1/94

WICHITA/SOUTH BEND (GERMANY)
Dietrich Schmitt — 375.33 kmh / 233.21 mph
Beechcraft Baron 58
2 Continental IO-550
2/1/94

WIGRAM/SYDNEY (NEW ZEALAND)
C. V. Tait — 145.62 kmh / 90.43 mph
Fletcher FU 24-950
1 Lycoming IO-720, 400 bhp
9/9-10/78

WINNIPEG/GOOSE BAY (UK)
Judith M. Chisholm — 317.97 kmh / 197.58 mph
Cessna T210 Centurion
1 Continental TSIO-520-H
2/2/80

WINNIPEG/LONDON (UK)
Judith M. Chisholm — 256.69 kmh / 159.50 mph
Cessna T210 Centurion
1 Continental TSIO-520-H
12/2-3/80

WINSTON SALEM/KITTY HAWK*
Danny Young — 220.57 mph
Janie Young
Beech Baron BE-58
12/17/83

YAKUTSK/ANADYR (USA)
Susan Nealey 317.14 kmh
Faith Hillman 197.06 mph
Cessna 310
2 Continental IO-520, 285 bhp
6/28/92

YAKUTSK/MAGADAN (GUATEMALA)
Frederico Bauer 441.22 kmh
Piper PA-60 274.16 mph
6/28/92

YAKUTSK/NOME (USA)
Susan Nealey 236.97 kmh
Faith Hillman 147.24 mph
Cessna 310
2 Continental IO-520, 285 bhp
6/28/92

YAKUTSK/VICTORIA (USA)
Susan Nealey 118.04 kmh
Faith Hillman 73.35 mph
Cessna 310
2 Continental IO-520, 285 bhp
6/27-30/92

ZURICH/ATHENS (USA)
Jeremie J. Kaelin 267.36 kmh
Piper Twin Comanche PA-30 166.12 mph
2 Lycoming TIO-320
6 hrs 3 min
6/13/69

ROUND TRIP

ANTWERP/BONN/ANTWERP (UK)
E. K. Conventry 316.62 kmh
Piper Navajo 196.74 mph
2 Lycoming TIO-540, 315 bhp
5/8/85

ANTWERP/COLOGNE/ANTWERP (UK)
E. K. Conventry 292.32 kmh
Piper Navajo 181.64 mph
2 Lycoming TIO-540, 315 bhp
5/8/85

BRUGES/BONN/BRUGES (UK)
E. K. Conventry 326.30 kmh
Piper Navajo 202.76 mph
2 Lycoming TIO-540, 315 bhp
5/8/85

BRUGES/COLGNE/BRUGES (UK)
E. K. Conventry 310.20 kmh
Piper Navajo 192.75 mph
2 Lycoming TIO-540, 315 bhp
5/8/85

CLEVELAND/KEY WEST/CLEVELAND (USA)
Jacob I. Rosenbaum 310.93 kmh
Cessna 210 193.2 mph
1 Continental IO-520-L
12 hrs 7 min 37 sec
5/20/77

COLUMBUS/SAN JUAN/COLUMBUS (USA)
Geraldine L. Mock 175.40 kmh
Cessna P-206, N155JM 108.99 mph
1 Continental IO-520
32 hrs 57 min 1.4 sec
6/24-25/68

LONDON/BONN/LONDON (UK)
E. K. Conventry 322.98 kmh
Piper Navajo 200.70 mph
2 Lycoming TIO-540, 315 bhp
5/8/85

LONDON/COLGNE/LONDON (UK)
E. K. Conventry 314.14 kmh
Piper Navajo 195.20 mph
2 Lycoming TIO-540, 315 bhp
5/8/85

LONGYEARBYEN/NORTH POLE/LONGYEARBYEN (SWITZERLAND)
Robert Plimpton 233.00 kmh
Piper PA-30 144.77 mph
8/2/91

MARAMBIO/SOUTH POLE/MARAMBIO (USA)
Richard Norton, Pilot 313.05 kmh
Calin Rosetti, Pilot 194.53 mph
Piper Malibu, N26033
1 Continental TSIO 520BE, 310 bhp
1/31-2/1/87

NORFOLK ISLAND/AUCKLAND/NORFOLK ISLAND (AUSTRALIA)
Claude Meunier 29.17 kmh
Denis Beresford 18.12 mph
Robyn Stewart
Piper 602P Aerostar
5/16/99

SAN DIEGO/BANGOR/SAN DIEGO (USA)
Stuart C. Goldberg 324.44 kmh
Beechcraft Bonanza V35B 201.60 mph
1 Continental IO-550, 300 bhp
10/2/99

SAN DIEGO/SYDNEY, NS/SAN DIEGO (USA)
Stuart C. Goldberg 323.02 kmh
Beechcraft Bonanza V35B 200.71 mph
1 Continental IO-550, 300 bhp
4/26/01

SAN JUAN/CARACAS/SAN JUAN (USA)
John F. X. Browne 234.72 kmh
Piper Aztec 145.85 mph
2 Lycoming IO-540C455, 250 bhp
3/4/85

CLASS C-1.E (3,000-6,000 Kg / 6,614-13,228 lbs)
GROUP I (PISTON ENGINE)

ABERDEEN/PARIS (USA)
Stephen A. Oster 432.72 kmh
Aerostar 601P 268.89 mph
2 Lycoming TIO-540-J2BD, 350 bhp
6/23/85

ABERDEEN/REYKJAVIK (USA)
William S. Kingson 406.80 kmh
Victor Grey 252.79 mph
Aerostar, N44XX
2 Lycoming TSIO-540, 350 bhp
8/19/88

ABU DHABI/DHAKA (USA)
James H. Knuppe, Pilot 392.31 kmh
Steve Picatti 243.78 mph
Marc Mosier
Cessna 421C
2 Continental TSIO-520L, 375 bhp
3/3/87

ADANA/IZMIR (CANADA)
Dawn Bartsch 342.00 kmh
Gordon Bartsch 212.51 mph
Cessna 421B
9/21/97

ADELAIDE/PERTH (AUSTRALIA)
George Gillespie Keirle 434.87 kmh
Cessna Chansillor C414A 270.23 mph
10/1/89

ALLENTOWN/KITTY HAWK (USA)
Peter Knox 343.08 kmh
Judy Dombroski 213.19 mph
North American T-28C
1 Wright R-1820, 1,425 bhp
12/16/89

AMMAN/ROME (USA)
James H. Knuppe, Pilot 271.28 kmh
Steve Picatti 168.57 mph
Marc Mosier
Cessna 421C
2 Continental TSIO-520L, 375 bhp
3/25/87

AMMAN/TRABZON (CANADA)
Dawn Bartsch 288.82 kmh
Gordon Bartsch 179.46 mph
Cessna 421B
9/19/97

AUCKLAND/CHRISTCHURCH (USA)
Jim York 284.02 kmh
Cessna 340A 176.50 mph
2 Continental TSIO-520-N, 310 bhp
10/2/84

BANDAR SERI BEGAWAN/SAN JOSE (QATAR)
Hamad Al-Thani 75.88 kmh
Piper Aerostar 601P 47.14 mph
2 Lycoming IO-540, 290 bhp
4/29/92

BANGOR/GOOSE BAY (USA)
William S. Kingson 438.84 kmh
Victor Grey 272.70 mph
Aerostar, N44XX
2 Lycoming TSIO-540, 350 bhp
8/3/88

BANGOR/LUXOR (USA)
Tim Mellon, Pilot 413.28 kmh
Tom Glista, Copilot 256.81 mph
Piper Aerostar 602P, N602PC
2 Lycoming IO-540
4/27-28/87

BANGOR/MANCHESTER (USA)
Tom Glista, Pilot 469.08 kmh
Tim Mellon, Copilot 291.49 mph
Piper Aerostar 602P, N602PC
2 Lycoming IO-540
4/27-28/87

BEIJING/GUANZHOU (USA)
James H. Knuppe, Pilot 309.33 kmh
Steve Picatti 192.22 mph
Marc Mosier
Cessna 421C
2 Continental TSIO-520L, 375 bhp
3/10-11/87

BRIDGEPORT/KITTY HAWK (USA)
Richard Hegenberger 337.68 kmh
Ryan Hegenberger 209.83 mph
North American T-28A
1 Wright R-1300, 800 bhp
12/16/89

BRISBANE/LONDON (AUSTRALIA)
D. N. Dalton 271.48 kmh
Beechcraft Duke 168.69 mph
61 hrs 12 min 46 sec
10/21-23/73

BRISBANE/NOUMEA (AUSTRALIA)
Claude Meunier 393.96 kmh
Robyn Stewart 244.79 mph
Denis Beresford
Piper PA-60-602P Aerostar
2 Lycoming IO-540, 350 bhp
6/7/00

BUDAPEST/STOCKHOLM (USA)
William S. Kingson 351.36 kmh
Victor Grey 218.34 mph
Aerostar, N44XX
2 Lycoming TSIO-540, 350 bhp
8/16/88

CALGARY/THE PAS (CANADA)
Dawn Bartsch 271.85 kmh
Gordon Bartsch 168.92 mph
Cessna 421B
9/5/97

CHARLESTON, SC/FT. LAUDERDALE (USA)
William S. Kingson 312.84 kmh
Victor Grey 194.40 mph
Aerostar, N44XX
2 Lycoming TSIO-540, 350 bhp
11/6/88

CHRISTCHURCH/SYDNEY (AUSTRALIA)
Adam Boyd Munro 325.76 kmh
Piper Navajo Panther 202.43 mph
2 TIO-540-J2BD, 350 bhp
3/31/87

CHICAGO/NEW YORK (USA)
William Kingson, Pilot 615.24 kmh
Victor Grey, Copilot 382.31 mph
Machen Super Star, N44XX
2 Lycoming TSIO-540 350 bhp
2/17/89

DARWIN/LONDON (AUSTRALIA)
D. N. Dalton 271.90 kmh
Beechcraft Duke 168.95 mph
51 hrs 14 min 25 sec
10/21-23/73

DENVER/ATLANTA (USA)
Peter J. Sones 407.97 kmh
Piper PA-31P-350 Mojave 253.50 mph
2 Lycoming IO-540, 350 bhp
12/30/95

DHAKA/BEIJING (USA)
James H. Knuppe, Pilot 323.04 kmh
Steve Picatti 200.74 mph
Marc Mosier
Cessna 421C
2 Continental TSIO-520L, 375 bhp
3/6/87

DOHA/SAN JOSE (QATAR)
Hamad Al-Thani 130.36 kmh
Piper Aerostar 601P 81.00 mph
2 Lycoming IO-540, 290 bhp
4/29/92

FROBISHER BAY/ALBANY (USA)
Stephen A. Oster 485.28 kmh
Aerostar 601P 301.55 mph
2 Lycoming TIO-540-J2BD, 350 bhp
6/30/85

FROBISHER BAY/BRIDGEPORT (USA)
Stephen A. Oster 457.23 kmh
Aerostar 601P 284.12 mph
2 Lycoming TIO-540-J2BD, 350 bhp
6/30-7/1/85

FROBISHER BAY/GOTHAB (USA)
Stephen A. Oster 362.88 kmh
Aerostar 601P 225.49 mph
2 Lycoming TIO-540-J2BD, 350 bhp
6/18/85

FROBISHER BAY/MONTREAL (USA)
Stephen A. Oster 488.88 kmh
Aerostar 601P 303.79 mph
2 Lycoming TIO-540-J2BD, 350 bhp
6/30/85

FROBISHER BAY/NEW YORK (USA)
Stephen A. Oster 483.84 kmh
Aerostar 601P 300.65 mph
2 Lycoming TIO-540-J2BD, 350 bhp
6/30/85

GANDER/LONDON (UK)
John J. A. Smith 426.71 kmh
Rockwell Commander 685 265.14 mph
2 Continental GTSIO-520-K1B, 435 bhp
8 hrs 47 min 32 sec
3/12/78

GOOSE BAY/BANGOR (USA)
William S. Kingson 379.44 kmh
Victor Grey 235.78 mph
Aerostar, N44XX
2 Lycoming TSIO-540, 350 bhp
8/20/88

GOOSE BAY/SONDERSTROM (USA)
William S. Kingson 406.80 kmh
Victor Grey 252.79 mph
Aerostar, N44XX
2 Lycoming TSIO-540, 350 bhp
8/4/88

GOTHAB/REYKJAVIK (USA)
Stephen A. Oster 466.20 kmh
Aerostar 601P 289.70 mph
2 Lycoming TIO-540-J2BD, 350 bhp
6/20/85

HILO/SANTA BARBARA (AUSTRALIA)
Claude Meunier 360.19 kmh
Piper PA-60-602P Aerostar 223.81 mph
2 Lycoming IO-540, 350 bhp
6/2/96

HOBART/CHRISTCHURCH (AUSTRALIA)
Adam Boyd Munro 492.05 kmh
Piper Navajo Panther 305.76 mph
2 TIO-540-J2BD, 350 bhp
3/26/87

HONG KONG/SINGAPORE (USA)
James H. Knuppe, Pilot 340.88 kmh
Steve Picatti 211.82 mph
Marc Mosier
Cessna 421C
2 Continental TSIO-520L, 375 bhp
3/15/87

HONOLULU/SAN FRANCISCO (AUSTRALIA)
D. N. Dalton, Pilot 342.56 kmh
K. Carmody, Copilot 212.85 mph
Beechcraft Duke 60
2 Lycoming TIO-540-E
11 hrs 23 min 27 sec
4/7-8/75

ISTANBUL/BUDAPEST (USA)
William S. Kingson 395.28 kmh
Victor Grey 245.63 mph
Aerostar, N44XX
2 Lycoming TSIO-540, 350 bhp
8/13/88

LAKELAND, FL/BEDFORD, MA (USA)
Christopher W. Peatridge, Pilot 298.12 kmh
Frank Hammerbacher, Copilot 185.24 mph
North American T-28B
1 Wright R1820-86B, 1,425 bhp
04/13/90

LONDON/BRISBANE (AUSTRALIA)
D. N. Dalton, Pilot 277.68 kmh
T. Gwynne-Jones, Copilot 172.54 mph
Beechcraft Duke 60
2 Lycoming TIO-540-E
59 hrs 51 min 17 sec
7/22-25/75

LONDON/CAIRO (USA)
David Long 127.50 kmh
Piper Chieftan 79.23 mph
2/19-2/20/83

LONDON/DARWIN (AUSTRALIA)
D. N. Dalton, Pilot 270.55 kmh
T. Gwynne-Jones, Copilot 168.11 mph
Beechcraft Duke 60
2 Lycoming TIO-540-E
51 hrs 29 min 38 se
7/22-24/75

LONDON/REYKJAVIK (USA)
Stephen A. Oster 549.00 kmh
Aerostar 601P 341.15 mph
2 Lycoming TIO-540-J2BD, 350 bhp
6/29/85

LONDON/SINGAPORE (AUSTRALIA)
D. N. Dalton, Pilot 324.62 kmh
K. Carmody, Copilot 201.70 mph
Beechcraft Duke 60
2 Lycoming TIO-540-E
33 hrs 38 min 54 sec
4/24-26/75

LONG BEACH/YOLO, CA (USA)
William Bronson 362.88 kmh
Beech Duke, N6731V 225.49 mph
2 Lycoming TIO-541
5/6/89

LOS ANGELES/NEW YORK (USA)
Russell H. Hancock 364.98 kmh
Piper Navajo Panther 226.80 mph
2 Lycoming TIO-540-J2BD, 350 bhp
7/31-8/1/79

MANCHESTER/LUXOR (USA)
Tim Mellon, Pilot 399.24 kmh
Tom Glista, Copilot 248.09 mph
Piper Aerostar 602P, N602PC
2 Lycoming IO-540
4/28/87

MELBOURNE/SYDNEY (AUSTRALIA)
Graham M. Thomas 615.08 kmh
PA 31-310 Colemill Conversion 382.21 mph
6/22/88

MEMPHIS/NEW YORK (USA)
William S. Kingson 471.60 kmh
Victor Grey 293.05 mph
Aerostar, N44XX
2 Lycoming TSIO-540, 350 bhp
11/23/88

MIAMI/GREENSBORO (USA)
William S. Kingson 365.76 kmh
Victor Grey 227.28 mph
Aerostar, N44XX
2 Lycoming TSIO-540, 350 bhp
11/9/88

MONTREAL/ALBANY (USA)
Stephen A. Oster
Aerostar 601P
2 Lycoming TIO-540-J2BD, 350 bhp
6/30/85
467.28 kmh
290.37 mph

MONTREAL/BRIDGEPORT (USA)
Stephen A. Oster
Aerostar 601P
2 Lycoming TIO-540-J2BD, 350 bhp
6/30-7/1/85
368.64 kmh
229.07 mph

MONTREAL/FROBISHER BAY (USA)
Stephen A. Oster
Aerostar 601P
2 Lycoming TIO-540-J2BD, 350 bhp
6/17/85
452.88 kmh
281.42 mph

MONTREAL/NEW YORK (USA)
Stephen A. Oster
Aerostar 601P
2 Lycoming TIO-540-J2BD, 350 bhp
6/30/85
469.44 kmh
291.71 mph

NADI/APIA (AUSTRALIA)
Claude Meunier
Robyn Stewart
Denis Beresford
Piper PA-60-602P Aerostar
2 Lycoming IO-540, 350 bhp
6/9/00
347.51 kmh
215.93 mph

NEW YORK/CHARLESTON, SC (USA)
William S. Kingson
Victor Grey
Aerostar, N44XX
2 Lycoming TSIO-540, 350 bhp
11/6/88
308.52 kmh
191.71 mph

NEW YORK/LONDON (AUSTRALIA)
A. Boyd Munro
Panther Navajo
2 Lycoming TIO-540-J2BD, 350 bhp
9/6/79
371.64 kmh
230.92 mph

NEW YORK/MONTREAL (USA)
Stephen A. Oster
Aerostar 601P
2 Lycoming TIO-540-J2BD, 350 bhp
6/15/85
459.00 kmh
285.22 mph

NORTH POLE/SOUTH POLE (USA)
Elgen M. Long
Piper Navajo, N9097Y
11/8-22/71
59.70 kmh
37.10 mph

NOUMEA/NADI (AUSTRALIA)
Claude Meunier
Denis Beresford
Robyn Stewart
Piper PA-60-602P Aerostar
2 Lycoming IO-540, 350 bhp
6/8/00
367.00 kmh
228.04 mph

OAKLAND/AUCKLAND (USA)
Jim York
Cessna 340A
2 Continental TSIO-520-N, 310 bhp
9/30-10/2/84
170.83 kmh
106.15 mph

OAKLAND/CHRISTCHURCH (USA)
Jim York
Cessna 340A
2 Continental TSIO-520-N, 310 bhp
9/30-10/2/84
173.70 kmh
107.94 mph

OAKLAND/HONOLULU (USA)
Jim York
Cessna 340A
2 Continental TSIO-520-N, 310 bhp
9/30-10/1/84
323.06 kmh
200.75 mph

OAKLAND/WELLINGTON (USA)
Jim York
Cessna 340A
2 Continental TSIO-520-N, 310 bhp
9/30-10/2/84
171.72 kmh
106.71 mph

FAGO PAGO/RAROTONGA (AUSTRALIA)
Claude Meunier
Robyn Stewart
Denis Beresford
Piper PA-60-602P Aerostar
2 Lycoming IO-540, 350 bhp
6/14/00
360.59 kmh
224.06 mph

PARIS/BEIJING (USA)
James H. Knuppe, Pilot
Steve Picatti
Marc Mosier
Cessna 421C
2 Continental TSIO-520L, 375 bhp
2/28-3/6/87
66.35 kmh
41.23 mph

PARIS/ISTANBUL (USA)
William S. Kingson
Victor Grey
Aerostar, N44XX
2 Lycoming TSIO-540, 350 bhp
8/8/88
467.64 kmh
290.59 mph

PARIS/LONDON (USA)
Stephen A. Oster
Aerostar 601P
2 Lycoming TIO-540-J2BD, 350 bhp
6/24/85
440.28 kmh
273.59 mph

PERTH/ADELAIDE (AUSTRALIA)
George Gillespie Keirle
Cessna Chancillor C414A
9/30/89
352.98 kmh
219.34 mph

RALEIGH/NEW YORK (USA)
William S. Kingson
Victor Grey
Aerostar, N44XX
2 Lycoming TSIO-540, 350 bhp
11/10/88
450.00 kmh
279.63 mph

RAROTONGA/TAHITI (AUSTRALIA)
Claude Meunier
Robyn Stewart
Denis Beresford
Piper PA-60-602P Aerostar
2 Lycoming IO-540, 350 bhp
6/14/00
342.89 kmh
213.06 mph

REYKJAVIK/ABDEREEN (USA)
Stephen A. Oster
Aerostar 601P
2 Lycoming TIO-540-J2BD, 350 bhp
6/22/85
413.64 kmh
257.04 mph

REYKJAVIK/ALBANY (USA)
Stephen A. Oster
Aerostar 601P
2 Lycoming TIO-540-J2BD, 350 bhp
6/30/85
372.96 kmh
231.76 mph

REYKJAVIK/BRIDGEPORT (USA)
Stephen A. Oster
Aerostar 601P
2 Lycoming TIO-540-J2BD, 350 bhp
6/30-7/1/85
360.72 kmh
224.15 mph

REYKJAVIK/FROBISHER BAY (USA)
Stephen A. Oster
Aerostar 601P
2 Lycoming TIO-540-J2BD, 350 bhp
6/30/85
456.12 kmh
283.43 mph

REYKJAVIK/MONTREAL (USA)
Stephen A. Oster — 373.68 kmh
Aerostar 601P — 232.20 mph
2 Lycoming TIO-540-J2BD, 350 bhp
6/30/85

REYKJAVIK/NEW YORK (USA)
Stephen A. Oster — 374.04 kmh
Aerostar 601P — 232.42 mph
2 Lycoming TIO-540-J2BD, 350 bhp
6/30/85

REYKJAVIK/PARIS (USA)
William S. Kingson — 402.48 kmh
Victor Grey — 250.10 mph
Aerostar, N44XX
2 Lycoming TSIO-540, 350 bhp
8/5/88

REYKJAVIK/SONDERSTROM (USA)
William S. Kingson — 406.80 kmh
Victor Grey — 252.78 mph
Aerostar, N44XX
2 Lycoming TSIO-540, 350 bhp
8/20/88

REYKJAVIK/STRASBOURG (CANADA)
Dawn Bartsch — 363.78 kmh
Gordon Bartsch — 226.04 mph
Cessna 421B
9/11/97

ROME/PARIS (USA)
James H. Knuppe, Pilot — 332.22 kmh
Steve Picatti — 206.44 mph
Marc Mosier
Cessna 421C
2 Continental TSIO-520, 375 bhp
3/27/87

ROME/TEL AVIV (CANADA)
Dawn Bartsch — 384.54 kmh
Gordon Bartsch — 238.94 mph
Cessna 421B
9/16/97

SAN JOSE/DOHA (QATAR)
Hamad Al-Thani — 77.73 kmh
Piper Aerostar 601P — 48.29 mph
2 Lycoming IO-540, 290 bhp
4/24/92

SAN JOSE/HONOLULU (QATAR)
Hamad Al-Thani — 296.15 kmh
Piper Aerostar 601P — 184.02 mph
2 Lycoming IO-540, 290 bhp
4/18/92

SEVILLA/ROME (CANADA)
Dawn Bartsch — 406.68 kmh
Gordon Bartsch — 252.70 mph
Cessna 421B
9/15/97

SHELBYVILLE/OSHKOSH (USA)
Gene Steven Sevier, Pilot — 171.53 kmh
Richard Tyler Wright, Copilot — 106.58 mph
Antonov An-2
1 Shevtson , 1,000 bhp
7/25/95

SINGAPORE/CHRISTCHURCH (AUSTRALIA)
Adam Boyd Munro — 368.84 kmh
Piper Navajo Panther — 229.20 mph
2 TIO-540-J2BD, 350 bhp
3/25-26/87

SINGAPORE/HOBART (AUSTRALIA)
Adam Boyd Munro — 360.77 kmh
Piper Navajo Panther — 224.18 mph
2 TIO-540-J2BD, 350 bhp
3/25-31/87

SONDERSTROM/GOOSE BAY (USA)
William S. Kingson — 396.72 kmh
Victor Grey — 246.52 mph
Aerostar, N44XX
2 Lycoming TSIO-540, 350 bhp
8/20/88

SONDERSTROM/REYKJAVIK (USA)
William S. Kingson — 410.76 kmh
Victor Grey — 255.25 mph
Aerostar, N44XX
2 Lycoming TSIO-540, 350 bhp
8/4/88

STOCKHOLM/ABERDEEN (USA)
William S. Kingson — 366.84 kmh
Victor Grey — 227.95 mph
Aerostar, N44XX
2 Lycoming TSIO-540, 350 bhp
8/19/88

STRASBOURG/SEVILLA (CANADA)
Dawn Bartsch — 318.49 kmh
Gordon Bartsch — 197.90 mph
Cessna 421B
9/13/97

SYDNEY/MELBOURNE (AUSTRALIA)
Graham M. Thomas — 374.78 kmh
PA 31-310 Colemill Conversion — 232.89 mph
6/21/88

THE PAS/IQALUIT (CANADA)
Dawn Bartsch — 356.95 kmh
Gordon Bartsch — 221.80 mph
Cessna 421B
9/5/97

TRABZON/ADANA (CANADA)
Dawn Bartsch — 309.38 kmh
Gordon Bartsch — 192.24 mph
Cessna 421B
9/20/97

VANCOUVER/OGDEN, UT (USA)
William G. Dilley — 392.59 kmh
Cessna 421B, N5957M — 243.95 mph
2 Continental GTSIO-520-H
3 hrs 7 min 34 sec
6/22/73

ROUND TRIP

LONGYEARBYEN/NORTH POLE/LONGYEARBYEN (CANADA)
Dave McCulloch, Pilot — 295.20 kmh
Alvin Solomon, Copilot — 183.44 mph
Milton Brown, Navigator
Cessna 414, C-FASB
2 Continental TSIO-520, 310 bhp
5/9/87

CLASS C-1.H (12,000-16,000 kg / 26,455-35,274lbs)
GROUP I (PISTON ENGINE)

BREMEM/SOUTHEND (UK)
Barbel C. Abela — 358.86 kmh
Douglas A-26 Invader — 222.98 mph
7/24/92

CANNES/LYDD (UK)
Barbel C. Abela — 348.44 kmh
Douglas A-26 Invader — 216.51 mph
7/15/92

GOOSE BAY/REYKJAVIK (UK)
Barbel C. Abela
Douglas A-26 Invader
7/9/92
420.72 kmh
261.42 mph

HELSINKI/BREMEN (UK)
Barbel C. Abela
Douglas A-26 Invader
7/23/92
360.03 kmh
223.71 mph

LYDD/CANNES (UK)
Barbel C. Abela
Douglas A-26 Invader
7/12/92
389.29 kmh
241.89 mph

REYKJAVIK/LIVERPOOL (UK)
Barbel C. Abela
Douglas A-26 Invader
7/10/92
404.21 kmh
251.16 mph

SOUTHEND/HELSINKI (UK)
Barbel C. Abela
Douglas A-26 Invader
7/19/92
427.94 kmh
265.90 mph

* * *

CLASS C-1.C (1,000-1,750 kg / 2,205 -3,858 lbs)
GROUP II (TURBOPROP)

ALBANY, GA/LAKELAND, FL (USA)
Jack Schweibold, Pilot
Sharon Schweibold, Copilot
Beech Bonanza A36, N34K
1 Allison 250-17B, 420 shp
4/13/88
422.28 kmh
262.41 mph

CARACAS/MARACAIBO (USA)
Jay Penner, CoCaptain
Henry Hoyas, CoCaptain
Beech AT-34, N4CN
1 Allison 250-B17D, 420 shp
4/29/88
359.28 kmh
223.26 mph

CHICAGO/BOSTON (USA)
Joe C. Boyd, Co-Captain
Dorothy Boyd, Co-Captain
Soloy Beech A-36
1 Allison 250-B17D, 420 shp
7/9/86
408.17 kmh
253.63 mph

ELMIRA, NY/BOSTON (USA)
Jack Schweibold
James Ernsting
Beech Bonanza, N34K
1 Allison 250-B17C, 420 shp
5/22/89
422.64 kmh
262.63 mph

INDIANAPOLIS/ATLANTA (USA)
Jack Schweibold, Pilot
Sharon Schweibold, Copilot
Beech Bonanza A36, N279WP
1 Allison 250 B-17, 420 shp
2/9-10/87
378.00 kmh
234.89 mph

INDIANAPOLIS/ATLANTIC CITY (USA)
Jack Schweibold, Pilot
Jay Penner, Co-Captain
Beech Bonanza A36
1 Allison 250B-17D, 420 bhp
11/18/85
405.05 kmh
251.70 mph

INDIANAPOLIS/BALTIMORE (USA)
Jack Schweibold, Pilot
Jay Penner, Co-Captain
Beech Bonanza A36
1 Allison 250B-17D, 420 bhp
11/18/85
408.62 kmh
253.90 mph

INDIANAPOLIS/BOSTON (USA)
Mark Schweibold, Co-Captain
Jack Schweibold, Co-Captain
Soloy Beeech A 36
1 Allison 250-C20S, 420 shp
7/6/86
389.52 kmh
242.05 mph

INDIANAPOLIS/BUFFALO (USA)
F. Jack Schweibold
Sharon Schweibold
Soloy Beech Bonanza A36, N34K
1 Allison 250-B17C, 420 shp
9/23/88
455.04 kmh
282.76 mph

INDIANAPOLIS/CLEVELAND (USA)
Brad Schweibold
Jack Schweibold
Beech Bonanza, N34K
1 Allison 250-B17C, 420 shp
7/1/89
362.16 kmh
225.05 mph

INDIANAPOLIS/ERIE (USA)
F. Jack Schweibold
Sharon Schweibold
Soloy Beech Bonanza A36, N34K
1 Allison 250-B17C, 420 shp
9/23/88
472.32 kmh
293.49 mph

INDIANAPOLIS/OSHKOSH (USA)
Jay Penner
Beech Bonanza
1 Allison 250-B17D, 420 shp
7/25/85
371.52 kmh
230.86 mph

INDIANAPOLIS/WASHINGTON (USA)
Jack Schweibold, Co-captain
Larrie Chambers, Co-captain
Beech Bonanza A36
1 Allison 250B-17D, 420 bhp
8/21/85
387.77 kmh
240.96 mph

MOLINE/INDIANAPOLIS (USA)
Robert Lark, Co-Captian
George Cummins, Co-Captain
Cessna 206
1 Allison 250-C20S, 420 shp
8/12/86
292.68 kmh
181.87 mph

MONTREAL/BOSTON (USA)
Dominic Sarazeno, Co-Captain
Jay Penner, Co-Captain
Soloy Beech A 36
1 Allison 250-B17D, 420 shp
7/11/86
391.32 kmh
243.17 mph

OSHKOSH/INDIANAPOLIS (USA)
Larrie Chambers
Soloy Beech A36
1 Allison 250-B17D, 420 shp
8/8-9/86
460.44 kmh
286.12 mph

RENO/INDIANAPOLIS (USA)
Larrie Chambers, Co-captain
Jack Schweibold, Co-captain
Beech Bonanza A36
1 Allison 250B-17D, 420 bhp
9/15-16/85
355.68 kmh
221.01 mph

SPRINGFIELD, MO/INDIANAPOLIS (USA)
Larrie Chambers, Co-Captain
Linda Chambers, Co-Captain
Soloy Beech A36
1 Allison 250-B17D, 420 shp
9/14/86
380.52 kmh
236.46 mph

TULSA/INDIANAPOLIS (USA)
Jack Schweibold, Pilot 428.73 kmh
Larrie Chambers, Co-Captain 266.43 mph
Harry Sutton, Co-Captain
Beech Bonanza A36
1 Allison 250B-17D, 420 bhp
11/6/85

WICHITA/INDIANAPOLIS (USA)
Jack Schweibold, Pilot 379.80 kmh
Jerry Molish II, Co-Captain 236.01 mph
Beech Bonanza A36
1 Allison 250B-17D, 420 bhp
9/19/85

ROUND TRIP

INDIANAPOLIS/CHARLESTON/INDIANAPOLIS (USA)
Don Richer, Co-Captain 351.72 kmh
Jay Penner, Co-Captain 218.56 mph
Soloy Beech A-36
1 Allison 250-B17C, 420 shp
8/12/86

CLASS C-1.D (1,750 -3,000 kg / 3,858-6,614 lbs) GROUP II (TURBOPROP)

WEST TO EAST TRANSCONTINENTAL*
Spencer Lane, Pilot 305.09 mph
Marc Mosier, Copilot
Socata TBM-700
1 P&W PT6A, 700 shp
9/26/92

***

BEDFORD/ASHEVILLE (USA)
Howard E. Cox, Jr. 397.36 kmh
Bruce R. MacGilvra 246.91 mph
Aerospatiale TBM-700-A
1 P&W PT6A-64, 700 shp
8/1/96

BEDFORD/FT. WAYNE (USA)
Howard E. Cox, Jr. 409.64 kmh
Bruce R. MacGilvra 254.53 mph
Aerospatiale TBM-700-A
1 P&W PT6A-64, 700 shp
7/30/96

BEDFORD/WHITE PLAINS*
Howard E. Cox, Jr. 206.62 mph
Mike S. Morvay
Aerospatiale TBM-700-A
1 P&W PT6A-64, 700 shp
7/30/96

CHADRON/JOHNSON CITY (USA)
Mike Smith 692.28 kmh
Smith Propjet 430.18 mph
1 P&W PT6A-41, 850 shp
12/18/85

CHARLESTON, SC/KITTY HAWK (USA)
John Clegg, Co-Captain 380.88 kmh
Peter Foster, Co-Captain 236.68 mph
Spirit 750, N731XH
1 P&W PT6-35, 450 shp
12/16/86

DENVER/LINCOLN (USA)
Mike Smith 794.34 kmh
Smith Propjet 493.60 mph
1 P&W PT6A-41, 850 shp
11/22/85

DULUTH/PARIS (FRANCE)
Jacques Lemaigre du Breuil 412.66 kmh
Nicolas Gorodiche 256.42 mph
Olivier Waisblat
Socata TBM 700
6/13-17/93

EDMONTON/INDIANAPOLIS (USA)
Jay D. Penner, Pilot 349.94 kmh
Larrie Chambers, Copilot 217.45 mph
Beech Bonanza A-36
1 Allison 250-B17D, 420 bhp
3/14/85

FT. WAYNE/WICHITA (USA)
Howard E. Cox, Jr. 421.83 kmh
Bruce R. MacGilvra 262.11 mph
Aerospatiale TBM-700-A
1 P&W PT6A-64, 700 shp
7/30/96

JOHNSON CITY/ATLANTA (USA)
Mike Smith 619.56 kmh
Smith Propjet 385.00 mph
1 P&W PT6A-41, 850 shp
9/30/84

KITTY HAWK/MIAMI (USA)
John Clegg, Co-Captain 255.96 kmh
Peter Foster, Co-Captain 159.05 mph
Spirit 750, N731XH
1 P&W PT6-35, 450 shp
12/17-18/86

KITTY HAWK/SAVANNAH (USA)
John Clegg, Co-Captain 308.88 kmh
Peter Foster, Co-Captain 191.94 mph
Spirit 750, N731XH
1 P&W PT6-35, 450 shp
12/17/86

MIAMI/CHARLESTON, SC (USA)
John Clegg, Co-Captain 336.60 kmh
Peter Foster, Co-Captain 209.16 mph
Spirit 750, N731XH
1 P&W PT6-35, 450 shp
12/16/86

MIAMI/KITTY HAWK (USA)
John Clegg, Co-Captain 293.40 kmh
Peter Foster, Co-Captain 182.32 mph
Spirit, 750, N731XH
1 P&W PT6-35, 450 shp
12/16/86

MILWAUKEE/BEDFORD (USA)
Howard E. Cox, Jr. 523.77 kmh
Bruce R. MacGilvra 325.45 mph
Aerospatiale TBM-700-A
1 P&W PT6A-64, 700 shp
8/1/96

PARIS/RHODOS (FRANCE)
Jacques Lemaigre du Breuil 491.75 kmh
Nicolas Gorodiche 305.60 mph
Olivier Waisblat
Socata TBM 700
6/13-17/93

PERTH/SYDNEY (AUSTRALIA)
Trevor J. Frost 594.13 kmh
Socata TBM 700 369.18 mph
1 P&WC PT6A-64, 700 shp
9/19/93

REYKJAVIK/PARIS (FRANCE)
Jacques Lemaigre du Breuil 540.40 kmh
Nicolas Gorodiche 335.79 mph
Olivier Waisblat
Socata TBM 700
6/13-17/93

SAN DIEGO/JACKSONVILLE (USA)
Spencer Lane, Pilot 491.00 kmh
Marc Mosier, Copilot 305.09 mph
Socata TBM-700
1 P&W PT6A, 700 shp
9/26/92

SAVANNAH/MIAMI (USA)
John Clegg, Co-Captain 358.20 kmh
Peter Foster, Co-Captain 222.58 mph
Spirit, 750, N731XH
1 P&W PT6-35, 450 shp
12/17-18/86

TAIPEI/SENDAI (FRANCE)
Jacques Lemaigre du Breuil 486.98 kmh
Nicolas Gorodiche 302.60 mph
Olivier Waisblat
Socata TBM 700
6/13-17/93

WICHITA/MILWAUKEE (USA)
Howard E. Cox, Jr. 490.53 kmh
Bruce R. MacGilvra 304.80 mph
Aerospatiale TBM-700-A
1 P&W PT6A-64, 700 shp
7/31/96

ROUND TRIP

INDIANAPOLIS/WINDSOR/INDIANAPOLIS (USA)
Larrie Chambers, Pilot 360.00 kmh
C. V. Glines, Copilot 223.70 mph
Soloy Beech A36, N7214D
1 Allison 250 B-17D, 420 shp
8/12/86

OAKLAND/LAS VEGAS/OAKLAND (USA)
Greg Presnell, Pilot 352.80 kmh
Eric Price, Copilot 219.23 mph
George Samson, Navigator
Beech Bonanza, N7214D
1 Allison 250-B17D, 420 shp
9/30/87

CLASS C-1.E (3,000-6,000 kg / 6,614-13,228 LBS)
GROUP II (TURBOPROP)

WEST TO EAST, TRANSCONTINENTAL * ★
Joseph J. Ritchie 880.01 kmh
Steve Fossett 546.81 mph
Piaggio P.180 Avanti
2 P&WC PT6A, 850 shp
2/6/03

---

* * *

AGANA/MAJURO (USA)
Robert E. Reinhold 420.41 kmh
Donald J. Grant 261.23 mph
Piper PA-42 Cheyenne III
2 P&W PT6A-41, 720 shp
3/22/82

AMMAN/AQABA (JORDAN)
Ziad Bilbeisi 382.52 kmh
Aero Commander 690A 237.68 mph
5/13/90

AMSTERDAM/BOSTON (USA)
Robert Thompson, Pilot 322.20 kmh
Bryant Jones, Copilot 200.21 mph
Cessna Conquest 441
2 Garrett TPE 331-8, 635.5 shp
4/11-12/84

AMSTERDAM/GOOSE BAY (USA)
Robert Thompson, Pilot 304.92 kmh
Bryant Jones, Copilot 189.47 mph
Cessna Conquest 441
2 Garrett TPE 331-8, 635.5 shp
4/11/84

AMSTERDAM/KEFLAVIK (USA)
Robert Thompson, Pilot 283.68 kmh
Bryant Jones, Copilot 176.27 mph
Cessna Conquest 441
2 Garrett TPE 331-8, 635.5 shp
4/11/84

AMSTERDAM/NEW YORK (USA)
Robert Thompson, Pilot 330.84 kmh
Bryant Jones, Copilot 205.58 mph
Cessna Conquest 441
2 Garrett TPE 331-8, 635.5 shp
4/11-12/84

AMSTERDAM/PRESTWICK (USA)
Robert Thompson, Pilot 375.48 kmh
Bryant Jones, Copilot 233.32 mph
Cessna Conquest 441
2 Garrett TPE 331-8, 635.5 shp
4/11/84

AMSTERDAM/WASHINGTON (USA)
Robert Thompson, Pilot 334.44 kmh
Bryant Jones, Copilot 207.82 mph
Cessna Conquest 441
2 Garrett TPE 331-8, 635.5 shp
4/11-12/84

ANADYR/PROVIDENIA (USA)
D. Lancy Allyn 290.88 kmh
Beechcraft A90 180.75 mph
2 P&W PT6A-20
07/15/89

ANCHORAGE/WICHITA (USA)
Gary Fiebach 522.54 kmh
Cessna Conquest 324.71 mph
7/29/83

ATLANTA/LINCOLN (USA)
Patrick McBride, Pilot 449.18 kmh
Thomas Smallen, Copilot 279.12 mph
Piper Cheyenne III
2 P&W PT6A-41, 720 shp
11/16/85

BANGOR/CANTON (USA)
J. Jeffrey Brausch 423.23 kmh
William H. Cox 262.99 mph
Piper PA-31T2 Cheyenne IIXL
2 P&WC PT6A, 620 shp
8/21/02

BANGOR/HALIFAX (USA)
J. Jeffrey Brausch 452.59 kmh
William H. Cox 281.23 mph
Piper PA-31T2 Cheyenne IIXL
2 P&WC PT6A, 620 shp
8/11/02

BARCELONA/NICE (USA)
James P. Hanson 280.23 kmh
Cessna 208 Caravan 174.12 mph
1 P&W PT-6A, 600 shp
6/6/96

BATON ROUGE/PANAMA CITY (USA)
Phil Cron 530.28 kmh
Beech King Air 200 329.50 mph
2 P&W PT6A-41, 850 shp
2/28/86

BOISE/DENVER (USA)
Robert Glaisyer, Pilot 695.88 kmh
Charles Oliver, Copilot 432.42 mph
Piper Cheyenne 400LS
2 Garrett TPE-331-14, 1000 shp
6/27/86

BORGER/SANTA ROSA (USA)
Edward B. Sleeper 200.02 kmh
Cessna 208B 124.29 mph
1 P&W PT6A, 675 shp
7/7/03

BOSTON/PARIS (USA)
Renald Davenport, Pilot 530.40 kmh
Calvin Arter, Copilot 329.57 mph
Piper Cheyenne IIIA
2 P&W PT6A-61, 720 shp
5/10/83

BOSTON/WASHINGTON (USA)
Charles "Chuck" Yeager, Pilot 512.28 kmh
Renald Davenport, Copilot 318.33 mph
Piper Cheyenne 400LS
2 Garrett TPE 331-14, 1,000 shp
7/31/85

BRATSK/MIRNY (USA)
D. Lancy Allyn 276.84 kmh
Beechcraft A90 172.03 mph
2 P&W PT6A-20
07/07/89

BRINDISI/IRAKLION (JORDAN)
Ziad Bilbeisi 349.03 kmh
Aero Commander 690A 216.87 mph
5/2/90

BURLINGTON/GOOSE BAY (AUSTRIA)
Ferdinand Mühlhofer 482.22 kmh
Mitsubishi MU-2 26A 299.64 mph
2 Garrett TPE 331, 665 shp
8/17/93

CAIRO/SHARJAH (USA)
Robert E. Reinhold 548.28 kmh
Donald J. Grant 340.69 mph
Piper PA-42 Cheyenne III
2 P&W PT6A-41, 720 shp
3/20/82

CANBERRA/MELBOURNE (AUSTRALIA)
William John McIntyre 456.77 kmh
Cessna Conquest II Model 441 283.82 mph
2 Garrett TPE 331-8-402, 635 shp
6/29/97

CANTON/BANGOR (USA)
J. Jeffrey Brausch 480.37 kmh
William H. Cox 298.49 mph
Piper PA-31T2 Cheyenne IIXL
2 P&WC PT6A, 620 shp
8/10/02

CANTON/ORLANDO (USA)
J. Jeffrey Brausch 469.40 kmh
Piper PA-31T2 Cheyenne IIXL 291.67 mph
2 P&WC PT6A, 620 shp
2/2/02

CHARLESTON/NEW YORK (USA)
Chuck Yeager, Pilot 830.52 kmh
Renald Davenport, Copilot 516.09 mph
Cheyenne 400LS
2 Garrett TPE 331-14, 1,000 shp
2/8/86

Pilots Joe Ritchie (l) and Steve Fossett (r), after setting a transcontinental speed record in a Piaggio Avanti.

CHICAGO/ORLANDO (USA)
Arthur St. Clair — 683.65 kmh
Piaggio P.180 Avanti — 424.80 mph
2 P&W Canada PT6A, 850 shp
10/5/03

CLEVELAND/TORONTO (USA)
Chuck Yeager, Pilot — 707.76 kmh
Renald Davenport, Copilot — 439.80 mph
Cheyenne 400LS
2 Garrett TPE 331-14, 1,000 shp
7/29/85

COLOMBO/SINGAPORE (USA)
Robert E. Reinhold, Pilot — 421.95 kmh
Frank X. Perullo — 262.19 mph
Donald J. Grant
Piper Cheyenne III PA-42
2 P&W PT6A-41, 720 shp
3/21/82

CONCORD/DAYTONA BEACH (USA)
Joseph S. Cornelius — 463.14 kmh
Steven E. Zahler — 287.78 mph
Beechcraft King Air 200
2 P&W PT6-41, 850 shp
7/3/96

DALLAS/JACKSONVILLE (USA)
Terri Jones — 705.96 kmh
Piper Cheyenne 400LS — 438.68 mph
2 Garrett TPE 331-14, 1,000 shp
2/8/86

DAYTON/KITTY HAWK (USA)
J. Jeffrey Brausch — 565.57 kmh
Piper PA-31T2 Cheyenne IIXL — 351.43 mph
2 P&WC PT6A, 620 shp
11/8/03

DAYTON/MEDINA *
J. Jeffrey Brausch — 321.41 mph
Piper PA-31T2 Cheyenne IIXL
2 P&WC PT6A, 620 shp
11/25/03

DENVER/CANTON (USA)
J. Jeffrey Brausch — 497.14 kmh
Piper PA-31T2 Cheyenne IIXL — 308.91 mph
2 P&WC PT6A, 620 shp
11/25/01

DENVER/DAYTON (USA)
J. Jeffrey Brausch — 535.51 kmh
Piper PA-31T2 Cheyenne IIXL — 332.75 mph
2 P&WC PT6A, 620 shp
10/11/02

DENVER/DES MOINES (USA)
Charles E. Yeager, Pilot — 742.32 kmh
Donald H. Jay, Copilot — 461.28 mph
Piper Cheyenne 400LS
2 Garrett TPE 331-14, 1,000 shp
6/22/86

DENVER/FORT WAYNE (USA)
J. Jeffrey Brausch — 504.38 kmh
Piper PA-31T2 Cheyenne IIXL — 313.41 mph
2 P&WC PT6A, 620 shp
11/25/01

DENVER/NEW ORLEANS (USA)
Chuck Yeager, Pilot — 767.88 kmh
Charles Oliver, Copilot — 477.16 mph
Piper Cheyenne 400LS
2 Garrett TPE 331-14, 1,000 shp
9/25/85

DETROIT/HOUSTON (USA)
Charles "Chuck" Yeager, Pilot — 594.72 kmh
Renald Davenport, Copilot — 369.56 mph
Piper Cheyenne 400LS
2 Garrett TPE 331-14, 1,000 shp
8/5/85

EDWARDS AFB/KITTY HAWK (USA)
Charles E. Yeager, Pilot — 723.44 kmh
Renald Davenport, Copilot — 449.55 mph
Piper Cheyenne 400LS, N9099U
2 Garrett TPE-331, 1,000 shp
12/17/86

EL PASO/CHARLESTON (USA)
Joseph J. Ritchie — 913.39 kmh
Steve Fossett — 567.56 mph
Piaggio P.180 Avanti
2 P&WC PT6A, 850 shp
2/6/03

EL PASO/FORT WORTH (USA)
Joseph J. Ritchie — 913.61 kmh
Steve Fossett — 567.69 mph
Piaggio P.180 Avanti
2 P&WC PT6A, 850 shp
2/6/03

FORT SMITH/PHOENIX (USA)
Robert Thompson — 460.00 kmh
Cessna Conquest — 285.85 mph
2 Garrett TPE-331-8, 635 shp
3/13-14/84

FORT WORTH/ATLANTA (USA)
Joseph J. Ritchie — 927.40 kmh
Steve Fossett — 576.26 mph
Piaggio P.180 Avanti
2 P&WC PT6A, 850 shp
2/6/03

FORT WORTH/ORLANDO (USA)
William A. Hauprich — 662.52 kmh
Piaggio P.180 Avanti — 411.67 mph
2 P&WC PT6A, 850 shp
9/8/02

GANDER/BERLIN (USA)
Knut Kramer — 423.95 kmh
Cessna Conquest — 263.44 mph
2 Garrett TPE 331-8-4025, 636 shp
3/27/83

GANDER/GOOSE BAY (USA)
J. Jeffrey Brausch — 307.14 kmh
William H. Cox — 190.85 mph
Piper PA-31T2 Cheyenne IIXL
2 P&WC PT6A, 620 shp
8/12/02

GANDER/LONDON (W. GERMANY)
Herman Krug — 597.87 kmh
Cheyenne III A — 371.54 mph
2 P&W PT6A-61, 720 shp
8/30/84

GANDER/PARIS (USA)
Charles "Chuck" Yeager, Pilot — 589.87 kmh
Renald Davenport, Copilot — 366.55 mph
Piper Cheyenne 400LS
2 TPE 331-14, 1,000 shp
5/28/85

GANDER/SHANNON (USA)
Douglas Smith — 595.32 kmh
Piper Cheyenne 400LS — 369.93 mph
2 TPE 331-14, 1,000 shp
6/8/85

GOOSE BAY/BANGOR (USA)
J. Jeffrey Brausch 434.08 kmh
William H. Cox 269.73 mph
Piper PA-31T2 Cheyenne IIXL
2 P&WC PT6A, 620 shp
8/21/02

GOOSE BAY/BOSTON (USA)
Robert Thompson, Pilot 554.20 kmh
Bryant Jones, Copilot 344.37 mph
Cessna Conquest 441
2 Garrett TPE 331-8, 635.5 shp
4/11-12/84

GOOSE BAY/IQALUIT (USA)
J. Jeffrey Brausch 437.49 kmh
William H. Cox 271.84 mph
Piper PA-31T2 Cheyenne IIXL
2 P&WC PT6A, 620 shp
8/12/02

GOOSE BAY/KEFLAVIK (USA)
Robert Thompson, Pilot 523.44 kmh
Bryant Jones, Copilot 325.26 mph
Cessna Conquest 441
2 Garrett TPE 331-8, 635.5 shp
4/5-6/84

GOOSE BAY/LONDON (USA)
Robert Thompson, Pilot 417.96 kmh
Bryant Jones, Copilot 259.72 mph
Cessna Conquest 441
2 Garrett TPE 331-8, 635.5 shp
4/5-6/84

GOOSE BAY/LUGANO (AUSTRIA)
Ferdinand Mühlhofer 196.45 kmh
Mitsubishi MU-2 26A 122.07 mph
2 Garrett TPE 331, 665 shp
8/18/93

GOOSE BAY/NEW YORK (USA)
Robert Thompson, Pilot 555.48 kmh
Bryant Jones, Copilot 345.17 mph
Cessna Conquest 441
2 Garrett TPE 331-8, 635.5 shp
4/11-12/84

GOOSE BAY/NARSSARSSUAQ (AUSTRIA)
Ferdinand Mühlhofer 483.63 kmh
Mitsubishi MU-2 26A 300.51 mph
2 Garrett TPE 331, 665 shp
8/17/93

GOOSE BAY/PARIS (USA)
Robert Thompson, Pilot 419.76 kmh
Bryant Jones, Copilot 260.83 mph
Cessna Conquest 441
2 Garrett TPE 331-8, 635.5 shp
4/5-6/84

GOOSE BAY/PRESTWICK (USA)
Robert Thompson, Pilot 419.40 kmh
Bryant Jones, Copilot 260.61 mph
Cessna Conquest 441
2 Garrett TPE 331-8, 635.5 shp
4/5-6/84

GOOSE BAY/WASHINGTON (USA)
Robert Thompson, Pilot 518.76 kmh
Bryant Jones, Copilot 322.34 mph
Cessna Conquest 441
2 Garrett TPE 331-8, 635.5 shp
4/11-12/84

HAGERSTOWN/BORGER (USA)
Edward B. Sleeper 224.64 kmh
Cessna 208B 139.58 mph
1 P&W PT6A, 675 shp
7/6/03

HALIFAX/ST. JOHN'S (USA)
J. Jeffrey Brausch 481.78 kmh
William H. Cox 299.36 mph
Piper PA-31T2 Cheyenne IIXL
2 P&WC PT6A, 620 shp
8/11/02

HELSINKI/LENINGRAD (USA)
D. Lancy Allyn 213.48 kmh
Beechcraft A90 132.66 mph
2 P&W PT6A-20
06/23/89

HELSINKI/NOME (USA)
D. Lancy Allyn 11.58 kmh
Beechcraft A90 7.19 mph
2 P&W PT6A-20
06/23-07/15/89

HONOLULU/SAN FRANCISCO (USA)
Robert E. Reinhold, Pilot 472.80 kmh
Frank X. Perullo and 293.78 mph
Donald J. Grant, Copilots
Piper Cheyenne III
2 P&W PT6A-41, 720 shp
3/23/82

HOUSTON/GRAND CAYMAN (USA)
Kimber Turner, Pilot 452.16 kmh
Al Dortenzio, Copilot 280.98 mph
Beech King Air
2 P&W PT6A-42, 850 shp
9/6/84

HOUSTON/LOS ANGELES (USA)
Charles "Chuck" Yeager, Pilot 629.64 kmh
Renald Davenport, Copilot 391.25 mph
Piper Cheyenne 400LS
2 Garrett TPE 331-14, 1,000 shp
8/7-8/85

HOUSTON/OSHKOSH (USA)
Charles "Chuck" Yeager, Pilot 624.24 kmh
Renald Davenport, Copilot 387.90 mph
Piper Cheyenne 400LS
2 Garrett TPE 331-14, 1,000 shp
7/26-27/85

HOUSTON/TAMPA (USA)
Glenn Walker, Pilot 645.48 kmh
Renald Davenport, Copilot 401.10 mph
Cheyenne 400LS
2 Garrett TPE 331-14, 1,000 shp
1/19-20/86

IDAHO FALLS/DENVER (USA)
J. Jeffrey Brausch 412.05 kmh
Piper PA-31T2 Cheyenne IIXL 256.04 mph
2 P&WC PT6A, 620 shp
10/11/02

INDIANAPOLIS/WASHINGTON (USA)
Charles E. Yeager, Pilot 694.28 kmh
John Beeson, Copilot 431.42 mph
Piper Cheyenne 400LS
2 Garrett TPE 331-14, 1,000 shp
6/25/86

IQALUIT/GOOSE BAY (USA)
J. Jeffrey Brausch 424.69 kmh
William H. Cox 263.89 mph
Piper PA-31T2 Cheyenne IIXL
2 P&WC PT6A, 620 shp
8/18/02

IQALUIT/KUUJJUAQ (USA)
James P. Hanson 248.11 kmh
Cessna 208 Caravan 154.16 mph
1 P&W PT-6A, 600 shp
5/16/96

IQALUIT/NORTH POLE (USA)
Michael K. Egan 510.41 kmh
James M. Conn 317.15 mph
Socata TBM700
1 P&WC PT6A, 700 shp
11/8/94

IQALUIT/SONDRE STROMFJORD (USA)
J. Jeffrey Brausch 447.87 kmh
William H. Cox 278.30 mph
Piper PA-31T2 Cheyenne IIXL
2 P&WC PT6A, 620 shp
8/13/02

IRAKLION/AMMAN (JORDAN)
Ziad Bilbeisi 283.09 kmh
Aero Commander 690A 175.90 mph
5/2/90

INVERNESS/LUGANO (AUSTRIA)
Ferdinand Mühlhofer 430.35 kmh
Mitsubishi MU-2 26A 267.41 mph
2 Garrett TPE 331, 665 shp
8/18/93

INVERNESS/REYKJAVIK (AUSTRIA)
Ferdinand Mühlhofer 456.77 kmh
Mitsubishi MU-2 26A 283.82 mph
2 Garrett TPE 331, 665 shp
7/24/93

KANSAS CITY/PHILADELPHIA (USA)
Robert E. Reinhold, Pilot 549.96 kmh
Frank X. Perullo and 341.73 mph
Donald J. Grant, Copilots
Piper PA-42 Cheyenne III
2 P&W PT6A-41, 720 shp
3/24/82

KEARNEY/DAYTON (USA)
J. Jeffrey Brausch 517.79 kmh
Piper PA-31T2 Cheyenne IIXL 321.74 mph
2 P&WC PT6A, 620 shp
10/18/03

KEFLAVIK/BOSTON (USA)
Robert Thompson, Pilot 423.00 kmh
Bryant Jones, Copilot 262.85 mph
Cessna Conquest 441
2 Garrett TPE 331-8, 635.5 shp
4/11-12/84

KEFLAVIK/GOOSE BAY (USA)
Robert Thompson, Pilot 418.68 kmh
Bryant Jones, Copilot 260.16 mph
Cessna Conquest 441
2 Garrett TPE 331-8, 635.5 shp
4/11/84

KEFLAVIK/LONDON (USA)
Robert Thompson, Pilot 480.60 kmh
Bryant Jones, Copilot 298.64 mph
Cessna Conquest 441
2 Garrett TPE 331-8, 635.5 shp
4/6/84

KEFLAVIK/MUNICH (USA)
Robert E. Reinhold, Pilot 491.28 kmh
Frank X. Perullo and 305.27 mph
Donald J. Grant, Copilots
Piper Cheyenne III
2 P&W PT6A-41, 720 shp
3/20/82

KEFLAVIK/NEW YORK (USA)
Robert Thompson, Pilot 431.37 kmh
Bryant Jones, Copilot 268.05 mph
Cessna Conquest 441
2 Garrett TPE 331-8, 635.5 shp
4/11-12/84

KEFLAVIK/PARIS (USA)
Robert Thompson, Pilot 478.08 kmh
Bryant Jones, Copilot 297.07 mph
Cessna Conquest 441
2 Garrett TPE 331-8, 635.5 shp
4/6/84

KEFLAVIK/PRESTWICK (USA)
Robert Thompson, Pilot 469.08 kmh
Bryant Jones, Copilot 291.48 mph
Cessna Conquest 441
2 Garrett TPE 331-8, 635.5 shp
4/6/84

KEFLAVIK/WASHINGTON (USA)
Robert Thompson, Pilot 427.68 kmh
Bryant Jones, Copilot 265.76 mph
Cessna Conquest 441
2 Garrett TPE 331-8, 653.5 shp
4/11-12/84

KEMEROVO/BRATSK (USA)
D. Lancy Allyn 306.00 kmh
Beechcraft A90 190.15 mph
2 P&W PT6A-20
07/05/89

KITTY HAWK/DAYTON (USA)
J. Jeffrey Brausch 314.54 kmh
Piper PA-31T2 Cheyenne IIXL 195.45 mph
2 P&WC PT6A, 620 shp
11/8/03

KUIBYSHEV/TUMEN (USA)
D. Lancy Allyn 296.64 kmh
Beechcraft A90 184.33 mph
2 P&W PT6A-20
07/02/89

KULUSUK/REYKJAVIK (USA)
J. Jeffrey Brausch 407.52 kmh
William H. Cox 253.22 mph
Piper PA-31T2 Cheyenne IIXL
2 P&WC PT6A, 620 shp
8/14/02

KULUSUK/SONDRE STROMFJORD (USA)
J. Jeffrey Brausch 391.53 kmh
William H. Cox 243.29 mph
Piper PA-31T2 Cheyenne IIXL
2 P&WC PT6A, 620 shp
8/17/02

LAGOS/RIO DE JANEIRO (USA)
Michael K. Egan 414.25 kmh
James M. Conn 257.40 mph
Socata TBM700
1 P&WC PT6A, 700 shp
11/10/94

LENINGRAD/MOSCOW (USA)
Garry W. Zinger, Pilot 212.04 kmh
Beechcraft A90 131.76 mph
2 P&W PT6A-20
06/26/89

LONDON/NEW YORK (USA)
Dominic Frontiere 374.50 kmh
Cessna Conquest 441 232.71 mph
2 Garrett TPE 331-3025
7/18/83

LONDON/PARIS (USA)
Robert Thompson, Pilot 447.48 kmh
Bryant Jones, Copilot 278.06 mph
Cessna Conquest 441
2 Garrett TPE 331-8, 635.5 shp
4/6/84

LONGYEARBYEN/TARBES (USA)
Michael K. Egan 494.82 kmh
James M. Conn 307.46 mph
Socata TBM700
1 P&WC PT6A, 700 shp
11/8/94

LOS ANGELES/CHARLESTON (USA)
Chuck Yeager, Pilot 728.28 kmh
Renald Davenport, Copilot 452.55 mph
Piper Cheyenne 400LS
2 Garrett TPE 331-14, 1,000 shp
2/8/86

LOS ANGELES/CHICAGO (USA)
Dominic Frontiere 502.56 kmh
Cessna Conquest 441 312.29 mph
2 Garrett TPE 331-402S, 635 shp
5/21/85

LOS ANGELES/DALLAS (USA)
Terri Jones 654.84 kmh
Piper Cheyenne 400LS 406.92 mph
2 Garrett TPE 331-14, 1,000 shp
2/8/86

LOS ANGELES/HOUSTON (USA)
Glenn Walker, Pilot 589.68 kmh
Renald Davenport, Copilot 366.43 mph
Piper Cheyenne 400LS
2 Garrett TPE 331-14, 1,000 shp
1/19-20/86

LOS ANGELES/JACKSONVILLE (USA)
Terri Jones 675.36 kmh
Piper Cheyenne 400LS 419.67 mph
2 Garrett TPE 331-14, 1,000 shp
2/8/86

LOS ANGELES/NEW YORK (USA)
Chuck Yeager, Pilot 731.88 kmh
Renald Davenport, Copilot 454.79 mph
Piper Cheyenne 400LS
2 Garrett TPE 331-14, 1,000 shp
2/8/86

LOS ANGELES/OAKLAND (USA)
V. C. Gardner 340.63 kmh
Volpar Turboliner 211.66 mph
2 Airesearch TPE-331-10E 620 shp
1 hrs 37 min 20 sec
3/28/70

LOS ANGELES/PHOENIX (USA)
James L. Badgett 609.08 kmh
Turbo Commander 690 378.47 mph
2 Garrett TPE-331
56 min 36 sec
4/7/72

LOS ANGELES/SACRAMENTO (USA)
Charles "Chuck" Yeager, Pilot 639.72 kmh
Renald Davenport, Copilot 397.72 mph
Piper Cheyenne 400LS
2 Garrett TPE 331-14, 1,000 shp
8/9/85

LOS ANGELES/TAMPA (USA)
Glenn Walker, Pilot 605.16 kmh
Renald Davenport, Copilot 376.05 mph
Piper Cheyenne 400LS
2 Garrett TPE 331-14, 1,000 shp
1/19-20/86

LOS ANGELES/WASHINGTON (USA)
Charles "Chuck" Yeager, Pilot 638.64 kmh
Renald Davenport, Copilot 396.85 mph
Piper Cheyenne 400LS
2 TPE 331-14, 1,000 shp
5/22/85

LUGANO/VIENNA (AUSTRIA)
Ferdinand Mühlhofer 452.10 kmh
Mitsubishi MU-2 26A 280.92 mph
2 Garrett TPE 331, 665 shp
7/23/93

MAGADAN/ANADYR (USA)
D. Lancy Allyn 338.76 kmh
Beechcraft A90 210.51 mph
2 P&W PT6A-20
07/13/89

MAJURO/HONOLULU (USA)
Robert E. Reinhold 486.36 kmh
Donald J. Grant 302.21 mph
Piper PA-42 Cheyenne III
2 P&W PT6A-41, 720 shp
3/22/82

MANILA/AGANA (USA)
Robert E. Reinhold, Pilot 430.67 kmh
Frank X. Perullo, 267.61 mph
Donald J. Grant, Copilots
Piper PA-42 Cheyenne III
2 P&W PT-6A-41, 720 shp
3/21-22/82

MEMPHIS/DALLAS (USA)
Frank Blazina 444.60 kmh
Piper Cheyenne 1A 276.27 mph
2 P&W PT6A-II, 500 shp
9/29/83

MIAMI/BOSTON (USA)
Mark O. Henry, Pilot 592.20 kmh
Renald Davenport, Copilot 367.98 mph
Piper Cheyenne 400LS
2 Garrett TPE 331-14, 1,500 shp
2/8/85

MIAMI/NEW YORK (USA)
Mark O. Henry, Pilot 612.36 kmh
Renald Davenport, Copilot 380.50 mph
Piper Cheyenne 400LS
2 Garrett TPE 331-14, 1,500 shp
2/8/85

MIRNY/YAKUTSK (USA)
D. Lancy Allyn 311.76 kmh
Beechcraft A90 193.73 mph
2 P&W PT6A-20
07/08/89

MONTEREY/SAN ANTONIO (USA)
Darrell A. Cottle 501.52 kmh
Beechcraft B200 311.63 mph
2 P&W PT6A, 850 shp
10/20/92

MOSCOW/IRKUTSK (FRANCE)
Claude Leprince 490.63 kmh
Beechcraft King Air 200 304.86 mph
6/23/92

MOSCOW/KUIBYSHEV (USA)
D. Lancy Allyn 333.36 kmh
Beechcraft A90 207.15 mph
2 P&W PT6A-20
7/1/89

MOSCOW/NOME (USA)
D. Lancy Allyn 19.44 kmh
Beechcraft A90 12.08 mph
2 P&W PT6A-20
7/1-15/89

MOSCOW/PROVIDENIA (USA)
D. Lancy Allyn 19.44 kmh
Beechcraft A90 12.08 mph
2 P&W PT6A-20
7/1-15/89

MUNICH/CAIRO (USA)
Robert E. Reinhold, Pilot 436.91 kmh
Frank X. Perullo, 271.48 mph
Donald J. Grant, Copilots
Piper PA-42 Cheyenne III
2 P&W PT6A-41, 720 shp
3/20/82

NARSSARSSUAQ/ILES DE LA MADELEINE (AUSTRIA)
Ferdinand Mühlhofer 495.15 kmh
Mitsubishi MU-2 26A 307.67 mph
2 Garrett TPE 331, 665 shp
7/24/93

NARSSARSSUAQ/REYKJAVIK (AUSTRIA)
Ferdinand Mühlhofer 433.86 kmh
Mitsubishi MU-2 26A 269.59 mph
2 Garrett TPE 331, 665 shp
8/17/93

NEW YORK/GANDER (W. GERMANY)
Herman Krug 568.86 kmh
Piper Cheyenne IIIA PA-42 353.49 mph
2 P&W PT6A-61, 720 shp
8/30/84

NEW YORK/GOOSE BAY (USA)
Robert Thompson, Pilot 552.24 kmh
Bryant Jones, Copilot 343.16 mph
Cessna Conquest 441
2 Garrett TPE 331-8, 635.5 shp
4/5/84

NEW YORK/KEFLAVIK (USA)
Robert Thompson, Pilot 502.56 kmh
Bryant Jones, Copilot 312.29 mph
Cessna Conquest 441
2 Garrett TPE 331-8, 635.5 shp
4/5-6/84

NEW YORK/LONDON (W. GERMANY)
Herman Krug 465.17 kmh
Cheyenne III A, PA-42 351.20 mph
2 P&W PT6A-61, 720 shp
8/30/84

NEW YORK/LOS ANGELES (USA)
Dominic Frontiere 434.52 kmh
Cessna Conquest 441 270.01 mph
2 Garrett
TPE 331-3025, 635 shp
7/18-19/83

NEW YORK/PARIS (USA)
Charles "Chuck" Yeager, Pilot 602.28 kmh
Renald Davenport, Copilot 374.26 mph
Piper Cheyenne 400LS
2 TPE 331-14, 1,000 shp
5/28/85

NEW YORK/PRESTWICK (USA)
Robert Thompson, Pilot 429.48 kmh
Bryant Jones, Copilot 266.87 mph
Cessna Conquest 441
2 Garrett TPE 331-8, 635.5 shp
4/5-6/84

NICE/VIENNA (AUSTRIA)
Ferdinand Mühlhofer 486.54 kmh
Mitsubishi MU-2 26A 302.32 mph
2 Garrett TPE 331, 665 shp
9/16/93

NORTH POLE/LONGYEARBYEN (USA)
Michael K. Egan 475.30 kmh
James M. Conn 295.34 mph
Socata TBM700
1 P&WC PT6A, 700 shp
11/8/94

OKHOTSK/MAGADAN (USA)
D. Lancy Allyn 258.84 kmh
Beechcraft A90 160.84 mph
2 P&W PT6A-20
7/11/89

OKLAHOMA CITY/GREENVILLE (USA)
Kevin G. Dunshee 523.65 kmh
David B. George 325.38 mph
Richard D. French
Beech C-12C
2 P&W Canada PT6A-41, 850 shp
4/25/98

OKLAHOMA CITY/KITTY HAWK (USA)
Tom A. Thomas, Jr. 553.54 kmh
Rockwell Commander, N711TT 343.97 mph
2 Garrett TPE-331-251K, 700 shp
12/16/86

OKLAHOMA CITY/ST. JOHN, NB (CANADA)
D. B. O'Connor, Pilot 590.01 kmh
D. Hall, Flight Engineer 366.63 mph
Gulfstream 1000
2 Garrett TPE 331-10-511K, 820 shp
11/7/85

OMSK/KEMEROVO (USA)
D. Lancy Allyn 251.28 kmh
Beechcraft A90 156.15 mph
2 P&W PT6A-20
7/4/89

ORLANDO/CANTON (USA)
J. Jeffrey Brausch 439.02 kmh
Piper PA-31T2 Cheyenne IIXL 272.80 mph
2 P&WC PT6A, 620 shp
4/7/02

ORLANDO/DALLAS (USA)
Andrew J. Cindric, Jr. 527.04 kmh
Piper Cheyenne PA-42A 327.50 mph
2 P&W PT6A-61, 720 shp
10/3/83

PALERMO/TUNIS (USA)
James P. Hanson 213.71 kmh
Cessna 208 Caravan 132.79 mph
1 P&W PT-6A, 600 shp
5/4/96

PALM BEACH/BALTIMORE (USA)
Tom Delantonas 609.69 kmh
Piper Cheyenne 400LS 378.86 mph
2 Garrett TPE 331-14, 1,000 shp
9/28/85

PARIS/AMSTERDAM (USA)
Robert Thompson, Pilot 414.97 kmh
Bryant Jones, Copilot 257.86 mph
Cessna Conquest 441
2 Garrett TPE 331-8, 635.5 shp
4/9/84

PERTH/ADELAIDE (AUSTRALIA)
Grant Hayden Kenny 548.42 kmh
Cessna 441 340.79 mph
3/12/88

PERTH/SYDNEY (AUSTRALIA)
Trevor James Frost 646.56 kmh
Piper Cheyenne 400LS 401.77 mph
2 Garrett TPE 331-14, 1,000 shp
4/25/86

PHILADELPHIA/GOOSE BAY (USA)
Robert E. Reinhold, Pilot 423.15 kmh
Frank X. Perullo, 262.93 mph
Donald J. Grant, Copilots
Piper PA-42 Cheyenne III
2 P&W PT6A-41, 720 shp
3/19/82

PHILADELPHIA/ORLANDO (USA)
Joe W. Kittinger, Pilot 533.88 kmh
Renald Davenport, Copilot 331.75 mph
Piper Cheyenne 400LS
2 Garrett TPE 331-14, 1,000 shp
2/9/86

PHOENIX/FORTH SMITH (USA)
Robert Thompson 578.30 kmh
Cessna Conquest 359.36 mph
2 Garrett TPE-331-8, 635 shp
3/13-14/84

PORTLAND/ATLANTA (USA)
Chuck Yeager, Pilot 625.68 kmh
Renald Davenport, Copilot 388.80 mph
Piper Cheyenne LS-400
2 Garrett TPE 331-14, 1,000 shp
10/1/84

PORTLAND/MINNEAPOLIS (USA)
Dan Vollum 711.14 kmh
Piper Cheyenne 400LS 442.08 mph
2 Garrett TPE 331-14, 1,000 shp
12/28/85

PORTLAND, OR/PORTLAND, ME (USA)
Randy Johnson 483.26 kmh
Gulfstream Commander 690C 300.28 mph
2 Garrett TPE 1331, 715 shp
2/9/84

PORTLAND/SACRAMENTO (USA)
Charles "Chuck" Yeager, Pilot 672.48 kmh
Renald Davenport, Copilot 417.88 mph
Piper Cheyenne 400LS
2 Garrett TPE 331-14, 1,000 shp
8/14-15/85

PRESTWICK/BOSTON (USA)
Robert Thompson, Pilot 346.32 kmh
Bryant Jones, Copilot 215.20 mph
Cessna Conquest 441
2 Garrett TPE 331-8, 635.5 shp
4/11-12/84

PRESTWICK/GOOSE BAY (USA)
Robert Thompson, Pilot 329.04 kmh
Bryant Jones, Copilot 204.46 mph
Cessna Conquest 441
2 Garrett TPE 331-8, 635.5 shp
4/11/84

PRESTWICK/KEFLAVIK (USA)
Robert Thompson, Pilot 342.00 kmh
Bryant Jones, Copilot 212.51 mph
Cessna Conquest 441
2 Garrett TPE 331-8, 635.5 shp
4/11/84

PRESTWICK/NEW YORK (USA)
Robert Thompson, Pilot 355.32 kmh
Bryant Jones, Copilot 220.79 mph
Cessna Conquest 441
2 Garrett TPE 331-8, 635.5 shp
4/11-12/84

PRESTWICK/PARIS (USA)
Robert Thompson, Pilot 493.10 kmh
Bryant Jones, Copilot 306.41 mph
Cessna Conquest 441
2 Garrett TPE 331-8, 635.5 shp
4/6/84

PRESTWICK/WASHINGTON (USA)
Robert Thompson, Pilot 358.56 kmh
Bryant Jones, Copilot 222.80 mph
Cessna Conquest 441
2 Garrett TPE 331-8, 635.5 shp
4/11-12/84

PROVIDENIA/NOME (USA)
D. Lancy Allyn 195.84 kmh
Beechcraft A90 121.70 mph
2 P&W PT6A-20
7/15/89

PUNTA ARENAS/ARICA (USA)
Michael K. Egan 391.92 kmh
James M. Conn 243.52 mph
Socata TBM700
1 P&WC PT6A, 700 shp
12/3/94

QUITO/SAN DIEGO (USA)
William F. Black 323.24 kmh
Turbo Commander 681B 200.85 mph
2 Garrett TPE-331-151-K, 675 shp
4/3-4/81

REYKJAVIK/INVERNESS (AUSTRIA)
Ferdinand Mühlhofer 424.00 kmh
Mitsubishi MU-2 26A 263.46 mph
2 Garrett TPE 331, 665 shp
8/18/93

REYKJAVIK/KULUSUK (USA)
J. Jeffrey Brausch 402.30 kmh
William H. Cox 249.98 mph
Piper PA-31T2 Cheyenne IIXL
2 P&WC PT6A, 620 shp
8/16/02

REYKJAVIK/LUGANO (AUSTRIA)
Ferdinand Mühlhofer 381.84 kmh
Mitsubishi MU-2 26A 237.26 mph
2 Garrett TPE 331, 665 shp
7/24/93

REYKJAVIK/NARSSARSSUAQ (AUSTRIA)
Ferdinand Mühlhofer 457.96 kmh
Mitsubishi MU-2 26A 284.56 mph
2 Garrett TPE 331, 665 shp
7/24/93

RIO DE JANEIRO/PUNTA ARENAS (USA)
Michael K. Egan 409.65 kmh
James M. Conn 254.54 mph
Socata TBM700
1 P&WC PT6A, 700 shp
11/10/94

ROME/ATHENS (USA)
James P. Hanson 267.80 kmh
Cessna 208 Caravan 166.40 mph
1 P&W PT-6A, 600 shp
4/28/96

SACRAMENTO/OKLAHOMA CITY (USA)
Kevin G. Dunshee 521.70 kmh
David B. George 324.17 mph
Richard D. French
Beech C-12C
2 P&W Canada PT6A-41, 850 shp
4/24/98

SAN DIEGO/CHARLESTON (USA)
Joseph J. Ritchie 880.01 kmh
Steve Fossett 546.81 mph
Piaggio P.180 Avanti
2 P&WC PT6A, 850 shp
2/6/03

SAN DIEGO/SACRAMENTO (USA)
Charles "Chuck" Yeager, Pilot 636.12 kmh
Renald Davenport, Copilot 395.29 mph
Piper Cheyenne 400LS
2 Garrett TPE 331-14, 1,000 shp
8/9/85

SAN FRANCISCO/CHARLESTON (USA)
Chuck Yeager, Pilot 666.53 kmh
Renald Davenport, Copilot 414.18 mph
Piper Cheyenne 400LS
2 Garrett TPE 331-14, 1,000 shp
10/13/85

SAN FRANCISCO/CINCINATTI (USA)
Chuck Yeager, Pilot 662.76 kmh
Renald Davenport, Copilot 411.84 mph
Piper Cheyenne 400LS
2 Garrett TPE 331-14, 1,000 shp
10/13/85

SAN FRANCISCO/HOUSTON (USA)
Darrell A. Cottle 616.11 kmh
Beechcraft B200 382.83 mph
2 P&W PT6A, 850 shp
11/28/92

SAN FRANCISCO/KANSAS CITY (USA)
Robert E. Reinhold, Pilot 436.62 kmh
Frank X. Perullo and 271.30 mph
Donald J. Grant, Copilots
Piper PA-42 Cheyenne III
2 P&W PT6A-41, 720 shp
3/23/82

SAN FRANCISCO/LOS ANGELES (USA)
Wayne A. Headlough, Pilot 732.37 kmh
Jean C. Ringwalt, Copilot 455.07 kmh
Piper Cheyenne 400LS
2 Garrett TPE-331, 1,000 shp
4/14/92

SAN FRANCISCO/NEW YORK (USA)
Charles "Chuck" Yeager, Pilot 622.44 kmh
Renald Davenport, Copilot 386.78 mph
Piper Cheyenne 400LS
2 TPE 331-14, 1,000 shp
4/19/85

SAN FRANCISCO/PORTLAND (USA)
Charles "Chuck" Yeager, Pilot 658.00 kmh
Renald Davenport, Copilot 408.88 mph
Piper Cheyenne 400LS
2 Garrett TPE 331-14, 1,000 shp
8/11/85

SAN FRANCISCO/POUGHKEEPSIE (USA)
Thomas W. Clements, Pilot 331.86 kmh
F. T. Elliot, Jr., Copilot 206.21 mph
Beechcraft King Air C-90
2 P&W PT6A-21, 550 shp
12 hrs 44 min 10 sec
5/21/77

SANTORINI/PALERMO (USA)
James P. Hanson 240.80 kmh
Cessna 208 Caravan 149.62 mph
1 P&W PT-6A, 600 shp
5/3/96

SAULT STE. MARIE/SEPT-ILES (USA)
James P. Hanson 250.72 kmh
Cessna 208 Caravan 155.79 mph
1 P&W PT-6A, 600 shp
4/8/96

SEATTLE/SACRAMENTO (USA)
Charles "Chuck" Yeager, Pilot 677.52 kmh
Renald Davenport, Copilot 421.01 mph
Piper Cheyenne 400LS
2 Garrett TPE 331-14, 1,000 shp
8/14-15/85

SEATTLE/PITTSBURGH (USA)
Leonard M. Greene, Pilot 415.17 kmh
Joyce Greene, Copilot 257.98 mph
Beechcraft King Aire B-90, N880X
2 P&W PT6A
8 hrs 44 min 30 sec
2/3/69

SEPT-ILES/GOOSE BAY (USA)
James P. Hanson 291.82 kmh
Cessna 208 Caravan 181.33 mph
1 P&W PT-6A, 600 shp
4/8/96

SHARJAH/COLOMBO (USA)
Robert E. Reinhold 483.66 kmh
Donald J. Grant 300.53 mph
Piper PA-42 Cheyenne III
2 P&W PT6A-41, 720 shp
3/20-21/82

SINGAPORE/MANILA (USA)
Robert E. Reinhold, Pilot 430.88 kmh
Frank X. Perullo and 267.74 mph
Donald J. Grant, Copilots
Piper PA-42 Cheyenne III
2 P&W PT6A-31, 720 shp
3/21/82

SONDRE STROMFJORD/IQALUIT (USA)
J. Jeffrey Brausch 414.71 kmh
William H. Cox 257.69 mph
Piper PA-31T2 Cheyenne IIXL
2 P&WC PT6A, 620 shp
8/17/02

SONDRE STROMFJORD/KULUSUK (USA)
J. Jeffrey Brausch 401.22 kmh
William H. Cox 249.31 mph
Piper PA-31T2 Cheyenne IIXL
2 P&WC PT6A, 620 shp
8/14/02

SONDRE STROMFJORD/REYKJAVIK (USA)
J. Jeffrey Brausch 268.63 kmh
William H. Cox 166.92 mph
Piper PA-31T2 Cheyenne IIXL
2 P&WC PT6A, 620 shp
8/14/02

STOCKHOLM/LENINGRAD (USA)
D. Lancy Allyn 14.40 kmh
Beechcraft A90 8.95 mph
2 P&W PT6A-20
6/21-23/89

STORNOWAY/REYKJAVIK (USA)
James P. Hanson 291.16 kmh
Cessna 208 Caravan 180.92 mph
1 P&W PT-6A, 600 shp
5/13/96

STORNOWAY/VAGAR (USA)
James P. Hanson 304.83 kmh
Cessna 208 Caravan 189.41 mph
1 P&W PT-6A, 600 shp
5/13/96

TARBES/LAGOS (USA)
Michael K. Egan 463.14 kmh
James M. Conn 287.78 mph
Socata TBM700
1 P&WC PT6A, 700 shp
11/8/94

TORONTO/BOSTON (USA)
Charles "Chuck" Yeager, Pilot 743.40 kmh
Renald Davenport, Copilot 461.95 mph
Piper Cheyenne 400LS
2 Garrett TPE 331-14, 1,000 shp
7/30/85

TUMEN/OMSK (USA)
D. Lancy Allyn 312.12 kmh
Beechcraft A90 193.95 mph
2 P&W PT6A-20
7/3/89

TUNIS/PALMA DE MAJORCA (USA)
James P. Hanson 254.08 kmh
Cessna 208 Caravan 157.88 mph
1 P&W PT-6A, 600 shp
5/5/96

USHUAIA/KING GEORGE ISLAND (USA)
James P. Hanson 269.32 kmh
Cessna 208 Caravan 167.34 mph
1 P&WC PT6A, 600 shp
11/20/99

USHUAIA/MARAMBIO (USA)
James P. Hanson 277.45 kmh
Cessna 208 Caravan 172.40 mph
1 P&WC PT6A, 600 shp
11/20/99

VAGAR/REYKJAVIK (USA)
James P. Hanson 347.09 kmh
Cessna 208 Caravan 215.67 mph
1 P&W PT-6A, 600 shp
5/13/96

VAIL/CANTON (USA)
J. Jeffrey Brausch 404.22 kmh
Piper PA-31T2 Cheyenne IIXL 251.17 mph
2 P&WC PT6A, 620 shp
7/8/02

VIENNA/ILES DE LA MADELEINE (AUSTRIA)
Ferdinand Mühlhofer 347.14 kmh
Mitsubishi MU-2 26A 215.70 mph
2 Garrett TPE 331, 665 shp
7/24/93

VIENNA/INVERNESS (AUSTRIA)
Ferdinand Mühlhofer 409.66 kmh
Mitsubishi MU-2 26A 254.55 mph
2 Garrett TPE 331, 665 shp
7/24/93

VIENNA/NARSSARSSUAQ (AUSTRIA)
Ferdinand Mühlhofer 354.06 kmh
Mitsubishi MU-2 26A 220.00 mph
2 Garrett TPE 331, 665 shp
7/24/93

VIENNA/REYKJAVIK (AUSTRIA)
Ferdinand Mühlhofer 385.89 kmh
Mitsubishi MU-2 26A 239.78 mph
2 Garrett TPE 331, 665 shp
7/24/93

VIENNA/RHODES (AUSTRIA)
Ferdinand Mühlhofer 454.62 kmh
Mitsubishi MU-2 26A 282.49 mph
2 Garrett TPE 331, 665 shp
9/18/93

WASHINGTON/DETROIT (USA)
Charles "Chuck" Yeager, Pilot 574.92 kmh
Renald Davenport, Copilot 357.26 mph
Piper Cheyenne 400LS
2 TPE 331-14, 1,000 shp
8/2/85

WASHINGTON/GANDER (W. GERMANY)
Herman Krug 564.01 kmh
Piper Cheyenne IIIA PA-42 350.84 mph
2 P&W PT6A-61, 720 shp
8/30/84

WASHINGTON/IQALUIT (USA)
Michael K. Egan 380.18 kmh
James M. Conn 236.23 mph
Socata TBM700
1 P&WC PT6A, 700 shp
11/7/94

WASHINGTON/LONDON (W. GERMANY)
Herman Krug 563.59 kmh
Piper Cheyenne IIIA PA-42 350.21 mph
2 P&W PT6A-61, 720 shp
8/30/84

WASHINGTON/PARIS (USA)
Charles "Chuck" Yeager, Pilot 598.51 kmh
Renald Davenport, Copilot 371.91 mph
Piper Cheyenne 400LS
2 TPE 331-14, 1,000 shp
5/27-28/85

YAKUTSK/OKHOTSK (USA)
D. Lancy Allyn 199.08 kmh
Beechcraft A90 123.27 mph
2 P&W PT6A-20
7/10/89

ROUND TRIP

DAYTON/KITTY HAWK/DAYTON (USA)
J. Jeffrey Brausch 404.25 kmh
Piper PA-31T2 Cheyenne IIXL 251.19 mph
2 P&WC PT6A, 620 shp
11/8/03

CLASS C-1.F (6,000-9,000 kg / 13,228-19,842 LBS)
GROUP II (TURBOPROP)

AGANA, GUAM/CEBU, P.I (USA)
V. C. Gardner 411.52 kmh
Volpar Turboliner, N353V 255.70 mph
2 Airesearch TPE-311-1-E, 620 shp
5 hrs 34 min 59.2 sec
4/1-2/70

ANADYR/FAIRBANKS (USA)
Allyn Caruso, Pilot 402.83 kmh
David Norgart, Copilot 250.30 mph
John Gallichon, Navigator
John Miller, Navigator
Dana Lovell, Mechanic
Fairchild Metroliner III
2 Garrett TPE331-11, 1,100 shp
6/5/91

ATLANTA/HOUSTON (USA)
Robert A. Brown 553.96 kmh
Colby W. Wilson 344.21 mph
Raytheon Beech 1900D
2 P&W PT6, 1,279 shp
8/27/97

BERLIN/MOSCOW (USA)
Allyn Caruso, Pilot 409.48 kmh
David Norgart, Copilot 254.44 mph
John Gallichon, Navigator
John Miller, Navigator
Dana Lovell, Mechanic
Fairchild Metroliner III
2 Garrett TPE331-11, 1,100 shp
6/1/91

CEBU, P.I/SINGAPORE (USA)
V. C. Gardner — 376.82 kmh, 234.15 mph
Volpar Turboliner, N353V
2 Airesearch TPE-311-1-E, 620 shp
6 hrs 27 min 54.7 sec
4/3/70

EDMONTON/MINNEAPOLIS (USA)
Allyn Caruso, Pilot — 413.79 kmh, 257.12 mph
David Norgart, Copilot
John Gallichon, Navigator
John Miller, Navigator
Dana Lovell, Mechanic
Fairchild Metroliner III
2 Garrett TPE331-11, 1,100 shp
6/6/91

FAIRBANKS/EDMONTON (USA)
Allyn Caruso, Pilot — 424.56 kmh, 263.81 mph
David Norgart, Copilot
John Gallichon, Navigator
John Miller, Navigator
Dana Lovell, Mechanic
Fairchild Metroliner III
2 Garrett TPE331-11, 1,100 shp
6/5-6/91

GANDER/ANKARA (USA)
Donald Holmes, Pilot — 533.16 kmh, 331.30 mph
Endogan Meneske, Copilot
Piper Cheyenne IIIA, TC-FAH
2 P&W PT6A-41, 720 shp
12/17-18/85

GUAM/WAKE ISLAND (USA)
Kevin G. Dunshee — 458.23 kmh, 284.73 mph
David B. George
Richard D. French
Beech C-12C
2 P&W Canada PT6A-41, 850 shp
4/22/98

HONOLULU/SACRAMENTO (USA)
Kevin G. Dunshee — 513.09 kmh, 318.81 mph
David B. George
Richard D. French
Beech C-12C
2 P&WC PT6A-41, 850 shp
4/23/98

HONOLULU/WAKE ISLAND (USA)
V. C. Gardner — 373.32 kmh, 231.97 mph
Volpar Turboliner, N353V
2 Airesearch TPE-311-1-E, 620 shp
9 hrs 56 min 27.6 sec
3/30-31/70

HOUSTON/ATLANTA (USA)
Allyn Caruso, Pilot — 410.03 kmh, 254.78 mph
David Norgart, Copilot
John Gallichon, Navigator
John Miller, Navigator
Dana Lovell, Mechanic
Fairchild Metroliner III
2 Garrett TPE331-11, 1,100 shp
6/7/91

HOUSTON/CHARLOTTE (USA)
Allyn Caruso, Pilot — 404.59 kmh, 251.40 mph
David Norgart, Copilot
John Gallichon, Navigator
John Miller, Navigator
Dana Lovell, Mechanic
Fairchild Metroliner III
2 Garrett TPE331-11, 1,100 shp
6/7/91

HOUSTON/NEW YORK (USA)
Robert A. Brown — 514.90 kmh, 319.94 mph
Colby W. Wilson
Raytheon Beech 1900D
2 P&W PT6, 1,279 shp
8/28/97

IRKUTSK/KHABAROVSK (USA)
Allyn Caruso, Pilot — 377.00 kmh, 234.25 mph
David Norgart, Copilot
John Gallichon, Navigator
John Miller, Navigator
Dana Lovell, Mechanic
Fairchild Metroliner III
2 Garrett TPE331-11, 1,100 shp
6/4/91

KEFLAVIK/BERLIN (USA)
Allyn Caruso, Pilot — 439.45 kmh, 273.06 mph
David Norgart, Copilot
John Gallichon, Navigator
John Miller, Navigator
Dana Lovell, Mechanic
Fairchild Metroliner III
2 Garrett TPE331-11, 1,100 shp
5/31/91

KHABAROVSK/ANADYR (USA)
Allyn Caruso, Pilot — 268.59 kmh, 166.89 mph
David Norgart, Copilot
John Gallichon, Navigator
John Miller, Navigator
Dana Lovell, Mechanic
Fairchild Metroliner III
2 Garrett TPE331-11, 1,100 shp
6/4-5/91

KHABAROVSK/MAGADAN (USA)
Allyn Caruso, Pilot — 402.05 kmh, 249.82 mph
David Norgart, Copilot
John Gallichon, Navigator
John Miller, Navigator
Dana Lovell, Mechanic
Fairchild Metroliner III
2 Garrett TPE331-11, 1,100 shp
6/4-5/91

LOS ANGELES/BOSTON (USA)
Richard Mitchell — 554.00 kmh, 344.26 mph
Fairchild 300
2 Garrett TPE 331 10,900 shp
4/3/84

LOS ANGELES/NEW YORK (USA)
Richard Mitchell — 574.56 kmh, 357.03 mph
Fairchild 300
2 Garrett TPE 331 10, 900 shp
4/3/84

MAGADAN/ANADYR (USA)
Allyn Caruso, Pilot — 381.14 kmh, 236.82 mph
David Norgart, Copilot
John Gallichon, Navigator
John Miller, Navigator
Dana Lovell, Mechanic
Fairchild Metroliner III
2 Garrett TPE331-11, 1,100 shp
6/5/91

MANILA/GUAM (USA)
Kevin G. Dunshee — 451.44 kmh, 280.51 mph
David B. George
Richard D. French
Beech C-12C
2 P&W Canada PT6A-41, 850 shp
4/14/98

MINNEAPOLIS/DALLAS-FORT WORTH (USA)
Allyn Caruso, Pilot 406.23 kmh
David Norgart, Copilot 252.42 mph
John Gallichon, Navigator
John Miller, Navigator
Dana Lovell, Mechanic
Fairchild Metroliner III
2 Garrett TPE331-11, 1,100 shp
6/6/91

MINNEAPOLIS/SAN ANTONIO (USA)
Robert A. Brown 571.08 kmh
Colby W. Wilson 354.85 mph
Raytheon Beech 1900D
2 P&W PT6, 1,279 shp
8/26/97

MINNEAPOLIS/TULSA (USA)
Robert A. Brown 554.69 kmh
Colby W. Wilson 344.66 mph
Raytheon Beech 1900D
2 P&W PT6, 1,279 shp
8/28/97

MOSCOW/EDMONTON (USA)
Allyn Caruso, Pilot 110.68 kmh
David Norgart, Copilot 68.77 mph
John Gallichon, Navigator
John Miller, Navigator
Dana Lovell, Mechanic
Fairchild Metroliner III
2 Garrett TPE331-11, 1,100 shp
6/3-6/91

MOSCOW/FAIRBANKS (USA)
Allyn Caruso, Pilot 110.17 kmh
David Norgart, Copilot 68.45 mph
John Gallichon, Navigator
John Miller, Navigator
Dana Lovell, Mechanic
Fairchild Metroliner III
2 Garrett TPE331-11, 1,100 shp
6/3-5/91

MOSCOW/IRKUTSK (USA)
Allyn Caruso, Pilot 338.35 kmh
David Norgart, Copilot 210.24 mph
John Gallichon, Navigator
John Miller, Navigator
Dana Lovell, Mechanic
Fairchild Metroliner III
2 Garrett TPE331-11, 1,100 shp
6/3/91

MOSCOW/MINNEAPOLIS (USA)
Allyn Caruso, Pilot 93.64 kmh
David Norgart, Copilot 58.19 mph
John Gallichon, Navigator
John Miller, Navigator
Dana Lovell, Mechanic
Fairchild Metroliner III
2 Garrett TPE331-11, 1,100 shp
6/3-6/91

MOSCOW/NOVOSIBIRSK (USA)
Allyn Caruso, Pilot 415.69 kmh
David Norgart, Copilot 258.30 mph
John Gallichon, Navigator
John Miller, Navigator
Dana Lovell, Mechanic
Fairchild Metroliner III
2 Garrett TPE331-11, 1,100 shp
6/3/91

MOSCOW/SAN ANTONIO (USA)
Allyn Caruso, Pilot 106.64 kmh
David Norgart, Copilot 66.26 mph
John Gallichon, Navigator
John Miller, Navigator
Dana Lovell, Mechanic
Fairchild Metroliner III
2 Garrett TPE331-11, 1,100 shp
6/3-6/91

MOSCOW/SVERDLOVSK (USA)
Allyn Caruso, Pilot 414.52 kmh
David Norgart, Copilot 257.57 mph
John Gallichon, Navigator
John Miller, Navigator
Dana Lovell, Mechanic
Fairchild Metroliner III
2 Garrett TPE331-11, 1,100 shp
6/3/91

NEW YORK/ANKARA (USA)
Donald Holmes, Pilot 533.52 kmh
Endogan Meneske, Copilot 331.53 mph
Piper Cheyenne IIIA, TC-FAH
2 P&W PT6A-41, 720 shp
12/17-18/85

NEW YORK/BERLIN (USA)
Allyn Caruso, Pilot 358.77 kmh
David Norgart, Copilot 222.93 mph
John Gallichon, Navigator
John Miller, Navigator
Dana Lovell, Mechanic
Fairchild Metroliner III
2 Garrett TPE331-11, 1,100 shp
5/30-31/91

NEW YORK/GOOSE BAY (USA)
Allyn Caruso, Pilot 405.37 kmh
David Norgart, Copilot 251.88 mph
John Gallichon, Navigator
John Miller, Navigator
Dana Lovell, Mechanic
Fairchild Metroliner III
2 Garrett TPE331-11, 1,100 shp
5/30/91

NEW YORK/MINNEAPOLIS (USA)
Robert A. Brown 496.32 kmh
Colby W. Wilson 308.39 mph
Raytheon Beech 1900D
2 P&W PT6, 1,279 shp
8/28/97

NEW YORK/MOSCOW (USA)
Allyn Caruso, Pilot 172.86 kmh
David Norgart, Copilot 107.41 mph
John Gallichon, Navigator
John Miller, Navigator
Dana Lovell, Mechanic
Fairchild Metroliner III
2 Garrett TPE331-11, 1,100 shp
5/30-6/1/91

NOVOSIBIRSK/IRKUTSK (USA)
Allyn Caruso, Pilot 370.82 kmh
David Norgart, Copilot 230.41 mph
John Gallichon, Navigator
John Miller, Navigator
Dana Lovell, Mechanic
Fairchild Metroliner III
2 Garrett TPE331-11, 1,100 shp
6/3/91

OAKLAND/HONOLULU (USA)
James H. Barker, Pilot 489.24 kmh
Robert J. Sutton, Copilot 304.00 mph
Fairchild Metro 23 (C-26B)
2 Garrett TPE331-12, 1,119 shp
6/8/92

SAN ANTONIO/ATLANTA (USA)
Robert A. Brown 539.21 kmh
Colby W. Wilson 335.05 mph
Raytheon Beech 1900D
2 P&W PT6, 1,279 shp
8/27/97

SAN ANTONIO/NEW YORK (USA)
Allyn Caruso, Pilot 395.18 kmh
David Norgart, Copilot 245.55 mph
John Gallichon, Navigator
John Miller, Navigator
Dana Lovell, Mechanic
Fairchild Metroliner III
2 Garrett TPE331-11, 1,100 shp
6/7/91

SAN ANTONIO/NORFOLK (USA)
Allyn Caruso, Pilot 446.35 kmh
David Norgart, Copilot 277.35 mph
John Gallichon, Navigator
John Miller, Navigator
Dana Lovell, Mechanic
Fairchild Metroliner III
2 Garrett TPE331-11, 1,100 shp
6/7/91

SAN LOUIS OBISPO/SIOUX CITY*
C. Martin Coddington 303.80 mph
Jetstream Aircraft, Jetstream 31
2 Garrett-AiResearch TPE331, 900 shp
6/15/94

TULSA/SAN ANTONIO (USA)
Robert A. Brown 540.62 kmh
Colby W. Wilson 335.92 mph
Raytheon Beech 1900D
2 P&W PT6, 1,279 shp
9/4/97

WAKE ISLAND/AGANA, GUAM (USA)
V. C. Gardner 345.30 kmh
Volpar Turboliner, N353V 214.56 mph
2 Airesearch TPE-311-1-E, 620 shp
7 hrs 37.8 sec
3/31-4/1/70

WAKE ISLAND/HONOLULU (USA)
Kevin G. Dunshee 492.79 kmh
David B. George 306.20 mph
Richard D. French
Beech C-12C
2 P&W Canada PT6A-41, 850 shp
4/22/98

WICHITA/MINNEAPOLIS (USA)
Robert A. Brown 522.02 kmh
Colby W. Wilson 324.37 mph
Raytheon Beech 1900D
2 P&W PT6, 1,279 shp
8/26/97

CLASS C-1.H (12,000-16,000 kg / 26,455-35,274 LBS)
GROUP II (TURBOPROP)

GOOSE BAY/MONTREAL (USA)
Benjamin C. White, Pilot 556.63 kmh
Meinhardt Feuersenger, Copilot 345.87 mph
Dornier 328
2 P&W PW119A, 2,180 shp
4/6/93

KANSAS CITY/PHOENIX (USA)
Benjamin C. White, Pilot 438.22 kmh
Meinhardt Feuersenger, Copilot 272.30 mph
Dornier 328
2 P&W PW119A, 2,180 shp
4/6/93

MONTREAL/KANSAS CITY (USA)
Benjamin C. White, Pilot 512.82 kmh
Meinhardt Feuersenger, Copilot 318.65 mph
Dornier 328
2 P&W PW119A, 2,180 shp
4/6/93

MUNICH/GOOSE BAY (USA)
Benjamin C. White, Pilot 465.80 kmh
Meinhardt Feuersenger, Copilot 289.43 mph
Dornier 328
2 P&W PW119A, 2,180 shp
4/4/93

MUNICH/PHOENIX (USA)
Benjamin C. White, Pilot 146.28 kmh
Meinhardt Feuersenger, Copilot 90.89 mph
Dornier 328
2 P&W PW119A, 2,180 shp
4/4-6/93

MUNICH/REYKJAVIK (USA)
Benjamin C. White, Pilot 520.56 kmh
Meinhardt Feuersenger, Copilot 323.46 mph
Dornier 328
2 P&W PW119A, 2,180 shp
4/4/93

REYKJAVIK/GOOSE BAY (USA)
Benjamin C. White, Pilot 592.74 kmh
Meinhardt Feuersenger, Copilot 368.31 mph
Dornier 328
2 P&W PW119A, 2,180 shp
4/4/93

CLASS C-1.J (20,000-25,000 kg / 44,092-55,116 LBS)
GROUP II (TURBOPROP)

TORONTO/NASSAU (CANADA)
Louis Sytsma 619.63 kmh
Kevin Spackman 385.02 mph
DHC-8 Dash 8 Q400
2 P&W PW150, 5100 shp
10/10/00

CLASS C-1.M (45,000-60,000 kg / 99,208-132,277 lbs)
GROUP II (TURBOPROP)

ATLANTA/SWINDON (USA)
Arlen D. Rens, Pilot 666.25 kmh
Jeffrey Bullen, Copilot 413.99 mph
Edward J. Delehant, III, Copilot
Timothy L. Gomez, Aircraft Systems Specialist
Lockheed Martin C-130J-30
4 RR Allison AE 2100, 4,591 shp
12/8/99

CLASS C-1.N (60,000-80,000 kg / 132,277-176,370 lbs)
GROUP II (TURBOPROP)

FAYETTEVILLE/CAMBRIDGE (USA)
Arlen D. Rens, Pilot 661.43 kmh
Edward J. Delehant, III, Copilot 410.99 mph
Richard L. White, Aircraft Systems Specialist
Lockheed Martin C-130J
4 R-R Allision AE2100, 4,591 shp
2/13/00

CLASS C-1.E (3,000-6,000 kg / 6,614-13,228 LBS) GROUP III (JET ENGINE)

WEST TO EAST TRANSCONTINENTAL*
Terry J. Kohler 432.60 mph
Cessna Citation, C-501-SP
2 P&W, 2,500 lbs
11/18/91

***

ADAK/ANCHORAGE (UK)
M. Naviede 575.41 kmh
Cessna 550 357.54 mph
5/12/91

ANCHORAGE/SEATTLE (UK)
M. Naviede 552.29 kmh
Cessna 550 343.17 mph
5/12/91

ATLANTA/PENSACOLA (USA)
Paul T. Entrekin 813.55 kmh
MiG-15bis 505.51 mph
1 Klimov VK-1, 6,000 lbs
7/6/93

BANGKOK/MANILA (UK)
M. Naviede 454.14 kmh
Cessna 550 282.18 mph
5/11/91

CALCUTTA/BANGKOK (UK)
M. Naviede 568.33 kmh
Cessna 550 353.14 mph
5/10/91

CHATTANOOGA/DAYTONA BEACH (USA)
Spencer L. Morgan 699.18 kmh
Raytheon 390 Premier I 434.45 mph
2 Williams FJ44, 2,300 lbs
2/14/03

GOOSE BAY/KEFLAVIK (UK)
M. Naviede 598.03 kmh
Cessna 550 371.59 mph
5/13/91

GUAM/WAKE ISLAND (UK)
M. Naviede 523.12 kmh
Cessna 550 325.05 mph
5/11/91

HONOLULU/OAKLAND (USA)
Mark Huffstutler 582.44 kmh
Cessna Citation 500 361.91 mph
2 P&W JT15D, 2,200 lbs
12/31/94

HOUSTON/WASHINGTON (USA)
Charles Hunter 708.48 kmh
Cessna Citation, N1UH 440.25 mph
2 P&W JT 15-D4, 2,500 lbs
1/14/89

ISTANBUL/DUBAI (USA)
H. James Knuppe 397.60 kmh
Lawrence M. Wilson 247.05 mph
Cessna 525 CitationJet
2 RR/Williams FJ44, 1,900 lbs
5/7/94

KARACHI/CALCUTTA (UK)
M. Naviede 595.08 kmh
Cessna 550 369.76 mph
5/10/91

KEFLAVIK/MANCHESTER (UK)
M. Naviede 670.95 kmh
Cessna 550 416.90 mph
5/13/91

LAS VEGAS, NV/WICHITA (UK)
M. Naviede 546.88 kmh
Cessna 550 339.81 mph
5/12/91

LOS ANGELES/NEW YORK (USA)
John M. Conroy 770.03 kmh
Lear Jet Model 23 N1965L 478.47 mph
2 GE CJ610-1
5 hrs 8 min 47 sec
5/21/65

LOS ANGELES/SAN FRANCISCO (USA)
Clay Lacy, Pilot 913.88 kmh
Harold Fishman, Copilot 567.86 mph
Lear Jet Model 24A
2 GE CJ-610-4
35 min 34 sec
11/4/71

LOS ANGELES/WASHINGTON (USA)
Charles Hunter 728.28 kmh
Cessna Citation 2SP, N1UH 452.55 mph
2 P&W JT 15 D-4, 2,500 lbs
12/21/87

LUXOR/MANAMA (UK)
M. Naviede 563.80 kmh
Cessna 550 350.32 mph
5/10/91

MANAMA/KARACHI (UK)
M. Naviede 577.11 kmh
Cessna 550 358.59 mph
5/10/91

MANCHESTER/VALETTA (UK)
M. Naviede 576.45 kmh
Cessna 550 358.18 mph
5/10/91

MANILA/GUAM (UK)
M. Naviede 513.93 kmh
Cessna 550 319.34 mph
5/11/91

MARRAKECH/VALLETTA (USA)
H. James Knuppe 567.96 kmh
Lawrence M. Wilson 352.91 mph
Cessna 525 CitationJet
2 RR/Williams FJ44, 1,900 lbs
5/5/94

MIAMI/WASHINGTON (USA)
Charles Hunter 752.76 kmh
Cessna Citation, N1UH 467.76 mph
2 P&W JT 15-D4, 2,500 lbs
1/22/89

MIDWAY/ADAK (UK)
M. Naviede 595.76 kmh
Cessna 550 370.18 mph
5/12/91

NEW YORK/LOS ANGELES (USA)
John M. Conroy 689.90 kmh
Lear Jet Model 23, N1965L 428.68 mph
2 GE CJ-610-1
5 hrs 44 min 38 sec
5/21/65

OKINAWA/SENDAI (USA)
H. James Knuppe 603.06 kmh
Lawrence M. Wilson 374.72 mph
Cessna 525 CitationJet
2 RR/Williams FJ44, 1,900 lbs
5/16/94

OSHKOSH/PEORIA, IL*
Michael P. Hare — 396.81 mph
Cessna OA-37
2 J-85, 3,500 lbs
8/1/91

PORTLAND/SAN FRANCISCO (USA)
Charles Kaady — 812.86 kmh
Art Ashton — 505.08 mph
Cessna Citation 500
2 P&W JT15D-1A, 2,200 lbs
2/18/91

ST. JOHN'S/MARRAKECH (USA)
H. James Knuppe — 511.50 kmh
Lawrence M. Wilson — 317.83 mph
Cessna 525 CitationJet
2 RR/Williams FJ44, 1,900 lbs
5/2/94

SAN ANTONIO/ALBANY (USA)
Carl Pascarell — 844.57 kmh
Swearingen SJ-30 — 524.79 mph
2 Williams/RR FJ44, 1,900 lbs
2/21/92

SAN DIEGO/BRUNSWICK (USA)
Terry J. Kohler — 696.20 kmh
Cessna Citation, C-501-SP — 432.60 mph
2 P&W, 2,500 lbs
11/18/91

SAN FRANCISCO/LOS ANGELES (USA)
Clay Lacy, Pilot — 1,007.86 kmh
Lear Jet Model 24A — 626.26 mph
2 GE CJ-610-4
32 min 15 sec
11/4/71

SEATTLE/LAS VEGAS, NV (UK)
M. Naviede — 585.88 kmh
Cessna 550 — 364.04 mph
5/12/91

SENDAI/PETROPAVLOVSK (USA)
H. James Knuppe — 534.13 kmh
Lawrence M. Wilson — 331.89 mph
Cessna 525 CitationJet
2 RR/Williams FJ44, 1,900 lbs
5/19/94

TORONTO/GOOSE BAY (UK)
M. Naviede — 574.85 kmh
Cessna 550 — 357.19 mph
5/12/91

VALLETTA/ISTANBUL (USA)
H. James Knuppe — 567.11 kmh
Lawrence M. Wilson — 352.38 mph
Cessna 525 CitationJet
2 RR/Williams FJ44, 1,900 lbs
5/6/94

VALLETTA/LUXOR (UK)
M. Naviede — 589.80 kmh
Cessna 550 — 366.48 mph
5/10/91

WAKE ISLAND/MIDWAY (UK)
M. Naviede — 560.68 kmh
Cessna 550 — 348.39 mph
5/11/91

WICHITA/NORFOLK (USA)
Mark A. Mills — 1,029.09 kmh
Raytheon 390 Premier I — 639.44 mph
2 Williams FJ44, 2,300 lbs
12/3/02

WICHITA/ORLANDO (USA)
Thomas O. Schiller — 894.26 kmh
Joseph S. Grubiak — 555.67 mph
Raytheon 390 Premier I
2 Williams FJ44, 2,300 lbs
10/5/03

WICHITA/OSHKOSH (USA)
Trevor E. Blackmer — 861.18 kmh
Joseph S. Grubiak — 535.11 mph
Raytheon 390 Premier I
2 Williams FJ44, 2,300 lbs
7/28/03

WICHITA/TORONTO (UK)
M. Naviede — 596.95 kmh
Cessna 550 — 370.92 mph
5/12/91

ROUND TRIP

LOS ANGELES/NEW YORK/LOS ANGELES (USA)
John M. Conroy — 682.66 kmh
Lear Jet Model 23 N1965L — 424.18 mph
2 GE CJ-610-1
11 hrs 32 min 40.8 sec
5/21/65

LOS ANGELES/SAN FRANCISCO/LOS ANGELES (USA)
Clay Lacy, Pilot — 937.84 kmh
Harold Fishman, Copilot — 582.75 mph
Lear Jet Model 24A
2 GE CJ-610-4
1 hr 9 min 19 sec
11/4/71

CLASS C-1.F (6,000-9,000 kg / 13,228-19,842 lbs)
GROUP III (JET ENGINE)

WEST TO EAST TRANSCONTINENTAL*
Nick A. Caporella, Pilot — 613.89 mph
Andy G. Grabnickas, Copilot
Dassault Falcon 10
2 Garrett TFE731, 3,230 lbs
Santa Ana to Fort Lauderdale
4/22/91

***

ANCHORAGE/SEATTLE (USA)
Henry G. Beaird, Pilot — 753.57 kmh
John O. Lear, — 468.25 mph
Robert R. King, Copilots
Lear Jet Model 24, N427LJ
2 GE CJ-610-4
3 hrs 6 min 54.9 sec
5/26/66

ASPEN/WASHINGTON (USA)
Peter T. Reynolds, Pilot — 812.33 kmh
James R. McClellan, Copilot — 504.75 mph
Kirk A. Vining, Copilot
Learjet 31A
2 Garrett TFE731-2, 3,500 lbs
8/25/94

ATLANTA/GANDER (USA)
Douglas Matthews, Pilot — 853.61 kmh
David Marco, Copilot — 530.43 mph
Learjet, N84DM
2 TFE-731-2-2B, 3,500 lbs
12/25-26/86

BAHRAIN/COLUMBO (USA)
Douglas Matthews — 830.52 kmh
Learjet 35A — 516.09 mph
2 Garrett TFE 731-B, 3,500 lbs
2/13/84

BARCELONA/ISTANBUL (USA)
Henry G. Beaird, Pilot
John O. Lear,
Robert R. King, Copilots
Lear Jet Model 24, N427LJ
2 GE CJ-610-4
3 hrs 2 min 37.8 sec
5/24/66
732.36 kmh
455.07 mph

BASSETERRE/SAL (USA)
Mark E. Calkins
Charles Conrad, Jr.
Paul Thayer
Daniel Miller
Learjet 35A
2 Garrett TFE731, 3,500 lbs
2/12/96
869.73 kmh
540.42 mph

BELFAST/PARIS (USA)
Rick E. Rowe
William A. Landrum
Learjet 31A
2 AlliedSignal TFE731, 3,500 lbs
6/14/01
824.16 kmh
512.11 mph

BIAK/WAKE (USA)
Douglas Matthews
Learjet 35A
2 Garrett TFE 731-B, 3,500 shp
2/14/84
717.12 kmh
445.61 mph

COLUMBO/SINGAPORE (USA)
Douglas Matthews
Learjet 35A
2 Garrett TFE 731-B, 3,500 lbs
2/13/84
791.02 kmh
491.54 mph

CONCORD/DAYTONA BEACH (USA)
Robert A. Combs, Jr.
Kenneth A. Combs
Mitsubishi Diamond I-A
2 P&WC JT15D, 2,500 lbs
7/3/96
712.90 kmh
442.98 mph

CORFU/BAHRAIN (USA)
Douglas Matthews
Learjet 35A
2 Garrett TFE 731-B, 3,500 lbs
2/13/84
745.92 kmh
463.51 mph

DENVER/BASSETERRE (USA)
Mark E. Calkins
Charles Conrad, Jr.
Paul Thayer
Daniel Miller
Learjet 35A
2 Garrett TFE731, 3,500 lbs
2/12/96
895.92 kmh
556.70 mph

DENVER/CHICAGO (USA)
Alan W. Zielinski
Verne Jobst
Cessna Citation V
2 P&W JT15D-5A, 2,900 lbs
4/21/96
869.10 kmh
540.03 mph

DENVER/NEW YORK (USA)
Brooke Knapp
Learjet 35A
2 TFE 731-2-2B, 3,500 lbs
1/16/84
948.68 kmh
589.51 mph

DENVER/OMAHA (USA)
Alan W. Zielinski
Verne Jobst
Cessna Citation V
2 P&W JT15D-5A, 2,900 lbs
4/21/96
861.61 kmh
535.37 mph

FT LAUDERDALE/OSHKOSH (USA)
Louis S. Beck
John Devlin
Learjet 35A
2 Garrett TFE731-2-2B, 3,500 lbs
7/31/97
719.70 kmh
447.20 mph

GANDER/KEFLAVIK (USA)
Rick E. Rowe
William A. Landrum
Learjet 31A
2 AlliedSignal TFE731, 3,500 lbs
6/12/01
755.99 kmh
469.75 mph

GANDER/LONDON (USA)
Douglas Matthews, Pilot
David Marco, Copilot
Learjet, N84DM
2 TFE-731-2-2B, 3,500 lbs
12/26/86
856.75 kmh
532.38 mph

GOOSE BAY/LONDON (USA)
Douglas Matthews
Learjet 35A
2 TFE 731-2-2B, 3,500 lbs
2/12-13/84
860.93 kmh
534.98 mph

HONOLULU/SAN FRANCISCO (USA)
Brooke Knapp, Pilot
James Topalian, Copilot
Paul Broyles, Copilot
James Magill, Crew Chief
Learjet 35A
2/18/83
988.89 kmh
614.50 mph

HONOLULU/SEATTLE (USA)
Douglas Matthews
Learjet 35A
2 TFE 731-2-2B, 3,500 lbs
2/14/84
855.72 kmh
531.75 mph

ISTANBUL/TEHERAN (USA)
Henry G. Beaird, Pilot
John O. Lear,
Robert R. King, Copilots
Lear Jet Model 24, N427LJ
2 GE CJ-610-4
2 hrs 52 min 24.7 sec
5/24/66
712.44 kmh
442.69 mph

KARACHI/COLOMBO (USA)
Henry G. Beaird, Pilot
John O. Lear,
Robert R. King, Copilots
Lear Jet Model 24, N427LJ
2 GE CJ-610-4
3 hrs 33 min 11.1 sec
6/24-25/66
680.75 kmh
423.03 mph

KING SALMON/DENVER (USA)
Mark E. Calkins
Charles Conrad, Jr.
Paul Thayer
Daniel Miller
Learjet 35A
2 Garrett TFE731, 3,500 lbs
2/14/96
854.75 kmh
531.11 mph

LAS VEGAS/CHICAGO (USA)
Alan W. Zielinski
Verne Jobst
Cessna Citation V
2 P&W JT15D-5A, 2,900 lbs
4/21/96
866.54 kmh
538.44 mph

LAS VEGAS/DENVER (USA)
Alan W. Zielinski
Verne Jobst
Cessna Citation V
2 P&W JT15D-5A, 2,900 lbs
4/21/96
866.01 kmh
538.12 mph

LAS VEGAS/OMAHA (USA)
Alan W. Zielinski — 862.88 kmh
Verne Jobst — 536.17 mph
Cessna Citation V
2 P&W JT15D-5A, 2,900 lbs
4/21/96

LONDON/CORFU (USA)
Brooke Knapp, Pilot — 750.70 kmh
James Topalian, Copilot — 466.48 mph
Paul Broyles, Copilot
James Magill, Crew Chief
Learjet 35A
2/17/83

LONDON/GANDER (USA)
Douglas Matthews, Pilot — 568.82 kmh
David Marco, Copilot — 353.47 mph
Learjet, N83DM
2 TFE-731-2-2B, 3,500 lbs
12/28/86

LONDON/LUXOR (USA)
Brooke Knapp, Pilot — 775.73 kmh
James Topalian, Copilot — 482.04 mph
Paul Broyles, Copilot
James Magill, Crew Chief
Learjet 35A
2/17/83

LONDON/NEWPORT NEWS, VA (USA)
Douglas Matthews, Pilot — 578.17 kmh
David Marco, Copilot — 359.28 mph
Learjet, N83DM
2 TFE-731-2-2B, 3,500 lbs
12/28/86

LONDON/SHANNON (USA)
Douglas Matthews, Pilot — 583.13 kmh
David Marco, Copilot — 362.36 mph
Learjet, N83DM
2 TFE-731-2-2B, 3,500 lbs
12/28/86

LOS ANGELES/KEY WEST (USA)
Donald J. Vecchie — 948.31 kmh
Leslie Bland — 589.25 mph
Learjet 35A
2 Garret TFE731, 3,500 lbs
3/24/92

LOS ANGELES/NEW YORK (USA)
Robert Berry — 782.59 kmh
Learjet 25C, N251GL — 486.28 mph
2 GE CJ-610-6
5 hrs 2 min 6.6 sec
8/30/70

LOS ANGELES/ORLANDO (USA)
Brooke Knapp — 1,034.28 kmh
Learjet 35A — 642.70 mph
2 TFE 731-2-2B, 3,500 lbs
3/23-24/83

LOS ANGELES/PARIS(USA)
Brooke Knapp — 693.62 kmh
Learjet 35A — 430.99 mph
2 TFE 731-2-2B, 3,500 lbs
5/26-27/83

LOS ANGELES/WASHINGTON (USA)
Brooke Knapp — 883.16 kmh
Learjet 35A — 548.80 mph
2 TFE 731-2-2B, 3,500 lbs
2/26/83

LOS ANGELES/WICHITA (USA)
Henry G. Beaird, Pilot — 837.00 kmh
John O. Lear, — 520.09 mph
Robert R. King, Copilots
Lear Jet Model 24, N427LJ
2 GE CJ-610-4
2 hrs 17 min 56.8 sec
5/26/66

MANILA/OSAKA (USA)
Henry G. Beaird, Pilot — 729.82 kmh
John O. Lear, — 453.49 mph
Robert R. King, Copilots
Lear Jet Model 24, N427LJ
2 GE CJ-610-4
3 hrs 39 min 24.7 sec
5/25/66

MANILA/YUZHNO-SAKHALINSK (USA)
Mark E. Calkins — 754.10 kmh
Charles Conrad, Jr. — 468.58 mph
Paul Thayer
Daniel Miller
Learjet 35A
2 Garrett TFE731, 3,500 lbs
2/14/96

MARSHALL/SAN ANTONIO (USA)
Jack Lenox — 531.06 kmh
James L. Holland — 329.98 mph
Cessna 560 Citation Excel
2 P&WC PW545A, 3,804 lbs
12/7/01

NEW YORK/LOS ANGELES (USA)
Robert Berry — 722.09 kmh
Learjet 25C, N251GL — 488.69 mph
2 GE CJ-610-6
5 hrs 27 min 25.9 sec
8/30/70

OLBIA/RIYADH (USA)
Mark E. Calkins — 794.34 kmh
Charles Conrad, Jr. — 493.58 mph
Paul Thayer
Daniel Miller
Learjet 35A
2 Garrett TFE731, 3,500 lbs
2/13/96

OMAHA/CHICAGO (USA)
Alan W. Zielinski — 877.76 kmh
Verne Jobst — 545.41 mph
Cessna Citation V
2 P&W JT15D-5A, 2,900 lbs
4/21/96

OSAKA/SAPPORO (USA)
Henry G. Beaird, Pilot — 662.61 kmh
John O. Lear, — 411.73 mph
Robert R. King, Copilots
Lear Jet Model 24, N427LJ
2 GE CJ-610-4
1 hr 36 min 2.6 sec
5/25/66

RIYADH/VARANASI (USA)
Mark E. Calkins — 865.52 kmh
Charles Conrad, Jr. — 537.81 mph
Paul Thayer
Daniel Miller
Learjet 35A
2 Garrett TFE731, 3,500 lbs
2/13/96

SAL/OLBIA (USA)
Mark E. Calkins — 820.84 kmh
Charles Conrad, Jr. — 510.04 mph
Paul Thayer
Daniel Miller
Learjet 35A
2 Garrett TFE731, 3,500 lbs
2/13/96

SAN FRANCISCO/DENVER (USA)
Clifford P. Crusan — 979.38 kmh
Benjamin C. Graves — 608.55 mph
Learjet 55
2 Garrett TFE731, 3,700 lbs
12/1/92

SAN FRANCISCO/GANDER (USA)
Brooke Knapp, Pilot — 814.44 kmh
James Topalian, Copilot — 506.09 mph
Paul Broyles, Copilot
James Magill, Crew Chief
Gates Learjet 35A
2 Garrett TFE 731-2-2B 3,500 lbs
2/16/83

SAN FRANCISCO/LONDON (USA)
Brooke Knapp, Pilot — 728.47 kmh
James Topalian, Copilot — 452.67 mph
Paul Broyles, Copilot
James Magill, Crew Chief
Gates Learjet 35A
2 Garrett TFE 731-2-2B 3,500 lbs
2/16-17/83

ST. JOHN'S, NF/SANTA MARIA (USA)
Henry G. Beaird, Pilot — 737.64 kmh
John O. Lear, — 458.35 mph
Robert R. King, Copilots
Lear Jet Model 24, N427LJ
2 GE CJ-610-4
3 hrs 26 min 27.5 sec
5/24/66

SANTA MARIA, AZORES/BARCELONA (USA)
Henry G. Beaird, Pilot — 722.46 kmh
John O. Lear, — 449.02 mph
Robert R. King, Copilots
Lear Jet Model 24, N427LJ
2 GE CJ-610-4
3 hrs 19 min 20.8 sec
5/24/66

SEATTLE/GOOSE BAY (USA)
Douglas Matthews — 854.07 kmh
Learjet 35A — 530.72 mph
2 Garrett TFE 731-2-2B, 3,500 lbs
2/12/84

SEATTLE/LONDON (USA)
Douglas Matthews — 762.84 kmh
Learjet 35A — 474.03 mph
2 Garrett TFE 731-2-2B, 3,500 lbs
2/12-13/84

SEATTLE/LOS ANGELES (USA)
Henry G. Beaird, Pilot — 758.49 kmh
John O. Lear, — 471.31 mph
Robert R. King, Copilots
Lear Jet Model 24, N427LJ
2 GE CJ-610-4
2 hrs 2 min 23.5 sec
5/26/66

SHANNON/GANDER (USA)
Douglas Matthews, Pilot — 686.92 kmh
David Marco, Copilot — 426.85 mph
Learjet, N83DM
2 TFE-731-2-2B, 3,500 lbs
12/28/86

SHANNON/NEWPORT NEWS, VA (USA)
Douglas Matthews, Pilot — 645.95 kmh
David Marco, Copilot — 401.39 mph
Learjet, N83DM
2 TFE-731-2-2B, 3,500 lbs
12/28/86

SINGAPORE/BIAK (USA)
Douglas Matthews — 631.46 kmh
Learjet 35A — 392.39 mph
2 Garrett TFE 731-2-2B, 3,500 lbs
2/13-14/84

SINGAPORE/HONOLULU (USA)
Douglas Matthews — 626.76 kmh
Learjet 35A — 389.47 mph
2 Garrett TFE 731-2-2B, 3,500 lbs
2/13-14/84

SINGAPORE/MANILA (USA)
Henry G. Beaird, Pilot — 605.43 kmh
John O. Lear, — 376.20 mph
Robert R. King, Copilots
Lear Jet Model 24, N427LJ
2 GE CJ-610-4
3 hrs 57 min 38.2 sec
5/25/66

TEHERAN/KARACHI (USA)
Henry G. Beaird, Pilot — 729.75 kmh
John O. Lear, — 453.45 mph
Robert R. King, Copilots
Lear Jet Model 24, N427LJ
2 GE CJ-610-4
2 hrs 38 min 4.7 sec
5/24/66

VARANASI/MANILA (USA)
Mark E. Calkins — 769.56 kmh
Charles Conrad, Jr. — 478.18 mph
Paul Thayer
Daniel Miller
Learjet 35A
2 Garrett TFE731, 3,500 lbs
2/13/96

YUZHNO-SAKHALINSK/KING SALMON (USA)
Mark E. Calkins — 814.43 kmh
Charles Conrad, Jr. — 506.06 mph
Paul Thayer
Daniel Miller
Learjet 35A
2 Garrett TFE731, 3,500 lbs
2/14/96

ROUND-TRIP

DAYTON/KITTY HAWK/DAYTON (USA)
Rex Anderson, Pilot — 769.18 kmh
Richard Smith, Copilot — 477.97 mph
Cessna Citation, N16QS
2 P&W JT15D-4B
7/24/87

LOS ANGELES/NEW YORK/LOS ANGELES (USA)
Robert Berry — 698.07 kmh
Lear Jet Model 25C, N251GL — 433.76 mph
2 GE CJ-610-6
11 hrs 17 min 22.7 sec
8/30/70

CLASS C-1.G (9,000-12,000 kg / 19,842-26,455 lbs)
GROUP III (JET ENGINE)

EAST TO WEST TRANSCONTINENTAL *
Rodney R. Lundy 468.75 mph
David S. Ryan
Learjet 45
2 AlliedSignal TFE731-20, 3,500 lbs
6/6/99

***

ANCHORAGE/SEATTLE (USA)
Arthur Godfrey 749.45 kmh
Richard Merrill 465.69 mph
Fred Austin & Karl Keller
Jet Commander, N1966J
2 GE CJ-610-6
3 hrs 6 min 17 sec
6/7/66

ATHENS/TEHERAN (USA)
Arthur Godfrey 672.64 kmh
Richard Merrill 417.96 mph
Fred Austin & Karl Keller
Jet Commander, N1966J
2 GE CJ-610-6
3 hrs 39 min 38 sec
6/5/66

AUCKLAND/SYDNEY (USA)
Donald M. Majors, Pilot 655.20 kmh
James Brown, Copilot 407.14 mph
Astra 1125, N40AJ
2 Garrett TFE 731-3-G-200, 3,700 lbs
8/9/89

BANGOR/PARIS (AUSTRALIA)
Bermar S. Stillwell 772.07 kmh
Learjet 55 LR 479.76 mph
2 TFE-731, 3,650 lbs
5/26/83

BOCA RATON/CHRISTIANSTED (USVI)
Svend A. Oversen, Jr. 769.50 kmh
Irvine Clifford 478.14 mph
IAI Astra AJ25
2 Garrett TFE-733, 3,700 lbs
4/20/95

BOMBAY/COLOMBO (USA)
Arthur Godfrey 643.89 kmh
Richard Merrill 400.10 mph
Fred Austin & Karl Keller
Jet Commander, N1966J
2 GE CJ-610-6
2 hrs 25 min 40 sec
6/6/66

BOSTON/GANDER (USA)
M. B. Burton 1,029.57 kmh
Falcon 20, N622R 639.75 mph
2 GE CF700-2C
1 hr 25 min 52.9 sec
9/25/66

BOSTON/LONDON (USA)
William S. Hallock, Captain 849.43 kmh
E. Judson Brandreth, 527.81 mph
Rockwell Sabreliner 65
2 Garrett TFE-731-3, 3,700 lbs
6 hrs 11 min 18 sec
2/22/80

BOSTON/LOS ANGELES (ISRAEL)
Adi Benaya, Pilot 633.24 kmh
Y. Geva, Copilot 393.49 mph
A. Hass, H. Oren, Flight Engineers
Westwind Astra, 4XWIA
2 Garrett TFE 731 3B 100
4/10/85

BOSTON/PARIS (USA)
William S. Hallock, Captain 832.42 kmh
E. Judson Brandreth, 517.24 mph
Rockwell Sabreliner 65
2 Garrett TFE-731-3, 3,700 lbs
6 hrs 39 min 55 sec
2/22/80

BRISBANE/PAGO PAGO (USA)
Donald M. Majors, Pilot 880.20 kmh
James Brown, Copilot 546.96 mph
Astra 1125, N40AJ
2 Garrett TFE 731-3-G-200, 3,700 lbs
8/27/89

CARACAS/JACKSONVILLE (USA)
William A. Landrum 740.63 kmh
Owen G. Zahnle 460.20 mph
Learjet 45
2 AlliedSignal TFE731-20, 3,500 lbs
6/5/99

CARACAS/NEW ORLEANS (USA)
Scott S. Evans 757.15 kmh
Jerry W. Blessing 470.47 mph
IAI 1125 Astra SPX
2 AlliedSignal TFE731, 4,250 lbs
9/1/01

CHRISTIANSTED/NEW YORK (USVI)
Svend A. Oversen, Jr. 616.66 kmh
Irvine Clifford 383.17 mph
IAI Astra AJ25
2 Garrett TFE-733, 3,700 lbs
4/17/95

COLOMBO/JAKARTA (USA)
Arnold Palmer 698.81 kmh
Learjet 36 434.22 mph
2 Garrett TFE 731-2-2B
4 hrs 45 min 45 sec
5/18/76

COLOMBO/MANILA (USA)
Arthur Godfrey 424.31 kmh
Richard Merrill 263.66 mph
Fred Austin & Karl Keller
Jet Commander, N1966J
2 GE CJ-610-6
10 hrs 46 min 39 sec
6/6/66

DENVER/BOSTON (USA)
Arnold Palmer 856.61 kmh
Learjet 36 532.27 mph
2 Garrett TFE 731-2-2B
3 hrs 18 min 51 sec
5/17/76

GANDER/GENEVA (USA)
Robert W. Agostino 774.67 kmh
Rick E. Rowe 481.36 mph
Learjet 60
2 P&WC PW305, 4,600 lbs
6/7/95

GANDER/PARIS (USA)
Robert W. Agostino 861.77 kmh
David S. Ryan 535.48 mph
Learjet 60
2 P&WC PW305A, 4,600 lbs
6/14/01

GANDER/REYKJAVIK (USA)
Charles E. Hall 725.96 kmh
Sabreliner 65 451.11 mph
2 Garrett TFE-731, 3,700 lbs
5/24/83

GANDER/SHANNON (USA)
Oakley H. Allen
Sabreliner 75A
2 GE CF700-2D2
4 hrs 11 min 13 sec
5/25/75
760.06 kmh
472.28 mph

HALIFAX/LOS ANGELES (AUSTRALIA)
Bermar S. Stillwell
Learjet 55LR
2 TFE-731, 3,650 lbs
6/6/83
671.04 kmh
416.98 mph

HONOLULU/DENVER (USA)
Arnold Palmer
Learjet 36
2 Garrett TFE 731-2-2B
6 hrs 54 min 28 sec
5/19/76
781.05 kmh
485.32 mph

HONOLULU/LAS VEGAS (USA)
Donald M. Majors, Pilot
James Brown, Copilot
Astra 1125, N40AJ
2 Garrett TFE 731-3-G-200, 3,700 lbs
8/28/89
805.68 kmh
500.65 mph

HONOLULU/PAGO PAGO (USA)
Donald M. Majors, Pilot
James Brown, Copilot
Astra 1125, N40AJ
2 Garrett TFE 731-3-G-200, 3,700 lbs
8/2/89
715.32 kmh
444.50 mph

JACKSONVILLE/LOS ANGELES (USA)
Rodney R. Lundy
David S. Ryan
Learjet 45
2 AlliedSignal TFE731-20, 3,500 lbs
6/6/99
754.39 kmh
468.75 mph

JAKARTA/MANILA (USA)
Arnold Palmer
Learjet 36
2 Garrett TFE 731-2-2B
3 hrs 50 min 50 sec
5/19/76
723.13 kmh
449.32 mph

KEFLAVIK/PARIS (USA)
Rodney R. Lundy
J. Shawn Christian
Learjet 45
2 AlliedSignal TFE731, 3,500 lbs
6/14/01
835.29 kmh
519.02 mph

KITTY HAWK/DAYTON (USA)
Donald E. Stroud, Pilot
Carl Day, Copilot
Jerald W. Dunn, Copilot
James V. Landolfi, Copilot
Jason A. Landolfi, Copilot
David W. Ross, Copilot
Sabreliner 60
2 P&W JT12, 3,300 lbs
7/22/95
669.17 kmh
415.80 mph

LAJES, AZORES/LISBON (USA)
Arthur D. Knapp
Sabreliner NA265-40, N788R
2 P&W JT12A-8
1 hr 44 min 35.8 sec
5/26/67
891.87 kmh
554.18 mph

LAS VEGAS/TOKYO (USA)
Niel Markussen
Westwind Astra, N716W
2 Garrett TFE 731-3A-200G, 23,500 lbs
6/28-29/87
645.03 kmh
400.82 mph

LISBON/PARIS (USA)
M. B. Burton
Falcon 20, N622R
2 GE CF700-2C
1 hr 40 min 34.8 sec
9/27/66
859.48 kmh
534.06 mph

LONDON/BONN (USA)
Jacqueline Cochran
Lockheed Jetstar, N172L
4 P&W JT12A-6
49 min 44.4 sec
4/22/62
620.49 kmh
385.56 mph

LONDON/BOSTON (ISRAEL)
Adi Benaya, Pilot
Y. Geva, Copilot
A. Hass, H. Oren, Flight Engineers
Westwind Astra, 4XWIA
2 Garrett TFE 731 3B 100
4/9/85
571.68 kmh
355.24 mph

LONDON/RENNES (UK)
CPT Denys Mobberley
HS 125
2 Garrett TFE 731-3R-1H, 3,700 lbs
11/12/85
651.10 kmh
404.59 mph

LOS ANGELES/BANGOR (AUSTRALIA)
Bermar S. Stillwell
Learjet 55LR
2 TFE-731, 3,650 lbs
5/26/83
795.23 kmh
494.16 mph

LOS ANGELES/GANDER (ISRAEL)
Adi Benaya, Pilot
Y. Geva, Copilot
A. Hass, H. Oren, Flight Engineers
Westwind Astra, 4XWIA
2 Garrett TFE 731 3B 100
5/14/85
771.12 kmh
479.17 mph

LOS ANGELES/HONOLULU (USA)
William Snell, Pilot
Robert McVey, Copilot
Randolph Burchett, II, Copilot
Westwind Astra
2 Garrett TFE 3B-1006, 3,650 lbs
11/23/85
711.57 kmh
442.17 mph

LOS ANGELES/NEW YORK (ISRAEL)
Haigai Koren, Copilot
Danny Shapiro, Copilot
Westwind Astra
2 Garrett TFE-731-3B-100G, 3,650 lbs
9/29/84
892.08 kmh
554.34 mph

LOS ANGELES/OKLAHOMA CITY (USA)
Arthur Godfrey
Richard Merrill
Fred Austin & Karl Keller
Jet Commander, N1966J
2 GE CJ-610-6
6/7/66
743.03 kmh
461.70 mph

LOS ANGELES/PARIS (AUSTRALIA)
Bermar S. Stillwell
Learjet 55LR
2 TFE-731, 3,650 lbs
5/26/83
721.82 kmh
448.54 mph

LOS ANGELES/PHILADELPHIA (USA)
John Olcott, Captain
Niel Markussen, Co-Captian
William Speaker, Co-Captain
Westwind Astra, N1125A
2 TFE 731-3A, 3,700 lbs
5/12/87
795.60 kmh
494.38 mph

MADRID/ATHENS (USA)
Arthur Godfrey
Richard Merrill
Fred Austin & Karl Keller
Jet Commander, N1966J
2 GE CJ-610-6
3 hrs 18 min 28.9 sec
6/5/66
716.02 kmh
444.92 mph

MANILA/TAIPEI (USA)
Arthur Godfrey
Richard Merrill
Fred Austin & Karl Keller
Jet Commander, N1966J
2 GE CJ-610-6
2 hrs 4 min 38 sec
6/6/66
566.66 kmh
352.11 mph

MANILA/TOKYO (USA)
Arthur Godfrey
Richard Merrill
Fred Austin & Karl Keller
Jet Commander, N1966J
2 GE CJ-610-6
6 hrs 38 min 38 sec
6/6/66
451.32 kmh
280.44 mph

MANILA/WAKE ISLAND (USA)
Arnold Palmer
Learjet 36
2 Garrett TFE 731-2-B
6 hrs 13 min 30 sec
5/19/76
783.00 kmh
486.56 mph

MELBOURNE/PERTH (USA)
Donald M. Majors, Pilot
James Brown, Copilot
Astra 1125, N40AJ
2 Garrett TFE 731-3-G-200, 3,700 lbs
8/16/89
707.04 kmh
439.35 mph

MIAMI/CARACAS (USA)
Jerry W. Blessing
Scott S. Evans
IAI 1125 Astra SPX
2 AlliedSignal TFE731, 4,250 lbs
9/1/01
765.84 kmh
475.87 mph

MIAMI/NEW ORLEANS (USA)
Jerry W. Blessing
Scott S. Evans
IAI 1125 Astra SPX
2 AlliedSignal TFE731, 4,250 lbs
8/31/01
832.12 kmh
517.05 mph

MILWAUKEE/PARIS (USA)
Charles E. Hall
Sabreliner 65
2 Garrett TE-731, 3,700 lbs
5/23-26/83
78.28 kmh
48.64 mph

MINNEAPOLIS/CARACAS (USA)
Robert McVey, Pilot
James Brown, Copilot
Westwind ASTRA, N1125A
2 Garrett TFE-731, 3,700 lbs
4/9/87
821.88 kmh
510.72 mph

MONTREAL/KEFLAVIK (USA)
Rodney R. Lundy
J. Shawn Christian
Learjet 45
2 AlliedSignal TFE731, 3,500 lbs
6/14/01
765.70 kmh
475.78 mph

MONTREAL/PARIS (USA)
Rodney R. Lundy
J. Shawn Christian
Learjet 45
2 AlliedSignal TFE731, 3,500 lbs
6/14/01
692.91 kmh
430.55 mph

NEW ORLEANS/MIAMI (USA)
Jerry W. Blessing
Scott S. Evans
IAI 1125 Astra SPX
2 AlliedSignal TFE731, 4,250 lbs
8/31/01
835.74 kmh
519.30 mph

NEW ORLEANS/SEATTLE (USA)
Jerry W. Blessing
Scott S. Evans
IAI 1125 Astra SPX
2 AlliedSignal TFE731, 4,250 lbs
9/2/01
794.16 kmh
493.47 mph

NEW YORK/ANCHORAGE (USA)
Niel Markussen, CoCaptain
Robert McVey, CoCaptain
Astra 1125, N30AJ
2 Garrett 731-3-G-200, 3,700 lbs
5/25-26/88
640.14 kmh
397.78 mph

NEW YORK/BOSTON (USA)
M. B. Burton
Falcon 20, N6224
2 GE CF700-2C
20 min 30.6 sec
9/25/66
886.13 kmh
550.62 mph

NEW YORK/CARACAS (USA)
Owen G. Zahnle
William A. Landrum
Learjet 45
2 AlliedSignal TFE731-20, 3,500 lbs
6/5/99
793.08 kmh
492.79 mph

NEW YORK/GANDER (USA)
M. B. Burton
Falcon 20, N622R
2 GE CF700-2C
1 hr 46 min 24.7 sec
9/25/66
1,001.72 kmh
622.44 mph

NEW YORK/LOS ANGELES (ISRAEL)
Haigai Koren, Pilot
Danny Shapiro, Copilot
Westwind Astra
2 Garrett TFE-731-3B-100G, 3,650 lbs
9/24/84
662.04 kmh
411.39 mph

NEW YORK/MADRID (USA)
Arthur Godfrey
Richard Merrill
Fred Austin & Karl Keller
Jet Commander, N1966J
2 GE CJ-610-6
11 hrs 46 sec
6/4-5/66
523.98 kmh
325.59 mph

NEW YORK/PARIS (ISRAEL)
Haigai Koren, Pilot
Danny Shapiro, Copilot
Westwind Astra
2 Garrett TFE-731-3B-100G, 3,650 lbs
10/7/84
747.72 kmh
464.63 mph

NEW YORK/ST. JOHN'S, NF (USA)
Arthur Godfrey
Richard Merrill
Fred Austin & Karl Keller
Jet Commander, N1966J
2 GE CJ-610-6
2 hrs 28 min 22 sec
6/4/66
745.64 kmh
463.32 mph

NEW YORK/SHANNON (USA)
James Brown
Astra 1125, N30AJ
2 Garrett TFE 731-3G-200, 3,700 lbs
9/3-4/88
816.84 kmh
507.58 mph

OKLAHOMA CITY/NEW YORK (USA)
Arthur Godfrey
Richard Merrill
Fred Austin & Karl Keller
Jet Commander, N1966J
2 GE CJ-610-6
2 hrs 56 min 15 sec
6/7/66
732.94 kmh
455.43 mph

PAGO PAGO/AUCKLAND (USA)
Donald M. Majors, Pilot
James Brown, Copilot
Astra 1125, N40AJ
2 Garrett TFE 731-3-G-200, 3,700 lbs
8/2/89
665.28 kmh
413.41 mph

PAGO PAGO/HONOLULU (USA)
Donald M. Majors, Pilot
James Brown, Copilot
Astra 1125, N40AJ
2 Garrett TFE 731-3-G-200, 3,700 lbs
8/27/89
774.72 kmh
481.41 mph

PARIS/HALIFAX (AUSTRALIA)
Bermar S. Stillwell
Learjet 55LR
2 TFE-731, 3,650 lbs
6/6/83
676.44 kmh
420.34 mph

PARIS/LOS ANGELES (AUSTRALIA)
Bermar S. Stillwell
Learjet 55LR
2 TFE-731, 3,650 lbs
6/6/83
623.88 kmh
387.68 mph

PARIS/TEHERAN (USA)
Arnold Palmer
Learjet 36
2 Garrett TFE 731-2-2B
5 hrs 38 min 15 sec
5/18/76
745.27 kmh
465.22 mph

PERTH/SYDNEY (USA)
Donald M. Majors, Pilot
James Brown, Copilot
Astra 1125, N40AJ
2 Garrett TFE 731-3-G-200, 3,700 lbs
8/19/89
920.16 kmh
571.79 mph

PITTSBURGH/GANDER (USA)
Robert W. Agostino
Damon K. Griebel
Learjet 60
2 P&W PW305A, 4,600 lbs
6/12/97
852.11 kmh
529.47 mph

PITTSBURGH/PARIS (USA)
Robert W. Agostino
Damon K. Griebel
Learjet 60
2 P&W PW305A, 4,600 lbs
6/12/97
782.48 kmh
486.21 mph

RENNES/LONDON (UK)
CPT Denys Mobberley
HS 125
2 Garrett TFE 731-3R-1H, 3,700 lbs
11/12/85
594.16 kmh
369.21 mph

REYKJAVIK/PARIS (USA)
Charles E. Hall
Sabreliner 65
2 Garrett TFE-731, 3,700 lbs
5/25/83
87.94 kmh
54.64 mph

REYKJAVIK/WILMINGTON (USA)
Robert McVey, Pilot
William Speaker, Copilot
James Erskine
Westwind Astra
2 Garrett TFE 731-3B-100G, 3,700 lbs
6/14/86
660.06 kmh
410.16 mph

ST. JOHN'S/LAJES, AZORES (USA)
Arthur D. Knapp
Sabreliner NA 265-40, N788R
2 P&W JT12A-8
2 hrs 29 min 17.9 sec
5/26/67
919.10 kmh
571.10 mph

ST. JOHN'S/LISBON (USA)
Arthur D. Knapp
Sabreliner NA 265-40 N788R
2 P&W JT12A-8
4 hrs 14 min 1.6 sec
5/26/67
855.30 kmh
531.40 mph

ST. JOHN'S/SANTA MARIA, AZORES (USA)
Arthur Godfrey
Richard Merrill
Fred Austin & Karl Keller
Jet Commander, N1966J
2 GE CJ-610-6
3 hrs 46 min 3 sec
6/4/66
674.60 kmh
419.18 mph

ST. LOUIS, MO/GANDER (USA)
Oakley H. Allen
Sabreliner 75A
2 GE CF700-2D2
3 hrs 59 min 31.3 sec
5/25/75
764.37 kmh
474.96 mph

ST. LOUIS, MO/PARIS (USA)
Oakley H. Allen
Sabreliner 75A
2 GE CF700-2D2
11 hrs 32 min 12.5 sec
5/25/75
611.95 kmh
380.25 mph

ST. LOUIS, MO/SHANNON (USA)
Oakley H. Allen
Sabreliner 75A
2 GE CF700-2D2
9 hrs 27 min 3.4 sec
5/25/75
652.17 kmh
405.24 mph

ST. LOUIS/WHITE PLAINS (USA)
William S. Hallock, Captain
E. Judson Brandreth, Jr.
Rockwell Sabreliner 65
2 Garrett TFE-731-3, 3,700 lbs
1 hr 38 min 45 sec
2/21/80
876.79 kmh
544.81 mph

SAN DIEGO/WASHINGTON (USA)
Robert W. Agostino
David S. Ryan
Learjet 45
2 AlliedSignal TFE731-20, 3,500 lbs
11/23/98
849.91 kmh
528.11 mph

SAN FRANCISCO/BOSTON (USA)
William S. Hallock, Captain
E. Judson Brandreth, Jr.
Rockwell Sabreliner 65
2 Garrett TFE-731-3, 3,700 lbs
5 hrs 1 min 7 sec.
2/22/80
860.99 kmh
534.99 mph

SAN FRANCISCO/HONOLULU (USA)
Donald M. Majors, Pilot
James Brown, Copilot
Astra 1125, N40AJ
2 Garrett TFE 731-3-G-200, 3,700 lbs
8/1/89
702.00 kmh
436.22 mph

SAN FRANCISCO/LONDON (USA)
William S. Hallock, Captain
E. Judson Brandreth, Jr.
Rockwell Sabreliner 65
2 Garrett TFE-731-3, 3,700 lbs
11 hrs 31 min 38 sec
2/22/80
748.17 kmh
464.89 mph

SAN FRANCISCO/NEW YORK (USA)
Clifford P. Crusan
Benjamin C. Graves
Learjet 55
2 Garrett TFE731, 3,700 lbs
12/7/92
886.02 kmh
550.55 mph

SAN FRANCISCO/PARIS (USA)
William S. Hallock, Captain
E. Judson Brandreth, Jr.
Rockwell Sabreliner 65
2 Garrett TFE-731-3, 3,700 lbs
12 hrs 15 sec
2/22/80
747.36 kmh
464.39 mph

SANTA MARIA, AZORES/MADRID (USA)
Arthur Godfrey
Richard Merrill, Fred Austin,
Karl Keller
Jet Commander, N1966J
2 GE CJ-610-6
3 hrs 2 min 16 sec
6/4/66
629.30 kmh
391.03 mph

SAPPORO, JAPAN/ANCHORAGE (USA)
Arthur Godfrey
Richard Merrill, Fred Austin
Karl Keller
Jet Commander, N1966J
2 GD CJ-610-6
9 hrs 13 min 40 sec
6/7/66
524.46 kmh
325.89 mph

SEATTLE/LOS ANGELES (USA)
George P. Eremea
Dassault Falcon 20
2 GE CF700-2C
1 hr 46 min 20.6 sec
9/16/66
867.88 kmh
539.28 mph

SEATTLE/NEW ORLEANS (USA)
Scott S. Evans
Jerry W. Blessing
IAI 1125 Astra SPX
2 AlliedSignal TFE731, 4,250 lbs
9/3/01
878.26 kmh
545.72 mph

SEATTLE/NEW YORK (USA)
Niel Markussen, CoCaptain
Robert McVey, CoCaptain
Astra 1125, N30AJ
2 Garrett 731-3-G-200, 3,700 lbs
6/4/88
829.00 kmh
515.14 mph

SHANNON/PARIS (USA)
Oakley H. Allen
Sabreliner 75A
2 GE CF00-2D2
1 hr 25 min 9.7 sec
5/25/75
632.74 kmh
393.17 mph

SHANNON/REYKJAVIK (USA)
Robert McVey, Pilot
William Speaker, Copilot
James Erskine
Westwind Astra
2 Garrett TFE 731-3B-100G, 3,700 lbs
6/14/86
725.76 kmh
450.99 mph

TAIPEI/TOKYO (USA)
Arthur Godfrey
Richard Merrill
Fred Austin & Karl Keller
Jet Commander, N1966J
2 GE CJ-610-6
2 hrs 47 min
6/6/66
753.97 kmh
468.50 mph

TEHERAN/BOMBAY (USA)
Arthur Godfrey
Richard Merrill
Fred Austin & Karl Keller
Jet Commander, N1966J
2 GE CJ/610/6
6 hrs 32 min 20 sec
6/5/66
429.24 kmh
266.72 mph

TEHRAN/COLOMBO (USA)
Arnold Palmer
Learjet 36
2 Garrett TFE-731-2-2B
5 hrs 54 min 39 sec
5/18/76
728.37 kmh
452.59 mph

TEL AVIV/COPENHAGEN *
Niel Markussen
Philip Talbott
Westwind Astra
2 Garrett TFE731, 3,650 lbs
7/13/86
456.36 mph

TEL AVIV/LONDON (ISRAEL)
Adi Benaya, Pilot
Y. Geva, Copilot
Westwind Astra
2 Garrett TFE 731-3B-100G, 3,650 lbs
4/8/85
629.64 kmh
391.26 mph

TEL AVIV/PARIS (ISRAEL)
Danny Shapiro, Pilot
Hagai Koren, Copilot
Westwind Astra
2 Garrett TFE-731-3B-1006, 3,700 lbs
5/28/85
722.88 kmh
449.92 mph

TEL AVIV/SHANNON (USA)
Robert McVey, Pilot
William Speaker, Copilot
James Erskine
Westwind Astra
2 Garrett TFE 731-3B-100G, 3,700 lbs
6/13/86
731.88 kmh
454.79 mph

TETERBORO/NEW ORLEANS (USA)
Scott S. Evans
Jerry W. Blessing
IAI 1125 Astra SPX
2 AlliedSignal TFE731, 4,250 lbs
8/31/01
806.99 kmh
501.44 mph

TOKYO/ANCHORAGE (USA)
Arthur Godfrey
Richard Merrill
Fred Austin & Karl Keller
Jet Commander, N1966J
2 GE CJ-610-6
11 hrs 35 min 15 sec
6/7/66
481.03 kmh
298.90 mph

TOKYO/LAS VEGAS (USA)
Niel Markussen
Westwind Astra, N716W
2 Garrett TFE 731-3A-200G, 3,700 lbs
7/9/87
741.52 kmh
460.78 mph

TOKYO/NEW YORK (USA)
Niel Markussen, CoCaptain 676.61 kmh
Robert McVey, CoCaptain 420.45 mph
Astra 1125, N30AJ
2 Garrett 731-3-G-200, 3,700 lbs
6/4/88

TOKYO/SAPPORO (USA)
Arthur Godfrey
Richard Merrill 628.14 kmh
Fred Austin & Karl Keller 390.31 mph
Jet Commander, N1966J
2 GE CJ-610-6
1 hr 18 min 40 sec
6/7/66

TOKYO/SEATTLE (USA)
Niel Markussen, CoCaptain 754.00 kmh
Robert McVey, CoCaptain 468.54 mph
Astra 1125, N30AJ
2 Garrett 731-3-G-200, 3,700 lbs
6/4/88

WAKE ISLAND/HONOLULU (USA)
Arnold Palmer 856.84 kmh
Learjet 36 532.42 mph
2 Garrett TFE 731-2-2B
4 hrs 19 min
5/19/76

WHITE PLAINS/SAN FRANCISCO (USA)
William S. Hallock, captain 689.19 kmh
E. Judson Brandreth, Jr. 428.24mph
Rockwell Sabreliner 65
2 Garrett TFE-731-3, 3,700 lbs
2/21-22/80

WICHITA/GENEVA (USA)
Robert W. Agostino 770.59 kmh
Rick E. Rowe 478.82 mph
Learjet 60
2 P&WC PW305, 4,600 lbs
6/7/95

WICHITA/PARIS (USA)
Robert W. Agostino 814.28 kmh
David S. Ryan 505.97 mph
Learjet 60
2 P&WC PW305A, 4,600 lbs
6/14/01

WILMINGTON, DE/CARACAS (USA)
Jim Brown, Pilot 712.44 kmh
Lee Herman, Copilot 442.71 mph
Westwind Astra IAI, N1125A
2 Garrett TFE 731-3A-200G, 3,700 lbs
9/19/87

WILMINGTON, DE/SAN FRANCISCO (USA)
Donald M. Majors, Pilot 678.60 kmh
James Brown, Copilot 421.68 mph
Astra 1125, N40AJ
2 Garrett TFE 731-3-G-200, 3,700 lbs
7/31/89

ROUND TRIP

LONDON/RENNES/LONDON (UK)
CPT Denys Mobberley 525.54 kmh
HS 125 226.57 mph
2 Garrett TFE 731-3R-1H, 3,700 lbs
11/12/85

CLASS C-1.H (12,000-16,000 kg / 26,455-35,274 lbs)
GROUP III (JET ENGINE)

EAST TO WEST, TRANSCONTINENTAL *
Steve Fossett, Pilot 591.96 mph
Darrin L. Adkins, Copilot
Cessna 750 Citation X
2 R-R AE 3007C, 6,442 lbs
9/17/00

WEST TO EAST, TRANSCONTINENTAL * ★
Steve Fossett 726.83 mph
Douglas A. Travis
Cessna 750 Citation X
2 R-R AE3007C, 6,442 lbs
2/5/03

TRANSCONTINENTAL, ROUND TRIP *
Steve Fossett, Pilot 537.36 mph
Darrin L. Adkins, Copilot
Cessna 750 Citation X
2 R-R AE 3007C, 6,442 lbs
9/17/00

***

ABIDJAN/JOHANNESBURG (UK)
Richard Britton 781.75 kmh
BAe 1000 485.75 mph
3/30/92

ANCHORAGE/SAPPORO (USA)
Stephen M. Johnson 767.95 kmh
Kevin A. Bixler 477.18 mph
IAI 1126 Galaxy
2 P&WC PW306A, 5,700 lbs
2/17/02

BANGKOK/TAIPEI (CANADA)
Edgar Kaiser, Captain 484.56 kmh
Willie Cove, Copilot 301.11 mph
Andy Anderson, Copilot
BAe 125-800A, C-GKRL
2 Garrett 731-5H, 4,300 lbs
2/21/88

BANGOR/PARIS (FRANCE)
Jean Pus 962.74 kmh
Jean-Louis Dumas 598.21 mph
Jerry Tritt
Falcon 2000
2 GE CFE-738, 5,725 lbs
11/1/94

BANGOR/WICHITA (USA)
Richard Trissell 914.07 kmh
William S. Dirks 567.97 mph
Cessna Model 750, Citation X
2 Allison GMA 3007C, 6,000 lbs
6/15/95

BORDEAUX/NEW YORK (USA)
Randolph M. Kennedy 692.94 kmh
Falcon Jet 50 430.57 mph
3 Garrett 731-3, 3,700 lbs
8 hrs 21 min
3/31/79

BORDEAUX/WASHINGTON (USA)
Randolph M. Kennedy 677.08 kmh
Falcon Jet 50 420.71 mph
3 Garrett 731-3, 3,700 lbs
9 hrs 3 min 35 sec
3/31/79

CALCUTTA/NAGASAKI (USA)
Steve Fossett, Pilot — 1,027.55 kmh / 638.49 mph
Darrin L. Adkins, Copilot
Alexander M. Tai, Copilot
Cessna 750 Citation X
2 R-R AE 3007C, 6,442 lbs
2/16/00

CHESTER/MONTREAL (UK)
J. C. Horscroft — 675.33 kmh / 419.63 mph
BAe 1000
10/13/92

DALLAS/ATLANTA (USA)
Steve Fossett — 1,180.28 kmh / 733.39 mph
Douglas A. Travis
Cessna 750 Citation X
2 R-R AE3007C, 6,442 lbs
2/5/03

DUBAI/BANGKOK (CANADA)
Edgar Kaiser, Captain — 857.16 kmh / 532.64 mph
Willie Cove, Copilot
Andy Anderson, Copilot
BAe 125-800A, C-GKRL
2 Garrett 731-5H, 4,300 lbs
2/21/88

DUBAI/SINGAPORE (UK)
J. C. Horscroft — 750.14 kmh / 466.11 mph
BAe 1000
2/17/92

FORT WORTH/ANCHORAGE (USA)
Scott S. Evans — 774.04 kmh / 480.96 mph
Bradley S. Robinson
IAI 1126 Galaxy
2 P&WC PW306A, 5,700 lbs
2/16/02

GANDER/NICE (CANADA)
Edgar Kaiser, Captain — 767.60 kmh / 476.71 mph
Willie Cove, Copilot
Andy Anderson, Copilot
BAe 125-800A, C-GKRL
2 Garrett 731-5H, 4,300 lbs
2/20-21/88

GANDER/PARIS (UK)
Alain F. George — 918.16 kmh / 570.52 mph
Dassault Mystère-Falcon 200
2 Garrett ATF 3-6A-4C, 5,200 lbs each
2/9/03

GOOSE BAY/BERLIN (USA)
Steve Fossett — 1,045.97 kmh / 649.93 mph
Douglas A. Travis
Cessna 750 Citation X
2 R-R AE 3007C, 6,442 lbs
10/8/03

GOOSE BAY/KRAKOW (USA)
Steve Fossett — 1,037.43 kmh / 644.63 mph
Douglas A. Travis
Cessna 750 Citation X
2 R-R AE 3007C, 6,442 lbs
10/8/03

GUAYAQUIL/SANTIAGO (USA)
Paulo S. Jancitsky — 870.65 kmh / 540.99 mph
Kevin A. Bixler
IAI 1126 Galaxy
2 P&WC PW306A, 5,700 lbs
3/31/02

HATFIELD/ABIDJAN (UK)
Richard Britton — 678.65 kmh / 421.69 mph
BAe 1000
3/30/92

HATFIELD/DUBAI (UK)
J. C. Horscroft — 784.97 kmh / 487.75 mph
BAe 1000
2/16/92

HATFIELD/MUSCAT (UK)
C. J. Shrimpton — 772.01 kmh / 479.70 mph
BAe 1000
10/16/92

HILO/VANCOUVER (CANADA)
Edgar Kaiser, Captain — 788.04 kmh / 489.69 mph
Willie Cove, Copilot
Andy Anderson, Copilot
BAe 125-800A, C-GKRL
2 Garrett 731-5H, 4,300 lbs
2/22/88

JAKARTA/MELBOURNE (USA)
Paulo S. Jancitsky — 898.79 kmh / 558.48 mph
Lanny C. Harwell
IAI 1126 Galaxy
2 P&WC PW306A, 5,700 lbs
3/6/02

LOS ANGELES/BANGOR (FRANCE)
Jean Pus — 931.00 kmh / 578.49 mph
Jean-Louis Dumas
Jerry Tritt
Falcon 2000
2 GE CFE-738, 5,725 lbs
11/1/94

LOS ANGELES/NEW YORK (USA)
Clay Lacy — 1,001.62 kmh / 622.37 mph
Jeffrey C. Brollier
Hal Fishman
Cessna 750 Citation X
2 RR Allison AE3007C, 6,400 lbs
2/12/99

LUXOR/CALCUTTA (USA)
Steve Fossett, Pilot — 988.02 kmh / 613.92 mph
Darrin L. Adkins, Copilot
Alexander M. Tai, Copilot
Cessna 750 Citation X
2 R-R AE 3007C, 6,442 lbs
2/16/00

MANILA/SINGAPORE (USA)
Stephen M. Johnson — 605.37 kmh / 376.16 mph
Kevin A. Bixler
IAI 1126 Galaxy
2 P&WC PW306A, 5,700 lbs
2/24/02

MONTREAL/CHESTER (UK)
J. C. Horscroft — 682.28 kmh / 423.94 mph
BAe 1000
10/13/92

MUNCIE/SARASOTA (USA)
Capt. J. T. Grange — 787.59 kmh / 489.38 mph
Capt. F. Wiley Grant
Falcon 20
2 Garrett TFE 731-5AR, 4,500 lbs
4/12/91

NAGASAKI/MIDWAY ISLAND (USA)
Steve Fossett, Pilot — 988.74 kmh / 614.37 mph
Darrin L. Adkins, Copilot
Alexander M. Tai, Copilot
Cessna 750 Citation X
2 R-R AE 3007C, 6,442 lbs
2/16/00

NAGOYA/HONG KONG (USA)
Stephen M. Johnson 591.05 kmh
Kevin A. Bixler 367.26 mph
IAI 1126 Galaxy
2 P&WC PW306A, 5,700 lbs
2/20/02

NEW YORK/LOS ANGELES (USA)
Clay Lacy 897.38 kmh
Jeffrey C. Brollier 557.60 mph
Hal Fishman
Cessna 750 Citation X
2 RR Allison AE3007C, 6,400 lbs
2/13/99

NICE/DUBAI (CANADA)
Edgar Kaiser, Captain 782.64 kmh
Willie Cove, Copilot 486.33 mph
Andy Anderson, Copilot
BAe 125-800A, C-GKRL
2 Garrett 731-5H, 4,300 lbs
2/21/88

PERTH/BRISBANE (USA)
Steve Fossett, Pilot 1,134.69 kmh
Alexander M. Tai, Copilot 705.06 mph
Cessna 750 Citation X
2 R-R AE 3007C, 6,442 lbs
7/28/01

PERTH/HOBART (USA)
Steve Fossett, Pilot 1,194.17 kmh
Alexander M. Tai, Copilot 742.02 mph
Cessna 750 Citation X
2 R-R AE 3007C, 6,442 lbs
7/30/01

PHOENIX/BORDEAUX (FRANCE)
Daniel Acton 846.05 kmh
Dominique Chenevier 525.71 mph
Dassault Falcon 2000
2 CFE
10/18/00

SAN DIEGO/CHARLESTON (USA)
Steve Fossett 1,169.71 kmh
Douglas A. Travis 726.83 mph
Cessna 750 Citation X
2 R-R AE3007C, 6,442 lbs
2/5/03

SAVANNAH/GUAYAQUIL (USA)
Paulo S. Jancitsky 829.78 kmh
Kevin A. Bixler 515.60 mph
IAI 1126 Galaxy
2 P&WC PW306A, 5,700 lbs
3/31/02

SHANNON/BANGOR (USA)
Richard Trissell 838.65 kmh
William S. Dirks 521.11 mph
Cessna Model 750, Citation X
2 Allison GMA 3007C, 6,000 lbs
6/15/95

TAIPEI/WAKE ISLAND (CANADA)
Edgar Kaiser, Captain 898.56 kmh
Willie Cove, Copilot 558.37 mph
Andy Anderson, Copilot
BAe 125-800A, C-GKRL
2 Garrett 731-5H, 4,300 lbs
2/21-22/88

TETERBORO/VAN NUYS (USA)
Benjamin M. Budzowski 819.39 kmh
Mark O. Schlegel 509.14 mph
Cessna Model 750, Citation X
2 Allison GMA 3007C, 6,000 lbs
5/11/95

TUCSON/MUNCIE (USA)
Capt. J. T. Grange 871.20 kmh
Capt. F. Wiley Grant 541.34 mph
Falcon 20
2 Garrett TFE 731-5AR, 4,500 lbs
4/28/91

VANCOUVER/GANDER (CANADA)
Edgar Kaiser, Captain 804.60 kmh
Willie Cove, Copilot 499.98 mph
Andy Anderson, Copilot
BAe 125-800A, C-GKRL
2 Garrett 731-5H, 4,300 lbs
2/20/88

VAN NUYS/WHITE PLAINS (USA)
Benjamin M. Budzowski 958.22 kmh
Mark O. Schlegel 595.40 mph
Cessna Model 750, Citation X
2 Allison GMA 3007C, 6,000 lbs
5/11/95

WAKE ISLAND/HILO (CANADA)
Edgar Kaiser, Captain 835.56 kmh
Willie Cove, Copilot 519.22 mph
Andy Anderson, Copilot
BAe 125-800A, C-GKRL
2 Garrett 731-5H, 4,300 lbs
2/22/88

WICHITA/BANGOR (USA)
Richard Trissell 944.41 kmh
Benjamin M. Budzowski 586.82 mph
Russell W. Meyer
Cessna Model 750, Citation X
2 Allison GMA 3007C, 6,000 lbs
6/6/95

ROUND TRIP

LOS ANGELES/NEW YORK/LOS ANGELES (USA)
Clay Lacy 792.36 kmh
Jeffrey C. Brollier 492.34 mph
Hal Fishman
Cessna 750 Citation X
2 RR Allison AE3007C, 6,400 lbs
2/13/99

CLASS C-1.I (16,000-20,000 kg / 35,274-44,092 lbs)
GROUP III (JET ENGINE)

ACAPULCO/BARCELONA (MEXICO)
José Alfredo Gastelum Arce 802.41 kmh
José Manuel Muradás Rodriguez 498.59 mph
Carlos Andrés López Herrera
Dassault Falcon 50
3 Garrett TFE731, 3,700 lbs
11/5/00

ACAPULCO/GANDER (MEXICO)
José Alfredo Gastelum Arce 825.91 kmh
Manuel Muradás Rodriguez 513.19 mph
Carlos Andrés López Herrera
Dassault Falcon 50
3 Garrett TFE731, 3,700 lbs
11/5/00

ANCHORAGE/TALLAHASSEE *
Nick A. Caporella
Charles D. Dague
Vincent T. Caporella
Dassault Falcon 2000
2 CFE738, 6,000 lbs
8/5/03
494.72 mph

BERLIN/PARIS (USA)
Warren Breeding
Fred J. Coffin
Canadair Challenger 604
2 GE CF34, 8,729 lbs
6/14/01
707.23 kmh
439.45 mph

BORDEAUX/LITTLE ROCK (USA)
Gene "Ed" Allen, Captain
Steven E. Bergner, Co-Captain
Dassault Falcon 900B
3 Garrett TFE-731, 4,750 lbs
6/20/92
768.74 kmh
477.67 mph

BOSTON/PARIS (USA)
James P. Dwyer
Robert W. Agostino
Rodney R. Lundy
Bombardier Challenger 300
2 Honeywell AS907, 6,700 lbs
6/12/03
911.74 kmh
566.53 mph

CARTAGENA/TOLUCA (MEXICO)
José Manuel Muradas Rodriguez
José Alfredo Gastelum Arce
Dassault Falcon 50
3 Garrett TFE 731, 3,700 lbs
6/4/00
736.68 kmh
457.75 mph

CHIHUAHUA/TOLUCA (MEXICO)
José Alfredo Gastelum Arce
Carlos Andrés López Herrera
Dassault Falcon 50
3 Garrett TFE731, 3,700 lbs
11/18/00
734.50 kmh
456.40 mph

GANDER/BARCELONA (MEXICO)
José Alfredo Gastelum Arce
José Manuel Muradás Rodriguez
Carlos Andrés López Herrera
Dassault Falcon 50
3 Garrett TFE731, 3,700 lbs
11/5/00
921.37 kmh
572.51 mph

HALIFAX/TOLUCA (MEXICO)
José Alfredo Gastelum Arce
Jose Manuel Muradas Rodriguez
Dassault Falcon 2000
2 GE/Honeywell CFE738, 5,888 lbs
11/14/02
671.28 kmh
415.62 mph

HAMILTON/TOLUCA (MEXICO)
José Manuel Muradás Rodriguez
José Alfredo Gastelum Arce
Carlos Andrés López Herrera
Dassault Falcon 50
3 Garrett TFE731, 3,700 lbs
11/11/00
692.28 kmh
430.16 mph

LONDON/HALIFAX (MEXICO)
José Alfredo Gastelum Arce
Jose Manuel Muradas Rodriguez
Dassault Falcon 2000
2 GE/Honeywell CFE738, 5,888 lbs
11/13/02
668.87 kmh
415.62 mph

Museum of Flight's Richard Beckerman, NAA Contest & Records Board Chairman Stan Nelson, pilots Bob Agostino and Jim Dwyer, and Apollo astronaut Gene Cernan (l to r) celebrate the record flight from Miami to Seattle of the Bombardier Challenger 300.

LONDON/NEW YORK (USA)
Ed Allen, Pilot — 817.56 kmh
Dave DeAngelis, Copilot — 508.03 mph
Falcon 900, F-GFJC

3 Garrett TFE-731-5A-1C, 4,500 lbs
9/5/86

LONDON/TOLUCA (MEXICO)
José Alfredo Gastelum Arce — 620.00 kmh
Jose Manuel Muradas Rodriguez — 385.25 mph
Dassault Falcon 2000
2 GE/Honeywell CFE738, 5,888 lbs
11/14/02

LOS ANGELES/WASHINGTON (MEXICO)
José Alfedo Gastelum Arce — 781.74 kmh
Carlos Andrés Lopez Herrera — 485.75 mph
Dassault Falcon 50
3 Garrett TFE 731-3-1C, 3,700 lbs
5/5/00

MADRID/LONDON (MEXICO)
José Alfredo Gastelum Arce — 780.00 kmh
Jose Manuel Muradas Rodriguez — 484.67 mph
Dassault Falcon 2000
2 GE/Honeywell CFE738, 5,888 lbs
11/11/02

MIAMI/SEATTLE (USA)
Robert W. Agostino — 752.92 kmh
James P. Dwyer — 467.84 mph
Eugene A. Cernan
Bombardier Challenger 300
2 Honeywell AS907, 6,700 lbs
6/10/03

NEW YORK/BORDEAUX (USA)
Dave DeAngelis, Pilot — 890.68 kmh
H. LePrince-Ringuet, Copilot — 553.47 mph
Falcon 900, FGFJC
3 Garrett TFE 731 5A, 4,500 lbs
4/2/87

NEW YORK/HANOVER (FRANCE)
Jerome Resal — 735.90 kmh
Falcon 50 — 457.27 mph
3 Garrett TFE-730, 3,700 lbs
4/24/80

NEW YORK/LONDON (USA)
Ed Allen, Pilot — 976.68 kmh
Rod Foster, Copilot — 609.91 mph
Randolph M. Kennedy, Copilot
Falcon 900, F-GFJC
3 Garrett TFE-731-5A-1C, 4,500 lbs
11/12/86

NEW YORK/MADRID (USA)
Ed Allen, Pilot — 915.12 kmh
Jan Williams, Copilot — 568.66 mph
Randolph M. Kennedy, Copilot
Falcon 900, F-GFJC
3 Garrett TFE-731-5A-1C, 4,500 lbs
10/31/86

NEW YORK/PARIS (FRANCE)
Herve LePrince-Ringuet, Pilot — 887.25 kmh
Gerard Joyeuse, Copilot — 551.31 mph
Falcon Jet 50
3 Garrett 731-3, 3,700 lbs
6 hrs 34 min 51 sec
10/9/77

PALMA DE MAJORCA/SANTA MARIA (MEXICO)
José Manuel Muradás Rodriguez — 708.41 kmh
José Alfredo Gastelum Arce — 440.18 mph
Carlos Andrés López Herrera
Dassault Falcon 50
3 Garrett TFE731, 3,700 lbs
11/11/00

PALMA DE MAJORCA/TOLUCA (MEXICO)
José Manuel Muradás Rodriguez — 629.80 kmh
José Alfredo Gastelum Arce — 430.16 mph
Carlos Andrés López Herrera
Dassault Falcon 50
3 Garrett TFE731, 3,700 lbs
11/11/00

PARIS/HOUSTON (USA)
Gene "Ed" Allen, Captain — 755.96 kmh
Herve LePrince-Ringuet, Co-Captain — 469.73 mph
Guy Mitaux Maurouard, 1st Officer
Dassault Falcon 900B
3 Garrett TFE731, 4,750 lbs
7/16/91

SANTA MARIA/HAMILTON (MEXICO)
José Manuel Muradás Rodriguez — 748.82 kmh
José Alfredo Gastelum Arce — 465.29 mph
Carlos Andrés López Herrera
Dassault Falcon 50
3 Garrett TFE731, 3,700 lbs
11/11/00

SEATTLE/KEY WEST (USA)
Donald J. Vecchie, Pilot — 832.75 kmh
Jon D. Vecchie, Copilot — 517.44 mph
Dassault Falcon 50
3 Garrett TFE731, 3,700 lbs
3/31/94

SEATTLE/WASHINGTON (USA)
Richard R. Severson — 1,033.22 kmh
Ronald L. Weight — 642.01 mph
LeRoy Herrman
Cessna 750
2 Allison AE 3007C, 6,400 lbs
5/5/97

STEPHENVILLE/MADRID (MEXICO)
Jose Manuel Muradas Rodriguez — 816.43 kmh
José Alfredo Gastelum Arce — 507.31 mph
Dassault Falcon 2000
2 GE/Honeywell CFE738, 5,888 lbs
11/8/02

TOLUCA/CARTAGENA (MEXICO)
José Alfredo Gastelum Arce — 717.37 kmh
José Manuel Muradas Rodriguez — 445.75 mph
Dassault Falcon 50
3 Garrett TFE 731, 3,700 lbs
5/31/00

TOLUCA/CHIHUAHUA (MEXICO)
José Alfredo Gastelum Arce — 680.09 kmh
Carlos Andrés López Herrera — 422.59 mph
Dassault Falcon 50
3 Garrett TFE731, 3,700 lbs
11/17/00

TOLUCA/MADRID (MEXICO)
Jose Manuel Muradas Rodriguez — 759.02 kmh
José Alfredo Gastelum Arce — 471.63 mph
Dassault Falcon 2000
2 GE/Honeywell CFE738, 5,888 lbs
11/8/02

TOLUCA/SAINT-MARTIN (MEXICO)
José Alfedo Gastelum Arce — 824.79 kmh
José Manuel Muradas Rodriguez — 512.50 mph
Carlos Andrés Lopez Herrera
Dassault Falcon 50
3 Garrett TFE 731-3-1C, 3,700 lbs
3/17/00

TOLUCA/STEPHENVILLE (MEXICO)
Jose Manuel Muradas Rodriguez — 833.08 kmh
José Alfredo Gastelum Arce — 517.65 mph
Dassault Falcon 2000
2 GE/Honeywell CFE738, 5,888 lbs
11/8/02

ROUND TRIP

RENO/GALVESTON/RENO (USA)
William S. Dirks
Jeffrey C. Brollier
Cessna Citation X
2 Allison AE 3007C, 6,400 lbs
Reno, NV - Galveston, TX
9/13/97
876.66 kmh
544.73 mph

CLASS C-1.J (20,000-25,000 kg / 44,092-55,116 lbs)
GROUP III (JET ENGINE)

BORDEAUX/LITTLE ROCK (USA)
Gene E. Allen, Pilot
Matthew A. Boyle, Copilot
Dassault Falcon 900EX
3 AlliedSignal TFE731-60, 5,000 lbs
8/1/00
802.29 kmh
498.52 mph

BUENOS AIRES/FT LAUDERDALE (USA)
Gene E. Allen
Frank G. Leskaukas
David M. Singer
Dassault Falcon 900EX
3 AlliedSignal TFE731, 4,750 lbs
1/4/98
693.53 kmh
430.94 mph

DUBLIN/DALIAN (FRANCE)
Jean-Michel Quentin
Jean-Louis Dumas
AMD-BA Falcon 900 EX
11/3/96
791.78 kmh
491.99 mph

KUALA LUMPUR/HONIARA (USA)
David J. Dwyer
David A. Montgomery
Dassault Falcon 900EX
3 Allied Signal TFE731-60, 5,000 lbs
7/12/98
730.49 kmh
453.90 mph

LONDON/EDINBURGH (UK)
JOINT RECORD
C. Spink
I. T. Gale
2 Phantom F4J
2 P&W J79/10B, 17,820 lbs
7/1/87
1,191.76 kmh
740.56 mph

NEW ORLEANS/PARIS (FRANCE)
Herve Leprince-Ringuet, Pilot
Olivier Dassault, Copilot
Dassault Falcon 900
9/25/87
851.63 kmh
529.20 mph

ORLANDO/PARIS (FRANCE)
Guy Mitaux-Maurouard
Dassault Falcon 900 EX
3 Allied Signal TFE-731, 4,750 lbs
11/22/96
925.60 kmh
575.14 mph

PALM SPRINGS/AMSTERDAM (FRANCE)
Jean Bongiraud
Daniel Acton
Dassault Falcon 900 EX
3 Allied Signal TFE-731, 4,250 lbs
2/1/97
815.35 kmh
506.64 mph

PARIS/ABU DHABI (FRANCE)
Guy Mitaux-Maurouard
Olivier Dassault
Patrick Experton
AMD-BA Falcon 900 EX
3 Allied Signal TFE731-60-1C
2/3/96
953.57 kmh
592.52 mph

PARIS/SINGAPORE (FRANCE)
Guy Mitaux-Maurouard
Olivier Dassault
Patrick Experton
AMD-BA Falcon 900 EX
3 Allied Signal TFE731-60-1C
2/4/96
844.08 kmh
524.49 mph

SAO PAULO/TETERBORO (USA)
Richard J. Iudice, Pilot
Robert D. Phillips, Copilot
Dassault Falcon 900EX
3 Allied Signal TFE731-60, 5,000 lbs
6/3/98
694.31 kmh
431.42 mph

SYDNEY/KAHULUI (FRANCE)
Guy Mitaux-Maurouard
Dassault Falcon 900 EX
3 Allied Signal TFE-731, 4,750 lbs
11/16/96
836.15 kmh
519.56 mph

WICHITA/PARIS (CANADA)
E. Bruce Robinson
Craig Kennedy
Canadair Challenger CL604
2 GE CF-343B, 9,220 lbs
6/9/95
778.86 kmh
483.96 mph

CLASS C-1.K (25,000-35,000 kg / 55,116-77,162 lbs)
GROUP III (JET ENGINE)

BEDFORD/MOSCOW (USA)
Robert S. McKenney
Ahmed M. Ragheb
Gulfstream IV-SP
2 R-R Tay 611, 13,850 lbs
8/13/01
860.64 kmh
534.77 mph

BEIJING/HONOLULU (USA)
Brooke Knapp
Gulfstream III
2 RR Spey MK511-8, 11,400 lbs
2/14-15/84
942.23 kmh
585.50 mph

BEIJING/LOS ANGELES (USA)
Brooke Knapp
Gulfstream III
2 RR Spey MK511-8, 11,400 lbs
2/14-15/84
931.79 kmh
579.01 mph

BEIJING/TOKYO (USA)
Brooke Knapp
Gulfstream III
2 RR Spey MK511-8, 11,400 lbs
2/14/84
887.92 kmh
551.75 mph

BEIJING/WASHINGTON (USA)
Brooke Knapp
Gulfstream III
2 RR Spey MK511-8, 11,400 lbs
2/14-15/84
928.66 kmh
577.07 mph

CHICAGO/BARCELONA (USA)
Ehab Hanna
Lynn Upham
Gulfstream IV
2 R-R Tay 611, 13,850 lbs
6/23/03
841.84 kmh
523.10 mph

CHICAGO/NEW YORK (USA)
Brooke Knapp
Gulfstream III
2 RR Spey MK511-8, 11,400 lbs
7/16-17/84
905.40 kmh
562.62 mph

CHICAGO/TETERBORO (USA)
Ehab Hanna 796.95 kmh
Lynn Upham 495.20 mph
Gulfstream IV
2 R-R Tay 611, 13,850 lbs
6/19/03

CHRISTCHURCH/MCMURDO STATION (USA)
Brooke Knapp 796.68 kmh
Gulfstream III 495.06 mph
2 RR Spey MK511-8, 11,400
11/15-16/83

FAIRBANKS/LOS ANGELES (USA)
Brooke Knapp 939.96 kmh
Gulfstream III 584.09 mph
2 RR Spey MK511-8, 11,400
11/18/83

GENEVA/BOSTON (USA)
William J. Hodde, Captain 803.40 kmh
Robert K. Smyth, Pilot 499.21 mph
Gulfstream III, N300GA
2 RR Spey 511-8, 11,400 lbs
7 hrs 22 min 20 sec
5/3/80

GENEVA/NEW YORK (USA)
William J. Hodde, Captain 801.99 kmh
Robert K. Smyth, Pilot 498.33 mph
Gulfstream III, N300GA
2 RR Spey 511.8, 11,400 lbs
5/3/80

GENEVA/WASHINGTON (USA)
William J. Hodde, Captain 788.02 kmh
Robert K. Smyth, Pilot 489.66 mph
Gulfstream III, N300GA
2 RR Spey 511.8, 11,400 lbs
8 hrs 20 min 35 sec

5/3/80

GOOSE BAY/LOS ANGELES (USA)
Ehab Hanna 743.00 kmh
Lynn Upham 461.68 mph
Gulfstream IV
2 R-R Tay 611, 13,850 lbs
7/8/03

GOOSE BAY/PARIS (USA)
Clay Lacy 870.92 kmh
Brian Kirkdoffer 541.16 mph
Gary Meermans
Hal Fishman
Gulfstream GIISP
2 RR Spey Mk 511-8, 11,400 lbs
6/9/95

HONOLULU/PAGO PAGO (USA)
Brooke Knapp 815.98 kmh
Gulfstream III 507.05 mph
2 RR Spey MK511-8, 11,400 lbs
11/15/83

HONOLULU/SOUTH POLE (USA)
Brooke Knapp 669.53 kmh
Gulfstream III 416.05 mph
2 RR Spey MK511-8, 11,400 lbs
11/15-16/83

Ehab Hanna in the cockpit of a Gulfstream IV, en route to one of eight speed records he set in 2003.

HONOLULU/WASHINGTON (USA)
Brooke Knapp 950.56 kmh
Gulfstream III 590.68 mph

2 RR Spey MK511-8, 11,400 lbs
2/15/84

JOHN O' GROATS/LANDS END (UK)
W/Cdr J. P. Brady 1,243.87 kmh
Phantom FG I 772.94 mph
2/24/88

KONA/CINCINNATI (USA)
Robert K. Smyth, Pilot 881.21 kmh
H. Ernest Rorer, Pilot 547.56 mph
Gulfstream II
2 RR Spey 511-8, 11,400 lbs
8 hrs 2 min 18 sec
2/15-16/81

KONA/WASHINGTON (USA)
Robert K. Smyth, Pilot 875.99 kmh
H. Ernest Rorer, Pilot 544.32 mph
Gulfstream III
2 RR Spey 511-8, 11,400 lbs
8 hrs 49 min 30 sec
2/15-16/81

KUSHIRO, JAPAN/LOS ANGELES (USA)
Robert K. Smyth, Captain 878.40 kmh
H. E. Rorer, Co-captain 545.84 mph
Thomas Goodwin, Flt Engineer
Gulfstream III, N300GA
2 RR Spey 511-8, 11,400 lbs
9 hrs 10 min 4 sec
10/26/81

LONDON/BEIJING (USA)
Brooke Knapp 688.85 kmh
Gulfstream III 428.05 mph
2 RR Spey MK511-8, 11,400 lbs
2/13-14/84

LONDON/MOSCOW (USA)
Brooke Knapp 733.50 kmh
Gulfstream III 455.80 mph
2 RR Spey MK511-8, 11,400 lbs
2/14/84

LONDON/NOVOSIBIRSK (USA)
Brooke Knapp 678.89 kmh
Gulfstream III 421.86 mph
2 RR Spey MK511-8, 11,400 lbs
2/14/84

LONDON/PARIS (USA)
Harold Curtis, Pilot 864.00 kmh
William L. Mack, Copilot 536.86 mph
Robert Dann Hardt, Copliot
Gulfstream III
2 RR Spey 511-8, 11,400 lbs
23 min 40 sec
1/8/82

LONDON/TOKYO (USA)
Brooke Knapp 731.09 kmh
Gulfstream III 454.30 mph
2 RR Spey MK511-8, 11,400 lbs
2/14/84

LOS ANGELES/CHICAGO (USA)
Ehab Hanna 786.71 kmh
Lynn Upham 488.84 mph
Gulfstream IV
2 R-R Tay 611, 13,850 lbs
6/18/03

LOS ANGELES/CHRISTCHURCH (USA)
Brooke Knapp 696.60 kmh
Gulfstream III 432.87 mph
2 RR Spey MK511-8, 11,400 lbs
11/15/83

LOS ANGELES/DALLAS (USA)
Clay Lacy 993.95 kmh
Alex A. Kvassay 617.61 mph
Gulfstream II SP
2 RR Spey MK 511-8, 11,400 lbs
9/22/97

LOS ANGELES/GOOSE BAY (USA)
Clay Lacy 1,126.57 kmh
Brian Kirkdoffer 700.02 mph
Gary Meermans
Hal Fishman
Gulfstream GIISP
2 RR Spey Mk 511-8, 11,400 lbs
6/8/95

LOS ANGELES/HONOLULU (USA)
Brooke Knapp 800.64 kmh
Gulfstream III 497.52 mph
2 RR Spey MK511-8, 11,400 lbs
11/15/83

LOS ANGELES/KITTY HAWK (USA)
Clay Lacy 1,019.22 kmh
Alex A. Kvassay 633.31 mph
Gulfstream IISP
2 R-R Spey Mk511-8, 11,400 lbs
12/15/03

LOS ANGELES/MCMURDO STATION (USA)
Brooke Knapp 627.48 kmh
Gulfstream III 389.91 mph
2 RR Spey MK511-8, 11,400 lbs
11/15-16/83

LOS ANGELES/PAGO PAGO (USA)
Brooke Knapp 724.68 kmh
Gulfstream III 450.32 mph
2 RR Spey MK511-8, 11,400 lbs
11/15/83

LOS ANGELES/PARIS (USA)
Clay Lacy 943.97 kmh
Brian Kirkdoffer 586.55 mph
Gary Meermans
Hal Fishman
Gulfstream GIISP
2 RR Spey Mk 511-8, 11,400 lbs
6/9/95

LOS ANGELES/SOUTH POLE (USA)
Brooke Knapp 574.92 kmh
Gulfstream III 357.25 mph
2 RR Spey MK511-8, 11,400 lbs
11/15-16/83

LOS ANGELES/TETERBORO (USA)
Ehab Hanna 812.20 kmh
John D. Gibson 504.68 mph
Gulfstream IV
2 R-R Tay 611, 13,850 lbs
7/29/03

LOS ANGELES/WICHITA (USA)
Clay Lacy 987.69 kmh
Alex A. Kvassay 613.72 mph
Gulfstream IISP
2 R-R Spey Mk511-8, 11,400 lbs
12/15/03

MCMURDO STATION/PUNTA ARENAS (USA)
Brooke Knapp 668.88 kmh
Gulfstream III 415.64 mph
2 RR Spey MK511-8, 11,400 lbs
11/16/83

MCMURDO STATION/RECIFE (USA)
Brooke Knapp 682.27 kmh
Gulfstream III 423.96 mph
2 RR Spey MK511-8, 11,400 lbs
11/16/83

MONCTON/GENEVA (FRANCE)
William Christophe 883.36 kmh
Gulfstream IV 548.89 mph
9/18/92

MOSCOW/BEIJING (USA)
Brooke Knapp 735.39 kmh
Gulfstream III 456.97 mph
2 RR Spey MK511-8, 11,400 lbs
2/14/84

MOSCOW/HONOLULU (USA)
Brooke Knapp 831.17 kmh
Gulfstream III 516.49 mph
2 RR Spey MK511-8, 11,400 lbs
2/14-15/84

MOSCOW/LOS ANGELES (USA)
Brian Kirkdoffer 699.33 kmh
Clay Lacy 434.54 mph
Gary Meermans
Gulfstream GIISP
2 RR Spey Mk 511-8, 11,400 lbs
6/18/95

MOSCOW/NOVOSIBIRSK (USA)
Brooke Knapp 769.05 kmh
Gulfstream III 477.89 mph
2 RR Spey MK511-8, 11,400 lbs
2/14/84

MOSCOW/TOKYO (USA)
Brooke Knapp 766.48 kmh
Gulfstream III 476.30 mph
2 RR Spey MK511-8, 11,400 lbs
2/14/84

NEW YORK/ZURICH (USA)
Brooke Knapp 888.12 kmh
Gulfstream III 551.88 mph
2 RR Spey MK511-8, 11,400 lbs
7/2/84

NORTH POLE/LOS ANGELES (USA)
Brooke Knapp 723.24 kmh
Gulfstream III 449.42 mph
2 RR Spey MK511-8, 11,400 lbs
11/18/83

NOVOSIBIRSK/BEIJING (USA)
Brooke Knapp 844.98 kmh
Gulfstream III 525.07 mph
2 RR Spey MK5111-8, 11,400 lbs
2/14/84

NOVOSIBIRSK/HONOLULU (USA)
Brooke Knapp 880.24 kmh
Gulfstream III 546.98 mph
2 RR Spey MK5111-8, 11,400 lbs
2/14-15/84

NOVOSIBIRSK/TOKYO (USA)
Brooke Knapp 815.11 kmh
Gulfstream III 506.50 mph
2 RR Spey MK5111-8, 11,400 lbs
2/14/84

PAGO PAGO/CHRISTCHURCH (USA)
Brooke Knapp 749.88 kmh
Gulfstream III 465.97 mph
2 RR Spey MK511-8, 11,400 lbs
11/15/83

PALMA DE MALLORCA/GOOSE BAY (USA)
Ehab Hanna 746.09 kmh
Lynn Upham 463.60 mph
Gulfstream IV
2 R-R Tay 611, 13,850 lbs
7/7/03

PALMA DE MALLORCA/LOS ANGELES (USA)
Ehab Hanna 715.30 kmh
Lynn Upham 444.47 mph
Gulfstream IV
2 R-R Tay 611, 13,850 lbs
7/8/03

PUNTA ARENAS/RECIFE (USA)
Brooke Knapp 851.18 kmh
Gulfstream III 528.92 mph
2 RR Spey MK511-8, 11,400 lbs
11/16/83

RECIFE/TENERIFE (USA)
Brooke Knapp 857.67 kmh
Gulfstream III 532.96 mph
2 RR Spey MK511-8, 11,400 lbs
11/17/83

RECIFE/TRONDHEIM (USA)
Brooke Knapp 728.28 kmh
Gulfstream III 452.55 mph
2 RR Spey MK511-8, 11,400 lbs
11/17-18/83

SAN DIEGO/NEW YORK (USA)
William L. Mack, Pilot 912.37 kmh
Harold Curtis, Pilot 566.92 mph
Gulfstream III
2 Spey RR 511-8, 11,400 lbs
1/22/82

SAN FRANCISCO/LAS VEGAS (USA)
Dwight V. Whitaker, III, Pilot 764.67 kmh
Harry G. Butler, III, Copilot 475.14 mph
Gulfstream IV
2 RR Tay Mk 611-8, 13,850 lbs
3/5/92

SAVANNAH/HANOVER (USA)
Robert K. Smyth, Captain 831.54 kmh
Allen Paulson, Pilot 516.69 mph
William J. Hodde, Pilot
Gulfstram III, N300GA
2 RR Spey 511.8, 11,400 lbs
8 hrs 50 min 10 sec
4/26/80

TENERIFE/NORTH POLE (USA)
Brooke Knapp 661.18 kmh
Gulfstream III 410.85 mph
2 RR Spey MK511-8, 11,400 lbs
11/17-18/83

TENERIFE/TRONDHEIM (USA)
Brooke Knapp 763.20 kmh
Gulfstream III 474.25 mph
2 RR Spey MK511-8, 11,400 lbs
11/17-18/83

TETERBORO/CHICAGO (USA)
Ehab Hanna 762.79 kmh
Lynn Upham 473.98 mph
Gulfstream IV
2 R-R Tay 611, 13,850 lbs
6/21/03

TOLUCA/GENEVA (FRANCE)
William Christophe 802.00 kmh
Gulfstream IV 498.34 mph
9/18/92

TOLUCA/MONCTON (FRANCE)
William Christophe 821.50 kmh
Gulfstream IV 510.46 mph
9/18/92

TOKYO/ALBUQUERQUE (USA)
Wayne Altman, Pilot 933.22 kmh
Robert Kirksey, Copilot 579.87 mph
Gulfstream IV
2 RR Tay Mk 611-8, 13,850 lbs
3/3/93

TOKYO/BOISE (USA)
Larry S. Mueller 867.97 kmh
James J. Keller 539.33 mph
Gulfstream IV
2 RR Tay Mk 611-8, 13,850 lbs
3/12/90

TOKYO/HONOLULU (USA)
Brooke Knapp 1,085.60 kmh
Gulfstream III 674.59 mph
2 RR Spey MK5111-8, 11,400 lbs
2/14-15/84

TOKYO/LAS VEGAS (USA)
Dwight V. Whitaker, III, Pilot 880.01 kmh
Harry G. Butler, III, Copilot 546.81 mph
Gulfstream IV
2 RR Tay Mk 611-8, 13,850 lbs
3/5/92

TOKYO/LOS ANGELES (USA)
Brooke Knapp 996.59 kmh
Gulfstream III 619.28 mph
2 RR Spey MK5111-8 11,400 lbs
2/14-15/84

TOKYO/SAN FRANCISCO (USA)
Dwight V. Whitaker, III, Pilot 895.35 kmh
Harry G. Butler, III, Copilot 556.34 mph
Gulfstream IV
2 RR Tay Mk 611-8, 13,850 lbs
3/5/92

TOKYO/SEATTLE (USA)
Larry S. Mueller 872.62 kmh
James J. Keller 542.21 mph
Gulfstream IV
2 RR Tay Mk 611-8, 13,850 lbs
3/12/90

TRONDHEIM/FAIRBANKS (USA)
Brooke Knapp 743.04 kmh
Gulfstream III 461.72 mph
2 RR Spey MK511-8, 11,400 lbs
11/18/83

TRONDHEIM/LOS ANGELES (USA)
Brooke Knapp 658.80 kmh
Gulfstream III 409.38 mph
2 RR Spey MK511-8, 11,400 lbs
11/18/83

WASHINGTON/MOSCOW (USA)
Brooke Knapp 795.57 kmh
Gulfstream III 494.36 mph
2 RR Spey MK5111-8, 11,400 lbs
2/13-14/84

WASHINGTON/NOVOSIBIRSK (USA)
Brooke Knapp 749.51 kmh
Gulfstream III 465.75 mph
2 RR Spey MK5111-8 11,400 lbs
2/13-14/84

WICHITA/KITTY HAWK (USA)
Clay Lacy 1,054.01 kmh
Alex A. Kvassay 654.93 mph
Gulfstream IISP
2 R-R Spey Mk511-8, 11,400 lbs
12/15/03

CLASS C-1.L (35,000-45,000 kg / 77,162-99,208 lbs)
GROUP III (JET ENGINE)

ANKARA/LANGKAWI (USA)
Raymond A. Wellington 870.97 kmh
William J. Watters 541.19 mph
Alberto L. Moros
Gulfstream G550
2 BMW R-R BR710, 15,385 lbs
10/1/03

BANDAR SERI BEGAWAN/SEOUL (USA)
Alberto L. Moros 788.95 kmh
William J. Watters 490.23 mph
Raymond A. Wellington
Gulfstream G550
2 BMW R-R BR710, 15,385 lbs
10/3/03

BEIJING/HOUSTON (USA)
Larry S. Mueller 869.32 kmh
Robert S. McKenney 540.17 mph
Rick B. Gowthrop
Gulfstream V
2 BMW-RR BR710-48, 14,750 lbs
9/30/99

BEIJING/NEW YORK (USA)
Robert S. McKenney 795.06 kmh
William J. Watters 494.03 mph
Raymond A. Wellington
Gulfstream G550
2 BMW R-R BR710, 15,385 lbs
12/14/03

BEIJING/TOKYO (USA)
James J. Keller 572.06 kmh
Dwight V. Whitaker, III 355.46 mph
Gulfstream V
2 BMW RR BR710-48, 14,750 lbs
4/13/97

BRISBANE/LOS ANGELES (USA)
Wilson A. Miles, Jr. 907.61 kmh
Alberto Moros 563.96 mph
Richard D. Robbins
Gulfstream V
2 BMW R-R BR710, 14,750 lbs
2/17/01

DALLAS/LONDON (USA)
Rick B. Gowthrop 978.02 kmh
H. Randy Gaston 607.71 mph
William J. Watters
Gulfstream V
2 BMW R-R BR710, 14,750 lbs
7/21/00

DALLAS/TOKYO (USA)
Andrew S. Milewski 751.16 kmh
Yves Tessier 466.75 mph
Durwood J. Heinrich
Bombardier BD-700 Global Express
2 BMW R-R BR710, 14,750 lbs
4/22/02

DAYTON/JOHNSTOWN (USA)
Robert F. Olszewski 1,139.86 kmh
DC-9-31, N942VJ 708.27 mph
2 P&W JT8D-7B, 14,000 lbs
1/5/88

DAYTON/PITTSBURGH*
Robert F. Olszewski 706.17 mph
DC-9-31, N942VJ
2 P&W JT8D-7B, 14,000 lbs
1/5/88

DUBAI/HONG KONG (USA)
Raymond A. Wellington 970.96 kmh
Robert S. McKenney 603.33 mph
William J. Watters
Gulfstream G550
2 BMW R-R BR710, 15,385 lbs
12/11/03

DUBLIN/SAN ANTONIO (USA)
Wilson A. Miles, Jr. 964.64 kmh
William J. Watters 599.40 mph
Gulfstream V
2 BMW R-R BR710, 14,750 lbs
7/2/01

FLORENCE/CINCINNATI (USA)
James P. Mayer 810.66 kmh
Charles B. Smith 503.72 mph
Bombardier BD-700 Global Express
2 BMW R-R BR710, 14,750 lbs
9/29/02

HAMILTON/CAPE TOWN (USA)
Bernard R. Ozbolt 834.56 kmh
Henry K. Gibson 518.57 mph
Christian M. Kennedy
Gulfstream V
2 BMW-RR BR710-48, 14,750 lbs
10/6/99

HAMILTON/LONDON (USA)
Franklin H. Davis 890.84 kmh
Max A. Dreckman 553.54 mph
Bombardier BD-700 Global Express
2 BMW R-R BR710, 14,750 lbs
6/16/00

HILTON HEAD/MAUI (USA)
Borden F. Schofield 807.15 kmh
E. Bruce Robinson 501.54 mph
Geoffrey M. Foster
Andrew Milewski
Bombardier BD 700
2 BMW RR BR-710, 14,690 lbs
5/5/99

HILTON HEAD/MILAN (USA)
Robert S. McKenney 892.62 kmh
Larry S. Mueller 554.65 mph
Gulfstream V
2 BMW R-R BR710, 14,750 lbs
1/31/01

HONG KONG/BEIJING (USA)
James J. Keller 848.28 kmh
Dwight V. Whitaker, III 527.10 mph
Gulfstream V
2 BMW RR BR710-48, 14,750 lbs
4/12/97

HOUSTON/CAIRO (USA)
Larry S. Mueller 926.35 kmh
John D. Mackay 575.60 mph
Gulfstream V
2 BMW R-R BR710, 14,750 lbs
3/18/01

JOHANNESBURG/DUBLIN (USA)
William J. Watters 820.26 kmh
Wilson A. Miles, Jr. 509.68 mph
Gulfstream V
2 BMW R-R BR710, 14,750 lbs
6/28/01

JOHANNESBURG/MANILA (USA)
Rick B. Gowthrop 805.96 kmh
Wilson A. Miles, Jr. 500.80 mph
Gulfstream V
2 BMW RR BR710-48, 14,750 lbs
11/10/98

LAS VEGAS/VENICE (USA)
William J. Watters 923.18 kmh
Richard D. Robbins 573.64 mph
Gulfstream V
2 BMW R-R BR710, 14,750 lbs
7/28/01

LIHUE/SINGAPORE (USA)
Bernard R. Ozbolt 740.58 kmh
Charles M. Barr II 460.17 mph
Gary M. Freeman
Gulfstream V
2 BMW Rolls-Royce BR710-48, 14,750 lbs
2/21/98

LONDON/DUBAI (USA)
Bernard R. Ozbolt 838.06 kmh
Henry K. Gibson 520.75 mph
Robert S. McKenney
Gulfstream V
2 BMW RR BR 710-48, 14,750 lbs
11/13/97

LONDON/HONG KONG (USA)
James J. Keller 794.63 kmh
Dwight V. Whitaker, III 493.76 mph
Gulfstream V
2 BMW RR BR710-48, 14,750 lbs
4/10/97

LONDON/LOS ANGELES (USA)
Peter T. Reynolds 846.66 kmh
E. Bruce Robinson 526.09 mph
Christopher D. Hall
Borden F. Schofield
Bombardier BD 700
2 BMW RR BR710, 14,690 lbs
5/3/99

LOS ANGELES/LONDON (USA)
James J. Keller 940.85 kmh
Dwight V. Whitaker, III 584.62 mph
Gulfstream V
2 BMW RR BR710-48, 14,750 lbs
4/8/97

MAUI/TOKYO (USA)
Geoffrey M. Foster 794.83 kmh
E. Bruce Robinson 493.88 mph
Andrew Milewski
Borden F. Schofield
Bombardier BD 700
2 BMW RR BR-710, 14,690 lbs
5/7/99

MEXICO CITY/MADRID (USA)
Peter T. Reynolds 938.72 kmh
E. Bruce Robinson 583.29 mph
Christopher D. Hall
Borden F. Schofield
Bombardier BD 700
2 BMW RR BR710, 14,690 lbs
5/2/99

MOSCOW/BEIJING (USA)
Andrew Milewski 857.62 kmh
E. Bruce Robinson 532.90 mph
Bombardier BD-700 Global Express
2 BMW-RR BR710, 14,750 lbs
9/22/99

MOSCOW/LOS ANGELES (USA)
Larry S. Mueller — 812.77 kmh
Henry K. Gibson — 505.03 mph
Robert S. McKenney
Gulfstream V
2 BMW RR BR710-48, 14,750 lbs
12/11/97

NASHVILLE/TOKYO (USA)
Gregory S. Sheldon — 778.69 kmh
Ahmed M. Ragheb — 483.85 mph
John A. Mullican
Gulfstream V
2 BMW R-R BR710, 14,750 lbs
11/6/02

NEW YORK/DUBAI (USA)
Timothy P. Freise — 853.36 kmh
Christian M. Kennedy — 530.25 mph
Henry K. Gibson
Gary M. Freeman
Gulfstream V
2 BMW-RR BR710-48, 14,750 lbs
11/13/99

NEW YORK/PARIS (USA)
John T. Race, Jr. — 958.30 kmh
Borden F. Schofield — 595.46 mph
Bombardier BD-700 Global Express
2 BMW R-R BR710, 14,750 lbs
6/14/01

PALM BEACH/TEL AVIV (USA)
Christian M. Kennedy — 949.79 kmh
James R. Cannon — 590.17 mph
Henry K. Gibson
Bruce A. Yates
Gulfstream V
2 BMW Rolls-Royce BR 710-48, 14,750 lbs
1/11/98

PARIS/SAVANNAH (USA)
Edward D. Mendenhall — 818.97 kmh
Dwight V. Whitaker, III — 508.88 mph
Gulfstream V
2 BMW RR BR710-48, 14,750 lbs
6/22/97

SAVANNAH/ANKARA (USA)
William J. Watters — 901.40 kmh
Raymond A. Wellington — 560.10 mph
Alberto L. Moros
Gulfstream G550
2 BMW R-R BR710, 15,385 lbs
9/29/03

SAVANNAH/DUBAI (USA)
William J. Watters — 883.84 kmh
Raymond A. Wellington — 549.19 mph
Ahmed M. Ragheb
Gulfstream G550
2 BMW R-R BR710, 15,385 lbs
12/6/03

SEOUL/ORLANDO (USA)
Wilson A. Miles, Jr. — 836.87 kmh
Anthony J. Briotta — 520.01 mph
Thomas C. Horne
William J. Watters
Gulfstream G550
2 BMW R-R BR710, 15,385 lbs
10/3/03

TEL AVIV/BOSTON (USA)
Christian M. Kennedy — 770.11 kmh
James R. Cannon — 478.52 mph
Henry K. Gibson
Bruce A. Yates
Gulfstream V
2 BMW Rolls-Royce BR 710-48, 14,750 lbs
1/11/98

TETERBORO/HAMILTON (USA)
Franklin H. Davis — 748.78 kmh
Max A. Dreckman — 465.27 mph
Bombardier BD-700 Global Express
2 BMW R-R BR710, 14,750 lbs
6/15/00

TETERBORO/LIHUE (USA)
Bernard R. Ozbolt — 825.65 kmh
Charles M. Barr II — 513.03 mph
Gary M. Freeman
Gulfstream V
2 BMW Rolls-Royce BR710-48, 14,750 lbs
2/20/98

TETERBORO/MOSCOW (USA)
Andrew Milewski — 922.92 kmh
E. Bruce Robinson — 573.47 mph
Bombardier BD-700 Global Express
2 BMW-RR BR710, 14,750 lbs
9/19/99

TETERBORO/TOKYO (USA)
E. Bruce Robinson — 826.69 kmh
Andrew Milewski — 513.68 mph
Roger Noble
Borden F. Schofield
Bombardier BD 700
2 BMW RR BR-710, 14,690 lbs
5/13/99

TETERBORO/VAN NUYS (USA)
Thomas W. Owens — 861.98 kmh
Christian M. Kennedy — 535.61 mph
Gulfstream V
2 BMW R-R BR710, 14,750 lbs
7/3/00

TETERBORO/WARSAW (USA)
Franklin H. Davis — 916.80 kmh
Gary Manchester — 569.67 mph
Bombardier BD-700 Global Express
2 BMW R-R BR710, 14,750 lbs
5/6/00

TOKYO/DALLAS (USA)
Andrew S. Milewski — 915.19 kmh
Yves Tessier — 568.67 mph
Durwood J. Heinrich
Bombardier BD-700 Global Express
2 BMW R-R BR710, 14,750 lbs
4/26/02

TOKYO/LOS ANGELES (USA) ★
Gregory S. Sheldon — 1,036.19 kmh
Ahmed M. Ragheb — 643.86 mph
John A. Mullican
Gulfstream V
2 BMW R-R BR710, 14,750 lbs
11/7/02

TOKYO/MONTREAL (USA)
E. Bruce Robinson — 911.29 kmh
Andrew Milewski — 566.24 mph
Roger Noble
Borden F. Schofield
Bombardier BD 700
2 BMW RR BR-710, 14,690 lbs
5/14/99

TOKYO/MOSCOW (USA)
Larry S. Mueller — 835.50 kmh
Christian M. Kennedy — 519.15 mph
Gulfstream V
2 BMW RR BR710-48, 14,750 lbs
8/22/97

TOKYO/PALM BEACH (USA)
Gregory S. Sheldon — 952.43 kmh
John D. Mackay — 591.81 mph
Neil E. Vernon
Gulfstream G550
2 BMW R-R BR710, 15,385 lbs
11/10/03

TOKYO/PARIS (USA)
Richard D. Robbins — 781.18 kmh
William J. Watters — 485.40 mph
Christian M. Kennedy
Gulfstream V
2 BMW RR BR710-48, 14,750 lbs
6/10/99

TOKYO/TETERBORO (USA)
Borden F. Schofield — 909.75 kmh
E. Bruce Robinson — 565.29 mph
Geoffrey M. Foster
Andrew Milewski
Bombardier BD 700
2 BMW RR BR-710, 14,690 lbs
5/9/99

TOKYO/WASHINGTON (USA)
Gregory S. Sheldon — 913.38 kmh
Robert S. McKenney — 567.55 mph
Gulfstream V
2 BMW R-R BR710, 14,750 lbs
3/5/02

ULAANBAATAR/CHICAGO (USA)
Larry S. Mueller — 766.13 kmh
William J. Watters — 476.05 mph
Kent R. Crenshaw
Gulfstream V
2 BMW R-R BR710, 14,750 lbs
6/27/00

WARSAW/SHANGHAI (USA)
Franklin H. Davis — 887.05 kmh
Gary Manchester — 551.19 mph
Bombardier BD-700 Global Express
2 BMW R-R BR710, 14,750 lbs
5/8/00

WASHINGTON/DUBAI (USA)
Bernard R. Ozbolt — 895.59 kmh
Henry K. Gibson — 556.49 mph
Robert S. McKenney
Gulfstream V
2 BMW RR BR 710-48, 14,750 lbs
11/13/97

WASHINGTON/LONDON (USA)
Bernard R. Ozbolt — 964.52 kmh
Henry K. Gibson — 599.32 mph
Robert S. McKenney
Gulfstream V
2 BMW RR BR 710-48, 14,750 lbs
11/13/97

WASHINGTON/PARIS (USA)
Edward D. Mendenhall — 919.46 kmh
Dwight V. Whitaker, III — 571.32 mph
Gulfstream V
2 BMW RR BR710-48, 14,750 lbs
6/12/97

WICHITA/PARIS (USA)
Peter T. Reynolds — 895.28 kmh
Alain Lacharite — 556.30 mph
Geoff M. Foster
Jeff Kirdeikis
Bombardier BD 700
2 BMW RR BR710, 14,970 lbs
6/12/97

ZURICH/BEIJING (USA)
Franklin H. Davis — 884.02 kmh
Charles L. Couture — 549.30 mph
Bombardier BD-700 Global Express
2 BMW R-R BR710, 14,750 lbs
3/26/00

CLASS C-1.M (45,000-60,000 kg / 99,208-132,277 lbs)
GROUP III (JET ENGINE)

HONOLULU/MAJURO (NEW ZEALAND)
Trevor J. Ancell — 731.50 kmh
Boeing 737-400 — 454.53 mph
4/30/92

HONOLULU/NOUMEA (AUSTRALIA)
Ian Haigh — 771.53 kmh
Boeing 737-400 — 479.40 mph
5/1/92

MAJURO/MELBOURNE (NEW ZEALAND)
Trevor J. Ancell — 732.38 kmh
Boeing 737-400 — 455.08 mph
5/1/92

NOUMEA/MELBOURNE (AUSTRALIA)
Ian Haig — 677.08 kmh
Australian Airlines — 420.72 mph
Boeing 737-400
7/24/92

PORT ANGELES/HILO (AUSTRALIA)
Ian Haig — 686.35 kmh
Australian Airlines — 426.48 mph
Boeing 737-400
7/21/92

SEATTLE/HONOLULU (NEW ZEALAND)
David A. Parrish — 638.80 kmh
Boeing 737-400 — 396.93 mph
4/24/92

CLASS C-1.N (60,000-80,000 kg / 132,277-176,370 lbs)
GROUP III (JET ENGINE)

HONOLULU/ORLANDO (USA)
Robert E. Siman — 782.20 kmh
Hassan A. Al-Kandry — 486.03 mph
McDonnell Douglas MD-83
2 P&W JT8D-219, 21,700 lbs
3/31/93

LONDON/ROME (UK)
Guy C. Westgate — 679.43 kmh
Boeing 737-400 — 422.18 mph
3/19/99

MONTREAL/BRUSSELS (BELGIUM)
Georges Rikir — 871.25 kmh
Boeing 737-300 — 541.39 mph
2 CFM 56-3, 20,000 lbs
3/22/86

WASHINGTON/ROME (USA)
Robert E. Siman — 785.80 kmh
Hassan A. Al-Kandry — 488.27 mph
McDonnell Douglas MD-83
2 P&W JT8D-219, 21,700 lbs
5/23/93

CLASS C-1.P (100,000-150,000 kg / 220,462-330,693 lbs)
GROUP III (JET ENGINE)

ASCENSION/PORT STANLEY (UK)
Ft/Lt John Halstead — 821.64 kmh
VC 10 C — 510.57 mph
4 RR Conway Mk 301, 22,500 lbs
12/20/87

ASCENSION/LONDON (UK)
Sqn/Ldr John Knapp — 827.55 kmh
VC 10 C — 514.24 mph
4 RR Conway Mk 301, 22,500 lbs
12/21-22/87

LONDON/ASCENSION (UK)
Ft/Lt John Halstead — 824.35 kmh
VC 10 C — 512.25 mph
4 RR Conway Mk 301, 22,500 lbs
12/19/87

LONDON/PORT STANLEY (UK)
Group Capt. Chris Lumb — 848.08 kmh
VC 10 C — 526.99 mph
4 RR Conway Mk 301, 22,500 lbs
12/19-20/87

PORT STANLEY/ASCENSION (UK)
Ft/Lt John Halstead — 912.03 kmh
VC 10 C — 566.74 mph
4 RR Conway Mk 301, 22,500 lbs
12/21/87

CLASS C-1.Q (150,000-200,000 kg / 330,693-440,924 lbs)
GROUP III (JET ENGINE)

AGANA/BANGKOK (UK)
Max Robinson — 1,469.09 kmh
BAe/Aérospatiale Concorde — 912.85 mph
10/30/99

AQABA/VENICE (UK)
Geoffrey Mussett — 1,172.05 kmh
BAe/Aérospatiale Concorde — 728.28 mph
11/11/99

BANGKOK/MADRAS (UK)
Max Robinson — 1,408.38 kmh
BAe/Aérospatiale Concorde — 875.12 mph
11/2/99

BERMUDA/NEW YORK (UK)
Capt. Peter Horton — 1,228.73 kmh
BAe/Aérospatiale Concorde — 763.49 mph
1/2/93

BIRMINGHAM/NEW YORK (UK)
Stuart Robertson — 1,100.05 kmh
BAe/Aérospatiale Concorde — 683.54 mph
9/27/93

BUENOS AIRES/USHUAIA (FRANCE)
Michel Rio — 1,316.38 kmh
Jean Marcot — 817.96 mph
Aérospatiale/BAC Concorde
1/14/99

CAIRO/LISBON (UK)
Richard Owen — 1,336.61 kmh
BAe/Aérospatiale Concorde — 830.53 mph
4 RR Olympus 593, 38,000 lbs
3/2/00

CAIRO/WASHINGTON (UK)
Richard Owen — 1,122.50 kmh
BAe/Aérospatiale Concorde — 697.49 mph
4 RR Olympus 593, 38,000 lbs
3/2/00

CAYENNE/RIO DE JANEIRO (FRANCE)
Michel Rio — 1,413.69 kmh
Jean Marcot — 878.42 mph
Aérospatiale/BAC Concorde
1/8/99

EASTER ISLAND/LIMA (FRANCE)
Michel Rio — 1,642.54 kmh
Jean Marcot — 1,020.62 mph
Aérospatiale/BAC Concorde
1/20/99

GRAND FORKS/MONROEVILLE (USA)
Capt. R. F. Lewandowski, Commander — 1,010.46 kmh
Capt. Kevin T. Kalen, Pilot — 627.87 mph
Lt. Col. Timothy W. Van Splunder, OSO
Capt. Jerrold A. Wangberg, DSO
Rockwell International B-1B
4 GE F101, 12,000 lbs
4/7/94

IGUAZU/BUENOS AIRES (FRANCE)
Michel Rio — 716.11 kmh
Jean Marcot — 444.97 mph
Aérospatiale/BAC Concorde
1/11/99

KAILUA/NADI (UK)
Max Robinson — 1,725.02 kmh
BAe/Aérospatiale Concorde — 1,071.87 mph
10/27/99

KAILUA/SYDNEY (UK)
Max Robinson — 1,237.75 kmh
BAe/Aérospatiale Concorde — 769.10 mph
10/27/99

LISBON/NEW YORK (UK)
Geoffrey Mussett — 1,668.37 kmh
BAe/Aérospatiale Concorde — 1,036.67 mph
11/14/99

LONDON/BOSTON (UK)
Mike Bannister — 1,689.47 kmh
BAe/Aérospatiale Concorde — 1,049.79 mph
4 RR Olympus, 38,000 lbs
10/8/03

LONDON/PERTH (UK)
W/CDR J. L. Upritchard — 914.07 kmh
VC 10K3 — 568.00 mph
4 Conway 550B, 22,500 lbs
4/8/87

LONDON/SHANNON (UK)
Capt. Peter Horton — 846.98 kmh
BAe/Aérospatiale Concorde — 526.29 mph
12/31/92

LONDON/WASHINGTON (UK)
Roger Mills — 1,537.76 kmh
Aérospatiale/BAC Concorde — 955.52 mph
2/11/00

MADRAS/NAIROBI (UK)
Max Robinson — 1,481.35 kmh
BAe/Aérospatiale Concorde — 920.47 mph
11/4/99

MEXICO CITY/MADRID (MEXICO)
Javier Heredia Viderique — 919.91 kmh
Manuel G. Hino-Josa Gomez — 571.60 mph
Carlos A. Ojeda Ruiz de la Pena
Boeing 767-200ER
10/1/98

MEXICO CITY/PARIS (MEXICO)
Alfonso Ponce Alcocer — 925.49 kmh
Jaime Soto Werschitz — 575.07 mph
Jorge O. Castillero Uribe
Boeing 767-200ER
9/24/98

MULLAN/MONROEVILLE (USA)
Capt. R. F. Lewandowski, Commander 976.48 kmh
Capt. Kevin T. Kalen, Pilot 606.76 mph
Lt. Col. Timothy W. Van Splunder, OSO
Capt. Jerrold A. Wangberg, DSO
Rockwell International B-1B
4 GE F101, 12,000 lbs
4/7/94

NADI/SYDNEY (UK)
Max Robinson 1,649.09 kmh
BAe/Aérospatiale Concorde 1,024.69 mph
10/27/99

NAIROBI/AQABA (UK)
Geoffrey Mussett 1,030.58 kmh
BAe/Aérospatiale Concorde 640.37 mph
11/9/99

NEW YORK/CAYENNE (FRANCE)
Michel Rio 1,747.29 kmh
Jean Marcot 1,085.71 mph
Aérospatiale/BAC Concorde
1/8/99

NEW YORK/LONDON (UK)
L. G. Scott 1,920.07 kmh
BAe/Aérospatiale Concorde 1,193.07 mph
4 RR Olympus, 38,000 lbs
2/7/96

NEW YORK/RIO DE JANEIRO (FRANCE)
Michel Rio 1,185.13 kmh
Jean Marcot 736.40 mph
BAe/Aérospatiale Concorde
1/8/99

NEW YORK/VANCOUVER (UK)
Roger Mills 1,004.67 kmh
BAe/Aérospatiale Concorde 624.27 mph
10/23/99

RIO DE JANEIRO/IGUAZU (FRANCE)
Michel Rio 948.74 kmh
Jean Marcot 589.52 mph
BAe/Aérospatiale Concorde
1/11/99

SAN JOSE, CR/NEW YORK (FRANCE)
Michel Rio 1,506.27 kmh
Jean Marcot 935.95 mph
BAe/Aérospatiale Concorde
1/27/99

SANTIAGO/EASTER ISLAND (FRANCE)
Michel Rio 1,707.38 kmh
Jean Marcot 1,060.91 mph
BAe/Aérospatiale Concorde
1/19/99

SEATTLE/CHRISTCHURCH (NEW ZEALAND)
Ian C. Varcoe 803.28 kmh
Gregory J. Vujcich 499.14 mph
Noel G. Farquhar
John Crosby
Boeing 767-300ER
10/28/97

SEATTLE/LUTON (UK)
Ivor Gibbs 870.94 kmh
Boeing 767-300 541.18 mph
2 CF6-80C, 61,500 lbs
5/1/96

SHANNON/BERMUDA (UK)
Capt. Peter Horton 1,757.42 kmh
BAe/Aérospatiale Concorde 1,092.01 mph
1/1/93

SHANNON/NEW YORK (UK)
Stuart Robertson 1,587.04 kmh
BAe/Aérospatiale Concorde 986.14 mph
9/27/93

SYDNEY/AGANA (UK)
Max Robinson 1,512.09 kmh
BAe/Aérospatiale Concorde 939.57 mph
10/30/99

SYDNEY/BANGKOK (UK)
Max Robinson 849.83 kmh
BAe/Aérospatiale Concorde 528.06 mph
10/30/99

USHUAIA/SANTIAGO (FRANCE)
Michel Rio 1,202.62 kmh
Jean Marcot 747.27 mph
BAe/Aérospatiale Concorde
1/17/99

VANCOUVER/KAILUA (UK)
Max Robinson 1,595.75 kmh
BAe/Aérospatiale Concorde 991.55 mph
10/24/99

VENICE/LISBON (UK)
Geoffrey Mussett 965.43 kmh
BAe/Aérospatiale Concorde 599.89 mph
11/14/99

VENICE/NEW YORK (UK)
Geoffrey Mussett 909.16 kmh
BAe/Aérospatiale Concorde 564.92 mph
11/14/99

CLASS C-1.R (200,000-250,000 kg / 440,924-551,155 lbs)
GROUP III (JET ENGINE)

MONROEVILLE/GRAND FORKS (USA)
Capt. Michael S. Menser, Commander 948.32 kmh
Capt. Brian P. Gallagher, Pilot 589.26 mph
Capt. Robert P. Boman, OSO
Capt. Matthew E. Grant, WSO
Rockwell International B-1B
4 GE F101, 12,000 lbs
4/8/94

MONROEVILLE/MULLAN (USA)
Capt. Michael S. Menser, Commander 920.72 kmh
Capt. Brian P. Gallagher, Pilot 572.11 mph
Capt. Robert P. Boman, OSO
Capt. Matthew E. Grant, WSO
Rockwell International B-1B
4 GE F101, 12,000 lbs
4/7/94

SINGAPORE/LOUISVILLE (USA)
John H. Ransom, Jr. 887.34 kmh
Michael L. Carlson 551.37 mph
Glenn J. Johnmeyer
Thomas A. Tonge
McDonnell Douglas MD-11F
3 GE CF6-80C2, 61,500 lbs
11/16/01

TAIPEI/SEATTLE (USA)
Susan P. Darcy 946.41 kmh
David C. Carbaugh 588.07 mph
Boeing 777-200
2 RR RB211, 76,000 lbs
3/1/96

TOULOUSE/MELBOURNE (AUSTRALIA)
Bruce Simpson 864.56 kmh
David Collier 537.21 mph
Bruce Van Eyle
James Peach
Airbus A330-200
2 GE CF6-80, 68,500 lbs
12/25/02

TOULOUSE/SANTIAGO (FRANCE)
Patrick Baudry 787.62 kmh
Airbus A330-222 489.40 mph
3/21/98

CLASS C-1.S (250,000-300,000 kg / 551,155-661,386 lbs)
GROUP III (JET ENGINE)

AUCKLAND/PARIS (FRANCE)
Bernard Ziegler 851.29 kmh
Pierre Baud 528.97 mph
Gérard Guyot
Airbus A340-211
6/16-18/93

DALLAS/SEOUL (USA)
Paul J. Shall, Captain 739.79 kmh
James A. Richmond, Co-captain 459.68 mph
McDonnell Douglas MD-11
3 GE CF6-80C2, 61,500 lbs
10/22/90

NORTH POLE/YUMA, AZ (USA)
John I. Miller, Captain 822.74 kmh
Paul J. Shall, Co-captain 511.23 mph
McDonnell Douglas MD-11
3 GE CF6-80C2, 61,500 lbs
10/25/90

PARIS/AUCKLAND (FRANCE)
Bernard Ziegler 861.56 kmh
Pierre Baud 535.35 mph
Gérard Guyot
Airbus A340-211
6/16-18/93

SEOUL/LONDON (USA)
James A. Richmond, Captain 663.03 kmh
Paul J. Shall, Co-captain 411.98 mph
McDonnell Douglas MD-11
3 GE CF6-80C2, 61,500 lbs
10/24/90

TOULOUSE/DJAKARTA (FRANCE)
Patrick Baudry 808.50 kmh
Gerard Guyot 502.38 mph
Airbus A340-313
4 CFM 56, 22,000 lbs
6/21/96

CLASS C-1.T (Over 300,000 kg / 661,386 lbs)
GROUP III (JET ENGINE)

LONDON/SYDNEY (AUSTRALIA)
Quantas Airways 979.20 kmh
Boeing 747-400 608.47 mph
8/16-17/89

NEW YORK/HONG KONG (UK)
Paul Horsting 830.96 kmh
Boeing 747-400 516.33 mph
7/5/98

SOUTH POLE/NORTH POLE (USSR)
Lev V. Kozlov, Captain 676.88 kmh
Yuri P. Resnitsky, Copilot 420.59 mph
Oleg I. Pripuskov, Copilot
Anatoly V. Andronov, Copilot
Antonov-124
12/1-4/90

Commercial Air Route

Like "Speed Over a Recognized Course" records, "Speed Over a Commercial Air Route" records are for speed between two cities. The difference is that these flights are made by commercial airliners on regularly scheduled commercial routes, and are timed from takeoff to touchdown. Accordingly, they are highly representative of the actual speeds that can be experienced by passengers for the price of a ticket.

There are two categories of world-class commercial records: those 6,500 kilometers (4,038.9 statute miles) or less and those over 6,500 kilometers. For routes over 6,500 kilometers, crew changes are permitted.

SPEED OVER A COMMERCIAL AIR ROUTE

WEST TO EAST TRANSCONTINENTAL*
(COMMERCIAL JET TRANSPORT)
Capt. Wylie H. Drummond — 680.90 mph
American Airlines
Boeing 707-123
4 P&W JT3D
Los Angeles Intl/Idlewild
Distance: 2,474 mi
3 hrs 38 min
4/10/63

EAST TO WEST TRANSCONTINENTAL*
(COMMERCIAL JET TRANSPORT)
Capt. Gene Kruse — 572.57 mph
American Airlines
Boeing 707-720B
4 P&W JT3D
Idlewild/Los Angeles Intl
Distance: 2,474 mi
4 hrs 19 min 15 sec
8/15/62

* * *

ACAPULCO/CHICAGO (USA)
Capt. Roger E. Ruch — 811.86 kmh / 504.47 mph
American Airlines
Boeing 707-123B, N757OA
4 P&W JT3D-1
3 hrs 43 min 50 sec
2/26-27/78

ACAPULCO/DALLAS*
Capt. Roger E. Ruch — 500.37 mph
American Airlines
Boeing 707-123B
4 P&W JT3D
General Juan N. Alvarez/Dallas-Ft. Worth Regional
2 hrs 12 min 35 sec
2/12/78

ACAPULCO/DETROIT*
Capt. Daniel J. Clisham — 501.23 mph
Evergreen Intl Airlines
Douglas DC-8-52
4 P&W JT3D
General Juan N. Alvarez/Detroit Metro
Distance: 2,030 mi
4 hrs 2 min 50 sec
5/9/77

ACAPULCO/LAS VEGAS (UK)
Capt. John Hutchinson — 1,046.95 kmh / 650.54 mph
British Airways
BAe/Aérospatiale Concorde
3/16/90

AGANA/HONOLULU (USA)
Richard Roberts — 683.28 kmh / 424.59 mph
Ozark Airlines
McDonnell Douglas DC-9-41
2 P&W JT8D-15, 15,000 lbs
8 hrs 57 min
2/5-6/83

ALBANY/BOSTON* *Group II*
Capt. William B. Moody — 325.00 mph
American Airlines
Lockheed Electra L-188
4 501D-13
Albany County/Logan Intl
Distance: 144 mi
26 min 35 sec
1/11/66

ALBUQUERQUE/CHICAGO*
Capt. Barry Schiff — 574.29 mph
Trans World Airlines
Boeing 727-231
3 P&W JT3D
Albuquerque Intl/O'Hare Intl
1 hr 56 min 42 sec
1/1/75

ALBUQUERQUE/PHOENIX*
David H. Koch, Captain — 533.99 mph
Officer Lyle T. Shelton, 1st Officer
Trans World Airlines
Boeing 727-231
3 P&W JT8D
Albuquerque Intl/Phoenix Sky Harbor Intl
36 min 58 sec
5/6/74

AMSTERDAM/LONDON*
Robert L. O'Neill, Captain — 336.26 mph
William R. Wallace, 1st Officer
United Airlines
Boeing 777-222
2 P&W PW4077, 77,000 lbs
41 min
8/13/95

ANCHORAGE/CHICAGO (USA)
Larry E. Steenstry, Captain
David L. Pond, 1st Officer
United Airlines
McDonnell Douglas DC-10-30
3 GE CF6-50C2, 52,500 lbs
4 hrs 45 min
8/13/98
961.74 kmh
597.59 mph

ANCHORAGE/HONG KONG (USA) §
Glenn W. Stickel, Captain
Thomas McCleary, 2nd Officer
United Parcel Service
Boeing 747-100
4 P&W JT9D, 45,000 lbs
11 hrs 32 min
12/22/99
707.96 kmh
439.90 mph

ANCHORAGE/LOUISVILLE (USA)
Glenn Stickel, Captain
Eric Steinbaugh, 1st Officer
C. Keith Haines, Jr., 2nd Officer
United Parcel Service
Boeing 747-100
4 P&W JT9D-7A, 46,100 lbs
4 hrs 55 min
12/10/95
1,019.58 kmh
633.53 mph

ANCHORAGE/MAGADAN (USA)
Capt. Steve Sanford
Co-Capt. Mike Swanigan
Flt Engineer Rob Spero
Alaska Airlines
Boeing 727-200
3 P&W JT8D-17, 16,000 lbs
4 hrs
6/18/91
783.33 kmh
486.73 mph

ANCHORAGE/ONTARIO (USA)
Glenn W. Stickel, Captain
Robert L. Peshkin, 1st Officer
William J. Borden, 2nd Officer
United Parcel Service
Boeing 747-200B
4 P&W JT9D, 53,000 lbs
3 hrs 38 min
12/15/01
1,044.10 kmh
648.77 mph

ANCHORAGE/PRESTWICK (USA) §
William A. Austin
Federal Express
DC-10-30
3 GE CF6-50C2, 52,500 lbs
7 hrs 31 min 45 sec
3/27/92
888.83 kmh
552.29 mph

ANCHORAGE/SEOUL (USA)
Capt. Glenn Stickel
United Parcel Service
Boeing 747-100
4 P&W JT9D, 46,100 lbs
7 hrs 56 min
3/10/96
765.38 kmh
475.58 mph

ANCHORAGE/TOKYO (USA)
Capt. George W. Murphy, Jr.
FedEx
McDonnell Douglas MD-11F
3 GE CF6-80C2, 61,500 lbs
6 hrs 7 min
7/22/00
901.58 kmh
560.21 mph

ARUBA/MIAMI (USA)
Capt. Roger E. Ruch
American Airlines
Boeing 727-100, N1964
3 P&W JT8D, 14,000 lbs
2 hrs 27 min
4/19/81
746.64 kmh
463.96 mph

Bob Falkins, Kellie Waddell, and Paul Entrekin (l to r) in the cockpit of a Delta Air Lines Boeing 727. Their record flight from Atlanta to Greensboro marked Delta's retirement of 727 service.

ATLANTA/BALTIMORE*
Capt. Peter C. Coxhead — 539.28 mph
Eastern Airlines
Boeing 727
3 P&W JT8D
Atlanta Intl/Friendship Airport
Distance: 576 mi
1 hr 4 min 5 sec
1/12/73

ATLANTA/BOSTON (USA)
Donald R. Champagne, Captain — 960.72 kmh
Robert E. Koski, 1st Officer — 596.96 mph
Delta Air Lines
Boeing 757
2 P&W PW2037, 38,200 lbs
1 hr 35 min
3/30/94

ATLANTA/CHICAGO* *Group I*
Capt. H. T. Merrill — 290.50 mph
Eastern Airlines
Lockheed Constellation
4 Wright R-3350
Atlanta Municipal/Chicago Municipal
Distance: 590 mi
2 hrs 1 min 55 sec
8/5/47

ATLANTA/GREENSBORO (USA)
Robert D. Falkins, Captain — 783.11 kmh
Paul T. Entrekin, 1st Officer — 486.60 mph
Kellie Waddell, 2nd Officer
Delta Air Lines
Boeing 727-200
3 P&W JT8D, 15,500 lbs
37 min
4/6/03

ATLANTA/LOS ANGELES*
Capt. L. L. Caruthers — 566.82 mph
Delta Air Lines
Douglas DC-8
4 P&W JT3D-1
Atlanta Municipal/Los Angeles Intl
Distance: 1,946 mi
3 hrs 25 min 59 sec
7/19/63

ALTANTA/MADRID (USA) §
Samuel Don Smith, Captain — 873.79 kmh
Bruce A. Doberstein, 1st Officer — 542.95 mph
Delta Air Lines
Boeing 767-300ER
2 GE CF6-80C2, 60,000 lbs
7 hrs 58 min
6/21/00

ATLANTA/MYRTLE BEACH*
Capt. Richard Y. Miller — 509.02 mph
Piedmont Airlines
Boeing 737-200, N754N
3 P&W JT8D-9A, 14,500 lbs
Hartsfield Intl/Myrtle Beach AFB
Distance: 322 mi
38 min
3/11/80

ATLANTA/NEW YORK*
Capt. Michael Winicki — 589.64 mph
Delta Air Lines
Douglas DC-8-51
4 P&W JT3D-1
Atlanta Municipal/Idlewild Intl
Distance: 760 mi
1 hr 17 min 20 sec
1/21/63

ATLANTA/ORLANDO (USA)
Patrick M. Broderick, Pilot — 749.88 kmh
G. Curt Ayers, 1st Officer — 465.98 mph
Binka Bone, 2nd Officer
Eastern Airlines
Boeing 727-225A, N885
3 P&W JT8D
52 min 5 sec
2/25/88

ATLANTA/PARIS (USA) §
Gordon R. Beck, Captain — 1,013.04 mph
Guy W. Davis, 1st Officer — 629.47 mph
Robert R. Leitzen, 2nd Officer
Delta Air Lines
Lockheed L-1011
3 RR RB211-524, 50,000 lbs
6 hrs 57 min 48 sec
12/1/93

ATLANTA/ST. PETERSBURG (USA)
Capt. Douglas Matthews — 658.99 kmh
Flight International — 409.50 mph
Boeing 727
3 P&W JT8D, 16,000 lbs
1 hr 2 min
7/5/85

ATLANTA/SAN FRANCISCO*
Capt. R. J. Nelson — 525.78 mph
Delta Air Lines
Douglas DC-8
4 P&W JT3D-1
Atlanta Municipal/San Francisco Intl
Distance: 2,139 mi
4 hrs 4 min 5 sec
4/27/64

ATLANTA/SEATTLE (USA)
Douglas R. Harper, Captain — 827.91 kmh
Robert R. Halstead, 1st Officer — 514.44 mph
Delta Air Lines
Boeing 767-300ER
2 GE CF6-80C2, 60,000 lbs
4 hrs 14 min
3/17/03

AUCKLAND/HONOLULU (USA) §
Capt. Robert M. McElrath — 864.16 kmh
United Airlines — 536.96 mph
Boeing 747-100
4 P&W JT9D-7A
8 hrs 12 min 22 sec
11/9/89

AUCKLAND/LONDON (UK)
Capt. J. W. Burton — 962.58 kmh
British Airways — 598.15 mph
BAe/Aérospatiale Concorde
4 Olympus 593/610, 38,000 lbs
4/7-8/86

AUCKLAND/LOS ANGELES (USA) §
Capt. N. B. Godlove — 963.55 kmh
United Airlines — 598.72 mph
Boeing 747-400
4 P&W PW4056, 56,750 lbs
10 hrs 53 min
4/6/94

AUCKLAND/MELBOURNE (USA)
Capt. N. B. Godlove — 865.18 kmh
United Airlines — 537.60 mph
Boeing 747-400
4 P&W PW4056, 56,750 lbs
3 hrs 3 min
1/28/94

AUCKLAND/SAN FRANCISCO (USA) §
Capt. Walter H. Mullikin — 910.11 kmh
Capt. Jake M. Marcum, — 565.52 mph
Chief of Relief Airman
Pan American World Airways, Inc.
Boeing 747 SP, N533 PA
4 P&W JT9D-7A
11 hrs 33 min 36 sec
10/30-31/77

AUCKLAND/SYDNEY (USA)
Capt. N. B. Godlove — 852.18 kmh
United Airlines — 529.52 mph
Boeing 747-400
4 P&W PW4056, 56,000 lbs
2 hrs 32 min
4/4/95

BALI/COLOMBO (UK)
Capt. Geoffrey Mussett — 1,600.68 kmh
British Airways — 994.61 mph
BAe/Aérospatiale Concorde
3/31/90

BALI/MOMBASA (UK)
Capt. Geoffrey Mussett — 1,300.63 kmh
British Airways — 808.17 mph
BAe/Aérospatiale Concorde
3/3/90

BALTIMORE/BERMUDA (USA)
Capt. Robert F. Olszewski — 841.76 kmh
USAir — 523.05 mph
Boeing 767-201ER
2 GE CF6-80C2, 52,000 lbs
1 hr 34 min
5/1/91

BALTIMORE/LOS ANGELES*
Capt. Howard U. Morton — 563.46 mph
American Airlines
Boeing 707
4 P&W JT3D
Friendship Intl/Los Angeles Intl
4 hrs 8 min
10/30/62

BANGKOK/HONG KONG (USA)
Capt. William A. Dias — 733.17 kmh
United Airlines — 455.57 mph
Boeing 747-400
4 P&W PW4056, 56,750 lbs
2 hrs 18 min
9/20/99

BANGKOK/TAIPEI (USA)
Capt. Robert E. Stanton — 842.55 kmh
United Airlines — 523.54 mph
Boeing 747-400
4 P&W PW4056, 56,750 lbs
2 hrs 57 min
4/29/93

BANGKOK/TOKYO (USA)
Capt. Jay C. Mallory — 925.69 kmh
United Airlines — 575.20 mph
Boeing 747-400
4 P&W PW4056, 56,750 lbs
5 hrs 1 min
3/15/99

BARBADOS/LONDON (UK) §
Capt. Norman Todd — 1,826.38 kmh
Brian Walpole, Copilot — 1,134.85 mph
Sydney Nolton, Engineering Officer
British Airways
BAe/Aérospatiale Concorde
4 RR Olympus, 38,000 lbs
3 hrs 42 min 5 sec
11/2/77

BEIJING/SAN FRANCISCO (USA) §
Capt. William A. Dias — 846.45 kmh
United Airlines — 525.96 mph
Boeing 747-400
4 P&W PW4056, 56,000 lbs
11 hrs 13 min
10/19/95

BEIJING/SHANGHAI (USA)
Capt. N. B. Godlove — 724.63 kmh
United Airlines — 450.26 mph
Boeing 747SP
4 P&W JT9D-7A, 46,500 lbs
1 hr 29 min
8/18/92

BEIJING/TOKYO (USA)
Capt. David F. Specht — 871.71 kmh
United Airlines — 541.65 mph
Boeing 747-400
4 P&W PW4056, 56,750 lbs
2 hrs 27 min
2/23/97

BERLIN/NEW YORK (USA) §
Don Binard, Captain — 868.87 kmh
Allan D. Gee, 1st Officer — 539.92 mph
Joseph P. Miller, 2nd Officer
Pan American Airlines
Airbus A310-300
2 P&W PW4152
8 hrs 48 min
12/26/88

BOGATA/MIAMI (USA)
Capt. Robert P. Gross — 807.09 kmh
American Airlines — 501.50 mph
Boeing 767-300ER
2 GE CF6-80C2, 61,500 lbs
3 hrs 1 min
8/18/94

BOISE/DENVER (USA)
Capt. Wallis E. Hubbell — 845.52 kmh
United Airlines — 525.38 mph
Boeing 737-300
2 CFM CFM56, 20,000 lbs
1 hr 14 min
10/19/01

BOSTON/ALBANY*
Capt. Robert F. Olszewski — 370.80 mph
Allegheny Airlines
McDonnell Douglas DC-9-31, N984VJ
2 P&W JT8D-7A, 14,000 lbs
Logan Intl/Albany County
Distance: 145 mi
23 min 15 sec
10/17/79

BOSTON/CHARLOTTE (USA)
Capt. Robert F. Olszewski — 832.23 kmh
1st Officer Russell L. Farris — 517.12 mph
USAir
Boeing 727-200A
3 P&W JT8D-17A, 16,000 lbs
1 hr 24 min 24 sec
07/04/90

BOSTON/CHICAGO (USA)
Robert F. Frank, Captain — 772.77 kmh
Larry E. Steenstry, 1st Officer — 480.17 mph
United Airlines
Boeing 757
2 P&W PW2037, 38,200 lbs
1 hr 48 min
6/1/94

BOSTON/LOS ANGELES*
Capt. George C. Dent — 562.57 mph
American Airlines
Boeing 707
4 P&W JT3D
Logan Intl/Los Angeles Intl
Distance: 2,610 mi
4 hrs 38 min 21 sec
4/29/65

BOSTON/NEW YORK*
Capt. Peter C. Bals — 414.61 mph
American Airlines
Convair 990 4 GE CJ805-23
Logan Intl/Idlewild Intl
Distance: 186 mi
26 min 55 sec
7/20/63

BOSTON/PARIS*
Capt. Michael G. Seay — 655.09 mph
Northwest Airlines
McDonnell Douglas DC10-40
3 P&W JT9D, 50,000 lbs
5 hrs 15 min
1/10/93

BOSTON/SAN FRANCISCO*
Capt. W. R. Swain — 508.64 mph
American Airlines
Boeing 707-123B
4 P&W JT3D
Logan Intl/San Francisco Intl
Distance: 2,704 mi
5 hrs 18 min 57 sec
1/7/62

BOSTON/SAN JOSE (USA)
Capt. Dennis D. Lyons — 742.09 kmh, 461.11 mph
American Airlines
Boeing 757-200
2 RR 535E4, 40,100 lbs
5 hrs 49 min
4/26/93

BOSTON/WASHINGTON*
Capt. Peter W. Williamson — 488.55 mph
Northeast Airlines
McDonnell Douglas DC-9-31
2 P&W JT8D-7
Logan Intl/Washington National
Distance: 399 mi
49 min 30 sec
3/8/71

BRISBANE/KOTA KINABALU (USA)
Capt. Leiland M. Duke, Jr. — 870.28 kmh, 540.77 mph
FedEx
McDonnell Douglas MD-11F
3 GE CF6-80C2, 61,500 lbs
6 hr 15 min
9/25/98

BRUSSELS/CHICAGO (USA) §
Capt. Thomas W. Mayer — 886.14 kmh, 550.63 mph
American Airlines
DC-10-30
3 GE CF6-50C2, 52,500 lbs
7 hrs 32 min
7/6/91

BUENOS AIRES/MIAMI (USA) §
Roy A. Berube, Captain — 942.87 kmh, 585.87 mph
Mario J. Lopes, 1st Officer
James A. Arnold, 2nd Officer
United Airlines
Boeing 747-100
4 P&W JT9D-7A, 46,950 lbs
7 hrs 33 min
1/5/97

BUENOS AIRES/MONTEVIDEO (USA)
William W. Critcher, Captain — 653.25 kmh, 405.91 mph
John M. Hogancamp, 1st Officer
United Airlines
Boeing 767-300ER
2 P&W PW4060, 60,000 lbs
21 min
9/17/99

BUENOS AIRES/NEW YORK (USA) §
Capt. Angela Masson — 911.17 kmh, 566.17 mph
American Airlines
Boeing 777-200
2 R-R Trent 892, 90,000 lbs
9 hrs 22 min
8/5/01

BUFFALO/NEW YORK*
Capt. Robert F. Olszewski — 529.09 mph
Allegheny Airlines
McDonnell Douglas DC-9-31
2 P&W JT8D-7A
Greater Buffalo Intl/LaGuardia
Distance: 291.29 mi
33 min
3/7/78

BUFFALO/SYRACUSE*
Capt. Robert F. Olszewski — 407.40 mph
Allegheny Airlines
McDonnell Douglas DC-9-51
2 P&W JT8D-17A
Greater Buffalo Intl/Syracuse Hancock Intl
Distance: 133 mi
19 min 35 sec
4/24/78

CAIRO/LOS ANGELES (USA) §
Donald L. Mullin — 775.95 kmh, 482.17 mph
Cammacorp
DC-8-72
4 CFM 56, 24,000 lbs
15 hrs 45 min 51 sec
3/29-30/83

CAPETOWN/LONDON (UK) §
Capt. Brian Walpole — 1,192.48 kmh, 740.99 mph
British Airways
BAe/Aérospatiale Concorde
4 RR Olympus, 38,000 lbs
8 hrs 7 min 39 sec
3/29/85

CARACAS/PARIS (FRANCE) §
Cdt. Pierre Chanoine — 1,833.78 kmh, 1,139.46 mph
Air France
BAe/Aérospatiale Concorde
4 RR Olympus, 38,000 lbs
9/24/91

CARACAS/SANTA MARIA (FRANCE)
Cdt. Jean-Paul Le Moel — 1,869.90 kmh, 1,161.96 mph
Air France
BAe/Aérospatiale Concorde
4 RR Olympus, 38,000 lbs
3/20/82

CEDAR RAPIDS/CHICAGO *
Capt. Martha Anne Collins Leeds — 345.43 mph
United Airlines
Boeing 727-200
3 P&W JT8D, 15,500 lbs
34 min
10/30/01

CHARLOTTE/DAYTON (USA)
Felix Jesus Villaverde — 714.91 kmh / 444.22 mph
Airlift International
McDonnell Douglas DC-8
4 P&W JT3D, 16,000 lbs
50 min
12/7/90

CHARLOTTE/DAYTONA BEACH (USA)
Stephen A. Clegg, Captain — 789.39 kmh / 490.50 mph
Scott A. Metz, 1st Officer
USAir
Boeing 737-200
2 P&W JT8D-15, 14,500 lbs
51 min
06/29/90

CHARLOTTE/FRANKFURT (USA) §
Don J. Rhynalds, Captain — 1,033.09 kmh / 641.93 mph
James S. Barbour, 1st Officer
J. Christopher Burkhart, 1st Officer
USAir Boeing 767-200
2 GE CF6-80C2, 52,000 lbs
6 hrs 50 min
1/9/93

CHARLOTTE/LONDON (USA)
Capt. Joseph W. Graham — 939.86 kmh / 584.00 mph
US Airways
Airbus A330-300
2 P&W PW4168A, 68,600 lbs
6 hrs 51 min
6/15/02

CHARLOTTE/SAN JUAN (USA)
Capt. Robert F. Olszewski — 913.89 kmh / 567.87 mph
USAir
Boeing 727-200A
3 P&W JT8D-17A, 16,000 lbs
2 hrs 36 min
10/17/90

CHARLOTTE/TAMPA (USA)
Felix Jesus Villaverde — 712.06 kmh / 442.45 mph
Airlift International
McDonnell Douglas DC-8
4 P&W JT3D, 16,000 lbs
1 hr 9 min
12/7/90

CHICAGO/ANCHORAGE (USA)
Larry E. Steenstry, Captain — 835.66 kmh / 519.25 mph
Colette P. Ireland, 1st Officer
United Airlines
McDonnell Douglas DC-10-30F
3 GE CF6-50C2, 52,500 lbs
5 hrs 28 min
2/22/99

CHICAGO/ATLANTA* *Group I*
Capt. H. T. Merrill — 326.93 mph
Eastern Airlines
Lockheed Constellation
4 Wright R-3350
Chicago Municipal/Atlanta Municipal
Distance: 590 mi
1 hr 48 min 20 sec
8/5/47

CHICAGO/BALTIMORE (USA)
Scott R. Lewis, Captain — 831.69 kmh / 516.78 mph
Martha Anne Collins Leeds, 1st Officer
R. Glenn Anderson, Flight Engineer
United Airlines
Boeing 727
3 P&W JT8D, 15,500 lbs
1 hr 12 min
2/27/99

CHICAGO/BERLIN (USA) §
Herbert L. Dabelow, Captain — 846.37 kmh / 525.91 mph
Robert E. Davies, 1st Officer
Robert S. Greer, 1st Officer
American Airlines
Boeing 767-200ER
2 GE CF6-80A, 48,000 lbs
8 hrs 22 min
8/31/93

CHICAGO/BOISE (USA)
Capt. Mario Jerez — 838.93 kmh / 521.28 mph
United Airlines
Boeing 737-500
2 CFM 56, 20,000 lbs
2 hrs 45 min
4/5/97

CHICAGO/BOSTON*
Capt. Alden Young — 692.28 mph
American Airlines
Convair 990
4 GE CT805-23
Chicago O'Hare/Logan Intl
Distance: 865 mi
1 hr 14 min 58 sec
2/16/65

CHICAGO/BRUSSELS (USA) §
Capt. Thomas W. Mayer — 940.24 kmh / 584.23 mph
American Airlines
DC-10-30
3 GE CF6-50C2, 52,500 lbs
7 hrs 6 min
9/21/91

CHICAGO/CHARLOTTE (USA)
Capt. Vaughn B. Cordle — 838.74 kmh / 521.17 mph
United Airlines
Boeing 737-300
2 CFM56, 20,000 lbs
1 hr 9 min
2/20/93

CHICAGO/COLUMBUS (USA)
Capt. Vaughn B. Cordle — 792.00 kmh / 492.12 mph
United Airlines
Boeing 737-300
2 CFM56, 20,000 lbs
36 min
3/14/93

CHICAGO/DALLAS (USA)
Capt. Carter W. Taylor — 790.43 kmh / 491.15 mph
American Airlines
McDonnell Douglas MD-11
3 GE CF6, 61,500 lbs
1 hr 38 min
9/16/97

CHICAGO/DAYTON*
Capt. Steven A. Shatto — 423.48 mph
United Airlines
Boeing 737-200
34 min
2 P&W JT8D, 14,000 lbs
6/23/96

CHICAGO/DENVER (USA)
Capt. David H. Friend — 806.97 kmh / 501.43 mph
United Airlines
Boeing 747-400
4 P&W PW4056, 56,750 lbs
1 hr 46 min
6/24/02

CHICAGO/DUSSELDORF (USA) §
Capt. Herbert L. Dabelow — 895.11 kmh
American Airlines — 556.20 kmh
Boeing 767-200ER
2 GE CF6-80A, 48,000 lbs
7 hrs 35 min
8/25/94

CHICAGO/EL PASO (USA)
Capt. Daniel J. Clisham — 784.61 kmh
American Airlines — 487.53 mph
McDonnell Douglas MD-82
2 P&W JT8D, 20,000 lbs
2 hrs 32 min
10/13/98

CHICAGO/FT. LAUDERDALE (USA)
Dan L. Kurt, Captain — 834.52 kmh
Steven A. Climer, 1st Officer — 518.54 mph
G. Robert Boller, 2nd Officer
United Airlines
Boeing 727
3 P&W JT8D-15, 15,500 lbs
2 hrs 17 min
4/30/94

CHICAGO/FT. MYERS (USA)
Randy S. Trull, Captain — 840.24 kmh
Larry E. Steenstry, 1st Officer — 522.09 kmh
United Airlines
Boeing 757
2 P&W PW2037, 38,200 lbs
2 hrs 9 min
11/19/94

CHICAGO/FORT WORTH*
Capt. Glen L. Stockwell — 548.85 mph
American Airlines
Boeing 727
3 P&W JT8D-1
Chicago O'Hare Intl/Greater Southwest Intl
Distance: 809 mi
1 hr 28 min 26 sec
4/20/65

CHICAGO/FRANKFURT (USA) §
Capt. Thomas W. Mayer — 992.16 kmh
Scott W. Russell, 1st Officer — 616.52 mph
Raymond Sleboda, Flt Engineer
American Airlines
McDonnell Douglas DC-10-30
3 GE CF6-50-C2, 52,500 lbs
7 hrs 2 min
1/2/88

CHICAGO/HARTFORD (USA)
Capt. William R. Wallace — 847.40 kmh
United Airlines — 526.54 mph
Airbus A319
2 IAE V2522, 22,000 lbs
1 hr 29 min
3/31/99

CHICAGO/HAYDEN (USA)
John H. Jenkins, Captain — 737.28 kmh
George Farrington, 1st Officer — 458.15 mph
Wayne Miller, Flt Engineer
American Airlines #829
Boeing 727-100, N1991
3 P&W JT8D-7A, 14,000 lbs
2 hrs 12 min
3/1/87

CHICAGO/HONG KONG (USA) §
Jay C. Mallory, Captain — 878.77 kmh
Mark A. Vogelsang, 1st Officer — 546.04 mph
Larry R. Palmer, 1st Officer
Dean A. Reily, 1st Officer
United Airlines
Boeing 747-400
4 P&W PW4056, 56,750 lbs
14 hrs 15 min
10/15/01

CHICAGO/HONOLULU (USA) §
Thomas W. Mayer, Captain — 888.48 kmh
Ken Weyler, 1st Officer — 552.10 mph
American Airlines
McDonnell Douglas DC-10-30
3 GE CF6-50-C2, 52,500 lbs
7 hrs 41 min
7/29-30/88

CHICAGO/INDIANAPOLIS*
Capt. P. G. Cook — 494.93 mph
American Airlines
Boeing 707
4 P&W JT3D
Chicago O'Hare Intl/Weir Cook Intl
Distance: 179 mi
21 min 42 sec
4/9/65

CHICAGO/LONDON (USA)
James O. Gross, Captain — 919.43 kmh
Andrew R. Tuson, 1st Officer — 571.31 mph
Dennis B. Walsh, 1st Officer
American Airlines
McDonnell Douglas MD-11
3 GE CF6-80C2, 61,500 lbs
6 hrs 54 min
4/2/98

CHICAGO/LOS ANGELES (USA)
Delmar J. Walker, Captain — 853.42 kmh
Steven C. Derebey, 1st Officer — 530.29 mph
Alton A. Coward, 2nd Officer
United Airlines
McDonnell Douglas DC-10-10
3 GE CF6-6, 40,000 lbs
3 hrs 17 min
10/20/94

CHICAGO/LOUISVILLE (USA)
Capt. Vaughn B. Cordle — 769.14 kmh
United Airlines — 477.92 mph
Boeing 737-300
2 CFM56, 20,000 lbs
36 min
3/8/93

CHICAGO/MANCHESTER (USA)
Capt. John G. Maxwell, Jr. — 887.69 kmh
American Airlines — 551.58 mph
Boeing 767-300ER
2 GE CF6-80C2, 61,500 lbs
6 hrs 55 min
7/29/02

CHICAGO/MEXICO CITY*
Capt. C. V. Jacobson — 882.97 kmh
American Airlines — 548.67 mph
Boeing 707-123
4 P&W JT3D
3 hrs 5 min 5 sec
12/12/62

CHICAGO/MIAMI*
Capt. Dean Clifton
Delta Air Lines
Convair 880
4 GE CJ805-3
Chicago Intl/Miami Intl
Distance: 1,184 mi
2 hrs 1 min
1/2-3/62
587.11 mph

CHICAGO/MINNEAPOLIS (USA)
David A. Ricketts, Captain
R. Glenn Anderson, Flt Engineer
United Airlines
Boeing 727
3 P&W JT8D, 15,500 lbs
52 min
10/24/98
618.94 kmh
384.59 mph

CHICAGO/MONTREAL (USA)
Capt. Daniel J. Clisham
American Airlines
McDonnell Douglas MD-82
2 P&W JT8D, 20,000 lbs
1 hr 28 min
10/18/98
818.50 kmh
508.59 mph

CHICAGO/MUNICH (USA) §
Capt. Hans L. Lagerloef
American Airlines
Boeing 767-300
2 GE CF6-80C2, 61,500 lbs
7 hrs 46 min
11/12/93
935.98 kmh
581.59 mph

CHICAGO/NEW HAVEN (USA)
Capt. Vaughn B. Cordle
United Airlines
Boeing 737-300
2 CFM56, 20,000 lbs
1 hr 33 min
11/19/92
805.94 kmh
500.78 mph

CHICAGO/NEW YORK*
Capt. W. J. Callahan
American Airlines
Boeing 727
3 P&W JT8D-1
O'Hare Intl/LaGuardia
Distance: 733 mi
1 hr 8 min 15 sec
10/28/65
644.36 mph

CHICAGO/NEWARK*
Capt. Louis L. Sokol
American Airlines
Boeing 707-720B
4 P&W JT3D
O'Hare Intl/Newark Intl
Distance: 719 mi
1 hr 8 min 45 sec
1/15/63
627.48 mph

CHICAGO/OMAHA (USA)
Capt. Vaughn B. Cordle
United Airlines
Boeing 737-300
2 CFM56, 20,000 lbs
50 min
10/6/92
801.44 kmh
497.99 mph

CHICAGO/PARIS (USA) §
Capt. Richard A. Ionata
United Airlines
Boeing 767-300
2 P&W PW4060, 60,000 lbs
6 hrs 33 min
3/2/95
1,017.52 kmh
632.26 mph

CHICAGO/PHILADELPHIA (USA)
Capt. Vaughn B. Cordle
United Airlines
Boeing 737-300
2 CFM56, 20,000 lbs
1 hr 15 min
3/15/93
870.41 kmh
540.84 mph

CHICAGO/PHOENIX*
Capt. Vernon M. Byrne
American Airlines
Boeing 720
4 P&W JT3D
O'Hare Intl/Sky Harbor
Distance: 1,442 mi
2 hrs 31 min 28 sec
2/12/63
571.21 mph

CHICAGO/PORTLAND (USA)
James E. Foley, Captain
Mo I. Morsy, 1st Officer
United Airlines
Boeing 757-200
2 P&W PW2037, 38,200 lbs
3 hrs 24 min
4/2/97
821.06 kmh
510.18 mph

CHICAGO/RALEIGH (USA)
Capt. James O. Gross
American Airlines
McDonnell Douglas MD-82
2 P&W JT8D, 21,000 lbs
1 hr 19 min
12/12/97
789.27 kmh
490.42 mph

CHICAGO/ROANOKE*
Capt. Richard Y. Miller
Piedmont Airlines
Boeing 737-200, N745N
2 P&W JT8D-9A
O'Hare Intl/Roanoke Municipal/Woodrum
1 hr 5 min
12/26/79
546.30 mph

CHICAGO/ROCHESTER*
Capt. Warren Julliard
American Airlines
Boeing 727
3 P&W JT8D-1
O'Hare Intl/Rochester Municipal
Distance: 528 mi
52 min 39 sec
1/29/65
601.70 mph

CHICAGO/ROME (USA) §
Capt. John G. Maxwell, Jr.
American Airlines
Boeing 767-300ER
2 GE CF6-80C2, 61,500 lbs
8 hrs 13 min
2/17/01
942.18 kmh
585.44 mph

CHICAGO/SAN FRANCISCO (USA)
Edward J. Chapman, Captain
David P. Culligan, 1st Officer
United Airlines
Airbus A319
2 IAE V2522, 22,000 lbs
3 hrs 25 min
11/28/02
867.50 kmh
539.04 mph

CHICAGO/SAO PAULO (USA) §
Capt. Martin K. Kemp
United Airlines
Boeing 767-300ER
2 P&W PW4060, 60,000 lbs
9 hrs 23 min
5/11/03
898.21 kmh
558.12 mph

CHICAGO/SEATTLE (USA)
James A. Zapata, Captain
Kevin J. Aunapu, 1st Officer
United Airlines
Airbus A320
2 IAE V2527, 26,500 lbs
3 hrs 30 min
6/27/01
789.07 kmh
490.31 mph

CHICAGO/SEOUL (USA) §
Paul V. Davis II, Captain
Shawn P. Buckley, 1st Officer
United Airlines
Boeing 747-400
4 P&W PW4056, 56,750 lbs
14 hrs 27 min
11/19/93
726.49 kmh
451.42 mph

CHICAGO/STOCKHOLM (USA) §
Capt. Herbert L. Dabelow
American Airlines
Boeing 767-200ER
2 GE CF6-80A, 48,000 lbs
7 hrs 46 min
4/27/93
882.68 kmh
548.47 mph

CHICAGO/TAIPEI (USA) §
David F. Specht, Captain
Anil J. Singh, 1st Officer
United Airlines
Boeing 747-400
4 P&W PW4056, 56,750 lbs
14 hrs 56 min
9/14/98
802.18 kmh
498.45 mph

CHICAGO/TAMPA (USA)
Jerry E. Soltis, Captain
Larry E. Steenstry, 1st Officer
United Airlines
Boeing 757
2 P&W PW2037, 38,200 lbs
1 hr 55 min
9/18/94
851.14 kmh
528.87 mph

CHICAGO/TOKYO (USA) §
Capt. David F. Specht
United Airlines
Boeing 747
4 P&W PW4056, 56,750 lbs
11 hrs 40 min
9/28/97
871.38 kmh
541.45 mph

CHICAGO/TORONTO (USA)
Capt. Daniel J. Clisham
American Airlines
McDonnell Douglas MD-83
2 P&W JT8D, 21,000 lbs
54 min
2/11/00
778.10 kmh
483.49 mph

CHICAGO/TUCSON*
Capt. C. W. Evans
American Airlines
Convair 990
4 GE CJ-805-23
O'Hare Intl/Tucson Municipal
Distance: 1,440 mi
2 hrs 28 min 05 sec
2/7/63
583.45 mph

CHICAGO/WASHINGTON*
Capt. David H. Friend
United Airlines
Boeing 777
2 P&W PW4077, 77,000 lbs
1 hr 11 min
1/8/00
496.41 mph

CHICAGO/WASHINGTON* *Group II*
Capt. A. B. Perriello
American Airlines
Lockheed Electra L-188
4 Allison 501-D-13
O'Hare Intl/Washington National
Distance: 612 mi
1 hr 11 min 5 sec
12/19/65
510.58 mph

CHICAGO/ZURICH (USA) §
Capt. John G. Maxwell, Jr.
American Airlines
Boeing 767-300ER
2 GE CF6-80C2, 61,500 lbs
7 hrs 26 min
7/16/01
959.28 kmh
596.07 mph

CHRISTCHURCH/SYDNEY (UK)
K. D. Leney
British Airways
BAe/Aérospatiale Concorde
4 RR Olympus 593, 38,050 lbs
4/5/89
1,516.32 kmh
942.20 mph

CINCINNATI/BOSTON (USA)
Carl R. Higginson, Captain
Carl D. Hendershot, 1st Officer
Delta Air Lines
Boeing 757-200
2 P&W PW2037, 37,500 lbs
1 hr 19 min
11/29/94
916.78 kmh
569.66 mph

CINCINNATI/FRANKFURT (USA) §
William D. McCarthy, Captain
Craig E. Stephens, 1st Officer
Delta Airlines
Boeing 767-300ER
2 P&W PW4060, 60,000 lbs
7 hrs 1 min
6/19/99
997.48 kmh
619.80 mph

CINCINNATI/NEW YORK*
Capt. Vincent R. O'Toole
American Airlines
Boeing 727
3 P&W JT8D-1
Greater Cincinnati/LaGuardia
Distance: 585 mi
53 min 53 sec
12/19/65
651.27 mph

CINCINNATI/TAMPA (USA)
Capt. Douglas R. Harper
Delta Air Lines Airlines
McDonnell Douglas MD-88
2 P&W JT8D-219, 21,000 lbs
1 hr 42 min
12/28/97
733.25 kmh
455.62 mph

CLEVELAND/BOSTON*
Capt. E. H. Riesh
American Airlines
Boeing 707-720
4 P&W JT3D
Hopkins Intl/Logan Intl
Distance: 561 mi
58 min 20 sec
9/28/65
577.00 mph

CLEVELAND/CINCINNATI*
Capt. Kenneth B. Nace
American Airlines
Boeing 727
3 P&W JT8D-1
Hopkins Intl/Greater Cincinnati
Distance: 221 mi
30 min 10 sec
4/5/65
439.52 mph

CLEVELAND/NEW YORK*
Capt. R. E. Bell — 599.50 mph
American Airlines
Boeing 727
3 P&W JT8D-1
Hopkins Intl/LaGuardia
Distance: 418 mi
41 min 50 sec
10/28/65

COLOGNE/LONDON (UK) *Group II*
Capt. A. S. Johnson — 422.78 kmh
Vickers Viscount, G-ALWE — 262.70 mph
4 Rolls Royce
1 hr 15 min 41 sec
1/22/53

COLOGNE/PHILADELPHIA (USA)
Capt. Glenn W. Stickel — 880.78 kmh
United Parcel Service — 547.29 mph
Boeing 747-200B
4 P&W JT9D, 53,000 lbs
7 hrs 3 min
12/18/02

COLUMBUS/LOS ANGELES (USA)
Barney Szymaniak, Pilot — 871.92 kmh
John C. Beatty, 1st Officer — 541.81 mph
William Allen, 2nd Officer
Flying Tigers DC 873F, N4869T
4 CFM-56
3 hrs 40 min 32 sec
3/28/88

COLUMBUS/MEMPHIS (USA)
Capt. Leiland M. Duke, Jr. — 754.45 kmh
FedEx — 468.79 mph
McDonnell Douglas MD-11F
3 GE CF6-80C2, 61,500 lbs
1 hr 5 min
8/26/98

COLUMBUS/PITTSBURGH*
Capt. Robert F. Olszewski — 457.14 mph
Allegheny Airlines
McDonnell Douglas DC-9-31
2 P&W JT8D-7A
Port Columbus Intl/Greater Pittsburgh Intl
Distance: 144 mi
18 min 54 sec
12/30/76

DAKAR/PARIS (FRANCE)
Cdt. M. Tardieu — 1,700.82 kmh
Air France — 1,056.89 mph
BAe/Aérospatiale Concorde
4 RR Olympus 593, 38,000 lbs
3/28-29/82

DAKAR/RIO DE JANEIRO (FRANCE)
Cdt. A. Gely — 1,721.57 kmh
Air France — 1,069.78 mph
BAe/Aérospatiale Concorde
4 RR Olympus 593, 38,000 lbs
3/28/82

DALLAS/ATLANTA*
Capt. Lloyd M. Damron — 644.54 mph
Delta Air Lines
Douglas DC-8-51
4 P&W JT3D-A
Love Field/Atlanta Municipal
Distance: 721 mi
1 hr 7 min 7 sec
2/7/64

DALLAS/BALTIMORE*
Capt. James B. Smay — 647.96 mph
American Airlines
Boeing 707/123
4 P&W JT3D
Love Field/Friendship Intl
Distance: 1,209 mi
1 hr 51 min 57 sec
3/12/65

DALLAS/CHICAGO (USA)
Capt. James O. Gross — 851.24 kmh
American Airlines — 528.93 mph
McDonnell Douglas MD-11
3 GE CF6-80C2, 61,500 lbs
1 hr 31 min
4/15/98

DALLAS/CINCINNATI*
Capt. I. E. Leach — 630.07 mph
American Airlines
Boeing 707
4 P&W JT3D
Love Field/Greater Cincinnati
Distance: 806 mi
1 hr 16 min 45 sec
12/14/65

DALLAS/DAYTON (USA)
Richard Y. Miller — 859.75 kmh
Piedmont Airlines — 534.25 mph
Boeing 727
3 P&W JT8D, 14,000 lbs
1 hr 41 min 17 sec
7/2/85

DALLAS/DENVER (USA)
Capt. James O. Gross — 687.43 kmh
American Airlines — 427.15 mph
McDonnell Douglas MD-82
2 P&W JT8D, 21,000 lbs
1 hr 30 min
12/3/97

DALLAS/EL PASO*
Capt. E. O. Medlin — 495.54 mph
American Airlines
Boeing 707-123
4 P&W JT3D-1
Love Field/El Paso Intl
Distance: 563 mi
1 hr 8 min 10 sec
2/25/65

DALLAS/FRANKFURT (USA) §
Capt. Roger E. Ruch — 888.00 kmh
American Airlines — 552.32 mph
DC-10-30
3 GE CF6-6, 40,000 lbs
9 hrs 18 min
5/24/86

DALLAS/HAYDEN (USA)
John H. Jenkins, Captain — 721.80 kmh
John Machamer, 1st Officer — 448.53 mph
Wayne Miller, Flt Engineer
American Airlines #847
Boeing 727-100, N1985
3 P&W JT8D-7A, 14,000 lbs
1 hr 43 min
2/24/87

DALLAS/HONOLULU (USA)
Thomas E. Travis, Captain — 800.28 kmh
Scott C. Davis, 1st Officer — 497.29 mph
James H. Fisher, 2nd Officer
American Airlines DC 1030
3 GE CF6-5062
7 hrs 36 min 27 sec
4/12/89

DALLAS/LITTLE ROCK*
Capt. A. B. DeSalvo — 504.50 mph
American Airlines
Boeing 727
3 P&W JT8D-1
Love Field/Adams Field
Distance: 295 mi
35 min 5 sec
7/6/65

DALLAS/LONDON (USA) §
Carter W. Taylor, Captain — 963.52 kmh
James O. Gross, 1st Officer — 598.70 mph
Michael H. Weber, 2nd Officer
American Airlines
McDonnell Douglas MD-11
3 GE CF6, 61,500 lbs
7 hrs 57 min
9/29/97

DALLAS/LOS ANGELES*
Capt. W. W. Gosnell — 614.89 mph
American Airlines
Boeing 707-123
4 P&W JT3D
Love Field/Los Angeles Intl
Distance: 1,245 mi
2 hrs 1 min 29 sec
2/18/65

DALLAS/MELBOURNE (USA)
Capt. C. Thad Kelly — 794.99 kmh
Atlantic Southeast Airlines — 493.98 mph
Canadair CRJ-200
2 GE CF34, 9,220 lbs
2 hrs 5 min
10/18/03

DALLAS/MIAMI (USA)
Capt. James O. Gross — 940.52 kmh
American Airlines — 584.41 mph

McDonnell Douglas MD-11
3 GE CF6-80C2, 61,500 lbs
1 hr 55 min
4/17/98

DALLAS/NEW YORK*
Capt. William F. Bonnell — 686.73 mph
American Airlines
Boeing 707/720
4 P&W JT3D
Love Field/John F. Kennedy Intl
Distance: 1,384
2 hrs 06 min
3/12/65

DALLAS/ONTARIO (USA)
Gary N. Graves, Captain — 689.63 kmh
Scott W. Miller, 1st Officer — 428.51 mph
American Airlines
Boeing 757-200
2 RR RB-211, 43,100 lbs
2 hrs 46 min
1/3/94

DALLAS/ORLANDO*
Capt. Earl R. Epperson — 664.59 mph
Delta Air Lines
Convair 880
4 CJ-805
Love Field/McCoy AFB
Distance: 972 mi
1 hr 27 min 45 sec
3/8/65

Captain Thad Kelly with the Canadair CRJ-200 he flies for Atlantic Southeast Airlines.

DALLAS/OSAKA (USA) §
James O. Gross, Captain — 831.38 kmh
Anthony A. Adams, 1st Officer — 516.59 mph
Walter J. Cappelli, 1st Officer
American Airlines
Boeing 777-200
2 R-R Trent 890, 90,000 lbs
12 hrs 56 min
12/26/99

DALLAS/RALEIGH (USA)
Capt. Dennis D. Lyons — 818.30 kmh
American Airlines — 508.47 mph
Boeing 757-200
2 RR 535E4, 40,100 lbs
2 hrs 5 min
4/6/93

DALLAS/SAN DIEGO*
Capt. Fred H. Braymer — 585.79 mph
American Airlines
Boeing 707-720 B
4 P&W JT3D
Love Field/Lindbergh Field
Distance: 1,182 mi
2 hrs 9 min 56 sec
1/31/62

DALLAS/SAN FRANCISCO*
Capt. W. R. Swain — 580.46 mph
American Airlines
Boeing 707-720
4 P&W JT3D
Love Field/San Francisco Intl
Distance: 1,477 mi
2 hrs 32 min 40 sec
2/18/65

DALLAS/SAO PAULO * §
Capt. Angela Masson — 537.30 mph
American Airlines
Boeing 777-200
2 R-R Trent 892, 91,300 lbs
9 hrs 32 min
6/5/02

DALLAS/TOKYO (USA) §
Thomas E. Travis, Captain — 868.32 kmh
James H. Fisher, Flt Engineer — 539.57 mph
American Airlines
Boeing 747 SP, N602AA
4 P&W JT9D-7ASP
11 hrs 54 min 10 sec
10/11-12/88

DALLAS/TORONTO (USA)
Capt. Dennis D. Lyons — 844.65 kmh
American Airlines — 524.84 mph
Boeing 757-200
2 RR 535E4, 40,100 lbs
2 hrs 14 min
4/18/93

DALLAS/WASHINGTON*
Capt. W. T. Fleming — 643.06 mph
American Airlines
Boeing 707-720
4 P&W JT3D
Love Field/Dulles Intl
Distance: 1,162 mi
1 hr 48 min 25 sec
3/12/65

DAYTON/CHARLOTTE (USA)
Felix Jesus Villaverde — 687.42 kmh
Airlift International — 427.14 mph
McDonnell Douglas DC-8
4 P&W JT3D, 16,000 lbs
52 min
12/7/90

DELHI/FRANKFURT (USA)
Capt. Keith J. Mackey — 758.82 kmh
Pan American Airways — 471.51 mph
Boeing 747
4 P&W JT9D
8 hrs 3 min 58 sec
8/21/91

DELHI/TOKYO (USA)
Walter H. Mullikin, Captain — 421.20 kmh
W. P. Monan, Chief of Relief Airmen — 261.72 mph
Pan American World Airways, Inc.
Boeing 747 SP
4 P&W JT9D-7A
14 hrs 1 min 23 sec
5/2/76

DENVER/ATLANTA (USA)
Capt. John E. Parsons — 995.47 kmh
1st Officer Eric E. Schoneberger — 618.55 mph
Delta Air Lines
Boeing 757-200
2 P&W PW2037, 38,250 lbs
1 hr 57 min
2/24/93

DENVER/BOSTON (USA)
Capt. Vaughn B. Cordle — 821.83 kmh
United Airlines — 510.66 mph
Boeing 737-300
2 CFM56, 20,000 lbs
3 hrs 27 min
9/7/92

DENVER/CHICAGO (USA)
David E. Gurney, Captain — 888.82 kmh
Patrick G. Nolta, 1st Officer — 552.29 mph
Frontier Airlines
Airbus A319-100
2 CFM CFM56, 22,000 lbs
1 hr 37 min
11/6/03

DENVER/CLEVELAND*
Capt. Jack W. Taffe — 660.73 mph
United Airlines
Boeing 727-222A
3 P&W JT8D-15
Stapleton/Hopkins
Distance: 1,211.3 mi
1 hr 50 min
1/20/82

DENVER/DALLAS (USA)
Capt. James O. Gross — 745.40 kmh
American Airlines — 463.17 mph
McDonnell Douglas MD-82
2 P&W JT8D, 21,000 lbs
1 hr 23 min
12/4/97

DENVER/KANSAS CITY (USA)
Larry E. Steenstry — 884.33 kmh
United Airlines — 549.49 mph
Boeing 757
2 P&W PW2037, 38,200 lbs
58 min
5/5/96

DENVER/LOS ANGELES (USA)
Capt. James C. Phillips — 903.55 kmh
United Airlines — 561.44 mph
Boeing 747-400
4 P&W PW4056, 56,750 lbs
1 hr 32 min
12/22/99

DENVER/MIAMI (USA)
Capt. Vaughn B. Cordle — 862.49 kmh
United Airlines — 535.93 mph
Boeing 737-500
2 CFM56, 20,000 lbs
3 hrs 12 min
12/20/92

DENVER/MINNEAPOLIS (USA)
Capt. Vaughn B. Cordle — 856.10 kmh
United Airlines — 531.96 mph
Boeing 737-300
2 CFM56, 20,000 lbs
1 hr 18 min
2/14/93

DENVER/NEW YORK (USA)
Capt. T. Eugene Scanlon — 927.01 kmh
United Airlines — 576.01 mph
Boeing 727
3 P&W JT8D-15, 15,500 lbs
2 hrs 48 min
3/22/94

DENVER/ST. LOUIS (USA)
David E. Gurney, Captain — 893.40 kmh
Patrick G. Nolta, 1st Officer — 555.13 mph
Frontier Airlines
Airbus A318-100
2 CFM CFM56, 22,000 lbs
1 hr 23 min
11/12/03

DENVER/SAN DIEGO (USA)
Capt. Larry Liguori — 748.14 kmh
United Airlines — 464.87 mph
Airbus A320
2 IAE V2527-A5, 26,500 lbs
1 hr 50 min
7/14/95

DENVER/SAN FRANCISCO (USA)
Capt. David H. Friend — 855.03 kmh
United Airlines — 531.29 mph
Boeing 777
2 P&W PW4090, 90,000 lbs
1 hr 49 min
3/3/00

DENVER/SAN JOSE (USA)
Edward J. Chapman, Captain — 845.99 kmh
Bridget Vietz, 2nd Officer — 525.67 mph
United Airlines
Boeing 727
3 P&W JT8D, 14,500 lbs
1 hr 48 min
12/26/99

DENVER/SPOKANE (USA)
Capt. Vaughn B. Cordle — 778.89 kmh
United Airlines — 483.98 mph
Boeing 737-300
2 CFM56, 20,000 lbs
1 hr 43 min
11/28/92

DENVER/TUCSON (USA)
Capt. Vaughn B. Cordle — 777.51 kmh
United Airlines — 483.12 mph
Boeing 737-300
2 CFM56, 20,000 lbs
1 hr 18 min
10/17/92

DENVER/WASHINGTON (USA)
Capt. Daniel M. Romcevich — 999.04 kmh
United Airlines — 620.77 mph
Boeing 727-200
3 P&W JT8D-15, 15,500 lbs
2 hrs 20 min
10/30/95

DES MOINES/DENVER (USA)
Capt. Vaughn B. Cordle — 782.97 kmh
United Airlines — 486.51 mph
Boeing 737-300
2 CFM56, 20,000 lbs
1 hr 14 min
11/27/92

DETROIT/BUFFALO* *Group II*
Capt. William B. Moody — 378.18 mph
American Airlines
Lockheed Electra L-188
4 501D-13
Detroit Metropolitan/Greater Buffalo Intl
Distance: 239 mi
37 min 55 sec
1/11/66

DETROIT/CHICAGO*
Capt. Peter W. Williamson — 431.62 mph
Northeast Airlines
McDonnell Douglas DC-9-31
2 P&W JT8D-7
Detroit Metropolitan/Chicago Midway
Distance: 229 mi
31 min 50 sec
10/1/71

DETROIT/DENVER (USA)
Edward J. Chapman, Captain — 819.22 kmh
Daniel C. Swanson, 1st Officer — 509.03 mph
United Airlines
Boeing 727
3 P&W JT8D, 14,500 lbs
2 hrs 12 min
12/27/99

DETROIT/LOS ANGELES*
Capt. P. F. Willis — 469.77 mph
American Airlines
Boeing 707-123
4 P&W JT3C-6
Detroit Metropolitan/Los Angeles Intl
Distance: 1,976 mi
4 hrs 12 min 23 sec
9/19/59

DETROIT/MIAMI* *Group I*
Capt. H. T. Merrill — 318.86 mph
Eastern Airlines
Lockheed Constellation
4 Wright R-3350
Willow Run/36th Street
Distance: 1,150 mi
3 hrs 36 min 29 sec
8/7/47

DETROIT/MIAMI (USA) *Group III*
Damon A. Powell, Captain — 780.41 kmh
Jeffrey D. Finch, 1st Officer — 484.92 mph
American Airlines
Boeing 737-800
2 CFM CFM56, 26,400 lbs
2 hrs 22 min
7/10/02

DETROIT/NEW YORK*
Capt. John Clark — 577.62 mph
American Airlines
Boeing 727
3 P&W JT8D-1
Detroit Metropolitan/JFK
Distance: 508 mi
52 min 46 sec
2/4/65

DETROIT/NEWARK*
Capt. Laurence B. Hauser — 607.89 mph
American Airlines
Boeing 727
3 P&W JT8D-1
Detroit Metropolitan/Newark Intl
Distance: 487 mi
48 min 4 sec
2/3/65

DETROIT/WASHINGTON*
Capt. Daniel J. Clisham — 484.04 mph
Evergreen Intl Airlines
DC-8-52
4 P&W JT3D
Detroit Metro/Dulles Intl
Distance: 455 mi
56 min 35 sec
4/27/77

EUGENE/DENVER (USA)
Capt. Vaughn B. Cordle — 828.43 kmh
United Airlines — 514.76 mph
Boeing 737-300
2 CFM56, 20,000 lbs
1 hr 55 min
12/19/92

FT LAUDERDALE/NEW YORK (USA)
Michael Piampiano, Captain — 830.16 kmh
Jim Kennedy, 1st Officer — 515.86 mph
Continental Airlines
Airbus, A 30 B-4
2 GE CF6-50-C2, 52,500 lbs
2 hrs 4 min 15 sec
2/17/89

FT MYERS/CHICAGO (USA)
David M. Harris, Captain — 821.14 kmh
Larry E. Steenstry, 1st Officer — 510.23 mph
United Airlines Boeing 757
2 P&W PW2037, 38,200 lbs
2 hrs 12 min
1/14/95

FRANKFURT/BANGKOK (USA) §
Daniel W. Dill, Captain — 926.56 kmh
Gordon R. Boettger, 1st Officer — 575.74 mph
Robert J. Schwartz, 1st Officer
FedEx
McDonnell Douglas MD-11F
3 GE CF6-80C2, 61,500 lbs
9 hrs 41 min
1/7/01

FRANKFURT/BOMBAY (USA) §
Capt. Roger A. Green — 866.27 kmh
Delta Air Lines — 538.27 mph
Boeing 767-300ER
2 P&W PW4060, 60,000 lbs
7 hrs 35 min
12/3/97

FRANKFURT/CHICAGO (USA)
Thomas W. Mayer, Captain — 863.32 kmh
Kenneth Weyler, 1st Officer — 536.47 mph
American Airlines
DC-10-30, N164AA
3 GE CF6-50C2, 52,000 lbs
8 hrs 5 min
5/28/88

FRANKFURT/DALLAS (USA) §
R. E. Miller, Captain — 873.98 kmh
D. M. Stubsten, 1st Officer — 543.07 mph
S. M. Provow, 2nd Officer
Delta Air Lines L-1011-500
3 RR RB211-524B, 50,000 lbs
9 hrs 26 min 57 sec
4/24/92

FRANKFURT/NEW YORK (USA)
William D. McCarthy, Captain — 836.20 kmh
Craig E. Stephens, 1st Officer — 519.59 mph
Delta Airlines
Boeing 767-300ER
2 GE CF6-80C2, 63,000 lbs
7 hrs 24 min
6/23/99

FRANKFURT/ORLANDO (USA) §
Capt. Keith J. Mackey — 825.88 kmh
Pan American — 513.20 mph
Boeing 747, N739PA
4 P&W JT9D
9 hrs 14 min 14 sec
6/16/88

FRANKFURT/PHILADELPHIA (USA)
Capt. Joseph W. Graham — 784.07 kmh
US Airways — 487.20 mph
Airbus A330-300
2 P&W PW4168A, 68,600 lbs
8 hrs 5 min
10/16/03

FRANKFURT/WASHINGTON (USA) §
Capt. Paul V. Davis II — 830.88 kmh
United Airlines — 516.28 mph
Boeing 767-300
2 P&W PW4000, 47,000 lbs
7 hrs 53 min
9/12/91

GUADALAJARA/LOS ANGELES (USA)
Capt. Daniel J. Clisham — 743.19 kmh
American Airlines — 461.80 mph
McDonnell Douglas MD-83
2 P&W JT8D-219, 21,000 lbs
2 hrs 50 min
11/6/97

GUATEMALA CITY/MIAMI*
Capt. Vaughn B. Cordle — 509.53 mph
United Airlines
Boeing 737-300
2 CFM56, 20,000 lbs
2 hrs
1/17/93

GLASGOW/CHICAGO*
Capt. Herbert L. Dabelow — 490.91 mph
American Airlines
Boeing 767-300ER
2 GE CF6-80C2, 61,500 lbs
7 hrs 28 min
9/30/93

HARTFORD/CHICAGO (USA)
Larry E. Steenstry, Captain — 732.18 kmh
Deborah G. Covlin, 1st Officer — 454.95 mph
United Airlines
Boeing 757
2 P&W PW2037, 38,200 lbs
1 hr 43 min
10/6/94

HAVANA/MADRID (SPAIN) *Group I*
Capt. R. de la Pena Moulie — 489.14 kmh
Iberian Airlines — 304.70 mph
Lockheed Super Constellation L-1049E
4 Wright R-3350
10/23/54

HAYDEN/CHICAGO (USA)
John H. Jenkins, Captain — 778.58 kmh
George Farrington, 1st Officer — 483.87 mph
Wayne Miller, Flt Engineer
American Airlines
Boeing 727-100, N1991
3 P&W JT8D-7A, 14,000 lbs
2 hrs 5 min
3/1/87

HAYDEN/DALLAS (USA)
John H. Jenkins, Captain — 721.80 kmh
George Farrington, 1st Officer — 448.53 mph
Wayne Miller, Flt Engineer
American Airlines
Boeing 727-100, N1987
3 P&W JT8D-7A, 14,000 lbs
1 hr 43 min
3/3/87

HONG KONG/ANCHORAGE* §
Glenn Stickel, Captain — 605.21 mph
Henry Kuhlman, 1st Officer
United Parcel Service
Boeing 747-100
4 P&W JT9D, 46,100 lbs
8 hrs 22 min
11/11/95

HONG KONG/ANCHORAGE (USA) §
Capt. Andrew W. Dulay — 958.74 kmh
United Parcel Service — 595.73 mph
Boeing 747-100
4 P&W JT9D, 48,000 lbs
8 hrs 30 min
4/26/95

HONG KONG/CHICAGO (USA) §
Robert H. Lamothe, Captain — 982.67 kmh
Paul M. Bristow, 1st Officer — 610.60 mph
Karen A. Anderson, 1st Officer
United Airlines
Boeing 747-400
4 P&W PW4056, 56,000 lbs
12 hrs 44 min
10/19/96

HONG KONG/LOS ANGELES (USA) §
N. B. Godlove, Captain — 1,086.52 kmh
Dan M. Romcevich, 1st Officer — 675.13 mph
United Airlines
Boeing 747-400
4 P&W PW4056, 56,750 lbs
10 hrs 43 min
1/29/95

HONG KONG/NEW DELHI (USA)
Robert E. Burnstein, Captain — 841.87 kmh
Gary R. McCaffrey, 1st Officer — 523.11 mph
United Airlines
Boeing 747-400
4 P&W PW4056, 56,750 lbs
4 hrs 27 min
7/11/01

HONG KONG/SAN FRANCISCO (USA) §
Capt. David L. Link — 1,064.65 kmh
United Airlines — 661.54 mph
Boeing 747-400
4 P&W PW4056, 56,750 lbs
10 hrs 26 min
2/18/97

HONG KONG/SINGAPORE (USA)
Capt. William A. Dias — 793.15 kmh
United Airlines — 492.84 mph
Boeing 747-400
4 P&W PW4056, 56,750 lbs
3 hrs 14 min
3/9/01

HONG KONG/TOKYO (USA)
Capt. N. B. Godlove — 995.64 kmh
United Airlines — 618.66 mph
Boeing 747-100
4 P&W JT9D-7A, 47,000 lbs
2 hrs 57 min
1/16/93

HONOLULU/AUCKLAND (USA) §
Capt. Robert M. McElrath — 943.99 kmh
United Airlines — 586.56 mph
Boeing 747-100
4 P&W JT9D-7A
7 hrs 30 min 44 sec
11/6/89

HONOLULU/CHICAGO (USA) §
Frederick C. Dauer, Captain — 1,049.11 kmh
Thomas W. Cannon, 1st Officer — 651.88 mph
Lee M. Olivas, 2nd Officer
American Airlines
McDonnell Douglas DC-10-30
3 GE CF6-50C2, 52,500 lbs
6 hrs 30 min
2/2/98

HONOLULU/DALLAS (USA)
Thomas E. Travis, Captain — 897.48 kmh
Scott C. Davis, 1st Officer — 557.69 mph
James H. Fisher, Flt Engineer
American Airlines DC 10-10H
3 CF6-6K2
6 hrs 47 min
4/13/89

HONOLULU/GREENSBORO (USA) §
James C. Sifford — 822.60 kmh
Piedmont Airlines — 511.16 mph
Boeing 727
3 P&W JT8D, 14,500 lbs
9 hrs 14 min
12/27/83

HONOLULU/JOHNSTON ISLAND (USA)
Bruce J. Mayes, Captain — 673.36 kmh
Glen Nakamura, 1st Officer — 418.40 mph
Aloha Airlines
Boeing 737-200
2 P&W JT8D, 14,500 lbs
1 hr 57 min 48 sec
1/24/96

HONOLULU/LOS ANGELES (USA)
Vance H. Weir, Captain — 1,082.84 kmh
Marvin C. Smith, 1st Officer — 672.84 mph
Trans World Airlines
Boeing 747-100
4 P&W JT9D, 46,150 lbs
3 hrs 47 min 38 sec
2/9/93

HONOLULU/MINNEAPOLIS (USA)
David J. Roth, Captain — 979.65 kmh
Patrick T. Duffy, 1st Officer — 608.73 mph
Paul T. Olson, 2nd Officer
Northwest Airlines
McDonnell Douglas DC-10-40
3 P&W JT9D-20, 49,400 lbs
6 hrs 31 min
1/10/95

HONOLULU/MONTREAL (CANADA) §
Capt. C. H. Simpson — 937.08 kmh
Capt. E. E. Doyle — 582.30 mph
Air Canada
Boeing 747-238B, VH-ECA
4 P&W JT9D-7A, 47,650 lbs
1/12/88

HONOLULU/NADI (USA)
Capt. Glenn W. Stickel — 914.63 kmh / 568.32 mph
United Parcel Service
Boeing 747-200B
4 P&W JT9D, 53,000 lbs
5 hrs 35 min
7/13/02

HONOLULU/OAKLAND (USA)
Capt. Steven M. Zeigler — 972.37 kmh / 604.21 mph
FedEx
McDonnell Douglas DC-10-30F
3 GE CF6-50C2, 54,000 lbs
3 hrs 59 min
5/7/03

HONOLULU/PAGO PAGO (USA)
Capt. Glenn W. Stickel — 860.65 kmh / 534.78 mph
United Parcel Service
Boeing 747-100
4 P&W JT9D, 46,950 lbs
4 hrs 53 min
11/11/00

HONOLULU/ST. LOUIS (USA) §
Gid Miller, Captain — 995.38 kmh / 618.50 mph
Thomas E. Vogel, 2nd Officer
Trans World Airlines
Boeing 747
4 P&W JT9D, 46,150 lbs
6 hrs 40 min
2/18/94

HONOLULU/SAN FRANCISCO (USA)
R. C. Bowman — 1,033.20 kmh / 642.03 mph
Western Airlines
McDonnell Douglas DC-10-10
3 GE GCS 6-10, 39,000 lbs
3 hrs 44 min 11 sec
2/1/83

HONOLULU/SEATTLE (USA)
Kendall J. Moeller, Captain — 982.80 kmh / 610.68 mph
Gary L. Sutton, 1st Officer
Northwest Airlines
McDonnell Douglas DC-10-30
3 GE CF6-50C, 51,000 lbs
4 hrs 23 min
2/5/02

HONOLULU/TOKYO (USA)
Jay C. Mallory, Captain — 785.07 kmh / 487.82 mph
Patrick E. Flanagan, 1st Officer
Valerie Scott, 1st Officer
United Airlines
Boeing 747-400
4 P&W PW4056, 56,750 lbs
7 hrs 49 min
1/28/00

HOUSTON/CHARLOTTE*
Capt. Richard Y. Miller — 610.24 mph
Piedmont Airlines
Boeing 727-100, N834N
3 P&W JT8D
Houston Intl/Douglas Municipal
1 hr 33 min
10/26/80

HOUSTON/LONDON (USA) §
Capt. David E. Moran — 979.20 kmh / 608.48 mph
Continental Airlines
DC 10-30, N68060
3 GE CF6, 52,000 lbs
7 hrs 58 min
9/13/89

HOUSTON/LOS ANGELES (USA)
William R. Wallace, Captain — 734.38 kmh / 456.32 mph
Joseph R. Bouley, 1st Officer
United Airlines
Airbus A319
2 IAE V2522, 22,000 lbs
3 hrs 1 min
9/19/99

HOUSTON/MIAMI (USA)
Capt. Damon A. Powell — 815.34 kmh / 506.63 mph
American Airlines
Boeing 737-800
2 CFM CFM56, 26,300 lbs
1 hr 54 min
6/19/03

HOUSTON/NEWARK*
Capt. John R. Brady — 561.85 mph
Eastern Airlines
Boeing 727
3 P&W JT8D
Houston Intl/Newark
Distance: 1,400 mi
2 hrs 29 min 30 sec
11/4/70

HOUSTON/SAN FRANCISCO (USA)
Charles E. Siebert, Captain — 852.15 kmh / 529.50 mph
Larry E. Steenstry, 1st Officer
United Airlines
Boeing 757
2 P&W PW2037, 38,200 lbs
3 hrs 15 min
7/19/94

IDAHO FALLS/ATLANTA*
Capt. Richard Y. Miller — 616.25 mph
Piedmont Airlines
Boeing 737-200, N736N
2 P&W JT8D-9A
Idaho Falls Municipal/Hartsfield Intl
2 hrs 50 min
1/13/80

INDIANAPOLIS/JACKSONVILLE (USA)
Felix Jesus Villaverde — 706.95 kmh / 439.27 mph
Airlift International
McDonnell Douglas DC-8
4 P&W JT3D, 16,000 lbs
1 hr 34 min
12/20/90

INDIANAPOLIS/PARIS (USA) §
Capt. Arnold H. Andrijeski — 970.18 kmh / 602.84 mph
FedEx
Boeing MD-11F
3 P&W PW4462, 62,000 lbs
6 hrs 56 min
6/21/00

INDIANAPOLIS/PHILADELPHIA (USA)
Robert F. Olszewski, Pilot — 975.03 kmh / 606.03 mph
Steven Wilson, 1st Officer
USAir DC 9-31, N935VJ
2 P&W JT8D-9B, 14,500 lbs
59 min 10 sec
12/20-21/87

INTERNATIONAL FALLS/MINNEAPOLIS (USA)
Group II
C. Martin Coddington, Captain — 522.08 kmh / 324.40 mph
Michael D. Monahan, 1st Officer
Express I
Jetstream International, Jetstream 31
2 Garrett TPE331, 900 shp
47 min 5 sec
4/4/95

JACKSON, MS/LOUISVILLE (USA)
Andrew Dulay, Pilot — 722.88 kmh
Greg Rassmussen, 1st Officer — 449.20 mph
Ryan International Airlines
Boeing 757-24, N402UP
2 P&W PW2040
1 hr 3 min 58 sec
5/20/88

JACKSONVILLE/BALTIMORE (USA)
Ward Anderson, Captain — 699.87 kmh
William Castle, 1st Officer — 434.97 mph
Piedmont Airlines #472
Boeing 737-200
2 P&W JT8D
1 hr 31 min 40 sec
4/30/87

JACKSONVILLE/INDIANAPOLIS (USA)
Felix Jesus Villaverde — 714.55 kmh
Airlift International — 444.00 mph
McDonnell Douglas DC-8
4 P&W JT3D, 16,000 lbs
1 hr 33 min
12/20/90

JOHNSTON ISLAND/HONOLULU (USA)
Bruce J. Mayes, Captain — 801.10 kmh
Glen Nakamura, 1st Officer — 497.78 mph
Aloha Airlines
Boeing 737-200
2 P&W JT8D, 14,500 lbs
1 hr 39 min 1 sec
1/24/96

JOPLIN/MEMPHIS (USA) *Group II*
C. Martin Coddington, Captain — 468.68 kmh
Paul J. Kegaly, 1st Officer — 291.22 mph
Express Airlines I
BAe Jetstream 31
2 Garrett TPE331, 900 shp
59 min 59 sec
2/24/98

KANSAS CITY/ST. LOUIS*
James J. Lawlor, Captain — 443.85 mph
James J. Lawlor, Jr., Flt Engineer
Trans World Airways
Boeing 727-231
3 P&W JT8D, 14,550 lbs
32 min
8/31/96

KEY WEST/ORLANDO (USA) *Group II*
Capt. Nicholas E. Mech — 412.26 kmh
Comair — 256.17 mph
Embraer EMB-120RT Brasilia
2 P&WC PW118, 1800 shp
59 min
3/17/01

KHABAROVSK/ANCHORAGE (USA)
Richard R. Walters, Captain — 849.37 kmh
Brett L. Church, 1st Officer — 527.77 mph
Polar Air Cargo
Boeing 747-132
4 P&W JT9D-7A, 46,950 lbs
5 hrs 31 min 57 sec
6/16/97

KONA/LOS ANGELES (USA)
Capt. David J. Passannante — 962.30 kmh
American Airlines — 597.94 mph
Boeing 757-200
2 R-R 535E4B, 43,100 lbs
4 hrs 11 min
3/3/02

Superseded
KONA/LOS ANGELES (USA)
Capt. L. Bruce Gilman — 918.39 kmh
American Airlines — 570.66 mph
Boeing 757-200
2 R-R 535E4B, 43,100 lbs
4 hrs 23 min
2/23/02

KONA/SAN FRANCISCO (USA)
Edward P. Akin, Captain — 972.40 kmh
Gaven C. Dunn, 1st Officer — 604.22 mph
United Airlines
McDonnell Douglas DC10-10
3 GE CF6, 40,000 lbs
3 hrs 55 min
2/16/94

LAS VEGAS/CHARLOTTE (USA)
Stephen A. Clegg, Captain — 928.10 kmh
William Gantt, 1st Officer — 576.69 mph
USAir
Boeing 737-300
2 CFM56, 22,000 lbs
3 hrs 18 min 53 sec
3/3/96

LAS VEGAS/CHICAGO (USA)
Larry E. Steenstry, Captain — 959.83 kmh
Randy S. Trull, 1st Officer — 596.40 kmh
United Airlines
Boeing 757
2 P&W PW2037, 38,200 lbs
2 hrs 32 min
11/25/94

LAS VEGAS/CLEVELAND (USA)
Jack W. Taffe, Pilot — 996.30 kmh
United Airlines — 600.46 mph
Boeing 727-100
3 P&W JT8D-7 of 14,000 lbs
2 hrs 56 min 36 sec
12/22-23/84

LAS VEGAS/DENVER (USA)
G.R. Madsen, Captain — 830.44 kmh
Julia Kerley, 1st Officer — 516.01 mph
Darren Chonko, 2nd Officer
United Airlines
McDonnell Douglas DC10-10
3 GE CF6-6D, 39,300 lbs
1 hr 13 min
10/12/96

LAS VEGAS/DETROIT (USA)
Patrick J. Casey, Captain — 920.74 kmh
Larry E. Steenstry, 1st Officer — 572.12 kmh
United Airlines
Boeing 757
2 P&W PW2037, 38,200 lbs
3 hrs 3 min
4/29/95

LAS VEGAS/HONOLULU (UK)
Capt. John Hutchinson — 1,258.55 kmh
British Airways — 782.02 mph
BAe/Aérospatiale Concorde
3/16/90

LAS VEGAS/PHOENIX*
Capt. Raymond H. Lutz — 498.91 mph
1st Officer Lyle T. Shelton
Trans World Airlines
Boeing 727
3 P&W JT8D
Las Vegas Intl/Phoenix Sky Harbor Intl
30 min 40 sec
7/1/74

LAS VEGAS/ST. LOUIS (USA)
Lyle T. Shelton, Captain — 958.68 kmh
Gary L. McDonald, 1st Officer — 595.72 mph
Trans World Airlines
McDonnell Douglas MD-80, N9054
2 P&W JT8D
2 hrs 18 min
2/14/89

LAS VEGAS/WASHINGTON (USA)
William R. Wallace, Captain — 912.95 kmh
Enrique Fernandez, 1st Officer — 567.28 mph
United Airlines
Airbus A-320
2 IAE V2527-A5, 26,500 lbs
3 hrs 38 min
2/18/97

LIHUE/LOS ANGELES (USA)
Capt. David J. Passannante — 977.44 kmh
American Airlines — 607.35 mph
Boeing 757-200
2 R-R RB211, 43,500 lbs
4 hrs 18 min
1/10/03

LIHUE/SAN FRANCISCO (USA)
Capt. Jamel Z. Abouremeleh — 914.91 kmh
United Airlines — 568.50 mph
Boeing 757-200
2 P&W PW2040, 41,700 lbs
4 hrs 18 min
1/8/01

LIMA/MIAMI (USA)
Paul J. Dunne, Captain — 884.98 kmh
Robert F. Frank, 1st Officer — 549.90 mph
United Airlines
Boeing 757-200
2 P&W PW2037, 38,200 lbs
4 hrs 46 min
1/16/95

LINCOLN/WASHINGTON*
Capt. Daniel J. Clisham — 491.28 mph
1st Officer Raymond Hoag
Evergreen Intl Airlines
McDonnel Douglas DC-8-52
4 P&W JT3D
Lincoln Municipal/Dulles Intl
2 hrs 8 min 18 sec
5/9/77

LITTLE ROCK/MEMPHIS*
Capt. A. B. DeSalvo — 384.22 mph
American Airlines
Boeing 727
3 P&W JT8D-1
Adams Field/Memphis Municipal
Distance: 130 mi
20 min 18 sec
7/6/75

LONDON/ACAPULCO (UK)
Capt. John Hutchinson — 1,094.57 kmh
British Airways — 680.13 mph
BAe/Aérospatiale Concorde
3/14/90

LONDON/AMSTERDAM*
Robert L. O'Neill, Captain — 328.26 mph
William R. Wallace, 1st Officer
United Airlines
Boeing 777-222
2 P&W PW4077, 77,000 lbs
42 min
8/12/95

LONDON/ATHENS (GREECE)
Capt. P. Ioanides — 814.80 kmh
Olympic Airways — 506.29 mph
Comet DH-106-4B, SX-DAK
4 RR Avon
3 hrs 13 min 42 sec
4/30/60

LONDON/AUCKLAND (UK) §
Capt. J. W. Burton — 983.22 kmh
British Airways — 610.97 mph
BAe/Aérospatiale Concorde
4 Olympus 593/610, 38,000 lbs
4/5-6/86

LONDON/BOSTON (UK)
Mike Bannister — 1,689.47 kmh
British Airways — 1,049.79 mph
BAe/Aérospatiale Concorde
4 RR Olympus, 38,000 lbs
10/8/03

LONDON/CAPETOWN (UK) §
Capt. Brian Walpole — 1,192.73 kmh
British Airways — 741.15 mph
BAe/Aérospatiale Concorde
4 RR Olympus, 38,000 lbs
8 hours 7 min 33 sec
3/28/85

LONDON/CHICAGO (USA)
James O. Gross, Captain — 865.10 kmh
Andrew R. Tuson, 1st Officer — 537.55 mph
Dennis B. Walsh, 1st Officer
American Airlines
McDonnell Douglas MD-11
3 GE CF6-80C2, 61,500 lbs
7 hrs 20 min
4/3/98

LONDON/COLOGNE (UK) *Group II*
Capt. A. S. Johnson — 455.26 kmh
British European Airways — 282.88 kmh
Vickers Viscount, G-ALWE
4 RR Dart
1 hr 10 min 17 sec
1/22/53

LONDON/COPENHAGEN (UK) *Group II*
Capt. R. F. Noden — 481.94 kmh
British European Airways — 299.46 mph
Vickers Viscout, G-AMOA
4 RR Dart
2 hrs 1 min 53 sec
7/28/53

LONDON/DALLAS (USA) §
Carter W. Taylor, Captain — 835.64 kmh
George E. Horn, 1st Officer — 519.24 mph
James M. Kennedy, 1st Officer
American Airlines
McDonnell Douglas MD-11
3 GE CF-6, 61,500 lbs
9 hrs 10 min
9/19/97

LONDON/GENEVA (UK) *Group II*
Capt. W. J. Wakelin — 491.74 kmh
Vickers Viscount, G-AMNY — 305.55 mph
4 RR Dart
1 hr 31 min 52 sec
6/1/53

LONDON/GENEVA (SWITZERLAND) *Group III*
Roger Pierre — 602.79 kmh
Gregory Fraschina — 374.55 mph
Crossair
BAe 146
4 Lycoming LF 507, 7,000 lbs
2/20/01

LONDON/KARACHI (PAKISTAN)
Capt. M. I. Baig — 938.78 kmh
Pakistan Intl Airlines — 583.30 mph
Boeing 720-B
4 P&W JT3D-1
6 hrs 43 min 51 sec
1/2/62

LONDON/LOS ANGELES (USA) §
Capt. Larry Liguori — 880.40 kmh
United Airlines — 547.05 mph
Boeing 777-200
2 P&W PW4090, 90,000 lbs
9 hrs 57 min
1/15/01

LONDON/MIAMI (USA) §
Capt. Angela Masson — 861.68 kmh
American Airlines — 535.42 mph
Boeing 777-200
2 R-R Trent 892, 91,300 lbs
8 hrs 15 min
10/26/03

LONDON/MINNEAPOLIS (USA)
Capt. Kendall J. Moeller — 863.78 kmh
Northwest Airlines — 536.72 mph
McDonnell Douglas DC-10-30
3 GE CF6-50C2, 54,000 lbs
7 hrs 30 min
6/4/03

Captain Angela Masson of American Airlines with the Boeing 777 she has flown on numerous speed records.

Photo: Dominique Berta Collins

LONDON/MOSCOW (UK)
Capt. Roger Mills — 898.71 kmh, 558.43 mph
British Airways
BAe/Aérospatiale Concorde
4 RR Olympus, 38,000 lbs

3/1/88

LONDON/NASSAU (UK) §
Capt. G. N. Henderson — 630.18 kmh, 391.57 mph
Capt. P. Wilson
Cunard Eagle Airways
Boeing 707-465
4 RR Conway 508
11 hrs 5 min 7 sec
5/5/62

LONDON/NEWARK (USA)
Dwight N. Bales, Captain — 842.74 kmh, 523.65 mph
William R. Wallace, 1st Officer
United Airlines
Boeing 777-222
2 P&W PW4077, 77,000 lbs
6 hrs 36 min
12/17/95

LONDON/NEW DELHI (USA) §
Jay C. Mallory, Captain — 903.51 kmh, 561.41 mph
Kent R. Cummins, 1st Officer
John F. Lisella, 1st Officer
United Airlines
Boeing 747-400
4 P&W PW4056, 56,750 lbs
7 hrs 27 min
6/8/01

LONDON/NEW YORK (UK)
Capt. Peter Horton — 1,644.02 kmh, 1,021.54 mph
British Airways
BAe/Aérospatiale Concorde
3/14/90

LONDON/PHILADELPHIA (USA)
Capt. Joseph W. Graham — 754.15 kmh, 468.61 mph
US Airways
Airbus A330-300
2 P&W PW4168A, 68,600 lbs
7 hrs 19 min
9/22/01

LONDON/RALEIGH (USA)
Stephen W. Whelan, Jr. — 720.47 kmh, 447.67 mph
Scott W. Miller
John P. Kane
American Airlines
Boeing 767-200ER
2 GE CF6-80A, 48,000 lbs
8 hrs 40 min
2/11/95

LONDON/ROME (UK)
Guy C. Westgate — 679.43 kmh, 422.18 mph
British Airways
Boeing 737-400
3/19/99

LONDON/ST. LOUIS (UK)
Capt. V. Trimble — 829.89 kmh, 515.67 mph
British Airways
DC 10-30
2/17/90

LONDON/SAN FRANCISCO (USA) §
Capt. Charles T. Kelly — 850.29 kmh, 528.34 mph
United Airlines
Boeing 747-200
4 P&W JT9D-7J, 52,000 lbs
10 hrs 8 min
3/30/95

LONDON/SYDNEY (UK) §
Capt. Hector McMullen — 998.59 kmh, 620.50 mph
British Airways
BAe/Aérospatiale Concorde
4 RR Snecma Olympus, 38,000 lbs
17 hrs 3 min 45 sec
2/13-14/85

LONDON/WASHINGTON (USA)
Capt. F. van Cortlandt de Peyster — 821.57 kmh, 510.50 mph
United Airlines
Boeing 747-400
4 P&W PW4056, 56,750 lbs
7 hrs 11 min
8/26/01

LONG BEACH/CHARLESTON*
Capt. W. B. Grubb — 520.27 mph
Southern Airways
Douglas DC-9
2 P&W JT8D-5
Long Beach Municipal/Charleston Municipal
Distance: 2,190 mi
4 hrs 12 min 33 sec
5/9/67

LONG BEACH/MIAMI*
Capt. F. E. Davis — 584.16 mph
Eastern Airlines
Douglas DC-8
4 P&W JT4A-3
Long Beach Municipal/Miami Intl
Distance: 2,326 mi
3 hrs 58 min 55 sec
1/3/60

LOS ANGELES/AUCKLAND (USA) §
Capt. N. B. Godlove — 901.44 kmh, 560.12 mph
United Airlines
Boeing 747-400
4 P&W PW4056, 56,750 lbs
11 hrs 38 min
5/3/95

LOS ANGELES/ATLANTA*
Capt. S. W. Hopkins — 658.94 mph
Delta Air Lines
Douglas DC-8
4 P&W JT3D
Los Angeles Intl/Atlanta
Distance: 1,946 mi
2 hrs 57 min 11 sec
3/8/62

LOS ANGELES/BALTIMORE*
Capt. Stan Smith — 657.68 mph
American Airlines
Boeing 707-123B
4 P&W JT3D
Los Angeles Intl/Friendship
Distance: 2,329 mi
3 hrs 32 min 28 sec
12/7/61

LOS ANGELES/BOSTON*
Capt. Simon F. Bittner — 666.61 mph
American Airlines
Boeing 707
4 P&W JT3D
Los Angeles Intl/Logan Intl
Distance: 2,610 mi
3 hrs 54 min 55 sec
11/26/65

LOS ANGELES/BRISBANE (USA) §
Capt. N. B. Godlove — 856.43 kmh, 532.16 mph
United Airlines
Boeing 747-400
4 P&W PW4056, 56,750 lbs
13 hrs 28 min
9/24/95

LOS ANGELES/CHARLESTON* *Group I*
Capt. T. P. Ball 344.19 mph
Delta Air Lines
Douglas DC-6
4 P&W R-2800 CA-15
Clover Field, Santa Monica/Charleston Municial
Distance: 2,203 mi
6 hrs 24 min 32 sec
11/6/48

LOS ANGELES/CHARLOTTE (USA)
Richard Y. Miller 853.77 kmh
Piedmont Airlines 530.54 mph
Boeing 727
3 P&W JT8D, 16,000 lbs
3 hrs 51 min 8 sec
7/5/85

LOS ANGELES/CHICAGO*
Capt. George R. Russon 672.80 mph
American Airlines
Boeing 707-123
4 P&W JT3D
Los Angeles Intl/O'Hare Intl
Distance: 1,746 mi
2 hrs 35 min 42 sec
4/9/63

LOS ANGELES/CINCINNATI (USA)
Capt. Michael H. Fields 1,052.23 kmh
Delta Air Lines 653.82 mph
Boeing 757
2 P&W PW2037, 38,200 lbs
2 hrs 54 min
11/21/94

LOS ANGELES/CLEVELAND*
Capt. Walter W. Gosnell 674.31 mph
American Airlines
Boeing 707
4 P&W JT3D
Los Angeles Intl/Hopkins
Distance: 2,053 mi
3 hrs 2 min 40 sec
11/23/65

LOS ANGELES/DALLAS*
Capt. W. R. Hunt 689.00 mph
American Airlines
Boeing 707-123
4 P&W JT3D
Los Angeles Intl/Love Field
Distance: 1,245 mi
1 hr 48 min 25 sec
3/9/65

LOS ANGELES/DAYTON (USA)
Richard Y. Miller 813.49 kmh
Piedmont Airlines 505.50 mph
Boeing 727-200
3 P&W JT8D-15A, 15,000 lbs
3 hrs 48 min 25 sec
4/5-6/85

LOS ANGELES/DENVER (USA)
Capt. David H. Friend 875.02 kmh
United Airlines 543.71 mph
Boeing 777
2 P&W PW4090, 90,000 lbs
1 hr 35 min
9/4/00

LOS ANGELES/DETROIT*
Capt. B. B. Bruce 571.76 mph
American Airlines
Boeing 707-123
4 P&W JT3C-6
Los Angeles Intl/Detroit Metropolitan
Distance: 1,976 mi
3 hrs 27 min 22.5 sec
9/17/59

LOS ANGELES/FORT WORTH*
Capt. H. D. Schmidt 660.85 mph
American Airlines
Boeing 707-720B
4 P&W JT3D
Los Angeles Intl/Amon Carter Field
Distance: 1,234 mi
1 hr 52 min 2 sec
3/7/62

LOS ANGELES/GUADALAJARA (USA)
Capt. Daniel J. Clisham 825.77 kmh
American Airlines 513.11 mph
McDonnell Douglas MD-83
2 P&W JT8D-219, 21,000 lbs
2 hrs 33 min
11/27/97

LOS ANGELES/HONG KONG (USA) §
Wayne P. Morgan, Captain 918.49 kmh
Gary R. McCaffrey, 1st Officer 570.72 mph
Jerry A. Porter, 1st Officer
United Airlines
Boeing 747-400
4 P&W PW4056, 56,750 lbs
12 hrs 42 min
7/25/00

LOS ANGELES/HONOLULU (USA)
J. D. Shagam, Captain 955.40 kmh
Brian Lapworth, 1st Officer 593.66 mph
Chris J. Miller, Flight Engineer
American Airlines
McDonnell Douglas DC-10-10
3 GE CF6-6K, 41,500 lbs
4 hrs 18 min
9/29/00

LOS ANGELES/HOUSTON* *Group I*
Capt. L. McBride 332.37 mph
Delta Air Lines
Douglas DC-7B
4 Wright R-3350
Clover Field, Santa Monica/Houston Intl
Distance: 1,390 mi
4 hrs 10 min 56.4 sec
8/14/57

LOS ANGELES/HOUSTON (USA) *Group III*
William R. Wallace, Captain 810.50 kmh
Daniel E. Smith, 1st Officer 503.62 mph
United Airlines
Airbus A319
2 IAE V2522, 22,000 lbs
2 hrs 44 min
9/25/99

LOS ANGELES/INDIANAPOLIS (USA)
Capt. Steven M. Zeigler 1,016.13 kmh
FedEx 631.39 mph
McDonnell Douglas DC-10-30
3 GE CF6-50C2, 52,500 lbs
2 hrs 52 min
5/1/02

LOS ANGELES/JACKSONVILLE* *Group I*
Capt. T. P. Ball 392.25 mph
Delta Air Lines
Douglas DC-7
4 Wright R-3350
Clover Field, Santa Monica/Jacksonville Municipal
Distance: 2,154 mi
5 hrs 29 min 33 sec
3/18/54

LOS ANGELES/KANSAS CITY*
Barry Schiff, Captain — 585.32 mph
James D. Bates, 1st Officer
John Meyers, Flt Engineer
Trans World Airlines
Boeing 727-231A
3 P&W JT8D-9A
Los Angeles Intl/Kansas Cty
2 hrs 19 min 43 sec
11/19/74

LOS ANGELES/LONDON (USA) §
Larry Liguori, Captain — 982.34 kmh
Trisha Speer, 1st Officer — 610.40 mph
Lynn Lee, 1st Officer
United Airlines
Boeing 777-200
2 P&W PW4084, 86,760 lbs
8 hrs 55 min
11/7/01

LOS ANGELES/MAUI (USA)
Capt. James E. Hamilton — 911.51 kmh
American Airlines — 566.38 mph
Boeing 757-200
2 R-R 535, 40,100 lbs
4 hrs 23 min
6/25/02

LOS ANGELES/MELBOURNE (USA) §
William A. Mullen, Captain — 898.44 kmh
Stephen E. Fitzgerald, 1st Officer — 558.26 mph
Greg N. Durgin, 1st Officer
United Airlines
Boeing 747-400
4 P&W PW4056, 56,750 lbs
14 hrs 12 min
3/1/00

LOS ANGELES/MEMPHIS*
Capt. P. F. Willis — 692.82 mph
American Airlines
Boeing 707-123
4 P&W JT3D
Los Angeles Intl/Memphis Municipal
Distance: 1,617 mi
2 hrs 20 min 2 sec
3/9/65

LOS ANGELES/MEXICO CITY (MEXICO)
Capt. Rafael T. Zapata — 962.19 kmh
Mexicana De Aviation — 597.88 mph
Boeing 727-64
3 P&W JT8D-1
2 hrs 36 min
12/1/66

LOS ANGELES/MIAMI (USA)
Capt. M. C. Wedge — 1,024.78 kmh
National Airlines — 636.77 mph
McDonnell Douglas DC-10
3 GD CF6-6
3 hrs 38 min
12/12/71

LOS ANGELES/NASHVILLE*
Capt. O. P. Brunsvold — 576.86 mph
American Airlines
Convair 990
4 G.E. CJ-805-23
Los Angeles Intl/Nashville
Distance: 1,797 mi
3 hrs 6 min 54 sec
2/21/65

LOS ANGELES/NEW ORLEANS*
Capt. G. O. Wright — 595.29 mph
Delta Air Lines Convair 880
4 G.E. CJ-805-3
Los Angeles Intl/Moisant Intl
Distance: 1,671 mi
2 hrs 48 min 25 sec
1/15/62

LOS ANGELES/NEW YORK*
Capt. Wylie H. Drummond — 680.90 mph
American Airlines
Boeing 707-123
4 P&W JT3D
Los Angeles Intl/Idlewild Intl
Distance: 2,474 mi
3 hrs 38 min
4/10/63

LOS ANGELES/NEWARK*
Capt. Conway F. Candler — 673.12 mph
American Airlines
Boeing 707-720
4 P&W JT3D
Los Angeles Intl/Newark Municipal
Distance: 2,454 mi
3 hrs 38 min 44 sec
11/25/65

LOS ANGELES/OKLAHOMA CITY*
Capt. Richard O. Robbins — 633.98 mph
American Airlines
Boeing 727
3 P&W JT8D-1
Los Angeles Intl/Will Rogers Field
Distance: 1,186 mi
1 hr 52 min 25 sec
9/28/65

LOS ANGELES/OSAKA (USA) §
Capt. David F. Specht — 831.13 kmh
United Airlines — 516.44 mph
Boeing 747-400
4 P&W PW4056, 56,750 lbs
11 hrs 3 min
3/9/95

LOS ANGELES/PAPEETE (FRANCE)
Cdt. Gilbert Demonceau — 925.81 kmh
Jacques Masson, Copilot — 575.30 mph
Air France, Boeing 747
4 GE CF6-50-E2, 52,500 lbs
9/26/86

LOS ANGELES/PARIS (FRANCE) §
Cdt. Rene Kimmel, Pilot — 962.22 kmh
Jean-Jacques Germain, Copilot — 597.92 mph
Alain Bac, Copilot
Air France Boeing 747
4 GE CF6-50-E2, 52,500 lbs
9/28/86

LOS ANGELES/PHILADELPHIA*
Capt. Ralph L. Johnson — 672.12 mph
American Airlines
Boeing 707-720
4 P&W JT3D
Los Angeles Intl/Philadelphia Intl
Distance: 2,402 mi
3 hrs 34 min 25 sec
4/19/63

LOS ANGELES/ST. LOUIS*
Capt. George J. Fell — 644.76 mph
American Airlines
Boeing 727
3 P&W JT8D-1
Los Angeles Intl/St. Louis Municipal
Distance: 1,591 mi
2 hrs 28 min 3 sec
4/13/65

LOS ANGELES/SAN SALVADOR*
William R. Wallace, Captain — 531.36 mph
Richard G. Johnson, Jr., 1st Officer
United Airlines
Airbus A320
2 IAE V2527, 26,500 lbs
4 hrs 22 min
2/25/00

LOS ANGELES/SEOUL (CANADA) §
Norman A. Macsween — 797.48 kmh
Frederick G. Braun — 495.53 mph
Boeing 747-400
4 GE CF6-80, 57,900 lbs
3/13/95

LOS ANGELES/SYDNEY (USA) §
Capt. N. B. Godlove — 896.72 kmh
United Airlines — 557.19 mph
Boeing 747-400
4 P&W PW4056, 56,750 lbs
13 hrs 27 min
3/24/95

LOS ANGELES/TAMPA* *Group I*
Capt. G. I. Baker — 354.41 mph
National Airlines
Douglas DC-6
4 P&W Double Wasp
Clover Field, Santa Monica/Drew Field
Distance: 2,157 mi
6 hrs 5 min 10 sec
6/3/47

LOS ANGELES/TAIPEI (USA) §
Capt. N. B. Godlove — 787.73 kmh
United Airlines — 489.47 mph
Boeing 747-400
4 P&W PW4056, 56,750 lbs
13 hrs 52 min
2/13/95

LOS ANGELES/TOKYO (USA) §
John B. Hilderbrant, Captain — 887.28 kmh
Stephen E. Fitzgerald, 1st Officer — 551.33 mph
United Airlines
Boeing 747-400
4 P&W PW4056, 56,750 lbs
9 hrs 52 min
8/19/98

LOS ANGELES/TUCSON (USA)
Edward J. Chapman — 820.63 kmh
United Airlines — 509.91 mph
Boeing 737-300
2 CFM56-3B1, 20,000 lbs
53 min
2/16/98

LOS ANGELES/WASHINGTON*
Capt. Howard Cady — 682.48 mph
American Airlines
Boeing 707-720 B
4 P&W JT3D
Los Angeles Intl/Dulles Intl
Distance: 2,286 mi
3 hrs 20 min 58 sec
1/25/65

LOUISVILLE/BUFFALO (USA)
Andrew Dulay, Pilot — 782.64 kmh
Greg Rassmussen, 1st Officer — 486.33 mph
Ryan International Airlines
Boeing 757-24, N402UP
2 P&W PW2040
1 hr 0 min 6 sec
5/18/88

LOUISVILLE/NEW YORK*
Capt. R. M. Sanderson — 567.90 mph
American Airlines
Boeing 707
4 P&W JT3D
Standiford Field/John F. Kennedy Intl
Distance: 611 mi
1 hr 9 min 50 sec
3/13/65

LOUISVILLE/PITTSBURGH (USA)
Robert Olszewski, Pilot — 779.04 kmh
Michael Kastrinelis, 1st Officer — 484.10 mph
USAir DC 9-31, N922VJ
2 P&W JT8D-9A, 14,500 lbs
41 min 33 sec
12/28/87

MADRID/MIAMI (USA) §
Capt. Steven P. Johnson — 800.05 kmh
American Airlines — 497.12 mph
Boeing 767-300ER
2 GE CF6-80C2, 61,500 lbs
8 hrs 53 min
5/17/99

MADRID/PHILADELPHIA (USA)
Capt. Joseph W. Graham — 784.62 kmh
US Airways — 487.54 mph
Airbus A330-300
2 P&W PW4168A, 68,600 lbs
7 hrs 32 min
7/2/03

MAGADAN/KHABAROVSK (USA)
Capt. Steve Sanford — 785.77 kmh
Co-Capt. Mike Swanigan — 488.25 mph
Flt Engineer Rob Spero
Alaska Airlines
Boeing 727-200
3 P&W JT8D-17, 16,000 lbs
2 hrs 3 min
6/18/91

MANCHESTER/PHILADELPHIA (USA)
Capt. Joseph W. Graham — 753.31 kmh
US Airways — 468.08 mph
Airbus A330-300
2 P&W PW4168A, 68,600 lbs
7 hrs 19 min
9/16/01

MANCHESTER/WASHINGTON (USA)
Capt. Vaughn B. Cordle — 720.26 kmh
United Airlines — 447.54 mph
Boeing 737-300
2 CFM56, 20,000 lbs
56 min
10/18/92

MANILA/OSAKA (USA)
Larry E. Steenstry, Captain — 877.49 kmh
Brad W. Kenwisher, 1st Officer — 545.25 mph
Jeffrey L. Snyder, 2nd Officer
United Airlines
McDonnell Douglas DC-10-30
3 GE CF6-50C2, 52,500 lbs
3 hrs 3 min
6/9/98

MANILA/SEOUL (USA)
Capt. David F. Specht — 807.40 kmh
United Airlines — 501.69 mph
Boeing 747-400
4 P&W PW4056, 56,750 lbs
3 hrs 15 min
11/27/97

MANILA/SINGAPORE (USA)
Capt. Maurice de Vas — 901.19 kmh / 560.00 mph
Boeing B747-312
4 P&W JT9D, 54,750 lbs
2 hrs 38 min 20 sec
6/28-29/84

MAUI/LOS ANGELES (USA)
Gordon Beck, Captain — 1,117.73 kmh / 694.52 mph
Jim Chapman, 1st Officer
Michael J. Fortanas, 2nd Officer
Delta Air Lines Lockheed L-1011
3 RR RB211-22B, 42,000 lbs
3 hrs 34 min 29 sec
2/8/92

MAUI/SAN FRANCISCO (USA)
Edward P. Akin, Captain — 948.01 kmh / 589.06 mph
Gaven C. Dunn, 1st Officer
Peter B. Letch, 2nd Officer
United Airlines
McDonnell Douglas DC10-10
3 GE CF6, 40,000 lbs
3 hrs 58 min
3/28/94

MEDFORD/OAKLAND (USA) *Group II*
Jeff D. Schrager, Captain — 421.55 kmh / 261.93 mph
Todd A. DiCello, 1st Officer
Sierra Expressway
BAe Jetstream Super 31
2 Garrett TPE331, 1,020 shp
1 hr 14 min 1 sec
9/3/95

MELBOURNE/AUCKLAND (USA)
Capt. Allan L. Holmes — 1,027.53 kmh / 638.47 mph
United Airlines
Boeing 747-400
4 P&W PW4056, 56,750 lbs
2 hrs 34 min
7/6/98

MELBOURNE/LOS ANGELES (USA) §
Lawrence A. Grihalva, Captain — 978.86 kmh / 608.23 mph
Scott E. Herman, 1st Officer
Richard R. Gordon, 1st Officer
Colin P. Winfield, 1st Officer
United Airlines
Boeing 747-400
4 P&W PW4056, 56,750 lbs
13 hrs 2 min
4/26/00

MELBOURNE/SYDNEY (USA)
David E. Seymour, Captain — 742.52 kmh / 461.38 mph
Gary R. McCaffrey, 1st Officer
United Airlines
Boeing 747-400
4 P&W PW4056, 56,750 lbs
57 min
12/20/03

MEMPHIS/ATLANTA (USA)
Capt. Leiland M. Duke, Jr. — 666.33 kmh / 414.03 mph
FedEx
McDonnell Douglas MD-11F
3 GE CF6-80C2, 61,500 lbs
48 min
8/11/98

MEMPHIS/HONOLULU (USA) §
Capt. Arnold H. Andrijeski — 840.19 kmh / 522.07 mph
FedEx
McDonnell Douglas MD-11F
3 GE CF6-80C2, 61,500 lbs
7 hrs 59 min
4/27/00

MEMPHIS/MINNEAPOLIS (USA)
Capt. Leiland M. Duke, Jr. — 743.70 kmh / 462.11 mph
FedEx
McDonnell Douglas MD-11F
3 GE CF6-80C2, 61,500 lbs
1 hr 31 min
8/26/98

MEMPHIS/NEW YORK*
Capt. A. B. DeSalvo — 566.28 mph
American Airlines
Boeing 707
3 P&W JT8D-1
Memphis Municipal/LaGuardia
Distance: 963 mi
1 hr 42 min 2 sec
7/6/65

MEMPHIS/OAKLAND (USA)
Capt. George W. Murphy, Jr. — 870.59 kmh / 540.96 mph
FedEx
McDonnell Douglas MD-11F
3 GE CF6-80C2, 61,500 lbs
3 hrs 19 min
7/8/99

MEMPHIS/PARIS (USA) §
Capt. Leiland M. Duke, Jr. — 980.05 kmh / 608.97 mph
FedEx
McDonnell Douglas MD-11F
3 GE CF6-80C2, 61,500 lbs
7 hrs 28 min
10/3/98

MEMPHIS/SEATTLE (USA)
Michael C. Healy, Captain — 808.39 kmh / 502.30 mph
James W. Dearborn, 1st Officer
Federal Express
McDonnell Douglas MD-11F
3 GE CF6-80C2D1F, 61,500 lbs
3 hrs 43 min
4/24/97

MEXICO CITY/CHICAGO*
Capt. O. D. Peterson — 1,004.46 kmh / 624.16 mph
American Airlines
Boeing 707-720B
4 P&W JT3D
2 hrs 42 min 42 sec
9/6/62

MEXICO CITY/LOS ANGELES (USA)
Diran L. Torigian, Jr., Captain — 824.11 kmh / 512.08 mph
Paul T. Brown, 1st Officer
United Airlines
Boeing 737-300
2 CFM56, 22,000 lbs
3 hrs 2 min
9/3/94

MEXICO CITY/MADRID (MEXICO) §
Javier Heredia Viderique — 919.91 kmh / 571.60 mph
Manuel G. Hino-Josa Gomez
Carlos A. Ojeda Ruiz de la Pena
Aeromexico
Boeing 767-200ER
10/1/98

MEXICO CITY/MIAMI (USA)
Capt. Keith J. Mackey — 917.08 kmh / 569.88 mph
Pan American Airways 747
4 P&W JT9D, 46,120 lbs
2 hrs 14 min 19 sec
12/29/88

MEXICO CITY/NEW YORK (FRANCE)
Commandant Pierre Debets — 1,366.17 kmh / 848.94 mph
Air France
BAe/Aérospatiale Concorde
4 RR Olympus 593, Mark 610
10/21/82

MEXICO CITY/PARIS (FRANCE)
Commandant Pierre Debets — 1,244.85 kmh
Air France — 773.55 mph
BAe/Aérospatiale Concorde
4 RR Olympus 593, Mark 610
10/21/82

MEXICO CITY/SAN ANTONIO (USA)
Capt. R. R. Newhouse — 900.62 kmh
American Airlines — 559.61 mph
Boeing 720B
4 P&W JT3D-1
1 hr 14 min 59.5 sec
9/6/62

MEXICO CITY/WASHINGTON (USA)
Capt. JohnGlen R. Fuentes — 953.09 kmh
United Airlines — 592.22 mph
Airbus A320
2 IAE V2527, 26,500 lbs
3 hrs 9 min
4/17/99

MIAMI/AMSTERDAM (USA) §
Capt. Kendall J. Moeller — 966.60 kmh
Northwest Airlines — 600.61 mph
McDonnell Douglas DC-10-30
3 GE CF6-50C, 51,000 lbs
7 hrs 42 min
7/7/02

MIAMI/ATLANTA*
Capt. T. P. Ball — 406.1 mph
Delta Air Lines
Douglas DC-8
4 P&W JT3C
Distance: 597 mi
1 hr 28 min 11 sec
2/24/62

MIAMI/BOSTON*
John B. Bennett — 474.96 mph
Northeast Airlines
Boeing 727
P&W JT8D-7
Distance: 1,258 mi
2 hrs 38 min 55 sec
4/18/71

MIAMI/BUENOS AIRES (USA) §
Capt. Mario J. Lopes — 944.96 kmh
United Airlines — 587.17 mph
Boeing 747-238B
4 P&W JT9D-7J, 48,000 lbs
7 hrs 32 min
11/9/96

MIAMI/CAMPINAS (USA)
Charles S. Tewksbury, Captain — 910.52 kmh
D. Rich Womack, 2nd Officer — 565.77 mph
Federal Express
McDonnell Douglas DC-10-30
3 GE CF6-50C2, 52,500 lbs
7 hrs 8 min
12/22/97

MIAMI/CAPE TOWN (SOUTH AFRICA) §
Karl Jensen — 1,016.49 kmh
Heinrich Gillen — 631.62 mph
Boeing 747-400
9/14/96

MIAMI/CHICAGO (USA)
Verne Jobst — 859.77 kmh
United Airlines — 534.26 mph
DC-10-10
3 GE CF-6 39,000 lbs
2 hrs 14 min 53 sec
12/19/83

MIAMI/DALLAS (USA)
Capt. Angela Masson — 777.87 kmh
American Airlines — 483.35 mph
Boeing 777-200
2 R-R Trent 892, 91,300 lbs
2 hrs 19 min
10/25/01

MIAMI/DENVER (USA)
Capt. David H. Friend — 739.40 kmh
United Airlines — 459.44 mph
Boeing 777-200
2 P&W PW4090, 90,000 lbs
3 hrs 43 min
3/7/01

MIAMI/FRANKFURT (USA) §
Capt. Robert P. Gross — 909.78 kmh
American Airlines — 565.31 mph
Boeing 767-300ER
2 GE CF6-80C2, 61,500 lbs
8 hrs 32 min
11/24/95

MIAMI/HOUSTON*
Capt. T. Outland — 477.82 mph
Delta Air Lines
Douglas DC 8-33
4 P&W JT3D
Miami Intl/Houston Intercontinental
Distance: 963 mi
2 hrs 0 min 55 sec
10/1/69

MIAMI/LIMA (USA)
Capt. Robert P. Gross — 855.10 kmh
American Airlines — 531.33 mph
Boeing 757-200
2 RR RB211, 48,000 lbs
4 hrs 56 min
4/16/95

MIAMI/LONDON (UK)
Brian Walpole — 1,179.44 kmh
British Airways — 732.90 mph
BAe/Aérospatiale Concorde
4 RR Olympus, 38,000 lbs
9/17/86

MIAMI/LOS ANGELES (USA)
Keith J. Mackey, Captain — 882.72 kmh
Pan American — 548.52 mph
Boeing 747, N7410U
4 P&W JT9D
4 hrs 16 min 3 sec
6/24/88

MIAMI/MADRID (USA) §
Capt. Steven P. Johnson — 875.62 kmh
American Airlines — 544.08 mph
Boeing 767-300ER
2 GE CF6-80C2, 61,500 lbs
8 hrs 7 min
5/16/99

MIAMI/MEXICO CITY (USA)
Daniel M. Siciliano, Captain — 799.00 kmh
Kenneth G. Brummage, 1st Officer — 496.47 mph
American Airlines
Boeing 757-200
2 RR 535-E4B, 40,100 lbs
2 hrs 34 min
6/21/97

MIAMI/MONTEGO BAY (USA)
Robert P. Gross, Captain — 715.36 kmh
Timothy Fountain, 1st Officer — 444.50 mph
American Airlines
Boeing 757-200
2 RR RB211, 48,000 lbs
1 hr 11 min
7/5/94

MIAMI/MONTREAL (USA)
Capt. James O. Gross — 937.00 kmh
American Airlines — 582.22 mph
McDonnell Douglas MD-82
2 P&W JT8D, 20,000 lbs
2 hrs 25 min
12/30/97

MIAMI/NEW YORK (USA)
Keith J. Mackey — 828.00 kmh
Pan American Airways — 514.52 mph
Boeing 747, 9674
4 P&W JT9D
2 hrs 7 min 30 sec
12/11-12/88

MIAMI/PANAMA CITY, PANAMA (USA)
Capt. Robert P. Gross — 839.89 kmh
American Airlines — 521.88 mph
Boeing 757-200
2 RR RB-211, 48,000 lbs
2 hrs 13 min
2/25/95

MIAMI/PARIS (FRANCE)
Cdt. Michel Brunet, Pilot — 1,013.21 kmh
Michel Launay, Copilot — 629.61 mph
Air France Boeing 747-228-B
4 GE, 23,800 lbs
10/27/86

MIAMI/PITTSBURGH (USA)
Capt. Robert F. Olszewski — 804.12 kmh
USAir — 499.65 mph
Boeing 767-201ER
2 GE CF6-80, 52,000 lbs
2 hrs 1 min 57 sec
6/20/91

MIAMI/RIO DE JANEIRO (USA) §
Capt. Keith J. Mackey — 911.52 kmh
PanAm — 566.42 mph
Boeing 747-S3R
7 hrs 22 min 35 sec
5/30/89

MIAMI/ST. JOHNS (USA)
Patrick M. Broderick, Captain — 897.12 kmh
Hobson Reynolds, 1st Officer — 557.47 mph
Joseph F. Basco, 2nd Officer
Eastern Airlines 727
3 P&W JT8D 7B
2 hrs 23 min
2/12-13/89

MIAMI/SAN JOSE, COSTA RICA (USA)
Daniel M. Siciliano, Captain — 839.35 kmh
Ken Kohlman, 1st Officer — 521.54 mph
American Airlines
Boeing 757-200
2 RR 535-EBA, 43,100 lbs
2 hrs 9 min
7/20/97

MIAMI/SAN JUAN (USA)
Capt. Robert P. Gross — 917.36 kmh
American Airlines — 570.02 mph
Boeing 767-300ER
2 GE CF6-80C2, 61,500 lbs
1 hr 50 min
4/9/95

MIAMI/SANTIAGO (USA) §
Capt. James R. Walker — 924.57 kmh
United Airlines — 574.50 mph
Boeing 777
2 P&W PW4090, 90,000 lbs
7 hrs 12 min
12/5/97

MIAMI/SAO PAULO (USA) §
James E. Goolsby, Captain — 921.52 kmh
Robert F. Frank, 1st Officer — 572.60 mph
Susan W. Harrison, 1st Officer
United Airlines
Boeing 777
2 P&W PW4077, 77,000 lbs
7 hrs 8 min
11/4/96

MIAMI/TEGUCIGALPA (USA)
Robert P. Gross, Captain — 813.28 kmh
Timothy Fountain, 1st Officer — 505.35 mph
American Airlines
Boeing 757-200
2 RR RB211, 48,000 lbs
1 hr 50 min
7/12/94

MILAN/NEW YORK (USA)
William A. Austin — 820.37 kmh
Federal Express DC-10-30 — 509.75 mph
3 GE CF6-50C2, 52,500 lbs
7 hrs 49 min
4/6/92

MILAN/WASHINGTON (USA) §
Capt. N. B. Godlove — 786.58 kmh
United Airlines — 488.75 mph
Boeing 747
4 P&W JT9D, 47,500 lbs
8 hrs 37 min
9/18/93

MINNEAPOLIS/ANCHORAGE (USA)
James R. Bramley, Captain — 767.49 kmh
Linda D. Larson, 1st Officer — 476.89 mph
Northwest Airlines
Boeing 757
2 P&W PW2037, 38,200 lbs
5 hrs 16 min
8/3/98

MINNEAPOLIS/BILLINGS*
Capt. Loran D. Gruman — 434.38 mph
Northwest Airlines
Boeing 727-2A
3 P&W JT8D, 14,000 lbs
Minneapolis-St. Paul Intl/Billings Logan Intl
Distance: 758 mi
1 hr 44 min 42 sec
12/17/78

MINNEAPOLIS/CHICAGO (USA)
Capt. Michael J. Kelly — 657.12 kmh
United Airlines — 408.32 mph
Boeing 727
3 P&W JT8D-15, 15,500 lbs
49 min
9/25/92

MINNEAPOLIS/COLUMBUS (USA)
Capt. Leiland M. Duke, Jr. — 834.21 kmh
FedEx — 518.35 mph
McDonnell Douglas MD-11F
3 GE CF6-80C2, 61,500 lbs
1 hr 13 min
8/26/98

MINNEAPOLIS/LAS VEGAS*
Capt. Gary Thompson 527.92 mph
Western Airlines
Boeing 727
3 P&W JT8D
Minneapolis-St. Paul Intl/Las Vegas McCarren
Distance: 1,300 mi
2 hrs 27 min 45 sec
5/28/47

MINNEAPOLIS/LOS ANGELES*
Capt. Robert C. Bowman 550.20 mph
Western Airlines
Boeing 720 B
4 P&W
Minneapolis-St. Paul Intl/Los Angeles Intl
Distance: 1,566.24 mi
2 hrs 50 min 42 sec
12/1/76

MINNEAPOLIS/TOKYO (USA) §
George D. Armstrong, Captain 835.52 kmh
Roger D. Break, Captain 519.16 mph
Gary E. Peterson, 1st Officer
Paul E. Ruohoniemi, 1st Officer
Northwest Airlines
Boeing 747-400
4 P&W PW4058, 58,000 lbs
11 hrs 26 min
12/28/96

MONTEVIDEO/BUENOS AIRES*
William W. Critcher, Captain 340.96 mph
Wilbur C. Biggin, III, 1st Officer
United Airlines
Boeing 767-300ER
2 P&W PW4060, 60,000 lbs
25 min
9/1/00

MONTREAL/CHICAGO (USA)
Capt. James O. Gross 654.80 kmh
American Airlines 406.87 mph
McDonnell Douglas MD-82
2 P&W JT8D, 21,000 lbs
1 hr 50 min
12/18/97

MONTREAL/LONDON (CANADA)
Capt. R. M. Smith 909.73 kmh
Trans-Canada Airlines 565.29 mph
Douglas DC-8-40
4 RR Conway
5 hrs 44 min 42 sec
5/28/60

MOSCOW/LONDON (UK)
Capt. Mike Bannister 772.09 kmh
British Airways 479.75 mph
BAe/Aérospatiale Concorde
4 RR Olympus, 38,000 lbs
3/2/88

MOSCOW/NEW YORK (USA) §
Capt. William D. McCarthy 834.22 kmh
Delta Airlines 518.36 mph
Boeing 767-300ER
2 P&W PW4060, 60,000 lbs
8 hrs 58 min
1/5/98

MUNICH/CHICAGO (USA) §
Capt. Hans L. Lagerloef 790.16 kmh
American Airlines 490.98 mph
Boeing 767-300
2 GE CF6-80C2, 61,500 lbs
9 hrs 12 min
9/19/93

MUNICH/WASHINGTON (USA) §
David H. Friend, Captain 758.93 kmh
Dale G. Goodrich, 1st Officer 471.57 mph
United Airlines
Boeing 777
2 P&W PW4077, 77,000 lbs
9 hrs 1 min
1/12/00

NADI/SYDNEY (USA)
Capt. Leiland M. Duke, Jr. 795.87 kmh
FedEx 494.53 mph
McDonnell Douglas MD-11F
3 GE CF6-80C2, 61,500 lbs
3 hrs 59 min
9/20/98

NASHVILLE/BOSTON (USA)
Capt. John H. Jenkins 900.62 kmh
American Airlines 559.64 mph
Boeing 727, N6824
1 hr 41 min
1/25/88

NASHVILLE/LOS ANGELES*
Capt. James W. Knight 491.40 mph
American Airlines
Boeing 707-123
4 P&W JT3D
Nashville Municipal/Los Angeles Intl
Distance: 1,797 mi
3 hrs 39 min 24 sec
2/14/65

NASHVILLE/MEMPHIS*
Capt. Robert F. Olszewski 453.6 mph
Allegheny Airlines
McDonnell Douglas DC-9-31
2 P&W JT8D-7A
Nashville Metropolitan/Memphis Intl
Distance: 200 mi
26 min 17 sec
7/9/79

NASHVILLE/NEW YORK*
Capt. Gordon Adams 639.12 mph
American Airlines
Boeing 707-720B
4 P&W JT3D
Berry Field/Idlewild Intl
Distance: 767 mi
1 hr 12 min
1/23/62

NASHVILLE/SAN FRANCISCO*
Capt. O. P. Brunsvold 492.80 mph
American Airlines
Convair 990
4 GE CJ805-23
Nashville Muni/San Francisco Intl
Distance: 1,970 mi
3 hrs 59 min 51 sec
2/22/65

NASHVILLE/TULSA*
Capt. J. Arthur Miller 499.03 mph
American Airlines
Boeing 727
3 P&W JT8D-1
Nashville Metro/Tulsa Intl
Distance: 515 mi
1 hr 1 min 55 sec
12/7/65

NASHVILLE/WASHINGTON* 583.81 mph
Capt. R. H. Gardner, Jr.
American Airlines
Boeing 707-720
4 P&W JT3D
Berry Field/Dulles Intl
Distance: 542 mi
55 min 42 sec
12/23/62

NASSAU/LONDON (ENGLAND) 759.85 kmh 471.14 mph
Capt. P. Wilson
Capt. M. Gudmundsson
Cunard Eagle Airways
Boeing 707/465
4 RR Conway
9 hrs 11 min 38 sec
5/6/62

NEWARK/CHICAGO (USA) 769.65 kmh 478.24 mph
Sather M. Ranum, Captain
Larry E. Steenstry, 1st Officer
United Airlines
Boeing 757-200
2 P&W PW2037, 38,200 lbs
1 hr 30 min
11/15/95

NEWARK/DENVER (USA) 792.60 kmh 492.50 mph
Capt. Larry Liguori
United Airlines
Airbus A320
2 IAE V2527-A5, 26,500 lbs
3 hrs 15 min
7/14/95

NEWARK/DETRIOT (USA) 585.36 kmh 363.74 mph
B. Michael Bristow, Capt
Continental Airlines
Boeing 737-200
2 P&W JT8D-7
1 hr 20 min 19 sec
6/13/89

NEWARK/FT. MYERS (USA) 720.72 kmh 447.85 mph
B. Michael Bristow
Continental Airlines
Boeing 737-200
2 P&W JT8D-9
2 hrs 23 min 30 sec
6/11/89

NEWARK/LONDON (USA) 969.83 kmh 602.62 mph
Capt. Arnold H. Andrijeski
FedEx
McDonnell Douglas MD-11F
3 GE CF6-80C2, 61,500 lbs
5 hrs 46 min
6/10/00

NEWARK/PARIS (USA) 958.35 mph 595.49 mph
Charles S. Tewksbury, Captain
R. Marshall Hodges, 1st Officer
Federal Express
McDonnell Douglas DC-10-30F
3 GE CF6-50C2, 52,500 lbs
6 hrs 6 min 35 sec
8/11/95

NEW DELHI/HONG KONG (USA) 899.11 kmh 558.68 mph
Owen V. Hafer, Captain
Gary R. McCaffrey, 1st Officer
United Airlines
Boeing 747-400
4 P&W PW4056, 56,750 lbs
4 hrs 10 min
11/13/98

NEW ORLEANS/BALTIMORE* 596.47 mph
Capt. T. Outland
Delta Air Lines
Convair 880
4 GE CJ-705-3
Moisant Intl/Friendship Intl
Distance: 999 mi
1 hrs 40 min 31 sec
10/30/60

NEW ORLEANS/LOS ANGELES (USA) 785.22 kmh 487.91 mph
Capt. Larry Liguori
United Airlines
Airbus A320
2 IAE V2527-A5, 26,500 lbs
3 hrs 25 min
8/8/95

NEW ORLEANS/NEW YORK* *Group I* 329.71 mph
Capt. H. T. Merrill
Eastern Airlines
Lockheed Constellation
4 Wright R-3350
Moisant Intl/LaGuardia
Distance: 1,182 mi
3 hrs 52 min 29.8 sec
7/23/47

NEW ORLEANS/SAN JUAN* 556.48 mph
Capt. H. E. Croft
Delta Air Lines
Convair 880
4 GE CJ-805
Moisant Intl/Puerto Rico Intl
Distance: 1,721 mi
3 hrs 5 min 33 sec
5/13/62

NEW YORK/ACAPULCO (UK) 1,307.60 kmh 812.50 mph
Capt. John Hutchinson
British Airways
BAe/Aérospatiale Concorde
3/14/90

NEW YORK/ATHENS (USA) § 944.82 kmh 587.08 mph
Douglas Boston, Captain
Barry Bogart, 1st Officer
Robert Vose, 2nd Officer
Delta Airlines
Boeing 767-ER
2 P&W PW4060, 60,000 lbs
8 hrs 23 min
11/21/97

NEW YORK/ATLANTA (USA) 707.71 kmh 439.75 mph
Richard Norton, Captain
Earl Ecklin, Flt Engineer
TWA Boeing 727-200
3 P&W JT8D, 14,500 lbs
1 hr 43 min 35 sec
7/26/91

NEW YORK/BERGEN (NORWAY) 857.99 kmh 533.13 mph
Capt. B. Bjornsta
Scandinavian Airlines System
Douglas DC-8
4 P&W JT4A-9
6 hrs 32 min 20 sec
6/2/63

NEW YORK/BERLIN (USA) § 1,017.36 kmh 632.19 mph
Don Binard, Captain
Alan Douglas Gee, 1st Officer
Joseph P. Miller, 2nd Officer
Pan American
Airbus A310-300
2 P&W PW4152
7 hrs 28 min
12/24/88

NEW YORK/BOSTON*
Capt. Fred E. Illston — 408.02 mph
American Airlines
Convair 990
4 GE CJ-805-23
Idlwewild Intl/Logan Intl
Distance: 186 mi
27 min 21 sec
7/20/63

NEW YORK/BUENOS AIRES (USA) §
Capt. Angela Masson — 907.99 kmh, 564.20 mph
American Airlines
Boeing 777-200
2 R-R Trent 892, 91,300 lbs
9 hrs 24 min
8/19/01

NEW YORK/CARACAS (USA)
Vaughn B. Cordle — 814.15 kmh, 505.89 mph
United Airlines
Boeing 757-200
2 P&W PW2037, 38,200 lbs
4 hrs 11 min
4/11/98

NEW YORK/CHICAGO*
Capt. Carl Jordan — 569.16 mph
American Airlines
Boeing 707/720
4 P&W JT3D-1
Idlewild Intl/O'Hare Intl
Distance: 740 mi
1 hr 18 min
8/13/61

NEW YORK/DELHI (USA) §
Capt. Walter H. Mullikin — 869.64 kmh, 540.36 mph
Capt. A. A. Frink
Pan American World Airways, Inc.
Boeing 747 SP
4 P&W JT9D-7A
13 hrs 31 min 20 sec
5/1/76

NEW YORK/EDINBURGH (UK)
Capt. John Cook — 1,729.38 kmh, 1,074.64 mph
British Airways
BAe/Aérospatiale Concorde
4 Olympus 593, 38,000 lbs
7/13/87

NEW YORK/FT LAUDERDALE (USA)
Michael Piampiano, Captain — 757.08 kmh, 470.45 mph
Jim Kennedy, 1st Officer
Continental Airlines
Airbus A 30 B-4
2 GE CF6-50-C2, 52,500 lbs
2 hrs 16 min 15 sec
2/17/89

NEW YORK/FRANKFURT (USA)
Capt. Don Gordon Loeschner — 845.93 kmh, 525.63 mph
Trans World Airlines
Boeing 767-200
2 P&W JT9D, 48,000 lbs
7 hrs 19 min
6/27/96

NEW YORK/HONG KONG (UK) §
Paul Horsting — 830.96 kmh, 516.33 mph
Cathay Pacific
Boeing 747-400
7/5/98

NEW YORK/HOUSTON*
Capt. J. W. Bishop — 570.70 mph
Delta Air Lines
Convair 880
4 GE CJ-805
JFK Intl/Houston Municipal
Distance: 1,430 mi
2 hrs 30 min 20 sec
7/23/65

NEW YORK/LISBON (USA)
Capt. Henry J. Stoecker — 939.92 kmh, 584.04 mph
Trans World Airlines
Boeing 757
2 P&W PW2037, 38,200 lbs
5 hrs 45 min
1/11/98

NEW YORK/LONDON (UK)
L. G. Scott — 1,920.07 kmh, 1,193.07 mph
BAe/Aérospatiale Concorde
4 RR Olympus, 38,000 lbs
2/7/96

NEW YORK/LOS ANGELES*
Capt. Eugene M. Kruse — 572.57 mph
American Airlines
Boeing 707-720B
4 P&W JT3D
4 hrs 19 min 15 sec
8/15/62

NEW YORK/LYON (USA)
Andrew J. Anderson, Captain — 927.80 kmh, 576.51 mph
Daniel S. Repasky, 1st Officer
Peter C. Baum, 1st Officer
Delta Air Lines
Boeing 767-300ER
2 P&W PW4060, 60,000 lbs
6 hrs 38 min
7/23/01

NEW YORK/MADRID (SPAIN) *Group I*
Capt. C. I. Batida — 610.90 kmh, 379.62 mph
Iberia Airlines
Lockheed Super Constellation L-1049E
4 Wright R-3350
9 hrs 29 min
11/26/54

NEW YORK/MADRID (USA) *Group III*
John J. Bauer, Captain — 1,002.08 kmh, 622.66 mph
Eric E. Schoneberger, 1st Officer
Robert Vose, 1st Officer
Delta Airlines
Boeing 767-300ER
2 P&W PW4060, 60,000 lbs
5 hrs 45 min
11/6/97

NEW YORK/MEMPHIS*
Capt. A. B. DeSalvo — 549.21 mph
American Airlines
Boeing 727
3 P&W JT8D-1
LaGuardia/Memphis Municipal
Distance: 963 mi
1 hr 45 min 12 sec
7/6/65

NEW YORK/MEXICO CITY (FRANCE)
Commandant Pierre Debets — 1,296.11 kmh, 805.40 mph
Air France
BAe/Aérospatiale Concorde
4 RR Olympus 593, Mark 610
10/20/82

NEW YORK/MIAMI* *Group I*
Capt. E. R. Brown — 275.62 mph
Eastern Airlines
Lockheed Constellation
4 Wright, 2100 hp
Laguardia/36th Street
5/28/47

NEW YORK/MIAMI (USA) *Group III*
Capt. Eugene W. Scholl — 780.94 kmh
Trans World Airlines — 485.25 mph
McDonnell Douglas MD-83
2 P&W JT8D, 20,850 lbs
2 hrs 15 min
8/16/00

NEW YORK/MUNICH (USA)
Capt. William D. McCarthy — 917.13 kmh
Delta Air Lines — 569.87 mph
Boeing 767-300ER
2 P&W PW4060, 60,000 lbs
7 hrs 4 min
12/23/97

NEW YORK/NEW ORLEANS* *Group I*
Capt. H. T. Merrill — 305.16 mph
Eastern Airlines
Lockheed Constellation
4 Wright R-3350
LaGuardia/Moissant Intl
7/23/47

NEW YORK/ORLANDO (USA)
Michael Piampiano, Captain — 700.92 kmh
Jim Kennedy, 1st Officer — 435.55 mph
Continental Airlines
Airbus A 30 B-4
2 GE CF6-50-C2, 52,500 lbs
2 hrs 9 min 30 sec
2/17/89

NEW YORK/PALM BEACH (USA)
C. Steve Snead, Captain — 850.87 kmh
William H. Dyer, Flt Engineer — 528.70 mph
Trans World Airlines
Boeing 727-200
3 P&W JT8D-9A, 14,500 lbs
1 hr 57 min 38 sec
10/29/92

NEW YORK/PARIS (FRANCE)
Commanndant Pierre Debets — 1,806.82 kmh
Air France — 1,122.76 mph
BAe/Aérospatiale Concorde
4 RR Olympus 593, Mark 610
10/21/82

NEW YORK/PHOENIX*
Capt. G. A. Ryan — 537.1 mph
American Airlines
Boeing 707-123B
4 P&W JT3D-1
Idlewild Intl Aiport/Sky Harbor
Distance: 2,154 mi
4 hrs 0 min 30 sec
10/15/61

NEW YORK/PROVIDENCE*
Capt. Robert F. Olszewski — 428.99 mph
Allegheny Airlines
McDonnell Douglas DC-9-31
2 P&W JT8D-7A
Distance: 143 mi
20 min 1 sec
3/7/78

NEW YORK/RIO DE JANEIRO (USA) §
Capt. Theodore C. Patecell — 956.40 kmh
Pan American World Airways — 594.28 mph
Boeing 707-300 B/A
4 P&W JT3D
8 hrs 5 min 30 sec
2/19/74

NEW YORK/SAN DIEGO*
Capt. Donald W. Fausner — 468.00 mph
American Airlines
McDonnell Douglas DC-10-10
3 GE CF-6, 40,000 lbs
JFK Intl/San Diego Intl
Distance: 2,494.35 mi
5 hrs 0 min 27 sec
9/28/79

NEW YORK/SAN FRANCISCO (USA)
Capt. D. W. Ledbetter — 898.50 kmh
American Airlines — 558.32 mph
Boeing 707-123B
4 P&W JT3D-1
4 hrs 38 min
8/18/61

NEW YORK/SAN JUAN (USA)
William W. Critcher, Captain — 827.12 kmh
Michael R. Dimento, 1st Officer — 513.94 mph
United Airlines
Boeing 757
2 P&W PW2037, 38,200 lbs
3 hrs 7 min
9/26/99

NEW YORK/SAO PAULO (USA) §
Capt. Angela Masson — 905.17 kmh
American Airlines — 562.45 mph
Boeing 777-200
2 R-R Trent 892, 91,300 lbs
8 hrs 28 min
12/12/01

NEW YORK/TEL AVIV (USA) §
Gid Miller, Captain — 1,000.03 kmh
Thomas E. Vogel, 2nd Officer — 621.39 kmh
Richard H. Gray, Reserve Officer
Trans World Airlines
Boeing 747
4 P&W JT9D, 46,150 lbs
9 hrs 7 min
9/6/95

NEW YORK/TOKYO (USA) §
Capt. John R. Deakin — 905.49 kmh
Japan Airlines — 562.64 mph
Boeing 747-200
4 P&W JT9D-7R4G2, 54,750 lbs
11 hrs 57 min 38 sec
05/17/90

NEW YORK/VENICE (USA) §
Capt. Andrew J. Anderson — 895.07 kmh
Delta Air Lines — 556.17 mph
Boeing 767-300ER
2 P&W PW4060, 60,000 lbs
7 hrs 27 min
12/25/00

NEW YORK/ZURICH (USA)
Capt. Dennis D. Lyons — 910.02 kmh
American Airlines — 565.46 mph
Boeing 767-300ER
2 GE CF6-80C2, 61,500 lbs
6 hrs 56 min
8/5/02

OAKLAND/HOUSTON (USA)
Capt. Wendell L. Moeller — 839.48 kmh
Continental Airlines — 521.62 mph
Boeing 737-300
2 CFM CFM56, 20,000 lbs
3 hrs 7 min
11/12/02

OAKLAND/MEDFORD (USA) *Group II*
Jeff D. Schrager, Captain — 497.50 kmh
Todd A. DiCello, 1st Officer — 309.13 mph
Sierra Expressway
BAe Jetstream Super 31
2 Garrett TPE331, 1,020 shp
1 hr 2 min 43 sec
9/4/95

OMAHA/ATLANTA (USA)
Samuel D. Smith — 826.92 kmh
Delta Air Lines — 513.84 mph
Boeing 737-232
1 hr 35 min 56 sec
5/28/89

ONTARIO/DALLAS (USA)
Gary N. Graves, Captain — 835.61 kmh
Scott W. Miller, 1st Officer — 519.22 mph
American Airlines
Boeing 757-200
2 RR 535, 43,100 lbs
2 hrs 17 min
12/31/93

ORLANDO/BOSTON (USA)
Capt. Vaughn B. Cordle — 814.88 kmh
United Airlines — 506.34 mph
Boeing 737-300
2 CFM56, 20,000 lbs
2 hrs 13 min
10/11/92

ORLANDO/CHICAGO (USA)
Richard R. Metler, Captain — 714.70 kmh
Larry E. Steenstry, 1st Officer — 444.09 mph
United Airlines
Boeing 757
2 P&W PW2037, 38,200 lbs
2 hrs 16 min
12/26/93

ORLANDO/FRANKFURT (USA) §
Capt. Keith J. Mackey — 873.37 kmh
Pan American — 542.71 mph
Boeing 747, N739PA
4 P&W JT9D
8 hrs 44 min 6 sec
6/15/88

ORLANDO/KEY WEST (USA) *Group II*
Capt. Nicholas E. Mech — 438.36 kmh
Comair — 272.38 mph
Embraer EMB-120RT Brasilia
2 P&WC PW118, 1800 shp
1 hr 3 min
3/7/01

ORLANDO/MEXICO CITY (USA)
Capt. Vaughn B. Cordle — 758.21 kmh
United Airlines — 471.13 mph
Boeing 737-300
2 CFM56, 20,000 lbs
2 hrs 43 min
12/12/92

ORLANDO/NEWARK (USA)
Capt. Vaughn B. Cordle — 768.43 kmh
United Airlines — 477.48 mph
Boeing 737-300
2 CFM56, 20,000 lbs
1 hr 58 min
12/2/92

ORLANDO/NEW YORK*
Capt. Daniel J. Clisham — 542.13 mph
Evergreen Intl Airlines
McDonnell Douglas DC-8-52
4 P&W JT3D
Orlando Intl/JFK Intl
Distance: 965 mi
1 hr 47 min 4 sec
4/25/77

ORLANDO/PARIS (USA)
Keith J. Mackey — 927.00 kmh
Pan American Airways — 576.01 mph
Boeing 747
4 P&W JT9D, 46,150 lbs
9/29/88

ORLANDO/PHILADELPHIA (USA)
Patrick M. Broderick, Captain — 785.52 kmh
G. Curt Ayers, 1st Officer — 488.12 mph
Binka Bone, 2nd Officer
Eastern Airlines
Boeing 727-200A
3 P&W JT8D
1 hr 46 min 12 sec
2/25/88

ORLANDO/WASHINGTON (USA)
Capt. Martha Anne Collins Leeds — 685.39 kmh
United Airlines — 425.88 mph
Boeing 727-200
3 P&W JT8D, 15,500 lbs
1 hr 47 min
10/30/01

OSAKA/ANCHORAGE (USA)
Larry E. Steenstry, Captain — 893.05 kmh
Michael J. McGarry, 1st Officer — 554.92 mph
R. W. Ashcraft, 2nd Officer
United Airlines
McDonnell Douglas DC-10-30F
3 GE CF6-50C2, 52,500 lbs
6 hrs 37 min
2/15/99

OSAKA/GREENSBORO (USA) §
James C. Sifford — 562.55 kmh
Piedmont Airlines — 349.55 mph
Boeing 727
3 P&W JT8D, 14,500 lbs
12/27/83

OSAKA/HONOLULU (USA) §
Robert H. Siteman, Captain — 1,117.32 kmh
Mark C. Schwing, 1st Officer — 694.27 mph
Paul Wessel, 2nd Officer
United Airlines
Boeing 747-100
4 P&W JT9D-7A, 46,950 lbs
5 hrs 54 min
1/31/95

OSAKA/LOS ANGELES (USA) §
Capt. Richard A. Saber — 1,045.60 kmh
United Airlines — 649.70 mph
Boeing 747-400
4 P&W PW4056, 57,000 lbs
8 hrs 47 min
2/5/96

OSAKA/MEMPHIS (USA) §
Arnold H. Andrijeski, Captain — 951.47 kmh
Michael K. Babin, Jr., 1st Officer — 591.21 mph
FedEx
McDonnell Douglas MD-11F
3 GE CF6-80C2, 61,500 lbs
11 hrs 32 min
4/8/00

OSAKA/OAKLAND (USA) §
Capt. George W. Murphy, Jr. 999.64 kmh
FedEx 621.14 mph
Boeing MD-11F
3 P&W PW4462, 62,000 lbs
8 hrs 42 min
3/1/00

OSAKA/SAN FRANCISCO (USA) §
Capt. F. van Cortlandt de Peyster 1,020.61 kmh
United Airlines 634.18 mph
Boeing 747-400
4 P&W PW4056, 56,750 lbs
8 hrs 31 min
5/13/01

OSAKA/SEOUL (USA)
Capt. David F. Specht 641.07 kmh
United Airlines 398.34 mph
Boeing 747-400
4 P&W PW4056, 56,750 lbs
1 hr 18 min
11/28/97

OTTAWA/LONDON (CANADA)
Robert Munro Smith 905.60 kmh
George Bayliss 562.71 mph
Trans Canada Airlines
McDonnell Douglas DC-8
4 R-R Conway, 17,500 lbs
5/28/1960

PAGO PAGO/SYDNEY (USA)
Capt. Glenn W. Stickel 788.41 kmh
United Parcel Service 489.90 mph
Boeing 747-100
4 P&W JT9D, 46,950 lbs
5 hrs 35 min
11/12/00

PAPEETE/LOS ANGELES (FRANCE)
Cdt. Gilbert Demonceau 932.33 kmh
Jacques Masson, Copilot 579.35 mph
Air France, Boeing 747
4 GE CF6-50-E2, 52,500 lbs
9/27/86

PARIS/ATHENS (GREECE)
Panajotis Kyratatos 830.00 kmh
Olympic Airways 515.74 mph
Comet DH-106 Mark 4B
4 RR Avon
2 hrs 39 min 29 sec
5/17/60

PARIS/BOMBAY (USA) §
Capt. William J. Wirth 840.92 kmh
Delta Air Lines 522.52 mph
McDonnell Douglas MD-11
3 P&W PW4460, 60,000 lbs
8 hrs 19 min
8/13/02

PARIS/CARACAS (FRANCE) §
Cdt. Jean-Paul Le Moel 1,392.19 kmh
Air France 865.11 mph
BAe/Aérospatiale Concorde
4 RR Olympus 593, 38,000 lbs
3/19/82

PARIS/CHARLOTTE (USA) §
Capt. Michael A. Creider 800.93 kmh
US Airways 497.67 mph
Boeing 767-200ER
2 GE CF6-80C2, 52,500 lbs
8 hrs 32 min
4/14/00

PARIS/CHICAGO (USA) §
Capt. Herbert L. Dabelow 858.60 kmh
American Airlines 533.50 mph
Boeing 767-200ER
2 GE CF6-80A, 48,000 lbs
7 hrs 46 min
7/19/94

PARIS/DAKAR (FRANCE)
Cdt. A. Gely 1,592.67 kmh
Air France 989.70 mph
BAe/Aérospatiale Concorde
4 RR Olympus 593 RR, 38,000 lbs
3/28/82

PARIS/DUBAI (USA)
Capt. Arnold H. Andrijeski 863.51 kmh
FedEx 536.55 mph
McDonnell Douglas MD-11F
3 GE CF6-80C2, 61,500 lbs
6 hrs 4 min
6/23/00

PARIS/LOS ANGELES (FRANCE) §
Cdt. Pierre Hersen, Pilot 890.16 kmh
Thierre Lacheretz, Copilot 553.14 mph
Francis Viala, Copilot
Air France Boeing 747
4 GE CF6-50-E2, 52,500 lbs
9/26/86

PARIS/MEMPHIS (USA) §
Capt. Leiland M. Duke, Jr. 837.90 kmh
FedEx 520.65 mph
McDonnell Douglas MD-11F
3 GE CF6-80C2, 61,500 lbs
8 hrs 44 min
10/6/98

PARIS/MEXICO CITY (FRANCE)
Commandant Pierre Debets 1,222.81 kmh
Air France 759.86 mph
BAe/Aérospatiale Concorde
4 RR Olympus 593, Mark 610
10/20/82

PARIS/MIAMI (FRANCE)
Cdt Yves Ros, Pilot 848.22 kmh
Jean-Louis Ferrand, Copilot 527.01 mph
Air France Boeing 747-228-B
10/27/86

PARIS/NEW YORK (FRANCE)
Pierre Chanoine 1,763.00 kmh
André Coloc 1,095.48 mph
André Blanc
Air France
BAe/Aérospatiale Concorde
4 Olympus 593 Mark 610, 38,000 lbs
4/1/81

PARIS/OLONGAPO (USA) §
Capt. Arnold H. Andrijeski 843.80 kmh
FedEx 524.31 mph
McDonnell Douglas MD-11F
3 GE CF6-80C2, 61,500 lbs
12 hrs 38 min
1/25/01

PARIS/PHILADELPHIA (USA)
Capt. Joseph W. Graham 819.74 kmh
US Airways 509.36 mph
Airbus A330-300
2 P&W PW4168A, 68,600 lbs
7 hrs 18 min
12/22/01

PARIS/RIO DE JANEIRO (FRANCE) §
Cdt. A. Gely — 1,382.18 kmh / 858.89 mph
Air France
BAe/Aérospatiale Concorde
4 RR Olympus 593, 38,000 lbs
3/28/82

PARIS/SAN FRANCISCO (USA) §
Richard A. Ionata, Captain — 861.68 kmh / 535.42 mph
Cari R. Davidson, 1st Officer
Scott E. Herman, 1st Officer
United Airlines
Boeing 767-300
2 P&W PW4060, 60,000 lbs
10 hrs 24 min
9/12/95

PARIS/SANTA MARIA (FRANCE)
Cdt. Jean-Paul Le Moel — 1,588.17 kmh / 986.89 mph
Air France
BAe/Aérospatiale Concorde
4 RR Olympus 593, 38,000 lbs
3/19/82

PARIS/WASHINGTON (FRANCE)
Capt. Pierre Chanoine — 1,725.00 kmh / 1,071.86 mph
Air France
BAe/Aérospatiale Concorde
4 Olympus 593 Mark 610, 38,000 lbs
3 hrs 35 min 15 sec
8/18/78

PERTH/MELBOURNE (AUSTRALIA)
Noel Patrick Mann, Pilot — 1,172.96 kmh / 728.88 mph
Boeing 727-77C
10/1/88

PHILADELPHIA/CANCUN (USA)
Ward R. Anderson, Captain — 670.13 kmh / 416.40 mph
Thomas T. Reynolds, 1st Officer
US Airways
Boeing 767-200ER
2 GE CF6-80C2, 52,500 lbs
3 hrs 32 min
3/27/03

PHILADELPHIA/CHICAGO*
Periklis Dervas, Captain — 494.67 mph
Larry E. Steenstry, 1st Officer
United Airlines
Boeing 757
2 P&W PW2037, 38,200 lbs
1 hr 22 min
6/23/95

PHILADELPHIA/DENVER (USA)
Capt. Larry Liguori — 773.08 kmh / 480.36 mph
United Airlines
Airbus A320-200
2 IAE V2527-A5, 26,500 lbs
3 hrs 14 min
7/28/95

PHILADELPHIA/FRANKFURT (USA)
Capt. Mitchell Cowan — 925.29 kmh / 574.95 mph
US Airways
Boeing 767-200
2 GE CF6-80C2, 52,500 lbs
6 hrs 51 min
2/20/00

PHILADELPHIA/LONDON (USA)
Capt. Joseph W. Graham — 886.66 kmh / 550.95 mph
US Airways
Airbus A330-300
2 P&W PW4168A, 68,600 lbs
6 hrs 22 min
9/21/01

PHILADELPHIA/LOS ANGELES*
Capt. William R. Hunt — 573.19 mph
American Airlines
Boeing 720B
4 P&W JT3D
Philadelphia Intl/Los Angeles Intl
Distance: 2,402 mi
4 hrs 11 min 30 sec
2/6/63

PHILADELPHIA/MADRID (USA)
Capt. Kenneth W. Pettigrew — 940.72 kmh / 584.53 mph
US Airways
Boeing 767-200
2 GE CF6-80C2, 55,000 lbs
6 hrs 17 min
1/31/02

PHILADELPHIA/MANCHESTER (USA)
Capt. Joseph W. Graham — 926.34 kmh / 575.60 mph
US Airways
Airbus A330-300
2 P&W PW4168A, 68,600 lbs
5 hrs 57 min
3/3/03

PHILADELPHIA/PARIS (USA)
Louis D. Pisane, Captain — 1,008.73 kmh / 626.79 mph
Thomas F. Delantonas, 1st Officer
Mark S. Nicholas, Relief Officer
USAir
Boeing 767-200
2 GE CF6-80C2, 52,000 lbs
5 hrs 56 min
2/6/96

PHILADELPHIA/ROME (USA) §
Paris Michaels, Captain — 990.57 kmh / 615.51 mph
Joseph W. Graham, 1st Officer
Carlisle C. Owen, 1st Officer
US Airways
Airbus A330-300
2 P&W PW4168A, 68,600 lbs
7 hrs 5 min
11/29/03

PHILADELPHIA/ST. LOUIS (USA)
Richard Norton, Captain — 802.08 kmh / 498.41 mph
Shari Thovson, 1st Officer
Cynthia Allen, 2nd Officer
Trans World Airlines
Boeing 727-200, N53213
1 hr 37 min 43 sec
9/23/89

PHILADELPHIA/SAN JUAN (USA)
Capt. Joseph W. Graham — 881.90 kmh / 547.99 mph
US Airways
Airbus A330-300
2 P&W PW4168A, 68,600 lbs
2 hrs 53 min
11/3/01

PHILADLEPHIA/TORONTO (USA)
Patrick M. Broderick, Pilot — 654.12 kmh / 406.47 mph
G. Curt Ayers, 1st Officer
Binka Bone, 2nd Officer
Eastern Airlines Boeing 727-225A, N888
3 P&W JT8D
51 min 13 sec
2/25/88

PHOENIX/ALBUQUERQUE*
Capt. David H. Koch — 561.32 mph
Trans World Airlines
Boeing 727-131
3 P&W JT8D
Phoenix Sky Harbor Intl/Albuquerque Intl
Distance: 329 mi
35 min 10 sec
3/5/75

PHOENIX/AMARILLO*
Barry Schiff, Captain 586.27 mph
Lyle T. Shelton, 1st Officer
Trans World Airlines
Boeing 727-231A
3 P&W JT8D
Phoenix Sky Harbor Intl/Amarillo Air Terminal
Distance: 598 mi
1 hr 1 min 18 sec
12/30/76

PHOENIX/CHICAGO*
Capt. W. H. Sumrall 653.22 mph
American Airlines
Boeing 707-123
4 P&W JT3D
Phoenix Sky Harbor Intl/O'Hare Intl
Distance: 1,442 mi
2 hrs 12 min 27 sec
1/1/66

PHOENIX/CLEVELAND (USA)
Robert F. Olszewski 1,059.07 kmh
USAir 658.07 mph
Boeing 727-100
3 P&W JT8D-7
2 hrs 38 min 16 sec
1/20/82

PHOENIX/DENVER*
Capt. Gary Thompson 552.9 mph
Western Airlines
Boeing 727
3 P&W JT8D
Phoenix Sky Harbor Intl/Denver Stapleton
Distance: 589 mi
1 hr 3 min 55 sec
8/30/74

PHOENIX/LOS ANGELES*
Capt. David H. Koch 486.72 mph
Mark McGargill, 1st Officer
Steve Holmes, Flt Eng
Trans World Airlines
Lockheed L-1011-1
3 Rolls Royce RB-211-22B
Phoenix Sky Harbor Intl/Los Angeles Intl
Distance: 380 mi
46 min 50 sec
1/2/76

PHOENIX/NEW YORK*
Capt. C. Walling 689.27 mph
American Airlines
Boeing 707-720B
4 P&W JT3D
Phoenix Sky Harbor Intl/JFK Intl
Distance: 2,154 mi
3 hrs 7 min 30 sec
12/17/64

PITTSBURGH/BALTIMORE*
Capt. Robert F. Olszewski 497.40 mph
Allegheny Airlines
McDonnell Douglas DC-9-31
2 P&W JT8D-7A
Greater Pittsburgh Intl/Baltimore-Washington Intl
25 min 18 sec
2/25/77

PITTSBURGH/BURLINGTON (USA)
Capt. Robert Olszewski 908.64 kmh
John Sabel, 1st Officer 564.63 mph
USAir, DC-9-31
2 P&W JT8D-7B, 14,000 lbs
48 min 13 sec
1/21/88

PITTSBURGH/FRANKFURT (USA) §
Paul L. Morell, Captain 980.00 kmh
James L. Hayhurst, 1st Officer 608.94 mph
Robert E. Tingey, 1st Officer
US Airways
Boeing 767-200ER
2 GE CF6-80C2, 52,500 lbs
6 hrs 44 min 50 sec
2/21/99

PITTSBURGH/LOS ANGELES*
Capt. Charles M. Davis 551.46 mph
Trans World Airlines
Lockheed L-1011 R
3 RR RB-211-22B
Pittsburgh Intl/Los Angeles Intl
Distance: 2,136 mi
3 hrs 52 min 24 sec
7/3/75

PITTSBURGH/PARIS (USA)
Paul L. Morell, Captain 933.23 kmh
John Barbas, 1st Officer 579.88 mph
James L. Hayhurst, 1st Officer
US Airways
Boeing 767-200ER
2 GE CF6-80C2, 52,500 lbs
6 hrs 43 min
1/19/99

PITTSBURGH/PHOENIX (USA)
Richard M. Nelson, Captain 713.34 kmh
William M. Tippins, 1st Officer 443.25 mph
USAir
Boeing 737-400
2 CFM56, 22,000 lbs
4 hrs 5 min
7/23/96

PITTSBURGH/SAN FRANCISCO (USA)
Capt. Don Ryhnalds 873.00 kmh
Gary Urban, 1st Officer 542.48 mph
Howell Finch, Flt Engineer
USAir
Boeing 727-200, N774AI
3 P&W JT8D-17, 16,000 lbs
4 hrs 9 min
8/31/87

PORTLAND, OR/CHICAGO (USA)
Larry E. Steenstry, Captain 946.27 kmh
Mo I. Morsy, 1st Officer 587.98 mph
United Airlines Boeing 757
2 P&W PW2037, 38,200 lbs
2 hrs 57 min
2/18/95

PORTLAND, OR/DENVER (USA)
Leslie Provow, Captain 884.91 kmh
Larry E. Steenstry, 1st Officer 549.85 mph
United Airlines
Boeing 757-200
2 P&W PW2037, 38,200 lbs
1 hr 48 min
10/18/96

PORTLAND, OR/LOUISVILLE (USA)
Glenn W. Stickel, Captain 1,004.55 kmh
Gary M. Cantara, 1st Officer 624.20 mph
United Parcel Service
Boeing 747-100
4 P&W JT9D-7A, 46,100 lbs
3 hrs 7 min
1/9/96

PUEBLA/DALLAS (USA)
Capt. Daniel J. Clisham — 800.01 kmh
American Airlines — 497.10 mph
McDonnell Douglas MD-83
2 P&W JT8D-219, 21,000 lbs
1 hr 55 min
12/21/97

RALEIGH/BOSTON (USA)
Capt. Dennis D. Lyons — 703.19 kmh
American Airlines — 436.94 mph
Boeing 757-200
2 RR 535E4, 40,100 lbs
1 hr 24 min
4/8/93

RALEIGH/LONDON (USA) §
Stephen W. Whelan, Jr. — 948.46 kmh
Scott W. Miller — 589.34 mph
John P. Kane
American Airlines
Boeing 767-200ER
2 GE CF6-80A, 48,000 lbs
6 hrs 35 min
2/13/95

RALEIGH/PITTSBURGH*
Capt. Robert Olszewski — 470.16 mph
USAir
McDonnell Douglas DC-9-31
2 P&W JT8D-7A, 14,000 lbs
Raleigh-Durham/Greater Pittsburgh Intl
Distance: 329 mi
41 min 59 sec
12/29/80

RENO/DENVER (USA)

Capt. Derek G. Mezo — 890.64 kmh
Frontier Airlines — 553.42 mph
Airbus A318-100
2 CFM CFM56, 22,000 lbs
1 hr 27 min
11/24/03

RIO DE JANEIRO/DAKAR (FRANCE)
Cdt. M. Tardieu — 1,794.32 kmh
Air France — 1,114.99 mph
BAe/Aérospatiale Concorde
4 RR Olympus 593, 38,000 lbs
3/28-29/82

RIO DE JANERIO/MIAMI (USA)
Keith J. Mackey — 932.76 kmh
PanAm — 579.62 mph
Boeing 747-100
7 hrs 12 min 30 sec
6/27/89

RIO DE JANEIRO/PARIS (FRANCE)
Cdt. M. Tardieu — 1,455.98 kmh
Air France — 904.75 mph
BAe/Aérospatiale Concorde
4 RR Olympus 593, 38,000 lbs
3/28-29/82

Paris Michaels set speed records from Philadelphia to Rome and back in a US Airways Airbus A330.

ROCHESTER/BOSTON*
Capt. Robert Olszewski
USAir
Boeing 727-200A
3 P&W JT8D-17A, 16,000 lbs
Monroe County/Logan Intl
Distance: 342.14 mi
554.78 mph

36 minutes 57 seconds
1/21/82

ROCHESTER/CHICAGO (USA)
Capt. Steven A. Shatto
United Airlines
Boeing 737-200
2 P&W JT8D, 14,000 lbs
1 hr 18 min
1/7/96
651.71 kmh
404.95 mph

ROME/CHICAGO (USA) §
Capt. John G. Maxwell, Jr.
American Airlines
Boeing 767-300ER
2 GE CF6-80C2, 61,500 lbs
9 hrs 20 min
9/18/03
829.45 kmh
515.40 mph

ROME/NEW YORK (USA) §
Capt. Richard B. Barnes
Delta Airlines
Boeing 767-300ER
2 P&W PW4060, 60,000 lbs
8 hrs 51 min
7/25/97
775.82 kmh
482.07 mph

ROME/PHILADELPHIA (USA) §
Joseph W. Graham, Captain
Paris Michaels, 1st Officer
US Airways
Airbus A330-300
2 P&W PW4168A, 68,600 lbs
8 hrs 33 min
11/10/03
820.65 kmh
509.93 mph

SACRAMENTO/CHICAGO (USA)
Kenneth L. Brown, Captain
Larry E. Steenstry, 1st Officer
United Airlines
Boeing 757
2 P&W PW2037, 38,250 lbs
3 hrs 20 min
8/10/93
857.91 kmh
533.08 mph

ST. LOUIS/CALGARY *
Edward L. Taylor, Captain
Michael W. Wyss, 1st Officer
Ozark Airlines
McDonnell Douglas DC-9-41
2 P&W JT8D, 16,000 lbs
5/17/85
490.58 mph

ST. LOUIS/CHICAGO* *Group II*
Capt. Ray E. Slaughter
Ozark Airlines
Fairchild-Hiller FH-227B
2 RR Dart 532-7
Lambert Intl/O'Hare Intl
Distance: 258 mi
1 hr 20 min 10 sec
1/12/72
296.74 mph

ST. LOUIS/COLORADO SPRINGS*
Capt. Lyle T. Shelton
Warren Nuffer, 1st Officer
James D. Speer, Flt Engineer
Trans World Airlines
Boeing 727
3 P&W JTD8
St Louis Intl/Colo Spgs Muni
1 hr 48 min
12/23/86
429.51 mph

ST. LOUIS/DALLAS (USA)
Lyle T. Shelton, Captain
Rodney Zapf, 1st Officer
Trans World Airlines
DC-9-31, N987Z
2 P&W JT8D, 14,500 lbs
1 hr 8 min 50 sec
6/3/88
772.56 kmh
480.07 mph

ST. LOUIS/HONOLULU (USA) §
Vance H. Weir, Captain
Gid Miller, 1st Officer
Trans World Airlines
Boeing 747-100
4 P&W JT9D, 46,150 lbs
7 hrs 55 min
12/19/93
838.21 kmh
520.84 mph

ST. LOUIS/KANSAS CITY *
Larry R. Warren, Captain
James J. Lawlor, Jr., 1st Officer
TWA
McDonnell Douglas MD-82
2 P&W JT8D, 20,850 lbs
36 min
12/2/01
394.60 mph

ST. LOUIS/LONDON (USA) §
Capt. Henry J. Stoecker
American Airlines
Boeing 767-300ER
2 P&W PW4060, 60,000 lbs
7 hrs 21 min
6/24/02
921.69 kmh
572.71 mph

ST. LOUIS/MEXICO CITY (USA)
Capt. Eugene W. Scholl
Trans World Airlines
McDonnell Douglas MD-80
2 P&W JT8D, 20,850 lbs
2 hrs 51 min
7/9/99
808.92 kmh
502.63 mph

ST. LOUIS/NEW YORK*
Capt. M. W. Baker
American Airlines
Boeing 707-123B
4 P&W JT3D-1
Lambert Intl/Idlewild Intl
Distance: 893 mi
1 hr 20 min 10 sec
12/8/61
667.58 mph

ST. LOUIS/PHOENIX (USA)
Capt. Don G. Loeschner
Trans World Airlines
Boeing 727
3 P&W JT8D-9A, 14,500 lbs
2 hrs 36 min
1/28/95
779.73 kmh
484.50 mph

ST. LOUIS/SALT LAKE CITY*
Capt. Lyle T. Shelton
Don Pittman, 1st Officer
Ed Streight, Flt Engineer
Trans World Airlines
Boeing 727, N64322
3 P&W JTD8
2 hrs 35 min
12/26/86
446.96 mph

ST. LOUIS/WICHITA (USA)
Richard Norton, Captain
Shari Thovson, 1st Officer
Cynthia Allen, 2nd Officer
Trans World Airlines
Boeing 727-200, N53213
47 min 48 sec
9/23/89
791.28 kmh
491.70 mph

SALT LAKE CITY/CHICAGO (USA)
Capt. Vaughn B. Cordle — 812.99 kmh, 505.17 mph
United Airlines
Boeing 737-300
2 CFM56, 20,000 lbs
2 hrs 28 min
3/8/93

SALT LAKE CITY/NEW ORLEANS (USA)
George T. Kyrazis, Captain — 867.82 kmh, 539.24 mph
John H. Baker, 1st Officer
Trevor R. John, 2nd Officer
Delta Air Lines
Boeing 727-200
3 P&W JT8D-15, 15,500 lbs
2 hrs 38 min 42 sec
7/27/95

SALT LAKE CITY/NEW YORK (USA)
Forrest W. Nelson, Captain — 943.70 kmh, 586.38 mph
Frederick L. Zimmerman, 1st Officer
Delta Air Lines
Boeing 767-200
2 GE CF6-80, 50,000 lbs
3 hrs 23 min
11/22/94

SAN ANTONIO/DALLAS*
Capt. James O. Gross — 411.98 mph
American Airlines
McDonnell Douglas MD-82
2 P&W JT8D, 21,000 lbs
36 min
12/11/97

SAN ANTONIO/MEXICO CITY (USA)
Capt. E. Basham — 892.09 kmh, 554.31 mph
Boeing 720B
4 P&W JT3D-1
1 hr 15 min 42.6 sec
9/6/62

SAN DIEGO/ATLANTA*
Capt. J. H. Longino — 555.75 mph
Delta Air Lines
Convair 880
4 GE CJ-805-5
Lindbergh Field/Atlanta
Distance: 1,889 mi
3 hrs 23 min 59 sec
5/11/60

SAN DIEGO/CHICAGO*
Robert A. Holt — 639.5 mph
American Airlines
Convair 990
GE CJ-805-23
Lindbergh Field/Chicago O'Hare Intl
Distance: 1,724 mi
2 hrs 41 min 45 sec
4/24/63

SAN DIEGO/CINCINNATI (USA)
Douglas R. Harper, Captain — 1,021.35 kmh, 634.64 mph
Wayne D. Erickson, First Officer
Delta Air Lines
Boeing 757-200
2 P&W PW2037, 38,200 lbs
2 hrs 56 min
11/3/02

SAN DIEGO/DALLAS (USA)
Capt. James O. Gross — 805.68 kmh, 500.63 mph
American Airlines
McDonnell Douglas MD-82
2 P&W JT8D, 21,000 lbs
2 hrs 20 min
12/3/97

SAN DIEGO/DENVER (USA)
Capt. Vaughn B. Cordle — 743.66 kmh, 462.09 mph
United Airlines
Boeing 737-300
2 CFM56, 20,000 lbs
1 hr 49 min
8/26/92

SAN DIEGO/MIAMI*
Capt. T. P. Ball — 641.77 mph
Delta Air Lines
Convair, 880
4 GE CJ-805
Lindbergh Field/Miami Intl
Distance: 2,266 mi
3 hrs 31 min 54 sec
2/10/60

SAN DIEGO/NEW YORK (USA)
Capt. Dennis D. Lyons — 832.79 kmh, 517.47 mph
American Airlines
Boeing 767-200ER
2 GE CF6-80A, 48,000 lbs
4 hrs 43 min
8/1/02

SAN DIEGO/ST. LOUIS (USA)
Capt. Eugene W. Scholl — 843.38 kmh, 524.05 mph
Trans World Airlines
Boeing MD-83
2 P&W JT8D, 20,850 lbs
2 hrs 58 min
7/17/00

SAN FRANCISCO/ANCHORAGE (USA)
Capt. George W. Murphy, Jr. — 922.47 kmh, 573.19 mph
FedEx
McDonnell Douglas MD-11F
3 GE CF6-80C2, 61,500 lbs
3 hrs 31 min
7/9/99

SAN FRANCISCO/ATLANTA*
Capt. G. M. Gradstreet — 578.95 mph
Delta Air Lines
Douglas DC-8
4 P&W JT3D-1
San Francisco Intl/Atlanta
Distance: 2,139 mi
3 hrs 40 min 38 sec
4/27/64

SAN FRANCISCO/BALTIMORE (USA)
Capt. William R. Wallace — 893.06 kmh, 554.92 mph
United Airlines
Airbus A-320-232
2 IAE V2527-A5, 26,500 lbs
4 hrs 25 min
9/18/96

SAN FRANCISCO/CHICAGO*
Capt. George F. Shores — 669.13 mph
American Airlines
Convair 990
4 GE CJ-805-23
San Francisco Intl/O'Hare Intl
2 hrs 45 min 37 sec
11/25/65

SAN FRANCISCO/CINCINNATI (USA)
Barry L. Watkins, Captain — 952.33 kmh, 591.75 mph
Robert E. Koski, 1st Officer
Delta Air Lines
Boeing 757
2 P&W PW2037, 38,200 lbs
3 hrs 26 min
10/29/94

SAN FRANCISCO/COLUMBUS (USA)
Fred Ferguson, Captain — 937.44 kmh
Rick Kingston, 1st Officer — 582.53 mph
John C. Beatty, 2nd Officer
Flying Tigers Line
Boeing 727-100F
3 P&W JT8D-7B, 14,000 lbs
3 hrs 38 min 2 sec
10/31/86

SAN FRANCISCO/DALLAS*
Capt. William R. Swain — 619.56 mph
American Airlines
Boeing 707-123
4 P&W JT3D
San Francisco Intl/Love Field
2 hrs 23 min 2 sec
4/10/63

SAN FRANCISCO/DENVER*
David H. Koch, Captain — 604.44 mph
Larry C. Aschcraft, 1st Officer
Ed McKee, Flt Engineer
Trans World Airlines
Boeing 727
3 P&W JT8D
San Francisco Intl/Denver
Distance: 955 mi
1 hr 34 min 46 sec
1/6/76

SAN FRANCISCO/DETROIT*
Capt. H. P. Luna — 555.51 mph
American Airlines
Boeing 707-123
4 P&W JT3D-6
San Francisco Intl/Detroit Metro, Wayne County
Distance: 2,075 mi
3 hrs 44 min 10 sec
9/17/59

SAN FRANCISCO/GREENSBORO (USA)
James C. Sifford — 918.36 kmh
Piedmont Airlines — 570.67 mph
Boeing 727
3 P&W JT8D, 14,500 lbs
4 hrs 5 min
12/27/83

SAN FRANCISCO/HONG KONG (USA) §
William R. Harrison, Captain — 883.19 kmh
David S. Valentine, 1st Officer — 548.79 mph
Sidney Maxey, 1st Officer
United Airlines
Boeing 747-400
4 P&W PW4056, 56,750 lbs
12 hrs 36 min
7/30/02

SAN FRANCISCO/HONOLULU (USA)
Capt. Jamel Z. Abouremeleh — 883.26 kmh
United Airlines — 548.83 mph
Boeing 757-200
2 P&W PW2040, 41,700 lbs
4 hrs 22 min
3/27/01

SAN FRANCISCO/KONA (USA)
Edward P. Akin, Captain — 862.32 kmh
Gaven C. Dunn, 1st Officer — 535.82 mph
Michael P. Anthony, 2nd Officer
United Airlines
McDonnell Douglas DC10-10
3 GE CF6, 40,000
4 hrs 25 min
2/1/94

SAN FRANCISCO/LAS VEGAS (USA)
Wallis E. Hubbell, Captain — 831.30 kmh
Kenneth Banks, 1st Officer — 516.54 mph
United Airlines
Boeing 737-300
2 CFM International CFM56, 20,000 lbs
48 min
2/2/98

SAN FRANCISCO/LONDON (USA) §
Capt. F. van Cortlandt de Peyster — 992.18 kmh
United Airlines — 616.51 mph
Boeing 747-400
4 P&W PW4056, 56,750 lbs
8 hrs 41 min
9/6/01

SAN FRANCISCO/LOS ANGELES (USA)
Capt. Vaughn B. Cordle — 775.96 kmh
United Airlines — 482.16 mph
Boeing 737-300
2 CFM56, 20,000 lbs
42 min
3/6/93

SAN FRANCISCO/MAUI (USA)
Edward P. Akin, Captain — 867.79 kmh
Gaven C. Dunn, 1st Officer — 539.22 mph
Peter B. Letch, 2nd Officer
United Airlines
McDonnell Douglas DC10-10
3 GE CF6, 40,000
4 hrs 20 min
3/7/94

SAN FRANCISCO/MEXICO CITY (USA)
Capt. Larry Liguori — 857.12 kmh
United Airlines — 532.59 mph
Airbus A320-232
2 IAE V2527-A5, 26,500 lbs
3 hrs 32 min
12/14/95

SAN FRANCISCO/MIAMI (USA)
Capt. Terry A. Lewis — 1,009.00 kmh
United Airlines — 626.96 mph
Boeing 777-200B
2 P&W PW4090, 90,000 lbs
4 hrs 7 min
3/5/98

SAN FRANCISCO/NEW ORLEANS*
Capt. Tom Bridges — 576.72 mph
Delta Air Lines
Convair 880
4 GE CJ-805
San Francisco Intl/Moisant
Distance: 1,913 mi
3 hrs 18 min 56 sec
10/5/61

SAN FRANCISCO/NEW YORK (USA)
Capt. T. E. Johnson — 1,059.00 kmh
Boeing 707B-123 — 658.00 mph
4 P&W JT3D-1
3 hrs 55 min 50 sec
12/8/61

SAN FRANCISCO/NEWARK (USA)
Michael Piampiano, Captain — 928.08 kmh
Jim Kennedy, 1st Officer — 576.71 mph
Continental Airlines
Airbus A 30 B-4
2 GE CF6-50-C2, 52,500 lbs
4 hrs 26 min 30 sec
2/19/89

SAN FRANCISCO/ORLANDO (USA)
Cathy Caseman Berdahl, Captain — 978.05 kmh
Larry E. Steenstry, 1st Officer — 607.73 mph
United Airlines
Boeing 757-200
2 P&W PW2037, 38,200 lbs
4 hrs 1 min
12/7/95

SAN FRANCISCO/OSAKA (USA) §
William R. Harrison, Captain — 862.02 kmh
Clark Hallam, 1st Officer — 535.64 mph
United Airlines
Boeing 747-400
4 P&W PW4056, 56,750 lbs
10 hrs 5 min
7/5/02

SAN FRANCISCO/PARIS (USA) §
Richard A. Ionata, Captain — 953.35 kmh
Derek S. Lucas, 1st Officer — 592.38 mph
United Airlines
Boeing 767-300
2 P&W PW4060, 60,000 lbs
9 hrs 24 min
8/27/95

SAN FRANCISCO/SEOUL (USA) §
David L. Link, Captain — 788.64 kmh
United Airlines — 490.04 mph
Boeing 747-400
4 P&W PW4056, 56,750 lbs
11 hrs 29 min
10/2/96

SAN FRANCISCO/SHANGHAI § *
Capt. William R. Harrison — 562.26 mph
United Airlines
Boeing 747-400
4 P&W PW4056, 56,750 lbs
10 hrs 55 min
8/11/01

SAN FRANCISCO/SYDNEY (USA) §
Graham H. Norris, Captain — 918.03 kmh
William R. Harrison, 1st Officer — 570.43 mph
Marvin L. Sparks, 1st Officer
C. Dave Tuck, 1st Officer
United Airlines
Boeing 747-400
4 P&W PW4056, 56,750 lbs
13 hrs 1 min
3/2/00

SAN FRANCISCO/TAIPEI (USA) §
Capt. David L. Link — 881.88 kmh
United Airlines — 547.98 mph
Boeing 747-400
4 P&W PW4056, 56,750 lbs
11 hrs 47 min
8/8/96

SAN FRANCISCO/TOKYO (USA) §
William R. Harrison, Captain — 845.39 kmh
Robert K. Spies, 1st Officer — 525.30 mph
United Airlines
Boeing 747-400
4 P&W PW4056, 56,750 lbs
9 hrs 44 min
3/6/01

SAN FRANCISCO/WASHINGTON* *Group I*
Capt. S. Flower — 382.34 mph
Pan American World Airways
Boeing 377 Stratocruiser
4 P&W Wasp Major R-4360
San Francisco/Washington Natl
Distance: 2,437 mi
6 hrs 22 min 25.4 sec
5/3/62

SAN FRANCISCO/WASHINGTON (USA) *Group III*
William J. Grossman, Captain — 931.96 kmh
Larry E. Steenstry, 1st Officer — 579.09 mph
United Airlines
Boeing 757
2 P&W PW2037, 38,200 lbs
4 hrs 10 min
6/14/94

SAN JOSE/SANTA ANA (USA)
Capt. Dennis D. Lyons — 687.94 kmh
American Airlines — 427.47 mph
Boeing 757-200
2 RR 535E4, 40,100 lbs
48 min
4/6/93

SAN JOSE/TOKYO (USA) §
James O. Gross, Captain — 820.86 kmh
Steven B. Jansen, 1st Officer — 510.06 mph
John E. Chambers, 1st Officer
American Airlines
Boeing 777
2 RR Trent 890, 90,000 lbs
10 hrs 5 min
5/22/99

SAN JOSE, COSTA RICA/MEXICO CITY (USA)
Capt. Vaughn B. Cordle — 785.08 kmh
United Airlines — 487.83 mph
Boeing 737-300
2 CFM56, 20,000 lbs
2 hrs 26 min
10/4/92

SAN JOSE DEL CABO/CHICAGO (USA)
Capt. Daniel J. Clisham — 851.40 kmh
American Airlines — 529.04 mph
McDonnell Douglas MD-83
2 P&W JT8D, 21,000 lbs
3 hrs 25 min
11/3/00

SAN JUAN/CHICAGO (USA)
Andrew E. Allen, Captain — 788.39 kmh
Larry E. Steenstry, 1st Officer — 489.88 mph
United Airlines
Boeing 757
2 P&W PW2037, 38,200 lbs
4 hrs 14 min
10/22/94

SAN JUAN/NEW ORLEANS*
Capt. H. E. Croft — 507.34 mph
Delta Air Lines
Convair 880
4 GE CJ-805
Puerto Rico Intl/Moisant Intl
Distance: 1,721 mi
3 hrs 23 min 31 sec
3/3/49

SAN JUAN/NEW YORK (USA)
William W. Critcher, Captain — 781.17 kmh
Michael R. Dimento, 1st Officer — 485.39 mph
United Airlines
Boeing 757
2 P&W PW2037, 38,200 lbs
3 hrs 18 min
9/26/99

SAN JUAN/PHILADELPHIA (USA)
Capt. Joseph W. Graham — 782.40 kmh
US Airways — 486.16 mph
Airbus A330-300
2 P&W PW4168A, 68,600 lbs
3 hrs 15 min
11/3/01

SAN JUAN/ST. LOUIS (USA)
Capt. Henry J. Stoecker
American Airlines
Boeing 757-200
2 P&W PW2037, 38,200 lbs
4 hrs 1 min
7/6/02
811.41 kmh
504.19 mph

SANTA ANA/DALLAS (USA)
Capt. Dennis D. Lyons
American Airlines
Boeing 757-200
2 RR 535E4, 40,100 lbs
2 hrs 14 min
4/12/93
866.19 kmh
538.22 mph

SANTA MARIA/CARACAS (FRANCE)
Cdt. Jean-Paul Le Moel
BAe/Aérospatiale Concorde
Air France
4 RR Olympus, 38,000 lbs
3/19/82
1,732.56 kmh
1,076.61 mph

SANTA MARIA/PARIS (FRANCE)
Cdt. Jean-Paul Le Moel
BAe/Aérospatiale Concorde
Air France
4 RR Olympus, 38,000 lbs
3/20/82
1,392.51 kmh
865.30 mph

SANTIAGO/BUENOS AIRES (USA)
Keith J. Mackey, Captain
Pam American Airways
Boeing 747, N4711U
4 P&W JT9D
1 hr 19 min 5 sec
7/9/88
949.11 kmh
589.78 mph

SANTIAGO/MIAMI (USA) §
Richard W. Wiley, Captain
Susan W. Harrison, 1st Officer
Enrique L. Valdes, 1st Officer
United Airlines
Boeing 777-200-IGW
2 P&W PW4090, 90,000 lbs
7 hrs 22 min
1/8/98
903.65 kmh
561.50 mph

SAO PAULO/CHICAGO (USA) §
Capt. David H. Friend
United Airlines
Boeing 777
2 P&W PW4090, 90,000 lbs
9 hrs 46 min
1/24/00
862.94 kmh
536.20 mph

SAO PAULO/MIAMI (USA) §
Capt. Robert J. Swain
United Airlines
Boeing 777-200
2 P&W PW4077, 77,000 lbs
7 hrs 19 min
10/18/97
898.43 kmh
558.25 mph

SAO PAULO/NEW YORK (USA) §
William W. Critcher, Captain
Shawn Werchan, 1st Officer
Gregory S. McDowell, 1st Officer
United Airlines
Boeing 767-300ER
2 P&W PW4060, 60,000 lbs
8 hrs 54 min
9/5/99
861.08 kmh
535.05 mph

SEATTLE/CHICAGO (USA)
Capt. Daniel J. Clisham
American Airlines
McDonnell Douglas MD-82
2 P&W JT8D, 20,000 lbs
2 hrs 54 min
1/14/00
952.32 kmh
591.74 mph

SEATTLE/DENVER*
Roland Schmidt, Captain
Wallis Hubbell, 1st Officer
United Airlines
Boeing 757-200
2 P&W PW2037, 38,200 lbs
1 hr 42 min
4/27/96
601.41 mph

SEATTLE/HOUSTON (USA)
Capt. Wendell L. Moeller
Continental Airlines
Boeing 737-800
2 CFM CFM56, 24,200 lbs
3 hrs 46 min
9/28/02
800.14 kmh
497.18 mph

SEATTLE/LONDON (UK)
Capt. G. A. C. Gray
British Airways
Boeing 757-236
2 RR RB211-535C, 37,400 lbs
5/15-16/86
844.96 kmh
525.06 mph

SEATTLE/LOS ANGELES (USA)
Stephen Lutz, Pilot
Charles King, Copilot
Western Airlines
Boeing 737-200
2 P&W JT8D-9, 14,000 lbs
1 hr 29 min 50 sec
12/13/84
1,028.12 kmh
638.84 mph

SEATTLE/MANILA (USA) §
Maurice de Vas
Boeing B747-312
4 P&W JT9D 54,750 lbs
12 hrs 33 min 40 sec
6/28-29/84
851.40 kmh
529.06 mph

SEATTLE/NEW YORK (USA)
Donald J. Taylor
Jared C. Brandt
United Airlines
Boeing 767-200,
2 P&W JT9D-7R4, 47,000 lbs
3 hrs 44 min 28 sec
5/12/84
1,033.56 kmh
642.24 mph

SEATTLE/SINGAPORE (USA) §
Capt. Maurice de Vas
Boeing B747-312
4 P&W JT9D, 54,750 lbs
15 hrs 12 min
6/28-29/84
853.00 kmh
530.05 mph

SEATTLE/TOKYO (USA) §
James O. Gross, Captain
Charles T. Coleman, 1st Officer
American Airlines
McDonnell Douglas MD-11
3 GE CF6-80C2, 61,500 lbs
9 hrs 7 min
8/13/98
839.79 kmh
521.82 mph

SEATTLE/WASHINGTON (USA)
Robert L. Hain III, Captain
Larry E. Steenstry, 1st Officer
United Airlines
Boeing 757
2 P&W PW2037, 38,200 lbs
3 hrs 54 min
5/9/94
949.09 kmh
589.73 mph

SEATTLE/WILMINGTON*
Capt. Lyle W. McNames
Piedmont Airlines
Boeing 737-200
2 P&W JT8D-7
Boeing Field/New Hanover County
Distance: 2,457 mi
4 hrs 47 min 44 sec
5/29/68
512.31 mph

SEOUL/ANCHORAGE (USA)
Glenn W. Stickel, Captain
Robert L. Peshkin, 1st Officer
Richard L. Levangie, 2nd Officer
United Parcel Service
Boeing 747-200B
4 P&W JT9D, 53,000 lbs
6 hrs 51 min
12/13/01
890.31 kmh
553.21 mph

SEOUL/CHICAGO* §
William T. Woolfolk, Captain
Paul V. Davis, II, Captain
United Airlines
Boeing 747-400
4 P&W PW4056, 56,750 lbs
12 hrs 6 min
6/24/94
539.09 mph

SEOUL/MANILA (USA)
Capt. David F. Specht
United Airlines
Boeing 747-400
4 P&W PW4056, 56,750 lbs
3 hrs 26 min
11/26/97
764.29 kmh
474.90 mph

SEOUL/OSAKA (USA)
Capt. David F. Specht
United Airlines
Boeing 747-400
4 P&W PW4056, 56,750 lbs
1 hr 13 min
11/28/97
684.98 kmh
425.63 mph

SEOUL/SAN FRANCISCO (USA) §
Capt. David L. Link
Thomas G. Parker, 1st Officer
Steve M. Zink, 1st Officer
United Airlines
Boeing 747-400
4 P&W PW4056, 56,750 lbs
9 hrs 4 min
4/3/96
998.72 kmh
620.57 mph

SEOUL/TOKYO (USA)
Capt. George W. Murphy, Jr.
FedEx
McDonnell Douglas MD-11F
3 GE CF6-80C2, 61,500 lbs
1 hr 29 min
12/8/01
847.70 kmh
526.74 mph

SHANGHAI/BEIJING (USA)
Capt. N. B. Godlove
United Airlines
Boeing 747SP
4 P&W JT9D-7A, 46,500 lbs
1 hr 21 min
8/31/92
796.20 kmh
494.73 mph

SHANGHAI/SAN FRANCISCO (USA) §
Capt. William A. Dias
United Airlines
Boeing 747-400
4 P&W PW4056, 56,750 lbs
9 hrs 35 min
1/13/01
1,033.55 kmh
642.22 mph

SHANGHAI/TOKYO (USA)
Irving S. Soble, Captain
Paul C. Mattson, 1st Officer
Kevin D. Garcia, 2nd Officer
United Airlines
Boeing 747-100
4 P&W JT9D-7A, 46,250 lbs
1 hr 59 min
2/4/95
925.43 kmh
575.03 mph

SINGAPORE/HONG KONG (USA)
William G. Winquist, Captain
William R. Harrison, 1st Officer
United Airlines
Boeing 747-400
4 P&W PW4056, 56,750 lbs
3 hrs 10 min
2/5/01
809.95 kmh
503.22 mph

SINGAPORE/OSAKA (USA)
Capt. Leiland M. Duke, Jr.
FedEx
McDonnell Douglas MD-11F
3 GE CF6-80C2, 61,500 lbs
5 hrs 25 min
9/29/98
904.96 kmh
562.31 mph

SINGAPORE/TOKYO (USA)
Capt. N. B. Godlove
United Airlines
Boeing 747-100
4 P&W JT9D-7A, 47,000 lbs
5 hrs 29 min
12/22/92
976.99 kmh
607.07 mph

SPOKANE/CHICAGO (USA)
Capt. Vaughn B. Cordle
United Airlines
Boeing 737-300
2 CFM56, 20,000 lbs
2 hrs 50 min
11/29/92
848.51 kmh
527.24 mph

STOCKHOLM/CHICAGO (USA) §
Capt. Herbert L. Dabelow
American Airlines
Boeing 767-200ER
2 GE CF6-80A, 48,000 lbs
8 hrs 7 min
7/23/93
844.62 kmh
524.82 mph

SYDNEY/BALI (UK)
Capt. G. Mussett
British Airways
BAe/Aérospatiale Concorde
3/19/90
1,331.01 kmh
827.05 mph

SYDNEY/BRISBANE (USA)
Capt. Leiland M. Duke, Jr.
FedEx
McDonnell Douglas MD-11F
3 GE CF6-80C2, 61,500 lbs
1 hr 1 min
9/25/98
740.67 kmh
460.23 mph

SYDNEY/HONG KONG (USA) §
Glenn W. Stickel, Captain
Mark A. Heaton, 1st Officer
United Parcel Service
Boeing 747-200
4 P&W JT9D, 53,000 lbs
8 hrs 1 min
3/14/00
922.21 kmh
573.03 mph

SYDNEY/LONDON (UK) §
Capt. Hector McMullen
British Airways
BAe/Aérospatiale Concorde
4 RR Snecma Olympus, 38,000 lbs
20 hrs 20 min 25 sec
2/15-16/85
837.67 kmh
520.50 mph

SYDNEY/LOS ANGELES (USA) §
Capt. David E. Seymour — 1,017.81 kmh / 632.44 mph
United Airlines
Boeing 747-400
4 P&W PW4056, 56,750 lbs
11 hrs 51 min
12/28/02

SYDNEY/MELBOURNE (USA)
Capt. Robert Z. Blue — 798.56 kmh / 496.20 mph
United Airlines
Boeing 747-400
4 P&W PW4056, 56,750 lbs
53 min
6/2/02

SYDNEY/PARIS (UK) §
Capt. Hector McMullen — 940.39 kmh / 584.35 mph
British Airways
BAe/Aérospatiale Concorde
4 RR Snecma Olympus, 38,000 lbs
18 hrs 2 min 15 sec
2/15/85

SYDNEY/SAN FRANCISCO (USA) §
Capt. Robert Z. Blue — 1,024.26 kmh / 636.45 mph
United Airlines
Boeing 747-400
4 P&W PW4056, 56,750 lbs
11 hrs 40 min
12/24/01

SYRACUSE/LOUISVILLE (USA)
Andrew Dulay, Pilot — 835.56 kmh / 519.22 mph
Greg Rassmussen, 1st Officer
Ryan International Airlines
Boeing 757-24, N402UP
2 P&W PW2040
1 hr 9 min 46 sec
5/18/88

TAIPEI/ANCHORAGE (USA) §
George W. Murphy, Jr., Captain — 978.64 kmh / 608.10 mph
David A. Swanson, 1st Officer
Robert J. Fielding, 1st Officer
FedEx
McDonnell Douglas MD-11F
3 GE CF6-80C2, 61,500 lbs
7 hrs 41 min
1/14/01

TAIPEI/HONG KONG (USA)
David F. Specht, Captain — 597.17 kmh / 371.06 mph
Anil J. Singh, 1st Officer
United Airlines
Boeing 747-400
4 P&W PW4056, 56,750 lbs
1 hr 21 min
9/14/98

TAIPEI/OSAKA (USA)
Truman Sterk, Captain — 1,001.63 kmh / 622.38 mph
Jim White, 1st Officer
United Airlines
McDonnell Douglas DC-10-30F
3 GE CF6-50C2, 52,500 lbs
1 hr 42 min
2/15/99

TAIPEI/SAN FRANCISCO (USA) § ✯
R. Z. Blue, Captain — 1,123.36 kmh / 698.02 mph
Edward R. Mieloch, 1st Officer
United Airlines
Boeing 747-400
4 P&W PW4056, 56,750 lbs
9 hrs 15 min
11/19/98

TAIPEI/TOKYO (USA)
John B. Horner, Captain — 1,081.07 kmh / 671.75 mph
Mark H. Buckner, 1st Officer
United Parcel Service
Boeing 747-100
4 P&W JT9D, 46,250 lbs
2 hrs 1 min
1/11/01

TAMPA/CHARLOTTE (USA)
Felix Jesus Villaverde — 599.17 kmh / 372.30 mph
Airlift International
McDonnell Douglas DC-8
4 P&W JT3D, 16,000 lbs
1 hr 22 min
12/7/90

TAMPA/CHICAGO (USA)
Oswaldo X. Sanchez, Captain — 714.46 kmh / 443.94 mph
Larry E. Steenstry, 1st Officer
United Airlines
Boeing 757
2 P&W PW2037, 38,200 lbs
2 hrs 17 min
2/27/94

TAMPA/LOS ANGELES (USA)
Capt. Harold L. Kennedy — 852.40 kmh / 529.66 mph
United Airlines
Airbus A320
2 IAE V2527-A5, 26,500 lbs
4 hrs 4 min
7/26/98

TAMPA/MIAMI* *Group I*
Capt. G. I. Baker — 312.77 mph
National Airlines
Douglas DC-6
4 P&W 2100 hp
Drew Field/36th Street
Distance: 204 mi
39 min 13 sec
6/3/47

TAMPA/WASHINGTON (USA)
Capt. Vaughn B. Cordle — 861.63 kmh / 535.39 mph
United Airlines
Boeing 737-300
2 CFM56, 20,000 lbs
1 hr 31 min
11/6/92

TOKYO/ANCHORAGE (USA) ✯
George W. Murphy, Jr., Captain — 1,008.78 kmh / 626.82 mph
Darrell D. Holmstrom, 1st Officer
FedEx
McDonnell Douglas MD-11F
3 GE CF6-80C2, 61,500 lbs
5 hrs 28 min
11/28/99

TOKYO/BANGKOK (USA)
Capt. William R. Wallace — 826.80 kmh / 513.75 mph
United Airlines
Boeing 747-400
4 P&W PW4056, 56,750 lbs
5 hrs 37 min
11/5/03

TOKYO/BEIJING (USA)
Jay C. Mallory, Captain — 723.69 kmh / 449.68 mph
J. Parker Merrill, 1st Officer
United Airlines
Boeing 747-400
4 P&W PW4056, 56,750 lbs
2 hrs 57 min
5/10/99

TOKYO/CHICAGO (USA) §
Jay C. Mallory, Captain — 1,047.63 kmh, 650.96 mph
P. Michael Bristow, 1st Officer
J. Bradford Lynn, 1st Officer
Ron L. Primrose, 1st Officer
United Airlines
Boeing 747-400
4 P&W PW4056, 56,750 lbs
9 hrs 37 min
9/23/98

TOKYO/DALLAS (USA) §
Dennis B. Walsh, Captain — 1,071.62 kmh, 665.87 mph
Edwin R. Bird, Jr., 1st Officer
Johnathan H. Greene, 1st Officer
American Airlines
McDonnell Douglas MD-11
3 GE CF6-80C2, 61,500 lbs
9 hrs 38 min
11/20/98

TOKYO/HONG KONG (USA)
Capt. David F. Specht — 773.26 kmh, 480.48 mph
United Airlines
Boeing 747-400
4 P&W PW4056, 56,000 lbs
3 hrs 48 min
7/28/95

TOKYO/HONOLULU (USA) ★
Irving S. Soble, Captain — 1,139.64 kmh, 708.17 mph
Paul C. Mattson, 1st Officer
Kevin D. Garcia, 2nd Officer
United Airlines
Boeing 747-100
4 P&W JT9D-7A, 46,250 lbs
5 hrs 23 min
2/6/95

TOKYO/LOS ANGELES (USA) § ★
Capt. James C. Phillips — 1,071.98 kmh, 666.10 mph
United Airlines
Boeing 747-400
4 P&W PW4056, 56,750 lbs
8 hrs 10 min
2/28/00

TOKYO/NEWARK (USA) §
Richard Moen, Captain — 945.47 kmh, 587.48 mph
William R. Wallace, 1st Officer
United Airlines
Boeing 747-400
4 P&W PW4056, 56,750 lbs
11 hrs 26 min
12/13/94

TOKYO/NEW YORK (USA) §
Capt. Philip P. Welsh — 986.07 kmh, 612.71 mph
United Airlines
Boeing 747-200
4 GE CF6-50E2, 52,500 lbs
10 hrs 59 min
12/12/97

TOKYO/OSAKA (USA)
Leiland M. Duke, Jr., Captain — 402.87 kmh, 250.33 mph
Jesse F. Roux, 1st Officer
FedEx
McDonnell Douglas MD-11F
3 GE CF6-80C2, 61,500 lbs
1 hr 13 min
9/12/98

TOKYO/PARIS (JAPAN) §
Taidi Oda — 669.56 kmh, 416.04 mph
Shigeru Nakano
Japan Airlines
McDonnell Douglas DC-8-62, JA8070
4 P&W JT3D-3B
14 hrs 17 min
3/28/70

TOKYO/SAN FRANCISCO (USA) §
Philip C. Ecklund, Captain — 1,066.36 kmh, 662.60 mph
Stephen E. Fitzgerald, 1st Officer
United Airlines
Boeing 747-400
4 P&W PW4056, 56,000 lbs
7 hrs 43 min
12/6/93

TOKYO/SAN JOSE (USA) §
James O. Gross, Captain — 1,054.40 kmh, 655.17 mph
Steve Schwartz, 1st Officer
John P. Romero, 1st Officer
American Airlines
McDonnell Douglas MD-11
3 GE CF6-80C2, 61,500 lbs
7 hrs 51 min
12/14/98

TOKYO/SEATTLE (USA) §
O. B. Phillips, Captain — 987.97 kmh, 613.89 mph
James A. Denton, 1st Officer
United Airlines
Boeing 747-400
4 P&W PW4056, 56,750 lbs
7 hrs 45 min
6/23/99

TOKYO/SINGAPORE (USA)
Capt. David H. Friend — 918.30 kmh, 570.60 mph
United Airlines
Boeing 747-400
4 P&W PW4056, 56,750 lbs
5 hrs 50 min
7/21/02

TOKYO/TAIPEI (USA)
Capt. William R. Wallace — 747.46 kmh, 464.45 mph
United Airlines
Boeing 747-400
4 P&W PW4056, 56,750 lbs
2 hrs 55 min
10/15/03

TORONTO/CHICAGO (USA)
Kenneth T. Firestone, Captain — 600.23 kmh, 372.96 mph
Larry E. Steenstry, 1st Officer
United Airlines
Boeing 757
2 P&W PW2038, 38,200 lbs
1 hr 10 min
3/25/95

TORONTO/NEW YORK*
Capt. C. P. Evans — 527.40 mph
American Airlines
Boeing 727
3 P&W JT8D-1
Toronto Intl/JFK Intl
Distance: 367 mi
41 min 45 sec
6/12/65

TUCSON/ALBUQUERQUE (USA)
Robert F. Olszewski, Captain — 867.60 kmh, 539.13 mph
J. B. Skiles, 1st Officer
USAir
Boeing 727-200A
36 min 12 sec
9/20/89

TUCSON/CHICAGO*
Capt. Jerry McMillan — 608.61 mph
Trans World Airlines
Boeing 727
3 P&W JT8D
Tucson Intl/O'Hare Intl
Distance: 1,437 mi
2 hrs 21 min 40 sec
2/1/75

TUCSON/DENVER (USA)
Capt. Vaughn B. Cordle — 689.16 kmh, 428.22 mph
United Airlines
Boeing 737-300
2 CFM56, 20,000 lbs
1 hr 28 min
8/25/92

TUCSON/LOS ANGELES*
Capt. E. O. Medlin — 486.10 mph
American Airlines
Boeing 707-123
P&W JT3D
Tucson Intl/Los Angeles Intl
Distance: 449 mi
55 min 25 sec
2/25/65

TULSA/CHARLOTTE (USA)
Richard Y. Miller — 949.74 kmh, 590.17 mph
Piedmont Airlines
Boeing 727
3 P&W JT8D, 16,000 lbs
1 hr 27 min 13 sec
7/5/85

TULSA/CHICAGO*
Capt. Frank B. Scott — 556.95 mph
American Airlines
Boeing 727
3 P&W JT8D-1
Tulsa Intl/O'Hare Intl
Distance: 587 mi
1 hr 3 min 4 sec
2/17/65

TULSA/NASHVILLE*
Capt. Robert L. Grove — 592.30 mph
American Airlines
Boeing 727
3 P&W JT8D-1
Tulsa Intl/Berry Field
Distance: 515 mi
52 min 10 sec
11/28/65

TULSA/NEW YORK*
Capt. Arthur Duffey, Jr. — 537.91 mph
American Airlines
Boeing 707-123B
4 P&W JT3D
Tulsa Municipal/Idlewild Intl
Distance: 1,235 mi
2 hrs 17 min 45 sec
4/1/65

WASHINGTON/ADDIS ABABA (USA) §
Capt. Zeleke Demissie — 870.79 kmh, 541.11 mph
Boeing 767-200 ER
2 P&W JT9D-7R4E, 54,750 lbs
13 hrs 17 min 39 sec
5/31-6/1/84

WASHINGTON/AMSTERDAM (USA)
Capt. David J. Roth — 993.13 kmh, 617.10 mph
Northwest Airlines
McDonnell Douglas DC10-40
3 P&W JT9D-20J, 49,000 lbs
6 hrs 15 min
2/17/96

WASHINGTON/ATLANTA (USA)
Capt. Vaughn B. Cordle — 705.65 kmh, 438.47 mph
United Airlines
Boeing 737-300
2 CFM56, 20,000 lbs
1 hr 13 min
10/16/92

WASHINGTON/BOSTON*
Capt. Peter W. Williamson — 536.56 mph
Northeast Airlines
McDonnell Douglas DC-9-31
2 P&W JT8D-7
Washington National/Logan Intl
Distance: 399 mi
44 min 37 sec
1/14/72

WASHINGTON/BUENOS AIRES (USA) §
Capt. Martin K. Kemp — 875.14 kmh, 543.79 mph
United Airlines
Boeing 767-300ER
2 P&W PW4060, 60,000 lbs
9 hrs 38 min
2/27/03

WASHINGTON/CHICAGO (USA)
Ellis Lea, Jr., Captain — 727.20 kmh, 451.86 mph
William R. Wallace, 1st Officer
United Airlines
Boeing 777-222
2 P&W PW4077, 77,000 lbs
1 hr 18 min
9/11/95

WASHINGTON/DALLAS*
Capt. Wendell T. Fleming — 567.97 mph
American Airlines
Boeing 707-123B
4 P&W JT3D
Dulles Intl/Love Field
Distance: 1,162 mi
2 hrs 2 min 45 sec
6/19/65

WASHINGTON/DETROIT (USA)
Capt. Vaughn B. Cordle — 626.50 kmh, 389.29 mph
United Airlines
Boeing 737-300
2 CFM56, 20,000 lbs
59 min
10/5/92

WASHINGTON/FRANKFURT (USA) §
Eugene R. Biscailuz, Captain — 926.77 kmh, 575.86 mph
Stephen E. Fitzgerald, 1st Officer
United Airlines
Boeing 747-400
4 P&W PW4056, 56,750 lbs
7 hrs 4 min
6/30/99

WASHINGTON/KANSAS CITY (USA)
James J. Lawlor, Captain — 870.93 kmh, 541.17 mph
James J. Lawlor, Jr., Flt Engineer
Trans World Airways
Boeing 727-231
3 P&W JT8D, 14,550 lbs
1 hr 45 min
7/13/96

WASHINGTON/LONDON (UK)
Brian Walpole — 1,712.49 kmh, 1,064.14 mph
British Airways
BAe/Aérospatiale Concorde
4 RR Olympus, 38,000 lbs
9/17/86

WASHINGTON/LOS ANGELES*
Capt. W. R. Hunt — 586.98 mph
American Airlines
Boeing 707-720
4 P&W JT3D
Dulles Intl/Los Angeles Intl
Distance: 2,286 mi
3 hrs 53 min 40 sec
7/25/63

WASHINGTON/MEXICO CITY (USA)
Capt. Vaughn B. Cordle — 720.56 kmh
United Airlines — 447.73 mph
Boeing 737-300
2 CFM56, 20,000 lbs
4 hrs 10 min
10/2/92

WASHINGTON/MIAMI (USA)
David H. Friend, Captain — 789.11 kmh
United Airlines — 490.33 mph
Boeing 777
2 P&W PW4090, 90,000 lbs
1 hr 53 min
3/5/00

WASHINGTON/MUNICH*
David H. Friend, Captain — 601.70 mph
Dale G. Goodrich, 1st Officer
United Airlines
Boeing 777
2 P&W PW4077, 77,000 lbs
7 hrs 4 min
1/11/00

WASHINGTON/NEW YORK*
Capt. Don Binard — 453.65 mph
Delta Airlines
Boeing 727-232
3 P&W JT8D-15A, 15,500 lbs
28 min 20 sec
11/23/95

WASHINGTON/ORLANDO (USA)
Capt. Vaughn B. Cordle — 797.10 kmh
United Airlines — 495.29 mph
Boeing 737-500
2 CFM56, 20,000 lbs
1 hr 32 min
12/11/92

WASHINGTON/PARIS (USA)
Capt. Don Binard — 923.38 kmh
Pan American Airways — 573.76 mph
Airbus A310-200
2 GE 7R4-E1, 50,000 lbs
6 hrs 43 min 10 sec
3/18/90

WASHINGTON/PHOENIX (USA)
Capt. Vaughn B. Cordle — 766.12 kmh
United Airlines — 476.04 mph
Boeing 737-300
2 CFM56, 20,000 lbs
4 hrs 6 min
8/21/92

WASHINGTON/SAN FRANCISCO (USA)
Grant C. Besley, Captain — 850.36 kmh
Larry E. Steenstry, 1st Officer — 528.39 mph
United Airlines
Boeing 747-400
4 P&W PW4056, 56,750 lbs
4 hrs 34 min
5/31/99

WASHINGTON/SEATTLE (USA)
Henry P. Kerr, Captain — 816.53 kmh
James R. Thompson, II, 1st Officer — 507.36 mph
United Airlines
Boeing 757
2 P&W PW2037, 38,200 lbs
4 hrs 32 min
1/30/00

WICHITA/AMARILLO*
Capt. Raymond H. Lutz — 534.96 mph
1st Officer Lyle T. Shelton
2nd Officer William J. O'Neill
Trans World Airlines
Boeing 727-231
3 P&W JT8D-9A
Wichita Mid-Continent/Amarillo Air Terminal
Distance: 292 mi
32 min 44 sec
8/19/76

ZURICH/GANDER (USA)
Noel H. Von Urff, Captain — 813.60 kmh
John C. Beatty, 1st Officer — 505.57 mph
Sharon D. Sowell, 2nd Officer
Federal Express DC-8-73F
4 CFM56
11/26/89

ZURICH/LONDON (UK) *Group II*
Capt. W. Baille — 441.70 kmh
British European Airways — 274.46 mph
Vickers Viscount, G-ALWE
4 RR Dart
1 hr 47 min 11 sec
3/19/60

VTOL AIRCRAFT

Vertical Take Off and Landing (VTOL) airplanes are one of the more recent aviation developments. The most practical example of this type, the British Hawker Sidley P.1127 "Kestrel," flew initially in October 1960. The Kestrel used swivelling exhaust outlets to deflect its thrust downward for vertical take-offs and landings.

Other lesser-known methods of jet lift have also been tried. In 1955, the Bell XV-3 used swivelling jet engines. The experimental German Dornier Do.31, built in 1967, used a cluster of small, vertically-mounted jet engines for takeoff and landing, and conventionally oriented jet engines for forward flight.

The Kestrel was later developed into the British Aerospace Harrier, a ground-support fighter-bomber which entered service with the Royal Air Force in 1969. One of the more interesting accomplishments of the Harrier was its winning of the East-West Speed Prize in the "Great Atlantic Air Race" of May, 1969. RAF Squadron Leader Leckie-Thompson traveled from downtown London to downtown New York City in just 6 hours, 12 minutes, using a helicopter and a sports car for the inner-city portions of his trip.

CLASS H (VTOL AIRCRAFT)

DISTANCE

DISTANCE WITHOUT LANDING (W. GERMANY)
Drury Wood — 681 km
Dieter Thomas — 423 mi
Dornier DO31 E-3
2 RR Pegasus, 7,000 kg
8 RR RB162, 2,000 kg
5/27/69

ALTITUDE

ALTITUDE (UK)
S/Ldr Bernard C. Scott, RAF — 15,499 m
Harrier DB6 — 50,852 ft
Pegasus MK II
Boscombe Down, UK
1/12/87

ALTITUDE WITH 1,000 KG PAYLOAD (USSR)
Andrei A. Sinitsine — 13,115 m
Yakovlev Yak-141 — 43,028 ft
Podmoskovnoe, USSR
4/25/91

ALTITUDE WITH 2,000 KG PAYLOAD (USSR)
Andrei A. Sinitsine — 13,115 m
Yakovlev Yak-141 — 43,028 ft
Podmoskovnoe, USSR
4/25/91

TIME TO CLIMB

TIME TO 3,000 METERS (UK)
A. J. Sephton — 36 sec
Harrier G-5
Filton, UK
8/14/89

TIME TO 3,000 METERS WITH 1,000 KG PAYLOAD (USSR)
Andrei A. Sinitsine — 1 min
YAK-141 — 2.41 sec
4/12/91

TIME TO 3,000 METERS WITH 2,000 KG PAYLOAD (USSR)
Andrei A. Sinitsine — 1 min
YAK-141 — 8.50 sec
Podmoskovnoe, USSR
4/25/91

TIME TO 6,000 METERS (UK)
H. Frick — 55 sec
Harrier G-5
Filton, UK
8/14/89

TIME TO 6,000 METERS WITH 1,000 KG PAYLOAD (USSR)
Andrei A. Sinitsine — 1 min
YAK-141 — 14.37 sec
4/12/91

TIME TO 6,000 METERS WITH 2,000 KG PAYLOAD (USSR)
Andrei A. Sinitsine — 1 min
YAK-141 — 29.00 sec
Podmoskovnoe, USSR
4/25/91

TIME TO 9,000 METERS (UK)
A. J. Sephton — 1 min
Harrier G-5 — 21 sec
Filton, UK
8/14/89

TIME TO 9,000 METERS WITH 1,000 KG PAYLOAD (USSR)
Andrei A. Sinitsine — 1 min 29.09 sec
YAK-141
4/12/91

TIME TO 9,000 METERS WITH 2,000 KG PAYLOAD (USSR)
Andrei A. Sinitsine — 1 min 50.00 sec
YAK-141
Podmoskovnoe, USSR
4/25/91

TIME TO 12,000 METERS (USSR)
Andrei A. Sinitsine — 1 min 56.15 sec
YAK-141
Podmoskovnoe, USSR
4/11/91

TIME TO 12,000 METERS WITH 1,000 KG PAYLOAD (USSR)
Andrei A. Sinitsine — 1 min 56.15 sec
YAK-141
Podmoskovnoe, USSR
4/11/91

TIME TO 12,000 METERS WITH 2,000 KG PAYLOAD (USSR)
Andrei A. Sinitsine — 2 min 10.50 sec
YAK-141
Podmoskovnoe, USSR
4/25/91

CLASS H (VTOL AIRCRAFT)
SPEED OVER A RECOGNIZED COURSE

MUNICH/PARIS (W. GERMANY)
Drury Wood — 513.96 kmh
Dieter Thomas — 319.36 mph
Dornier DO31 E-3
2 RR Pegasus, 7,000 kgs
8 RR RB162, 2,000 kgs
5/27/69

STOL Aircraft

STOL, the abbreviation for Short Take-Off and Landing, describes the potential of this special class of aircraft. These aircraft resemble normal aircraft in every way except for their amazing performance. The most common methods of obtaining this increased performance are: increasing wing area, installing more powerful engines, and, particularly in the case of larger aircraft, installing complex high-lift devices.

CLASS N (STOL AIRCRAFT) GROUP I (PISTON ENGINE)

SPEED

SPEED OVER A 3 KM STRAIGHT COURSE (USA)
Robert F. Olszewski — 429.33 kmh
Glasair III, N540RG — 266.78 mph
1 Lycoming IO-540, 300 hp
Everett, WA
11/30/88

SPEED OVER A 15/25 KM STRAIGHT COURSE (USA)
Robert F. Olszewski — 421.96 kmh
Glasair III, N540RG — 262.21 mph
1 Lycoming IO-540, 300 hp
Arlington, WA
12/1/88

CLASS N (STOL AIRCRAFT) GROUP II (TURBOPROP)

ALTITUDE

ALTITUDE (USA)
Lyle H. Schaefer, Pilot — 12,309 m
Arlen D. Rens, Copilot — 40,386 ft
Timothy L. Gomez, Aircraft Systems Specialist
Lockheed Martin C-130J
4 RR Allison AE 2100, 4,591 shp
Marietta, GA
5/14/99

ALTITUDE IN HORIZONTAL FLIGHT (USA)
Lyle H. Schaefer, Pilot — 11,915 m
Arlen D. Rens, Copilot — 39,092 ft
Timothy L. Gomez, Aircraft Systems Specialist
Lockheed Martin C-130J
4 RR Allison AE 2100, 4,591 shp
Marietta, GA
5/14/99

ALTITUDE WITH 1,000 KG PAYLOAD (USA)
Lyle H. Schaefer, Pilot — 12,309 m
Arlen D. Rens, Copilot — 40,386 ft
Timothy L. Gomez, Aircraft Systems Specialist
Lockheed Martin C-130J
4 RR Allison AE 2100, 4,591 shp
Marietta, GA
5/14/99

ALTITUDE WITH 2,000 KG PAYLOAD (USA)
Lyle H. Schaefer, Pilot — 12,309 m
Arlen D. Rens, Copilot — 40,386 ft
Timothy L. Gomez, Aircraft Systems Specialist
Lockheed Martin C-130J
4 RR Allison AE 2100, 4,591 shp
Marietta, GA
5/14/99

ALTITUDE WITH 5,000 KG PAYLOAD (USA)
Lyle H. Schaefer, Pilot — 12,309 m
Arlen D. Rens, Copilot — 40,386 ft
Timothy L. Gomez, Aircraft Systems Specialist
Lockheed Martin C-130J
4 RR Allison AE 2100, 4,591 shp
Marietta, GA
5/14/99

ALTITUDE WITH 10,000 KG PAYLOAD (USA) ✯
Lyle H. Schaefer, Pilot — 12,309 m
Arlen D. Rens, Copilot — 40,386 ft
Timothy L. Gomez, Aircraft Systems Specialist
Lockheed Martin C-130J
4 RR Allison AE 2100, 4,591 shp
Marietta, GA
5/14/99

GREATEST PAYLOAD CARRIED TO 2,000 METERS (USA)
Lyle H. Schaefer, Pilot — 10,217 kg
Arlen D. Rens, Copilot — 22,525 lbs
Timothy L. Gomez, Aircraft Systems Specialist
Lockheed Martin C-130J
4 RR Allison AE 2100, 4,591 shp
Marietta, GA
5/14/99

SPEED

SPEED OVER A 1,000 KM CLOSED CIRCUIT WITHOUT PAYLOAD (USA)
Arlen D. Rens, Pilot — 598.04 kmh
Lyle H. Schaefer, Copilot — 371.60 mph
Timothy L. Gomez, Aircraft Systems Specialist
Lockheed Martin C-130J
4 RR Allison AE 2100, 4,591 shp
Marietta, GA
5/14/99

SPEED OVER A 1,000 KM CLOSED CIRCUIT WITH 1,000 KG PAYLOAD (USA)
Arlen D. Rens, Pilot — 598.04 kmh
Lyle H. Schaefer, Copilot — 371.60 mph
Timothy L. Gomez, Aircraft Systems Specialist
Lockheed Martin C-130J
4 RR Allison AE 2100, 4,591 shp
Marietta, GA
5/14/99

SPEED OVER A 1,000 KM CLOSED CIRCUIT WITH 2,000 KG PAYLOAD (USA)
Arlen D. Rens, Pilot — 598.04 kmh
Lyle H. Schaefer, Copilot — 371.60 mph
Timothy L. Gomez, Aircraft Systems Specialist
Lockheed Martin C-130J
4 RR Allison AE 2100, 4,591 shp
Marietta, GA
5/14/99

SPEED OVER A 1,000 KM CLOSED CIRCUIT WITH 5,000 KG PAYLOAD (USA)
Arlen D. Rens, Pilot 598.04 kmh
Lyle H. Schaefer, Copilot 371.60 mph
Timothy L. Gomez, Aircraft Systems Specialist
Lockheed Martin C-130J
4 RR Allison AE 2100, 4,591 shp
Marietta, GA
5/14/99

SPEED OVER A 1,000 KM CLOSED CIRCUIT WITH 10,000 KG PAYLOAD (USA)
Arlen D. Rens, Pilot 598.04 kmh
Lyle H. Schaefer, Copilot 371.60 mph
Timothy L. Gomez, Aircraft Systems Specialist
Lockheed Martin C-130J
4 RR Allison AE 2100, 4,591 shp
Marietta, GA
5/14/99

SPEED OVER A 2,000 KM CLOSED CIRCUIT WITHOUT PAYLOAD (USA)
Arlen D. Rens, Pilot 598.61 kmh
Lyle H. Schaefer, Copilot 371.96 mph
Timothy L. Gomez, Aircraft Systems Specialist
Lockheed Martin C-130J
4 RR Allison AE 2100, 4,591 shp
Marietta, GA
5/14/99

SPEED OVER A 2,000 KM CLOSED CIRCUIT WITH 1,000 KG PAYLOAD (USA)
Arlen D. Rens, Pilot 598.61 kmh
Lyle H. Schaefer, Copilot 371.96 mph
Timothy L. Gomez, Aircraft Systems Specialist
Lockheed Martin C-130J
4 RR Allison AE 2100, 4,591 shp
Marietta, GA
5/14/99

SPEED OVER A 2,000 KM CLOSED CIRCUIT WITH 2,000 KG PAYLOAD (USA)
Arlen D. Rens, Pilot 598.61 kmh
Lyle H. Schaefer, Copilot 371.96 mph
Timothy L. Gomez, Aircraft Systems Specialist
Lockheed Martin C-130J
4 RR Allison AE 2100, 4,591 shp
Marietta, GA
5/14/99

SPEED OVER A 2,000 KM CLOSED CIRCUIT WITH 5,000 KG PAYLOAD (USA)
Arlen D. Rens, Pilot 598.61 kmh
Lyle H. Schaefer, Copilot 371.96 mph
Timothy L. Gomez, Aircraft Systems Specialist
Lockheed Martin C-130J
4 RR Allison AE 2100, 4,591 shp
Marietta, GA
5/14/99

SPEED OVER A 2,000 KM CLOSED CIRCUIT WITH 10,000 KG PAYLOAD (USA)
Arlen D. Rens, Pilot 598.61 kmh
Lyle H. Schaefer, Copilot 371.96 mph
Timothy L. Gomez, Aircraft Systems Specialist
Lockheed Martin C-130J
4 RR Allison AE 2100, 4,591 shp
Marietta, GA
5/14/99

TIME TO CLIMB

TIME TO 3,000 METERS (USA)
Robert Olszewski, Pilot 3 min
Robert Gavinsky, Copilot 7 sec
Glasair III, N253LC
1 Allison 250-B17D, 420 shp
Arlington, WA
12/14/88

TIME TO 3,000 METERS WITH 1,000 KG PAYLOAD (USA)
Lyle H. Schaefer, Pilot 3 min
Arlen D. Rens, Copilot 49 sec
Timothy L. Gomez, Aircraft Systems Specialist
Lockheed Martin C-130J
4 RR Allison AE 2100, 4,591 shp
Marietta, GA
5/14/99

TIME TO 3,000 METERS WITH 2,000 KG PAYLOAD (USA)
Lyle H. Schaefer, Pilot 3 min
Arlen D. Rens, Copilot 49 sec
Timothy L. Gomez, Aircraft Systems Specialist
Lockheed Martin C-130J
4 RR Allison AE 2100, 4,591 shp
Marietta, GA
5/14/99

TIME TO 3,000 METERS WITH 5,000 KG PAYLOAD (USA)
Lyle H. Schaefer, Pilot 3 min
Arlen D. Rens, Copilot 49 sec
Timothy L. Gomez, Aircraft Systems Specialist
Lockheed Martin C-130J
4 RR Allison AE 2100, 4,591 shp
Marietta, GA
5/14/99

TIME TO 3,000 METERS WITH 10,000 KG PAYLOAD (USA)
Lyle H. Schaefer, Pilot 3 min
Arlen D. Rens, Copilot 49 sec
Timothy L. Gomez, Aircraft Systems Specialist
Lockheed Martin C-130J
4 RR Allison AE 2100, 4,591 shp
Marietta, GA
5/14/99

TIME TO 6,000 METERS (USA)
Frank D. Hadden 7 min
Lockheed 382 E (Modified) 21 sec
4 Allison 501-M71K, 5250 shp
Palmdale, CA
5/18/89

TIME TO 6,000 METERS WITH 1,000 KG PAYLOAD (USA)
Lyle H. Schaefer, Pilot 8 min
Arlen D. Rens, Copilot
Timothy L. Gomez, Aircraft Systems Specialist
Lockheed Martin C-130J
4 RR Allison AE 2100, 4,591 shp
Marietta, GA
5/14/99

TIME TO 6,000 METERS WITH 2,000 KG PAYLOAD (USA)
Lyle H. Schaefer, Pilot 8 min
Arlen D. Rens, Copilot
Timothy L. Gomez, Aircraft Systems Specialist
Lockheed Martin C-130J
4 RR Allison AE 2100, 4,591 shp
Marietta, GA
5/14/99

TIME TO 6,000 METERS WITH 5,000 KG PAYLOAD (USA)
Lyle H. Schaefer, Pilot 8 min
Arlen D. Rens, Copilot
Timothy L. Gomez, Aircraft Systems Specialist
Lockheed Martin C-130J
4 RR Allison AE 2100, 4,591 shp
Marietta, GA
5/14/99

TIME TO 6,000 METERS WITH 10,000 KG PAYLOAD (USA)
Lyle H. Schaefer, Pilot 8 min
Arlen D. Rens, Copilot
Timothy L. Gomez, Aircraft Systems Specialist
Lockheed Martin C-130J
4 RR Allison AE 2100, 4,591 shp
Marietta, GA
5/14/99

TIME TO 9,000 METERS (USA)
Frank D. Hadden 13 min 20 sec
Lockheed 382 E (Modified)
4 Allison 501-M17K, 5250 sph
Palmdale, CA
5/18/89

TIME TO 9,000 METERS WITH 1,000 KG PAYLOAD (USA)
Lyle H. Schaefer, Pilot 15 min 12 sec
Arlen D. Rens, Copilot
Timothy L. Gomez, Aircraft Systems Specialist
Lockheed Martin C-130J
4 RR Allison AE 2100, 4,591 shp
Marietta, GA
5/14/99

TIME TO 9,000 METERS WITH 2,000 KG PAYLOAD (USA)
Lyle H. Schaefer, Pilot 15 min 12 sec
Arlen D. Rens, Copilot
Timothy L. Gomez, Aircraft Systems Specialist
Lockheed Martin C-130J
4 RR Allison AE 2100, 4,591 shp
Marietta, GA
5/14/99

TIME TO 9,000 METERS WITH 5,000 KG PAYLOAD (USA)
Lyle H. Schaefer, Pilot 15 min 12 sec
Arlen D. Rens, Copilot
Timothy L. Gomez, Aircraft Systems Specialist
Lockheed Martin C-130J
4 RR Allison AE 2100, 4,591 shp
Marietta, GA
5/14/99

TIME TO 9,000 METERS WITH 10,000 KG PAYLOAD (USA)
Lyle H. Schaefer, Pilot 15 min 12 sec
Arlen D. Rens, Copilot
Timothy L. Gomez, Aircraft Systems Specialist
Lockheed Martin C-130J
4 RR Allison AE 2100, 4,591 shp
Marietta, GA
5/14/99

CLASS N (STOL AIRCRAFT) GROUP III (JET ENGINE)

ALTITUDE

ALTITUDE IN HORIZONTAL FLIGHT (USSR)
Nikolai Sadovnikov 19,335 m 63,435 ft
P-42
2 P-32, 13,600 kg
Podmoscovnoe
6/10/87

GREATEST PAYLOAD CARRIED TO 2,000 METERS (USA) ★
Maj. Andre A. Gerner, Aircraft Commander 19,997 kg 44,088 lbs
John D. Burns, Pilot
Craig S. Johnson, Loadmaster
McDonnell Douglas C-17A
4 P&W F117-PW-100, 41,700 lbs
Edwards AFB, CA
6/3/94

TIME TO CLIMB

TIME TO 3,000 METERS (USSR)
Nikolai Sadovnikov 25 sec
Sukhoi P-42
2 P-32, 13,600 kg
Podmoscovnoe
4/11/87

TIME TO 3,000 METERS WITH 1,000 KG PAYLOAD (USSR)
Oleg Tsoi 28 sec
Sukhoi P-42
Podmoscovnoe, USSR
5/17/88

TIME TO 6,000 METERS WITH 1,000 KG PAYLOAD (USSR)
Oleg Tsoi 38 sec
Sukhoi P-42
Podmoscovnoe, USSR
4/19/88

TIME TO 9,000 METERS WITH 1,000 KG PAYLOAD (USSR)
Oleg Tsoi 48 sec
Sukhoi P-42
Podmoscovnoe, USSR
5/17/88

TIME TO 12,000 METERS (USSR)
Nikolai Sadovnikov 57 sec
Sukhoi P-42
2 P-32, 13,600 kg
Podmoscovnoe
4/11/87

TIME TO 12,000 METERS WITH 1,000 KG PAYLOAD (USSR)
Oleg Tsoi 59 sec
Sukhoi P-42
Podmoscovnoe, USSR
5/17/88

TIME TO 15,000 METERS (USSR)
Nikolai Sadovnikov 1 min 16 sec
Sukhoi P-42
2 P-32, 13,600 kg
Podmoscovnoe
4/11/87

SEAPLANES AND AMPHIBIANS

The first seaplane flight was made on March 28, 1910, when Henri Fabre flew briefly near Marseilles, France. He later explained that he built a seaplane rather than one of the more popular landplanes because he assumed it would be gentler on the pilot to crash on water than on land. Glenn Curtiss built a "hydroaeroplane" 10 months later and made his first flight at San Diego, California.

The primary difference between an amphibian and a seaplane is that an amphibian has retractable wheels that enable it to operate from land as well as water, while the seaplane is limited to water takeoffs and landings.

In 1913, the Jacques Schneider Trophy Race became a part of the aviation scene. The technical progress stimulated by this prestigious series of races had a major impact on the development of airplanes and engines. The trophy was captured by the British in 1931.

From 1927 to 1939, the world's fastest airplanes were seaplanes. The first airplanes to exceed both 300 mph and 400 mph were seaplanes. The first airplane to carry paying passengers on a regularly-scheduled flight was a small seaplane in 1914.

Waterborne aircraft dominated important phases of aviation from racing to long-distance passenger carrying until World War II because there were no long, hard-surfaced runways that could accommodate high-performance or heavily-laden airplanes. Large bodies of water provided the distances needed for gradual acceleration to takeoff speed. The demise of the seaplane came with the invention of variable-pitch propellers and effective wing flaps, making it possible for land-based airplanes to accelerate rapidly, and to operate well at both low and high speeds. From World War II onward, their use has been primarily limited to maritime patrol and rescue functions and recreational-use homebuilt craft.

Most seaplane records date back to 1939, while most records for amphibians are held by more modern American and Soviet aircraft.

CLASS C-2 (SEAPLANES) GROUP I (PISTON ENGINE)

DISTANCE

DISTANCE WITHOUT LANDING (UK)
CAPT D.C.T. Benneti 9,652.00 km
4 Napier Rapier 5,997.46 m
J.I.Short-Mayo "Mercury"
Dundee, Scotland to
near Fort Niloth, South Africa
10/6-8/38

DISTANCE OVER A CLOSED CIRCUIT WITHOUT LANDING (ITALY)
Mario Stoppani 5,200 km
Cant Z-506 I-LERO 3,231 mi
3 Alfa-Romeo 126RC-34
Confalcone, Italy
5/27-28/37

ALTITUDE

ALTITUDE (ITALY)
COL Nicola di Mauro 13,542 m
Caproni 161 44,429 ft
Piaggio SI PC 100
Vigna di Valle, Italy
9/25/39

ALTITUDE IN HORIZONTAL FLIGHT (USA)
Robert G. Mann, Jr. 8,260 m
Lake Renegade 27,100 ft
Sanford, ME
11/2/89

ALTITUDE WITH 1,000 KG PAYLOAD (ITALY)
Nicola di Mauro, 10,389 m
Mario Stoppani 34,086 ft
Cant Z-506 B.
Monfalcone, Italy
11/12/37

ALTITUDE WITH 2,000 KG PAYLOAD (ITALY)
Mario Stoppani, 8,951 m
Nicola di Mauro 29,368 ft
Cant Z-506-B
3 Alfa-Romeo
11/3/37

ALTITUDE WITH 5,000 KG PAYLOAD (ITALY)
Mario Stoppani, 7,410 m
Nicola di Mauro 24,311 ft
Cant Z-506-B
3 Alfa-Romeo
Monfalcone, Italy
11/7/37

ALTITUDE WITH 10,000 KG PAYLOAD (ITALY)
Mario Stoppani 4,863 m
Cant Z-508 15,955 ft
3 Isotta-Franchini Asso 11 RD 40
Monfalcone, Italy
4/13/37

ALTITUDE WITH 15,000 KG PAYLOAD (FRANCE)
Guillaumet 3,508 m
Latecoere 521 13,509 ft
6 Hispano-Suiza
Biscarosse, France
12/30/37

GREATEST PAYLOAD CARRIED TO 2,000 METERS (FRANCE)
Guillaumet 18,040 kg
Latecoere 521 39,771 lbs
6 Hispano-Suiza
Biscarosse, France
12/30/37

SPEED

SPEED OVER A 3 KM STRAIGHT COURSE (ITALY)
Francesco Agello 709.21 kmh
Macchi-Castoldi M.C. 72 440.68 mph
Fiat AS 6
Lake Garda, Italy
10/23/34

SPEED OVER A 100 KM CLOSED CIRCUIT WITHOUT PAYLOAD (ITALY)
Guglielmo Cassinelli 629.37 kmh
Macchi-Castoldi M.C. 72 391.07 mph
Fiat AS. 6
Falconara, Italy
10/8/33

SPEED OVER A 1,000 KM CLOSED CIRCUIT WITHOUT PAYLOAD (ITALY)
M. Stoppani 403.42 kmh
G. Gorini 250.68 mph
Cant Z 509
3 Fiat A-80 RC 41
3/30/38

SPEED OVER A 1,000 KM CLOSED CIRCUIT WITH 1,000 KG PAYLOAD (ITALY)
M. Stoppani 403.42 kmh
G. Gorini 250.68 mph
Cant Z-509
3 Fiat A-80 PC 41
3/30/38

SPEED OVER A 1,000 KM CLOSED CIRCUIT WITH 2,000 KG PAYLOAD (ITALY)
M. Stoppani 403.42 kmh
G. Gorini 250.68 mph
Cant Z-509
3 Fiat A-80 PC 41
3/30/38

SPEED OVER A 1,000 KM CLOSED CIRCUIT WITH 5,000 KG PAYLOAD (ITALY)
M. Stoppani 251.89 kmh
Ing. A. Maiorana 156.52 mph
Cant Z-508
3 Isotta-Frachini Asso 11 RC
Grado, Italy
5/1/37

SPEED OVER A 1,000 KM CLOSED CIRCUIT WITH 10,000 KG PAYLOAD (FRANCE)
Guillaumet, 211.00 kmh
Leclaire, Comet, 131.11 mph
Leduff, Le Morvan and Chapaton
Latecoere 521, "LT de Vaisseau Paris"
6 Hispano-Suiza
Luco-Aureilhan, France
12/27/37

SPEED OVER A 1,000 KM CLOSED CIRCUIT WITH 15,000 KG PAYLOAD (FRANCE)
Guillaumet 189.74 kmh
Latecoere 521 117.90 mph
6 Hispano-Suiza
Lucon, France
12/29/37

SPEED OVER A 2,000 KM CLOSED CIRCUIT WITHOUT PAYLOAD (ITALY)
M. Stoppani 396.46 kmh
G. Gorini 246.35 mph
Cant Z 509
3 Fiat A-80 RC 41
3/30/38

SPEED OVER A 2,000 KM CLOSED CIRCUIT WITH 1,000 KG PAYLOAD (ITALY)
M. Stoppani — 396.46 kmh
G. Gorini — 246.35 mph
Cant Z-509 Seaplane
3 Fiat A-80 PC 41 Engines
3/30/38

SPEED OVER A 2,000 KM CLOSED CIRCUIT WITH 2,000 KG PAYLOAD (ITALY)
M. Stoppani — 396.46 kmh
G. Gorini — 246.35 mph
Cant Z-509
3 Fiat A-80 RC 41
3/30/38

SPEED OVER A 2,000 KM CLOSED CIRCUIT WITH 5,000 KG PAYLOAD (ITALY)
M. Stoppani — 248.41 kmh
Ing. A. Maiorana — 154.36 mph
Cant Z-508
3 Isotta-Frachini Asso 11 RC
Grado, Italy
5/1/37

SPEED OVER A 5,000 KM CLOSED CIRCUIT WITHOUT PAYLOAD (ITALY)
Mario Stoppani — 308.42 kmh
Carlo Tonini — 191.65 mph
Cant Z-506, I-LERO
3 Alfa-Romeo 126 RC-34
Monfalcone, Italy
5/27-28/37

SPEED OVER A 5,000 KM CLOSED CIRCUIT WITH 1,000 KG PAYLOAD (ITALY)
Mario Stoppani — 308.24 kmh
Carlo Tonini — 191.53 mph
Cant Z-506
3 Alfa-Romeo 126 RC 34
5/28/37

CLASS C-2.A (300-600 kg / 661-1,323 lbs) GROUP I (PISTON ENGINE)

DISTANCE

DISTANCE WITHOUT LANDING (USA)
Elvis W. Cruz — 296.89 km
Modified Eagle Ultralight — 184.49 mi
1 Zenoah, 20 bhp
Paradise Island, Bahamas to Miami, FL
10/6/82

DISTANCE OVER A CLOSED CIRCUIT WITHOUT LANDING (FRANCE)
Louis-Yvon Schmitz — 546.00 km
Mistral Twin Hydro — 339.28 mi
Rotax 462, 50 hp, Rotax 503, 46 hp
Lac de Roquebrune France
12/14/88

ALTITUDE

ALTITUDE (USA)
Charles L. Davis — 7,467 m
Piper Super Cub PA-18 — 24,498 ft
Lycoming O-290, 125 bhp
Detroit, MI
6/18/52

SPEED

SPEED OVER A 3 KM STRAIGHT COURSE (USA)
Verne Jobst — 151.11 kmh
Piper J3S, N31067 — 93.88 mph
1 Continental C-90-8F, 90 hp
McHenry, IL
5/21/88

SPEED OVER A 15/25 KM STRAIGHT COURSE (USA)
Verne Jobst — 154.17 kmh
Piper J3S, N31067 — 95.80 mph
1 Continental C-90-8F, 90 hp
McHenry, IL
5/21/88

SPEED OVER A 100 KM CLOSED CIRCUIT WITHOUT PAYLOAD (USA)
Charles L. Davis — 175.1 kmh
Piper Super Cub PA-18 — 108.8 mph
Lycoming O-290, 125 hp
8/29/52

SPEED OVER A 500 KM CLOSED CIRCUIT WITHOUT PAYLOAD (USA)
Charles L. Davis — 169.56 kmh
Piper Super Cub PA-18 — 105.35 mph
Lycoming O-290, 125 hp
8/29/52

CLASS C-2.B (600-1,200 kg / 1,323-2,646 lbs) GROUP I (PISTON ENGINE)

DISTANCE

DISTANCE WITHOUT LANDING (USA)
Dan E. Pariseau — 1,589.88 km
Christen A-1 Husky — 987.89 mi
Lycoming O-360, 180 bhp
Seeley Lake, MT to St. Paul, MN
8/15/94

ALTITUDE

ALTITUDE (USA)
Robert G. Mann, Jr. — 8,321 m
Lake Renegade — 27,300 ft
1 Lycoming IO-540, 250 hp
Sanford, ME
11/2/89

ALTITUDE IN HORIZONTAL FLIGHT (USA)
Robert G. Mann, Jr. — 8,260 m
Lake Renegade — 27,100 ft
1 Lycoming IO-540, 250 hp
Sanford, ME
11/2/89

SPEED

SPEED OVER A 15/25 KM STRAIGHT COURSE (USA)
Charles I. Fitzsimmons — 221.42 kmh
Lake LA 4-200 — 137.59 mph
1 Lycoming IO-360, 200 hp
10/19/86

SPEED OVER A 100 KM CLOSED CIRCUIT WITHOUT PAYLOAD (USA)
Dan E. Pariseau — 180.40 kmh
Christen A-1 Husky — 112.09 mph
Lycoming O-360, 180 bhp
Renton, WA to Everett, WA
12/13/94

SPEED OVER A 500 KM CLOSED CIRCUIT WITHOUT PAYLOAD (USA)
Dan E. Pariseau — 180.06 kmh
Christen A-1 Husky — 111.88 mph
Lycoming O-360, 180 bhp
Deer Park, WA to Elensburg, WA
7/13/94

SPEED OVER A 1,000 KM CLOSED CIRCUIT WITHOUT PAYLOAD (USA)
Dan E. Pariseau — 163.10 kmh
Christen A-1 Husky — 101.34 mph
Lycoming O-360, 180 bhp
Deer Park, WA to Elensburg, WA
8/14/94

CLASS C-2.C (1,200-2,100 kg / 2,646-4,630 lbs)
GROUP I (PISTON ENGINE)

ALTITUDE

ALTITUDE (USA)
Robert G. Mann, Jr. 7,772 m
Lake Renegade 25,500 ft
1 Lycoming TIO-540, 250 hp
Sanford, ME
11/2/89

ALTITUDE IN HORIZONTAL FLIGHT (USA)
Robert G. Mann, Jr. 7,772 m
Lake Renegade 25,500 ft
1 Lycoming TIO-540, 250 hp
Sanford, ME
11/2/89

SPEED

SPEED OVER A 15/25 KM STRAIGHT COURSE (USA)
Robert F. Olszewski, Pilot 244.40 kmh
Paul Babcock, Copilot 151.87 mph
Lake LA-250
Lake Winnisquam, NH
11/8/84

SPEED OVER A 100 KM CLOSED CIRCUIT WITHOUT PAYLOAD (ITALY)
Giuseppe Alesini 271.78 kmh
SIAI Marchetti FN 333 I-RAID 168.88 mph
Sesona, Italy
7/22/60

SPEED OVER A 500 KM CLOSED CIRCUIT WITHOUT PAYLOAD (ITALY)
Giuseppe Alesini 268.95 kmh
SIAI Marchetti FN 333 I-RAID 167.12 mph
Continental 0-470-H
Sesona, Italy
7/22/60

CLASS C-2 (SEAPLANES)
GROUP II (TURBOPROP)

DISTANCE

DISTANCE WITHOUT LANDING (USSR)
Nickolai Shlikov, Pilot 2,647.63 km
Sergei Tyukavkin, Copilot 1,645.24 mi
M-12 Chaika; 2 AI-20, 4000 SHP
Severny - Morskoy
10/31/83

DISTANCE OVER A CLOSED CIRCUIT WITHOUT LANDING (USSR)
Gennadi Effimov 2,581.62 km
M-12 Chaika 1,604.14 mi
2 AI-20
11/20/73

ALTITUDE

ALTITUDE (USSR)
Edward Kolhov, Chief Pilot 9,159 m
Vladimir Belov, Copilot 30,049 ft
Guennadi Chetkive, navigator
Boris O. Sine, radio operator
M-12 Chaika
2 AI-20, 4,000 shp
Morskoy, USSR
11/2/77

ALTITUDE IN HORIZONTAL FLIGHT (USSR)
Vladimir Belov 9,850 m
M-12 Chaika 32,316 ft
2 AI-20, 4,000 shp
Morskoy, USSR
8/11/81

ALTITUDE WITH 1,000 KG PAYLOAD (USSR)
Nikolai Chkourha, Pilot 9,043 m
Youri Ouryadov, Copilot 29,669 ft
Leonid Korkine, navigator
Pavil Kousine, radio operator
M-12 Chaika
2 AI-20, 4,000 shp
11/2/77

ALTITUDE WITH 2,000 KG PAYLOAD (USSR)
N. Chkourko 9,407 m
M-12 Chaika 30,863 ft
2 AI-20, 4,000 shp
1/16/78

ALTITUDE WITH 5,000 KG PAYLOAD (USSR)
Vasily Avershin 8,312 m
M-12 Chaika 27,270 ft
2 AI-20, 4,000 shp
Morskoy
8/18/81

GREATEST PAYLOAD CARRIED TO 2,000 METERS (USSR)
Vladimir Effimov 7,029.20 kg
M-12 Chaika 15,492 lbs
2 AI-20, 4,000 shp
Morskoy
9/23/81

SPEED

SPEED OVER A 15/25 KM STRAIGHT COURSE (USA)
F.J. "Jack" Schweibold, Pilot 288.97 kmh
Roscoe Morton, Copilot 179.60 mph
Cessna 206 Floatplane
1 Allison C-250, 420 shp
John Brown Seaport, FL
3/18/86

SPEED OVER A 100 KM CLOSED CIRCUIT WITHOUT PAYLOAD (USSR)
Gennadi Effimov 597.00 kmh
M-12 Chaika 370.96 mph
4/19/76

SPEED OVER A 500 KM CLOSED CIRCUIT WITHOUT PAYLOAD (USSR)
E. Nikitine 565.35 kmh
M-12 Chaika 351.29 mph
Morskoi, USSR
4/25/68

SPEED OVER A 1,000 KM CLOSED CIRCUIT WITHOUT PAYLOAD (USSR)
E. Nikitine 551.87 kmh
M-12 Chaika 342.91 mph
10/12/68

SPEED OVER A 1,000 KM CLOSED CIRCUIT WITH 1,000 KG PAYLOAD (USSR)
A. Zakharov 536.07 kmh
A. Smirnov 333.10 mph
M-12 Chaika
Evpatoria, USSR
4/21/70

SPEED OVER A 1,000 KM CLOSED CIRCUIT WITH 2,000 KG PAYLOAD (USSR)
P. Iakoushin 535.23 kmh
M-12 Chaika 332.58 mph
2 AI-20 4,000 shp
Evpatoria, USSR
7/8/70

SPEED OVER A 1,000 KM CLOSED CIRCUIT WITH 5,000 KG PAYLOAD (USSR)
E. Nikitine — 529.00 kmh / 328.70 mph
M-12 Chaika
2 AI-20
Evpatoria, USSR
7/9/70

SPEED OVER A 2,000 KM CLOSED CIRCUIT WITHOUT PAYLOAD (USSR)
A. Zakharov — 555.98 kmh / 345.47 mph
M-12 Chaika
10/30/72

SPEED OVER A 2,000 KM CLOSED CIRCUIT WITH 1,000 KG PAYLOAD (USSR)
A. Zakharov — 555.98 kmh / 345.47 mph
M-12 Chaika
10/30/72

SPEED OVER A 2,000 KM CLOSED CIRCUIT WITH 2,000 KG PAYLOAD (USSR)
Vassily Averchine — 548.54 kmh / 340.84 mph
M-12 Chaika
10/28/73

SPEED OVER A 2,000 KM CLOSED CIRCUIT WITH 5,000 KG PAYLOAD (USSR)
E. Nikitine — 479.47 kmh / 297.92 mph
M-12 Chaika
10/29/73

Time to Climb

TIME TO 3,000 METERS (USSR)
V. Belov — 5 min 10 sec
M-12 Chaika
1/5/74

TIME TO 6,000 METERS (USSR)
E. Nikitine — 11 min 57 sec
M-12 Chaika
11/14/74

TIME TO 9,000 METERS (USSR)
Vassily Averchine — 22 min 10 sec
M-12 Chaika
Morskoi, USSR
4/28/75

CLASS C-2 (SEAPLANES) GROUP III (JET ENGINE)

Altitude

ALTITUDE (USSR)
Gueorgui Bourianov — 14,961 m / 49,085 ft
Beriev G.M. M-10
2 AL-7PB
Sea of Azov, USSR
9/9/61

ALTITUDE WITH 1,000 KG PAYLOAD (USSR)
Gueorgui Bourianov — 14,062 m / 46,135 ft
Beriev G.M. M-10
2 AL-7PB
Sea of Azov, USSR
9/8/61

ALTITUDE WITH 2,000 KG PAYLOAD (USSR)
Gueorgui Bourianov — 14,062 m / 46,135 ft
Beriev G.M. M-10
2 AL-7PB
Sea of Azov, USSR
9/8/61

ALTITUDE WITH 5,000 KG PAYLOAD (USSR)
Gueorgui Bourianov — 14,062 m / 46,135 ft
Beriev G.M. M-10
2 AL-7PB
Sea of Azov, USSR
9/8/61

ALTITUDE WITH 10,000 KG PAYLOAD (USSR)
Gueorgui Bourianov — 12,733 m / 41,774 ft
Beriev G.M. M-10
2 AL-7PB
Sea of Azov, USSR
9/11/61

ALTITUDE WITH 15,000 KG PAYLOAD (USSR)
Gueorgui Bourianov — 11,997 m / 39,360 ft
Beriev G.M. M-10
2 AL-7PB
Sea of Azov, USSR
9/12/61

GREATEST PAYLOAD CARRIED TO 2,000 METERS (USSR)
Gueorgui Bourianov — 15,206 kg / 33,523 lbs
Beriev G.M. M-10
2 AL-7PB
Sea of Azov, USSR
9/12/61

Speed

SPEED OVER 15/25 KM STRAIGHT COURSE (USSR)
Nicolai Andrievsky — 911.98 kmh / 566.68 mph
Beriev G.M. M-10
2 AL-7PB
Joukovsky-Petrovskoe, USSR
8/7/61

SPEED OVER A 1,000 KM CLOSED CIRCUIT WITHOUT PAYLOAD (USSR)
Gueorgui Bourianov — 875.86 kmh / 543.23 mph
Beriev G.M. M-10
2 AL-7PB
Sea of Azov, USSR
9/3/61

SPEED OVER A 1,000 KM CLOSED CIRCUIT WITH 1,000 KG PAYLOAD (USSR)
Gueorgui Bourianov — 875.86 kmh / 543.23 mph
Beriev G.M. M-10
2 AL-7PB
Sea of Azov, USSR
9/3/61

SPEED OVER A 1,000 KM CLOSED CIRCUIT WITH 2,000 KG PAYLOAD (USSR)
Gueorgui Bourianov — 875.86 kmh / 543.23 mph
Beriev G.M. M-10
2 AL-7PB
Sea of Azov, USSR
9/3/61

SPEED OVER A 1,000 KM CLOSED CIRCUIT WITH 5,000 KG PAYLOAD (USSR)
Gueorgui Bourianov — 875.86 kmh / 543.23 mph
Beriev G.M. M-10
2 AL-7PB
Sea of Azov, USSR
9/3/61

Time to Climb

TIME TO 3,000 METERS (RUSSIA)
Bogdan I. Lissak, Pilot — 3 min 6.3 sec
Gennadi Kalyuzhny, Copilot
Albatros
11/21/91

TIME TO 3,000 METERS WITH 1,000 KG PAYLOAD (RUSSIA)
Bogdan I. Lissak, Pilot — 3 min
Gennadi Kalyuzhny, Copilot — 6.3 sec
Albatros
11/21/91

TIME TO 3,000 METERS WITH 2,000 KG PAYLOAD (RUSSIA)
Bogdan I. Lissak, Pilot — 3 min
Gennadi Kalyuzhny, Copilot — 6.3 sec
Albatros
11/21/91

TIME TO 3,000 METERS WITH 5,000 KG PAYLOAD (RUSSIA)
Bogdan I. Lissak, Pilot — 3 min
Gennadi Kalyuzhny, Copilot — 6.3 sec
Albatros
11/21/91

TIME TO 3,000 METERS WITH 10,000 KG PAYLOAD (RUSSIA)
Bogdan I. Lissak, Pilot — 3 min
Gennadi Kalyuzhny, Copilot — 6.3 sec
Albatros
11/21/91

TIME TO 3,000 METERS WITH 15,000 KG PAYLOAD (RUSSIA)
Genaddy Kalujniy, Pilot — 4 min
Gennady Parchin, Copilot — 10 sec
Beriev A-40 Albatros
Gilindzhik-Black Sea, Russia
7/3/98

TIME TO 6,000 METERS (RUSSIA)
Bogdan I. Lissak, Pilot — 7 min
Gennadi Kalyuzhny, Copilot — 40.7 sec
Albatros
11/21/91

TIME TO 6,000 METERS WITH 1,000 KG PAYLOAD (RUSSIA)
Bogdan I. Lissak, Pilot — 7 min
Gennadi Kalyuzhny, Copilot — 40.7 sec
Albatros
11/21/91

TIME TO 6,000 METERS WITH 2,000 KG PAYLOAD (RUSSIA)
Bogdan I. Lissak, Pilot — 7 min
Gennadi Kalyuzhny, Copilot — 40.7 sec
Albatros
11/21/91

TIME TO 6,000 METERS WITH 5,000 KG PAYLOAD (RUSSIA)
Bogdan I. Lissak, Pilot — 7 min
Gennadi Kalyuzhny, Copilot — 40.7 sec
Albatros
11/21/91

TIME TO 6,000 METERS WITH 10,000 KG PAYLOAD (RUSSIA)
Bogdan I. Lissak, Pilot — 7 min
Gennadi Kalyuzhny, Copilot — 40.7 sec
Albatros
11/21/91

TIME TO 6,000 METERS WITH 15,000 KG PAYLOAD (RUSSIA)
Genaddy Kalujniy, Pilot — 9 min
Gennady Parchin, Copilot — 21 sec
Beriev A-40 Albatros
Gilindzhik-Black Sea, Russia
7/3/98

TIME TO 9,000 METERS (RUSSIA)
Bogdan I. Lissak, Pilot — 12 min
Gennadi Kalyuzhny, Copilot — 14 sec
Albatros
11/21/91

TIME TO 9,000 METERS WITH 1,000 KG PAYLOAD (RUSSIA)
Bogdan I. Lissak, Pilot — 12 min
Gennadi Kalyuzhny, Copilot — 14 sec
Albatros
11/21/91

TIME TO 9,000 METERS WITH 2,000 KG PAYLOAD (RUSSIA)
Bogdan I. Lissak, Pilot — 12 min
Gennadi Kalyuzhny, Copilot — 14 sec
Albatros
11/21/91

TIME TO 9,000 METERS WITH 5,000 KG PAYLOAD (RUSSIA)
Bogdan I. Lissak, Pilot — 12 min
Gennadi Kalyuzhny, Copilot — 14 sec
Albatros
11/21/91

TIME TO 9,000 METERS WITH 10,000 KG PAYLOAD (RUSSIA)
Bogdan I. Lissak, Pilot — 12 min
Gennadi Kalyuzhny, Copilot — 14 sec
Albatros
11/21/91

TIME TO 9,000 METERS WITH 15,000 KG PAYLOAD (RUSSIA)
Genaddy Kalujniy, Pilot — 17 min
Gennady Parchin, Copilot — 14 sec
Beriev A-40 Albatros
Gilindzhik-Black Sea, Russia
7/3/98

CLASS C-2.H (20,000-25,000 kg / 44,092-55,116 lbs) GROUP III (JET ENGINE)

ALTITUDE

ALTITUDE WITHOUT PAYLOAD (USSR)
Bogdan I. Lissak, Pilot — 13,070 m
K.V. Babitch, Copilot — 42,881 ft
Beriev " Albatros"
2 Soloviev D-30 KPV 12,000 shp
9/14/89

SPEED

SPEED OVER A 1,000 KM CLOSED CIRCUIT WITHOUT PAYLOAD (USSR)
Vladimir Demianovsky — 743.40 kmh
Albatros — 461.92 mph
7/23/91

SPEED OVER A 1,000 KM CLOSED CIRCUIT WITH 1,000 KG PAYLOAD (USSR)
Vladimir Demianovsky — 743.40 kmh
Albatros — 461.92 mph
7/23/91

SPEED OVER A 1,000 KM CLOSED CIRCUIT WITH 2,000 KG PAYLOAD (USSR)
Vladimir Demianovsky — 743.40 kmh
Albatros — 461.92 mph
7/23/91

SPEED OVER A 1,000 KM CLOSED CIRCUIT WITH 5,000 KG PAYLOAD (USSR)
Vladimir Demianovsky — 743.40 kmh
Albatros — 461.92 mph
7/23/91

SPEED OVER A 1,000 KM CLOSED CIRCUIT WITH 10,000 KG PAYLOAD (USSR)
Vladimir Demianovsky — 743.40 kmh
Albatros — 461.92 mph
7/23/91

SPEED OVER A 2,000 KM CLOSED CIRCUIT WITHOUT PAYLOAD (USSR)
Gennadi Kalyuzhny — 746.90 kmh
Albatros — 464.10 mph
7/22/91

SPEED OVER A 2,000 KM CLOSED CIRCUIT WITH 1,000 KG PAYLOAD (USSR)
Gennadi Kalyuzhny — 746.90 kmh
Albatros — 464.10 mph
7/22/91

SPEED OVER A 2,000 KM CLOSED CIRCUIT WITH 2,000 KG PAYLOAD (USSR)
Gennadi Kalyuzhny — 746.90 kmh
Albatros — 464.10 mph
7/22/91

SPEED OVER A 2,000 KM CLOSED CIRCUIT WITH 5,000 KG PAYLOAD (USSR)
Gennadi Kalyuzhny — 746.90 kmh
Albatros — 464.10 mph
7/22/91

SPEED OVER A 2,000 KM CLOSED CIRCUIT WITH 10,000 KG PAYLOAD (USSR)
Gennadi Kalyuzhny — 746.90 kmh
Albatros — 464.10 mph
7/22/91

TIME TO CLIMB

TIME TO 3,000 METERS (RUSSIA)
Bogdan I. Lissak, Pilot — 3 min
Gennadi Kalyuzhny, Copilot — 6.3 sec
Albatros
11/21/91

TIME TO 3,000 METERS WITH 1,000 KG PAYLOAD (RUSSIA)
Bogdan I. Lissak, Pilot — 3 min
Gennadi Kalyuzhny, Copilot — 6.3 sec
Albatros
11/21/91

TIME TO 3,000 METERS WITH 2,000 KG PAYLOAD (RUSSIA)
Bogdan I. Lissak, Pilot — 3 min
Gennadi Kalyuzhny, Copilot — 6.3 sec
Albatros
11/21/91

TIME TO 3,000 METERS WITH 5,000 KG PAYLOAD (RUSSIA)
Bogdan I. Lissak, Pilot — 3 min
Gennadi Kalyuzhny, Copilot — 6.3 sec
Albatros
11/21/91

TIME TO 3,000 METERS WITH 10,000 KG PAYLOAD (RUSSIA)
Bogdan I. Lissak, Pilot — 3 min
Gennadi Kalyuzhny, Copilot — 6.3 sec
Albatros
11/21/91

TIME TO 6,000 METERS (RUSSIA)
Bogdan I. Lissak, Pilot — 7 min
Gennadi Kalyuzhny, Copilot — 40.7 sec
Albatros
11/21/91

TIME TO 6,000 METERS WITH 1,000 KG PAYLOAD (RUSSIA)
Bogdan I. Lissak, Pilot — 7 min
Gennadi Kalyuzhny, Copilot — 40.7 sec
Albatros
11/21/91

TIME TO 6,000 METERS WITH 2,000 KG PAYLOAD (RUSSIA)
Bogdan I. Lissak, Pilot — 7 min
Gennadi Kalyuzhny, Copilot — 40.7 sec
Albatros
11/21/91

TIME TO 6,000 METERS WITH 5,000 KG PAYLOAD (RUSSIA)
Bogdan I. Lissak, Pilot — 7 min
Gennadi Kalyuzhny, Copilot — 40.7 sec
Albatros
11/21/91

TIME TO 6,000 METERS WITH 10,000 KG PAYLOAD (RUSSIA)
Bogdan I. Lissak, Pilot — 7 min
Gennadi Kalyuzhny, Copilot — 40.7 sec
Albatros
11/21/91

TIME TO 9,000 METERS (RUSSIA)
Bogdan I. Lissak, Pilot — 12 min
Gennadi Kalyuzhny, Copilot — 14 sec
Albatros
11/21/91

TIME TO 9,000 METERS WITH 1,000 KG PAYLOAD (RUSSIA)
Bogdan I. Lissak, Pilot — 12 min
Gennadi Kalyuzhny, Copilot — 14 sec
Albatros
11/21/91

TIME TO 9,000 METERS WITH 2,000 KG PAYLOAD (RUSSIA)
Bogdan I. Lissak, Pilot — 12 min
Gennadi Kalyuzhny, Copilot — 14 sec
Albatros
11/21/91

TIME TO 9,000 METERS WITH 5,000 KG PAYLOAD (RUSSIA)
Bogdan I. Lissak, Pilot — 12 min
Gennadi Kalyuzhny, Copilot — 14 sec
Albatros
11/21/91

TIME TO 9,000 METERS WITH 10,000 KG PAYLOAD (RUSSIA)
Bogdan I. Lissak, Pilot — 12 min
Gennadi Kalyuzhny, Copilot — 14 sec
Albatros
11/21/91

TIME TO 12,000 METERS (RUSSIA)
Vladimir Demianovsky — 25 min
Albatros — 8 sec
3/26/92

TIME TO 12,000 METERS WITH 1,000 KG PAYLOAD (RUSSIA)
Vladimir Demianovsky — 25 min
Albatros — 8 sec
3/26/92

TIME TO 12,000 METERS WITH 2,000 KG PAYLOAD (RUSSIA)
Vladimir Demianovsky — 25 min
Albatros — 8 sec
3/26/92

TIME TO 12,000 METERS WITH 5,000 KG PAYLOAD (RUSSIA)
Vladimir Demianovsky — 25 min 8 sec
Albatros
3/26/92

TIME TO 12,000 METERS WITH 10,000 KG PAYLOAD (RUSSIA)
Vladimir Demianovsky — 25 min 8 sec
Albatros
3/26/92

CLASS C-2.I (25,000-35,000 kg / 55,116-77,162 lbs)
GROUP III (JET ENGINE)

TIME TO CLIMB

TIME TO 3,000 METERS (RUSSIA)
Konstantin Babich — 3 min 38 sec
Dmitry Morozov
Nikolai Kuznetsov
Nikolai Okhotnikov
Beriev Be-200
2 Progress D-436T, 16,535 lbs
Gelendzhik, Russia
9/7/02

TIME TO 3,000 METERS WITH 1,000 KG PAYLOAD (RUSSIA)
Konstantin Babich — 3 min 38 sec
Dmitry Morozov
Nikolai Kuznetsov
Nikolai Okhotnikov
Beriev Be-200
2 Progress D-436T, 16,535 lbs
Gelendzhik, Russia
9/7/02

TIME TO 3,000 METERS WITH 2,000 KG PAYLOAD (RUSSIA)
Konstantin Babich — 3 min 38 sec
Dmitry Morozov
Nikolai Kuznetsov
Nikolai Okhotnikov
Beriev Be-200
2 Progress D-436T, 16,535 lbs
Gelendzhik, Russia
9/7/02

TIME TO 3,000 METERS WITH 5,000 KG PAYLOAD (RUSSIA)
Konstantin Babich — 3 min 35 sec
Nikolai Kuleshov
Beriev Be-200
Gelendzhik, Russia
9/8/00

TIME TO 6,000 METERS (RUSSIA)
Konstantin Babich — 7 min 54 sec
Dmitry Morozov
Nikolai Kuznetsov
Nikolai Okhotnikov
Beriev Be-200
2 Progress D-436T, 16,535 lbs
Gelendzhik, Russia
9/7/02

TIME TO 6,000 METERS WITH 1,000 KG PAYLOAD (RUSSIA)
Konstantin Babich — 7 min 54 sec
Dmitry Morozov
Nikolai Kuznetsov
Nikolai Okhotnikov
Beriev Be-200
2 Progress D-436T, 16,535 lbs
Gelendzhik, Russia
9/7/02

TIME TO 6,000 METERS WITH 2,000 KG PAYLOAD (RUSSIA)
Konstantin Babich — 7 min 54 sec
Dmitry Morozov
Nikolai Kuznetsov
Nikolai Okhotnikov
Beriev Be-200
2 Progress D-436T, 16,535 lbs
Gelendzhik, Russia
9/7/02

TIME TO 6,000 METERS WITH 5,000 KG PAYLOAD (RUSSIA)
Konstantin Babich — 7 min 43 sec
Nikolai Kuleshov
Beriev Be-200
Gelendzhik, Russia
9/8/00

TIME TO 9,000 METERS (RUSSIA)
Konstantin Babich — 17 min 29 sec
Dmitry Morozov
Nikolai Kuznetsov
Nikolai Okhotnikov
Beriev Be-200
2 Progress D-436T, 16,535 lbs
Gelendzhik, Russia
9/7/02

TIME TO 9,000 METERS WITH 1,000 KG PAYLOAD (RUSSIA)
Konstantin Babich — 17 min 29 sec
Dmitry Morozov
Nikolai Kuznetsov
Nikolai Okhotnikov
Beriev Be-200
2 Progress D-436T, 16,535 lbs
Gelendzhik, Russia
9/7/02

TIME TO 9,000 METERS WITH 2,000 KG PAYLOAD (RUSSIA)
Konstantin Babich — 17 min 29 sec
Dmitry Morozov
Nikolai Kuznetsov
Nikolai Okhotnikov
Beriev Be-200
2 Progress D-436T, 16,535 lbs
Gelendzhik, Russia
9/7/02

TIME TO 9,000 METERS WITH 5,000 KG PAYLOAD (RUSSIA)
Konstantin Babich — 15 min 39 sec
Nikolai Kuleshov
Beriev Be-200
Gelendzhik, Russia
9/8/00

CLASS C-2.L (More Than 60,000 kg / 132,277 lbs)
GROUP III (JET ENGINE)

ALTITUDE

GREATEST PAYLOAD CARRIED TO 2,000 METERS (RUSSIA)
Genaddy Kalujniy, Pilot — 15,040 kg 33,158 lbs
Gennady Parchin, Copilot
Beriev A-40 Albatros
Gilindzhik-Black Sea, Russia
7/3/98

TIME TO CLIMB

TIME TO 3,000 METERS WITH 15,000 KG PAYLOAD (RUSSIA)
Genaddy Kalujniy, Pilot — 4 min 10 sec
Gennady Parchin, Copilot
Beriev A-40 Albatros
Gilindzhik-Black Sea, Russia
7/3/98

TIME TO 6,000 METERS WITH 15,000 KG PAYLOAD (RUSSIA)
Genaddy Kalujniy, Pilot — 9 min
Gennady Parchin, Copilot — 21 sec
Beriev A-40 Albatros
Gilindzhik-Black Sea, Russia
7/3/98

TIME TO 9,000 METERS WITH 15,000 KG PAYLOAD (RUSSIA)
Genaddy Kalujniy, Pilot — 17 min
Gennady Parchin, Copilot — 14 sec
Beriev A-40 Albatros
Gilindzhik-Black Sea, Russia
7/3/98

CLASS C-3 (AMPHIBIANS)
GROUP I (PISTON ENGINE)

DISTANCE

DISTANCE WITHOUT LANDING (USA)
CDR W. Fenlon, USCG — 5,748.04 km
Grumman UF-2G Albatross — 3,571.65 mi
2 Wright R-1820-76B
Kodiak, AK - Pensacola, FL
10/25/62

ALTITUDE

ALTITUDE (USA)
LTC Charles H. Manning — 10,023 m
Grumman HU-16B Albatross — 32,883 ft
2 Wright R-1820-76D
Homestead, FL
7/4/73

ALTITUDE WITH 1,000 KG PAYLOAD (USA)
LTCDR Don Moore, USN — 8,984 m
Grumman UF-2G Albatross — 29,475 ft
2 Wright R-1820-76B
Floyd Bennett Field, NY
9/12/62

ALTITUDE WITH 2,000 KG PAYLOAD (USA)
LTCDR Fred Franke, USN — 8,353 m
Grumman UF-2G Albatross — 27,405 ft
2 Wright R-1820-76B
Floyd Bennett Field, NY
9/12/62

ALTITUDE WITH 5,000 KG PAYLOAD (USA)
CAPT Henry E. Erwin, Jr., USAF — 6,119 m
Grumman HU-16B Albatross — 19,747 ft
2 Wright R-1820-76B
Eglin AFB, FL
3/20/63

GREATEST PAYLOAD CARRIED TO 2,000 METERS (USA)
CAPT Henry E. Erwin, Jr. — 5,517.00 kg
Grumman HU-16B Albatross — 12,162.90 lbs
2 Wright R-1820 76B
Eglin AFB
3/20/63

SPEED

SPEED OVER A 3 KM STRAIGHT COURSE (USA)
MAJ Alexander P. de Seversky — 370.81 kmh
Seversky N3PB — 230.41 mph
Wright Cyclone, 710 hp
Detroit, MI
9/15/35

SPEED OVER A 15/25 KM STRAIGHT COURSE (USA)
Arthur C. Stifel, III — 265.58 kmh
Grumman G-44 Widgeon — 165.03 mph
2 Lycoming GO-480-G2D6, 295 hp
Leesburg, VA to Annapolis, MD
12/17/82

SPEED OVER A 100 KM CLOSED CIRCUIT WITHOUT PAYLOAD (UK)
L.R. Colquhoun — 389.27 kmh
Vickers Supermarine — 241.88 mph
Seagull I, PA 147
Rolls Royce Griffon Mark 29
Marston Moor, England
7/22/50

SPEED OVER A 1,000 KM CLOSED CIRCUIT WITHOUT PAYLOAD (USA)
CDR Wallace C. Dahlgren — 373.32 kmh
Grumman UF-2G Albatross — 231.96 mph
2 Wright R-1820-76BFloyd Bennett, NY
8/13/62

SPEED OVER A 1,000 KM CLOSED CIRCUIT WITH 1,000 KG PAYLOAD (USA)
CDR Wallace C. Dahlgren — 373.32 kmh
Grumman UF-2G Albatross — 231.96 mph
2 Wright R-1820-76B
Floyd Bennett, NY to
Elizabeth City, NC
8/13/62

SPEED OVER A 1,000 KM CLOSED CIRCUIT WITH 2,000 KG PAYLOAD (USA)
CDR Wallace C. Dahlgren — 373.32 kmh
Grumman UF-2G Albatross — 231.96 mph
2 Wright R-1820-76B
Floyd Bennett, NY to
Elizabeth City, NC
8/13/62

SPEED OVER A 1,000 KM CLOSED CIRCUIT WITH 5,000 KG PAYLOAD (USA)
CAPT Glenn A. Higginson — 247.28 kmh
Grumman HU-16B Albatross — 153.65 mph
2 Wright R-1820-76B
Eglin AFB, FL - Albany, GA
3/19/63

SPEED OVER A 2,000 KM CLOSED CIRCUIT WITHOUT PAYLOAD (ITALY)
Giuseppe Burei — 248.97 kmh
Enrica Rossaldi — 154.70 mph
Macchi C-94 I-NAPI
2 Wright Cyclone
Rovine Ansedonia, Italy
5/6/37

SPEED OVER A 5,000 KM CLOSED CIRCUIT WITH 1,000 KG PAYLOAD (USA)
LTCDR Richard A. Hoffman — 243.64 kmh
Grumman UF-2G Albatross — 151.39 mph
2 Curtis-Wright R-1820-76B
Floyd Bennett Field, Plattsburg, NY, Dupress, SD, Floyd Bennett, NY
9/16/62

CLASS C-3a/0 (Less Than 300 kg / 661 lbs)
GROUP I (PISTON ENGINE)

SPEED

SPEED OVER A 3 KM STRAIGHT COURSE (USA)
Matthew Naylor — 93.85 kmh
Quicksilver GT 400, N18MN — 58.32 mph
1 Rotax, 50 hp
San Marcos, TX
6/5/87

SPEED OVER A 15/25 KM STRAIGHT COURSE (USA)
Matthew Naylor — 120.86 kmh
Quicksilver GT 400, N18MN — 75.10 mph
1 Rotax, 50 hp
San Marcos, TX
6/6/87

CLASS C-3.A (300-600 kg / 661-1,323 lbs) GROUP I (PISTON ENGINE)

ALTITUDE

ALTITUDE (FRANCE)
Benard d'Ottreppe — 5,457 m / 17,904 ft
Mistral Twin Hydro
Rotax 462, 50 hp
Rotax 503, 46 hp
La Palud Frejus, France
12/15/88

SPEED

SPEED OVER A 3 KM STRAIGHT COURSE (USA)
Matthew Naylor — 128.12 kmh / 79.61 mph
Eipper GT400, N18MN
2 Rotax Bombardier 377, 35 hp ea
Canyon Lake Arpt., TX
3/27/88

SPEED OVER A 15/25 KM STRAIGHT COURSE (USA)
Matthew Naylor — 131.39 kmh / 81.65 mph
Eipper GT400, N18MN
2 Rotax Bomadier 377, 35 hp ea
Canyon Lake Arpt., TX
4/2/88

SPEED OVER A 100 KM CLOSED CIRCUIT WITHOUT PAYLOAD (CANADA)
Claude Roy — 110.51 kmh / 68.67 mph
Challenger II Ultralight
Ottawa, Canada
5/30/99

CLASS C-3.B (600-1,200 kg / 1,323-2,646 lbs) GROUP I (PISTON ENGINE)

SPEED

SPEED OVER A 3 KM STRAIGHT COURSE (USA)
Ronald Lueck, Pilot — 323.55 kmh / 201.05 mph
Mark Woodbury, Engineer
Air Shark I, N204FM
1 Lycoming IO-360-A3B, 200 hp
4/14/88

SPEED OVER A 15/25 KM STRAIGHT COURSE (USA)
Charles I. Fitzsimmons — 231.19 kmh / 143.66 mph
Lake LA 4-200
1 Lycoming IO-360, 200 hp
10/19/86

CLASS C-3.C (1,200-2,100 kg / 2,646-4,630 lbs) GROUP I (PISTON ENGINE)

DISTANCE

DISTANCE WITHOUT LANDING (W. GERMANY)
Dietrich Schmitt — 3,165.10 km / 1,966.79 mi
Lake Buccaneer
1 Lycoming I0-360-A1B, 200 hp
Kulusuk, Greenland to Heidelberg, W. Germany
7/2/81

ALTITUDE

ALTITUDE (USA)
Peter L. Foster, Pilot — 7,498 m / 24,600 ft
Robert G. Mann, Copilot
Lake Renegade, N250L
1 Lycoming TIO-540-AA1AD
Sanford, ME
8/31/88

ALTITUDE IN HORIZONTAL FLIGHT (USA)
Peter L. Foster, Pilot — 7,468 m / 24,500 ft
Robert Mann, Copilot
Lake Renegade, N250L
1 Lycoming TIO-540-AA1AD
Sanford, ME
8/31/88

SPEED

SPEED OVER A 15/25 KM STRAIGHT COURSE (USA)
Robert Olszewski, Pilot — 242.28 kmh / 150.55 mph
Paul Babcock, Copilot
Lake LA-250
Laconia to Lake Winnisquam, NH
11/8/84

SPEED OVER A 100 KM CLOSED CIRCUIT WITHOUT PAYLOAD (ITALY)
Giuseppe Alesini — 270.27 kmh / 167.94 mph
SIAI Marchetti FN 333
Continental 0-470-H
Sesona, Italy
7/21/60

SPEED OVER A 500 KM CLOSED CIRCUIT WITHOUT PAYLOAD (ITALY)
Giuseppe Alesini — 268.91 kmh / 167.09 mph
SIAI Marchetti FN 333
Continental 0-470-H
Sesona, Italy
7/21/60

CLASS C-3.D (2,100-3,400 kg / 4,630-7,496 lbs) GROUP I (PISTON ENGINE)

SPEED

SPEED OVER A 3 KM STRAIGHT COURSE (USA)
Arthur C. Stifel, III — 270.43 kmh / 168.05 mph
Grumman G-44 Widgeon
2 Lycoming GO-480-G2D6, 295 hp
Dulles Intl Airport, VA
12/17/82

SPEED OVER A 15/25 KM STRAIGHT COURSE (USA)
Arthur C. Stifel, III — 265.58 kmh / 165.03 mph
Grumman G-44 Widgeon
2 Lycoming GO-480-G2D6, 295 hp
Leesburg, VA to Annapolis, MD
12/17/82

CLASS C-3 (AMPHIBIANS) GROUP II (TURBOPROP)

DISTANCE

DISTANCE WITHOUT LANDING (USSR)
Nickolai Shlikov, — 2,647.63 km / 1,645.24 mi
Sergei Tyukavkin, Copilot
M-12 Chaika; 2 AI-20, 4000 5 SHP
Morskoy - Sakhoputny
10/30/83

DISTANCE OVER A CLOSED CIRCUIT WITHOUT LANDING (USSR)
Vladimir Sviatochnuk — 2,562.90 km / 1,592.51 mi
M-12 Chaika
2 AI-20
10/25/73

ALTITUDE

ALTITUDE (USSR)
M. Mikhailov — 12,185 m / 39,977 ft
M-12 Chaika
10/23/64

ALTITUDE IN HORIZONTAL FLIGHT (USSR)
Victor Toroubarov 9,970 m
M-12 Chaika 32,709 ft
2 AI-20, 4,000 hp
8/12/81

ALTITUDE WITH 1,000 KG PAYLOAD (USSR)
M. Mikhailov 11,366 m
M-12 Chaika 37,290 ft
2 AI-20D
10/23/64

ALTITUDE WITH 2,000 KG PAYLOAD (USSR)
M. Mikhailov 11,366 m
M-12 Chaika 37,290 ft
2 AI-20D
10/23/64

ALTITUDE WITH 5,000 KG PAYLOAD (USSR)
M. Mikhailov 10,685 m
M-12 Chaika 35,056 ft
2 AI-20D
10/24/64

ALTITUDE WITH 10,000 KG PAYLOAD (USSR)
M. Mikhailov 9,352 m
M-12 Chaika 30,682 ft
2 AI-20D
10/27/64

GREATEST PAYLOAD CARRIED TO 2,000 METERS (USSR)
M. Mikhailov 10,100.00 kg
M-12 Chaika 22,266.69 lbs
2 AI-20D
10/27/64

SPEED

SPEED OVER A 3 KM STRAIGHT COURSE (USA)
Jay Penner, Pilot 288.17 kmh
William Cocker, Copilot 179.07 mph
Cessna 206 Floatplane
1 Allison C-250, 420 shp
John Brown Seaport to Lakeland Arpt. FL
3/21/86

SPEED OVER A 15/25 KM STRAIGHT COURSE (USSR)
Serguei Andreev, Pilot 573.27 kmh
Nikolay Shylykov, Copilot 356.23 mph
M-12 Chayka
Podmoskovnoya, USSR
10/30/87

SPEED OVER A 100 KM CLOSED CIRCUIT WITHOUT PAYLOAD (USSR)
Valadimir Sviatochnuk 596.00 kmh
M-12 Chaika 370.34 mph
4/19/76

SPEED OVER A 500 KM CLOSED CIRCUIT WITHOUT PAYLOAD (USSR)
A. Souchko 552.28 kmh
M-12 Chaika 343.17 mph
4/24/68

SPEED OVER A 1,000 KM CLOSED CIRCUIT WITHOUT PAYLOAD (USSR)
A. Souchko 544.69 kmh
M-12 Chaika 338.45 mph
10/31/72

SPEED OVER A 1,000 KM CLOSED CIRCUIT WITH 1,000 KG PAYLOAD (USSR)
A. Suchkov, A. Zakharov, 526.01 kmh
M. Moskalenko, F. Mogak 326.85 mph
M-12 Chaika 2 AI-20
4/21/70

SPEED OVER A 1,000 KM CLOSED CIRCUIT WITH 2,000 KG PAYLOAD (USSR)
A. Smirnov 530.50 kmh
M-12 Chaika 329.64 mph
7/8/70

SPEED OVER A 1,000 KM CLOSED CIRCUIT WITH 5,000 KG PAYLOAD (USSR)
A. Zakharov 526.61 kmh
M-12 Chaika 327.22 mph
2 AI-20D
Evpatoria, USSR
7/9/70

SPEED OVER A 2,000 KM CLOSED CIRCUIT WITHOUT PAYLOAD (USSR)
P. Yakouchine 556.79 kmh
M-12 Chaika 345.97 mph
10/9/68

SPEED OVER A 2,000 KM CLOSED CIRCUIT WITH 1,000 KG PAYLOAD (USSR)
P. Yakouchine 556.79 kmh
M-12 Chaika 345.97 mph
10/31/72

SPEED OVER A 2,000 KM CLOSED CIRCUIT WITH 2,000 KG PAYLOAD (USSR)
P. Yakouchine 556.79 kmh
M-12 Chaika 345.97 mph
10/31/72

SPEED OVER A 2,000 KM CLOSED CIRCUIT WITH 5,000 KG PAYLOAD (USSR)
A. Souchko 488.72 kmh
M-12 Chaika 303.67 mph
10/30/73

TIME TO CLIMB

TIME TO 3,000 METERS (USA)
Bobby K. Wilkes 4 min
Maule MX-7-420 23 sec
1 Allison 250-B17C
Lakeland, FL
4/21/98

TIME TO 6,000 METERS (USSR)
Guennadi Effimov 11 min
M-12 Chaika 21 sec
2 AI-20
7/15/81

TIME TO 9,000 METERS (USSR)
Nikolai Shlykov 25 min
M-12 Chaika 32 sec
2 AI-20
7/21/81

CLASS C-3.B (600-1,200 kg / 1,323-2,646 lbs) GROUP II (TURBOPROP)

TIME TO CLIMB

TIME TO 3,000 METERS (USA)
Bobby K. Wilkes 4 min
Maule MX-7-420 23 sec
1 Allison 250-B17C
Lakeland, FL
4/21/98

CLASS C-3.C (1,200-2,100 kg / 2,646-4,630 lbs) GROUP II (TURBOPROP)

ALTITUDE

ALTITUDE (USA)
Robert F. Harvey — 6,707 m / 22,005 ft
Cessna U206
Allison 350-B17D, 420 hp
Lakeland - Winterhaven, FL
3/23/85

SPEED

SPEED OVER A 3 KM STRAIGHT COURSE (USA)
F. J. "Jack" Schweibold — 286.92 kmh / 178.75 mph
Phil A. Cron, Co-Captain
Cessna U206U
Allison 250 C205, 420 hp
Acadiana, LA
4/10/85

SPEED OVER A 15/25 KM STRAIGHT COURSE (USA)
F. J. "Jack" Schweibold — 285.18 kmh / 177.21 mph
Cessna U260G
Allison 250-C30S, 420 hp
Lakeland - Winterhaven, FL
3/22/85

SPEED OVER A 100 KM CLOSED CIRCUIT WITHOUT PAYLOAD (USA)
C. V. Glines, Co-Captain — 264.24 kmh / 164.20 mph
Jay Penner, Co-Captain
Cessna 206 Floatplane
1 Allison 250 C20S, 420 shp
Bloomington, IN
8/12/86

TIME TO CLIMB

TIME TO 3,000 METERS (USA)
Robert F. Harvey — 7 min 48 sec
Cessna U206G
Allison 350-B17D, 420 hp
Lakeland - Winterhaven, FL
3/23/85

TIME TO 6,000 METERS (USA)
Robert F. Harvey — 23 min 25 sec
Cessna U206G
Allison 350-B17D, 420 hp
Lakeland - Winterhaven, FL
3/23/85

CLASS C-3.H (More Than 20,000 kg / 44,092 lbs) GROUP II (TURBOPROP)

SPEED

SPEED OVER A 15/25 KM STRAIGHT COURSE (USSR)
Serguei Andreev, Pilot — 573.27 kmh / 356.23 mph
Nikolay Shylykov, Copilot
M-12 Chayka
Podmoskovnoya, USSR
10/30/87

CLASS C-3 (AMPHIBIANS) GROUP III (JET ENGINE)

ALTITUDE

ALTITUDE (USSR)
Bogdan I. Lissak, Pilot — 13,367 m / 43,855 ft
K.V. Babitch, Copilot
Beriev "Albatros"
2 Soloviev D-30 NKD 12,000 shp
9/13/89

ALTITUDE IN HORIZONTAL FLIGHT (USSR)
Bogdan I. Lissak, Pilot — 12,857 m / 42,184 ft
K.V. Babitch, Copilot
Beriev "Albatros"
2 Soloviev D-30 NKD 12,000ehp
9/13/89

ALTITUDE WITH 1,000 KG PAYLOAD (USSR)
Bogdan I. Lissak, Pilot — 13,367 m / 43,855 ft
K.V. Babitch, Copilot
Beriev "Albatros"
2 Soloviev D-30 NKD 12,000ehp
9/13/89

ALTITUDE WITH 2,000 KG PAYLOAD (USSR)
Bogdan I. Lissak, Pilot — 13,367 m / 43,855 ft
K.V. Babitch, Copilot
Beriev "Albatros"
2 Soloviev D-30 NKD 12,000ehp
9/13/89

ALTITUDE WITH 5,000 KG PAYLOAD (USSR)
Bogdan I. Lissak, Pilot — 13,367 m / 43,855 ft
K.V. Babitch, Copilot
Beriev "Albatros"
2 Soloviev D-30 NKD 12,000ehp
9/13/89

ALTITUDE WITH 10,000 KG PAYLOAD (USSR)
Bogdan I. Lissak, Pilot — 13,281 m / 43,573 ft
K.V. Babitch, Copilot
Beriev "Albatros"
2 Soloviev D-30 NKD 12,000ehp
9/14/89

SPEED

SPEED OVER A 500 KM CLOSED CIRCUIT WITHOUT PAYLOAD (RUSSIA)
Nikolai Okhotnikov — 731.70 kmh / 454.66 mph
Sergei Parkhaev
Beriev A-40 Albatross
Gelendzhik, Russia
9/9/00

SPEED OVER A 2,000 KM CLOSED CIRCUIT WITHOUT PAYLOAD (USSR)
Gennadi Kalyuzhny — 761.93 kmh / 473.44 mph
Albatros
7/19/91

SPEED OVER A 2,000 KM CLOSED CIRCUIT WITH 1,000 KG PAYLOAD (USSR)
Gennadi Kalyuzhny — 761.93 kmh / 473.44 mph
Albatros
7/19/91

SPEED OVER A 2,000 KM CLOSED CIRCUIT WITH 2,000 KG PAYLOAD (USSR)
Gennadi Kalyuzhny — 761.93 kmh / 473.44 mph
Albatros
7/19/91

SPEED OVER A 2,000 KM CLOSED CIRCUIT WITH 5,000 KG PAYLOAD (USSR)
Gennadi Kalyuzhny — 761.93 kmh / 473.44 mph
Albatros
7/19/91

SPEED OVER A 2,000 KM CLOSED CIRCUIT WITH 10,000 KG PAYLOAD (USSR)
Gennadi Kalyuzhny — 761.93 kmh / 473.44 mph
Albatros
7/19/91

TIME TO CLIMB

TIME TO 3,000 METERS (RUSSIA)
Konstantin V. Babitch, Pilot 3 min
Vladimir Demianovsky, Copilot 9.5 sec
Albatros
11/19/91

TIME TO 3,000 METERS WITH 1,000 KG PAYLOAD (RUSSIA)
Konstantin V. Babitch, Pilot 3 min
Vladimir Demianovsky, Copilot 9.5 sec
Albatros
11/19/91

TIME TO 3,000 METERS WITH 2,000 KG PAYLOAD (RUSSIA)
Konstantin V. Babitch, Pilot 3 min
Vladimir Demianovsky, Copilot 9.5 sec
Albatros
11/19/91

TIME TO 3,000 METERS WITH 5,000 KG PAYLOAD (RUSSIA)
Konstantin V. Babitch, Pilot 3 min
Vladimir Demianovsky, Copilot 9.5 sec
Albatros
11/19/91

TIME TO 3,000 METERS WITH 10,000 KG PAYLOAD (RUSSIA)
Konstantin V. Babitch, Pilot 3 min
Vladimir Demianovsky, Copilot 9.5 sec
Albatros
11/19/91

TIME TO 3,000 METERS WITH 15,000 KG PAYLOAD (RUSSIA)
Gennady Parchin, Pilot 3 min
Genaddy Kalujniy, Copilot 55 sec
Beriev A-40 Albatros
Gilindzhik-Black Sea, Russia
7/3/98

TIME TO 6,000 METERS (RUSSIA)
Konstantin V. Babitch, Pilot 6 min
Vladimir Demianovsky, Copilot 54.2 sec
Albatros
11/19/91

TIME TO 6,000 METERS WITH 1,000 KG PAYLOAD (RUSSIA)
Konstantin V. Babitch, Pilot 6 min
Vladimir Demianovsky, Copilot 54.2 sec
Albatros
11/19/91

TIME TO 6,000 METERS WITH 2,000 KG PAYLOAD (RUSSIA)
Konstantin V. Babitch, Pilot 6 min
Vladimir Demianovsky, Copilot 54.2 sec
Albatros
11/19/91

TIME TO 6,000 METERS WITH 5,000 KG PAYLOAD (RUSSIA)
Konstantin V. Babitch, Pilot 6 min
Vladimir Demianovsky, Copilot 54.2 sec
Albatros
11/19/91

TIME TO 6,000 METERS WITH 10,000 KG PAYLOAD (RUSSIA)
Konstantin V. Babitch, Pilot 6 min
Vladimir Demianovsky, Copilot 54.2 sec
Albatros
11/19/91

TIME TO 6,000 METERS WITH 15,000 KG PAYLOAD (RUSSIA)
Gennady Parchin, Pilot 9 min
Genaddy Kalujniy, Copilot 8 sec
Beriev A-40 Albatros
Gilindzhik-Black Sea, Russia
7/3/98

TIME TO 9,000 METERS WITH 5,000 KG PAYLOAD (RUSSIA)
Konstantin Babich 15 min
Yuri Sheffer 19 sec
Beriev Be-200
Gelendzhik, Russia
9/9/00

TIME TO 9,000 METERS WITH 15,000 KG PAYLOAD (RUSSIA)
Gennady Parchin, Pilot 15 min
Genaddy Kalujniy, Copilot 35 sec
Beriev A-40 Albatros
Gilindzhik-Black Sea, Russia
7/3/98

CLASS C-3.H (20,000-25,000 kg / 44,092-55,116 lbs)
GROUP III (JET ENGINE)

ALTITUDE

ALTITUDE (USSR)
Bogdan I. Lissak, Pilot 13,367 m
K.V. Babitch, Copilot 43,855 ft
Beriev "Albatros"
2 Soloviev D-30 NKD 12,000 shp
9/13/89

ALTITUDE IN HORIZONTAL FLIGHT (USSR)
Bogdan I. Lissak, Pilot 12,857 m
K.V. Babitch, Copilot 42,184 ft
Beriev "Albatros"
2 Soloviev D-30 NKD 12,000 shp
9/13/89

ALTITUDE WITH 1,000 KG PAYLOAD (USSR)
Bogdan I. Lissak, Pilot 13,367 m
K.V. Babitch, Copilot 43,855 ft
Beriev "Albatros"
2 Soloviev D-30 NKD 12,000 shp
9/13/89

ALTITUDE WITH 2,000 KG PAYLOAD (USSR)
Bogdan I. Lissak, Pilot 13,367 m
K.V. Babitch, Copilot 43,855 ft
Beriev "Albatros"
2 Soloviev D-30 NKD 12,000 shp
9/13/89

ALTITUDE WITH 5,000 KG PAYLOAD (USSR)
Bogdan I. Lissak, Pilot 13,367 m
K.V. Babitch, Copilot 43,855 ft
Beriev "Albatros"
2 Soloviev D-30 NKD 12,000 shp
9/13/89

ALTITUDE WITH 10,000 KG PAYLOAD (USSR)
Bogdan I. Lissak, Pilot 13,281 m
K.V. Babitch, Copilot 43,573 ft
Beriev "Albatros"
2 Soloviev D-30 NKD 12,000 shp
9/14/89

TIME TO CLIMB

TIME TO 3,000 METERS (RUSSIA)
Konstantin V. Babitch, Pilot 3 min
Vladimir Demianovsky, Copilot 9.5 sec
Albatros
11/19/91

TIME TO 3,000 METERS WITH 1,000 KG PAYLOAD (RUSSIA)
Konstantin V. Babitch, Pilot 3 min
Vladimir Demianovsky, Copilot 9.5 sec
Albatros
11/19/91

TIME TO 3,000 METERS WITH 2,000 KG PAYLOAD (RUSSIA)
Konstantin V. Babitch, Pilot 3 min
Vladimir Demianovsky, Copilot 9.5 sec
Albatros
11/19/91

TIME TO 3,000 METERS WITH 5,000 KG PAYLOAD (RUSSIA)
Konstantin V. Babitch, Pilot 3 min
Vladimir Demianovsky, Copilot 9.5 sec
Albatros
11/19/91

TIME TO 3,000 METERS WITH 10,000 KG PAYLOAD (RUSSIA)
Konstantin V. Babitch, Pilot 3 min
Vladimir Demianovsky, Copilot 9.5 sec
Albatros
11/19/91

TIME TO 6,000 METERS (RUSSIA)
Konstantin V. Babitch, Pilot 6 min
Vladimir Demianovsky, Copilot 54.2 sec
Albatros
11/19/91

TIME TO 6,000 METERS WITH 1,000 KG PAYLOAD (RUSSIA)
Konstantin V. Babitch, Pilot 6 min
Vladimir Demianovsky, Copilot 54.2 sec
Albatros
11/19/91

TIME TO 6,000 METERS WITH 2,000 KG PAYLOAD (RUSSIA)
Konstantin V. Babitch, Pilot 6 min
Vladimir Demianovsky, Copilot 54.2 sec
Albatros
11/19/91

TIME TO 6,000 METERS WITH 5,000 KG PAYLOAD (RUSSIA)
Konstantin V. Babitch, Pilot 6 min
Vladimir Demianovsky, Copilot 54.2 sec
Albatros
11/19/91

TIME TO 6,000 METERS WITH 10,000 KG PAYLOAD (RUSSIA)
Konstantin V. Babitch, Pilot 6 min
Vladimir Demianovsky, Copilot 54.2 sec
Albatros
11/19/91

TIME TO 9,000 METERS (RUSSIA)
Gennadi Kalyuzhniy 13 min
Albatros 29 sec
3/26/92

TIME TO 9,000 METERS WITH 1,000 KG PAYLOAD (RUSSIA)
Gennadi Kalyuzhniy 13 min
Albatros 29 sec
3/26/92

TIME TO 9,000 METERS WITH 2,000 KG PAYLOAD (RUSSIA)
Gennadi Kalyuzhniy 13 min
Albatros 29 sec
3/26/92

TIME TO 9,000 METERS WITH 5,000 KG PAYLOAD (RUSSIA)
Gennadi Kalyuzhniy 13 min
Albatros 29 sec
3/26/92

TIME TO 9,000 METERS WITH 10,000 KG PAYLOAD (RUSSIA)
Gennadi Kalyuzhniy 13 min
Albatros 29 sec
3/26/92

TIME TO 12,000 METERS (RUSSIA)
Gennadi Kalyuzhniy 24 min
Albatros 48 sec
3/26/92

TIME TO 12,000 METERS WITH 1,000 KG PAYLOAD (RUSSIA)
Gennadi Kalyuzhniy 24 min
Albatros 48 sec
3/26/92

TIME TO 12,000 METERS WITH 2,000 KG PAYLOAD (RUSSIA)
Gennadi Kalyuzhniy 24 min
Albatros 48 sec
3/26/92

TIME TO 12,000 METERS WITH 5,000 KG PAYLOAD (RUSSIA)
Gennadi Kalyuzhniy 24 min
Albatros 48 sec
3/26/92

TIME TO 12,000 METERS WITH 10,000 KG PAYLOAD (RUSSIA)
Gennadi Kalyuzhniy 24 min
Albatros 48 sec
3/26/92

CLASS C-3.I (25,000-35,000 kg / 55,116-77,162 lbs)
GROUP III (JET ENGINE)

TIME TO CLIMB

TIME TO 3,000 METERS (RUSSIA)
Nikolai Kuleshov 3 min
Nikolai Okhotnikov 35 sec
Yuri Gerasimov
Grigory Kaliuzhny
Beriev Be-200
2 Progress D-436T, 16,535 lbs
Gelendzhik, Russia
9/7/02

TIME TO 3,000 METERS WITH 1,000 KG PAYLOAD (RUSSIA)
Nikolai Kuleshov 3 min
Nikolai Okhotnikov 35 sec
Yuri Gerasimov
Grigory Kaliuzhny
Beriev Be-200
2 Progress D-436T, 16,535 lbs
Gelendzhik, Russia
9/7/02

TIME TO 3,000 METERS WITH 2,000 KG PAYLOAD (RUSSIA)
Nikolai Kuleshov 3 min
Nikolai Okhotnikov 35 sec
Yuri Gerasimov
Grigory Kaliuzhny
Beriev Be-200
2 Progress D-436T, 16,535 lbs
Gelendzhik, Russia
9/7/02

TIME TO 3,000 METERS WITH 5,000 KG PAYLOAD (RUSSIA)
Konstantin Babich — 3 min
Yuri Sheffer — 46 sec
Beriev Be-200
Gelendzhik, Russia
9/9/00

TIME TO 6,000 METERS (RUSSIA)
Nikolai Kuleshov — 7 min
Nikolai Okhotnikov — 39 sec
Yuri Gerasimov
Grigory Kaliuzhny
Beriev Be-200
2 Progress D-436T, 16,535 lbs
Gelendzhik, Russia
9/7/02

TIME TO 6,000 METERS WITH 1,000 KG PAYLOAD (RUSSIA)
Nikolai Kuleshov — 7 min
Nikolai Okhotnikov — 39 sec
Yuri Gerasimov
Grigory Kaliuzhny
Beriev Be-200
2 Progress D-436T, 16,535 lbs
Gelendzhik, Russia
9/7/02

TIME TO 6,000 METERS WITH 5,000 KG PAYLOAD (RUSSIA)
Konstantin Babich — 7 min
Yuri Sheffer — 39 sec
Beriev Be-200
Gelendzhik, Russia
9/9/00

TIME TO 9,000 METERS (RUSSIA)
Nikolai Kuleshov — 17 min
Nikolai Okhotnikov — 8 sec
Yuri Gerasimov
Grigory Kaliuzhny
Beriev Be-200
2 Progress D-436T, 16,535 lbs
Gelendzhik, Russia
9/7/02

TIME TO 9,000 METERS WITH 1,000 KG PAYLOAD (RUSSIA)
Nikolai Kuleshov — 17 min
Nikolai Okhotnikov — 8 sec
Yuri Gerasimov
Grigory Kaliuzhny
Beriev Be-200
2 Progress D-436T, 16,535 lbs
Gelendzhik, Russia
9/7/02

TIME TO 9,000 METERS WITH 2,000 KG PAYLOAD (RUSSIA)
Nikolai Kuleshov — 17 min
Nikolai Okhotnikov — 8 sec
Yuri Gerasimov
Grigory Kaliuzhny
Beriev Be-200
2 Progress D-436T, 16,535 lbs
Gelendzhik, Russia
9/7/02

TIME TO 9,000 METERS WITH 5,000 KG PAYLOAD (RUSSIA)
Konstantin Babich — 15 min
Yuri Sheffer — 19 sec
Beriev Be-200
Gelendzhik, Russia
9/9/00

CLASS C-3.L (More Than 60,000 kg / 132,277 lbs) GROUP III (JET ENGINE)

ALTITUDE

GREATEST PAYLOAD CARRIED TO 2,000 METERS (RUSSIA)
Gennady Parchin, Pilot — 15,040 kg
Genaddy Kalujniy, Copilot — 33,158 lbs
Beriev A-40 Albatros
Gilindzhik-Black Sea, Russia
7/3/98

SPEED

SPEED OVER A 500 KM CLOSED CIRCUIT WITHOUT PAYLOAD (RUSSIA)
Nikolai Okhotnikov — 731.70 kmh
Sergei Parkhaev — 454.66 mph
Beriev A-40 Albatross
Gelendzhik, Russia
9/9/00

TIME TO CLIMB

TIME TO 3,000 METERS WITH 15,000 KG PAYLOAD (RUSSIA)
Gennady Parchin, Pilot — 3 min
Genaddy Kalujniy, Copilot — 55 sec
Beriev A-40 Albatros
Gilindzhik-Black Sea, Russia
7/3/98

TIME TO 6,000 METERS WITH 15,000 KG PAYLOAD (RUSSIA)
Gennady Parchin, Pilot — 9 min
Genaddy Kalujniy, Copilot — 8 sec
Beriev A-40 Albatros
Gilindzhik-Black Sea, Russia
7/3/98

TIME TO 9,000 METERS WITH 15,000 KG PAYLOAD (RUSSIA)
Gennady Parchin, Pilot — 15 min
Genaddy Kalujniy, Copilot — 35 sec
Beriev A-40 Albatros
Gilindzhik-Black Sea, Russia
7/3/98

ROTORCRAFT

Man has long wanted to fly. The Chinese made toys of wood and feathers that went straight up. They still exist today. Leonardo da Vinci understood the theory of vertical flight but lacked the materials, manufacturing technology, and a power plant to build a machine. Many inventors of the modern era believed that vertical flight was the proper approach for man to take to achieve flight.

First to fly a helicopter was either M. Volumard or Paul Cornu, both of whom made serious attempts in France in 1907, although historians differ as to which deserves credit for having been first. M. Volumard made a flight with a tethered craft, the Bregeut-Richet Gyroplane #1 at Douai on September 29. Cornu made a free flight of his tandem-rotor device at Lisieux on November 13.

In the early 1920s, progress resumed. Henry Berliner flew one of his father's helicopters at College Park, Maryland, in early 1922. On December 18, 1922, George de Bothezat flew a large, complicated four-rotor craft for one and a half minutes, at an altitude of six feet, over a distance of three hundred feet.

Many other pioneers continued the early attempts, especially in Europe. The only successful one was Juan de la Cierva of Spain. His autogyros, although incapable of hover, were in production for many years between the World Wars.

A number of inventors built machines that actually got off the ground for short hops, but none were successful in sustained, full helicopter flight, except the Folke-Wolf, which flew in Germany in the 1930s. Hannah Reich, the world's first woman test pilot, made many flights in it, but it never went into full production.

In the United States, Igor Sikorsky never gave up his ambition to build a helicopter and finally succeeded in 1939. Three different models were produced for the Army Air Force in WW II. In June 1943, Lawrence D. Bell's Model 30 was flown on its first untethered flight. This led the way for the certification of the world's first commercial helicopter, the Bell 47, in March 1946. Since then, the U.S. has led the world in rotorcraft technology and operational applications.

The U.S. has set more National and World rotorcraft records than any other country; the first around-the-world flight by Ross Perot,Jr., in a Bell Jet Ranger, the first non-stop flight across the United States in a Piasecki H-21 by the U.S. Army, the first flight across the Atlantic in a Sikorsky HH-3 by the U. S. Air Force, and the first non-stop, unrefuelled flight across the United States by Bob Ferry in a Hughes OH-6A.

CLASS E-1, HELICOPTERS, GENERAL

SUBCLASS E-1.A (Less Than 500 kg / 1,102 lbs) GROUP I (PISTON ENGINE)

DISTANCE

DISTANCE WITHOUT LANDING (FRANCE)
Bruno Guimbal 481.34 km
Cabri 299.09 mi
1 Lycoming O-320, 150 bhp
Paris, France
5/21/96

DISTANCE OVER A CLOSED CIRCUIT WITHOUT LANDING (FRANCE)
Richard Fenwick 315.88 km
Robinson R22 196.29 mi
1 Lycoming O-320, 160 hp
Toussus - Deauville - Toussus
10/19/85

ALTITUDE

ALTITUDE (USA)
Wayne H. Mulgrew 5,938 m
Robinson R22, N8398A 19,480 ft
1 Lycoming O-320-B2C, 160 hp
Redding, CA
3/4/89

SPEED

SPEED OVER A 3 KM STRAIGHT COURSE (USA)
Wayne H. Mulgrew 231.12 kmh
Robinson R22 143.61 mph
1 Lycoming O-320, 160 hp
Red Bluff, CA
9/22/90

SPEED OVER A 15/25 KM STRAIGHT COURSE (USA)
Wayne H. Mulgrew 213.91 kmh
Robinson R22 132.91 mph
1 Lycoming O-320, 160 hp
Red Bluff, CA
9/22/90

100 KM CLOSED CIRCUIT WITHOUT PAYLOAD (USA)
Wayne H. Mulgrew 204.90 kmh
Robinson R22 127.32 mph
1 Lycoming O-320, 160 hp
Red Bluff, CA
9/22/90

TIME TO CLIMB

TIME TO 3,000 METERS (FRANCE)
Richard Fenwick, Pilot 7 min
Robinson R22 45 sec
1 Lycoming O-320, 160 hp
5/29/90

SUBCLASS E-1.B (500-1,000 kg / 1,102-2,205 lbs) GROUP I (PISTON ENGINE)

DISTANCE

DISTANCE WITHOUT LANDING (USA)
A. P. Averill 1,176.79 km
Bell 47-G 731.15 mi
Franklin O-335
El Paso to Texarkana, TX
2/8/61

ALTITUDE

ALTITUDE (USA)
James E. Bowman 9,076 m
Cessna YH-41 29,777 ft
1 Continental FSO-526A, 270 shp
Wichita, KS
12/28/57

SUBCLASS E-1.C (1,000-1,750 kg / 2,205-3,858 lbs) GROUP I (PISTON ENGINE)

ALTITUDE

ALTITUDE (USA)
James E. Bowman 8,562 m
Cessna YH-41 28,090 ft
1 Continental FSO-526A, 270 shp
Wichita, KS
12/28/57

SPEED

SPEED OVER A 3 KM STRAIGHT COURSE (USA)
Bertram G. Leach 198.68 kmh
Hiller OH-23G 123.45 mph
1 Lycoming VO-540-C2A, 305 shp
10/29/63

SPEED OVER A 15/25 KM STRAIGHT COURSE (USA)
Bertram G. Leach 199.18 kmh
Hiller OH-23G 123.76 mph
1 Lycoming VO-540-C2A, 305 shp
10/29/63

SPEED OVER A 100 KM CLOSED CIRCUIT WITHOUT PAYLOAD (USA)
Bertram G. Leach 192.72 kmh
Hiller OH-23G 119.75 mph
1 Lycoming VO-540-C2A, 305 shp
10/29/63

SPEED OVER A 500 KM CLOSED CIRCUIT WITHOUT PAYLOAD (USSR)
Vsevolod Vinitsky 170.45 kmh
Kamov KA-15 105.91 mph
1 AI-14, 260 shp
Mikhailovskaia Sloboda, USSR
5/6/59

SPEED AROUND THE WORLD, EASTBOUND (UK)
Jennifer Murray, Pilot 22.44 kmh
Quentin Smith, Copilot 13.94 mph
Robinson R44 Astro
Denham, UK
5/10-8/8/97

CLASS E-1 (HELICOPTERS) GROUP II (TURBINE ENGINE)

DISTANCE

DISTANCE WITHOUT LANDING (USA)
Robert G. Ferry 3,561.55 km
Hughes YOH-6A No. 24213 2,213.04 mi
1 Allison T-63-A-5
Culver City, CA to
Ormond Beach, FL
4/6-7/66

DISTANCE OVER A CLOSED CIRCUIT WITHOUT LANDING (USA)
F.J. "Jack" Schweibold 2,800.20 km
Hughes YOH-6A No. 24213 1,739.96 mi
1 Allison T-63-A-5
Edwards AFB, CA
3/26/66

ALTITUDE

ALTITUDE (FRANCE)
Jean Boulet 12,442 m
Alouette SA 315-001 "Lama" 40,820 ft
Artouste IIIB 735 KW
Istres, France
6/21/72

ALTITUDE IN HORIZONTAL FLIGHT (USA)
CWO James K. Church, US Army 11,010 m
Sikorsky CH-54B No. 18488 36,122 ft
2 P&W JFTD-12
Stratford, CT
11/4/71

ALTITUDE WITH 1,000 KG PAYLOAD (USA)
CAPT B.P. Blackwell, USA 9,499 m
Sikorsky CH-54B, No. 18488 31,165 ft
2 P&W JFTD-12
Stratford, CT
10/26/71

ALTITUDE WITH 2,000 KG PAYLOAD (USA)
CWO Eugene E. Price, USA 9,595 m
Sikorsky CH-54B No. 18488 31,480 ft
2 P&W JFTD-12
Stratford, CT
10/29/71

ALTITUDE WITH 5,000 KG PAYLOAD (USA)
CWO Eugene E. Price, USA 7,778 m
Sikorsky CH-54B, No. 18488 25,518 ft
2 P&W JFTD-12
Stratford, CT
10/27/71

ALTITUDE WITH 10,000 KG PAYLOAD (USSR)
G. P. Karapetjane, Pilot 6,400 m
J. F. Tchapaev, Pilot 20,992 ft
Mil MI-26
2 D-136, 11,400 shp
Podmoscovnoe, USSR
2/2/82

ALTITUDE WITH 15,000 KG PAYLOAD (USSR)
S. V. Petrov, Pilot 5,600 m
A. I. Tchetverik, Pilot 18,368 ft
Mil MI-26
2 D-136, 11,400 shp
Podmoscovnoe, USSR
2/4/82

ALTITUDE WITH 20,000 KG PAYLOAD (USSR)
A. P. Kholoupov, Pilot 4,600 m
V. I. Kostine, Pilot 15,088 ft
Mil MI-26
2 D-136, 11,400 shp
Podmoscovnoe, USSR
2/4/82

ALTITUDE WITH 25,000 KG PAYLOAD (USSR)
G. V. Alfeurov, Pilot 4,100 m
L. A. Indeev, Pilot 13,448 ft
Mil MI-26
2 D-136, 11,400 shp
Podmoscovnoe, USSR
2/3/82

ALTITUDE WITH 30,000 KG PAYLOAD (USSR)
Vassili P. Kolochenko 2,951 m
Mil V-12, 4D-25-VF 9,682 ft
Podmoscovnoe, USSR
2/22/69

ALTITUDE WITH 35,000 KG PAYLOAD (USSR)
Vassili P. Kolochenko 2,255 m
Mil V-12, 4 D-25-VF 7,398 ft
Podmoscovnoe, USSR
8/6/69

ALTITUDE WITH 40,000 KG PAYLOAD (USSR)
Vassili P. Kolochenko 2,255 m
Mil V-12, 4 D-25-VF 7,398 ft
Podmoscovnoe, USSR
8/6/69

GREATEST PAYLOAD CARRIED TO 2,000 METERS (USSR)
G. V. Alfeurov, Pilot 40,204.50 kg
L. A. Indeev, Pilot 88,635.80 lbs
Mil MI-26
2 D-136, 11,400 shp
Podmoscovnoe, USSR
2/3/82

SPEED

SPEED OVER A 3 KM STRAIGHT COURSE (FRANCE)
Guy Dabadie, Pilot 372.00 kmh
I. B. Fouques, Copilot 231.15 mph
Aérospatiale SA 365 Dauphin
11/19/91

SPEED OVER A 15/25 KM STRAIGHT COURSE (UK)
John Trevor Egginton, Pilot 400.87 kmh
Derek J. Clews, Copilot 249.10 mph
Westland Lynx
2 RR GEM Mk. 531
Glastonbury, Somerset
8/11/86

SPEED OVER A 100 KM CLOSED CIRCUIT WITHOUT PAYLOAD (USSR)
Boris Galitsky 340.15 kmh
Mil MI-6 211.35 mph
2 TB-2BM
Podmoscovnoe, USSR
8/26/64

SPEED OVER A 500 KM CLOSED CIRCUIT WITHOUT PAYLOAD (USA)
Thomas Doyle 345.74 kmh
Sikorsky S-76 214.84 mph
2 Allison 250-C-30, 650 shp
West Palm Beach, FL
2/8/82

SPEED OVER A 1,000 KM CLOSED CIRCUIT WITHOUT PAYLOAD (USSR)
Galina Rastorgoueva 322.65 kmh
Mil A-10 200.48 mph
2 TV2 117A
8/13/75

SPEED OVER A 1,000 KM CLOSED CIRCUIT WITH 1,000 KG PAYLOAD (USSR)
Boris Galitsky 300.37 kmh
Mil MI-6; 2 TB-2BM 186.64 mph
Touchino, Kalouga, Tikhonova
Poustnyi, Touchino, USSR
9/15/62

SPEED OVER A 1,000 KM CLOSED CIRCUIT 2,000 KG PAYLOAD (USSR)
Boris Galitsky 300.38 kmh
Mil MI-6; 2 TB-2BM 186.64 mph
Touchino, Kalouga, Tikhanova
Poustnyi, Touchino, USSR
9/15/62

SPEED OVER A 1,000 KM CLOSED CIRCUIT 5,000 KG PAYLOAD (USSR)
Vassili P. Kolochenko 284.35 kmh
Mil MI-6; 2 TB-2BM 176.69 mph
Touchino, USSR
9/11/62

SPEED OVER A 2,000 KM CLOSED CIRCUIT WITHOUT PAYLOAD (USSR)
Anatoly Razbegaev — 278.96 kmh
MI-26 — 173.35 mph
8/17/88

SPEED AROUND THE WORLD, EASTBOUND (USA) ★
Joe Ronald Bower — 65.97 kmh
Bell JetRanger III — 40.99 mph
1 Allison 250-C20J, 317 shp
Hurst, TX
6/28-7/22/94

SPEED AROUND THE WORLD, WESTBOUND (USA)
Joe Ronald Bower — 91.75 kmh
John W. Williams — 57.01 mph
Bell 430
2 Allison 250-C40, 811 shp
Hurst, TX
8/17-9/3/96

TIME TO CLIMB

TIME TO 3,000 METERS (USA)
MAJ John C. Henderson, USA — 1 min
Sikorsky CH-54B — 22 sec
2 P&W JFTD-12
Stratford, CT
4/12/72

TIME TO 6,000 METERS (USA)
MAJ John C. Henderson, USA — 2 min
Sikorsky CH-54B — 59 sec
2 P&W JFTD-12
Stratford, CT
4/12/72

TIME TO 9,000 METERS (USA)
CWO Delbert W. Hunt, USA — 5 min
Sikorsky CH-54B, No. 18488 — 58 sec
2 P&W JFTD-12
Stratford, CT
11/4/71

SUBCLASS E-1.A (Less Than 500 kg / 1,102 lbs) GROUP II (TURBINE ENGINE)

ALTITUDE

ALTITUDE (FRANCE)
Jean Dabos — 4,789 m
SO 1221 — 15,712 ft
1 Turboméca Palouste
Villacoublay, France
12/29/53

SUBCLASS E-1.B (500-1,000 kg / 1,102-2,205 lbs) GROUP II (TURBINE ENGINE)

DISTANCE

DISTANCE OVER A CLOSED CIRCUIT WITHOUT LANDING (USA)
Jack L. Zimmerman — 1,700.12 km
Hughes YOH-6A — 1,056.41 mi
1 Allison T-63-A5
Edwards AFB, CA
3/21/66

ALTITUDE

ALTITUDE (FRANCE)
Jean Boulet — 12,442 m
Alouette SA315-001 Lama — 40,820 ft
Artouste III B 735
Istres, France
6/21/72

ALTITUDE IN HORIZONTAL FLIGHT (USA)
Terrell R. Clark — 8,503 m
Hiller UH12-E — 27,889 ft
AVCO Lycoming
TiV-540-A2A, 315 HP
Grand Junction, CO
3/16/83

SPEED

SPEED OVER A 3 KM STRAIGHT COURSE (USA)
LTC Richard T. Heard — 274.73 kmh
Hughes YOH-6A — 170.70 mph
1 Allison T-63-A-5
Edwards AFB, CA
3/24/66

SPEED OVER A 15/25 KM STRAIGHT COURSE (USA)
LTC Richard T. Heard — 276.51 kmh
Hughes YOH-6A — 171.85 mph
1 Allison T-63-A-5
Edwards AFB, CA
3/24/66

SPEED OVER A 100 KM CLOSED CIRCUIT WITHOUT PAYLOAD (USA)
MAJ A.L. Darling — 259.47 kmh
Hughes YOH-6A — 161.22 mph
1 Allison T-63-A-5
Edwards AFB, CA
3/13/66

SPEED OVER A 500 KM CLOSED CIRCUIT WITHOUT PAYLOAD (USA)
MAJ A.L. Darling — 254.59 kmh
Hughes YOH-6A — 158.19 mph
1 Allison T-63-A-5
Edwards AFB, CA
3/13/66

SPEED OVER A 1,000 KM CLOSED CIRCUIT WITHOUT PAYLOAD (USA)
MAJ A.L. Darling — 249.76 kmh
Hughes YOH-6A — 155.19 mph
1 Allison T-63-A-5
Edwards AFB, CA
3/13/66

TIME TO CLIMB

TIME TO 3,000 ME TERS (USA)
Steve Hanvey — 3 min
Hughes 530 F — 15 sec
1 Allison 250-C30, 650 hp
Thermal, CA
8/30/84

TIME TO 6,000 METERS (USA)
Steve Hanvey — 6 min
Hughes 530 F — 34 sec
1 Allison 250-C30, 650 HP
Thermal, CA
8/30/84

TIME TO 9,000 METERS (FRANCE)
Jean Boulet — 17 min
Alouette SA315-001 Lama — 44 sec
Artouste III B 735
Istres, France
6/13/58

SUBCLASS E-1.C (1,000-1,750 kg / 2,205-3,858 lbs) GROUP II (TURBINE ENGINE)

DISTANCE

DISTANCE WITHOUT LANDING (USA)
Robert G. Ferry — 3,561.55 km / 2,213.04 mi
Hughes YOH-6A
1 Allison T-63-A-5
Culver City, CA to Ormond Beach, FL
4/6-7/66

DISTANCE OVER A CLOSED CIRCUIT WITHOUT LANDING (USA)
F. J. "Jack" Schweibold — 2,800.20 km / 1,739.96 mi
Hughes YOH-6A
1 Allison T-63-A-5
Edwards AFB, CA
3/26/66

ALTITUDE

ALTITUDE (FRANCE)
Jean Boulet — 10,856 m / 35,617 ft
Alouette SA-351-001
Artouste IIIB 735 KW
Istres, France
6/19/72

ALTITUDE IN HORIZONTAL FLIGHT (USA)
Terrell R. Clark — 8,184 m / 26,843 ft
Hiller UH12-E
AVCO Lycoming TIV-540-A2A, 315 HP
Grand Junction, CO
3/16/83

SPEED

SPEED OVER A 3 KM STRAIGHT COURSE (FRANCE)
Dennis Prost — 310.00 kmh / 192.60 mph
SA 341-01 Gazelle
Istres, France
5/13/71

SPEED OVER A 15/25 KM STRAIGHT COURSE (FRANCE)
Dennis Prost — 312.00 kmh / 193.80 mph
SA 341-01 Gazelle
Istres, France
5/14/71

SPEED OVER A 100 KM CLOSED CIRCUIT WITHOUT PAYLOAD (FRANCE)
Dennis Prost — 296.00 kmh / 183.90 mph
SA 341-01 Gazelle
Istres, France
5/13/71

SPEED OVER A 500 KM CLOSED CIRCUIT WITHOUT PAYLOAD (USA)
COL David M. Kyle — 249.84 kmh / 155.24 mph
Hughes YOH-6A
1 Allison T-63-A-5
Edwards AFB, CA
3/12/66

SPEED OVER A 1,000 KM CLOSED CIRCUIT WITHOUT PAYLOAD (USA)
COL David M. Kyle — 246.38 kmh / 153.09 mph
Hughes YOH-6A
1 Allison T-63-A-5
Edwards AFB, CA
3/12/66

SPEED OVER A 2,000 KM CLOSED CIRCUIT WITHOUT PAYLOAD (USA)
CWO Richard D. Szczepanski — 227.74 kmh / 141.52 mph
Hughes YOH-6A
1 Allison T-63-A-5
Edwards AFB, CA
3/20/66

SPEED AROUND THE WORLD, EASTBOUND (USA) ★
Joe Ronald Bower — 65.97 kmh / 40.99 mph
Bell JetRanger III
1 Allison 250-C20J, 317 shp
Hurst, TX
6/28-7/22/94

TIME TO CLIMB

TIME TO 3,000 METERS (FRANCE)
Pierre Loranchet — 2 min 59 sec
Bernard Certain
AS 350 B1 "Ecureuil"
1 Turbomeca Arriel 1 D
Istres
5/14/85

TIME TO 6,000 METERS (FRANCE)
Pierre Loranchet — 6 min 55 sec
Bernard Certain
AS 350 B1 "Ecureuil"
1 Turbomeca Arriel 1 D
Istres
5/14/85

TIME TO 9,000 METERS (FRANCE)
Pierre Loranchet — 13 min 51 sec
Bernard Certain
AS 350 B1 "Ecureuil"
1 Turbomeca Arriel 1 D
Istres
5/14/85

SUBCLASS E-1.D (1,750-3,000 kg / 3,858-6,614 lbs) GROUP II (TURBINE ENGINE)

DISTANCE

DISTANCE WITHOUT LANDING (W. GERMANY)
Siegfried Hoffman, Pilot — 1,714.84 km / 1,065.55 mi
Adam Teleki, Flight Engineer
BO 105-C, D-HABV
2 Allison 250-C20, 640 shp
Ottobrun, W. Germany to Hinojosa de Duque, Spain
4/20/74

DISTANCE OVER A CLOSED CIRCUIT WITHOUT LANDING (USSR)
Alexay Anossov — 1,528.77 km / 949.93 mi
Mil MI-1; AL-26B
4/17/68

ALTITUDE

ALTITUDE (USA)
MAJ E.F. Sampson, USA — 10,714 m / 35,150 ft
Bell UH-1D
1 Lycoming T-52-L-13
Forth Worth, TX
12/11/64

ALTITUDE IN HORIZONTAL FLIGHT (USA)
Byron Graham — 7,940 m / 26,049 ft
Sikorsky S-76
2 Allison 250-C-30, 650 shp
West Palm Beach, FL
2/5/82

SPEED

SPEED OVER A 3 KM STRAIGHT COURSE (USA)
Nicholas Lappos 335.50 kmh
Sikorsky S-76 208.48 mph
West Palm Beach, FL
2/4/82

SPEED OVER A 15/25 KM STRAIGHT COURSE (USA)
Nicholas Lappos 342.61 kmh
Sikorsky S-76 212.90 mph
West Palm Beach, FL
2/4/82

40 years ago - Major John Johnston with the Bell UH-1D in which he set a closed circuit speed record in September of 1964.

SPEED OVER A 100 KM CLOSED CIRCUIT WITHOUT PAYLOAD (USA)
David Wright 331.22 kmh
Sikorsky S-76 205.82 mph
West Palm Beach, FL
2/6/82

SPEED OVER A 500 KM CLOSED CIRCUIT WITHOUT PAYLOAD (USA)
MAJ Billy L. Odneal, USA 286.83 kmh
Bell UH-1D 178.22 mph
1 Lycoming T-53-L-13
Forth Worth, TX
11/23/64

SPEED OVER A 1,000 KM CLOSED CIRCUIT WITHOUT PAYLOAD (USA)
MAJ John A. Johnston, USA 235.10 kmh
Bell UH-1D 178.09 mph
1 Lycoming T-53-L-13
Forth Worth, TX
9/15/64

SPEED AROUND THE WORLD, EASTBOUND (USA)
H. Ross Perot, Jr., Pilot 56.97 kmh
Jay W. Coburn, Copilot 35.40 mph
Bell 206 L-II Long Ranger, N3911Z
1 Allison 250-C28B, 435 hp
Dallas, Indianapolis, Montreal, Sept Iles, Shefferville, Koartac, Frobisher Bay, Sondrestrom AB, Angmagssalik, Keflavik, Sumburgh, Stansted, Farnborough, Tilford, London, Marseille, Naples, Athens, Cairo, Luxor, Jeddah, Riyadh, Bahrein, Muscat, Karachi, Delhi, Calcutta, Rangoon, Mergui, Phuket, Singapore, Kuching, Kota, Kinabaul, Manila, Clark AFB, Laoag, Taipei, Naha, Kagoshima, Fukuoka, Niigata, Kushiro, SS President McKinley, Shemya AFB, Adak, Cold Bay, Anchorage, Whitehorse Yukon, Fort Nelson, Fort St. John, Simonette, Calgary, Billings, Wray, Renner Field, Garden City, Dallas
29 days, 3 hrs, 8 min, 13 sec
9/1-30/82

Time to Climb

TIME TO 3,000 METERS (FRANCE)
Max Jot, Pilot 2 min
Pierre Rougier, Copilot 53 sec
SA 365M "Panther"
2 Turbomeca TM3331A, 837 hp & 750 hp
Istres - Le Tube
9/15/87

TIME TO 6,000 METERS (FRANCE)
Max Jot, Pilot 6 min
Pierre Rougier, Copilot 13 sec
SA 365M "Panther"
2 Turbomeca TM3331A, 837 hp & 750 hp
Istres - Le Tube
9/15/87

SUBCLASS E-1.E (3,000-4,500 kg / 6,614-9,921 lbs)
GROUP II (TURBINE ENGINE)

Distance

DISTANCE WITHOUT LANDING (USA) ★
Ron Williamson 3,180.15 km
Bell UH-1H 1,976.05 mi
1 LHTEC T800, 1,350 shp
Oxnard, CA - Marietta, GA
4/22/93

DISTANCE OVER A CLOSED CIRCUIT WITHOUT LANDING (USA)
CWO Joseph C. Watts, USA 2,000.15 km
Bell UH-1D 1,242.83 mi
1 Lycoming T-53-L-11
Edwards AFB, CA
9/23/64

Altitude

ALTITUDE (IRAN)
MGEN M. Khosrowdad 9,071 m
Bell Model 214A 29,760 ft
1 Lycoming LTC4B-8D
Ahwaz, Iran
4/29/75

ALTITUDE IN HORIZONTAL FLIGHT (IRAN)
MGEN M. Khosrowdad 9,071 m
Bell Model 214A 29,760 ft
1 Lycoming LTC4B-8D
Ahwaz, Iran
4/29/75

ALTITUDE WITH 1,000 KG PAYLOAD (RUSSIA)
Nikolay Grigoriev 4,760 m
MIL Mi-2 15,617 ft
2 GTD-350 IV, 400 shp
Moscow, Russia
1/27/97

GREATEST PAYLOAD CARRIED TO 2,000 METERS (RUSSIA)
Nikolay Grigoriev 1,734 kg
MIL Mi-2 3,823 lbs
2 GTD-350 IV, 400 shp
Moscow, Russia
3/24/97

Speed

SPEED OVER A 3 KM STRAIGHT COURSE (FRANCE)
Guy Dabadie, Pilot 372.00 kmh
I. B. Fouques, Copilot 231.15 mph
Aérospatiale SA 365 Dauphin
11/19/91

SPEED OVER A 15/25 KM STRAIGHT COURSE (UK)
John Trevor Egginton, Pilot 400.87 kmh
Derek J. Clews, Copilot 249.10 mph
Westland Lynx
2 RR GEM Mk. 531
Gastonbury, Somerset
8/11/86

SPEED OVER A 100 KM CLOSED CIRCUIT WITHOUT PAYLOAD (USA)
David Wright 334.69 kmh
Sikorsky S-76 207.97 mph
2 Allison 250-C-30, 650 shp
West Palm Beach, FL
2/6/82

SPEED OVER A 500 KM CLOSED CIRCUIT WITHOUT PAYLOAD (USA)
Thomas Doyle 345.74 kmh
Sikorsky S-76 214.84 mph
2 Allison 250-C-30, 650 shp
West Palm Beach, FL
2/8/82

SPEED OVER A 1,000 KM CLOSED CIRCUIT WITHOUT PAYLOAD (USA)
David Wright 305.10 kmh
Sikorsky S-76 189.59 mph
2 Allison 250-C-30, 650 shp
West Palm Beach, FL
2/9/82

SPEED OVER A 2,000 KM CLOSED CIRCUIT WITHOUT PAYLOAD (USA)
CWO Joseph C. Watts, USA 215.63 kmh
Bell UH-1D 133.98 mph
1 Lycoming T-53-L-11
Edwards AFB, CA
9/23/64

SPEED AROUND THE WORLD, WESTBOUND (USA)
Joe Ronald Bower 91.75 kmh
John W. Williams 57.01 mph
Bell 430
2 Allison 250-C40, 811 shp
Hurst, TX
8/17-9/3/96

TIME TO CLIMB

TIME TO 3,000 METERS (IRAN)
MGEN M. Khosrowdad 1 min
Bell Model 214A 58 sec
1 Lycoming LTC4B-8D
Ahwaz, Iran
4/29/75

TIME TO 3,000 METERS WITH 1,000 KG PAYLOAD (RUSSIA)
Nikolay Grigoriev 7min
MIL Mi-2 55 sec
2 GTD-350 IV, 400 shp
Moscow, Russia
2/20/97

TIME TO 6,000 METERS (IRAN)
MGEN M. Khosrowdad 5 min
Bell Model 214A 13 sec
1 Lycoming LTC4B-8D
Ahwaz, Iran
4/29/75

TIME TO 9,000 METERS (IRAN)
MGEN M. Khosrowdad 15 min
Bell Model 214A 5 sec
1 Lycoming LTC4B-1D
Ahwaz, Iran
4/29/75

SUBCLASS E-1.F (4,500-6,000 kg / 9,921-13,228 lbs) GROUP II (TURBINE ENGINE)

DISTANCE

DISTANCE WITHOUT LANDING (USA)
F. J. "Jack" Schweibold, Pilot 1,508.91 km
John B. Silva, Copilot 937.59 mi
Sikorsky S-76
2 Allison 250-C30S
Dallas, TX to Toledo OH
4/12-13/85

ALTITUDE

ALTITUDE IN HORIZONTAL FLIGHT (USA)
Gerald T. Golden, Pilot 4,590 m
Harry B. Sutton, Copilot 15,060 ft
Sikorsky S-76
2 Allison 250-C30S
Lafayette, LA
4/11/85

SPEED

SPEED OVER A 3 KM STRAIGHT COURSE (USA)
Leslie E. White, Pilot 312.15 kmh
F. J. "Jack" Schweibold, Copilot 193.97 mph
Sikorsky S-76
2 Allison 250-C30S
Lafayette, LA
4/9-10/85

SPEED OVER A 15/25 KM STRAIGHT COURSE (USA)
Vernon Albert, Pilot 304.73 kmh
F. J. "Jack" Schweibold, Copilot 189.35 mph
Arthur Chadbourne, Crew Chief
Sikorsky S-76
2 Allison 250-C30S
Lafayette, LA
4/9/85

TIME TO CLIMB

TIME TO 3,000 METERS (USA)
Joseph R. Bolen, Pilot 6 min
Harry B. Sutton, Copilot 15 sec
Bruce A. Schneider, Crew Chief
Sikorsky S-76
2 Allison 250-C30S
Lafayette, LA
4/9/85

SUBCLASS E-1.G (6,000-10,000 kg / 13,228-22,046 lbs) GROUP II (TURBINE ENGINE)

ALTITUDE

ALTITUDE (USSR)
Igor V. Piliai 6,230 m
Mil Mi-8 20,440 ft
9/17/91

ALTITUDE WITH 1,000 KG PAYLOAD (USSR)
Igor V. Piliai 6,230 m
Mil Mi-8 20,440 ft
9/17/91

SPEED

SPEED OVER A 100 KM CLOSED CIRCUIT WITHOUT PAYLOAD (USSR)
Yuri V. Jutchkov 246.94 kmh
Mil Mi-8 153.44 mph
7/18/91

SUBCLASS E-1.H (10,000-20,000 kg / 22,046-44,092 lbs) GROUP II (TURBINE ENGINE)

ALTITUDE

ALTITUDE (USSR)
Vladimir A. Pukhvatov 5,930 m
Mil Mi-8 19,455 ft
9/18/91

ALTITUDE WITH 1,000 KG PAYLOAD (USSR)
Vladimir A. Pukhvatov 5,930 m
Mil Mi-8 19,455 ft
9/18/91

ALTITUDE WITH 2,000 KG PAYLOAD (USSR)
Vladimir A. Pukhvatov 5,685 m
Mil Mi-8 18,652 ft
9/19/91

GREATEST PAYLOAD CARRIED TO 2,000 METERS (USSR)
Valery Kalashnikov 4,014.00 kg
Mil Mi-8 8,849.35 lbs
9/19/91

SPEED

SPEED OVER A 500 KM CLOSED CIRCUIT WITHOUT PAYLOAD (USSR)
Valery G. Kotchine 259.40 kmh
Mil Mi-8 161.18 mph
7/22/91

CLASS E-1, HELICOPTER, FEMININE

CLASS E-1 (HELICOPTERS) GROUP I (PISTON ENGINE)

SPEED

SPEED AROUND THE WORLD, EASTBOUND (UK)
Jennifer Murray 16.99 kmh
Robinson R44 Astro 10.55 mph
Brooklands, UK
9/6/00

SUBCLASS E-1.A (Less Than 500 kg / 1,102 lbs) GROUP I (PISTON ENGINE)

ALTITUDE

ALTITUDE (UK)
Georgina Hunter-Jones 4,475 m
Robinson R-22 14,681 ft
2/27/94

SUBCLASS E-1.C (1,000-1,750 kg / 2,205-3,858 lbs) GROUP I (PISTON ENGINE)

SPEED

SPEED AROUND THE WORLD, EASTBOUND (UK)
Jennifer Murray 16.99 kmh
Robinson R44 Astro 10.55 mph
Brooklands, UK
9/6/00

CLASS E-1 (HELICOPTERS) GROUP II (TURBINE ENGINE)

DISTANCE

DISTANCE WITHOUT LANDING (USSR)
Inna Kopets 2,232.22 km
Mil MI-8 1,387.03 mi
2 "V" type turbines
8/15/69

DISTANCE OVER A CLOSED CIRCUIT WITHOUT LANDING (USSR)
Inna Kopets 2,082.22 km
Mil MI-8 1,293.80 mi
9/14/67

ALTITUDE

ALTITUDE (USSR)
Tatiana Zueva, Pilot 8,250 m
Nadezhda Eremina, CoPilot 27,067 ft
Ka-32
2 TV3 - 117V, 2,225 hp
Podmoskovnoe
1/29/85

ALTITUDE IN HORIZONTAL FLIGHT (USSR)
Tatiana Zueva, Pilot 8,215 m
Nadezhda Eremina, CoPilot 26,952 ft
Ka-32
2 TV3 - 117V, 2,225 hp
Podmoskovnoe
1/29/85

ALTITUDE WITH 1,000 KG PAYLOAD (USSR)
Nadezhda Eremina, Pilot 7,305 m
Tatiana Zueva, CoPilot 23,967 ft
Ka-32
2 TV3 - 117V, 2,225 hp
Podmoskovnoe
1/29/85

ALTITUDE WITH 2,000 KG PAYLOAD (USSR)
Nadezhda Eremina, Pilot 6,400 m
Tatiana Zueva, CoPilot 20,997 ft
Ka-32
2TV3 - 117V, 2,225 hp
Podmoskovnoe
1/29/85

ALTITUDE WITH 5,000 KG PAYLOAD (USSR)
Inna Kopets, Pilot 5,750 m
Valentina Volkova, CoPilot 18,860 ft
MI-26
2 D-136, 11,400 hp
Podmoscovnoe
12/1/82

ALTITUDE WITH 10,000 KG PAYLOAD (USSR)
Inna Kopets, Pilot 5,750 m
Valentina Volkova, CoPilot 18,860 ft
MI-26
2 D-136, 11,400 hp
Podmoscovnoe
12/1/82

ALTITUDE WITH 15,000 KG PAYLOAD (USSR)
Inna Kopets, Pilot 4,800 m
Valentina Volkova, CoPilot 15,744 ft
MI-26
2 D-136, 11,400 hp
Podmoscovnoe
12/2/82

ALTITUDE WITH 20,000 KG PAYLOAD (USSR)
Inna Kopets, Pilot 4,050 m
Valentina Volkova, CoPilot 13,284 ft
MI-26
2 D-136, 11,400 hp
Podmoscovnoe
12/3/82

ALTITUDE WITH 25,000 KG PAYLOAD (USSR)
Inna Kopets, Pilot 3,750 m
Valentina Volkova, CoPilot 12,300 ft
MI-26
2 D-136, 11,400 hp
Podmoscovnoe
12/3/82

GREATEST PAYLOAD CARRIED TO 2,000 METERS (USSR)
Inna Kopets, Pilot 25,110.70 kg
Valentina Volkova, CoPilot 55,344.00 lbs
MI-26
2 D0136, 11,400 hp
Podmoscovnoe
12/3/82

SPEED

SPEED OVER A 15/25 KM STRAIGHT COURSE (USSR)
Galina Rastorgoueva 341.32 kmh
Mil A-10 212.07 mph
2 TV2-117A, 1500 shp
Ramenskoye, USSR
7/16/75

SPEED OVER A 100 KM CLOSED CIRCUIT WITHOUT PAYLOAD (USSR)
Galina Rastorgoueva 334.46 kmh
Mil A-10 207.82 mph
2 TV-2117A, 1,500 shp
Ramenskoye, USSR
7/18/75

SPEED OVER A 500 KM CLOSED CIRCUIT WITHOUT PAYLOAD (USSR)
Galina Rastorgoueva 331.02 kmh
Mil A-10 205.69 mph
8/1/75

SPEED OVER A 1,000 KM CLOSED CIRCUIT WITHOUT PAYLOAD (USSR)
Galina Rastorgoueva 322.65 kmh
Mil A-10 200.48 mph
8/13/76

SPEED OVER A 2,000 KM CLOSED CIRCUIT WITHOUT PAYLOAD (USSR)
Inna Kopets 235.12 kmh
Mil MI-8 146.09 mph
9/14/67

TIME TO CLIMB

TIME TO 3,000 METERS (USSR)
Nadezhda Eremina 2 min
Ka-32 11 sec
2 TV3-117V of 2,225 HP
Podmoscovnoe
5/12/83

TIME TO 6,000 METERS (USSR)
Tatiana Zueva 4 min
Ka-32 47 sec
2 TV3-117V 2,225 HP
Podmoscovnoe
5/11/83

SUBCLASS E-1.D (1,750-3,000 kg / 3,858-6,614 lbs) GROUP II (TURBINE ENGINE)

ALTITUDE

ALTITUDE (USSR)
Tatiana Soueva 5,626 m
KA-26 18,453 ft
2 M-14 B26, 325 hp
Podmoscovnoe
3/11/82

ALTITUDE IN HORIZONTAL FLIGHT (USSR)
Tatiana Soueva 5,602 m
KA-26 18,374 ft
2 M-14 B26 325 hp
Podmoscovnoe
3/11/82

TIME TO CLIMB

TIME TO 3,000 METERS (USSR) 8 min
Nadezhda Eremina 19 sec
KA-26
2 M-14 B26 325 hp
3/11/82

CLASS E-1, HELICOPTERS, SPEED OVER A RECOGNIZED COURSE

ANCHORAGE/DALLAS (USA)
H. Ross Perot, Jr., Pilot 80.35 kmh
Jay W. Coburn, Copilot 49.93 mph
Bell 206L Long Ranger
Allison 250-C28B, 435 shp
9/28-30/82

ATHENS/DALLAS (USA)
H. Ross Perot, Jr., Pilot 56.88 kmh
Jay W. Coburn, Copilot 35.35 mph
Bell 206L Long Ranger
Allison 250-C28B, 435 shp
9/10-30/82

BERMUDA/WASHINGTON (AUSTRALIA)
Adam Boyd Munro 189.01 kmh
Bell Jet Ranger 117.45 mph
Allison T-36-A5, 250 shp
10/31-11/1/82

BILLINGS/DALLAS (USA)
H. Ross Perot, Jr., Pilot 106.64 kmh
Jay W. Coburn, Copilot 66.27 mph
Bell 206L Long Ranger
Allison 250-C28B, 435 shp
9/29-30/82

BOSTON/NEW YORK (USA)
Robert W. Hawes 283.79 kmh
Sikorsky S-76 176.04 mph
1 hr 5 min 35 sec
2/27/79

CALCUTTA/DALLAS (USA)
H. Ross Perot, Jr., Pilot 51.62 kmh
Jay W. Coburn, Copilot 32.08 mph
Bell 206L Long Ranger
Allison 250-C28B, 435 shp
9/14-30/82

CALGARY/DALLAS (USA)
H. Ross Perot, Jr., Pilot 118.38 kmh
Jay W. Coburn, Copilot 70.45 mph
Bell 206L Long Ranger
Allison 250-C28B, 435 shp
9/29-30/82

CHICAGO/NEW YORK (USA)
Nicholas Lappos 331.39 kmh
Sikorsky S-76 205.91 mph
2 DDA 250-C30, 650 shp
3 hrs 27 min 3 sec
9/26/80

DALLAS/ATHENS (USA)
H. Ross Perot, Jr., Pilot 59.91 kmh
Jay W. Coburn, Copilot 37.23 mph
Bell 206L Long Ranger
Allison 250-C28B, 435 shp
9/1-9/82

DALLAS/CAIRO (USA)
H. Ross Perot, Jr., Pilot 60.76 kmh
Jay W. Coburn, Copilot 37.76 mph
Bell 206L Long Ranger
Allison 250-C28B, 435 shp
9/1-10/82

DALLAS/CALCUTTA (USA)
H. Ross Perot, Jr., Pilot 64.24 kmh
Jay W. Coburn, Copilot 39.92 mph
Bell 206L-11 Long Ranger
Allison 250/C28B, 435 shp
9/1-14/82

DALLAS/INDIANAPOLIS (USA)
F. J. "Jack" Schweibold 268.56 kmh
John B. Silva, Jr. Copilot 166.88 mph
Sikorsky S-76
2 Allison 250-C30S
4/12/85

DALLAS/KARACHI (USA)
H. Ross Perot, Jr., Pilot 65.26 kmh
Jay W. Coburn, Copilot 40.55 mph
Bell 206L-11 Long Ranger
Allison 250/C28B, 435 shp
9/1-12/82

DALLAS/KUSHIRO (USA)
H. Ross Perot, Jr., Pilot 59.36 kmh
Jay W. Coburn, Copilot 36.89 mph
Bell 206L-11 Long Ranger
Allison 250/C28B, 435 shp
9/1-22/82

DALLAS/LONDON (USA)
H. Ross Perot, Jr., Pilot 63.08 kmh
Jay W. Coburn, Copilot 39.20 mph
Bell 206L-11 Long Ranger
Allison 250/C28B, 435 shp
9/1-7/82

DALLAS/LOS ANGELES (USA)
Arthur J. Hill 69.62 kmh
Aerospatiale AS350D 43.26 mph
1 LTS-101, 616 shp
1/19-20/86

DALLAS/MARSEILLE (USA)
H. Ross Perot, Jr., Pilot 66.16 kmh
Jay W. Coburn, Copilot 41.11 mph
Bell 206L-11 Long Ranger
Allison 250/C28B, 435 shp
9/1-7/82

DALLAS/MONTREAL (USA)
F. J. "Jack" Schweibold 255.96 kmh
John B. Silva, Jr. Copilot 159.05 mph
Sikorsky S-76
2 Allison 250-C30S
4/12-13/85

DALLAS/NAPLES (USA)
H. Ross Perot, Jr., Pilot 57.63 kmh
Jay W. Coburn, Copilot 35.81 mph
Bell 206L-11 Long Ranger
Allison 250/C28B, 435 shp
9/1-9/82

DALLAS/NEW DELHI (USA)
H. Ross Perot, Jr., Pilot 64.23 kmh
Jay W. Coburn, Copilot 39.91 mph
Bell 206L-11 Long Ranger
Allison 250/C28B, 435 shp
9/1-13/82

DALLAS/REYKJAVIK (USA)
H. Ross Perot, Jr., Pilot 83.76 kmh
Jay W. Coburn, Copilot 52.05 mph
Bell 206L-11 Long Ranger
Allison 250/C28B, 435 shp
9/1-4/82

DENVER/KANSAS CITY (USA)
William Kramer 292.49 kmh
Sikorsky S-76 181.74 mph
2 DDA 250-C30, 650 shp
3 hrs 4 min 14 sec
9/24/80

FISHERS ISLAND/NORTH DUMPLING ISLAND (USA)
Dean Kamen, Pilot 116.08 kmh
Richard Rutan, Copilot 101.34 mph
Enstrom 280FX, N36DK
1 HIO-360
6/29/88

HOUSTON/LAFAYETTE, LA (USA)
Vernon Albert, CoCaptain 312.12 kmh
Dave Harvey, CoCaptain 193.95 mph
Sikorsky S76, N31217
1 Allison 250-C30S
8/4/89

INDIANAPOLIS/ALBUQUERQUE (USA)
F. J. "Jack" Schweibold 225.36 kmh
R. Frederick Harvey and 140.04 mph
Harry B. Sutton, Copilots
Sikorsky S-76
2 Allison 250-C30S
4/13-14/85

INDIANAPOLIS/LAFAYETTE (USA)
John Connolly, Co-Captain 237.13 kmh
Paul Nielson, Co-Captain 147.35 mph
Astar 350
1 Allison C-30M, 600 shp
8/2/85

INDIANAPOLIS/LOS ANGELES (USA)
F. J. "Jack" Schweibold 202.68 kmh
R. Frederick Harvey and 125.95 mph
Harry B. Sutton, Copilots
Sikorsky S-76
2 Allison 250-C30S
4/13-14/85

INDIANAPOLIS/MEMPHIS (USA)
John Connolly, Co-Captain 259.66 kmh
Paul Nielson, Co-Captain 161.35 mph
Astar 350
1 Allison C-30M, 600 shp
8/2/85

INDIANAPOLIS/MONTREAL (USA)
F. J. "Jack" Schweibold 244.44 kmh
John B. Silva, Jr. Copilot 151.90 mph
Sikorsky S-76
2 Allison 250-C30S
4/12-13/85

INDIANAPOLIS/NASHVILLE (USA)
F. J. Schweibold, Co-Captain 244.80 kmh
C. V. Glines, Co-Captain 152.12 mph
Bell Long Ranger
1 Allison 250-C30R, 420 shp
8/11/86

INDIANAPOLIS/TRAVERSE CITY (USA)
F. J. "Jack" Schweibold, Pilot 263.16 kmh
Daniel Black, Copilot 163.53 mph
Bell 206L
1 Allison 250C30R, 650 shp
6/21/85

INDIANAPOLIS/WICHITA (USA)
F. J. "Jack" Schweibold 254.88 kmh
R. Frederick Harvey and 159.05 mph
Harry B. Sutton, Copilots
Sikorsky S-76
2 Allison 250-C30S
4/13-14/85

KANSAS CITY/INDIANAPOLIS (USA)
F. J. "Jack" Schweibold 276.12 kmh
Thomas Needham, Copilot 171.58 mph
Hughes 530F
2 Allison 250-C30S, 650 shp
4/25/85

KARACHI/DALLAS (USA)
H. Ross Perot, Jr., Pilot 53.75 kmh
Jay W. Coburn, Copilot 33.40 mph
Bell 206L-11 Long Ranger
Allison 250/C28B, 435 shp
9/13-30/82

KUSHIRO/ANCHORAGE (USA)
H. Ross Perot, Jr., Pilot 93.94 kmh
Jay W. Coburn, Copilot 58.37 mph
Bell 206L-11 Long Ranger
Allison 250/C28B, 435 shp
9/25-27/82

KUSHIRO/DALLAS (USA)
H. Ross Perot, Jr., Pilot 85.77 kmh
Jay W. Coburn, Copilot 53.30 mph
Bell 206L-11 Long Ranger
Allison 250/C28B, 435 shp
9/25-30/82

LEWISTON/OSHKOSH (USA)
Jack Schweibold, Co-captain 226.23 kmh
John Connolly, Co-captain 140.57 mph
Astar 350
1 Allison 250-C30M, 650 shp
7/26-27/85

LONDON/ATHENS (USA)
H. Ross Perot, Jr., Pilot 53.28 kmh
Jay W. Coburn, Copilot 33.11 mph
Bell 206L-11 Long Ranger
Allison 250/C28B, 435 shp
9/7-9/82

LONDON/CAIRO (USA)
H. Ross Perot, Jr., Pilot 57.69 kmh
Jay W. Coburn, Copilot 35.85 mph
Bell 206L-11 Long Ranger
Allison 250/C28B, 435 shp
9/7-10/82

LONDON/MARSEILLE (USA)
H. Ross Perot, Jr., Pilot 179.39 kmh
Jay W. Coburn, Copilot 111.47 mph
Bell 206L-11 Long Ranger
Allison 250/C28B, 435 shp
9/7/82

LONDON/NAPLES (USA)
H. Ross Perot, Jr., Pilot 41.77 kmh
Jay W. Coburn, Copilot 25.96 mph
Bell 206L-11 Long Ranger
Allison 250/C28B, 435 shp
9/7-9/82

LONDON/PARIS (FRANCE)
Bernard Pasquet 281.05 kmh
Dauphin II 174.63 mph
2 Turbomeca Arriel 486 KW
2/8/80

LONDON/SYDNEY (UK)
Jennifer Murray 26.39 kmh
Colin Bodill 16.40 mph
Robinson R44
1 Lycoming O-540, 205 bhp
1/12/01

LOS ANGELES/LAS VEGAS (USA)
Lawrence Graves 274.26 kmh
Augusta 109A 170.43 mph
2 Allison C-20-B, 420 shp
1 hr 23 min 37 sec
9/13/81

MORGAN CITY, LA/HOUSTON (USA)
Vernon Albert, CoCaptain — 278.28 kmh
Dave Harvey, CoCaptain — 172.92 mph
Skiorsky S76, N31217
1 Allison 250-C30S
8/4/89

NEW DELHI/DALLAS (USA)
H. Ross Perot, Jr., Pilot — 53.85 kmh
Jay W. Coburn, Copilot — 33.46 mph
Bell 206L Long Ranger
Allison 250/C28B, 435 shp
9/14-30/82

NEW ORLEANS/INDIANAPOLIS (USA)
Harry B. Sutton, Pilot — 231.84 kmh
Jerry Marlette, Copilot — 144.07 mph
George Cummins, Crew Chief
Sikorsky S-76
2 Allison 250-C30S
4/17/85

NEW YORK/BOSTON (USA)
Larry Graves, Pilot — 305.28 kmh
Agusta 109 MKII, N1PQ — 189.70 mph
1 Allison 250C20R, 450 shp
6/1/88

NEW YORK/DALLAS (USA)
Arthur J. Hill — 75.96 kmh
Aerospatiale AS350D — 47.20 mph
1 LTS-101, 616 shp
1/18-19/86

NEW YORK/INDIANAPOLIS (USA)
F. J. "Jack" Schweibold — 261.36 kmh
R. Frederick Harvey and — 162.41 mph
Harry B. Sutton, Copilots
Sikorsky S-76
2 Allison 250-C30S
4/13/85

NEW YORK/LONDON (USA)
MAJ Donald B. Maurras, USAF — 190.12 kmh
Sikorsky HH-3E — 118.14 mph
2 GE T58-5
29 hrs 13 min 35 sec
6/1/67

NEW YORK/LOS ANGELES (USA)
F. J. "Jack" Schweibold — 209.52 kmh
R. Frederick Harvey and — 130.20 mph
Harry B. Sutton, Copilots
Sikorsky S-76
2 Allison 250-C30S
4/13/85

NEW YORK/PARIS (USA)
MAJ Herbert Zehnder, USAF — 189.95 kmh
Sikorsky HH-3E — 118.03 mph
2 GE T58-5
30 hrs 46 min 10.8 sec
6/1/67

NEW YORK/PHILADELPHIA (USA)
Larry Graves, Pilot — 253.44 kmh
John McLeod, Copilot — 157.49 mph
Agusta 109 MKII, N1PQ
1 Allison 250C20R, 450 shp
5/31/88

NEW YORK/WASHINGTON (USA)
Robert W. Hawes — 279.46 kmh
Sikorsky S-76 — 173.52 mph
DDA 250-C30, 650 shp
1 hr 11 min
12/27/79

OSHKOSH/INDIANAPOLIS (USA)
F. J. "Jack" Schweibold, Pilot — 278.71 kmh
Junior Van der Vleit, Copilot — 173.20 mph
Hughes 530F
1 Allison 250-C-30, 650 shp
6/13/85

PARIS/LONDON (FRANCE)
Bernard Pasquet — 321.91 kmh
Dauphin II — 200.02 mph
2 Turbomeca Arriel, 486 shp
2/8/80

PHILADELPHIA/BOSTON (USA)
Larry Graves, Pilot — 303.20 kmh
Agusta 109 MKII, N1PQ — 188.41 mph
1 Allison 250C20R, 450 shp
6/1/88

PHILADELPHIA/NEW YORK (USA)
Larry Graves, Pilot — 317.16 kmh
Agusta 109 MKII, N1PQ — 197.08 mph
1 Allison 250C20R, 450 shp
6/1/88

PHILADELPHIA/WASHINGTON (USA)
Larry Graves, Pilot — 280.44 kmh
Gary DiVincenzo, Copilot — 174.27 mph
Agusta 109 MKII, N1PQ
1 Allison 250C20R, 450 shp
5/31/88

SPOKANE/MINNEAPOLIS (USA)
Jack Schweibold, Co-captain — 230.22 kmh
John Connolly, Co-captain — 143.06 mph
Astar 350
1 Allison 250-C30M, 650 shp
7/26/85

SPOKANE/OSHKOSH (USA)
Jack Schweibold, Co-captain — 226.14 kmh
John Connolly, Co-captain — 140.52 mph
Astar 350
1 Allison 250-C30M, 650 shp
7/26-27/85

SYDNEY/FORT WORTH (AUSTRALIA)
R. H. Smith — 9.81 kmh
Bell Jetranger 206B — 6.09 mph
5/25-7/22/83

SYDNEY/HONG KONG (AUSTRALIA)
R. H. Smith — 18.79 kmh
Bell Jetranger 206B — 11.68 mph
5/25-6/10/83

SYDNEY/TOKYO (AUSTRALIA)
R. H. (Dick) Smith — 11.98 kmh
Bell Jetranger 206B — 7.44 mph
1 Allison
5/25-6/21/83

WASHINGTON/BOSTON (USA)
Nicholas Micskey — 298.36 kmh
Sikorsky S-76 — 185.04 mph
2 DDA 250/C30, 650 shp
2 hrs 8 min 53 sec
12/27/79

WASHINGTON/NEW YORK (USA)
Lawrence Graves — 312.23 kmh
Agusta 109A — 194.02 mph
2 Allison C-20-B, 420 shp
1 hr 3 min 33 sec
10/7/81

WASHINGTON/PHILADELPHIA (USA)
Larry Graves, Pilot 287.28 kmh
Gary DiVincenzo, Copilot 178.52 mph
Agusta 109 MK11, N1PQ
1 Allison 250C20R, 450 shp
5/31/88

WICHITA/LOS ANGELES (USA)
F. J. "Jack" Schweibold 197.28 kmh
R. Frederick Harvey and 122.59 mph
Harry B. Sutton, Copilots
Sikorsky S-76
Allison 250-C30S
4/14/85

ROUND TRIP

INDIANAPOLIS/OSHKOSH/INDIANAPOLIS (USA)
F. J. "Jack" Schweibold, Pilot 237.81 kmh
Junior Van der Vleit, Copilot 147.77 mph
Hughes 530F
1 Allison 250-C-30, 650 shp
6/13/85

PARIS/LONDON/PARIS (FRANCE)
Bernard Pasquet 294.26 kmh
Dauphin II 182.84 mph
2 Turbomeca Arriel 486 KW
2/6/80

SUB CLASS E-1.B (500-1,000 kg / 1,102-2,205 lbs) GROUP I (PISTON ENGINE)

ESPERANCE/NORTHAM (AUSTRALIA)
Claude Meunier 188.31 kmh
Robinson R44 Astro 117.01 mph

1 Lycoming O-540, 210 hp
4/29/03

NORTHAM/ESPERANCE (AUSTRALIA)
Claude Meunier 175.3 kmh
Robinson R44 Astro 108.9 mph
1 Lycoming O-540, 210 hp
4/29/03

PERTH/SYDNEY (AUSTRALIA)
Claude Meunier 97.82 kmh
Robinson R22 60.78 mph
1 Lycoming O-320, 160 bhp
1/12/01

SYDNEY/PERTH (AUSTRALIA)
Claude Meunier 103.34 kmh
Robinson R22 64.21 mph
1 Lycoming O-320, 160 bhp
1/19/01

ROUND TRIP

NORTHAM/ESPERANCE/NORTHAM (AUSTRALIA)
Claude Meunier 174.18 kmh
Robinson R44 Astro 108.23 mph
1 Lycoming O-540, 210 hp
4/29/03

Randy Waldman (l) and Dave Riggs (r) prepare to leave Fresno before their record-setting flight to Los Angeles.

KAITAIA/NORFOLK (NEW ZEALAND)
Charles Bernard Lewis, Pilot 287.06 kmh
Grant Biel, Copilot 178.37 mph
Aerospatiale SA 365N Dauphin
5/29/91

LONDON/MONTREAL (USA)
Joe Ronald Bower 108.58 kmh
John W. Williams 67.47 mph
Bell 430
2 Allison 250-C40, 811 shp
8/17-19/96

LORD HOWE/SYDNEY (NEW ZEALAND)
Charles Bernard Lewis, Pilot 219.66 kmh
Grant Biel, Copilot 136.49 mph
Aerospatiale SA 365N Dauphin
5/29/91

NORFOLK/LORD HOWE (NEW ZEALAND)
Charles Bernard Lewis, Pilot 246.27 kmh
Grant Biel, Copilot 153.02 mph
Aerospatiale SA 365N Dauphin
5/29/91

OXNARD/MARIETTA (USA)
Ron Williamson 242.75 kmh
Bell UH-1H 150.84 mph
1 LHTEC T800, 1,350 shp
4/22/93

PROVIDENIYA/ST. PETERSBURG (USA)
Joe Ronald Bower 58.14 kmh
John W. Williams 36.13 mph
Bell 430
2 Allison 250-C40, 811 shp
8/28-9/2/96

STOCKHOLM/LONDON (USA)
Joe Ronald Bower 177.34 kmh
John W. Williams 110.19 mph
Bell 430
2 Allison 250-C40, 811 shp
9/3/96

SUBCLASS E-1.F (4,500-6,000 kg / 9,921-13,228 lbs)
GROUP II (TURBINE ENGINE)

SAVANNAH/RALEIGH DURHAM (USA)
Frank Candela, Pilot 266.40 kmh
Arthur Tobey, CoCaptain 165.54 mph
J. Michael Peters, Crew Chief
Sikorsky SK76A, N202SK
2 Allison 250-C30, 650 shp
12/11/87

WEST PALM BEACH/WASHINGTON (USA)
Frank Candela, Pilot 212.04 kmh
Arthur Tobey, CoCaptain 131.76 mph
J. Michael Peters, Crew Chief
Sikorsky SK76A, N202SK
2 Allison 250-C30, 650 shp
12/11/87

SUBCLASS E-1.G (6,000-10,000 kg / 13,228-22,046 lbs)
GROUP II (TURBINE ENGINE)

HOUMA/MEXICO CITY (USA)
Andrew A. Evans 239.83 kmh
Sergio Delgadillo Guadarrama 149.02 mph
Pedro Navarrete Alvarez
Sikorsky S-70A-24
2 GE T700-GE-701C, 1,900 shp
10/20/94

STRATFORD/BIRMINGHAM (USA)
Kevin L. Bredenbeck 199.56 kmh
Juan Diego Solano Aguayo 124.00 mph
Jose A. Meneces
Sikorsky S-70A-24
2 GE T700-GE-701C, 1,900 shp
10/19/94

SUBCLASS E-1.J (30,000-40,000 kg / 66,139-88,185 lbs)
GROUP II (TURBINE ENGINE)

PATUXENT RIVER/JUPITER (USA)
John L. Carson, Pilot 238.94 kmh
Richard Muldoon, Copilot 148.47 mph
Sikorsky MH-53E
3 GE T64-GE-416, 4,380 shp
6/2/93

CLASS E-3, AUTOGYROS

CLASS E-3 (AUTOGYROS) GROUP I (PISTON ENGINE)

DISTANCE

DISTANCE WITHOUT LANDING (UK)
Barry Jones — 941.25 km
Magni M 16 — 584.86 mi
1 Rotax 912 ULS, 100 bhp
Culdrose - Wick
2/24/03

DISTANCE OVER A CLOSED CIRCUIT WITHOUT LANDING (UK)
Kenneth H. Wallis — 1,002.75 km
Wallis Autogyro, WA-116/F/S — 623.11 mi
8/5/88

ALTITUDE

ALTITUDE (USA) ★
William B. Clem, III — 7,456 m
Dominator — 24,463 ft
1 Rotax 914, 115 bhp
Wauchula, FL
4/17/98

SPEED

SPEED OVER A 3 KM STRAIGHT COURSE (UK)
Kenneth H. Wallis — 207.70 kmh
Wallis WA-121/Mc — 129.06 mph
1 McCulloch, 90 bhp
Norfolk, England
11/16/02

SPEED OVER A 15/25 KM STRAIGHT COURSE (UK)
Kenneth H. Wallis — 189.58 kmh
Wallis WA-116/FS — 117.80 mph
1 Franklin 2A0120B, 60 hp
RAF Wyton
10/17/84

SPEED OVER A 100 KM CLOSED CIRCUIT WITHOUT PAYLOAD (UK)
Kenneth H. Wallis — 190.41 kmh
Wallis Q-116/F/S — 118.32 mph
1 Franklin 2A-120B, 60 hp
Waterbeach Airfield
4/17/85

SPEED OVER A 500 KM CLOSED CIRCUIT WITHOUT PAYLOAD (UK)
Kenneth H. Wallis — 134.04 kmh
Wallis Autogyro — 83.29 mph
8/5/88

SPEED OVER A 1,000 KM CLOSED CIRCUIT WITHOUT PAYLOAD (UK)
Kenneth H. Wallis — 130.80 kmh
Wallis Autogyro — 81.28 mph
8/5/88

TIME TO CLIMB

TIME TO 3,000 METERS (UK)
Kenneth H. Wallis — 7 min
Wallis WA-121 — 20 sec
1 McCulloch, 90 hp
Norfolk, UK
3/19/98

SUBCLASS E-3.A (Less Than 500 kg / 1,102 lbs) GROUP I (PISTON ENGINE)

DISTANCE

DISTANCE WITHOUT LANDING (UK)
Barry Jones — 941.25 km
Magni M 16 — 584.86 mi
1 Rotax 912 ULS, 100 bhp
Culdrose - Wick
2/24/03

DISTANCE OVER A CLOSED CIRCUIT WITHOUT LANDING (UK)
Kenneth H. Wallis — 1,002.75 km
Wallis Autogyro, WA-116/F/S — 623.11 mi
8/5/88

ALTITUDE

ALTITUDE (USA)
William B. Clem, III — 7,456 m
Dominator — 24,463 ft
1 Rotax 914, 115 bhp
Wauchula, FL
4/17/98

SPEED

SPEED OVER A 3 KM STRAIGHT COURSE (UK)
Kenneth H. Wallis — 207.70 kmh
Wallis WA-121/Mc — 129.06 mph
1 McCulloch, 90 bhp
Norfolk, England
11/16/02

SPEED OVER A 15/25 KM STRAIGHT COURSE (UK)
Kenneth H. Wallis — 189.58 kmh
WA-116/F/S — 117.80 mph
1 Franklin 2A-120-B, 60 hp
RAF Wyton
10/14/84

100 KM CLOSED CIRCUIT WITHOUT PAYLOAD (UK)
Kenneth H. Wallis — 190.41 kmh
Wallis WA-116/F/S — 118.32 mph
1 Franklin 2A-120B, 60 hp
Waterbeach Airfield
4/17/85

TIME TO CLIMB

TIME TO 3,000 METERS (UK)
Kenneth H. Wallis — 7 min
Wallis WA-121 — 20 sec
1 McCulloch, 90 hp
Norfolk, UK
3/19/98

SUBCLASS E-3.B (500-1,000 kg / 1,102-2,205 lbs) GROUP I (PISTON ENGINE)

DISTANCE

DISTANCE WITHOUT LANDING (USA)
Frank Anders — 612.34 km
Farrington Air and Space — 380.50 mi
Lycoming, 180 hp
Paducah, KY to Columbus, GA
4/16/89

ALTITUDE

ALTITUDE (USA)
Charles I. Fitzsimmons — 3,317 m
Farrington Air and Space 18A — 10,880 ft
1 Lycoming O-360, 180 hp
Paducah, KY
6/22/86

SPEED OVER A RECOGNIZED COURSE

EAST TO WEST TRANSCONTINENTAL*
Andrew C. Keech — 16.45 kmh
Little Wing LW-5 — 10.22 mph
1 Rotax 914, 115 bhp
10/12/03

WEST TO EAST TRANSCONTINENTAL*
Andrew C. Keech — 31.89 kmh
Little Wing LW-5 — 19.82 mph
1 Rotax 914, 115 bhp
10/22/03

***

DARWIN/ADELAIDE (AUSTRALIA)
John MacQueen — 12.34 kmh
Homebuilt — 7.66 mph
1 Rotax
6/8/94

KITTY HAWK/SAN DIEGO (USA)
Andrew C. Keech — 16.45 kmh
Little Wing LW-5 — 10.22 mph
1 Rotax 914, 115 bhp
10/12/03

PADUCAH, KY/COLUMBUS, GA (USA)
Frank Anders — 102.60 kmh
Farrington Air and Space — 93.75 mph
Lycoming, 180 bhp
4/16/89

SAN DIEGO/KITTY HAWK (USA)
Andrew C. Keech — 31.89 kmh
Little Wing LW-5 — 19.82 mph
1 Rotax 914, 115 bhp
10/22/03

ROUND TRIP

ROUND TRIP TRANSCONTINENTAL*
Andrew C. Keech — 16.42 kmh
Little Wing LW-5 — 10.20 mph
1 Rotax 914, 115 bhp
10/22/03

KITTY HAWK/SAN DIEGO/KITTY HAWK (USA)
Andrew C. Keech — 16.42 kmh
Little Wing LW-5 — 10.20 mph
1 Rotax 914, 115 bhp
10/22/03

Andy Keech, with his Little Wing autogyro, at Kitty Hawk before departing on a transcontinental flight.

Photo: A.W. Greenfield

Tilt-Wing/Tilt-Engine Airplanes

In 1990, the tilt-wing/tilt-engine class (Class M) was added to the FAI Sporting Code. The Class M airplane is defined as "an airplane capable of both horizontal and vertical flight which, in forward horizontal flight, derives most of its lift from fixed wings, and which achieves vertical or hovering flight by tilting the wings or engines upward to a position substantially vertical." During record attempts, all takeoffs and landings must be made vertically.

CLASS M (TILT-WING / TILT-ENGINE) GROUP II (TURBINE ENGINE)

Altitude

ALTITUDE (USA)
Ronald G. Erhart, Pilot — 6,907 m
Thomas L. Warren, Copilot — 22,660 ft
Bell XV-15
2 Textron/Lycoming LTC 1K-4K, 1,550 shp
Arlington, TX
03/15/90

ALTITUDE IN HORIZONTAL FLIGHT (USA)
Ronald G. Erhart, Pilot — 6,876 m
Thomas L. Warren, Copilot — 22,560 ft
Bell XV-15
2 Textron/Lycoming LTC 1K-4K, 1,550 shp
Arlington, TX
03/15/90

ALTITUDE WITH 1,000 KG PAYLOAD (USA)
Ronald G. Erhart, Pilot — 6,879 m
Thomas L. Warren, Copilot — 22,570 ft
Bell XV-15
2 Textron/Lycoming LTC 1K-4K, 1,550 shp
Arlington, TX
03/15/90

Time to Climb

TIME TO 3,000 METERS (USA)
Ronald G. Erhart, Pilot — 4 min 24 sec
Thomas L. Warren, Copilot
Bell XV-15
2 Textron/Lycoming LTC 1K-4K, 1,550 shp
Arlington, TX
03/15/90

TIME TO 6,000 METERS (USA)
Ronald G. Erhart, Pilot — 8 min 28 sec
Thomas L. Warren, Copilot
Bell XV-15
2 Textron/Lycoming LTC 1K-4K, 1,550 shp
Arlington, TX
03/15/90

SPEED OVER A RECOGNIZED COURSE

ARLINGTON, TX/BATON ROUGE (USA)
Thomas L. Warren, Pilot — 456.02 kmh
Ronald G. Erhart, Copilot — 283.36 mph
Bell XV-15
2 Textron/Lycoming LTC 1K-4K, 1,550 shp
4/6/90

Balloons

The first recorded free flight of any man-carrying aerial vehicle occurred on November 21, 1783, in Paris, France. François Pilatre de Rozier and the Marquis d'Arlandes flew for 25 minutes, at an altitude of about 1,500 feet, and landed some five miles from where they took off.

Thus, aeronautics began in a 49-foot hot-air balloon, after a surprisingly short period of development which included unmanned flights, at least one flight with animals as passengers, and then a tethered man-carrying flight before the first voyage across country. Less than two weeks later, the first hydrogen-filled balloon flew 27 miles with two men on board. The seeds of aerial competition were planted, and balloon ascents soon became a popular pastime and money-making attraction all over Europe.

In January, 1785, a Frenchman and an American crossed the English Channel in a hydrogen balloon, making the first international journey by air. The potential of flying from one place to another was awakening interest around the world, and less than 10 years later, a balloon was used for military reconnaissance by the French Army.

Ballooning was the first aeronautical sport, with pleasure flying and competition becoming popular in the final years of the 19th century. The first National Aero Clubs grew up around ballooning, and were already in existence when powered airplanes appeared on the scene a few years later.

The first major international aviation contest—the James Gordon Bennett Race—was for gas-filled balloons. It premiered in Paris in 1906 and continued until 1938. With the rebirth of hot air balloons, the race was resurrected in 1983 and three U.S. teams compete annually.

Records for ballooning are of three major types: Altitude (above mean sea level); Distance (nonstop); and Duration (from takeoff to landing, regardless of distance and direction). The Sub-Classes for lighter-than-air records include the following:

AA – non-pressurized gas balloons
AM – gas / hot-air balloons (Roziere)
AN – ammonia gas
AS – pressurized gas balloons
AX – hot-air balloons
BA – gas airships (without rigid framework)
BR – gas airships (with rigid framework)
BX – hot-air airships

The FAI oversees the sport through its International Ballooning Commission (C.I.A.). Ballooning competitions are supervised in the U.S. by NAA through its affiliate, the Balloon Federation of America, P.O. Box 400, Indianola, IA 50125, (515) 961-8809.

ABSOLUTE WORLD RECORDS - BALLOONS

ALTITUDE (USA)
CDR M. D. Ross, USNR
LTCDR V. A. Prather
Lee Lewis Memorial
Gulf of Mexico
5/4/61
34,668 m
113,740 ft

DISTANCE (SWITZERLAND/UK)
Bertrand Piccard
Brian Jones
Cameron Balloons R-650
Chateau d'Oex, Switzerland to Dâkhla, Egypt
3/1-21/99
40,814 km
25,360 mi

DURATION (SWITZERLAND/UK)
Bertrand Piccard
Brian Jones
Cameron Balloons R-650
Chateau d'Oex, Switzerland to Dâkhla, Egypt
3/1-21/99
477 hrs
47 min

SHORTEST TIME AROUND THE WORLD (USA)
Steve Fossett
Cameron/Cole R-550
Northam, Australia
6/19-7/3/02
320 hrs
33 min

U.S. NATIONAL AND WORLD CLASS RECORDS, GENERAL

SUBCLASS AA-1 (less than 250 cubic meters)

ALTITUDE (USA)
Dr. Coy Foster
OZ A-165
Plano, TX
7/14/84
5,516 m
18,097 ft

DISTANCE (USA)
Dr. Coy Foster
OZ A-165
Plano, TX
9/8/84
695.74 km
432.32 mi

DURATION (USA)
Dr. Coy Foster
OZ A-165
Plano, TX
5/30/83
8 hrs
12 min

SUBCLASS AA-2 (250 - 400 cubic meters)

ALTITUDE (USA)
Donald L. Piccard
"Sioux City Sue"
Sioux City, IA
8/24/62
5,409 m
17,747 ft

DISTANCE (USA)
Dr. Coy Foster
OZ A-165
Plano, TX
9/8/84
695.74 km
432.32 mi

DURATION (FRANCE)
Vincent Leys
Leys VL 248
Lille, France to Ransel, Germany
6/5/94
13 hrs
55 min

DURATION*
Timothy S. Cole
West Wind 600M
Kersey, CO
9/26/91
8 hrs
47 min
20 sec

SUBCLASS AA-3 (400 - 600 cubic meters)

ALTITUDE (USA)
Tracy Barnes
Barnes 14-A
Rosemount, MN
5/10/64
11,780 m
38,650 ft

DISTANCE (USA)
Troy Bradley
homebuilt
Amarillo, TX
1/3-5/02
1,971.82 km
1,225.23 mi

DURATION (USSR)
Serge Sinoveev
URSS-VR 80
Dolgoproudnaia, USSR
3/30/41
46 hrs
10 min

SUBCLASS AA-4 (600 - 900 cubic meters)

ALTITUDE (USA)
Tracy Barnes
Barnes 14-A
Rosemount, MN
5/10/64
11,780 m
38,650 ft

DISTANCE (USA)
Troy Bradley
homebuilt
Amarillo, TX
1/3-5/02
1,971.82 km
1,225.23 mi

DURATION (USSR)
F. Bourlouski
Moscow to Charaboulski, USSR
4/3/39
61 hrs
30 min

SUBCLASS AA-5 (900 - 1,200 cubic meters)

ALTITUDE (USA)
Tracy Barnes
Barnes 14-A
Rosemount, MN
5/10/64
11,780 m
38,650 ft

DISTANCE (USA)
Troy Bradley 1,971.82 km
homebuilt 1,225.23 mi
Amarillo, TX
1/3-5/02

DURATION (USSR)
F. Bourlouski 61 hrs
Moscow to Charaboulski, USSR 30 min
4/3/39

SUBCLASS AA-6 (1,200 - 1,600 cubic meters)

ALTITUDE (USA)
Tracy Barnes 11,780 m
Barnes 14-A 38,650 ft
Rosemount, MN
5/10/64

DISTANCE (USA)
Richard Abruzzo 3,338.25 km
homebuilt GROM-1 2,074.29 mi
San Diego, CA to Waverly, GA
2/2-5/03

DURATION (USA)
Richard Abruzzo 80 hrs
homebuilt Grom-1 18 min
Albuquerque, NM to Crawfordville, GA
2/19/01

SUBCLASS AA-7 (1,600 - 2,200 cubic meters)

ALTITUDE (USA)
Tracy Barnes 11,780 m
Barnes 14-A 38,650 ft
Rosemount, MN
5/10/64

DISTANCE (USA)
Richard Abruzzo 3,338.25 km
homebuilt GROM-1 2,074.29 mi
San Diego, CA to Waverly, GA
2/2-5/03

Richard Abruzzo gets his balloon ready for liftoff from Torrey Pines Park near San Diego, California on a flight that will take him all the way across the United States.

DURATION (GERMANY)
Wilhelm Eimers 92 hrs
Bernd Landmann 11 min

Columbus II
Wil, Switzerland to Ludzasraj, Latvia
9/9-13/95

DURATION*
Richard Abruzzo 80 hrs
homebuilt Grom-1 18 min
Albuquerque, NM to Crawfordville, GA
2/19/01

SUBCLASS AA-8 (2,200 - 3,000 cubic meters)

ALTITUDE (USA)
Tracy Barnes 11,780 m
Barnes 14-A 38,650 ft
Rosemount, MN
5/10/64

DISTANCE (USA)
Edward Yost 3,983.18 km
Silver Fox GB-47 2,475.03 mi
Milbridge, ME to 37°11'N 020°52'W
10/5-10/76

DURATION (USA)
Edward Yost 107 hrs
Silver Fox GB-47 37 min
Milbridge, ME to 37°11'N 020°52'W
10/5-10/76

SUBCLASS AA-9 (3,000 - 4,000 cubic meters)

ALTITUDE (USA)
Tracy Barnes 11,780 m
Barnes 14-A 38,650 ft
Rosemount, MN
5/10/64

DISTANCE (USA)
Edward Yost 3,983.18 km
Silver Fox GB-47 2,475.03 mi
Milbridge, ME to 37°11'N 020°52'W
10/5-10/76

DURATION (USA)
Edward Yost 107 hrs
Silver Fox GB-47 37 min
Milbridge, ME to 37°11'N 020°52'W
10/5-10/76

SUBCLASS AA-10 (4,000 - 6,000 cubic meters)

ALTITUDE (USA)
Tracy Barnes 11,780 m
Barnes 14-A 38,650 ft
Rosemount, MN
5/10/64

DISTANCE (USA)
Joseph W. Kittinger 5,703.03 km
Yost GB55 3,544.25 mi
Caribou, ME to Cairo Montenotte, Italy
9/15-18/84

DURATION (USA)
Edward Yost 107 hrs
Silver Fox GB-47 37 min
Milbridge, ME to 37°11'N 020°52'W
10/5-10/76

SUBCLASS AA-11 (6,000 - 9,000 cubic meters)

ALTITUDE (USA)
Tracy Barnes 11,780 m
Barnes 14-A 38,650 ft
Rosemount, MN
5/10/64

DISTANCE (USA)
Joseph W. Kittinger 5,703.03 km
Yost GB55 3,544.25 mi
Caribou, ME to Cairo Montenotte, Italy
9/15-18/84

DURATION (USA)
Ben L. Abruzzo, Commander 137 hrs
Maxie L. Anderson, Co-Commander 5 min
Larry M. Newman, Radio Operator 50 sec
Double Eagle II
Presque Isle, ME, USA to Miserey, France
8/12-17/78

SUBCLASS AA-12 (9,000 - 12,000 cubic meters)

ALTITUDE (USA)
Tracy Barnes 11,780 m
Barnes 14-A 38,650 ft
Rosemount, MN
5/10/64

DISTANCE (USA)
Joseph W. Kittinger 5,703.03 km
Yost GB55 3,544.25 mi
Caribou, ME to Cairo Montenotte, Italy
9/15-18/84

DURATION (USA)
Ben L. Abruzzo, Commander 137 hrs
Maxie L. Anderson, Co-Commander 5 min
Larry M. Newman, Radio Operator 50 sec
Double Eagle II
Presque Isle, ME, USA to Miserey, France
8/12-17/78

SUBCLASS AA-13 (12,000 - 16,000 cubic meters)

ALTITUDE (USA)
Tracy Barnes 11,780 m
Barnes 14-A 38,650 ft
Rosemount, MN
5/10/64

DISTANCE (USA)
Joseph W. Kittinger 5,703.03 km
Yost GB55 3,544.25 mi
Caribou, ME to Cairo Montenotte, Italy
9/15-18/84

DURATION (USA)
Ben L. Abruzzo, Commander 137 hrs
Maxie L. Anderson, Co-Commander 5 min
Larry M. Newman, Radio Operator 50 sec
Double Eagle II
Presque Isle, ME, USA to Miserey, France
8/12-17/78

SUBCLASS AA-14 (16,000 - 22,000 cubic meters)

ALTITUDE (SWITZERLAND)
August Piccard 16,201 m
Max Cosyns 53,153 ft
Riedinger OO-BFH
Zurich, Switzerland
8/18/32

ALTITUDE*
Tracy Barnes 11,780 m
Barnes 14-A 38,650 ft
Rosemount, MN
5/10/64

DISTANCE (USA)
Ben L. Abruzzo, pilot 8,382.54 km
Larry M. Newman, Copilot; 5,208.67 mi
Rocky Aoki, pilot
Ron Clark, pilot
Raven Experimental
Nagashima, Japan to Covello, CA
11/9-12/81

DURATION (USA)
Ben L. Abruzzo, Commander — 137 hrs
Maxie L. Anderson, Co-Commander — 5 min
Larry M. Newman, Radio Operator — 50 sec
Double Eagle II
Presque Isle, ME, USA to Miserey, France
8/12-17/78

SUBCLASS AA-15 (more than 22,000 cubic meters)

ALTITUDE (USA)
CDR M. D. Ross, USNR — 34,668 m
LTCDR V. A. Prather — 113,740 ft
Lee Lewis Memorial
Gulf of Mexico
5/4/61

DISTANCE (USA)
Ben L. Abruzzo, Pilot — 8,382.54 km
Larry M. Newman, Copilot — 5,208.67 mi
Rocky Aoki, Pilot
Ron Clark, Pilot
Raven Experimental
Nagashima, Japan to Covello, CA
11/9-12/81

DURATION (USA)
Ben L. Abruzzo, Commander — 137 hrs
Maxie L. Anderson, Co-Commander — 5 min
Larry M. Newman, Radio Operator — 50 sec
Double Eagle II
Presque Isle, ME, USA to Miserey, France
8/12-17/78

SUBCLASS AX-1 (less than 250 cubic meters)

ALTITUDE (AUSTRIA)
Günter Schabus — 2,308 m
Lindstrand LBL9A — 7,572 ft
Saurachberg, Austria to Pichlerrs, Austria
2/28/95

ALTITUDE*
Dr. Coy Foster — 5,379 ft
Colt Balloon, N-BMAM
Plano, TX
1/10/88

DISTANCE (USA)
Dr. Coy Foster — 47.44 km
Colt Balloon, N-BMAM — 29.48 mi
Plano, TX
2/7/88

DURATION (AUSTRIA)
Günter Schabus — 2 hrs
Lindstrand LBL9A — 50 min
Tigring-Moosburg, Austria — 39 sec
2/11/95

DURATION*
Dr. Coy Foster — 1 hr
Colt Balloon, N-BMAM — 37 min
Plano, TX — 20 sec
2/7/88

SUBCLASS AX-2 (250 - 400 cubic meters)

ALTITUDE (SWEDEN)
Oscar Lindström — 4,371 m
Lindstrand LBL-14 — 14,341 ft
Longtora, Sweden
3/15/98

ALTITUDE*
Dr. Coy Foster — 3,671 m
Colt Cloudhopper — 12,044 ft
Plano, TX
1/23/83

DISTANCE (SWEDEN)
Oscar Lindström — 138.01 km
Lindstrand LBL-14 — 85.76 mi
Storvik to Sandviken-Strängnäs, Sweden
2/16/97

DISTANCE*
Dr. Coy Foster — 56.67 km
Colt Cloudhopper — 35.21 mi
Plano, TX to Pilot Point, TX
3/22/83

DURATION (SWEDEN)
Oscar Lindström — 4 hrs
Lindstrand LBL-14 — 8 min
Storvik to Sandviken-Strängnäs, Sweden — 20 sec
2/16/97

DURATION*
Dr. Coy Foster — 2 hrs
Colt Cloudhopper — 44 min
Plano, TX to Pilot Point, TX — 20 sec
3/22/83

SUBCLASS AX-3 (400 - 600 cubic meters)

ALTITUDE (USA)
Dr. Coy Foster — 6,163 m
Colt 21A — 20,220 ft
Plano, TX to Jackass Flats, TX
2/15/86

DISTANCE (SWEDEN)
Oscar Lindström — 341.36 km
LBL-21 — 212.11 mi
Årjäng, Sweden to Löstabruk, Sweden
3/4/01

DISTANCE*
Dr. Coy Foster — 186.52 km
Colt Cloudhopper AX-3 — 115.90 mi
Plano, TX to Donie, TX
2/3/83

DURATION (SWEDEN)
Oscar Lindström — 7 hrs
LBL-21 — 27 min
Årjäng, Sweden to Löstabruk, Sweden — 31 sec
3/4/01

SUBCLASS AX-4 (600 - 900 cubic meters)

ALTITUDE (USA)
Dr. Coy Foster — 8,992 m
Colt 31A — 29,500 ft
Plano, TX to Brashear, TX
2/23/86

DISTANCE (SWEDEN)
Oscar Lindström — 341.36 km
LBL-21 — 212.11 mi
Årjäng, Sweden to Löstabruk, Sweden
3/4/01

DISTANCE*
Michael C. Emich — 295.85 km
homebuilt — 183.83 mi
Galesburg, IL to Centralia, IL
2/13/99

DURATION (USA)
Dr. Coy Foster — 8 hrs
Colt AX4 — 39 min
Plano, TX to Bellfalls, TX
3/20/86

SUBCLASS AX-5 (900 - 1,200 cubic meters)

ALTITUDE (USA)
Carol R. Davis — 9,540 m
Firefly — 31,300 ft
Moriarty, NM
12/8/79

DISTANCE (USA)
Michael C. Emich — 434.70 km
homebuilt — 270.11 mi
Washington, IL to St. Marys, OH
1/10/98

DURATION (USA)
William J. Cloninger — 12 hrs
Aerostar XP-3011 — 21 min
Tea, SD
2/10/01

SUBCLASS AX-6 (1,200 - 1,600 cubic meters)

ALTITUDE (AUSTRIA)
Josef Starkbaum — 12,326 m
Colt 56 A — 40,440 ft
Heilbrunn, Austria
6/12/96

ALTITUDE*
Edward J. Chapman — 10,084 m
Barnes Firefly — 33,085 ft
Prior Lake, MN
12/12/82

DISTANCE (USA)
William E. Bussey — 1,434.24 km
Thunder and Colt, "Colt 56A" — 891.19 mi
Chanute, KS to Fleming, GA
2/6/95

DURATION (USA)
John A. Cayton — 23 hrs
Aerostar XP-3010 — 55 min
Tea, SD — 51 sec
12/13/00

SUBCLASS AX-7 (1,600 - 2,200 cubic meters)

ALTITUDE (AUSTRIA)
Josef Starkbaum — 14,018 m
Thunder and Colt "Colt 77A" — 45,991 ft
Dimbach, Austria
7/8/95

ALTITUDE*
Edward J. Chapman — 11,857 m
Barnes Firefly — 38,900 ft
Prior Lake, MN
2/3/81

DISTANCE (USA)
William E. Bussey — 1,434.24 km
Thunder and Colt, "Colt 56A" — 891.19 mi
Chanute, KS to Fleming, GA
2/6/95

DURATION (USA)
John R. Petrehn — 24 hrs
Barnes Firefly — 11 min
Huron, SD to Middletown, IL — 54 sec
1/17-19/84

SUBCLASS AX-8 (2,200 - 3,000 cubic meters)

ALTITUDE (AUSTRIA)
Josef Starkbaum — 15,011 m
Kubicek BB-30 — 49,249 ft
Stolzalpe to Knittelfeld, Austria
7/21/98

DISTANCE (USA)
William E. Bussey — 1,434.24 km
Thunder and Colt, "Colt 56A" — 891.19 mi
Chanute, KS to Fleming, GA
2/6/95

DURATION (USA) ✯
William E. Bussey — 29 hrs
Thunder and Colt, Colt 105A — 14 min
Amarillo, TX to Milbank, SD — 35 sec
1/24-25/93

SUBCLASS AX-9 (3,000 - 4,000 cubic meters)

ALTITUDE (USA)
Chauncey M. Dunn — 16,154 m
Raven S66A N5682C — 53,000 ft
Indianola, IA
8/1/79

DISTANCE (USA)
Troy A. Bradley — 1,636.84 km
Aerostar S-66A — 1,017.09 mi
Rockerville, SD to Shelbyville, KY
1/20/03

DURATION (USA)
William E. Bussey — 29 hrs
Thunder and Colt, Colt 105A — 14 min
Amarillo, TX to Milbank, SD — 35 sec
1/24-25/93

SUBCLASS AX-10 (4,000 - 6,000 cubic meters)

ALTITUDE (USA)
Chauncey M. Dunn — 16,154 m
Raven S66A N5682C — 53,000 ft
Indianola, IA
8/1/79

DISTANCE (JAPAN)
Michio Kanda — 2,366.10 km
Chikatsu Kakinuma — 1,470.22 mi
Cameron N-210
Mullewa to Frome Downs, Australia
6/23/94

DISTANCE*
Troy A. Bradley — 1,017.09 mi
Aerostar S-66A
Rockerville, SD to Shelbyville, KY
1/20/03

DURATION (JAPAN)
Michio Kanda — 50 hrs
Hirosuke Takezawa — 38 min
Cameron N-210
Chestemere Lake, AB, Canada to Jordan, MT
2/1-3/97

DURATION*
William E. Bussey — 29 hrs
Thunder and Colt, Colt 105A — 14 min
Amarillo, TX to Milbank, SD — 35 sec
1/24-25/93

SUBCLASS AX-11 (6,000 - 9,000 cubic meters)

ALTITUDE (USA)
Chauncey M. Dunn — 16,154 m
Raven S66A N5682C — 53,000 ft
Indianola, IA
8/1/79

DISTANCE (JAPAN)
Michio Kanda — 2,366.10 km
Chikatsu Kakinuma — 1,470.22 mi
Cameron N-210
Mullewa to Frome Downs, Australia
6/23/94

DISTANCE*
Troy A. Bradley 1,017.09 mi
Aerostar S-66A
Rockerville, SD to Shelbyville, KY
1/20/03

DURATION (JAPAN)
Michio Kanda 50 hrs
Hirosuke Takezawa 38 min
Cameron N-210
Chestemere Lake, AB, Canada to Jordan, MT
2/1-3/97

DURATION*
William E. Bussey 29 hrs
Thunder and Colt, Colt 105A 14 min
Amarillo, TX to Milbank, SD 35 sec
1/24-25/93

SUBCLASS AX-12 (9,000 - 12,000 cubic meters)

ALTITUDE (UK)
Julian R.P. Nott 16,805 m
Cameron 55,134 ft
Longmont, CO to Woodrow, CO
10/31/80

DISTANCE (JAPAN)
Michio Kanda 2,366.10 km
Chikatsu Kakinuma 1,470.22 mi
Cameron N-210
Mullewa to Frome Downs, Australia
6/23/94

DISTANCE*
Troy A. Bradley 1,017.09 mi
Aerostar S-66A
Rockerville, SD to Shelbyville, KY
1/20/03

DURATION (JAPAN)
Michio Kanda 50 hrs
Hirosuke Takezawa 38 min
Cameron N-210
Chestemere Lake, AB, Canada to Jordan, MT
2/1-3/97

DURATION*
William E. Bussey 29 hrs
Thunder and Colt, Colt 105A 14 min
Amarillo, TX to Milbank, SD 35 sec
1/24-25/93

SUBCLASS AX-13 (12,000 - 16,000 cubic meters)

ALTITUDE (UK)
Julian R.P. Nott 16,805 m
Cameron 55,134 ft
Longmont, CO to Woodrow, CO
10/31/80

DISTANCE (JAPAN)
Michio Kanda 2,366.10 km
Chikatsu Kakinuma 1,470.22 mi
Cameron N-210
Mullewa to Frome Downs, Australia
6/23/94

DISTANCE*
Troy A. Bradley 1,017.09 mi
Aerostar S-66A
Rockerville, SD to Shelbyville, KY
1/20/03

DURATION (JAPAN)
Michio Kanda 50 hrs
Hirosuke Takezawa 38 min
Cameron N-210
Chestemere Lake, AB, Canada to Jordan, MT
2/1-3/97

DURATION*
William E. Bussey 29 hrs
Thunder and Colt, Colt 105A 14 min
Amarillo, TX to Milbank, SD 35 sec
1/24-25/93

SUBCLASS AX-14 (16,000 - 22,000 cubic meters)

ALTITUDE (UK)
Per Lindstrand19,812 m
Thunder Colt Balloon 65,000 ft
Laredo, TX
6/6/88

DISTANCE (JAPAN)
Michio Kanda 2,366.10 km
Chikatsu Kakinuma 1,470.22 mi
Cameron N-210
Mullewa to Frome Downs, Australia
6/23/94

DISTANCE*
Troy A. Bradley 1,017.09 mi
Aerostar S-66A
Rockerville, SD to Shelbyville, KY
1/20/03

DURATION (JAPAN)
Michio Kanda 50 hrs
Hirosuke Takezawa 38 min
Cameron N-210
Chestemere Lake, AB, Canada to Jordan, MT
2/1-3/97

DURATION*
William E. Bussey 29 hrs
Thunder and Colt, Colt 105A 14 min
Amarillo, TX to Milbank, SD 35 sec
1/24-25/93

SUBCLASS AX-15 (more than 22,000 cubic meters)

ALTITUDE (UK)
Per Lindstrand 19,812 m
Thunder Colt Balloon 65,000 ft
Laredo, TX
6/6/88

DISTANCE (UK)
Per Lindstrand, Pilot 7,671.91 km
Richard Branson, Copilot 4,767.10 mi
Thunder and Colt, Colt 2500A
1/15-17/91

DISTANCE*
Troy A. Bradley 1,017.09 mi
Aerostar S-66A
Rockerville, SD to Shelbyville, KY
1/20/03

DURATION (JAPAN)
Michio Kanda 50 hrs
Hirosuke Takezawa 38 min
Cameron N-210
Chestemere Lake, AB, Canada to Jordan, MT
2/1-3/97

DURATION*
William E. Bussey 29 hrs
Thunder and Colt, Colt 105A 14 min
Amarillo, TX to Milbank, SD 35 sec
1/24-25/93

SUBCLASS AM-1 (less than 250 cubic meters)

ALTITUDE (AUSTRIA)
Günter Schabus 5,816 m
Cameron RN-9 19,081 ft
Semley, UK
10/13/94

ALTITUDE*
Troy Bradley 18,143 ft
Pandammonia Partners Gas-008
Albuquerque, NM to Moriarty, NM
6/15/94

DISTANCE (USA)
Troy Bradley 568.63 km
Pandammonia Partners Gas-008 353.33 mi
Ozark, MO to Spring Hill, TN
10/20-22/99

DURATION (USA) ✯
Troy Bradley 27 hrs
Pandammonia Partners Gas-008 25 min
Ozark, MO to Spring Hill, TN
10/20-22/99

SUBCLASS AM-2 (250 - 400 cubic meters)

ALTITUDE (AUSTRIA)
Günter Schabus 5,816 m
Cameron RN-9 19,081 ft
Semley, UK
10/13/94

ALTITUDE*
Troy Bradley 18,143 ft
Pandammonia Partners Gas-008
Albuquerque, NM to Moriarty, NM
6/15/94

DISTANCE (USA)
Troy Bradley 568.63 km
Pandammonia Partners Gas-008 353.33 mi
Ozark, MO to Spring Hill, TN
10/20-22/99

DURATION (USA)
Troy Bradley 27 hrs
Pandammonia Partners Gas-008 25 min
Ozark, MO to Spring Hill, TN
10/20-22/99

SUBCLASS AM-3 (400 - 600 cubic meters)

ALTITUDE (UK)
Janet Folkes 6,866 m
Lindstrand LBL-14M 22,526 ft
Reno, NV
11/4/98

ALTITUDE*
Troy Bradley 18,143 ft
Pandammonia Partners Gas-008
Albuquerque, NM to Moriarty, NM
6/15/94

DISTANCE (USA)
Troy Bradley 568.63 km
Pandammonia Partners Gas-008 353.33 mi
Ozark, MO to Spring Hill, TN
10/20-22/99

DURATION (USA)
Troy Bradley 27 hrs
Pandammonia Partners Gas-008 25 min
Ozark, MO to Spring Hill, TN
10/20-22/99

SUBCLASS AM-4 (600 - 900 cubic meters)

ALTITUDE (UK)
Janet Folkes 6,866 m
Lindstrand LBL-14M 22,526 ft
Reno, NV
11/4/98

ALTITUDE*
Troy Bradley 18,143 ft
Pandammonia Partners Gas-008
Albuquerque, NM to Moriarty, NM
6/15/94

DISTANCE (USA)
Troy Bradley 568.63 km
Pandammonia Partners Gas-008 353.33 mi
Ozark, MO to Spring Hill, TN
10/20-22/99

DURATION (USA)
Troy Bradley 27 hrs
Pandammonia Partners Gas-008 25 min
Ozark, MO to Spring Hill, TN
10/20-22/99

SUBCLASS AM-5 (900 - 1,200 cubic meters)

ALTITUDE (UK)
Janet Folkes 6,866 m
Lindstrand LBL-14M 22,526 ft
Reno, NV
11/4/98

ALTITUDE*
Troy Bradley 18,143 ft
Pandammonia Partners Gas-008
Albuquerque, NM to Moriarty, NM
6/15/94

DISTANCE (USA)
Troy Bradley 568.63 km
Pandammonia Partners Gas-008 353.33 mi
Ozark, MO to Spring Hill, TN
10/20-22/99

DURATION (USA)
Troy Bradley 27 hrs
Pandammonia Partners Gas-008 25 min
Ozark, MO to Spring Hill, TN
10/20-22/99

SUBCLASS AM-6 (1,200 - 1,600 cubic meters)

ALTITUDE (UK)
Janet Folkes 6,866 m
Lindstrand LBL-14M 22,526 ft
Reno, NV
11/4/98

ALTITUDE*
Troy Bradley 18,143 ft
Pandammonia Partners Gas-008
Albuquerque, NM to Moriarty, NM
6/15/94

DISTANCE (USA)
Troy Bradley 568.63 km
Pandammonia Partners Gas-008 353.33 mi
Ozark, MO to Spring Hill, TN
10/20-22/99

DURATION (USA)
Troy Bradley 27 hrs
Pandammonia Partners Gas-008 25 min
Ozark, MO to Spring Hill, TN
10/20-22/99

SUBCLASS AM-7 (1,600 - 2,200 cubic meters)

ALTITUDE (UK)
Janet Folkes 6,866 m
Lindstrand LBL-14M 22,526 ft
Reno, NV
11/4/98

ALTITUDE*
Troy Bradley 18,143 ft
Pandammonia Partners Gas-008
Albuquerque, NM to Moriarty, NM
6/15/94

DISTANCE (SPAIN)
Tomas Feliu Rius, Pilot 5,046.00 km
Jesus Gonzalez Green, Copilot 3,135.44 mi
Cameron R-60
Hierro Island, Spain to Maturin, Venezuela
2/8-14/92

DISTANCE*
Troy Bradley 568.63 km
Pandammonia Partners Gas-008 353.33 mi
Ozark, MO to Spring Hill, TN
10/20-22/99

DURATION (SPAIN)
Tomas Feliu Rius, Pilot 129 hrs
Jesus Gonzalez Green, Copilot 10 min
Cameron R-60
Hierro Island, Spain to Maturin, Venezuela
2/8-14/92

DURATION*
Troy Bradley 27 hrs
Pandammonia Partners Gas-008 25 min
Ozark, MO to Spring Hill, TN
10/20-22/99

SUBCLASS AM-8 (2,200 - 3,000 cubic meters)

ALTITUDE (UK)
Per Lindstrand 10,589 m
Lindstrand AM-2200 34,741 ft
Reno, NV
11/25/96

ALTITUDE*
Steve Fossett 27,485 ft
Bruce P. Comstock
Cameron R-77
Erie, CO
9/22/96

DISTANCE (USA)
Richard Abruzzo 5,340.18 km
Troy Bradley 3,318.23 mi
Cameron R-77
Bangor, ME, USA to Ben Slimane, Morocco
9/16-22/92

DURATION (FRANCE)
Laurent Lajoye 146 hrs
Christophe Houver 48 min
Lindstrand LBL 77M 45 sec
St John, Canada to Saône, France
9/5/00

DURATION*
Richard Abruzzo 144 hrs
Troy Bradley 16 min
Cameron R-77
Bangor, ME, USA to Ben Slimane, Morocco
9/16-22/92

SUBCLASS AM-9 (3,000 - 4,000 cubic meters)

ALTITUDE (UK)
Per Lindstrand 10,589 m
Lindstrand AM-2200 34,741 ft
Reno, NV
11/25/96

ALTITUDE*
Steve Fossett 27,485 ft
Bruce P. Comstock
Cameron R-77
Erie, CO
9/22/96

DISTANCE (USA)
Richard Abruzzo 5,340.18 km
Troy Bradley 3,318.23 mi
Cameron R-77
Bangor, ME, USA to Ben Slimane, Morocco
9/16-22/92

DURATION (FRANCE)
Laurent Lajoye 146 hrs
Christophe Houver 48 min
Lindstrand LBL 77M 45 sec
St John, Canada to Saône, France
9/5/00

DURATION*
Richard Abruzzo 144 hrs
Troy Bradley 16 min
Cameron R-77
Bangor, ME, USA to Ben Slimane, Morocco
9/16-22/92

SUBCLASS AM-10 (4,000 - 6,000 cubic meters)

ALTITUDE (UK)
Per Lindstrand 10,589 m
Lindstrand AM-2200 34,741 ft
Reno, NV
11/25/96

ALTITUDE*
Steve Fossett 27,485 ft
Bruce P. Comstock
Cameron R-77
Erie, CO
9/22/96

DISTANCE (USA)
Steve Fossett 8,748.11 km
Cameron R-150 5,435.82 mi
Seoul, Korea to Mendham, SK
2/17-22/95

DURATION (FRANCE)
Laurent Lajoye 146 hrs
Christophe Houver 48 min
Lindstrand LBL 77M 45 sec
St John, Canada to Saône, France
9/5/00

DURATION*
Richard Abruzzo 144 hrs
Troy Bradley 16 min
Cameron R-77
Bangor, ME, USA to Ben Slimane, Morocco
9/16-22/92

SUBCLASS AM-11 (6,000 - 9,000 cubic meters)

ALTITUDE (UK)
Per Lindstrand 10,589 m
Lindstrand AM-2200 34,741 ft
Reno, NV
11/25/96

ALTITUDE*
Steve Fossett 27,485 ft
Bruce P. Comstock
Cameron R-77
Erie, CO
9/22/96

DISTANCE (USA)
Steve Fossett — 16,673.81 km / 10,360.61 mi
Cameron R-210
St. Louis, MO to Sultanpur, India
1/14-20/97

DURATION (USA)
Steve Fossett — 146 hrs 44 min
Cameron R-210
St. Louis, MO to Sultanpur, India
1/14-20/97

SUBCLASS AM-12 (9,000 - 12,000 cubic meters)

ALTITUDE (UK)
Per Lindstrand — 10,589 m / 34,741 ft
Lindstrand AM-2200
Reno, NV
11/25/96

ALTITUDE*
Steve Fossett — 27,485 ft
Bruce P. Comstock
Cameron R-77
Erie, CO
9/22/96

DISTANCE (USA)
Steve Fossett — 16,673.81 km / 10,360.61 mi
Cameron R-210
St. Louis, MO to Sultanpur, India
1/14-20/97

DURATION (USA)
Steve Fossett — 146 hrs 44 min
Cameron R-210
St. Louis, MO to Sultanpur, India
1/14-20/97

SUBCLASS AM-13 (12,000 - 16,000 cubic meters)

ALTITUDE (UK)
Per Lindstrand — 10,589 m / 34,741 ft
Lindstrand AM-2200
Reno, NV
11/25/96

ALTITUDE*
Steve Fossett — 27,485 ft
Bruce P. Comstock
Cameron R-77
Erie, CO
9/22/96

DISTANCE (USA)
Steve Fossett — 22,909.53 km / 14,235.33 mi
Cameron R-450
Mendoza, Argentina to Coral Sea
8/7-16/98

DURATION (USA)
Kevin J. Uliassi — 243 hrs 28 min
homebuilt
Loves Park, IL to Nyaungu, Myanmar
2/22-3/3/00

SUBCLASS AM-14 (16,000 - 22,000 cubic meters)

ALTITUDE (UK)
Per Lindstrand — 10,589 m / 34,741 ft
Lindstrand AM-2200
Reno, NV
11/25/96

ALTITUDE*
Steve Fossett — 27,485 ft
Bruce P. Comstock
Cameron R-77
Erie, CO
9/22/96

DISTANCE (USA)
Steve Fossett — 33,195.1 km / 20,626.4 mi
Cameron/Cole R-550
Northam, Australia
6/19-7/3/02

DURATION (USA)
Steve Fossett — 355 hrs 50 min
Cameron/Cole R-550
Northam, Australia
6/19-7/3/02

SHORTEST TIME AROUND THE WORLD (USA)
Steve Fossett — 320 hrs 33 min
Cameron/Cole R-550
Northam, Australia
6/19-7/3/02

SUBCLASS AM-15 (more than 22,000 cubic meters)

ALTITUDE (SWITZERLAND/UK)
Bertrand Piccard — 11,737 m / 38,507 ft
Brian Jones
Cameron Balloons R-650
Chateau d'Oex, Switzerland to Dâkhla, Egypt
3/1-21/99

ALTITUDE*
Steve Fossett — 27,485 ft
Bruce P. Comstock
Cameron R-77
Erie, CO
9/22/96

DISTANCE (SWITZERLAND/UK)
Bertrand Piccard — 40,814 km / 25,360 mi
Brian Jones
Cameron Balloons R-650
Chateau d'Oex, Switzerland to Dâkhla, Egypt
3/1-21/99

DISTANCE*
Steve Fossett — 20,626.4 mi
Cameron/Cole R-550
Northam, Australia
6/19-7/3/02

DURATION (SWITZERLAND/UK)
Bertrand Piccard — 477 hrs 47 min
Brian Jones
Cameron Balloons R-650
Chateau d'Oex, Switzerland to Dâkhla, Egypt
3/1-21/99

DURATION*
Steve Fossett — 355 hrs 50 min
Cameron/Cole R-550
Northam, Australia
6/19-7/3/02

SHORTEST TIME AROUND THE WORLD (USA)
Steve Fossett — 320 hrs 33 min
Cameron/Cole R-550
Northam, Australia
6/19-7/3/02

SUBCLASS AS-2 (250 - 400 cubic meters)

ALTITUDE (USA)
Coy Foster — 2,111 m / 6,926 ft
Lindstrand AS-2
Plano, TX to Clinton, MO
3/8/03

DISTANCE (USA)
Coy Foster — 647.21 km / 402.15 mi
Lindstrand AS-2
Plano, TX to Clinton, MO
3/8/03

DURATION (USA)
Coy Foster — 12 hrs 25 min
Lindstrand AS-2
Plano, TX to Clinton, MO
3/8/03

SUBCLASS AS-3 (400 - 600 cubic meters)

ALTITUDE (USA)
Coy Foster — 2,111 m 6,926 ft
Lindstrand AS-2
Plano, TX to Clinton, MO
3/8/03

DISTANCE*
Tom F. Heinsheimer — 571.63 mi
Peter C. Neushul
Atmostat America
San Angelo, TX to Ruleton, KS
4/18-19/76

DURATION*
Tom F. Heinsheimer — 32 hrs 30 min
Atmostat America
Rancho Palos Verdes, CA
7/2-4/76

SUBCLASS AS-4 (600 - 900 cubic meters)

ALTITUDE (USA)
Coy Foster — 2,111 m 6,926 ft
Lindstrand AS-2
Plano, TX to Clinton, MO
3/8/03

DISTANCE*
Tom F. Heinsheimer — 571.63 mi
Peter C. Neushul
Atmostat America
San Angelo, TX to Ruleton, KS
4/18-19/76

DURATION*
Tom F. Heinsheimer — 32 hrs 30 min
Atmostat America
Rancho Palos Verdes, CA
7/2-4/76

SUBCLASS AS-5 (900 - 1,200 cubic meters)

ALTITUDE (USA)
Coy Foster — 2,111 m 6,926 ft
Lindstrand AS-2
Plano, TX to Clinton, MO
3/8/03

DISTANCE*
Tom F. Heinsheimer — 571.63 mi
Peter C. Neushul
Atmostat America
San Angelo, TX to Ruleton, KS
4/18-19/76

DURATION*
Tom F. Heinsheimer — 32 hrs 30 min
Atmostat America
Rancho Palos Verdes, CA
7/2-4/76

SUBCLASS AS-6 (1,200 - 1,600 cubic meters)

ALTITUDE (UK)
Julian Nott, Pilot — 5,415 m 17,766 ft
Spider Andersen, Copilot
Balloon G-BLHF
Pearce RAFF, Western Australia
11/20-22/84

ALTITUDE*
Coy Foster — 6,926 ft
Lindstrand AS-2
Plano, TX to Clinton, MO
3/8/03

DISTANCE (UK)
Julian Nott, Pilot — 2,391.46 km 1,486.03 mi
Spider Andersen, Copilot
Balloon G-BLHF
Pearce RAFF, Western Australia
11/20-22/84

DISTANCE*
Tom F. Heinsheimer — 571.63 mi
Peter C. Neushul
Atmostat America
San Angelo, TX to Ruleton, KS
4/18-19/76

DURATION (UK)
Julian Nott, Pilot — 33 hrs 8 min 42 sec
Spider Andersen, Copilot
Balloon G-BLHF
Pearce RAFF, Western Australia
11/20-22/84

DURATION*
Tom F. Heinsheimer — 32 hrs 30 min
Atmostat America
Rancho Palos Verdes, CA
7/2-4/76

SUBCLASS AS-7 (1,600 - 2,200 cubic meters)

ALTITUDE (UK)
Julian Nott, Pilot — 5,415 m 17,766 ft
Spider Andersen, Copilot
Balloon G-BLHF
Pearce RAFF, Western Australia
11/20-22/84

ALTITUDE*
Coy Foster — 6,926 ft
Lindstrand AS-2
Plano, TX to Clinton, MO
3/8/03

DISTANCE (UK)
Julian Nott, Pilot — 2,391.46 km 1,486.03 mi
Spider Andersen, Copilot
Balloon G-BLHF
Pearce RAFF, Western Australia
11/20-22/84

DISTANCE*
Tom F. Heinsheimer — 571.63 mi
Peter C. Neushul
Atmostat America
San Angelo, TX to Ruleton, KS
4/18-19/76

DURATION (UK)
Julian Nott, Pilot — 33 hrs 8 min 42 sec
Spider Andersen, Copilot
Balloon G-BLHF
Pearce RAFF, Western Australia
11/20-22/84

DURATION*
Tom F. Heinsheimer — 32 hrs 30 min
Atmostat America
Rancho Palos Verdes, CA
7/2-4/76

SUBCLASS AS-8 (2,200 - 3,000 cubic meters)

ALTITUDE (UK)
Julian Nott, Pilot — 5,415 m
Spider Andersen, Copilot — 17,766 ft
Balloon G-BLHF
Pearce RAFF, Western Australia
11/20-22/84

ALTITUDE*
Coy Foster — 6,926 ft
Lindstrand AS-2
Plano, TX to Clinton, MO
3/8/03

DISTANCE (UK)
Julian Nott, Pilot — 2,391.46 km
Spider Andersen, Copilot — 1,486.03 mi
Balloon G-BLHF
Pearce RAFF, Western Australia
11/20-22/84

DISTANCE*
Tom F. Heinsheimer — 571.63 mi
Peter C. Neushul
Atmostat America
San Angelo, TX to Ruleton, KS
4/18-19/76

DURATION (UK)
Julian Nott, Pilot — 33 hrs
Spider Andersen, Copilot — 8 min
Balloon G-BLHF — 42 sec
Pearce RAFF, Western Australia
11/20-22/84

DURATION*
Tom F. Heinsheimer — 32 hrs
Atmostat America — 30 min
Rancho Palos Verdes, CA
7/2-4/76

SUBCLASS AS-9 (3,000 - 4,000 cubic meters)

ALTITUDE (UK)
Julian Nott, Pilot — 5,415 m
Spider Andersen, Copilot — 17,766 ft
Balloon G-BLHF
Pearce RAFF, Western Australia
11/20-22/84

ALTITUDE*
Coy Foster — 6,926 ft
Lindstrand AS-2
Plano, TX to Clinton, MO
3/8/03

DISTANCE (UK)
Julian Nott, Pilot — 2,391.47 km
Spider Andersen, Copilot — 1,486.03 mi
Balloon G-BLHF
Pearce RAFF, Western Australia
11/20-22/84

DISTANCE*
Tom F. Heinsheimer — 571.63 mi
Peter C. Neushul
Atmostat America
San Angelo, TX to Ruleton, KS
4/18-19/76

DURATION (UK)
Julian Nott, Pilot — 33 hrs
Spider Andersen, Copilot — 8 min
Balloon G-BLHF — 42 sec
Pearce RAFF, Western Australia
11/20-22/84

DURATION*
Tom F. Heinsheimer — 32 hrs
Atmostat America — 30 min
Rancho Palos Verdes, CA
7/2-4/76

SUBCLASS AS-10 (4,000 - 6,000 cubic meters)

ALTITUDE (UK)
Julian Nott, Pilot — 5,415 m
Spider Andersen, Copilot — 17,766 ft
Balloon G-BLHF
Pearce RAFF, Western Australia
11/20-22/84

ALTITUDE*
Coy Foster — 6,926 ft
Lindstrand AS-2
Plano, TX to Clinton, MO
3/8/03

DISTANCE (UK)
Julian Nott, Pilot — 2,391.47 km
Spider Andersen, Copilot — 1,486.03 mi
Balloon G-BLHF
Pearce RAFF, Western Australia
11/20-22/84

DISTANCE*
Tom F. Heinsheimer — 571.63 mi
Peter C. Neushul
Atmostat America
San Angelo, TX to Ruleton, KS
4/18-19/76

DURATION (UK)
Julian Nott, Pilot — 33 hrs
Spider Andersen, Copilot — 8 min
Balloon G-BLHF — 42 sec
Pearce RAFF, Western Australia
11/20-22/84

DURATION*
Tom F. Heinsheimer — 32 hrs
Atmostat America — 30 min
Rancho Palos Verdes, CA
7/2-4/76

SUBCLASS AS-11 (6,000 - 9,000 cubic meters)

ALTITUDE (UK)
Julian Nott, Pilot — 5,415 m
Spider Andersen, Copilot — 17,766 ft
Balloon G-BLHF
Pearce RAFF, Western Australia
11/20-22/84

ALTITUDE*
Coy Foster — 6,926 ft
Lindstrand AS-2
Plano, TX to Clinton, MO
3/8/03

DISTANCE (UK)
Julian Nott, Pilot — 2,391.47 km
Spider Andersen, Copilot — 1,486.03 mi
Balloon G-BLHF
Pearce RAFF, Western Australia
11/20-22/84

DISTANCE*
Tom F. Heinsheimer — 571.63 mi
Peter C. Neushul
Atmostat America
San Angelo, TX to Ruleton, KS
4/18-19/76

DURATION (UK)
Julian Nott, Pilot — 33 hrs 8 min 42 sec
Spider Andersen, Copilot
Balloon G-BLHF
Pearce RAFF, Western Australia
11/20-22/84

DURATION*
Tom F. Heinsheimer — 32 hrs 30 min
Atmostat America
Rancho Palos Verdes, CA
7/2-4/76

SUBCLASS AS-12 (9,000 - 12,000 cubic meters)

ALTITUDE (UK)
Julian Nott, Pilot — 5,415 m 17,766 ft
Spider Andersen, Copilot
Balloon G-BLHF
Pearce RAFF, Western Australia
11/20-22/84

ALTITUDE*
Coy Foster — 6,926 ft
Lindstrand AS-2
Plano, TX to Clinton, MO
3/8/03

DISTANCE (UK)
Julian Nott, Pilot — 2,391.47 km 1,486.03 mi
Spider Andersen, Copilot
Balloon G-BLHF
Pearce RAFF, Western Australia
11/20-22/84

DISTANCE*
Tom F. Heinsheimer — 571.63 mi
Peter C. Neushul
Atmostat America
San Angelo, TX to Ruleton, KS
4/18-19/76

DURATION (UK)
Julian Nott, Pilot — 33 hrs 8 min 42 sec
Spider Andersen, Copilot
Balloon G-BLHF
Pearce RAFF, Western Australia
11/20-22/84

DURATION*
Tom F. Heinsheimer — 32 hrs 30 min
Atmostat America
Rancho Palos Verdes, CA
7/2-4/76

SUBCLASS AS-13 (12,000 - 16,000 cubic meters)

ALTITUDE (UK)
Julian Nott, Pilot — 5,415 m 17,766 ft
Spider Andersen, Copilot
Balloon G-BLHF
Pearce RAFF, Western Australia
11/20-22/84

ALTITUDE*
Coy Foster — 6,926 ft
Lindstrand AS-2
Plano, TX to Clinton, MO
3/8/03

DISTANCE (UK)
Julian Nott, Pilot — 2,391.47 km 1,486.03 mi
Spider Andersen, Copilot
Balloon G-BLHF
Pearce RAFF, Western Australia
11/20-22/84

DISTANCE*
Tom F. Heinsheimer — 571.63 mi
Peter C. Neushul
Atmostat America
San Angelo, TX to Ruleton, KS
4/18-19/76

DURATION (UK)
Julian Nott, Pilot — 33 hrs 8 min 42 sec
Spider Andersen, Copilot
Balloon G-BLHF
Pearce RAFF, Western Australia
11/20-22/84

DURATION*
Tom F. Heinsheimer — 32 hrs 30 min
Atmostat America
Rancho Palos Verdes, CA
7/2-4/76

SUBCLASS AS-14 (16,000 - 22,000 cubic meters)

ALTITUDE (UK)
Julian Nott, Pilot — 5,415 m 17,766 ft
Spider Andersen, Copilot
Balloon G-BLHF
Pearce RAFF, Western Australia
11/20-22/84

ALTITUDE*
Coy Foster — 6,926 ft
Lindstrand AS-2
Plano, TX to Clinton, MO
3/8/03

DISTANCE (UK)
Julian Nott, Pilot — 2,391.47 km 1,486.03 mi
Spider Andersen, Copilot
Balloon G-BLHF
Pearce RAFF, Western Australia
11/20-22/84

DISTANCE*
Tom F. Heinsheimer — 571.63 mi
Peter C. Neushul
Atmostat America
San Angelo, TX to Ruleton, KS
4/18-19/76

DURATION (UK)
Julian Nott, Pilot — 33 hrs 8 min 42 sec
Spider Andersen, Copilot
Balloon G-BLHF
Pearce RAFF, Western Australia
11/20-22/84

DURATION*
Tom F. Heinsheimer — 32 hrs 30 min
Atmostat America
Rancho Palos Verdes, CA
7/2-4/76

SUBCLASS AS-15 (more than 22,000 cubic meters)

ALTITUDE (UK)
Julian Nott, Pilot — 5,415 m 17,766 ft
Spider Andersen, Copilot
Balloon G-BLHF
Pearce RAFF, Western Australia
11/20-22/84

ALTITUDE*
Coy Foster — 6,926 ft
Lindstrand AS-2
Plano, TX to Clinton, MO
3/8/03

DISTANCE (UK)
Julian Nott, Pilot — 2,391.47 km / 1,486.03 mi
Spider Andersen, Copilot
Balloon G-BLHF
Pearce RAFF, Western Australia
11/20-22/84

DISTANCE*
Tom F. Heinsheimer — 571.63 mi
Peter C. Neushul
Atmostat America
San Angelo, TX to Ruleton, KS
4/18-19/76

DURATION (UK)
Julian Nott, Pilot — 33 hrs 8 min 42 sec
Spider Andersen, Copilot
Balloon G-BLHF
Pearce RAFF, Western Australia
11/20-22/84

DURATION*
Tom F. Heinsheimer — 32 hrs 30 min
Atmostat America
Rancho Palos Verdes, CA
7/2-4/76

SUBCLASS AN-3 (400 - 600 cubic meters)

DURATION* ★
Timothy S. Cole — 8 hrs 47 min 20 sec
West Wind 600M
Kersey, CO
9/26/91

SUBCLASS AN-4 (600 - 900 cubic meters)

DURATION*
Timothy S. Cole — 8 hrs 47 min 20 sec
West Wind 600M
Kersey, CO
9/26/91

SUBCLASS AN-5 (900 - 1,200 cubic meters)

DURATION*
Timothy S. Cole — 8 hrs 47 min 20 sec
West Wind 600M
Kersey, CO
9/26/91

SUBCLASS AN-6 (1,200 - 1,600 cubic meters)

DISTANCE*
Carl W. Eidsness — 1,124.28 mi
homebuilt
Reserve, MT to Tablerville, OK
10/20/95

DURATION*
Timothy S. Cole — 8 hrs 47 min 20 sec
West Wind 600M
Kersey, CO
9/26/91

SUBCLASS AN-7 (1,600 - 2,200 cubic meters)

DISTANCE*
Carl W. Eidsness — 1,124.28 mi
homebuilt
Reserve, MT to Tablerville, OK
10/20/95

DURATION*
Timothy S. Cole — 8 hrs 47 min 20 sec
West Wind 600M
Kersey, CO
9/26/91

SUBCLASS AN-8 (2,200 - 3,000 cubic meters)

DISTANCE*
Carl W. Eidsness — 1,124.28 mi
homebuilt
Reserve, MT to Tablerville, OK
10/20/95

DURATION*
Timothy S. Cole — 8 hrs 47 min 20 sec
West Wind 600M
Kersey, CO
9/26/91

SUBCLASS AN-9 (3,000 - 4,000 cubic meters)

DISTANCE*
Carl W. Eidsness — 1,124.28 mi
homebuilt
Reserve, MT to Tablerville, OK
10/20/95

DURATION*
Timothy S. Cole — 8 hrs 47 min 20 sec
West Wind 600M
Kersey, CO
9/26/91

SUBCLASS AN-10 (4,000 - 6,000 cubic meters)

DISTANCE*
Carl W. Eidsness — 1,124.28 mi
homebuilt
Reserve, MT to Tablerville, OK
10/20/95

DURATION*
Timothy S. Cole — 8 hrs 47 min 20 sec
West Wind 600M
Kersey, CO
9/26/91

SUBCLASS AN-11 (6,000 - 9,000 cubic meters)

DISTANCE*
Carl W. Eidsness — 1,124.28 mi
homebuilt
Reserve, MT to Tablerville, OK
10/20/95

DURATION*
Timothy S. Cole — 8 hrs 47 min 20 sec
West Wind 600M
Kersey, CO
9/26/91

SUBCLASS AN-12 (9,000 - 12,000 cubic meters)

DISTANCE*
Carl W. Eidsness — 1,124.28 mi
homebuilt
Reserve, MT to Tablerville, OK
10/20/95

DURATION*
Timothy S. Cole — 8 hrs 47 min 20 sec
West Wind 600M
Kersey, CO
9/26/91

SUBCLASS AN-13 (12,000 - 16,000 cubic meters)

DISTANCE*
Carl W. Eidsness — 1,124.28 mi
homebuilt
Reserve, MT to Tablerville, OK
10/20/95

DURATION*
Timothy S. Cole 8 hrs
West Wind 600M 47 min
Kersey, CO 20 sec
9/26/91

SUBCLASS AN-14 (16,000 - 22,000 cubic meters)

DISTANCE*
Carl W. Eidsness 1,124.28 mi
homebuilt
Reserve, MT to Tablerville, OK
10/20/95

DURATION*
Timothy S. Cole 8 hrs
West Wind 600M 47 min
Kersey, CO 20 sec
9/26/91

SUBCLASS AN-15 (more than 22,000 cubic meters)

DISTANCE*
Carl W. Eidsness 1,124.28 mi
homebuilt
Reserve, MT to Tablerville, OK
10/20/95

DURATION*
Timothy S. Cole 8 hrs
West Wind 600M 47 min
Kersey, CO 20 sec
9/26/91

U.S. NATIONAL AND WORLD CLASS RECORDS, FEMININE

SUBCLASS AA-1 (less than 250 cubic meters)

DISTANCE (USA)
Wilma Piccard — 28.33 km / 17.60 mi
Piccard S-10
Indianola, IA
8/12/72

SUBCLASS AA-2 (250 - 400 cubic meters)

DISTANCE (USA)
Wilma Piccard — 28.33 km / 17.60 mi
Piccard S-10
Indianola, IA
8/12/72

SUBCLASS AA-3 (400 - 600 cubic meters)

DISTANCE (FRANCE)
Paulette Weber — 511.978 km / 318.128 mi
F-AMAQ
5/8/53

DURATION (USSR)
A. Kondratyeva — 22 hrs 40 min
Moscow to Loukino Polie
SSR BP-31
5/14/39

SUBCLASS AA-4 (600 - 900 cubic meters)

ALTITUDE (GERMANY)
Astrid Gerhardt — 4,692 m / 15,393 ft
Wörner NL-510
Stuttgart, Germany
10/13/96

DISTANCE (GERMANY)
Astrid Gerhardt — 789.78 km / 490.74 mi
Wörner NL-510/STU
Albuquerque, NM to Hutchinson, KS
10/7/01

DURATION (GERMANY)
Astrid Gerhardt — 23 hrs 17 min
Wörner NL-510/STU
Albuquerque, NM to Hutchinson, KS
10/7/01

SUBCLASS AA-5 (900 - 1,200 cubic meters)

ALTITUDE (GERMANY)
Astrid Gerhardt — 4,692 m / 15,393 ft
Wörner NL-510
Stuttgart, Germany
10/13/96

DISTANCE (GERMANY)
Astrid Gerhardt — 789.78 km / 490.74 mi
Wörner NL-510/STU
Albuquerque, NM to Hutchinson, KS
10/7/01

DURATION (GERMANY)
Astrid Gerhardt — 23 hrs 17 min
Wörner NL-510/STU
Albuquerque, NM to Hutchinson, KS
10/7/01

CLASS AA-6 (1,200 - 1,600 cubic meters)

ALTITUDE (W. GERMANY)
Renate Peter — 6,176 m / 20,262 ft
D. Trevira Balloon
Augsburg, W. Germany
7/31/75

DISTANCE (USA)
Lesley Pritchard Davies, Pilot — 2,331.17 km / 1,448.52 mi
Dr. Carol Rymer Davis, Copilot
Aero Vail AA-1000
Albuquerque, NM to Lisbon, OH
10/8-10/95

DURATION (USA)
Lesley Pritchard Davies, Pilot — 60 hrs 12 min
Dr. Carol Rymer Davis, Copilot
Aero Vail AA-1000
Albuquerque, NM to Lisbon, OH
10/8-10/95

SUBCLASS AA-7 (1,600 - 2,200 cubic meters)

ALTITUDE (W. GERMANY)
Renate Peter — 6,176 m / 20,262 ft
D. Trevira balloon
Augsburg, W. Germany
7/31/75

DISTANCE (USA)
Lesley Pritchard Davies, Pilot — 2,331.17 km / 1,448.52 mi
Dr. Carol Rymer Davis, Copilot
Aero Vail AA-1000
Albuquerque, NM to Lisbon, OH
10/8-10/95

DURATION (USA)
Lesley Pritchard Davies, Pilot — 60 hrs 12 min
Dr. Carol Rymer Davis, Copilot
Aero Vail AA-1000
Albuquerque, NM to Lisbon, OH
10/8-10/95

SUBCLASS AA-8 (2,200 - 3,000 cubic meters)

ALTITUDE (W. GERMANY)
Renate Peter — 6,176 m / 20,262 ft
D. Trevira Balloon
Augsburg, W. Germany
7/13/75

DISTANCE (USA)
Lesley Pritchard Davies, Pilot — 2,331.17 km / 1,448.52 mi
Dr. Carol Rymer Davis, Copilot
Aero Vail AA-1000
Albuquerque, NM to Lisbon, OH
10/8-10/95

DURATION (USA)
Lesley Pritchard Davies, Pilot — 60 hrs 12 min
Dr. Carol Rymer Davis, Copilot
Aero Vail AA-1000
Albuquerque, NM to Lisbon, OH
10/8-10/95

SUBCLASS AA-9 (3,000 - 4,000 cubic meters)

ALTITUDE (W. GERMANY)
Renate Peter — 6,176 m / 20,262 ft
D. Trevira balloon
Augsburg, W. Germany
7/31/75

DISTANCE (USA)
Lesley Pritchard Davies, Pilot — 2,331.17 km / 1,448.52 mi
Dr. Carol Rymer Davis, Copilot
Aero Vail AA-1000
Albuquerque, NM to Lisbon, OH
10/8-10/95

DURATION (USA)
Lesley Pritchard Davies, Pilot — 60 hrs 12 min
Dr. Carol Rymer Davis, Copilot
Aero Vail AA-1000
Albuquerque, NM to Lisbon, OH
10/8-10/95

SUBCLASS AA-10 (4,000 - 6,000 cubic meters)

ALTITUDE (W. GERMANY)
Renate Peter — 6,176 m 20,262 ft
D. Trevira balloon
Augsburg, W. Germany
7/31/75

DISTANCE (USA)
Lesley Pritchard Davies, Pilot — 2,331.17 km 1,448.52 mi
Dr. Carol Rymer Davis, Copilot
Aero Vail AA-1000
Albuquerque, NM to Lisbon, OH
10/8-10/95

DURATION (USA)
Lesley Pritchard Davies, Pilot — 60 hrs 12 min
Dr. Carol Rymer Davis, Copilot
Aero Vail AA-1000
Albuquerque, NM to Lisbon, OH
10/8-10/95

SUBCLASS AA-11 (6,000 - 9,000 cubic meters)

ALTITUDE (W. GERMANY)
Renate Peter — 6,176 m 20,262 ft
D. Trevira balloon
Augsburg, W. Germany
7/31/75

DISTANCE (USA)
Lesley Pritchard Davies, Pilot — 2,331.17 km 1,448.52 mi
Dr. Carol Rymer Davis, Copilot
Aero Vail AA-1000
Albuquerque, NM to Lisbon, OH
10/8-10/95

DURATION (USA)
Lesley Pritchard Davies, Pilot — 60 hrs 12 min
Dr. Carol Rymer Davis, Copilot
Aero Vail AA-1000
Albuquerque, NM to Lisbon, OH
10/8-10/95

SUBCLASS AA-12 (9,000 - 12,000 cubic meters)

ALTITUDE (W. GERMANY)
Renate Peter — 6,176 m 20,262 ft
D. Trevira balloon
Augsburg, W. Germany
7/31/75

DISTANCE (USA)
Lesley Pritchard Davies, Pilot — 2,331.17 km 1,448.52 mi
Dr. Carol Rymer Davis, Copilot
Aero Vail AA-1000
Albuquerque, NM to Lisbon, OH
10/8-10/95

DURATION (USA)
Lesley Pritchard Davies, Pilot — 60 hrs 12 min
Dr. Carol Rymer Davis, Copilot
Aero Vail AA-1000
Albuquerque, NM to Lisbon, OH
10/8-10/95

SUBCLASS AA-13 (12,000 - 16,000 cubic meters)

ALTITUDE (W. GERMANY)
Renate Peter — 6,176 m 20,262 ft
D. Trevira balloon
Augsburg, W. Germany
7/31/75

DISTANCE (USA)
Lesley Pritchard Davies, Pilot — 2,331.17 km 1,448.52 mi
Dr. Carol Rymer Davis, Copilot
Aero Vail AA-1000
Albuquerque, NM to Lisbon, OH
10/8-10/95

DURATION (USA)
Lesley Pritchard Davies, Pilot — 60 hrs 12 min
Dr. Carol Rymer Davis, Copilot
Aero Vail AA-1000
Albuquerque, NM to Lisbon, OH
10/8-10/95

SUBCLASS AA-14 (16,000 - 22,000 cubic meters)

ALTITUDE (W. GERMANY)
Renate Peter — 6,176 m 20,262 ft
D. Trevira balloon
Augsburg, W. Germany
7/31/75

DISTANCE (USA)
Lesley Pritchard Davies, Pilot — 2,331.17 km 1,448.52 mi
Dr. Carol Rymer Davis, Copilot
Aero Vail AA-1000
Albuquerque, NM to Lisbon, OH
10/8-10/95

DURATION (USA)
Lesley Pritchard Davies, Pilot — 60 hrs 12 min
Dr. Carol Rymer Davis, Copilot
Aero Vail AA-1000
Albuquerque, NM to Lisbon, OH
10/8-10/95

SUBCLASS AA-15 (more than 22,000 cubic meters)

ALTITUDE (W. GERMANY)
Renate Peter — 6,176 m 20,262 ft
D. Trevira balloon
Augsburg, W. Germany
7/31/75

DISTANCE (USA)
Lesley Pritchard Davies, Pilot — 2,331.17 km 1,448.52 mi
Dr. Carol Rymer Davis, Copilot
Aero Vail AA-1000
Albuquerque, NM to Lisbon, OH
10/8-10/95

DURATION (USA)
Lesley Pritchard Davies, Pilot — 60 hrs 12 min
Dr. Carol Rymer Davis, Copilot
Aero Vail AA-1000
Albuquerque, NM to Lisbon, OH
10/8-10/95

SUBCLASS AX-1 (less than 250 cubic meters)

ALTITUDE (USA)
Katherine E. Boland — 143.95 m 472.17 ft
Boland Balloon
Farmington, CT
7/29/78

DISTANCE (USA)
Katherine E. Boland — 4.81 km 2.99 mi
Boland Balloon
Farmington, CT
7/29/78

DURATION (USA)
Stephanie Shinn — 1 hr 11 min 12 sec
Avian Balloon
Moses Lake, WA
1/16/83

SUBCLASS AX-2 (250 - 400 cubic meters)

ALTITUDE (UK)
Janet Folkes — 2,293 m / 7,523 ft
Colt Cloudhopper 14A
Villamartin, Spain
1/28/95

ALTITUDE*
Donna Wiederkehr — 595 m / 1,952 ft
Modified Raven
St. Paul, MN
3/13/75

DISTANCE (USA)
Donna Wiederkehr — 18.01 km / 11.19 min
Modified Raven
St. Paul, MN
3/13/75

DURATION (USA)
Donna Wiederkehr — 2 hrs 40 min
St. Paul, MN
Modified Raven
3/13/75

SUBCLASS AX-3 (400 - 600 cubic meters)

ALTITUDE (USA)
Brenda Jo Bogan — 2,978 m / 9,770 ft
Barnes Firefly 18
9/3/65

DISTANCE (UK)
Julia Bayly — 59.57 km / 37.01 mi
Lindstrand Cloudhopper
Great Eccleston to Todmorden, UK
6/5/94

DURATION (USA)
Donna Wiederkehr — 2 hrs 40 min
St. Paul, MN
Modified Raven
3/13/75

SUBCLASS AX-4 (600 - 900 cubic meters)

ALTITUDE (USA)
Connie Lee March — 3,858 m / 12,659 ft
Raven Balloon
Moriarty, NM
3/2/80

DISTANCE (USA)
Susan A. Harwell — 74.74 km / 46.44 mi
BRET 31A
Oil City, LA to Williams, LA
1/24/98

DURATION (USA)
Donna Wiederkehr — 2 hrs 40 min
Modified Raven
St. Paul, MN
3/13/75

SUBCLASS AX-5 (900 - 1,200 cubic meters)

ALTITUDE (USA)
Carol R. Davis — 9,540 m / 31,300 ft
Firefly 5
Moriarty, NM
12/8/79

DISTANCE (USA)
Deborah L. Young — 288.29 km / 179.13 mi
Aerostar XP-3011
Ottumwa, IA
2/26/01

DURATION (USA)
Deborah L. Young — 9 hrs 32 min
Aerostar XP-3011
Ottumwa, IA
2/26/01

SUBCLASS AX-6 (1,200 - 1,600 cubic meters)

ALTITUDE (USA)
Carol R. Davis — 9,540 m / 31,300 ft
Firefly 5
Moriarty, NM
12/8/79

DISTANCE (USA)
Denise Wiederkehr — 366.99 km / 228.04 mi
Raven S-50A
St. Paul, MN to Waupun, WI
3/23/74

DURATION (JAPAN)
Toshiko Ichiyoshi — 12 hrs 18 min
Cameron N-56
Merredin, Australia
8/22/97

DURATION*
Denise Wiederkehr — 11 hrs 10 min
Raven S-50A
St. Paul, MN to Waupun, WI
3/23/74

SUBCLASS AX-7 (1,600 - 2,200 cubic meters)

ALTITUDE (USA) ★
Jetta Miller Schantz — 9,954 m / 32,657 ft
Aerostar Rally
China Lake, CA
8/19/94

DISTANCE (USA) ★
Karen L. Gould — 736.79 km / 457.82 mi
Balloon Works Firefly
Amarillo, TX to Beatrice, NE
2/19/94

DURATION (USA)
Jetta Miller Schantz — 15 hrs 11 min
Aerostar Rally
McAlester, OK to Wellborn, TX
2/4/96

SUBCLASS AX-8 (2,200 - 3,000 cubic meters)

ALTITUDE (USA)
Jetta Miller Schantz — 9,954 m / 32,657 ft
Aerostar Rally
China Lake, CA
8/19/94

DISTANCE (USA)
Karen L. Gould — 736.79 km / 457.82 mi
Balloon Works Firefly
Amarillo, TX to Beatrice, NE
2/19/94

DURATION (USA)
Jetta Miller Schantz — 15 hrs 11 min
Aerostar Rally
McAlester, OK to Wellborn, TX
2/4/96

SUBCLASS AX-9 (3,000 - 4,000 cubic meters)

ALTITUDE (USA)
Jetta Miller Schantz — 9,954 m / 32,657 ft
Aerostar Rally
China Lake, CA
8/19/94

DISTANCE (USA)
Karen L. Gould — 736.79 km / 457.82 mi
Balloon Works Firefly
Amarillo, TX to Beatrice, NE
2/19/94

DURATION (USA)
Jetta Miller Schantz — 15 hrs / 11 min
Aerostar Rally
McAlester, OK to Wellborn, TX
2/4/96

SUBCLASS AX-10 (4,000 - 6,000 cubic meters)

ALTITUDE (USA)
Jetta Miller Schantz — 9,954 m / 32,657 ft
Aerostar Rally
China Lake, CA
8/19/94

DISTANCE (USA)
Karen L. Gould — 736.79 km / 457.82 mi
Balloon Works Firefly
Amarillo, TX to Beatrice, NE
2/19/94

DURATION (UK)
Lindsay Muir — 19 hrs / 7 min / 55 sec
Lindstrand LBL-210 A
Paddock Wood, UK - Lisbourg, France
5/21/00

DURATION*
Jetta Miller Schantz — 15 hrs / 11 min
Aerostar Rally
McAlester, OK to Wellborn, TX
2/4/96

SUBCLASS AX-11 (6,000 - 9,000 cubic meters)

ALTITUDE (USA)
Jetta Miller Schantz — 9,954 m / 32,657 ft
Aerostar Rally
China Lake, CA
8/19/94

DISTANCE (USA)
Karen L. Gould — 736.79 km / 457.82 mi
Balloon Works Firefly
Amarillo, TX to Beatrice, NE
2/19/94

DURATION (UK)
Lindsay Muir — 19 hrs / 7 min / 55 sec
Lindstrand LBL-210 A
Paddock Wood, UK - Lisbourg, France
5/21/00

DURATION*
Jetta Miller Schantz — 15 hrs / 11 min
Aerostar Rally
McAlester, OK to Wellborn, TX
2/4/96

SUBCLASS AX-12 (9,000 - 12,000 cubic meters)

ALTITUDE (USA)
Jetta Miller Schantz — 9,954 m / 32,657 ft
Aerostar Rally
China Lake, CA
8/19/94

DISTANCE (USA)
Karen L. Gould — 736.79 km / 457.82 mi
Balloon Works Firefly
Amarillo, TX to Beatrice, NE
2/19/94

DURATION (UK)
Lindsay Muir — 19 hrs / 7 min / 55 sec
Lindstrand LBL-210 A
Paddock Wood, UK - Lisbourg, France
5/21/00

DURATION*
Jetta Miller Schantz — 15 hrs / 11 min
Aerostar Rally
McAlester, OK to Wellborn, TX
2/4/96

SUBCLASS AX-13 (12,000 - 16,000 cubic meters)

ALTITUDE (USA)
Jetta Miller Schantz — 9,954 m / 32,657 ft
Aerostar Rally
China Lake, CA
8/19/94

DISTANCE (USA)
Karen L. Gould — 736.79 km / 457.82 mi
Balloon Works Firefly
Amarillo, TX to Beatrice, NE
2/19/94

DURATION (UK)
Lindsay Muir — 19 hrs / 7 min / 55 sec
Lindstrand LBL-210 A
Paddock Wood, UK - Lisbourg, France
5/21/00

DURATION*
Jetta Miller Schantz — 15 hrs / 11 min
Aerostar Rally
McAlester, OK to Wellborn, TX
2/4/96

SUBCLASS AX-14 (16,000 - 22,000 cubic meters)

ALTITUDE (USA)
Jetta Miller Schantz — 9,954 m / 32,657 ft
Aerostar Rally
China Lake, CA
8/19/94

DISTANCE (USA)
Karen L. Gould — 736.79 km / 457.82 mi
Balloon Works Firefly
Amarillo, TX to Beatrice, NE
2/19/94

DURATION (UK)
Lindsay Muir — 19 hrs / 7 min / 55 sec
Lindstrand LBL-210 A
Paddock Wood, UK - Lisbourg, France
5/21/00

DURATION*
Jetta Miller Schantz — 15 hrs / 11 min
Aerostar Rally
McAlester, OK to Wellborn, TX
2/4/96

SUBCLASS AX-15 (more than 22,000 cubic meters)

ALTITUDE (USA)
Jetta Miller Schantz — 9,954 m / 32,657 ft
Aerostar Rally
China Lake, CA
8/19/94

DISTANCE (USA)
Karen L. Gould — 736.79 km / 457.82 mi
Balloon Works Firefly
Amarillo, TX to Beatrice, NE
2/19/94

DURATION (UK)
Lindsay Muir — 19 hrs
Lindstrand LBL-210 A — 7 min
Paddock Wood, UK - Lisbourg, France — 55 sec
5/21/00

DURATION*
Jetta Miller Schantz — 15 hrs
Aerostar Rally — 11 min
McAlester, OK to Wellborn, TX
2/4/96

SUBCLASS AM-3 (400 - 600 cubic meters)

ALTITUDE (UK)
Janet Folkes — 6,866 m
Lindstrand LBL-14M — 22,526 ft
Reno, NV
11/4/98

DISTANCE (UK)
Janet Folkes — 69.7 km
Lindstrand LBL-14M — 43.3 mi
Reno, NV
11/4/98

DURATION (UK)
Janet Folkes — 1 hr
Lindstrand LBL-14M — 38 min
Reno, NV
11/4/98

SUBCLASS AM-4 (600 - 900 cubic meters)

ALTITUDE (UK)
Janet Folkes — 6,866 m
Lindstrand LBL-14M — 22,526 ft
Reno, NV
11/4/98

DISTANCE (UK)
Janet Folkes — 69.7 km
Lindstrand LBL-14M — 43.3 mi
Reno, NV
11/4/98

DURATION (UK)
Janet Folkes — 1 hr
Lindstrand LBL-14M — 38 min
Reno, NV
11/4/98

SUBCLASS AM-5 (900 - 1,200 cubic meters)

ALTITUDE (UK)
Janet Folkes — 6,866 m
Lindstrand LBL-14M — 22,526 ft
Reno, NV
11/4/98

DISTANCE (UK)
Janet Folkes — 69.7 km
Lindstrand LBL-14M — 43.3 mi
Reno, NV
11/4/98

DURATION (UK)
Janet Folkes — 1 hr
Lindstrand LBL-14M — 38 min
Reno, NV
11/4/98

SUBCLASS AM-6 (1,200 - 1,600 cubic meters)

ALTITUDE (UK)
Janet Folkes — 6,866 m
Lindstrand LBL-14M — 22,526 ft
Reno, NV
11/4/98

DISTANCE (UK)
Janet Folkes — 69.7 km
Lindstrand LBL-14M — 43.3 mi
Reno, NV
11/4/98

DURATION (UK)
Janet Folkes — 1 hr
Lindstrand LBL-14M — 38 min
Reno, NV
11/4/98

SUBCLASS AM-7 (1,600 - 2,200 cubic meters)

ALTITUDE (UK)
Janet Folkes — 6,866 m
Lindstrand LBL-14M — 22,526 ft
Reno, NV
11/4/98

DISTANCE (UK)
Janet Folkes — 69.7 km
Lindstrand LBL-14M — 43.3 mi
Reno, NV
11/4/98

DURATION (UK)
Janet Folkes — 1 hr
Lindstrand LBL-14M — 38 min
Reno, NV
11/4/98

SUBCLASS AM-8 (2,200 - 3,000 cubic meters)

ALTITUDE (UK)
Janet Folkes — 6,866 m
Lindstrand LBL-14M — 22,526 ft
Reno, NV
11/4/98

DISTANCE (UK)
Janet Folkes — 69.7 km
Lindstrand LBL-14M — 43.3 mi
Reno, NV
11/4/98

DURATION (UK)
Janet Folkes — 1 hr
Lindstrand LBL-14M — 38 min
Reno, NV
11/4/98

SUBCLASS AM-9 (3,000 - 4,000 cubic meters)

ALTITUDE (UK)
Janet Folkes — 6,866 m
Lindstrand LBL-14M — 22,526 ft
Reno, NV
11/4/98

DISTANCE (UK)
Janet Folkes — 69.7 km
Lindstrand LBL-14M — 43.3 mi
Reno, NV
11/4/98

DURATION (UK)
Janet Folkes — 1 hr
Lindstrand LBL-14M — 38 min
Reno, NV
11/4/98

SUBCLASS AM-10 (4,000 - 6,000 cubic meters)

ALTITUDE (UK)
Janet Folkes — 6,866 m
Lindstrand LBL-14M — 22,526 ft
Reno, NV
11/4/98

DISTANCE (UK)
Janet Folkes 69.7 km
Lindstrand LBL-14M 43.3 mi
Reno, NV
11/4/98

DURATION (UK)
Janet Folkes 1 hr
Lindstrand LBL-14M 38 min
Reno, NV
11/4/98

SUBCLASS AM-11 (6,000 - 9,000 cubic meters)

ALTITUDE (UK)
Janet Folkes 6,866 m
Lindstrand LBL-14M 22,526 ft
Reno, NV
11/4/98

DISTANCE (UK)
Janet Folkes 69.7 km
Lindstrand LBL-14M 43.3 mi
Reno, NV
11/4/98

DURATION (UK)
Janet Folkes 1 hr
Lindstrand LBL-14M 38 min
Reno, NV
11/4/98

SUBCLASS AM-12 (9,000 - 12,000 cubic meters)

ALTITUDE (UK)
Janet Folkes 6,866 m
Lindstrand LBL-14M 22,526 ft
Reno, NV
11/4/98

DISTANCE (UK)
Janet Folkes 69.7 km
Lindstrand LBL-14M 43.3 mi
Reno, NV
11/4/98

DURATION (UK)
Janet Folkes 1 hr
Lindstrand LBL-14M 38 min
Reno, NV
11/4/98

SUBCLASS AM-13 (12,000 - 16,000 cubic meters)

ALTITUDE (UK)
Janet Folkes 6,866 m
Lindstrand LBL-14M 22,526 ft
Reno, NV
11/4/98

DISTANCE (UK)
Janet Folkes 69.7 km
Lindstrand LBL-14M 43.3 mi
Reno, NV
11/4/98

DURATION (UK)
Janet Folkes 1 hr
Lindstrand LBL-14M 38 min
Reno, NV
11/4/98

SUBCLASS AM-14 (16,000 - 22,000 cubic meters)

ALTITUDE (UK)
Janet Folkes 6,866 m
Lindstrand LBL-14M 22,526 ft
Reno, NV
11/4/98

DISTANCE (UK)
Janet Folkes 69.7 km
Lindstrand LBL-14M 43.3 mi
Reno, NV
11/4/98

DURATION (UK)
Janet Folkes 1 hr
Lindstrand LBL-14M 38 min
Reno, NV
11/4/98

SUBCLASS AM-15 (more than 22,000 cubic meters)

ALTITUDE (UK)
Janet Folkes 6,866 m
Lindstrand LBL-14M 22,526 ft
Reno, NV
11/4/98

DISTANCE (UK)
Janet Folkes 69.7 km
Lindstrand LBL-14M 43.3 mi
Reno, NV
11/4/98

DURATION (UK)
Janet Folkes 1 hr
Lindstrand LBL-14M 38 min
Reno, NV
11/4/98

ABSOLUTE WORLD RECORDS - AIRSHIPS

ALTITUDE (CANADA)
Hokan Colting
Tim Buss
21st Century SPAS-62.5
Drumheller, Alberta to Gull Lake, Sasketchawan
6/12/03

6,234 m
20,452 ft

DISTANCE IN STRAIGHT LINE (GERMANY)
Dr. Hugo Eckner
L. Z. 127, Graf Zeppelin
5 Maybach 450-550 hp
Lakehurst, NJ to Friedrichshafen, Germany
10/29-11/1/28

6,384.50 km
3,966.80 mi

DURATION (GERMANY)
Dr. Hugo Eckner
L. Z. 127, Graf Zeppelin
5 Maybach 450-550 hp
Lakehurst, NJ to Friedrichshafen, Germany
10/29-11/1/28

71 hrs

SPEED OVER A 3 KILOMETER COURSE (UK)
James Dexter
Michael Kendrick
ABC Lightship A-60 Plus
Staffordshire, England
1/19/00

92.80 kmh
57.66 mph

WORLD CLASS RECORDS, GENERAL

SUBCLASS BA-1 (less than 400 cubic meters)

DISTANCE (UK)
Donald Cameron
Cameron DG/14
10/12/90

94.86 km
58.94 mi

DURATION (USA)
Bryan Allen
Raven Helium "White Dwarf"
Thermal, CA to Brawley, CA
2/12/85

8 hrs
50 min
12 sec

SUBCLASS BA-2 (400 - 900 cubic meters)

ALTITUDE (CANADA)
Hokan Colting
21st Century SPAS
Holt, ON
11/8/92

1,898 m
6,227 ft

DISTANCE (UK)
Donald Cameron
Cameron DG/14
10/12/90

94.86 km
58.94 mi

DURATION (USA)
Bryan Allen
Raven Helium "White Dwarf"
Thermal, CA to Brawley, CA
2/12/85

8 hrs
50 min
12 sec

SUBCLASS BA-3 (900 - 1,600 cubic meters)

ALTITUDE (CANADA)
Hokan Colting
21st Century SPAS
Holt, ON
11/8/92

1,898 m
6,227 ft

DISTANCE (USA)
Paul Woessner, Pilot
Dr. Coy Foster, Copilot
Thunder and Colt GA-42
Dallas, TX to Marion, TX
10/25/90

374.71 km
232.83 mi

DURATION (USA)
Bryan Allen
Raven Helium "White Dwarf"
Thermal, CA to Brawley, CA
2/12/85

8 hrs
50 min
12 sec

SPEED OVER A 3 KILOMETER COURSE (USA)
Paul Woessner, Pilot
Dr. Coy Foster, Copilot
Thunder and Colt GA-42
Red Oak, TX
10/24/90

77.50 kmh
48.16 mph

SUBCLASS BA-4 (1,600 - 3,000 cubic meters)

ALTITUDE (CANADA)
Hokan Colting
21st Century SPAS
Holt, ON
11/8/92

1,898 m
6,227 ft

DISTANCE (USA)
Paul Woessner, Pilot
Dr. Coy Foster, Copilot
Thunder and Colt GA-42
Dallas, TX to Marion, TX
10/25/90

374.71 km
232.83 mi

DURATION (USA)
Bryan Allen
Raven Helium "White Dwarf"
Thermal, CA to Brawley, CA
2/12/85

8 hrs
50 min
12 sec

SPEED OVER A 3 KILOMETER COURSE (UK)
James Dexter
Michael Kendrick
ABC Lightship A-60 Plus
Staffordshire, England
1/19/00

92.80 kmh
57.66 mph

SPEED OVER A 3 KILOMETER COURSE*
Paul Woessner, Pilot — 77.50 kmh
Dr. Coy Foster, Copilot — 48.16 mph
Thunder and Colt GA-42
Red Oak, TX
10/24/90

SUBCLASS BA-5 (3,000 - 6,000 cubic meters)

ALTITUDE (CANADA)
Hokan Colting — 6,234 m
Tim Buss — 20,452 ft
21st Century SPAS-62.5
Drumheller, Alberta to Gull Lake, Sasketchawan
6/12/03

DISTANCE (USA)
Paul Woessner, Pilot — 374.71 km
Dr. Coy Foster, Copilot — 232.83 mi
Thunder and Colt GA-42
Dallas, TX to Marion, TX
10/25/90

DURATION (USA)
Bryan Allen — 8 hrs
Raven Helium "White Dwarf" — 50 min
Thermal, CA to Brawley, CA — 12 sec
2/12/85

SPEED OVER A 3 KILOMETER COURSE (UK)
James Dexter — 92.80 kmh
Michael Kendrick — 57.66 mph
ABC Lightship A-60 Plus
Staffordshire, England
1/19/00

SPEED OVER A 3 KILOMETER COURSE*
Paul Woessner, Pilot — 77.50 kmh
Dr. Coy Foster, Copilot — 48.16 mph
Thunder and Colt GA-42
Red Oak, TX
10/24/90

SUBCLASS BA-6 (6,000 - 12,000 cubic meters)

ALTITUDE (CANADA)
Hokan Colting — 6,234 m
Tim Buss — 20,452 ft
21st Century SPAS-62.5
Drumheller, Alberta to Gull Lake, Sasketchawan
6/12/03

DISTANCE (USA)
Paul Woessner, Pilot — 374.71 km
Dr. Coy Foster, Copilot — 232.83 mi
Thunder and Colt GA-42
Dallas, TX to Marion, TX
10/25/90

DURATION (USA) ✯
James O. Gross — 14 hrs
John L. McHugh — 9 min
Kenneth W. Petschow
H. Mark Pinsky
Airship Industries Skyship 600
2 Porsche 930, 255 hp
White Plains, NY
9/6/98

SPEED OVER A 3 KILOMETER COURSE (UK)
James Dexter — 92.80 kmh
Michael Kendrick — 57.66 mph
ABC Lightship A-60 Plus
Staffordshire, England
1/19/00

SPEED OVER A 3 KILOMETER COURSE*
Paul Woessner, Pilot — 77.50 kmh
Dr. Coy Foster, Copilot — 48.16 mph
Thunder and Colt GA-42
Red Oak, TX
10/24/90

SUBCLASS BA-7 (12,000 - 25,000 cubic meters)

ALTITUDE (CANADA)
Hokan Colting — 6,234 m
Tim Buss — 20,452 ft
21st Century SPAS-62.5
Drumheller, Alberta to Gull Lake, Sasketchawan
6/12/03

DISTANCE (USA)
Paul Woessner, Pilot — 374.71 km
Dr. Coy Foster, Copilot — 232.83 mi
Thunder and Colt GA-42
Dallas, TX to Marion, TX
10/25/90

DURATION (USA)
James O. Gross — 14 hrs
John L. McHugh — 9 min
Kenneth W. Petschow
H. Mark Pinsky
Airship Industries Skyship 600
2 Porsche 930, 255 hp
White Plains, NY
9/6/98

SPEED OVER A 3 KILOMETER COURSE (UK)
James Dexter — 92.80 kmh
Michael Kendrick — 57.66 mph
ABC Lightship A-60 Plus
Staffordshire, England
1/19/00

SPEED OVER A 3 KILOMETER COURSE*
Paul Woessner, Pilot — 77.50 kmh
Dr. Coy Foster, Copilot — 48.16 mph
Thunder and Colt GA-42
Red Oak, TX
10/24/90

SUBCLASS BA-8 (25,000 - 50,000 cubic meters)

ALTITUDE (CANADA)
Hokan Colting — 6,234 m
Tim Buss — 20,452 ft
21st Century SPAS-62.5
Drumheller, Alberta to Gull Lake, Sasketchawan
6/12/03

DISTANCE (USA)
Paul Woessner, Pilot — 374.71 km
Dr. Coy Foster, Copilot — 232.83 mi
Thunder and Colt GA-42
Dallas, TX to Marion, TX
10/25/90

DURATION (USA)
James O. Gross — 14 hrs
John L. McHugh — 9 min
Kenneth W. Petschow
H. Mark Pinsky
Airship Industries Skyship 600
2 Porsche 930, 255 hp
White Plains, NY
9/6/98

SPEED OVER A 3 KILOMETER COURSE (UK)
James Dexter — 92.80 kmh
Michael Kendrick — 57.66 mph
ABC Lightship A-60 Plus
Staffordshire, England
1/19/00

SPEED OVER A 3 KILOMETER COURSE*
Paul Woessner, Pilot 77.50 kmh
Dr. Coy Foster, Copilot 48.16 mph
Thunder and Colt GA-42
Red Oak, TX
10/24/90

SUBCLASS BA-9 (50,000 - 100,000 cubic meters)

ALTITUDE (CANADA)
Hokan Colting 6,234 m
Tim Buss 20,452 ft
21st Century SPAS-62.5
Drumheller, Alberta to Gull Lake, Sasketchawan
6/12/03

DISTANCE (USA)
Paul Woessner, Pilot 374.71 km
Dr. Coy Foster, Copilot 232.83 mi
Thunder and Colt GA-42
Dallas, TX to Marion, TX
10/25/90

DURATION (USA)
James O. Gross 14 hrs
John L. McHugh 9 min
Kenneth W. Petschow
H. Mark Pinsky
Airship Industries Skyship 600
2 Porsche 930, 255 hp
White Plains, NY
9/6/98

SPEED OVER A 3 KILOMETER COURSE (UK)
James Dexter 92.80 kmh
Michael Kendrick 57.66 mph
ABC Lightship A-60 Plus
Staffordshire, England
1/19/00

SPEED OVER A 3 KILOMETER COURSE*
Paul Woessner, Pilot 77.50 kmh
Dr. Coy Foster, Copilot 48.16 mph
Thunder and Colt GA-42
Red Oak, TX
10/24/90

SUBCLASS BA-10 (more than 100,000 cubic meters)

ALTITUDE (CANADA)
Hokan Colting 6,234 m
Tim Buss 20,452 ft
21st Century SPAS-62.5
Drumheller, Alberta to Gull Lake, Sasketchawan
6/12/03

DISTANCE (USA)
Paul Woessner, Pilot 374.71 km
Dr. Coy Foster, Copilot 232.83 mi
Thunder and Colt GA-42
Dallas, TX to Marion, TX
10/25/90

DURATION (USA)
James O. Gross 14 hrs
John L. McHugh 9 min
Kenneth W. Petschow
H. Mark Pinsky
Airship Industries Skyship 600
2 Porsche 930, 255 hp
White Plains, NY
9/6/98

SPEED OVER A 3 KILOMETER COURSE (UK)
James Dexter 92.80 kmh
Michael Kendrick 57.66 mph
ABC Lightship A-60 Plus
Staffordshire, England
1/19/00

SPEED OVER A 3 KILOMETER COURSE*
Paul Woessner, Pilot 77.50 kmh
Dr. Coy Foster, Copilot 48.16 mph
Thunder and Colt GA-42
Red Oak, TX
10/24/90

SUBCLASS BR-10 (more than 100,000 cubic meters)

DISTANCE IN STRAIGHT LINE (GERMANY)
Dr. Hugo Eckner 6,384.50 km
L. Z. 127, Graf Zeppelin 3,966.80 mi
5 Maybach 450-550 hp
Lakehurst, NJ to Friedrichshafen, Germany
10/29-11/1/28

DURATION (GERMANY)
Dr. Hugo Eckner 71 hrs
L. Z. 127, Graf Zeppelin
5 Maybach 450-550 hp
Lakehurst, NJ to Friedrichshafen, Germany
10/29-11/1/28

SUBCLASS BX-3 (900 - 1,600 cubic meters)

ALTITUDE (AUSTRALIA)
R.W. Taaff 3,159 m
Cameron D-38 10,361 ft
Cunderdin, W.A. to Kitkittering Creek
8/27/82

DISTANCE (SWEDEN)
Oscar Lindström 93.03 km
Colt Airship 57.81 mi
3/20/88

DURATION (SWEDEN)
Oscar Lindström 3 hrs
Colt Airship 41 min
3/20/88 55 sec

SUBCLASS BX-4 (1,600 - 3,000 cubic meters)

ALTITUDE (USA)
Brian Boland 5,059 m
Boland Rover 16,598 ft
Luxemburg
8/5/88

DISTANCE (SWEDEN)
Oscar Lindström 93.03 km
Colt Airship 57.81 mi
3/20/88

DURATION (LUX)
Guy Moyano 5 hrs
Cameron DP 80 6 min
1/25/92 42 sec

SPEED (UK)
David Hempleman-Adams 25.71 kmh
Cameron DP 70 15.97 mph
Mondovì, Italy
12/17/03

SUBCLASS BX-5 (3,000 - 6,000 cubic meters)

ALTITUDE (USA)
Brian Boland 5,059 m
Boland Rover 16,598 ft
Luxemburg
8/5/88

DISTANCE (SWEDEN)
Oscar Lindström 93.03 km
Colt Airship 57.81 mi
3/20/88

DURATION (LUX)
Guy Moyano 5 hrs
Cameron DP 80 6 min
1/25/92 42 sec

SPEED (UK)
David Hempleman-Adams 25.71 kmh
Cameron DP 70 15.97 mph
Mondovì, Italy
12/17/03

SUBCLASS BX-6 (6,000 - 12,000 cubic meters)

ALTITUDE (USA)
Brian Boland 5,059 m
Boland Rover 16,598 ft
Luxemburg
8/5/88

DISTANCE (SWEDEN)
Oscar Lindström 93.03 km
Colt Airship 57.81 mi
3/20/88

DURATION (LUX)
Guy Moyano 5 hrs
Cameron DP 80 6 min
1/25/92 42 sec

SPEED (UK)
David Hempleman-Adams 25.71 kmh
Cameron DP 70 15.97 mph
Mondovì, Italy
12/17/03

SUBCLASS BX-7 (12,000 - 25,000 cubic meters)

ALTITUDE (USA)
Brian Boland 5,059 m
Boland Rover 16,598 ft
Luxemburg
8/5/88

DISTANCE (SWEDEN)
Oscar Lindström 93.03 km
Colt Airship 57.81 mi
3/20/88

DURATION (LUX)
Guy Moyano 5 hrs
Cameron DP 80 6 min
1/25/92 42 sec

SPEED (UK)
David Hempleman-Adams 25.71 kmh
Cameron DP 70 15.97 mph
Mondovì, Italy
12/17/03

SUBCLASS BX-8 (25,000 - 50,000 cubic meters)

ALTITUDE (USA)
Brian Boland 5,059 m
Boland Rover 16,598 ft
Luxemburg
8/5/88

DISTANCE (SWEDEN)
Oscar Lindström 93.03 km
Colt Airship 57.81 mi
3/20/88

DURATION (LUX)
Guy Moyano 5 hrs
Cameron DP 80 6 min
1/25/92 42 sec

SPEED (UK)
David Hempleman-Adams 25.71 kmh
Cameron DP 70 15.97 mph
Mondovì, Italy
12/17/03

SUBCLASS BX-9 (50,000 - 100,000 cubic meters)

ALTITUDE (USA)
Brian Boland 5,059 m
Boland Rover 16,598 ft
Luxemburg
8/5/88

DISTANCE (SWEDEN)
Oscar Lindström 93.03 km
Colt Airship 57.81 mi
3/20/88

DURATION (LUX)
Guy Moyano 5 hrs
Cameron DP 80 6 min
1/25/92 42 sec

SPEED (UK)
David Hempleman-Adams 25.71 kmh
Cameron DP 70 15.97 mph
Mondovì, Italy
12/17/03

SUBCLASS BX-10 (more than 100,000 cubic meters)

ALTITUDE (USA)
Brian Boland 5,059 m
Boland Rover 16,598 ft
Luxemburg
8/5/88

DISTANCE (SWEDEN)
Oscar Lindström 93.03 km
Colt Airship 57.81 mi
3/20/88

DURATION (LUX)
Guy Moyano 5 hrs
Cameron DP 80 6 min
1/25/92 42 sec

SPEED (UK)
David Hempleman-Adams 25.71 kmh
Cameron DP 70 15.97 mph
Mondovì, Italy
12/17/03

GLIDERS

Once man learned that he could not only survive in the air, but even travel where the winds blew his balloon, the next step was to achieve some control over his flying. Just 21 years after the first balloons lifted off in France, George Cayley launched the first unmanned, heavier-than-air craft—a glider—in England.

Cayley was the father of scientific thought in the basics of heavier-than-air flight. He first put down in writing the principles of the wing's curved airfoil, the man-carrying fuselage, and the cruciform tail. He designed both monoplanes and biplanes during his long career. In 1804, Cayley flew an unmanned glider down a hill, and in 1849 he built one that carried a small local boy as its history-making passenger. Four years later, a Cayley glider flew down a hill with a grown man—a flight long enough to qualify as "sustained" flight.

The first known efforts at gliding in the United States were those of John Montgomery, whose experiments began around 1883. Ten years later, Otto Lilienthal was gliding in Germany, eventually making many hundreds of flights in both monoplanes and biplanes. In 1896, Octave Chanute was building gliders in Chicago, which were flown by Augustus Herring. In 1897, Percey Pilcher, an Englishman, made a glide of more than 2,000 feet in his Hawk.

The development of gliders was seen as a step toward powered airplanes, rather than a movement of its own. The Wrights, in particular, spent several years building gliders and learning to fly them in preparation for flying their first powered airplane.

Despite considerable gliding activity in the early part of the 20th century, the FAI did not record its first official gliding records until 1923. Then, on February 7, E. Descamps, of France, flew a Dewoitine glider to a height of 1,788 feet above his launching altitude. On August 26, Lt. Thoret covered a distance of 5 miles at Vauville, France, in a Bardin glider.

The real emergence of gliding and its more sophisticated outgrowth, soaring, began in the 1930s in Germany, where the Treaty of Versailles prohibited Germans from flying most forms of powered airplanes. The Wasserkuppe, a hill near Berlin, became the center of gliding.

A series of flights by German pilots raised the record for altitude gained to more than 22,400 feet before the start of World War II. By 1935, other German pilots increased the distance record to more than 310 miles. The Russians then took over and increased it to almost 465 miles by the start of World War II. After the war, the development of fiberglass and composite construction led to even greater achievements.

American soaring activity is governed by the Soaring Society of America, P.O. Box E, Hobbs, NM 88241, (505) 392-1177. The FAI's I.G.C. (International Gliding Commission) is responsible for records and international competition.

CLASS D, GLIDERS, WORLD RECORDS, GENERAL

DO, OPEN CLASS GLIDERS

DISTANCE

FREE DISTANCE (GERMANY)
Klaus Ohlmann 2,174.50 km
Hervé Lefranc 1,351.17 mi
Schempp-Hirth Nimbus 4DM
Calafate, Argentina to San Juan, Argentina
11/23/03

Superseded
FREE DISTANCE (GERMANY)
Klaus Ohlmann 1,687.00 km
Hervé Lefranc 1,048.25 mi
Schempp-Hirth Nimbus 4DM
Malrgue, Argentina to Calafate, Argentina
11/21/03

Superseded
FREE DISTANCE (GERMANY)
Klaus Ohlmann 1,501.90 km
Schempp-Hirth Nimbus 4DM 933.24 mi
Calafate, Argentina
1/9/03

STRAIGHT DISTANCE TO A GOAL (GERMANY)
Klaus Ohlmann 2,123.00 km
Hervé Lefranc 1,319.17 mi
Schempp-Hirth Nimbus 4DM
Calafate, Argentina to San Juan, Argentina
11/23/03

Superseded
STRAIGHT DISTANCE TO A GOAL (GERMANY)
Klaus Ohlmann 1,449.30 km
Schempp-Hirth Nimbus 4DM 900.55 mi
Calafate, Argentina
1/9/03

OUT AND RETURN DISTANCE (GERMANY)
Klaus Ohlmann 1,708.40 km
Schempp-Hirth Nimbus 4DM 1,061.55 mi
San Martin de los Andes, Argentina
12/20/02

TRIANGLE DISTANCE (USA)
Steve Fossett 1,502.55 km
Terrence R. Delore 933.64 mi
Schleicher ASH 25 Mi
San Carlos de Bariloche, Argentina
12/13/03

FREE TRIANGLE DISTANCE (USA)
Steve Fossett 1,508.42 km
Terrence R. Delore 937.29 mi
Schleicher ASH 25 Mi
San Carlos de Bariloche, Argentina
12/13/03

FREE THREE TURN POINT DISTANCE (GERMANY)
Klaus Ohlmann 3,009.00 km
Schempp-Hirth Nimbus 4DM 1,869.70 mi
San Martin de los Andes, Argentina
1/21/03

Superseded
FREE THREE TURN POINT DISTANCE (GERMANY)
Klaus Ohlmann 2,624.10 km
Schempp-Hirth Nimbus 4DM 1,630.54 mi
San Martin de los Andes, Argentina
11/12/02

FREE OUT AND RETURN DISTANCE (NEW ZEALAND)
Terrence R. Delore 2,008.40 km
Steve Fossett 1,247.96 mi
Schleicher ASH 25 Mi
Esquel, Argentina
11/14/03

Superseded
FREE OUT AND RETURN DISTANCE (GERMANY)
Klaus Ohlmann 1,715.50 km
Schempp-Hirth Nimbus 4DM 1,065.96 mi
San Martin de los Andes, Argentina
12/20/02

SPEED

SPEED OVER A 100 KM TRIANGLE (ARGENTINA)
Horacio Miranda 249.09 kmh
SZD Jantar Standard 2 154.78 mph
Chos Malal, Argentina
12/1/03

Superseded
SPEED OVER A 100 KM TRIANGLE (USA)
Tom K. Serkowski 243.40 kmh
Schleicher ASH 26 E 151.25 mph
Canon City, CO
11/9/03

SPEED OVER A 300 KM TRIANGLE (NEW ZEALAND)
Terrence R. Delore 179.30 kmh
Rolladen-Schneider LS6-C 111.41 mph
Omarama, New Zealand
12/17/97

SPEED OVER A 500 KM TRIANGLE (USA)
Steve Fossett 187.13 kmh
Terrence R. Delore 116.28 mph
Schleicher ASH 25
Omarama, New Zealand
11/15/02

SPEED OVER A 750 KM TRIANGLE (USA)
Steve Fossett 171.29 kmh
Terrence R. Delore 106.43 mph
Schleicher ASH 25 Mi
Ely, NV
7/29/03

SPEED OVER A 1,000 KM TRIANGLE (GERMANY)
Helmut H. Fischer 169.72 kmh
Schempp-Hirth Ventus 105.45 mph
Hendrik Verwoerd Dam, South Africa
1/5/95

SPEED OVER 1,250 KM TRIANGLE (W. GERMANY)
Hans Werner Grosse 143.46 kmh
Hans H. Kohlmeier 89.14 mph
Schleicher ASH 25
Alice Springs, Australia
1/10/87

SPEED OVER A 1,500 KM TRIANGLE (USA)
Steve Fossett 119.11 kmh
Terrence R. Delore 74.01 mph
Schleicher ASH 25 Mi
San Carlos de Bariloche, Argentina
12/13/03

SPEED OVER AN OUT AND RETURN COURSE OF 500 KM (USA)
James M. Payne 247.49 kmh
Thomas S. Payne 153.78 mph
Schleicher ASH 25
California City, CA
3/4/99

SPEED OVER AN OUT AND RETURN COURSE OF 1,000 KM (NEW ZEALAND)
Terrence Raymond Delore 166.53 kmh
Steve Fossett 103.48 mph
Schleicher ASH 25
Omarama, New Zealand
12/12/02

SPEED OVER AN OUT AND RETURN COURSE OF 1,500 KM (GERMANY)
Klaus Ohlmann 136.82 kmh
Schempp-Hirth Nimbus 4DM 85.02 mph
San Martin de los Andes, Argentina
12/20/02

ALTITUDE (available in Open Class only)

GAIN OF HEIGHT (USA)
Paul F. Bikle 12,894 m
Schweizer SGS 1-23E 42,303 ft
Lancaster, CA
2/25/61

ABSOLUTE ALTITUDE (USA)
Robert Harris 14,938 m
Grob G 102 49,009 ft
California City, CA
2/17/86

D15, 15 METER CLASS GLIDERS

DISTANCE

FREE THREE TURN POINT DISTANCE (GERMANY)
Klaus Ohlmann 2,029.7 km
Schempp-Hirth Ventus 2cM 1,261.20 mi
San Martin de los Andes, Argentina
1/1/02

SPEED

SPEED OVER A 100 KM TRIANGLE (ARGENTINA)
Horacio Miranda 249.09 kmh
SZD Jantar Standard 2 154.78 mph
Chos Malal, Argentina
12/1/03

SPEED OVER A 300 KM TRIANGLE (NEW ZEALAND)
Terrence R. Delore 179.30 kmh
Rolladen-Schneider LS6-c 111.41 mph
Omarama, New Zealand
12/17/97

SPEED OVER A 500 KM TRIANGLE (NEW ZEALAND)
Terrence R. Delore 158.43 kmh
Rolladen-Schneider LS6-c 98.44 mph
Mt. Chudleigh, New Zealand
10/18/98

SPEED OVER A 750 KM TRIANGLE (USA)
Philippe Athuil 149.71 kmh
Rolladen-Schneider LS6-b 93.02 mph
Ely, NV
8/1/02

SPEED OVER A 1,000 KM TRIANGLE (POLAND)
Janusz Centka 144.95 kmh
SZD-56 90.07 mph
Ely, NV
7/31/02

SPEED OVER AN OUT AND RETURN COURSE OF 500 KM (USA)
Martin J. Eiler 205.92 kmh
Schempp-Hirth Discus b 127.95 mph
California City, CA
1/12/00

DW, WORLD CLASS GLIDERS

DISTANCE

FREE DISTANCE (USA)
William B. Snead 741.57 km
PW-5 460.78 mi
Briggs, TX
7/22/01

STRAIGHT DISTANCE TO A GOAL (USA)
William B. Snead 602.26 km
PW-5 374.22 mi
Briggs, TX
7/21/01

OUT AND RETURN DISTANCE (GERMANY)
Diether Memmert 591.80 km
PW-5 367.72 mi
Heuberg, Germany
6/24/01

TRIANGLE DISTANCE (GERMANY)
Diether Memmert 551.6 km
PW-5 342.75 mi
Vogtareuth, Germany
6/14/02

FREE THREE TURN POINT DISTANCE (USA)
William B. Snead 771.74 km
PW-5 479.53 mi
Briggs, TX
7/22/01

FREE OUT AND RETURN DISTANCE (USA)
William B. Snead 591.12 km
PW-5 367.30 mi
Briggs, TX
7/30/00

SPEED

SPEED OVER A 100 KM TRIANGLE (USA)
Larry Pardue 115.38 kmh
PW-5 71.69 mph
Hobbs, NM
8/2/03

SPEED OVER A 300 KM TRIANGLE (GERMANY)
Guenther Jacobs 108.53 kmh
PW-5 67.44 mph
Parowan, UT
6/21/01

SPEED OVER A 500 KM TRIANGLE (GERMANY)
Axel Reich 82.00 kmh
Thomas Wartha 50.95 mph
PW-5
Lillo, Spain
6/21/01

SPEED OVER AN OUT AND RETURN COURSE OF 500 KM (GERMANY)
Guenther Jacobs 97.87 kmh
PW-5 60.81 mph
Parowan, UT
7/2/01

DU, ULTRALIGHT GLIDERS

DISTANCE

FREE DISTANCE (USA) ★
William G. Osoba, Jr. 508.10 km
Woodstock One 315.71 mi
Hutchinson, KS
4/21/98

STRAIGHT DISTANCE TO A GOAL (USA)
William G. Osoba, Jr. 508.10 km
Woodstock One 315.71 mi
Hutchinson, KS
4/21/98

OUT AND RETURN DISTANCE (SLOVENIA)
Bostjan Pristavec 603.0 km
Albastar APIS WR 374.69 mi
Lesce, Slovenia
7/21/01

TRIANGLE DISTANCE (USA)
William G. Osoba, Jr. 501.69 km
homebuilt SparrowHawk 311.73 mi
Big Spring, TX
8/5/02

FREE THREE TURN POINT DISTANCE (SLOVENIA)
Bostjan Pristavec 799.50 km
Albastar Apis WR 496.79 mi
Lesce, Slovenia
5/22/03

FREE OUT AND RETURN DISTANCE (SLOVENIA)
Bostjan Pristavec 627.2 km
Albastar APIS WR 389.72 mi
Lesce, Slovenia
7/21/01

SPEED

SPEED OVER A 100 KM TRIANGLE (SLOVENIA)
Bostjan Pristavec 118.2 kmh
Albastar APIS WR 73.4 mph
Lesce, Slovenia
8/22/00

SPEED OVER A 300 KM TRIANGLE (USA)
William G. Osoba, Jr. 83.82 kmh
homebuilt SparrowHawk 52.08 mph
Ulysses, KS
7/20/02

SPEED OVER A 500 KM TRIANGLE (USA)
William G. Osoba, Jr. 79.74 kmh
homebuilt SparrowHawk 49.55 mph
Big Spring, TX
8/5/02

SPEED OVER AN OUT AND RETURN COURSE OF 500 KM (SLOVENIA)
Andrej Kolar 82.10 kmh
Albastar Apis WR 51.01 mph
Lesce, Slovenia
6/5/01

CLASS D, GLIDERS, WORLD RECORDS, FEMININE

DO, OPEN CLASS GLIDERS

DISTANCE

FREE DISTANCE (UK)
Pamela Kurstjens-Hawkins 1,078.20 km
Schempp-Hirth Nimbus 4T 669.96 mi
Kingaroy Airfield, Australia
1/5/03

STRAIGHT DISTANCE TO A GOAL (UK)
Pamela Kurstjens-Hawkins 965.3 km
Schempp-Hirth Nimbus 4T 599.81 mi
Kingaroy Airfield, Australia
1/5/03

OUT AND RETURN DISTANCE (USA)
Doris F. Grove 1,127.68 km
Schempp-Hirth Nimbus 2 700.71 mi
Lock Haven, PA
9/28/81

TRIANGLE DISTANCE (UK)
Pamela Hawkins 1,036.56 km
Schempp-Hirth Nimbus 4T 644.09 mi
Tocumwal, Australia
12/25/98

FREE TRIANGLE DISTANCE (USA)
Steve Fossett 1,508.42 km
Terrence R. Delore 937.29 mi
Schleicher ASH 25 Mi
San Carlos de Bariloche, Argentina
12/13/03

FREE THREE TURN POINT DISTANCE (UK)
Pamela Kurstjens-Hawkins 1,081.20 km
Schempp-Hirth Nimbus 4T 671.82 mi
Kingaroy Airfield, Australia
1/5/03

FREE OUT AND RETURN DISTANCE (CZECH REPUBLIC)
Hana Zejdova 1,042.55 km
SZD-56 647.81 mi
Tocumwal, Australia
12/25/98

SPEED

SPEED OVER A 100 KM TRIANGLE (UK)
Pamela Kurstjens-Hawkins 159.06 kmh
Schempp-Hirth Nimbus 4T 98.84 mph
McCaffrey's Airfield, Australia
12/14/02

SPEED OVER A 300 KM TRIANGLE (UK)
Pamela Kurstjens-Hawkins 153.83 kmh
Schempp-Hirth Nimbus 4T 95.59 mph
Tocumwal, Australia
1/3/00

SPEED OVER A 500 KM TRIANGLE (UK)
Pamela Hawkins 151.37 kmh
Schempp-Hirth Nimbus 4T 94.06 mph
Tocumwal, Australia
1/4/99

SPEED OVER A 750 KM TRIANGLE (UK)
Pamela Kurstjens-Hawkins 150.75 kmh
Schempp-Hirth Nimbus 4T 93.67 mph
Ely, NV
7/26/01

SPEED OVER A 1,000 KM TRIANGLE (GERMANY)
Angelika Machinek 126.09 kmh
Schleicher ASH 25 E 78.35 mph
Bitterwasser, Namibia
1/5/99

SPEED OVER 1,250 KM TRIANGLE
(No record registered)

SPEED OVER AN OUT AND RETURN COURSE OF 500 KM (UK)
Pamela Kurstjens-Hawkins 156.91 kmh
Schempp-Hirth Nimbus 4T 97.50 mph
Parowan, UT
7/20/01

SPEED OVER AN OUT AND RETURN COURSE OF 1,000 KM (UK)
Pamela Kurstjens-Hawkins 133.89 kmh
Schempp-Hirth Nimbus 4T 83.19 mph
Corowa Airfield, Australia
1/7/03

SPEED OVER AN OUT AND RETURN COURSE OF 1,500 KM
(No record registered)

ALTITUDE (available in Open Class only)

GAIN OF HEIGHT (NEW ZEALAND)
Yvonne M. Loader 10,213 m
Schempp-Hirth Nimbus 2 33,507 ft
Omarama, New Zealand
1/12/88

ABSOLUTE ALTITUDE (USA)
Sabrina Jackintell 12,637 m
Grob Astir CS 41,460 ft
Colorado Springs, CO
2/14/79

D15, 15 METER CLASS GLIDERS

DISTANCE

OUT AND RETURN DISTANCE (CZECH REPUBLIC)
Hana Zejdova 1,042.55 km
SZD-56 647.81 mi
Tocumwal, Australia
12/25/98

TRIANGLE DISTANCE (CZECH REPUBLIC)
Hana Zejdova 1,012.33 km
SZD-55 629.03 mi
Tocumwal, Australia
12/1/99

FREE THREE TURN POINT DISTANCE (CZECH REPUBLIC)
Hana Zejdova 1,042.55 km
SZD-56 647.81 mi
Tocumwal, Australia
12/25/98

FREE OUT AND RETURN DISTANCE (CZECH REPUBLIC)
Hana Zejdova 1,042.55 km
SZD-56 647.81 mi
Tocumwal, Australia
12/25/98

SPEED

SPEED OVER A 100 KM TRIANGLE (GERMANY)
Angelika Machinek 142.4 kmh
Schempp-Hirth Ventus 2cM 88.48 mph
Bitterwasser, Namibia
12/4/02

SPEED OVER A 300 KM TRIANGLE (GERMANY)
Angelika Machinek 153.8 kmh
Schempp-Hirth Ventus 2cM 95.57 mph
Bitterwasser, Namibia
12/6/02

SPEED OVER A 500 KM TRIANGLE (UK)
Gillian Spreckley 141.6 kmh
Rolladen-Schneider LS-8 87.99 mph
Tswalu, South Africa
11/26/02

SPEED OVER A 750 KM TRIANGLE (CZECH REPUBLIC)
Hana Zejdova 132.5 kmh
Schempp-Hirth Ventus 82.3 mph
Tocumwal, Australia
2/17/00

SPEED OVER A 1,000 KM TRIANGLE (CZECH REPUBLIC)
Hana Zejdova 116.10 kmh
SZD-55 72.14 mph
Tocumwal, Australia
12/1/99

SPEED OVER AN OUT AND RETURN COURSE OF 500 KM (GERMANY)
Angelika Machinek 136.59 kmh
Schempp-Hirth Discus 84.87 mph
Bitterwasser, Namibia
12/18/98

SPEED OVER AN OUT AND RETURN COURSE OF 1,000 KM (CZECH REPUBLIC)
Hana Zejdova 109.44 kmh
SZD-56 68.00 mph
Tocumwal, Australia
12/25/98

DW, WORLD CLASS GLIDERS

DISTANCE

OUT AND RETURN DISTANCE (AUSTRALIA)
Kerrie Claffey 503.22 km
PW-5 312.68 mi
Flying M Ranch, NV
7/14/00

FREE THREE TURN POINT DISTANCE (AUSTRALIA)
Kerrie Claffey 517.38 km
PW-5 321.48 mi
Flying M Ranch, NV
7/14/00

FREE OUT AND RETURN DISTANCE (AUSTRALIA)
Kerrie Claffey 507.97 km
PW-5 315.63 mi
Flying M Ranch, NV
7/14/00

SPEED

SPEED OVER A 100 KM TRIANGLE (POLAND)
Adela Dankowska 87.75 kmh
PW-5 54.53 mph
Leszno, Poland
5/16/00

SPEED OVER A 300 KM TRIANGLE (POLAND)
Adela Dankowska 73.22 kmh
PW-5 45.50 mph
Leszno, Poland
7/28/99

SPEED OVER AN OUT AND RETURN COURSE OF 500 KM (AUSTRALIA)
Kerrie Claffey 68.82 kmh
PW-5 42.76 mph
Flying M Ranch, NV
7/14/00

DU, ULTRALIGHT GLIDERS

DISTANCE

FREE DISTANCE (SLOVENIA)
Tanja Pristavec 156.10 km
Albastar Apis WR 97.00 mi
Lesce, Slovenia
5/24/03

OUT AND RETURN DISTANCE (CZECH REPUBLIC)
Jana Koutna 204.50 km
TeST TST-3 Alpin T 127.07 mi
Krizanov, Czech Republic
8/26/01

FREE THREE TURN POINT DISTANCE (SLOVENIA)
Tanja Pristavec 357.40 km
Albastar Apis WR 222.08 mi
Lesce, Slovenia
5/24/03

FREE OUT AND RETURN DISTANCE (SLOVENIA)
Tanja Pristavec 312.20 km
Albastar Apis WR 193.99 mi
Lesce, Slovenia
5/24/03

SPEED

SPEED OVER A 100 KM TRIANGLE (SLOVENIA)
Tanja Pristavec 76.90 kmh
Albastar Apis WR 47.78 mph
Lesce, Slovenia
8/8/01

CLASS D, GLIDERS, U.S. NATIONAL RECORDS, GENERAL

GLIDERS, OPEN CLASS, SINGLE PLACE

DISTANCE

STRAIGHT DISTANCE*
Michael S. Koerner — 1,452.80 km
Kestrel 19 — 902.95 mi
California City, CA
4/19/84

STRAIGHT DISTANCE TO A GOAL* ★
Karl H. Striedieck — 1,288.79 km
Schleicher ASW 27 — 800.82 mi
Mill Hall, PA to Selma, AL
4/18/97

OUT AND RETURN DISTANCE*
Thomas L. Knauff — 1,646.68 km
Schempp-Hirth Nimbus 3 — 1,023.25 mi
Williamsport, PA
4/25/83

TRIANGLE DISTANCE*
JOINT RECORD
Karl Striedieck — 1362.68 km
Schleicher ASW 20 B — 846.77 mi
and
John C. Seymour
Schleicher ASW 20 B
and
Leonard R. McMaster
Schleicher ASW 20 B
and
Thomas L. Knauff
Schempp-Hirth Nimbus 3
Julian, PA
5/2/86

FREE THREE TURN POINT DISTANCE* ★
Karl H. Striedieck — 1,434.99 km
Schleicher ASW 20 B — 891.66 mi
Julian, PA
5/12/94

FREE OUT AND RETURN DISTANCE*
John F. Good — 1,261.65 km
Schempp-Hirth Discus — 783.95 mi
Mill Hall, PA
4/18/97

SPEED

SPEED OVER A 100 KM TRIANGLE*
James M. Payne — 234.95 kmh
Schempp-Hirth Discus a — 145.99 mph
Rosamond, CA
5/7/00

SPEED OVER A 300 KM TRIANGLE*
Gregory W. Chaffee — 153.21 kmh
Schempp-Hirth Ventus cA 15m — 95.20 mph
California City, CA
8/6/97

SPEED OVR A 500 KM TRIANGLE*
Carl D. Herold — 159.79 kmh
Schempp-Hirth Nimbus 3 — 99.29 mph
Minden, NV
7/9/88

SPEED OVER A 750 KM TRIANGLE*
Philippe Athuil — 149.71 kmh
Rolladen-Schneider LS6-b — 93.02 mph
Ely, NV
8/1/02

SPEED OVER A 1,000 KM TRIANGLE*
Philippe Athuil — 144.64 kmh
Rolladen-Schneider LS6-b — 89.87 mph
Ely, NV
7/31/02

SPEED OVER A 1,250 KM TRIANGLE*
Thomas L. Knauff — 124.05 kmh
Schempp-Hirth Nimbus 3 — 77.08 mph
Julian, PA
5/2/86

SPEED OVER AN OUT AND RETURN COURSE OF 300 KM*
James M. Payne — 231.79 kmh
Schempp-Hirth Discus a — 144.03 mph
Owens Peak, CA
5/18/96

SPEED OVER AN OUT AND RETURN COURSE OF 500 KM*
Martin J. Eiler — 205.92 kmh
Schempp-Hirth Discus b — 127.95 mph
California City, CA
1/12/00

SPEED OVER AN OUT AND RETURN COURSE OF 750 KM*
Peter Deane — 152.46 kmh
Rolladen-Schneider LS8-a — 94.73 mph
Minden, NV
9/3/02

SPEED OVER AN OUT AND RETURN COURSE OF 1,000 KM*
Thomas L. Knauff — 146.85 kmh
Schempp-Hirth Discus b — 91.25 mph
Julian, PA
4/23/93

SPEED OVER AN OUT AND RETURN COURSE OF 1,250 KM*
John F. Good — 131.14 kmh
Schempp-Hirth Discus — 81.48 mph
Mill Hall, PA
4/18/97

ALTITUDE

GAIN OF HEIGHT*
Paul F. Bikle — 12,894 m
Schweizer SGS 1-23E — 42,303 ft
Lancaster, CA
2/25/61

ABSOLUTE ALTITUDE*
Robert Harris — 14,938 m
Grob G 102 — 49,009 ft
California City, CA
2/17/86

GLIDERS, OPEN CLASS, MULTI-PLACE

DISTANCE

STRAIGHT DISTANCE*
Lazlo Horvath — 901.92 km
Schempp-Hirth Janus — 560.43 mi
Estrella, AZ
5/10/80

STRAIGHT DISTANCE TO A GOAL*
Kenneth Arterburn — 657.90 km
Lark — 408.90 mi
Woodsboro, TX
8/17/78

OUT AND RETURN DISTANCE*
James M. Payne 1,016.93 km
Thomas S. Payne 631.89 mi
Schleicher ASH 25
Rosamond, CA
4/2/01

TRIANGLE DISTANCE*
James M. Payne, Pilot 1,015.03 km
Thomas S. Payne, Passenger 630.71 mi
Schleicher ASH 25 E
Inyokern, CA
6/8/96

FREE THREE TURN POINT DISTANCE*
Doug Haluza 1,134.73 km
John D. Godfrey 705.08 mi
Schempp-Hirth Janus C
Julian, PA
4/23/03

FREE OUT AND RETURN DISTANCE*
James M. Payne 1,017.88 km
Thomas S. Payne 632.48 mi
Schleicher ASH 25
Rosamond, CA
4/2/01

SPEED

SPEED OVER A 100 KM TRIANGLE*
James M. Payne, Pilot 201.61 kmh
Thomas S. Payne, Passenger 125.28 mph
Schleicher ASH 25 E
California City, CA
3/28/98

SPEED OVER A 300 KM TRIANGLE*
James M. Payne, Pilot 138.01 kmh
Jacqueline Payne, Passenger 85.75 mph
Schleicher ASH 25
Minden, NV
7/29/98

SPEED OVER A 500 KM TRIANGLE*
Steve Fossett 187.13 kmh
Terrence R. Delore 116.28 mph
Schleicher ASH 25
Omarama, New Zealand
11/15/02

SPEED OVER A 750 KM TRIANGLE*
James M. Payne, Pilot 123.72 kmh
Thomas S. Payne, Passenger 76.88 mph
Schleicher ASH 25 E
Minden, NV
5/29/95

SPEED OVER AN OUT AND RETURN COURSE OF 300 KM* ✯
James M. Payne 269.36 kmh
Thomas S. Payne 167.37 mph
Schleicher ASH 25
California City, CA
3/3/99

SPEED OVER AN OUT AND RETURN COURSE OF 500 KM*
James M. Payne 247.49 kmh
Thomas S. Payne 153.78 mph
Schleicher ASH 25
California City, CA
3/4/99

SPEED OVER AN OUT AND RETURN COURSE OF 750 KM *
Doug Haluza 108.07 kmh
John D. Godfrey 67.15 mph
Schempp-Hirth Janus C
Julian, PA
4/23/03

SPEED OVER AN OUT AND RETURN COURSE OF 1,000 KM*
James M. Payne 147.85 kmh
Thomas S. Payne 91.87 mph
Schleicher ASH 25
Rosamond, CA
4/2/01

ALTITUDE

GAIN OF HEIGHT*
Laurence E. Edgar, Pilot 10,493 m
Harold E. Klieforth, Passenger 34,426 ft
Pratt Read PR-GI
Bishop, CA
3/19/52

ABSOLUTE ALTITUDE*
Laurence E. Edgar, Pilot 13,489 m
Harold E. Klieforth, Passenger 44,255 ft
Pratt Read PR-GI
Bishop, CA
3/19/52

MOTOR GLIDERS, SINGLE PLACE

DISTANCE

STRAIGHT DISTANCE *
James S. Ketcham 1,029.59 km
Schleicher ASH 26 E 639.76 mi
Agua Dulce, CA to American Falls, ID
5/29/03

STRAIGHT DISTANCE TO A GOAL*
Bernard R. Gross 714.04 km
Schempp-Hirth Ventus bT 443.68 mi
Alamogordo, NM
6/12/00

OUT AND RETURN DISTANCE *
Richard A. Howell 1,027.32 km
Schempp-Hirth Ventus bT 638.34 mi
Julian, PA
10/2/03

TRIANGLE DISTANCE*
JOINT RECORD
Stephen S. Dashew 1,008.54 km
Glaser-Dirks DG-800B 626.68 mi
and
Michael N. Parker
Schleicher ASH 26 E
Tonopah, NV
7/18/03

FREE THREE TURN POINT DISTANCE *
Philippe Athuil 1,194.11 km
Schleicher ASH 25 Mi 741.99 mi
Ely, NV
7/31/03

Superseded
*FREE THREE TURN POINT DISTANCE **
Philippe Athuil 1,100.22 km
Schleicher ASH 25 Mi 683.64 mi
Llano, CA
6/19/03

Superseded
*FREE THREE TURN POINT DISTANCE**
James S. Ketcham 1,089.90 km
Schleicher ASH 26 E 677.23 mi
Agua Dulce, CA to American Falls, ID
5/29/03

FREE OUT AND RETURN DISTANCE*
Richard A. Howell 1,027.32 km
Schempp-Hirth Ventus bT 638.34 mi
Julian, PA
10/2/03

SPEED

SPEED OVER A 100 KM TRIANGLE*
Tom K. Serkowski — 243.40 kmh
Schleicher ASH 26 E — 151.25 mph
Canon City, CO
11/9/03

SPEED OVER A 300 KM TRIANGLE*
Gerald Kaufman — 152.22 kmh
Schempp-Hirth Ventus bT — 94.59 mph
Ely, NV
8/14/98

SPEED OVER A 500 KM TRIANGLE*
Gerald Kaufman — 137.59 kmh
Schempp-Hirth, Ventus bT — 85.49 mph
Parowan, UT
7/24/91

SPEED OVER A 750 KM TRIANGLE*
Gerald Kaufman — 139.37 kmh
Schempp-Hirth Ventus bT — 86.60 mph
Minden, NV
8/6/98

SPEED OVER A 1,000 KM TRIANGLE*
Michael N. Parker — 122.23 kmh
Schleicher ASH 26 E — 75.95 mph
Tonopah, NV
7/18/03

SPEED OVER AN OUT AND RETURN COURSE OF 300 KM*
William H. Seed, Jr. — 168.93 kmh
Glaser-Dirks DG-600M — 104.97mph
Tonopah, NV
7/14/94

SPEED OVER AN OUT AND RETURN COURSE OF 500 KM*
Gerald Kaufman — 141.96 kmh
Schempp-Hirth Ventus bT — 88.21 mph
Salida, CO
6/17/02

SPEED OVER AN OUT AND RETURN COURSE OF 750 KM*
Gerald Kaufman — 119.27 kmh
Schempp-Hirth Ventus bT — 74.11 mph
Salida, CO
6/25/99

SPEED OVER AN OUT AND RETURN COURSE OF 1,000 KM*
Richard A. Howell — 117.45 kmh
Schempp-Hirth Ventus bT — 72.98 mph
Julian, PA
10/2/03

ALTITUDE

GAIN OF HEIGHT*
Malcolm D. Stevenson — 9,935 m
Glaser-Dirks DG-400 — 32,595 ft
North Conway, NH
10/25/85

ABSOLUTE ALTITUDE*
Malcolm D. Stevenson — 10,282 m
Glaser-Dirks DG-400 — 33,733 ft
North Conway, NH
10/25/85

MOTOR GLIDERS, MULTI-PLACE

DISTANCE

STRAIGHT DISTANCE*
Jerry A. Wenger — 969.75 km
Donald W. Aitken — 602.57 mi
Schempp-Hirth Nimbus 3DM
Minden, NV
7/8/89

STRAIGHT DISTANCE TO A GOAL*
Jerry A. Wenger, Pilot — 777.81 km
Donald W. Aitken, Passenger — 483.31 mi
Schempp-Hirth Nimbus 3DM
Minden, NV to Grand Junction, CO
7/7/89

OUT AND RETURN DISTANCE*
Stanley F. Nelson — 760.91 km
Anthony H. Johnson — 472.80 mi
Schleicher ASH 25 E
Alamogordo, NM
5/31/01

TRIANGLE DISTANCE*
Steve Fossett — 1,502.55 km
Terrence R. Delore — 933.64 mi
Schleicher ASH 25 Mi
San Carlos de Bariloche, Argentina
12/13/03

FREE TRIANGLE DISTANCE*
Steve Fossett — 1,508.42 km
Terrence R. Delore — 937.29 mi
Schleicher ASH 25 Mi
San Carlos de Bariloche, Argentina
12/13/03

FREE THREE TURN POINT DISTANCE *
Thomas L. Knauff — 1,057.94 km
Doris F. Grove — 657.38 mi
Schempp-Hirth Duo Discus T
Julian, PA
4/23/03

FREE OUT AND RETURN DISTANCE*
Stanley F. Nelson — 762.87 km
Anthony H. Johnson — 474.02 mi
Schleicher ASH 25 E
Alamogordo, NM
5/31/01

SPEED

SPEED OVER A 100 KM TRIANGLE*
Thomas L. Knauff — 154.09 kmh
Michael R. Robison — 95.75 mph
Schempp-Hirth Duo Discus T
Julian, PA
10/7/02

SPEED OVER A 300 KM TRIANGLE*
Allan Martini — 121.35 kmh
Ed Peerens — 75.40 mph
Schempp-Hirth Nimbus 3DM
Ely, NV
7/22/01

SPEED OVER A 500 KM TRIANGLE*
Deborah B. Kutch — 109.43 kmh
Allan Martini — 68.00 mph
Stemme S 10-VT
Ely, NV
7/24/03

SPEED OVER A 750 KM TRIANGLE*
Steve Fossett 171.29 kmh
Terrence R. Delore 106.43 mph
Schleicher ASH 25 Mi
Ely, NV
7/29/03

SPEED OVER A 1,000 KM TRIANGLE*
James M. Payne, Pilot 125.38 kmh
Thomas S. Payne, Passenger 78.10 mph
Schleicher ASH 25 E
Inyokern, CA
6/8/96

SPEED OVER A 1,250 KM TRIANGLE*
Steve Fossett 143.47 kmh
Terence R. Delore 89.15 mph
Schleicher ASH 25
Ely, NV
7/31/03

SPEED OVER A 1,500 KM TRIANGLE*
Steve Fossett 119.11 kmh
Terrence R. Delore 74.01 mph
Schleicher ASH 25 Mi
San Carlos de Bariloche, Argentina
12/13/03

SPEED OVER AN OUT AND RETURN COURSE OF 300 KM*
Thomas L. Knauff 167.27 kmh
Daniel Scott 103.93 mph
Schempp-Hirth Duo Discus T
Julian, PA
10/7/02

SPEED OVER AN OUT AND RETURN COURSE OF 500 KM*
James M. Payne, Pilot 139.10 kmh
Thomas S. Payne, Passenger 86.44 mph
Schleicher ASH 25 E
Rosamond, CA
6/18/97

SPEED OVER AN OUT AND RETURN COURSE OF 750 KM *
Thomas L. Knauff 151.93 kmh
Doris F. Grove 94.35 mph
Schempp-Hirth Duo Discus T
Julian, PA
4/23/03

GLIDERS, 15-METER CLASS

DISTANCE

STRAIGHT DISTANCE*
Karl H. Striedieck 1,288.79 km
Schleicher ASW 27 800.82 mi
Mill Hall, PA to Selma, AL
4/18/97

STRAIGHT DISTANCE TO A GOAL* ✯
Karl H. Striedieck 1,288.79 km
Schleicher ASW 27 800.82 mi
Mill Hall, PA to Selma, AL
4/18/97

OUT AND RETURN DISTANCE*
Leonard R. McMaster 1,298.97 km
Std. Cirrus 807.14 mi
Lock Haven, PA
3/17/76

TRIANGLE DISTANCE*
Leonard R. McMaster 1,362.68 km
John C. Seymour 846.73 mi
Karl Striedieck
Schleicher ASW 20 B
Port Matilda, PA
5/2/86

FREE THREE TURN POINT DISTANCE* ✯
Karl H. Striedieck 1,434.99 km
Schleicher ASW 20 B 891.66 mi
Julian, PA
5/12/94

FREE OUT AND RETURN DISTANCE*
John F. Good 1,261.65 km
Schempp-Hirth Discus 783.95 mi
Mill Hall, PA
4/18/97

SPEED

SPEED OVER A 100 KM TRIANGLE*
James M. Payne 234.95 kmh
Schempp-Hirth Discus a 145.99 mph
Rosamond, CA
5/7/00

SPEED OVER A 300 KM TRIANGLE*
Gregory W. Chaffee 153.21 kmh
Schempp-Hirth Ventus cA 15m 95.20 mph
California City, CA
8/6/97

SPEED OVER A 500 KM TRIANGLE*
Alan K. Reeter 142.03 kmh
Rolladen-Schneider LS8 88.26 mph
Tucson, AZ
5/21/00

SPEED OVER A 750 KM TRIANGLE*
Philippe Athuil 149.71 kmh
Rolladen-Schneider LS6-b 93.02 mph
Ely, NV
8/1/02

SPEED OVER A 1,000 KM TRIANGLE*
Philippe Athuil 144.64 kmh
Rolladen-Schneider LS6-b 89.87 mph
Ely, NV
7/31/02

SPEED OVER A 1,250 KM TRIANGLE*
Leonard R. McMaster 109.01 kmh
John C. Seymour 67.74 mph
Karl Striedieck
Schleicher ASW 20 B
Port Matilda, PA
5/2/86

SPEED OVER AN OUT AND RETURN COURSE OF 300 KM*
James M. Payne 231.79 kmh
Schempp-Hirth Discus a 144.03 mph
Owens Peak, CA
5/18/96

SPEED OVER AN OUT AND RETURN COURSE OF 500 KM (USA)
Martin J. Eiler 205.92 kmh
Schempp-Hirth Discus b 127.95 mph
California City, CA
1/12/00

SPEED OVER AN OUT AND RETURN COURSE OF 750 KM*
Peter Deane 152.46 kmh
Rolladen-Schneider LS8-a 94.73 mph
Minden, NV
9/3/02

SPEED OVER AN OUT AND RETURN COURSE OF 1,000 KM*
Thomas L. Knauff 146.85 kmh
Schempp-Hirth Discus b 91.25 mph
Julian, PA
4/23/93

SPEED OVER AN OUT AND RETURN COURSE OF 1,250 KM*
John F. Good — 131.14 kmh / 81.48 mph
Schempp-Hirth Discus
Mill Hall, PA
4/18/97

ALTITUDE

GAIN OF HEIGHT*
Harold Katinszky — 9,784m / 32,100 ft
Jantar Standard 2
California City, CA
3/27/85

ABSOLUTE ALTITUDE*
Harold Katinszky — 11,582 m / 38,000 ft
Jantar Standard 2
California City, CA
3/27/85

GLIDERS, STANDARD CLASS

DISTANCE

STRAIGHT DISTANCE*
John F. Good — 1,082.90 km / 672.88 mi
Schempp-Hirth Discus
Mill Hall, PA
3/22/97

STRAIGHT DISTANCE TO A GOAL*
Fritz Seger — 721.37 km / 448.24 mi
Grob Astir CS/77
Crystalaire, CA
8/29/87

OUT AND RETURN DISTANCE*
John F. Good — 1,261.65 km / 783.95 mi
Schempp-Hirth Discus
Mill Hall, PA
4/18/97

TRIANGLE DISTANCE*
Sergio Colacevich — 1,025.25 km / 637.06 mi
Schempp-Hirth Discus a
Truckee, CA
7/27/01

FREE THREE TURN POINT DISTANCE* ★
Thomas L. Knauff — 1,394.04 km / 866.22 mi
Schempp-Hirth Discus b
Julian, PA
6/1/93

FREE OUT AND RETURN DISTANCE*
John F. Good — 1,261.65 km / 783.95 mi
Schempp-Hirth Discus
Mill Hall, PA
4/18/97

SPEED

SPEED OVER A 100 KM TRIANGLE*
James M. Payne — 234.95 kmh / 145.99 mph
Schempp-Hirth Discus a
Rosamond, CA
5/7/00

SPEED OVER A 300 KM TRIANGLE*
Alan K. Reeter — 147.48 kmh / 91.64 mph
Rolladen-Schneider LS8
Tucson, AZ
5/20/00

SPEED OVER A 500 KM TRIANGLE*
Alan K. Reeter — 142.03 kmh / 88.26 mph
Rolladen-Schneider LS8
Tucson, AZ
5/21/00

SPEED OVER A 750 KM TRIANGLE*
Richard R. Wagner — 130.46 kmh / 81.06 mph
Rolladen-Schneider LS4
Tehachapi, CA
7/4/91

SPEED OVER A 1,000 KM TRIANGLE*
James M. Payne — 135.06 kmh / 83.92 mph
Schempp-Hirth Discus a
Inyokern, CA
8/11/96

SPEED OVER AN OUT AND RETURN COURSE OF 300 KM*
James M. Payne — 231.79 kmh / 144.03 mph
Schempp-Hirth Discus a
Owens Peak, CA
5/18/96

SPEED OVER AN OUT AND RETURN COURSE OF 500 KM*
Martin J. Eiler — 205.52 kmh / 127.70 mph
Schempp-Hirth Discus b
California City, CA
1/12/00

SPEED OVER AN OUT AND RETURN COURSE OF 750 KM*
Peter Deane — 152.46 kmh / 94.73 mph
Rolladen-Schneider LS8-a
Minden, NV
9/3/02

SPEED OVER AN OUT AND RETURN COURSE OF 1,000 KM*
Thomas L. Knauff — 146.85 kmh / 91.25 mph
Schempp-Hirth Discus b
Julian, PA
4/23/93

SPEED OVER AN OUT AND RETURN COURSE OF 1,250 KM*
John F. Good — 131.14 kmh / 81.48 mph
Schempp-Hirth Discus
Mill Hall, PA
4/18/97

ALTITUDE

GAIN OF HEIGHT*
Robert R. Harris — 12,162 m / 39,900 ft
Grob G 102
California City, CA
2/17/86

ABSOLUTE ALTITUDE*
Robert R. Harris — 14,938 m / 49,009 ft
Grob G 102
California City, CA
2/17/86

DW, WORLD CLASS GLIDERS

DISTANCE

FREE DISTANCE*
William B. Snead — 741.57 km / 460.78 mi
PW-5
Briggs, TX
7/22/01

STRAIGHT DISTANCE TO A GOAL*
William B. Snead — 602.26 km / 374.22 mi
PW-5
Briggs, TX
7/21/01

OUT AND RETURN DISTANCE*
John C. Downing — 324.78 km / 201.81 mi
PW-5
Parowan, UT
8/9/03

FREE THREE TURN POINT DISTANCE*
William B. Snead — 771.74 km / 479.53 mi
PW-5
Briggs, TX
7/22/01

FREE OUT AND RETURN DISTANCE*
William B. Snead — 591.12 km / 367.30 mi
PW-5
Briggs, TX
7/30/00

SPEED

SPEED OVER A 100 KM TRIANGLE*
Larry Pardue — 115.38 kmh / 71.69 mph
PW-5
Hobbs, NM
8/2/03

SPEED OVER A 300 KM TRIANGLE*
Patrick L. Tuckey — 91.71 kmh / 56.98 mph
PW-5
Waxahachie, TX
8/19/99

SPEED OVER AN OUT AND RETURN COURSE OF 300 KM*
John C. Downing — 72.04 kmh / 44.76 mph
PW-5
Parowan, UT
8/9/03

DU, ULTRALIGHT GLIDERS

DISTANCE

STRAIGHT DISTANCE* ✯
William G. Osoba, Jr. — 508.10 km / 315.71 mi
Woodstock One
Hutchinson, KS
4/21/98

STRAIGHT DISTANCE TO A GOAL*
William G. Osoba, Jr. — 508.10 km / 315.71 mi
Woodstock One
Hutchinson, KS
4/21/98

OUT AND RETURN DISTANCE*
William G. Osoba, Jr. — 260.87 km / 162.09 mi
Woodstock One
Hearne, TX
8/15/00

TRIANGLE DISTANCE*
William G. Osoba, Jr. — 501.69 km / 311.73 mi
homebuilt SparrowHawk
Big Spring, TX
8/5/02

FREE THREE TURN POINT DISTANCE*
William G. Osoba, Jr. — 508.10 km / 315.71 mi
Woodstock One
Hutchinson, KS
4/21/98

FREE OUT AND RETURN DISTANCE*
William G. Osoba, Jr. — 262.88 km / 163.35 mi
Woodstock One
Hearne, TX
8/15/00

SPEED

SPEED OVER A 100 KM TRIANGLE*
William G. Osoba Jr. — 84.32 kmh / 52.39 mph
Woodstock One
Hearne, TX
8/17/00

SPEED OVER A 300 KM TRIANGLE*
William G. Osoba, Jr. — 83.82 kmh / 52.08 mph
homebuilt SparrowHawk
Ulysses, KS
7/20/02

SPEED OVER A 500 KM TRIANGLE*
William G. Osoba, Jr. — 79.74 kmh / 49.55 mph
homebuilt SparrowHawk
Big Spring, TX
8/5/02

CLASS D, GLIDERS, U.S. NATIONAL RECORDS, FEMININE

GLIDERS, OPEN CLASS, SINGLE PLACE

DISTANCE

STRAIGHT DISTANCE*
Joann B. Shaw — 951.43 km / 591.19 mi
Schempp-Hirth Nimbus 2
Hobbs, NM
7/2/90

STRAIGHT DISTANCE TO A GOAL*
Joann B. Shaw — 951.43 km / 591.19 mi
Schempp-Hirth Nimbus 2
Hobbs, NM - North Platte, NE
07/02/90

OUT AND RETURN DISTANCE*
Doris F. Grove — 1,127.68 km / 700.74 mi
Schempp Hirth Nimbus 2
Lock Haven, PA
9/28/81

TRIANGLE DISTANCE*
Joann B. Shaw — 847.27 km / 526.47 mi
Schempp-Hirth Nimbus 2
Snyder, Texas
8/5/84

FREE THREE TURN POINT DISTANCE*
Joann B. Shaw — 1,015.25 km / 630.84 mi
Schempp-Hirth Nimbus 3
Hobbs, NM
8/27/00

FREE OUT AND RETURN DISTANCE*
Joann B. Shaw — 823.80 km / 511.88 mi
Schempp-Hirth Nimbus 3
Hobbs, NM
8/27/00

SPEED

SPEED OVER A 100 KM TRIANGLE*
Joann B. Shaw — 151.12 kmh / 93.90 mph
Schempp-Hirth Nimbus 3
Hobbs, NM
7/17/97

SPEED OVER A 300 KM TRIANGLE*
Joann B. Shaw — 129.71 kmh / 80.61 mph
Schempp-Hirth Nimbus 2
Snyder, TX
8/15/83

SPEED OVER A 500 KM TRIANGLE*
Joann B. Shaw — 112.77 kmh / 70.09 mph
Schempp-Hirth Nimbus 2
Snyder, TX
7/30/83

SPEED OVER A 750 KM TRIANGLE*
Joann B. Shaw — 100.55 kmh / 62.48 mph
Schempp-Hirth Nimbus 2
Snyder, TX
8/5/84

SPEED OVER AN OUT AND RETURN COURSE OF 300 KM*
Iris Mittendorf — 184.81 kmh / 114.83 mph
AS-W 24
Port Matilda, PA
11/17/90

SPEED OVER AN OUT AND RETURN COURSE OF 500 KM*
Joann B. Shaw — 113.82 kmh / 70.74 mph
Schempp-Hirth Nimbus 2
Snyder, TX
8/13/83

ALTITUDE

GAIN OF HEIGHT*
Betsy W. Proudfit — 8,533 m / 27,994 ft
Pratt Read
Bishop, CA
4/14/55

ABSOLUTE ALTITUDE*
Sabrina Jackintell — 12,637 m / 41,460 ft
Grob Astir CS
Colorado Springs, CO
2/14/79

GLIDERS, OPEN CLASS, MULTI-PLACE

DISTANCE

STRAIGHT DISTANCE*
Betsy W. Proudfit — 274.10 km / 170.32 mi
Pratt Read
El Mirage, CA
7/11/52

STRAIGHT DISTANCE TO A GOAL*
Betsy W. Proudfit — 274.10 km / 170.32 mi
Pratt Read
El Mirage, CA
7/11/52

OUT AND RETURN DISTANCE*
Cynthia Brickner, Pilot — 246.07 km / 152.90 mi
Susan Schionning, Copilot
Schleicher ASK 21
California City, CA
8/17/92

SPEED

SPEED OVER A 100 KM TRIANGLE*
Karol A. Hines, Pilot — 89.32 kmh / 55.50 mph
Alexa Heimbach, Passenger
Schempp-Hirth Janus a
Minden, NV
9/16/92

ALTITUDE

GAIN OF HEIGHT*
Babbs Nutt — 7,481 m / 24,545 ft
Schweizer SGS 2-32
Colorado Springs, CO
3/5/75

ABSOLUTE ALTITUDE*
Babs Nutt — 10,809 m / 35,463 ft
Schweizer SGS 2-32
Colorado Springs, CO
3/5/75

MOTOR GLIDERS, SINGLE PLACE

DISTANCE

OUT AND RETURN DISTANCE*
Deborah B. Kutch — 504.99 km / 313.79 mi
Stemme S10-VT
Ely, NV
7/25/01

TRIANGLE DISTANCE*
Deborah B. Kutch 303.06 km
Stemme S10-VT 188.31 mi
Ely, NV
7/27/01

FREE OUT AND RETURN DISTANCE*
Deborah B. Kutch 504.99 km
Stemme S10-VT 313.79 mi
Ely, NV
7/25/01

SPEED

SPEED OVER A 300 KM TRIANGLE*
Deborah B. Kutch 124.02 kmh
Stemme S10-VT 77.06 mph
Ely, NV
7/27/01

SPEED OVER AN OUT AND RETURN COURSE OF 500 KM*
Deborah B. Kutch 104.22 kmh
Stemme S10-VT 64.76 mph
Ely, NV
7/25/01

MOTOR GLIDERS, MULTI-PLACE

DISTANCE

OUT AND RETURN DISTANCE*
Deborah B. Kutch 306.05 km
Reba Coombs 190.17 mi
Stemme S10-VT
Ely, NV
7/27/00

TRIANGLE DISTANCE*
Deborah B. Kutch 304.18 km
Ruthann Povinelli 189.01 mi
Stemme S10-VT
Ely, NV
7/31/02

FREE OUT AND RETURN DISTANCE*
Deborah B. Kutch 309.41 km
Reba Coombs 192.26 mi
Stemme S10-VT
Ely, NV
7/27/00

SPEED

SPEED OVER A 300 KM TRIANGLE*
Deborah B. Kutch 124.99 kmh
Ruthann Povinelli 77.67 mph
Stemme S10-VT
Ely, NV
7/31/02

SPEED OVER AN OUT AND RETURN COURSE OF 300 KM*
Deborah B. Kutch 117.45 kmh
Reba Coombs 72.98 mph
Stemme S10-VT
Ely, NV
7/27/00

GLIDERS, 15-METER CLASS

DISTANCE

TRIANGLE DISTANCE*
Cynthia Brickner 300.09 km
Schempp-Hirth Discus b 186.47 mi
California City, CA
6/29/00

FREE THREE TURN POINT DISTANCE*
Cynthia Brickner 369.13 km
Schempp-Hirth Discus b 229.36 mi
California City, CA
6/29/00

FREE OUT AND RETURN DISTANCE*
Cynthia Brickner 281.44 km
Schempp-Hirth Discus b 174.88 mi
California City, CA
6/29/00

SPEED

SPEED OVER A 300 KM TRIANGLE*
Cynthia Brickner 78.34 kmh
Schempp-Hirth Discus b 48.68 mph
California City, CA
6/29/00

DW, WORLD CLASS GLIDERS

DISTANCE

STRAIGHT DISTANCE*
Valeria Paget 194.07 km
PW-5 120.59 mi
Waller, TX
9/10/00

PARACHUTES

There are reports of parachute-like devices being used as far back as 12th century China and 16th century Venice, but the first confirmed intentional parachute jump was in 1783, the same year that men began to fly in balloons. Sebastian Lenormand, in France, used a crude parachute to demonstrate his plan for escaping from burning buildings.

In 1797, Andre Garnerin stepped from a balloon at 2,000 feet over Paris and arrived safely on the ground, thanks to a parachute. The first emergency use of a parachute was in 1808, when Jodaki Kuparento bailed out of his flaming balloon over Warsaw and survived.

Parachuting as a sport began just before World War I, when Georgia "Tiny" Broadwick made freefall jumps for exhibition purposes. After the war, parachuting became a popular attraction at fairs and air shows.

In 1930, the first parachute competition was staged by the Russians, who graded the participants on their nearness to a ground target upon landing. The first NAA-sanctioned contest was in conjunction with the 1932 National Air Races, at Cleveland, Ohio.

It was not until after World War II that the sport began to come of age. In 1948, techniques for stabilized freefall were developed by Leo Valentin, in France. This led to today's highly controlled descents in freefall, with intricate maneuvering and the forming of linked formations of dozens of jumpers.

Also in 1948, FAI first recognized the sport of parachuting at the recommendation of pioneer Joe Crane on behalf of NAA. The first FAI-sanctioned World Parachuting Championships were held in 1951 in Yugoslavia.

The sport of parachuting is directed in the U.S. by the U.S. Parachute Association, 1440 Duke St., Alexandria, VA 22314, (703) 836-3495, and on the world level by the FAI's International Parachuting Commission (I.P.C.).

PERFORMANCE RECORDS

LARGEST FREEFALL FORMATION (G-2-C)

GENERAL CATEGORY (INTERNATIONAL) ★
Christoph Aarns, Geoffro Abrahams, 300 persons
Sean Adams, Tim Addison,
Marcel Aeby, Alberto Alibrandi, Mark Allen, Edward Anderson, Ron Andre, John Appleton, Peter Arthen, Eric Asendorf, Lise Aune, Mary Pat Avery, Guy Banal, Jeannie Bankston, Orla Bannan, Theresa Baron, Mary Bauer, Jennifer Behrens, Alexandre Beloglazov, Pål Bergan, Cherie Berke, John Berke III, Robyn Best, Gary Beyer, Christian Blau, Christine Blohme, Hans Blohme, Steven Boekel, Robert Bonitz, Dan Brodsky-Chenfeld, Mark Brown, James Browning, Sally Burgess, David Burton, Richard Calledare, William Campbell, Marco Carrara, Bruce Chapman, Andrew Cliff, John Coffman, Philip Colmer, Rob Colpus, Kate Cooper, Melanie Conaster, Santiago Corella, Michael Cotter, Ivan Coufal, David Craigmile, Donat Curti, Chris Dales, Sandro D'Aloia, Simon Darch, Carl Daugherty, Richard Davis, Julian Dawson, Breno de Assis, Marinus DeHeij, Ilario DeMarchi, Jean-marc Desfourneaux, Beatrice Dodet, Sherry Dodson, Tony Domenico, Stepehn Dossenback, Barbara Duke, Gretchen Dunn, Kelly Dunn, Richard Duran, Gilles Dutrisac, John Eagle, Albert Edwards, Lara Eisenberg, Jerry Elkind, Don Ellisor, Lawrence Elmore, DeLayne Etheridge, Esben Evensen, Rod Fairweather, Thomas Falzone, Marcia Farkouh, John Farrell, Mark Farrell, Raymond Ferrell, Douglas Forth, Gregory Frank, Scott Franklin, James Freegard, Craig Fronk, Bill Furlong, Cris Fucci, Nick Furchner, Christopher Gade, E. Fernando Gallegos, Catherine Gallet, Peter Galli, Malcolm Gareau, Paul Gemmell, David Gifford, Ken Gillespie, Pablo Gimenez, Eric Gin, Craig Girard, Viktor Gorbunkov, James Graham, Sandy Grillet, Paul Grisoni, Kathleen Grix, Amy Haass, Gary Haass, Gregory Habermann, John Hamilton, Ken Hansen, Mark Harrington, Bruce Heiler, Beverly Hein, Eric Heinsheimer, Joseph Helfer, Debra Henry, Jay Henry, Howard Hensen, Charles Hoare, Marc Hogue, John Hoover, Neal M. Houston, Alexander Huber, Justo Ibarra, Jack Jefferies, Thomas Jenkins, Carsten Jensen, Geroge Jicha, Peter Johansen, Michael Johnston, Martin Jones, R. Joey Jones, James Keery, Andrey Kharitonov, Mark Kirkby, Dieter Kirsch, Peter Knight, Mariann Kramer, Brian Krause, Linda Kretzler, Mark Kruse, Helena Kubler, Yasuhiro Kubo, Douglas Kuhn, Josep Lagares, Herman Landsman, Michael Lange, Robin Larsen, Lu Lastra, Ari Laukkanen, Gilles Leboeuf, William Legard, Michel Lemay, Fred Leslie, Kathy Leslie, Thomas Lewetz, Stephan Lipp, Laurent Lobjoit, David Loncasty, Elizabeth Loncasty, Antonio J. Lopez, Julien Losantos, Philip Ludwig, Guy Manos, Pamela Manos, Jennifer Martin, Stephane Mattoni, Mario Mauro, Cheryl McClure, James McCormick, Christopher McDougal, Colleen McGrath, Gary McGuinness, Malcolm McKee, Sandy McRobbie, Jeff W. McVey, Ben McWilliam, Jon McWilliam, William Mendenhall, Alexandre Merts, Jan Meyer, Truett Miley, Robin Mills, Alexey Minaev, Mikhail Mineev, Ellen Monsees, Timothy Monsees, Alan Moss, Dave Mudge, Terrance Murphy, Pierre Nadeau, Douglas Nintzel, Mickey Nuttal, Dave Osbon, Daniel Paquette, Timothy Patterson, Patrick Passe, Edward Pawlowski, Carey Peck, Robert Pecnik Jonathan Perl, John Peschio, Ricardo Pettena, Paula Philbrook, Roger Ponce De Leon, Robert Ponzi, Mario Prevost, Mark Procos, Russell Racey, Paul Rafferty, Raider Ramstad, James Rees, Carsten Reese, Michael Reinert, Klaus Renz, Eli Rhodes, Julie Richter, Eliana Rodriguez, John Ronalter, David Rucker, Thomas Sætren, Karin Sako, Robert Sammis, Kirill Samotsvetov, Umberto Sampieri, Jeff Sands, Reiner Schuchmann, Martin Schweinoch, Juliana Se Raidel, Francesco Sforza, Oleg Shalamihin, Mike Sheerin, Andrey Shemetov, Sergey Shenin, Eiko Shimura, Susanne Siedler, Gail Sims, Greg Sitkowski, Angus Smith, Catherine Smith, Paul Smith, Phil Smith, Ryan Smith, Scott Smith, Nikolay Soukharnikov, Billy Somerville, Scott Stainforth, Dave Stephens, Tim Stevens, Martin Stromeyer, Christopher Talbert, Steve Tambosso, Mark Tennison, John TerBeek, Jorn Thiele, Max Thiele, Wolfgang Thiele, Bryan Thivierge, Derek Thomas, Stephen Thomas, Bekie Thompson, Paula Thues, Louis Tommaso, Mary Traub, Bruce R. Travis, Michael Treman, Joe Trinko, Michael Truffer, Rossella Tura, Claude Tzifkansky, Klaus Unterhofer, Laurens Van Der Post, Steven Vargas, John Verley, Kirk Verner, Elena Vertiprakhova, Rusty Vest, William Von Novak, Ted Wagner, Elizabeth Wagner-Gantzer, Lisa Walker, Scott Wall, James Wallace, Steven Webb, Peter Weber-Heinz, Joseph Weber, Mary Ellen Weber, William Weber, Richard Whitehill, Fred W. Whitsitt, Tyree Wilde, Solly Williams, Kelly Wolf, Stephen Woodford, Stephen Wood, Blaine Wright, Bob Wright
Eloy, AZ
12/13/02

FEMININE CATEGORY (INTERNATIONAL)
Nicole Angelides, Jeannie Bankston, 131 persons
Theresa Baron, Dana Bartels,
Mary Bauer, Jennifer Behrens, Betty Bennett, Tone Bergan, Robyn Best, Karen Bilder, Lisa Briggs, Marta Brooks, Carol Bufalino, Betsy Burkey, Laurel Callen, Janice Chandler, Marie Chapman, Nesta Chapman, Yong Chisholm, Loraine Clark, Carolyn Clay, Jennifer Clayton, Kate Cooper, Melanie Conaster, Laticia Craig, Shelly Crowell, Brenda Culver, Monica Dean, Jennifer DeLong, Sherry Dodson, Shauna Dorsett, Barbara Duke, Tracey Eckersley, Joanne Eichorn, Lara Eisenberg, Christiane Enard, Jessie Farrington, Keri Farrington, Mary Farwell, Heidi Flagg, Nicole Fouché, Kelley Fredrickson, Deborah Frost, Mary Garrison, Nathalie Gaudreault, Cynthia Gibson, Lilian Goodin, Amy Goriesky, Wendell Graham, Kathleen Grix, Linda Groarke, Amy Haass, Joanne Haberl, Linda Hardesty, Debra Henry, Tracie Henry, Cynthia Jardine, Debbie Jasek, Jaimie Johnson, Rhonda Joyce, Angela Jung, Charlene Kerr, Claudia Ketterer, Lynda Kies, Bambi Knight, Barbara Kobzik, Cecile Kohrs, Alexandra Kolacio, Brenda Kramar, Mariann Kramer, Linda Kretzler, Virginia Kuhlmann, Nancy LaRiviere, Janet Lundquist, Jennifer Martin, Libbi Mathews, Sarah McCarthy, Josepha McComb, Kim Meijer, Patricia Menke, Linda Merricks, Lisa Methvin, Christian Meyer, Jan Meyer, Ellen Monsees, Carmen Mullenix, Julie Nichol, Molly Osborne, Anita Paul-Oliver, Arlene Rapkin, Pamela Rhodes, Julie Richter, Pamela Riggs, Audrey Rowe, Alison Sahl, Karin Sako, Chigusa Sansen, Mary SantAngelo, Robin Saunders, Amanda Schaeffer, Jill Scheidel, Eva Schumann, Elizabeth Settel, Amy Shreve, Eiko Shimura, Sally Sidman, Susanne Siedler, Gail Sims, Catherine Smith, Wendy Smith, Jayna Sommer, Lynne Stewart, Susan Sweetman, Tammy Tatum, Valerie Thal-Slocum, Sandra Thiede, Bekie Thompson, Paula Thues, Rossella Tura, June Urschel, Elena Vertiprakhova, Elizabeth Wagner-Gantzer, Lisa Walker, Cornelia Waymouth, Judy Welker, Gabriele Welte, Birgit Weyer, Kathryn Wilson, Marie Winther, Sarah Withers, Janna Wynne
Perris, CA
10/19/02

LARGEST CANOPY FORMATION (G-2-D)

GENERAL CATEGORY (INTERNATIONAL)
Paulo Assis, Christophe Balisky, 70 persons
Claudia Balisky, Bruce Barnett,
David Basinger, Vern Bates, Remko Bolt, Charles Bunch, Scott Chew, Bill Clement, Martin Clennon, Jeffrey Cornelius, James Cowan, Wendy Fauljner, Michael Fedak, Bob Feisthamel, Scott Fiore, Francis Fowler, Petra Gatti, Chris Gay, Mark Gregory, Eduardo Guillen, Allen Gutshall, Lyn Hannah, Victor Herrick, Brett Higgins, Mauri Hujanen, Kevin Ingley, Eric Johnson, Kevin Keenan, Michael Lewis, Robert Lievsay, Robert Lyon, Keith Mac Beth, Shahin Mahmoudzadeh, Frank Matrone, Jerome McCauley, Roger McClelland, Cheryl Michaels, Kenneth Oka, Brian Pangburn, Luis Perez, Russel Pinney, Ernie Pliscott, Andrew Preston, Paul Quandt, Raul Ramirez, Jim Rasmussen, Timothy Raup, Robert Reger, David Richardson, Thomas Rohde-Seelbinder, Raymond Rumpf, William Runyon, Scott Shumway, Eward Slot, Herman Slot, Keith Thivierge, Michael Tomaselli, Kees Tops, Johan Verstegan, William Vest, Thomas Vetter, Shaun Vineyard, Lyal Waddel, Aidan Walters, James Walton, Chris Warnock, Henny Wiggers, Joel Zane
Lake Wales, FL
11/29/03

Superseded
GENERAL CATEGORY (INTERNATIONAL)
David Anderson, Paulo Assis, 65 parachutists
Christophe Balisky, David Basinger,
Vern Bates, Remko Bolt, Charles Bunch, Scott Chew, Bill Clement, Martin Clennon, Jeffrey Cornelius, James Cowan, Wendy Fauljner, Bob Feisthamel, Scott Fiore, Francis Fowler, Petra Gatti, Chris Gay, Holger Gnoth, Jenifer Gordon, Mark Gregory, Eduardo Guillen, Allen Gutshall, Victor Herrick, Mauri Hujanen, Kevin Ingley, Eric Johnson, Kevin Keenan, Robert Lievsay, Robert Lyon, Keith Mac Beth, Shahin Mahmoudzadeh, Frank Matrone, Jerome McCauley, Roger McClelland, Cheryl Michaels, Craig Nadler, Brian Pangburn, Luis Perez, Pasi Pirttikoski, Andrew Preston, Paul Quandt, Raul Ramirez, Jim Rasmussen, Robert Reger, David Richardson, Eber Amaral Rodrigues, Thomas Rohde-Seelbinder, Raymond Rumpf, William Runyon, Eward Slot, Herman Slot, Keith Thivierge, Michael Tomaselli, Kirk Vanzandt, Johan Verstegan, William Vest, Thomas Vetter, Shaun Vineyard, Lyal Waddel, Aidan Walters, James Walton, Chris Warnock, Henny Wiggers, Joel Zane
Lake Wales, FL
11/26/03

Superseded
GENERAL CATEGORY (INTERNATIONAL)
Christophe Balisky, David Basinger, 64 persons
Vern Bates, Remko Bolt,
Charles Bunch, Scott Chew, Bill Clement, Martin Clennon, Jeffrey Cornelius, James Cowan, Wendy Faulkner, Michael Fedak, Bob Feisthamel, Scott Fiore, Petra Gatti, Chris Gay, Mark Gregory, Eduardo Guillen, Allen Gutshall, Vitor Herrick, Mauri Hujanen, Kevin Ingley, Eric Johnson, Kevin Keenan, Michael Lewis, Robert Lievsay, Robert Lyon, Keith Macbeth, Shahin Mahmoudzadeh, Frank Matrone, Jerome McCauley, Roger McClelland, Cheryl Michaels, Kenneth Oka, Brian Pangburn, Russel Pinney, Pasi Pirttikoski, Andrew Preston, Paul Quandt, Rraul Ramirez, Jim Rasmussen, Timothy Raup, Robert Reger, David Richardson, Thomas Rohde-Seelbinder, Raymond Rumpf, William Runyon, Scott Shumway, Sharon Shumway, Eward Slot, Herman Slot, Ketih Thivierger, Michael Tomaselli, Kees Tops, Johan Verstegan, William Vest, Thomas Vetter, Shaun Vineyard, Lyal Waddell, Aidan Walters, James Walton, Chris Warnock, Henny Wiggers, Joel Zane
Lake Wales, FL
11/25/03

Superseded
GENERAL CATEGORY (INTERNATIONAL)
David Anderson, Paulo Assis, 64 persons
Christophe Balisky, Claudia Balisky,
Bruce Barnett, David Basinger, Vern Bates, Remko Bolt, Charles Bunch, Ricardo Castillo, Scott Chew, Martin Clennon, Jeffrey Cornelius, James Cowan, Wendy Faulkner, Bob Feisthamel, Scott Fiore, Petra Gatti, Chris Gay, Elizabeth Godwin, Mark Gregory, Eduardo Guillen, Allen Gutshall, Lyn Hannah, Vitor Herrick, Kevin Ingley, Robert Lievsay, Robert Lyon, Keith Macbeth, Shahin Mahmoudzadeh, Frank Matrone, Jerome McCauley, Roger McClelland, Cheryl Michaels, Craig Madler, Brian Pangburn, Pasi Pirttikoski, Ernie Pliscott, Andrew Preston, Paul Quandt, Raul Ramirez, Jim Rasmussen, Robert Reger, David Richardson, Thomas Rohde-Seelbinder, Raymond Rumpf, William Runyon, Juan Ventura Sanchez-Finger, John Scott, Eward Slot, Herman Slot, Ketih Thivierger, Michael Tomaselli, Michael Trudell, Johan Terstegan, William Vest, Thomas Vetter, Shaun Vineyard, Lyal Waddell, Aidan Walters, James Walton, Chris Warnock, Henny Wiggers, Joel Zane
Lake Wales, FL
11/25/03

FEMININE CATEGORY (USA) ✯
Audrey Alexander, Beatrice Baudequin, 25 persons
Fran Bowen, Claudia Buss,
Marie-Helen Colinet, Wendy Faulkner, Peggy Fleming, Danielle Ferroni, Petra Gatti, Lillian Goodin, Susanne Krause, Janine Marcacci, Terina McMichael, Dawn Menard, Cheryl Michaels, Anne Mykkanen, Christine Offen, Francoise Paule, Shirley Pelland, Germana Proletta, Claudia Turkiewicz, Lucienne Tranchant, Renate Wieler, Sandra Williams, Karen Young
Perris, CA
10/17/97

LARGEST HEAD-DOWN FORMATION

GENERAL CATEGORY (INTERNATIONAL)
Omar Alhegelan, Manuel Basso, 24 persons
Stephen Boyd, David Brown,
Emmanuelle Celicourt, Amy Chmelecki, Maxwell Cohn, Kenneth Cosgrove, Gabriele Dassori, Filippo Fabbi, Chris Fiala, Nathan Gilbert, Stanley Gray, Anthony Landgren, Chris Lynch, Gabriel Mata, Matt Nelson, Bobby Page, Timothy Porter, Adam Rosen, Mike Swanson, Claude Tzifkansky, Steve Utter, Sven Zimmermann
Pahokee, FL
2/10/03

FEMININE CATEGORY (INTERNATIONAL)
Gigliola Borgnis, Cathy Bouette, 16 persons
Julie Christensen, Amy Chmelecki,
Sara Commandeur, Joi Falk, Sayaka Ihara, Michelle Israel, Jennifer Key, Shannon Kidd, Melissa Nelson, Yoko Okazaki, Gillian Parker, Carine Pausset, Maria Laura Vago, Ursula Wagner
Eloy, AZ
11/29/03

Superseded
FEMININE CATEGORY (USA)
Michele M. Bish, Kate Cooper, 4 persons
Blue McGowan, Kimm A. Wakefield
Perris, CA
5/10/03

ALTITUDE/FREEFALL DISTANCE (G-1-A)

GENERAL CATEGORY
E. Andreev (USSR) 24,500 m
11/1/62 80,360 ft

FEMININE CATEGORY
E. Fomitcheva (USSR) 14,800 m
Odessa Aerodrome 48,556 ft
10/26/77

LANDING ACCURACY WITH .03 METER DISC (G-1-B)

GENERAL CATEGORY
Serguei Vertiprakhov (USSR) 14 landings
Kolomma, Russia (on dead center)
8/23-27/1/97

FEMININE CATEGORY
Cheryl Stearns (USA) 6 landings
Marana, AZ (on dead center)
5/10-13/98 plus .01 meter landing

COMPETITION RECORDS

FREEFALL STYLE (G-1-C)

GENERAL CATEGORY
Franck Bernachot (FRANCE) — 5.40 sec
Grenade, Spain
6/8/92

FEMININE CATEGORY
Tatiana Osipova (RUSSIA) — 6.10 sec
Bekescsaba, Hungary
9/19/96

LANDING ACCURACY WITH .03 METER DISC (G-2-B)

GENERAL CATEGORY (CZECH REPUBLIC)
Jiri Gecnuk — 2 landings (on dead center) plus .03 m landing
Ivan Hovorka
Milos Jurca
Cesatmir Zitka
Prostejov, Czech Republic
8/6/00

FEMININE CATEGORY (CZECH REPUBLIC)
Martina Benesova — 2 landings (on dead center) plus .07 m landing
Blanka Vojtkova
Katerina Papezikova
Ivana Hubalkova
Jindrichuv Hradec Airfield, Czech Republic
6/24/00

FORMATION/LONGEST SEQUENCE (G-2-C)

4-WAY

GENERAL CATEGORY (FRANCE)
Marin Ferre, Julien Losantos, Davide Moy, Erwan Pouliquen — 42 formations
Pujaut, France
7/07/02

FEMININE CATEGORY
(No record registered.)

8-WAY

GENERAL CATEGORY (USA)
Charles Brown, Eric Heinsheimer, John Hoover, Trevor McCarthy, Vernon Miller, Carey Mills, Paul Raspino, Scott Rhodes — 31 formations
Efes, Turkey
9/17/97

FEMININE CATEGORY (USA)
W. Neustrup, D. Rowland, B. Chalfant, V. Thal, T. McCormick, C. Deli, D. Arbogast, D. Morrison — 5 formations
Perris Valley, CA
3/1/80

16-WAY

GENERAL CATEGORY (USA)
Gary Beyer, Christopher Bobo, Dan Brodsky-Chenfeld, Kyle Collins, John Eagle, Craig Girard, Neal Houston, Christopher Irwin, Jack Jeffries, Mark Kirkby, Alan Metni, Steve Nowak, Shannon Pilcher, Chad Smith, David van Greuningen, Kirk Verner — 19 formations
Eloy, AZ
10/25/01

CANOPY FORMATION/CANOPY RELATIVE WORK (G-2-D)

4-WAY ROTATION

GENERAL CATEGORY (RUSSIA)
Oleg Baleev, Denis Dodonov, Sergei Filippov, Alexei Volynskiy — 22 stacks
Granada, Spain
6/28/01

FEMININE CATEGORY (USA)
Sharon Allen, Lori Bartlett, Lilian Goodin, Cheryl Lane — 2 stacks
Eloy, AZ
10/7/91

LONGEST SEQUENCE

GENERAL CATEGORY (RUSSIA)
Jean-Michel Poulet, Alban Mathe, Stéphane Correas, Stéphane Menoux, Werner Norenberg — 12 formations
Immola, Finland
7/4/00

FEMININE CATEGORY (USA)
Sharon Allen, Lori Bartlett, Lilian Goodin, Cheryl Lane — 1 formation
Eloy, AZ
10/11/91

8-WAY SPEED FORMATION

GENERAL CATEGORY (RUSSIA)
Yuri Arifulin, Sergei Kulakov, Guennadi Goriaev, Oleg Pavljuk, Dmitry Khjuppenen, Vladimir Andreev, Sergei Urvantsev, Andrei Kuznetsov — 24.65 sec
Immola, Finland
7/6/00

FEMININE CATEGORY (USSR)
L. Diachek, L. Gusakovskaya, E. Logvinenko, T. Majara, M. Savelieva, O. Seriapova, S. Stchelkunova, L. Yokovleva — 94.05 sec
Grozni, USSR
10/16/90

FREEFLYING - LONGEST SEQUENCE (G-2-E)

GENERAL CATEGORY (USA)
Trenton E. Alkek — 13 formations
Jed A. Lloyd
Stephen R. Boyd
Lake Wales, FL
10/22/03

Hang Gliders and Paragliders

The first hang glider was built and flown by Otto Lilienthal in Germany in 1893. He used a series of monoplane and biplane designs to make the first organized effort to learn to fly. Since the hang glider was the simplest form of heavier-than-air craft, one could be built and then repaired with relative ease. As the powered airplane was perfected and more efficient forms of gliders designed, the simpler types, like hang gliders, faded away.

It was not until the National Aeronautics and Space Administration (NASA) began its Gemini two-man orbital spacecraft project in the early 1960s that the immediate forerunner of the modern hang glider was born. In an effort to enable the returning Gemini spacecraft to maneuver and then land without a parachute, Francis and Gertrude Rogallo invented the flexible, triangular device which led to the first sport hang gliders. That flexible, triangular device is known today as the Rogallo Wing.

After the flex-wing hang glider came the rigid-wing style which offered a considerable increase in handling and performance. More recently, double surface flex-wings have increased performance to the extent that most flex-wing world records are almost double ridge wing records. Both rigid and flex wings are regularly flown for record setting and in competition. Aerodynamic improvements which increase performance include using faired kingposts and tubing and a prone flying position with cocoon-type harness. In addition, new electronics allow pilots the ability to carry an onboard variometer, altitude readout, temperature gauge, barometer and more. This, along with improvements in structural materials and coverings continue to make the sport safer as well as more competitive.

The United States Hang Gliding Association, P.O. Box 1330, Colorado Springs, CO 80901, (719) 632-8300, has been delegated the responsibility of governing sporting activities regarding hang gliding and paragliding in the U.S. The FAI's hang gliding and paragliding committee is the Commission Internationale de Vol Libre (C.I.V.L.).

CLASS 1 - SINGLE PLACE - GENERAL

DISTANCE

STRAIGHT DISTANCE (AUSTRIA)
Manfred Ruhmer — 700.6 km
Icaro Laminar MRx — 435.3 mi
Zapata, TX
7/17/01

DISTANCE OVER A TRIANGULAR COURSE (CZECH REPUBLIC)
Tomas Suchanek — 357.12 km
Moyes Litespeed 4 — 221.90 mi
Riverside, Australia
12/16/00

STRAIGHT DISTANCE TO A DECLARED GOAL (USA)
JOINT RECORD
Lawrence Lehmann — 517.23 km
Wills Wing Talon 150 — 321.39 mi
Michael Barber
Moyes Litespeed LS4
Zapata, TX
6/20/02

OUT AND RETURN DISTANCE (AUSTRALIA)
Rohan Holtkamp — 330.6 km
Moyes CSX-5 — 205.4 mi
Eucla, Australia
2/12/98

SPEED

SPEED OVER AN OUT AND RETURN 100 KM COURSE (CZECH REPUBLIC)
Tomas Suchanek — 75.73 kmh
Moyes CSX-5 181M — 47.06 mph
Eucla, Australia
1/9/98

SPEED OVER AN OUT AND RETURN 200 KM COURSE (CZECH REPUBLIC)
Tomas Suchanek — 73.05 kmh
Moyes CSX-5 181M — 45.39 mph
Eucla, Australia
1/10/98

SPEED OVER AN OUT AND RETURN 300 KM COURSE (AUSTRALIA)
Rohan Holtkamp — 56.59 kmh
Moyes CSX-5 — 35.16 mph
Eucla, Australia
2/12/98

SPEED OVER A 25 KM TRIANGULAR COURSE (CZECH REPUBLIC)
Tomas Suchanek — 50.81 kmh
Moyes Litespeed 4 — 31.57 mph
Riverside, Australia
12/15/00

SPEED OVER A 50 KM TRIANGULAR COURSE (CZECH REPUBLIC)
Tomas Suchanek — 48.84 kmh
Moyes Litespeed 4 — 30.34 mph
Riverside, Australia
12/15/00

SPEED OVER A 100 KM TRIANGULAR COURSE (CANADA)
Martin Henry — 44.1 kmh
Aeros Combat 2-14 — 27.4 mph
Mansfield, WA
7/24/02

SPEED OVER A 150 KM TRIANGULAR COURSE (CZECH REPUBLIC)
Tomas Suchanek — 44.68 kmh
Moyes CSX-4 CSX 017M — 27.76 mph
Hay, Australia
1/2/97

SPEED OVER A 200 KM TRIANGULAR COURSE (HUNGARY)
Attila Bertok — 42.40 kmh
Moyes Litespeed 5 — 26.34 mph
Riverside, Australia
12/15/00

SPEED OVER A 300 KM TRIANGULAR COURSE (CZECH REPUBLIC)
Tomas Suchanek — 45.10 kmh
Moyes Litespeed 4 — 28.02 mph
Riverside, Australia
12/16/00

GAIN OF HEIGHT

GAIN OF HEIGHT (USA)
Larry Tudor — 4,344 m
G2-155 — 14,251 ft
Horseshoe Meadows, CA
South of Gabbs
4/8/85

CLASS 1 - SINGLE PLACE - FEMININE

DISTANCE

STRAIGHT DISTANCE (USA)
Kari Castle — 403.5 km
Wills Wing Talon — 250.7 mi
Zapata, TX
7/28/01

DISTANCE OVER A TRIANGULAR COURSE (UK)
Nichola Hamilton — 167.2 km
Moyes SX-4 — 103.9 mi
Hay, Australia
1/2/97

STRAIGHT DISTANCE TO A DECLARED GOAL (USA)
Kari Castle — 353.4 km
Wills Wing Talon — 219.6 mi
Zapata, TX
7/21/01

OUT AND RETURN DISTANCE (NEW ZEALAND)
Tascha McLelland — 143.85 km
Moyes CSX-4 — 89.38 mi
Forbes, Australia
1/3/99

SPEED

SPEED OVER AN OUT AND RETURN 100 KM COURSE (UK)
Nichola Hamilton — 63.59 kmh
Moyes CSX-4 — 39.51 mph
Eucla, Australia
1/9/98

SPEED OVER A 25 KM TRIANGULAR COURSE (AUSTRALIA)
Jenny Ganderton — 26.00 kmh
Foil 152 C — 16.15 mph
Forbes, Australia
2/14/90

SPEED OVER A 50 KM TRIANGULAR COURSE (CANADA)
Mia Schokker — 27.63 kmh
Aeros Stealth 14 KPL3 — 17.17 mph
Mansfield, WA
7/10/01

SPEED OVER A 100 KM TRIANGULAR COURSE (NEW ZEALAND)
Tascha McLelland — 30.81 kmh
Moyes CSX-4 — 19.14 mph
Forbes, Australia
12/31/98

SPEED OVER A 150 KM TRIANGULAR COURSE (GERMANY)
Rosmarie Brams — 31.45 kmh
Guggenmos Bullet RCS — 19.54 mph
Croyten, Australia
1/2/97

GAIN OF HEIGHT

GAIN OF HEIGHT (UK)
Judy Leden — 3,970 m
Wills Wing HP AT 145 — 13,025 ft
Kuruman, S. Africa
12/1/92

CLASS 1 - MULTI PLACE

DISTANCE

STRAIGHT DISTANCE (CZECH REPUBLIC)
Tomas Suchanek — 368.8 km
Corinna Swiegershausen — 229.2 mi
Moyes Xtralite 164T
Hillston, NSW, Australia
12/7/94

DISTANCE TO A GOAL (USA)
Peter Debellis — 99.2 km
Moyes X2 — 61.6 mi
Lone Pine, CA
7/7/97

OUT AND RETURN DISTANCE (USA)
Kevin Klinefelter — 131.96 km
Tom Klinefelter — 81.99 mi
Moyes Delta Glider
Bishop Airport, CA
7/6/89

SPEED

SPEED OVER A 25 KM TRIANGULAR COURSE (NETHERLANDS)
J.C. de Keijzer — 12.90 kmh
Shelley Smith — 8.02 mph
Icaro XL 5
Plaine Joux, France
4/22/03

GAIN OF HEIGHT

GAIN OF HEIGHT (CZECH REPUBLIC)
Tomas Suchanek — 3,500 m
Corinna Swiegershausen — 11,483 ft
Moyes Xtralite 164T
Hillston, NSW, Australia
12/7/94

CLASS 2 - SINGLE PLACE - GENERAL

DISTANCE

STRAIGHT DISTANCE (USA) ✯
Davis Straub — 655.0 km
AIR ATOS — 407.0 mi
Zapata, TX
7/18/01

DISTANCE OVER A TRIANGULAR COURSE (GERMANY)
Marcus Hoffmann-Guben — 258.9 km
AIR ATOS — 160.9 mi
St-André-des-Alpes, France
8/15/00

STRAIGHT DISTANCE TO A GOAL (USA)
David Glover — 354.6 km
AIR ATOS — 220.4 mi
Zapata, TX
7/28/01

OUT AND RETURN DISTANCE (CANADA)
James M. Neff — 330.6 km
Flight Design Exxtacy-160 — 205.4 mi
Eucla, Australia
2/12/98

SPEED

SPEED OVER AN OUT AND RETURN 100 KM COURSE (USA)
Davis Straub — 51.42 kmh
AIR ATOS — 31.95 mph
Big Spring, TX
8/2/03

SPEED OVER AN OUT AND RETURN 200 KM COURSE (CANADA)
James M. Neff — 65.15 km
Flight Design Exxtacy-160 — 40.48 mi
Eucla, Australia
2/15/98

SPEED OVER AN OUT AND RETURN 300 KM COURSE (CANADA)
James M. Neff — 61.53 km
Flight Design Exxtacy-160 — 38.23 mi
Eucla, Australia
2/12/98

SPEED OVER A 25 KM TRIANGULAR COURSE (USA)
Davis Straub — 50.4 kmh
AIR ATOS — 31.3 mph
Davenport, FL
5/20/01

SPEED OVER A 50 KM TRIANGULAR COURSE (UK)
Robin Hamilton — 45.46 kmh
Swift — 28.25 mph
Hearne, TX
9/22/02

SPEED OVER A 100 KM TRIANGULAR COURSE (UK)
Robin Hamilton — 53.73 kmh
Swift — 33.39 mph
Hearne, TX
8/30/02

SPEED OVER A 150 KM TRIANGULAR COURSE (UK)
Robin Hamilton — 47.0 kmh
Swift — 29.2 mph
Hearne, TX
8/15/01

SPEED OVER A 200 KM TRIANGULAR COURSE (UK)
Robin Hamilton — 45.50 kmh
Swift — 28.27 mph
Hearne, TX
9/29/02

GAIN OF HEIGHT

GAIN OF HEIGHT (SOUTH AFRICA)
Rainer Scholl — 3,820 m
5/8/85 — 12,533 ft

CLASS 4 - SINGLE PLACE - GENERAL

DISTANCE

STRAIGHT DISTANCE (UK)
Nick Chitty — 86.4 km
US Aviation Corp. Super Floater — 53.7 mi
Oxford, UK
7/20/97

SPEED

(No records registered)

GAIN OF HEIGHT

(No records registered)

CLASS 5 - SINGLE PLACE - GENERAL

DISTANCE

STRAIGHT DISTANCE (USA) ★
Davis Straub 655.0 km
AIR ATOS 407.0 mi
Zapata, TX
7/18/01

DISTANCE OVER A TRIANGULAR COURSE (GERMANY)
Marcus Hoffmann-Guben 258.9 km
AIR ATOS 160.9 mi
St-André-des-Alpes, France
8/15/00

STRAIGHT DISTANCE TO A GOAL (USA)
David Glover 354.6 km
AIR ATOS 220.4 mi
Zapata, TX
7/28/01

OUT AND RETURN DISTANCE (CANADA)
James M. Neff 330.6 km
Flight Design Exxtacy-160 205.4 mi
Eucla, Australia
2/12/98

SPEED

SPEED OVER AN OUT AND RETURN 100 KM COURSE (USA)
Davis Straub 51.42 kmh
AIR ATOS 31.95 mph
Big Spring, TX
8/2/03

SPEED OVER AN OUT AND RETURN 200 KM COURSE (CANADA)
James M. Neff 65.15 km
Flight Design Exxtacy-160 40.48 mi
Eucla, Australia
2/15/98

SPEED OVER AN OUT AND RETURN 300 KM COURSE (CANADA)
James M. Neff 61.53 km
Flight Design Exxtacy-160 38.23 mi
Eucla, Australia
2/12/98

SPEED OVER A 25 KM TRIANGULAR COURSE (USA)
Davis Straub 50.4 kmh
AIR ATOS 31.3 mph
Davenport, FL
5/20/01

SPEED OVER A 50 KM TRIANGULAR COURSE (USA)
Davis Straub 40.8 kmh
AIR ATOS 25.4 mph
Davenport, FL
5/17/01

SPEED OVER A 100 KM TRIANGULAR COURSE (USA)
Davis Straub 34.47 kmh
AIR ATOS 21.42 mph
Hearne, TX
8/16/00

SPEED OVER A 150 KM TRIANGULAR COURSE (No record registered)

SPEED OVER A 200 KM TRIANGULAR COURSE (GERMANY)
Marcus Hoffmann-Guben 43.44 kmh
Flight Design GhostBuster 26.99 mph
St-André-les-Alpes, France
8/22/01

GAIN OF HEIGHT

GAIN OF HEIGHT (SOUTH AFRICA)
Rainer Scholl 3,820 m
5/8/85 12,533 ft

PARAGLIDERS - SINGLE PLACE - GENERAL

DISTANCE

STRAIGHT DISTANCE (CANADA)
Will Gadd — 423.4 km
Gin Boomerang Superfly — 263.1 mi
Zapata, TX
6/21/02

DISTANCE OVER A TRIANGULAR COURSE (FRANCE)
Pierre Bouilloux — 237.1 km
Gin Boomerang — 147.3 mi
Pralognan-la-Vanoise, France
8/10/03

DISTANCE TO A GOAL (USA)
Josh Cohn — 285.17 km
Windtech Nitro — 177.19 mi
Carrizo Springs, TX
6/22/03

OUT AND RETURN DISTANCE (SLOVENIA)
Primoz Susa — 213.8 km
Gradient Avax RSE — 132.8 mi
Soriska Planina, Slovenia
6/12/03

Superseded
OUT AND RETURN DISTANCE (SLOVENIA)
Primoz Susa — 204.9 km
Gradient Avax RSE — 127.3 mi
Soriska Planina, Slovenia
5/24/03

Superseded
OUT AND RETURN DISTANCE (SLOVENIA)
Marko Novak — 200.9 km
Gradient Avax RSE — 124.8 mi
Ratitovec, Slovenia
5/16/03

Superseded
OUT AND RETURN DISTANCE (SLOVENIA)
Primoz Susa — 189.2 km
Gradient Avax RS — 117.6 mi
Soriska Planina, Slovenia
5/4/03

SPEED

SPEED OVER AN OUT AND RETURN 100 KM COURSE (UK)
Howard Travers — 28.04 kmh
Airwave XXX — 17.42 mph
Eucla, Australia
1/24/99

SPEED OVER AN OUT AND RETURN 200 KM COURSE (SLOVENIA)
Primoz Susa — 28.8 kmh
Gradient Avax RSE — 17.9 mph
Soriska Planina, Slovenia
6/12/03

Superseded
SPEED OVER AN OUT AND RETURN 200 KM COURSE (SLOVENIA)
Primoz Susa — 26.6 kmh
Gradient Avax RSE — 16.5 mph
Soriska Planina, Slovenia
5/24/03

Superseded
SPEED OVER AN OUT AND RETURN 200 KM COURSE (SLOVENIA)
Marko Novak — 25.11 kmh
Gradient Avax RSE — 15.60 mph
Ratitovec, Slovenia
5/16/03

SPEED OVER A 25 KM TRIANGULAR COURSE (FRANCE)
Patrick Berod — 28.26 kmh
Edel Energy 30 — 17.55 mph
Albertville, France
6/27/95

SPEED OVER A 50 KM TRIANGULAR COURSE (AUSTRALIA)
Enda Murphy — 23.60 kmh
Advance Omega 4/28 — 14.66 mph
Mt. Borah, Australia
12/5/99

SPEED OVER A 100 KM TRIANGULAR COURSE (GERMANY)
Burkhard Martens — 19.47 kmh
Gin Boomerang — 12.10 mph
Brauneck, Germany
7/28/01

SPEED OVER A 200 KM TRIANGULAR COURSE (AUSTRIA)
Klaus Heimhofer — 23.50 kmh
Gin Boomerang — 14.60 mph
Stubnerkogel, Austria
6/19/00

GAIN OF HEIGHT

GAIN OF HEIGHT (UK)
Robbie Whittall — 4,526 m
Firebird Navajo — 14,849 ft
Brandvlei, S. Africa
1/6/93

PARAGLIDERS - SINGLE PLACE - FEMININE

DISTANCE

STRAIGHT DISTANCE (UK)
Kat Thurston — 285.0 km
Nova Xyon 22 — 177.1 mi
Kuruman Airfield, South Africa
12/25/95

DISTANCE OVER A TRIANGULAR COURSE (AUSTRIA)
Karin Wimmer — 92.8 km
Edel Saber S — 57.7 mi
St. Veit, Austria
5/11/98

DISTANCE TO A GOAL (DENMARK)
Louise Crandal — 213.7 km
Gin Boomerang — 132.8 mi
Quixada, Brazil
11/28/01

OUT AND RETURN DISTANCE (AUSTRIA)
Karin Wimmer — 100.6 km
Gin Bonanza S — 62.51 mi
Schmittenhöhe, Austria
7/18/99

SPEED

SPEED OVER A 25 KM TRIANGULAR COURSE (UK)
Fiona Macaskill — 22.75 kmh
Aerodine Shamam — 14.10 mph
Plaine Joux, France
4/24/03

SPEED OVER A 50 KM TRIANGULAR COURSE (UK)
Judy Leden — 15.80 kmh
Edel Rainbow 26 — 9.81 mph
Piedrahita, Spain
7/20/94

GAIN OF HEIGHT

GAIN OF HEIGHT (UK)
Kat Thurston — 4,325 m
Nova Xyon 22 — 14,190 ft
Kuruman Airfield, South Africa
1/1/96

PARAGLIDERS - MULTIPLACE

DISTANCE

STRAIGHT DISTANCE (BRAZIL)
André Luis Grosso Fleury — 299.7 km
Claudia Otilia Guimaraes Ribeiro — 186.2 mi
Sol Kangaroo
Varzea da Cacimba, Brazil
10/17/03

DISTANCE OVER A TRIANGULAR COURSE (AUSTRIA)
Jürgen Stock — 101.9 km
Sabine Kröll — 63.3 mi
Flight Design - Twin 2
Melchboden Kiosk, Austria
6/20/00

STRAIGHT DISTANCE TO A DECLARED GOAL (UK)
Richard Westgate
Jim Coutts — 215.0 km
Edel Prime — 133.6 mi
Poranga, Brazil
11/30/00

OUT AND RETURN DISTANCE (AUSTRIA)
Jürgen Stock — 129.60 km
Manuela Konold — 80.53 mi
Sub-class O-3 (Paragliders)
Melchboden Kiosk, Austria
Flight Design - Twin 2
5/15/00

SPEED

SPEED OVER AN OUT AND RETURN 100 KM COURSE (UK)
Howard Travers — 24.59 kmh
Apco Futura Tandem 42 — 15.28 mph
Eucla, Australia
1/25/99

SPEED OVER A 25 KM TRIANGULAR COURSE (GERMANY)
Burkhard Martens
Renate Bruemmer — 27.1 kmh
Airea Cargo — 16.84 mph
Oetz, Austria
6/27/01

SPEED OVER A 50 KM TRIANGULAR COURSE (SWITZERLAND)
Roland Würgler, Pilot — 18.91 kmh
Jacqueline Gubler, Passenger — 11.75 mph
UP Pick-Up XL
Bellinzona, Switzerland
5/23/99

SPEED OVER A 100 KM TRIANGULAR COURSE (AUSTRIA)
Jürgen Stock — 19.41 kmh
Sabine Kröll — 12.06 mph
Flight Design - Twin 2
Melchboden Kiosk, Austria
6/20/00

GAIN OF HEIGHT

GAIN OF HEIGHT (UK)
Richard Westgate — 4,380 m
Guy Westgate — 14,370 ft
UP Pick-Up 43
Kuruman Airfield, South Africa
1/1/96

GAIN OF HEIGHT*
Robert B. Schick, Pilot — 6,700 ft
Mike Hanrahan, Passenger
Edel Space
Orem, UT
5/22/94

HUMAN POWERED AIRCRAFT

One of man's earliest dreams of flight was of propelling himself through the sky unassisted--using only his own muscles. Until the late 19th century, the idea of human-powered flight was closely tied to ornithopters. But once an airplane had flown with an engine-driven propeller, both of these ideas were relegated to the "crackpot" category.

In 1959, British industrialist Henry Kremer offered a prize of 5,000 pounds Sterling to the first person to fly a mile-long, figure-8 course in a truly human-powered airplane. Several groups of English university students took up the challenge and began to build long-winged and amazingly light airplanes with pedal-driven propellers.

The first of these to fly was Southampton University's "SUMPAC," which covered 50 yards at an altitude of about 6 feet on November 9, 1961. Within a year, both it and the rival "Puffin" were flying as far as 2,000 feet in a straight line and making gentle turns. But neither made an official attempt for the Kremer Prize, being unable to manage the required 180 degree turns.

Seeing interest wane, Kremer increased his prize to 10,000 pounds and then to 50,000 pounds in 1973. But it wasn't until a veteran American soaring pilot and engineer, Paul MacReady, discarded all previous design philosophies and struck out on his own, that the seemingly impossible goal came within reach. On August 23, 1977, his Gossamer Condor, with racing bicyclist Bryan Allen at the pedals, traversed a 1.15-mile figure-8 course in just under six and a half minutes to win the more than $200,000 prize.

Kremer then re-stoked the intellectual fires with a new challenge: 100,000 pounds Sterling for the first flight across the English Channel, more than 20 miles at its narrowest point. MacReady applied his experience to this test and built three new aircraft known as Gossamer Albatrosses in an attempt to win the prize and to set the first FAI records for human-powered flying.

On June 12, 1979, Bryan Allen took off from the Channel Coast near Deal, where Caesar's Roman Legions had landed for their first invasion of England in 79 B.C. Almost three hours later, Allen landed in France after a 22.3 mile journey that completely overshadowed all previous human-powered flights in its distance and duration.

In January 1987, the Michelob Light Eagle, a 92-pound human-powered aircraft (HPA) with a 114-foot wingspan designed by a research team from the Massachusetts Institute of Technology, shattered the 22-mile Allen/MacCready record with a closed circuit flight of 36.45 miles in 2 hours, 13 minutes, at a dry lake bed at Edwards AFB, California. The pilot-powerplant was 26-year-old amateur triathlete, Glenn Tremml. He averaged 16.4 mph for the flight. During the same month, Lois McCallin, 25, the team's first official pilot, set feminine records of 9.60 miles (closed circuit), 4.24 miles (straight line), and 37 minutes, 38 seconds (duration).

CLASS I, HUMAN-POWERED AIRCRAFT, WORLD RECORDS

SUBCLASS I-C, HUMAN-POWERED AIRPLANES, GENERAL

STRAIGHT DISTANCE (GREECE)
Kanellos Kanellopoulos — 115.10 km
Daedalus 88 — 71.52 mi
Crete - Santorini Island
4/23/88

DISTANCE IN A CLOSED CIRCUIT (USA)
P. Glenn Tremml — 58.66 km
Michelob Light Eagle — 36.45 mi
Edwards AFB, CA
1/22/87

DURATION (GREECE)
Kanellos Kanellopoulos — 3 hrs 54 min
Daedalus 88 — 59 sec
Crete - Santorini Island
4/23/88

SPEED OVER A CLOSED CIRCUIT (W. GERMANY)
Holger Rochelt — 44.32 kmh
Musculair II — 27.54 mph
Oberschleissheim Airport
10/2/85

SUBCLASS I-D, HUMAN-POWERED AIRPLANES WITH STORED ENERGY, GENERAL

(No records registered)

SUBCLASS I-E, HUMAN-POWERED ROTORCRAFT, GENERAL

(No records registered)

SUBCLASS I-C, HUMAN-POWERED AIRPLANES, FEMININE

STRAIGHT DISTANCE (USA)
Lois McCallin — 6.83 km
Michelob Light Eagle — 4.24 mi
Edwards AFB, CA
1/21/87

DISTANCE IN A CLOSED CIRCUIT (USA)
Lois McCallin — 15.44 km
Michelob Light Eagle — 9.60 mi
Edwards AFB, CA
1/21/87

DURATION (USA)
Lois McCallin — 37 min
Michelob Light Eagle — 38 sec
Edwards AFB, CA
1/21/87

SUBCLASS I-D, HUMAN-POWERED AIRPLANES WITH STORED ENERGY, FEMININE

(No records registered)

SUBCLASS I-E, HUMAN-POWERED ROTORCRAFT, FEMININE

(No records registered)

Ultralights / Microlights

John Moody is credited with beginning the ultralight movement when he strapped a 10 hp motor to his rigid wing hang glider and left the flatlands of the Midwest below. From that time ultralights have sprouted into all shapes and configurations of flying machine from canard to conventional, trike to flying wing. Most present day ultralights are manufactured from aircraft quality components powered by two cycle engines ranging from 30–65 hp.

Ultralights are defined as fixed wing craft having an empty weight of no more than 254 pounds, maximum airspeed of 55 knots, and fuel capacity of no more than 5 gallons. For record attempts, ultralights fit into a broader class known as microlights as defined by FAI. This class allows for a single place landplane with a maximum gross weight 661 pounds, and a two place landplane with a maximum weight of 992 pounds. In addition, the stall speed at maximum gross weight must be less than 35 knots. Any plane that would fit the microlight class and be above the ultralight limits would fall into the amateur-built experimental category.

The relatively slow and docile ultralight has captured the essence of powered flying for many individuals. Current government regulations allow for minimal qualifications of pilot and aircraft through industry programs. Pilot and vehicle registration is administered by the U.S. Ultralight Association (USUA). USUA is the newest NAA air sport affiliate, accepted into the organization in 1986. The association operates the Colibri badge program (an international flight achievement program), and a national competition program.

World records for ultralights have been set in distance, time to climb, and altitude. Many more records will be set as more of these lightweight aircraft take to the skies.

For further information about ultralights in the U.S., contact USUA at P.O. Box 667, Frederick, MD 21705, Phone (301) 695-9100, Fax (301) 695-0763.

CLASS R MICROLIGHTS, WORLD RECORDS

Subclass RWL1
Landplanes, Weightshift, Solo

DISTANCE

DISTANCE IN A STRAIGHT LINE WITHOUT LANDING (FRANCE)
Patricia Taillebresse 811.64 km
Air Création/Rotax 447 504.33 mi
1 Rotax 447 cc, 40 hp
Leopoldsburg, Belgium - Nîmes-Courbessac, France
8/20/87

DISTANCE IN A STRAIGHT LINE WITHOUT ENGINE POWER
(No record registered)

DISTANCE IN A STRAIGHT LINE WITH LIMITED FUEL (FRANCE)
Yves Rousseau 397.00 km
Tecma - Colt Ascender 246.68 mi
1 Solo 210, 13 hp
Le Davier, Etriché, France
10/22/89

DISTANCE IN A CLOSED CIRCUIT WITHOUT LANDING (FRANCE)
Michel Serane 1,071.16 km
Air Creation Racer 665.59 mi
1 Rotax 462, 52 hp
Besançon-Thise, France
8/5/91

DISTANCE IN A CLOSED CIRCUIT WITHOUT ENGINE POWER
(No record registered)

DISTANCE IN A CLOSED CIRCUIT WITH LIMITED FUEL (FRANCE)
Yves Rousseau 213.60 km
Tecma - Colt Ascender 132.72 mi
1 Solo 210, 13 hp
Argentan, France
7/16/89

ALTITUDE

ALTITUDE (FRANCE)
Serge Zin 9,720 m
Air Création Norgil 31,890 ft
1 Rotax 582, 65 hp
Saint Auban sur Durance, France
9/18/94

TIME TO CLIMB

TIME TO CLIMB TO 3,000 METERS (BELGIUM)
Roland Coddens 5 min 40 sec
Air Creation Racer
1 Arrow GP 1000 90-123, 120 hp
Baisy-Thy, Belgium
4/2/97

TIME TO CLIMB TO 6,000 METERS (BELGIUM)
Roland Coddens 14 min 54 sec
Air Creation Racer
1 Arrow GP 1000 90-123, 120 hp
Baisy-Thy, Belgium
4/2/97

SPEED

SPEED OVER A STRAIGHT 15/25 KM COURSE (UK)
Richard Meredith-Hardy 128.00 kmh
Air Creation Racer SX12 II 79.54 mph
1 Rotax 462, 52 hp
Hethel, UK
8/25/91

SPEED OVER A CLOSED CIRCUIT OF 50 KM WITHOUT LANDING (GERMANY)
Helmut Großklaus 131.52 kmh
Moyes Silent Racer 81.72 mph
Geschendorf, Germany
9/24/02

SPEED OVER A CLOSED CIRCUIT OF 100 KM WITHOUT LANDING (GERMANY)
Helmut Großklaus 129.97 kmh
Moyes Silent Racer 80.76 mph
Geschendorf, Germany
8/21/02

SPEED OVER A CLOSED CIRCUIT OF 500 KM WITHOUT LANDING (HUNGARY)
Bela Dull 116.00 kmh
homebuilt SX 12 II 72.08 mph
Dunaújváros, Hungary
9/29/01

Subclass RWL2
Landplanes, Weightshift, 2-place

(Availability of Records per R1, Group 1, Solo Category)

DISTANCE

DISTANCE IN A STRAIGHT LINE WITH LIMITED FUEL (GERMANY)
Robert Mair 144.82 km
Dietmar Spekking 89.99 mi
Enduro Diesel
Vilshofen, Germany - Altomünster, Germany
6/22/03

DISTANCE IN A CLOSED CIRCUIT WITHOUT LANDING (HUNGARY)
Guti Gabor 698.86 km
Pal Matuska 434.25 mi
Apollo C15 Delta Jet
1 Rotax 582, 65 hp
Bekescsaba, Hungary
5/22/99

DISTANCE IN A CLOSED CIRCUIT WITH LIMITED FUEL (GERMANY)
Robert Mair 126.8 km
Ernst A. Graf 78.8 mi
Enduro Diesel
Altomünster, Germany
6/13/03

ALTITUDE

ALTITUDE (GERMANY)
Robert Mair 6,245 m
Werner von Zeppelin 20,489 ft
Enduro XP
Griesau, Germany
5/23/98

TIME TO CLIMB

TIME TO CLIMB TO 3,000 METERS (HUNGARY)
Gabor Guti 9 min
Mihaly Huszar 19 sec
Apollo C-15 Racer GT
Békéscsaba, Hungary
10/13/01

TIME TO CLIMB TO 6,000 METERS (HUNGARY)
Gabor Guti 42 min
Mihaly Huszar 13 sec
Apollo C-15 Racer GT
Békéscsaba, Hungary
11/15/01

SPEED

SPEED OVER A STRAIGHT 15/25 KM COURSE (HUNGARY)
Vince Ferinc 132.00 kmh
Laszlo Toth 82.02 mph
Apollo C15 Delta Jet
Totszerdahely - Nagykanizsa, Hungary
9/26/99

SPEED OVER A CLOSED CIRCUIT OF 50 KM WITHOUT LANDING (GERMANY)
Helmut Großklaus 132.9 kmh
Dalia Alewi 82.58 mph
Moyes Silent Racer
Geschendorf, Germany
10/2/02

SPEED OVER A CLOSED CIRCUIT OF 100 KM WITHOUT LANDING (HUNGARY)
Jozsef Szurma 129.11 kmh
Vince S. Varga 80.22 mph
ESO 13
Dunaujváros, Hungary
4/26/98

SPEED OVER A CLOSED CIRCUIT OF 500 KM WITHOUT LANDING (HUNGARY)
Guti Gabor 100.33 kmh
Pal Matuska 62.34 mph
Apollo C15 Delta Jet
1 Rotax 582, 65 hp
Bekescsaba, Hungary
5/22/99

Subclass RAL1
Landplanes, Aerodynamic Controls, Solo

(Availability of Records per R1, Group 1, Solo Category)

DISTANCE

DISTANCE IN A STRAIGHT LINE WITHOUT LANDING (BELGIUM)
Bernard d'Otreppe 1,369.00 km
Aviasud Engineering - Albatros 850.66 mi
1 Modified Rotax 377, 35 hp
Fréjus La Palud, France
9/6/88

DISTANCE IN A STRAIGHT LINE WITH LIMITED FUEL (USA) ✯
Jon A. Jacobs 274.61 km
Mitchell Wing b-10 170.63 mi
1 Volkswagen, 14 bhp
Burnettsville, IN - Petersburg, MI
11/30/03

DISTANCE IN A CLOSED CIRCUIT WITHOUT LANDING (USA)
Thomas E. Pratt 805.40 km
Mitchell Wing A-10 Silver Eagle 500.45 mi
1 Zenoah, 23 hp
Elyria, OH - West Salem, OH
9/2/86

DISTANCE IN A CLOSED CIRCUIT WITH LIMITED FUEL (USA)
Michael Brawner 197.96 km
Mitchell Wing A-10 Silver Eagle 123.01 mi
1 Rotax 277 cc, 26 hp
Houston, TX
9/20/87

ALTITUDE

ALTITUDE (AUSTRALIA)
Eric S. Winton 9,144 m
Facet Opal 30,000 ft
1 Rotax 447, 40 hp
Tyagarah, NSW, Australia
3/11/89

TIME TO CLIMB

TIME TO CLIMB TO 3,000 METERS (AUSTRALIA)
Eric S. Winton 6 min 46 sec
Facet Opal
1 Rotax 447 cc, 40 hp
Evans Head, NSW, Australia
3/5/89

TIME TO CLIMB TO 6,000 METERS (BELGIUM)
Louis Y. Schmitz 18 min 51 sec
Aviasud Engineering - Albatros
1 Rotax 503 Free Air, 50 hp
Fréjus La Palud, France
7/23/89

SPEED

SPEED OVER A STRAIGHT 15/25 KM COURSE (CZECH REPUBLIC)
Pavel Skarytka 194.2 kmh
B-612 120.6 mph
Bubovice, Czech Republic
10/11/03

SPEED OVER A CLOSED CIRCUIT OF 50 KM WITHOUT LANDING (FRANCE)
Serge Ferrari 157.44 kmh
Micro B Proto 97.83 mph
1 Rotax 447, 42 hp
Belley-Peyrieu, France
6/30/95

SPEED OVER A CLOSED CIRCUIT OF 100 KM WITHOUT LANDING (FRANCE)
Serge Ferrari 163.61 kmh
Micro B Proto 101.66 mph
1 Rotax 447, 42 hp
Belleville sur Saône, France
10/10/95

Subclass RAL2
Landplanes, Aerodynamic Controls, 2-place

(Availability of Records per R1, Group 1, Solo Category)

DISTANCE

DISTANCE IN A STRAIGHT LINE WITHOUT LANDING (GERMANY)
Johannes Kessler 1,033.15 km
Margarete Krawiel 641.97 mi
MCR 01 UL
Mainz Finthen, Germany
9/3/99

DISTANCE IN A STRAIGHT LINE WITH LIMITED FUEL (SWITZERLAND)
Patrick Watermann 202.00 km
Nathalie Meyer 125.52 mi
Ikarus Comco C-42
1 Rotax 912, 90 bhp
Tannheim, Germany
8/23/00

DISTANCE IN A CLOSED CIRCUIT WITHOUT LANDING (SWITZERLAND)
Patrick Watermann 616.00 km
Nathalie Meyer 382.76 mi
Ikarus Comco C-42
Rotax 912, 90 bhp
Herten, Germany
8/23/00

ALTITUDE

ALTITUDE (ITALY)
Walter Mauri 7,143 m
Heike Goettlicher 23,435 ft
Rodaro Storch
1 Rotax 582, 64 hp
Udine, Italy
4/16/93

TIME TO CLIMB

TIME TO CLIMB TO 3,000 METERS (FRANCE)
Philippe Zen 9 min 20 sec
Gilbert Huguet
Rans Coyotte II S-6 ES
1 Rotax 912, 80 hp
Yenne, France
6/3/94

SPEED

SPEED OVER A STRAIGHT 15/25 KM COURSE (GERMANY)
Johannes Kessler 265.00 kmh
Margarete Krawiec 164.66 mph
MCR 01 UL
Osthofen - Freimersheim, Germany
8/23/99

SPEED OVER A CLOSED CIRCUIT OF 50 KM WITHOUT LANDING (UK)
Julian Harris 186.00 kmh
Robert Sharp 115.57 mph
Jabiru
Isle of Sheppey, UK
4/13/02

SPEED OVER A CLOSED CIRCUIT OF 100 KM WITHOUT LANDING (UK)
Julian Harris 187.00 kmh
Robert Sharp 116.20 mph
Jabiru
Isle of Sheppey, UK
4/14/02

SPEED OVER A CLOSED CIRCUIT OF 500 KM WITHOUT LANDING (SWITZERLAND)
Patrick Watermann 162.16 kmh
Ewald Ritter 100.76 mph
Ikarus Comco C-42
Herten-Rheinfelden, Germany
8/13/99

Subclass RAS1
Seaplanes, Aerodynamic Controls, Solo

DISTANCE

DISTANCE IN A STRAIGHT LINE WITHOUT LANDING
(No record registered)

DISTANCE IN A CLOSED CIRCUIT WITHOUT LANDING (FRANCE)
Richard Fenwick 247.50 km
Ultrastar Zodiac/Rotax 377 153.79 mi
1 Rotax 377, 92 hp
Meaux, France
10/22/85

ALTITUDE

ALTITUDE
(No record registered)

TIME TO CLIMB

TIME TO CLIMB TO 3,000 METERS
(No record registered)

TIME TO CLIMB TO 6,000 METERS
(No record registered)

SPEED OVER A 15/25 KM STRAIGHT COURSE
(No record registered)

SPEED OVER A CLOSED CIRCUIT
(No record registered)

Subclass RAS2
Seaplanes, Aerodynamic Controls, 2-place

DISTANCE

(No records registered)

ALTITUDE

(No records registered)

TIME TO CLIMB

(No records registered)

SPEED

SPEED OVER A STRAIGHT 15/25 KM COURSE (FRANCE)
Albert Ukena 102.25 kmh
Danièle Mazzier 63.54 mph
Chickinox
1 Rotax 582, 65 hp
Gosier, Guadeloupe, France
4/6/93

SPEED OVER A CLOSED CIRCUIT OF 50 KM WITHOUT LANDING (FRANCE)
Albert Ukena 97.53 kmh
Danièle Mazzier 60.60 mph
Chickinox
1 Rotax 582, 65 hp
Gosier, Guadeloupe, France
4/6/93

Subclass RWF1
Foot Launched, Powered Hang Gliders, Solo

DISTANCE

(No records registered)

ALTITUDE

ALTITUDE (FRANCE)
Yves Rousseau 5,230 m
Medium Ropuleim 17,159 ft
1 Rotax 447, 42 hp
Bégrolles en Mauges, France
7/18/92

TIME TO CLIMB

TIME TO CLIMB TO 3,000 METERS (FRANCE)
Yves Rousseau 24 min
Medium Ropuleim
1 Rotax 447, 42 hp
Bégrolles en Mauges, France
7/18/92

SPEED

(No records registered)

Subclass RPF1
Foot Launched, Powered Paragliders, Solo

DISTANCE

DISTANCE IN A STRAIGHT LINE WITHOUT LANDING (SPAIN)
Ramon Morillas Salmeron 644.00 km
Nova Vertex 24 400.16 mi
Almonte, Spain
6/21/98

DISTANCE IN A STRAIGHT LINE WITH LIMITED FUEL (JAPAN)
Etsushi Matsuo 153.65 km
La Paz, Spain 95.47 mi
7/3/01

DISTANCE IN A CLOSED CIRCUIT WITHOUT LANDING (MONACO)
Frédéric Jacques 231.6 km
Korsair 28 143.9 mi
Saint-André-les-Alpes, France
8/25/01

ALTITUDE

ALTITUDE (MONACO)
Frédéric Jacques 5,243 m
Back-Bone/Airwave Explorer 17,201 ft
Sisteron, France
10/5/02

TIME TO CLIMB

TIME TO CLIMB TO 3,000 METERS (FRANCE)
Pierre Allet 31 min 20 sec
Technic'air Koyott
1 Hirth F34 (modified), 28 hp
Bergerac, France
7/20/96

SPEED

SPEED OVER A STRAIGHT 15/25 KM COURSE (SLOVENIA)
Igor Marentic 42.55 kmh
Swing Arcus 26.44 mph
1 Solo 220
Moravske Toplice, Slovenia
8/8/03

SPEED OVER A CLOSED CIRCUIT OF 50 KM WITHOUT LANDING (FRANCE)
Jean-Luc Thuin 37.13 kmh
Marbella Parapente PAP 23.07 mph
Soissons, France
10/13/01

Subclass RPF2
Foot Launched, Powered Paragliders, 2-place

ALTITUDE

ALTITUDE (MONACO)
Frédéric Jacques 3,036 m
Back-Bone/Airwave Scenic 9,960 ft
Savines-le-Lac, France
8/18/02

Subclass RPL1
Landplane, Powered Paraglider, Solo

DISTANCE

DISTANCE IN A STRAIGHT LINE WITHOUT LANDING (FRANCE)
Pierre Allet 740.60 km
Trekking 42 m2 460.19 mi
JPX D320, 20 bhp
Fréjus, France
5/22/99

ALTITUDE

ALTITUDE (USA) ✯
Howard M. Gish, Jr. 5,386 m
Six Chuter SR2 17,671 ft
Rotax 503, 52 bhp
Birchwood, AK
9/9/00

TIME TO CLIMB

TIME TO CLIMB TO 3,000 METERS (USA)
Howard M. Gish, Jr. 22 min 30 sec
Six Chuter SR2
Rotax 503, 52 bhp
Birchwood, AK
9/9/00

SPEED

SPEED OVER A STRAIGHT 15/25 KM COURSE (SLOVENIA)
Roman Zelenko 43.42 kmh
Team Paratrike BWP2000 26.98 mph
Murska Sobota, Slovenia
2/20/03

SPEED OVER A CLOSED CIRCUIT OF 50 KM WITHOUT LANDING (SLOVENIA)
Miro Drobec 41.21 kmh
Team Paratrike BWP2000 25.61 mph
Verzej, Airfield Krapje (Slovenia)
3/25/03

Subclass RPL2
Landplane, Powered Paraglider, 2-place

ALTITUDE

ALTITUDE (SLOVENIA)
Alenka Prah 3,409 m
Vojko Prah 11,184 ft
Air System Big Berta
Murska Sobota, Slovenia
3/30/03

TIME TO CLIMB

TIME TO CLIMB TO 3,000 METERS (SLOVENIA)
Alenka Prah 39 min
Vojko Prah 6 sec
Air System Big Berta
Murska Sobota, Slovenia
3/30/03

SPEED

SPEED OVER A STRAIGHT 15/25 KM COURSE (SLOVENIA)
Vojko Prah 50.85 kmh
Anton Kos-Toncek 31.60 mph
Pipistrel Plus
Murska Sobota, Slovenia
6/9/01

CLASS R, MICROLIGHTS, U.S. NATIONAL RECORDS

POINT-TO-POINT

CENTERVILLE, TN/LITTLE ROCK, AR*
Joseph M. Clinard, Jr. 38.58 mph
Quicksilver MX Super
1 Rotax 503, 46 hp
9/5/93

LITTLE ROCK, AR/CENTERVILLE, TN*
Joseph M. Clinard, Jr. 36.32 mph
Quicksilver MX Super
1 Rotax 503, 46 hp
9/6/93

AIRPLANE RACING

Airplane racing is almost as old as the airplane. The first airplane meet was held in Reims, France in 1909; Glenn Curtiss won the Gordon Bennett Cup at a speed of 47 mph. Along with the growth in numbers of aircraft, such public events became the proving grounds for aircraft performance. Pylon (closed-course) racing and cross-country contests, especially those offering generous prizes, pushed aviation development prior to World War I. After the war, the Pulitzer Trophy races in the U.S. and the international Schneider Cup races (for seaplanes) provided arenas for gaining official speed records and recognition.

In the 1920s and '30s, air racing grew with the National Air Races series at Cleveland, Ohio, Los Angeles, California, and other sites. The "Golden Age of Air Racing" produced Thompson, Greve, and Bendix Trophy race winners who became national heroes. The Gee Bee racers of the Granville Brothers, the Weddell-Williams racers, the speedsters produced by Benny Howard and Gordon Israel, and many other outstanding airplanes proved themselves against the cream of aviation. The custom-built racers outflew the best the Army and Navy had. The attraction of speed and daring created the fastest men in the world—the civilian air racers.

As national events, the Cleveland and Los Angeles air races far exceeded the on-site crowds of present-day Indianapolis 500 races. Even in the Depression, the Cleveland Air Races drew hundreds of thousands of spectators at week-long extravaganzas. NAA was present at major U.S. races as observers and race officials, and often in a sanctioning capacity.

Suspended during WWII, air racing resumed in 1946 in the U.S. Many of racing's designers and builders had gone to work for Uncle Sam, and the feature attractions were surplus WWII fighter planes, now in private hands and modified for racing. The Goodyear class of homebuilt midget racers was introduced in 1947.

Propeller-driven raceplanes were no longer the absolute fastest aircraft—the Jet Age had arrived. However, jets proved to be unsuitable for pylon racing, their speed outstripping the turning radiuses required to keep the planes in sight on the old courses. In 1949, an airplane accident at the Cleveland races ended major racing events for what was thought to be "forever." The promoters retired, the suburbs overran the race sites, and the public found other forms of entertainment, including television. Only the Goodyear racers survived, calling themselves "Formula 1" and holding races at airshows through the 1950s and into the 1960s.

The memory and lure of air racing never died. In 1964, Bill Stead invited a number of friends to take part in an organized air race on his ranch outside of Reno, Nevada. Stock and modified WWII fighter planes arrived, along with Formula racers, small sport biplanes, and former military trainers (T-6s). This air racing event, moved to a former military airfield north of Reno, has been held in September every year since.

While the thunderous speed of the Unlimiteds is the main attraction for racers and fans, air racing continues to offer opportunities to tinker with existing designs and to break new ground. The frontrunners in Unlimited and Formula I racing in the 1990s, like their counterparts in decades past, are those who work hard for incremental increases in speed.

The Formula racers, with 500-pound fixed-gear airplanes powered by "stock" Continental O-200s, are far from rag-and-tube or plywood antiques. Working within the Formula requirements, the leaders in this highly competitive group utilize present-day composite materials, cutting-edge wing and airfoil designs, blueprinted engines, and exceptional pilot skills to reach race speeds exceeding 250 mph.

Additional classes of racing, while not in the developmental forefront with the Unlimiteds and Formulas, are also exciting and competitive. The AT-6/SNJ class is primarily "stock" and therefore very closely matched. Competing on a 5-mile course, these raceplanes are colorful, loud, and fun to watch. The Sport Biplanes included both "racing" and "sport" versions some years ago, leading to domination by the small number of the much faster "racing" types. This mismatch was eventually settled by eliminating the "racing" versions and restoring closer competition on the racecourse. The Formula Vee class, resembling Formula I aircraft but powered by Volkswagen engines, was introduced in the 1970s as an entry-level form of air racing which, because of the slower speed, could race on courses as small as two miles around.

In 1999, the Reno Air Races added the Sport Class. This class is for kit built planes with reciprocating engines of less than 650 cubic inches displacement. These planes race on a course that is about five miles long and achieve speeds around the pylons in the low to mid 300 mph range.

Real raceplanes, large or small, are a handful to control. Concentration, skill, and nerve are requirements for those who fly these unforgiving airplanes to their limits. Preparing and maintaining them generally falls to dedicated (and usually unpaid) crews who work for the love of the sport. The days of gigantic prizes are long gone: hearts, hands, brains, and guts keep modern air racing alive. The owners, builders, crews, and pilots—the air racers—are responding to the challenge that racing engenders: to think and work to improve, to go a little faster a little longer, and ultimately to demonstrate superiority in the sheer excitement of head-to-head competition.

For more information about pylon air racing in the U.S., please contact: National Air-racing Group, % Frank Ronco, 731 St. James Dr., Morgan Hill, CA 95037.

UNLIMITED CLASS

All-Time Qualifying Record*
Skip Holm — 497.78 mph
Modified P-51D Mustang
#4 "Dago Red," N5410V
8.269 mile course, 1 lap
Reno, NV
9/10/02

All-Time Race Record
Lyle Shelton — 481.62 mph
Modified Grumman F8F Bearcat
#77 "Rare Bear," N777L
9.128 mile course, 8 laps
Reno, NV
9/15/91

FORMULA 1 CLASS (International Formula 1)

All-Time Best Qualifying Speed*
Jon Sharp — 263.19 mph
Sharp DR90
#3 "Nemesis," N18JS
3.110 mile course, 1 lap
Reno, NV
9/14/99

All-Time Race Record*
Jon Sharp — 255.26 mph
Sharp DR90
#3 "Nemesis," N18JS
3.110 mile course, 6 laps
Reno, NV
9/16/93

AT-6/SNJ CLASS

All-Time Qualifying Record
Eddie Van Fossen — 235.22 mph
SNJ-4 Texan
#27 "Miss TNT," N127VF
5.033 mile course, 1 lap
Reno, NV
9/15/92

All-Time Race Record
Jack Frost — 235.38 mph
SNJ-6 Texan
#47 "Frostbite," N611F
4.990 mile course, 5 laps
Reno, NV
9/17/98

SPORT BIPLANE CLASS

All-Time Qualifying Record*
David Rose — 226.91 mph
Rose Peregrine
#3 "Rags," N3NG
3.119 mile course, 1 lap
Reno, NV
9/10/02

All-Time Race Record*
David Rose — 225.65 mph
Rose Peregrine
#3 "Rags," N3NG
3.119 mile course, 5 laps
Reno, NV
9/14/02

SPORT CLASS

All-Time Qualifying Record
Darryl G. Greenamyer — 347.77 mph
Lancair Legacy
#33, N33XP
7.799/6.369 mile course, 1 lap
Reno, NV
9/10/02

All-Time Race Record
Darryl G. Greenamyer — 335.98 mph
Lancair Legacy
#33, N33XP
7.799/6.369 mile course, 6 laps
Reno, NV
9/14/02

T-28 CLASS

All-Time Qualifying Record
Bruce Wallace — 287.19 mph
T-28B
#33, "The Bear," N137NA
6.39 mile course, 1 lap
Reno, NV
9/15/98

All-Time Race Record
Rick Raesz — 273.02 mph
T-28B
#29, "Monster," N28NE
6.39 mile course, 6 laps
Reno, NV
9/20/98

FORMULA V CLASS

All-Time Qualifying Record
Brian Dempsey — 175.78 mph
#8 "Miss Annapolis," N8FV
2.0 mile course, 1 lap
Virginia Beach, VA
9/9/94

All-Time Race Record
Brian Dempsey — 168.94 mph
#8 "Miss Annapolis," N8FV
2.0 mile course, 8 laps
Virginia Beach, VA
9/11/94

AEROMODELS

The flying model is the oldest form of flying machine, predating man-carrying aircraft by several centuries. Toy helicopters were sold on the streets of Paris in the 1400's, and a steam-powered model plane may have flown in 1871. We know for certain that at the turn of the century Samuel Langley successfully flew the Smithsonian's Aerodrome steam-powered model. The first model with a gasoline engine probably flew within a few years of the first flights by the Wright brothers.

Before the flying of model airplanes became a hobby, it was an inexpensive and practical method for inventors to try out new ideas they hoped would lead to man-carrying airplanes. Once that was achieved, models became the rage with those whose interest in aviation inspired them to become involved in any way possible.

Official interest in model aircraft competition and records began on a national level not long after World War I. The first U.S. National Model Championships to be sanctioned by the National Aeronautic Association were at St. Louis, Mo., in 1923, as part of that year's National Air Races. Formal records were kept by national aero clubs during the 1930s in the U.S., Great Britain, Germany, Australia and elsewhere.

The first sign of interest from FAI came during its annual General Conference in 1935, when a proposal was offered to add model records to those of full-size aircraft. The first official record was set on September 13, 1936, when a rubber powered model flew in France for 7 minutes, 36 seconds, after takeoff from the ground. Models with gasoline engines entered the international record book in 1938 with a thirteen-and-a-half-minute flight by a Russian modeler. In 1934, a U.S. modeler flew his gasoline engine model for 2 hours, 35 minutes. The Italians set the first model glider record in 1934, the year those records were first recognized.

Currently there are scores of categories of model records to fit the many types of flying models: free-flight models with no power (gliders), rubber power, electric power, and piston engines; radio controlled models with no motors, with piston engines and with electric motors. Powered models are additionally divided into airplanes, seaplanes and helicopters. Rubber-powered models can also set records for indoor flying. Most categories include records for duration, straight-line distance, height and speed. Radio controlled models can also set records for closed circuit distance and closed circuit speed.

Modeling records and competition are controlled in the U.S. by the Academy of Model Aeronautics (AMA), 5151 E. Memorial Drive, Muncie, IN 47302, (765) 741-0072. The FAI governs model activities through its International Commission for Model Aircraft (C.I.A.M.).

CLASS F, AEROMODELS, WORLD RECORDS

FREE FLIGHT

GLIDERS: OPEN CLASS

DURATION (YUGOSLAVIA)
No. 101 — 4 hrs 58 min 10 sec
M. Milutinovic
5/15/60

DISTANCE IN A STRAIGHT LINE (CZECHOSLOVAKIA)
No. 102 — 310.33 km / 192.71 mi
Z. Taus
3/31/62

GAIN IN ALTITUDE (HUNGARY)
No. 103 — 2,364 m / 7,755 ft
G. Benedek
5/23/48

EXTENSIBLE MOTOR AIRPLANE: OPEN CLASS

DURATION (USSR)
No. 104 — 1 hr 41 min 32 sec
V. Fiortorov
6/19/64

DISTANCE IN A STRAIGHT LINE (USSR)
No. 105 — 371.19 km / 230.50 mi
G. Tchiglitsev
7/1/64

GAIN IN ALTITUDE (USSR)
No. 106 — 1,732 m / 5,682 ft
V. Fiortorov
6/19/64

SPEED (USSR)
No. 107 — 187.68 kmh / 116.62 mph
Andrew Belanov
Zhovtnevoye
9/6/87

PISTON MOTOR AIRPLANE: OPEN CLASS

DURATION (USSR)
No. 108 — 6 hrs 1 min
Koulakovsky
8/6/52

DISTANCE IN A STRAIGHT LINE (USSR)
No. 109 — 378.76 km / 235.20 mi
E. Boricevitch
8/15/52

GAIN IN ALTITUDE (CHINA)
No. 110 — 6,467 m / 21,218 ft
Yin Chengbai
Jiangsu
8/8/82

SPEED (RUSSIA)
No. 111 — 189.30 kmh / 117.63 mph
Artur Shaginian
Orel, Russia
7/3/99

INDOOR AIRPLANE, EXTENSIBLE MOTOR: OPEN CLASS

(Category 1: Less Than 8 meters)
DURATION (USA)
No. 115(a) — 39 min 19 sec
Robert Randolph
Loma Linda, CA
1/21/96

(Category 2: 8 - 15 meters)
DURATION (USA)
No. 115(b) — 45 min 32 sec
Robert Randolph
San Bernardino, CA
10/17/93

(Category 3: 15 - 30 Meters)
DURATION (USA) ✯
No. 115(c) — 47 min 19 sec
James Richmond
West Baden, IN
8/4/02

(Category 4: Over 30 Meters)
DURATION (USA)
No. 115(d) — 60 min 1 sec
Stephen Brown
Tustin, CA
6/1/97

INDOOR AIRPLANE, EXTENSIBLE MOTOR: CLASS FID

(Category 1: Less than 8 meters)
DURATION (UKRAINE)
No. 125(a) — 26 min 45 sec
Oleg Korniychuk
Energodar, Ukraine
5/20/02

(Category 2: 8 - 15 meters)
DURATION (UKRAINE)
No. 125(b) — 29 min 8 sec
Oleg Korniychuk
Kharkiv, Ukraine
8/24/02

(Category 3: 15 - 30 meters)
DURATION (USA)
No. 125(c) — 33 min 47 sec
James Richmond
West Baden, IN
8/4/02

(Category 4: Over 30 meters)
DURATION (ROMANIA)
No. 125(d) — 41 min 42 sec
Popa Aurel
Berlin, Germany
6/28/03

INDOOR AIRPLANE, EXTENSIBLE MOTOR: CLASS FID

(Category 1: Less Than 8 meters)
DURATION (JAPAN)
No. 116(a) — 19 min 39 sec
Akihiro Danjo
Tokyo, Japan
5/6/01

(Category 2: 8 - 15 meters)
DURATION (JAPAN)
No. 116(b) — 20 min 36 sec
Akihiro Danjo
Kanagawa, Japan
4/26/01

(Category 3: 15 - 30 meters)
DURATION (JAPAN)
No. 116(c) — 21 min 23 sec
Akihiro Danjo
Kanagawa, Japan
10/20/01

(Category 4: Over 30 meters)
DURATION (SWEDEN)
No. 116(d) 23 min 47 sec
Jonas Romblad
Berlin, Germany
3/3/01

INDOOR AIRPLANE, EXTENSIBLE MOTOR: CLASS F1M

(Category 1: Less Than 8 meters)
DURATION (JAPAN)
No. 117(a) 16 min 44 sec
Akihiro Danjo
Tokyo, Japan
5/6/01

(Category 2: 8 - 15 meters)
DURATION (JAPAN)
No. 117(b) 18 min 14 sec
Akihiro Danjo
Kanagawa, Japan
4/26/01

(Category 3: 15 - 30 meters)
DURATION (JAPAN)
No. 117(c) 18 min 20 sec
Akihiro Danjo
Kanagawa, Japan
5/27/01

(Category 4: Over 30 meters)
DURATION (JAPAN)
No. 117(d) 19 min 40 sec
Akihiro Danjo
Tokyo, Japan
8/11/02

INDOOR GLIDER: CLASS F1N

(Category 1: Less Than 8 meters)
DURATION (AUSTRALIA)
No. 118(a) 39.6 sec
Leonard George Surtees
Tamworth, Australia
11/24/01

(Category 2: 8 - 15 meters)
DURATION (JAPAN)
No. 118(b) 1 min 0 sec
Akihiro Danjo
Mitsuru Ishii
Tokyo, Japan
10/28/03

(Category 3: 15 - 30 meters)
DURATION (USA)
No. 118(c) 1 min 2 sec
Jim Buxton
Flint, MI
5/4/03

(Category 4: Over 30 meters)
DURATION (USA)
No. 118(d) 1 min 23 sec
Jim Buxton
Johnson City, TN
5/28/03

CONTROL LINE CIRCULAR FLIGHT

REACTION MOTOR AIRPLANE

SPEED (USSR)
No. 135 395.64 kmh
Leonid Lipinski 245.69 mph
12/6/71

PISTON MOTOR AIRPLANE

(Cubic Capacity Less Than 1.00 cc)
SPEED (CHINA)
No. 130 (up to 1.00 cm3) 251.67 kmh
Zhao Jihe 138.69 mph
Chengdon City
8/22/84

(Cubic Capacity 1.01 - 2.50 cc)
SPEED (UK)
No. 131 335 kmh
Paul F. Eisner 208 mph
Esher, England
10/13/01

(Cubic Capacity 2.51 - 5.00 cc)
SPEED (UK)
No. 132 312.29 kmh
P. A. Halman 194.06 mph
Fairmile Common Esher Surrey
10/12/86

(Cubic Capacity 5.01 - 10.00 cc)
SPEED (UK)
No. 133 345.0 kmh
Ken Morrissey 214.3 mph
Oakington, UK
8/25/03

SPEED IN COMPETITION (SPAIN)
No. 134 302.5 kmh
Luis Parramon 188.0 mph
Sebnitz, Germany
7/19/02

PISTON MOTOR AIRPLANE, TEAM RACING: CLASS F2C

SPEED IN CIRCULAR FLIGHT, 100 LAPS (UKRAINE)
No. 57 3 min 10.5 sec
Valeriy Kramarenko
Yuriy Chayka
Rouille, France
7/15/03

SPEED IN CIRCULAR FLIGHT, 200 LAPS (FRANCE)
No. 58 6 min 28.9 sec
Jean Maret
Jean-Paul Perret
Landres, France
7/18/00

RADIO-CONTROLLED FLIGHT PISTON MOTOR AIRPLANE

DURATION (USA)
No. 141 38 hrs 52 min 19 sec
Maynard L. Hill
Barrett J. Foster
David G. Brown
Cape Spear, Newfoundland to Mannin Beach, Ireland
8/9-11/03

DISTANCE IN A STRAIGHT LINE (USA)
No. 142 3,030.0 km
Maynard L. Hill 1,882.8 mi
Barrett J. Foster
David G. Brown
Cape Spear, Newfoundland to Mannin Beach, Ireland
8/9-11/03

GAIN IN ALTITUDE (USA)
No. 143 8,208 m
Maynard L. Hill 26,929 ft
9/6/70

SPEED (USSR)
No. 145 343.92 kmh
V. Goukoune and V. Myakinine 213.57 mph
9/21/71

DISTANCE IN A CLOSED CIRCUIT (USA) ★
No. 146 1,301 km
Maynard S. Hill 808 mi
Hagerstown, MD
8/3/98

SPEED ON A CLOSED CIRCUIT (USA)
No. 147 241.80 kmh
Maynard L. Hill 150.25 mph
11/26/84

PISTON MOTOR SEAPLANE

DURATION (ESTONIA)
No. 148 16 hrs 14 min
Toomas Mardna 11 sec
Tallinn, Estonia
8/14/99

DISTANCE IN A STRAIGHT LINE (ITALY)
No. 149 308.84 km
Gian Maria Aghem 191.90 mi
3/26/94

GAIN IN ALTITUDE (USA)
No. 150 5,651 m
Maynard L. Hill 18,540 ft
9/3/67

DISTANCE GOAL AND RETURN (USA)
No. 151 60.20 km
George W. Finch, 37.40 mi
Ronald C. Clem, Donald Westergren
Lake Elsinore, CA
6/20/02

SPEED (USSR)
No. 152 294.98 kmh
V. Goukoune and V. Myakinine 183.18 mph
9/25/71

DISTANCE IN A CLOSED CIRCUIT (FRANCE)
No. 153 601.00 km
Daniel Coince 373.44 mi
9/15/91

SPEED IN A CLOSED CIRCUIT (UKRAINE)
No. 154 192.90 kmh
Michael Ischenko 119.86 mph
6/20/93

GLIDER

DURATION (UK)
No. 155 36 hrs 3 min
Nicholas Shaw 19 sec
Ivinghoe, England
9/8/01

DISTANCE IN A STRAIGHT LINE (USA)
No. 156 226.44 km
Joseph M. Wurts 140.71 mi
Lancaster, CA
5/28/88

GAIN IN ALTITUDE (MONACO)
No. 157 2,068 m
Frédéric Jacques 6,784 ft
Thierry Regis
St. Vincent-les-Forts, France
7/19/03

DISTANCE GOAL AND RETURN (USA)
No. 158 1.90 km
David L. Hall 1.18 mi
El Mirage Lake, CA
5/27/03

SPEED (FRG)
No. 159 239.00 kmh
Klaus Kowalski 148.50 mph
7/20/91

DISTANCE IN A CLOSED CIRCUIT (CZECHO-SLOVAKIA)
No. 160 716.10 km
E. Svoboda 444.96 mi
7/23/79

SPEED IN A CLOSED CIRCUIT (RUSSIA)
No. 161 129.70 kmh
Zufar Vakkasov 80.59 mph
Oriol, Russia
6/23/97

PISTON MOTOR HELICOPTER

DURATION (RUSSIA)
No. 162 5 hr 59 min
Vladimir Bulatnikov 51 sec
Yuri Bazanov
Alexander Orlov
Moscow, Russia
7/2/01

DISTANCE IN A STRAIGHT LINE (AUSTRALIA)
No. 163 134.10 km
Michael Farnan 83.33 mi
Yatpool, Australia
6/2/96

GAIN IN ALTITUDE (FRANCE)
No. 164 2,940 m
J. P. Allogne 9,646 ft
4/12/92

SPEED (RUSSIA)
No. 166 144.67 kmh
Vladimir Bulatnikov 89.89 mph
Orel, Russia
7/2/98

DISTANCE IN A CLOSED CIRCUIT (FRANCE)
No. 167 101.00 km
Jean Philippe Allogne 62.76 mi
Blagnac, Haute-Garonne
6/18/88

SPEED IN A CLOSED CIRCUIT (UK)
No. 168 112.72 kmh
David W. Whitney 70.04 mph
RAF Elvington
11/2/86

PISTON MOTOR AIRPLANE: CLASS F3D

SPEED IN A CLOSED CIRCUIT (10 LAPS) (AUSTRALIA)
No. 83 57.7 sec
Christopher Callow
Melnik, Czech Republic
8/25/03

RADIO-CONTROLLED ELECTRIC POWERED FLIGHT

RECHARGEABLE (S) ELECTRIC MOTOR AIRPLANE

DURATION (SWITZERLAND)
No. 171 10 hours 38 min
Emil Hilber 30 sec
Dübendorf, Switzerland
6/21/98

DISTANCE IN A STRAIGHT LINE (USSR)
No. 172 102.40 km
Anatoly Dubinetsky 63.62 mi
Evdokievka, USSR
8/25/90

GAIN IN ALTITUDE (AUSTRALIA)
No. 173 2,573 m
Raymond Cooper 8,441 ft
Boralma, Australia
11/9/03

DISTANCE GOAL AND RETURN (USA)
No. 174 6.21 km
Gary B. Fogel 3.86 mi
El Mirage Lake, CA
5/27/03

SPEED IN A STRAIGHT LINE (GERMANY)
No. 175 282.60 kmh
Werner Vauth 175.59 mph
Versmold, Germany
8/8/92

DISTANCE IN A CLOSED CIRCUIT (SWITZERLAND)
No. 176 315.50 km
Emil Hilber 196.04 mi
Dübendorf, Switzerland
6/21/98

SPEED IN A CLOSED CIRCUIT (W. GERMANY)
No. 177 163.68 kmh
Franz Weissgerber 101.70 mph
Bremen
10/31/85

NON-RECHARGEABLE (P) ELECTRIC MOTOR AIRPLANE

DURATION (SWITZERLAND)
No. 178 15 hrs 12 min
Walter Engel 30 sec
Dübendorf, Switzerland
6/21/98

DISTANCE IN A STRAIGHT LINE (ITALY)
No. 179 135.00 km
Gian Maria Aghem 83.88 mi
6/19/91

GAIN IN ALTITUDE (ITALY)
No. 180 2,200 m
Gian Maria Aghem 7,218 ft
9/19/88

SPEED IN A STRAIGHT LINE (GERMANY)
No. 182 168.51 kmh
H. J. Hackstein 104.70 mph
Markdorf, Germany
6/10/90

DISTANCE IN A CLOSED CIRCUIT (FRG)
No. 183 424.50 km
Walter Engel 263.77 mi
Dübendorf, Switzerland
6/21/98

SPEED IN A CLOSED CIRCUIT (SWITZERLAND)
No. 184 97.20 kmh
H. J. Hackstein 60.40 mph
Markdorf, Germany
6/10/90

SOLAR POWERED (SOL) ELECTRIC MOTOR AIRPLANE

DURATION (GERMANY)
No. 185 11 hrs 34 min
Wolfgang Schäper 18 sec
Markdorf, Germany
7/13/97

DISTANCE IN A STRAIGHT LINE (GERMANY)
No. 186 48.21 km
Wolfgang Schäper 29.96 mi
Markdorf, Germany
7/13/97

GAIN IN ALTITUDE (GERMANY)
No. 187 2,065 m
Wolfgang Schäper 6,775 ft
Markdorf, Germany
5/30/99

SPEED IN A STRAIGHT LINE (GERMANY)
No. 189 80.63 kmh
Wolfgang Schäper 50.10 mph
Markdorf, Germany
6/21/98

DISTANCE IN A CLOSED CIRCUIT (GERMANY)
No. 190 190.00 km
Wolfgang Schäper 118.06 mi
Markdorf, Germany
6/17/90

SPEED IN A CLOSED CIRCUIT (GERMANY)
No. 191 62.15 kmh
Wolfgang Schäper 38.61 mph
Markdorf, Germany
6/17/90

ALL SOURCES (COMB) ELECTRIC MOTOR AIRPLANE

DURATION (FRG)
No. 192 15 hrs 36 min
Wolfgang Schäper 55 sec
6/22/91

DISTANCE IN A STRAIGHT LINE (ITALY)
No. 193 109.00 km
Gian Maria Aghem 67.72 mi
6/19/91

GAIN IN ALTITUDE (ITALY)
No. 194 4,539 m
Gina Maria Aghem 14,892 ft
Torino, Italy
2/20/95

SPEED IN A STRAIGHT LINE (GERMANY)
No. 196 274.28 kmh
Wolfgang Kuppers 170.43 mph
Delmenhorst, Germany
8/9/98

DISTANCE IN A CLOSED CIRCUIT (FRG)
No. 197 490.00 km
Wolfgang Schäper 304.47 mi
6/22/91

SPEED IN A CLOSED CIRCUIT (WEST GERMANY)
No. 198 160.44 kmh
Franz Weissgerber 99.69 mph
Bremen
11/25/89

ELECTRIC MOTOR HELICOPTER

DURATION (UK)
No. 199 1 hr 10 min
Mark Hopkins 50 sec
Rye, UK
3/16/03

ELECTRIC MOTOR AIRPLANE: CLASS F5D
(No record registered)

CLASS S, SPACE MODELS, WORLD RECORDS

ALTITUDE

(S-1-A) 0.00 –2.50 Total Impulse (Slovenia)
No. 040 — 310 m / 1,017 ft
Miha Cuden
Sremska Mitrovica, Yugoslavia
9/12/03

(S-1-B) 2.51–5.00 Total Impulse
No. 141
No record registered

(S-1-C) 5.01–10.00 Total Impulse (Slovenia)
No. 102 — 507 m / 1,663 ft
Anton Sijanec
Sremska Mitrovica, Yugoslavia
9/12/03

(S-1-D) 10.01–20.00 Total Impulse
No. 142
No record registered

(S-1-E) 20.01–40.00 Total Impulse
No. 143
No record registered

(S-1-F) 40.01–80.00 Total Impulse
No. 104
No record registered

ALTITUDE WITH PAYLOAD

(S-2-C) 0.00–10.00 Total Impulse
No. 105
No record registered

(S-2-E) 10.01–40.00 Total Impulse
No. 106
No record registered

(S-2-F) 40.01–80.00 Total Impulse
No. 107
No record registered

PARACHUTE DURATION

(S-3-A) 0.00–2.50 Total Impulse (JAPAN)
No. 008 — 1 hr 11 min 10 sec
Takashi Suzuki
Ljublijana, Slovenia
9/10/96

(S-3-B) 2.51–5.00 Total Impulse (SLOVENIA)
No. 109 — 22 min
Drago Perc
Sentjernej, Slovenia
5/12/02

(S-3-C) 5.01–10.00 Total Impulse
No. 110
No record registered

(S-3-D) 10.01–20.00 Total Impulse
No. 111
No record registered

BOOST GLIDER DURATION

(S-4-A) 0.00–2.50 Total Impulse (USSR)
No. 012 — 48 min 15 sec
Alexandre Stakhovsky
Planerskoe
10/15/88

(S-4-B) 2.51–5.00 Total Impulse (USSR)
No. 013 — 32 min 19 sec
Valery Miakinin
Planerskoe
10/15/81

(S-4-C) 5.01–10.00 Total Impulse (USSR)
No. 014 — 2 hrs 22 min
Valery Miakinin
Planerskoe
10/13/81

(S-4-D) 10.01–20.00 Total Impulse (POLAND)
No. 044 — 18 min 2 sec
Ireneusz Pudelko
Krakow, Poland
7/3/94

(S-4-E) 20.01–40.00 Total Impulse (POLAND)
No. 045 — 4 min 21 sec
Leszek Pienkowski
Kosice, Slovakia
5/30/98

(S-4-F) 40.01–80.00 Total Impulse (USSR)
No. 016 — 3 hrs 28 min
Valery Miakinin
Planerskoe
10/13/81

SCALE ALTITUDE

(S-5-A) 0.00–2.50 Total Impulse
No. 117
No record registered

(S-5-B) 2.51–5.00 Total Impulse (USSR)
No. 018 — 772.00 m / 2,532.93 ft
Alexandre Mitiurev
Bolnisi
4/21/87

(S-5-C) 5.01–10.00 Total Impulse
No. 119
No record registered

(S-5-D) 10.01–20.00 Total Impulse
No. 146
No record registered

(S-5-E) 20.01–40.00 Total Impulse
No. 147
No record registered

(S-5-F) 40.01–80.00 Total Impulse
No. 121
No record registered

STREAMER DURATION

(S-6-A) 0.00–2.50 Total Impulse (ROMANIA)
No. 022 — 16 min 48 sec
Dumitru Tudorel
Sibiu
8/15/84

(S-6-B) 2.51–5.00 Total Impulse (YUGOSLAVIA)
No. 123 — 14 min 53 sec
Zoran Katanic
Sremska Mitrovica, Yugoslavia
6/16/01

(S-6-C) 5.01–10.00 Total Impulse
No. 124
No record registered

(S-6-D) 10.01–20.00 Total Impulse
No. 125
No record registered

ROCKET GLIDER DURATION

(S-8-A) 0.00–2.50 Total Impulse (POLAND)
No. 026 22 min
Piotr Malczyyk 04 sec
5/3/91

(S-8-B) 2.51–5.00 Total Impulse (BULGARIA)
No. 027 11 min
Krasimir Tasev 43 sec
Jambol
5/13/84

(S-8-C) 5.01–10.00 Total Impulse (BULGARIA)
No. 028 14 min
Emil Petrov 26 sec
Jambol
8/30/85

(S-8-D) 10.01–20.00 Total Impulse (USSR)
No. 029 27 min
S. Iljin 40 sec
4/24/88

(S-8-E) 20.01–40.00 Total Impulse (GERMANY)
No. 030 36 min
Gunther Gschwilm 55 sec
Streitheim, Germany
6/20/92

(S-8-F) 40.01–80.00 Total Impulse (USSR)
No. 031 24 min
A. Koryapin 8 sec
4/24/88

GYROCOPTER DURATION

(S-9-A) 0.00–2.50 Total Impulse (YUGOSLAVIA)
No. 032 3 min
Radojica Katanic 23 sec
Bratislava-Vajnory, Slovakia
5/9/98

(S-9-B) 2.51–5.00 Total Impulse (Yugoslavia)
No. 133 3 min
Radojica Katanic 1 sec
Sremska Mitrovica, Yugoslavia
6/17/01

(S-9-C) 5.01-10.00 Total Impulse (Yugoslavia)
No. 134 1 min
Andrija Ducak 5 sec
Ruma, Yugoslavia
9/29/02

(S-9-D) 10.01-20.00 Total Impulse (UK)
No. 135 1 min
Stuart Lodge 33 sec
Ruma, Yugoslavia
9/29/02

FLEX-WING DURATION

(S-10-A) 0.00–2.50 Total Impulse
No. 036 9 min
Andrei Angelov 7 sec
Krdgali
4/16/89

(S-10-B) 2.51–5.00 Total Impulse
No. 137
No record registered

(S-10-C) 5.01–10.00 Total Impulse
No. 138
No record registered

(S-10-D) 10.01–20.00 Total Impulse
No. 139
No record registered

CLASS F, AEROMODELS, NATIONAL RECORDS

INDOOR AMA CEILING CATEGORY I

Age	Min/Sec	Held By	Date Set
HL Stick			
Jr.	14:18	Doug Schaefer	11-18-01
Sr.	13:15	Don Slusarczyk	12-11-88
Op.	39:19	Robert Randolph	01-26-96
Intermediate Stick			
Jr.	11:49	Don Slusarczyk	08-12-86
Sr.	22:08	Don Slusarczyk	12-09-90
Op.	30:40	Stan Chilton	08-05-95
F1D (FAI)			
Jr.	14:18	Doug Schaefer	11-18-01
Sr.	12:20	Doug Schaefer	01-04-02
Op.	22:37	John Kagan	08-19-01
ROG Cabin			
Jr.	10:01	Jonathan Harlan	10-01-83
Sr.	05:40	Don Deloach	04-21-89
Op.	16:15	Robert Randolph	04-05-91
Manhattan Cabin			
Jr.	03:12	Don Slusarczyk	08-19-86
Sr.	05:40	Don Deloach	04-21-89
Op.	08:31	Walter VanGorder	12-09-90
Easy B			
Jr.	08:26	Don Slusarczyk	08-19-86
Sr.	13:04	Don Slusarczyk	12-09-90
Op.	21:44	Larry Coslick	02-11-96
Pennyplane			
Jr.	07:43	David Rigotti	08-19-01
Sr.	09:56	Don DeLoach	05-19-90
Op.	13:45	Jim Clem	08-06-95
Limited Pennyplane			
Jr.	07:43	David Rigotti	08-19-01
Sr.	09:56	Don DeLoach	05-19-90
Op.	16:14	Warren J. Williams	05-10-96
Helicopter			
Jr.	06:24	Jon Harlan	06-29-85
Sr.	04:32	Ronnie Stransky	03-31-73
Op.	11:46	Larry Loucka	12-21-03
Ornithopter			
Jr.	03:22	Scott T. Robbins	03-23-91
Sr.	05:16	Don Slusarczyk	12-11-88
Op.	09:01	Roy White	10-01-95
Autogiro			
Jr.	02:43	Don Slusarczyk	08-21-86
Sr.	06:52	Don Slusarczyk	12-09-90
Op.	10:10	Don Slusarczyk	12-22-96
HLG			
Jr.	01:01.4	Jon Chancey	03-31-84
Sr.	01:03.3	Don DeLoach	05-11-91
Op.	01:34.8	Stan Buddenbohm	03-13-91
ROG Stick			
Jr.	08:09	Mike Clem	12-14-80
Sr.	05:51.05	James Buxton	04-14-91
Op.	17:17	Larry Coslick	10-20-98
Standard Class Catapult Glider			
Jr.	00:44.2	Ryan Naccarato	11-13-98
Sr.	00:47.1	Matthew Chalker	05-24-01
Op.	01:24.8	Wayne Trivin	07-13-97
Unlimited Class Catapult Glider			
Jr.	00:43.5	Ryan Naccarato	11-13-98
Sr.	00:46.9	Jeff Daulton	05-24-01
Op.	01:24.8	Wayne Trivin	07-13-97
Mini Stick			
Jr.	06:55	David Rigotti	11-18-01
Sr.	06:04	Jonathan Sayre	04-26-98
Op.	11:13	Stan Chilton	10-05-02
FF Electric Powered			
Jr.		No record established	
Sr.		No record established	
Op.	20:54	Don Slusarczyk	12-14-03

INDOOR AMA CEILING CATEGORY II

Age	Min/Sec	Held By	Date Set
HL Stick			
Jr.	18:13	Doug Schaefer	08-04-01
Sr.	23:16	Don Slusarczyk	07-12-87
Op.	45:32	Robert Randolph	10-17-93
Intermediate Stick			
Jr.	10:45	Parker Parrish	11-25-00
Sr.	21:04	Don Slusarczyk	07-13-87
Op.	31:37	Larry Coslick	09-12-95
F1D (FAI)			
Jr.	18:13	Doug Schaefer	08-04-01
Sr.	23:16	Don Slusarczyk	07-12-87
Op.	35:58	Robert Randolph	10-24-93
ROG Cabin			
Jr.	08:26	Paul Loucka	04-05-81
Sr.	17:37	Don Slusarczyk	07-13-87
Op.	21:57	William Shailor	07-13-87
Manhattan Cabin			
Jr.	04:19	Don Slusarczyk	04-27-86
Sr.	07:23	Don DeLoach	03-20-88
Op.	09:32	Walt VanGorder	07-12-87
Easy B			
Jr.	11:45	Doug Schaefer	03-25-01
Sr.	15:44	Benjamin Saks	03-25-01
Op.	21:14	Larry Coslick	05-08-97
Pennyplane			
Jr.	08:35	Alex Johnson	04-16-00
Sr.	10:10	David Rigotti	04-07-02
Op.	16:41	Raymond B Harlan	10-11-97
Limited Pennyplane			
Jr.	08:35	Alex Johnson	04-16-00
Sr.	10:10	David Rigotti	04-07-02
Op.	12:28	William Pavek	11-17-96
Helicopter			
Jr.	06:39	Jonathan Harlan	04-15-84
Sr.	02:28	Thomas Norell	04-14-85
Op.	11:00	Larry Coslick	04-27-95
Ornithopter			
Jr.	03:13	Scott Robbins	04-11-91
Sr.	04:27	Jonathon Sayre	04-24-99
Op.	10:01	Roy White	12-30-95
Autogiro			
Jr.	04:40	Chris Lu	11-03-01
Sr.		No record established	
Op.	08:38	Ronald Ganser	02-13-93
HLG			
Jr.	01:16	Aaron Markos	09-30-84
Sr.	01:21.8	Bryan Fulmer	03-20-83
Op.	01:39.8	Daniel Domina	06-13-82
ROG Stick			
Jr.	07:04	Don Slusarczyk	11-30-86
Sr.	12:14	Mark Drela	06-12-77
Op.	17:06	Larry Coslick	10-24-98
Standard Class Catapult Glider			
Jr.	01:01.7	Kenny Krempetz	11-16-97
Sr.	00:20.5	Kenny Krempetz	11-10-02
Op.	01:47.9	Wayne R. Trivin	10-26-96
Unlimited Class Catapult Glider			
Jr.	01:07.0	Kenny Krempetz	11-22-98
Sr.	01:10.6	Kenny Krempetz	11-10-02
Op.	01:51.0	James Buxton	04-06-03
Mini Stick			
Jr.	07:23	Ethan Aaron	02-14-04
Sr.	07:45	Jonathan Sayre	04-26-98
Op.	11:00	Stan Chilton	03-26-00
FF Electric Powered			
Jr.		No record established	
Sr.		No record established	
Op.	26:28	Raymond Harlan	04-19-03

INDOOR AMA CEILING CATEGORY III

Age	Min/Sec	Held By	Date Set
HL Stick			
Jr.	23:46	Joshua Merseal	08-16-03
Sr.	32:14	Doug Schaefer	08-16-03
Op.	47:19	James Richmond	08-04-02
Intermediate Stick			
Jr.	24:44	Joshua Merseal	08-17-03
Sr.	32:14	Doug Schaefer	08-16-03
Op.	39:54	Lawrence Coslick	08-16-03

FID (FAI)

Age	Min/Sec	Held By	Date Set
Jr.	14:26	David Rigotti	05-06-01
Sr.	29:31	Richard Whitten	07-30-76
Op.	20:39	Richard Doig	05-06-01

ROG Cabin

Age	Min/Sec	Held By	Date Set
Jr.	15:07	Paul Loucka	06-25-81
Sr.	15:42	Tom Sova	08-06-73
Op.	27:45	Larry Loucka	08-04-02

Manhattan Cabin

Age	Min/Sec	Held By	Date Set
Jr.	03:15	Don Slusarczyk	04-13-86
Sr.	08:32	Don Slusarczyk	08-21-88
Op.	11:43	Larry Loucka	08-16-03

Easy B

Age	Min/Sec	Held By	Date Set
Jr.	24:13	Joshua Merseal	08-17-03
Sr.	24:00	Doug Schaefer	05-04-03
Op.	27:03	Alan Cohen	08-15-03

Pennyplane

Age	Min/Sec	Held By	Date Set
Jr.	12:36	Jeni Jaecks	06-14-83
Sr.	14:36	Dave Lindley	06-17-82
Op.	17:45	Gordon Wisniewski	08-17-03

Limited Pennyplane

Age	Min/Sec	Held By	Date Set
Jr.	12:24	Aaron Markos	06-15-83
Sr.	13:30	Brian Johnson	08-03-02
Op.	15:13	James Richmond	08-16-03

Helicopter

Age	Min/Sec	Held By	Date Set
Jr.	06:18	Jonathan Harlan	06-18-84
Sr.	08:03	Doug Schaefer	08-16-03
Op.	10:18	James Richmond	08-15-03

Ornithopter

Age	Min/Sec	Held By	Date Set
Jr.	05:21	Scott T. Robbins	12-09-90
Sr.	09:11	Don Slusarczyk	08-21-88
Op.	15:20	Ray Harlan	08-16-03

Autogiro

Age	Min/Sec	Held By	Date Set
Jr.	04:08	Dave Lindley	06-26-80
Sr.	08:36	Don Slusarczyk	10-30-90
Op.	10:58	Ronald Ganser	05-01-94

HLG

Age	Min/Sec	Held By	Date Set
Jr.	02:10.4	Darryl Stevens	08-07-77
Sr.	02:32.6	Gary Stevens	08-07-77
Op.	02:35.1	James C. Buxton	10-03-93

ROG Stick

Age	Min/Sec	Held By	Date Set
Jr.	10:02	Dave Lindley	06-28-80
Sr.	11:09	Richard Whitten	06-02-77
Op.	20:03	Lawrence Coslick	05-03-98

Standard Class Catapult Glider

Age	Min/Sec	Held By	Date Set
Jr.	01:27.5	Kenny Krempetz	05-02-99
Sr.	01:59.1	Kenny Krempetz	08-18-03
Op.	02:48.2	Stan Buddenbohm	12-27-98

Unlimited Class Catapult Glider

Age	Min/Sec	Held By	Date Set
Jr.	01:31.3	Kenny Krempetz	05-02-99
Sr.	02:15.4	Kenny Krempetz	08-19-03
Op.	03:08	Stan Buddenbohm	12-27-98

Mini Stick

Age	Min/Sec	Held By	Date Set
Jr.	11:30	Joshua Merseal	08-16-03
Sr.	11:34	Doug Schaefer	08-16-03
Op.	13:23	Walter VanGorder	08-15-03

FF Electric Powered

Age	Min/Sec	Held By	Date Set
Jr.		No record established	
Sr.		No record established	
Op.	31:00	Ray Harlan	08-15-03

INDOOR AMA CEILING CATEGORY IV

HL Stick

Age	Min/Sec	Held By	Date Set
Jr.	34:44	Jonathan Harlan	09-01-85
Sr.	41:25	Don Slusarczyk	09-02-90
Op.	60:01	Stephen Brown	06-01-97

Intermediate Stick

Age	Min/Sec	Held By	Date Set
Jr.	15:54	Mike VanGorder	07-15-79
Sr.	36:31	Doug Schaefer	05-25-03
Op.	41:48	Larry Coslick	09-03-95

FID (FAI)

Age	Min/Sec	Held By	Date Set
Jr.	29:42	Doug Schaefer	09-01-01
Sr.	36:12	Doug Schaefer	10-11-02
Op.	37:59	John Kagan	08-31-03

ROG Cabin

Age	Min/Sec	Held By	Date Set
Jr.	14:35	Jonathan Harlan	08-14-83
Sr.	27:10	Don Slusarczyk	06-08-90
Op.	33:20	Robert Randolph	06-05-82

Manhattan Cabin

Age	Min/Sec	Held By	Date Set
Jr.	04:31	Jim Buxton	06-07-87
Sr.	10:28	Don Slusarczyk	06-03-89
Op.	15:17	James Grant	07-03-95

Easy B

Age	Min/Sec	Held By	Date Set
Jr.	18:07	Andrew Tagliafico	08-05-88
Sr.	26:29	Doug Schaefer	05-30-03
Op.	34:13	Larry Coslick	05-18-97

Pennyplane

Age	Min/Sec	Held By	Date Set
Jr.	13:56	Chris Lu	09-02-01
Sr.	16:05	Brian Johnson	07-07-01
Op.	20:52	Anthony D'Alessandro	09-13-03

Limited Pennyplane

Age	Min/Sec	Held By	Date Set
Jr.	15:19	Vito Iacobellis	07-05-02
Sr.	16:05	Brian Johnson	07-07-01
Op.	18:34	Lawrence Cailliau	08-17-97

Helicopter

Age	Min/Sec	Held By	Date Set
Jr.	05:47	Nick Leonard, Jr.	08-01-96
Sr.	08:15	Benjamin Saks	06-02-01
Op.	12:00	Larry Coslick	08-02-96

Ornithopter

Age	Min/Sec	Held By	Date Set
Jr.	04:28	Scott Robbins	07-23-89
Sr.	09:02	Don Slusarczyk	07-03-88
Op.	21:44	Roy White	09-02-95

Autogiro

Age	Min/Sec	Held By	Date Set
Jr.	05:22	Chris Lu	09-01-01
Sr.	12:11	Don Slusarczyk	09-01-90
Op.	19:53	Anthony D'Alessandro	07-19-03

HLG

Age	Min/Sec	Held By	Date Set
Jr.	02:07.2	Steve Wittman	01-27-74
Sr.	02:39.44	Charles Gagliano	09-05-88
Op.	02:58.6	Ron Wittman	02-18-73

ROG Stick

Age	Min/Sec	Held By	Date Set
Jr.	12:47	Robert Skrjanc	05-16-82
Sr.	15:51	Mark Drela	09-18-77
Op.	24:51	Larry Coslick	07-05-98

Standard Class Catapult Glider

Age	Min/Sec	Held By	Date Set
Jr.	01:29.5	Jason Gomes	05-15-99
Sr.	01:50	Christopher Sydor	09-03-94
Op.	03:25.4	Robert De Shields	02-02-97

Unlimited Class Catapult Glider

Age	Min/Sec	Held By	Date Set
Jr.	01:26.5	Jason Gomes	05-15-99
Sr.	01:55.5	Christopher Sydor	09-04-94
Op.	04:11.9	Stan Buddenbohm	10-04-98

Mini Stick

Age	Min/Sec	Held By	Date Set
Jr.	09:49	David Rigotti	06-01-01
Sr.	13:32	Max Zaluska	08-31-03
Op.	15:06	Robert Romash	09-01-01

FF Electric Powered

Age	Min/Sec	Held By	Date Set
Jr.		No record established	
Sr.		No record established	
Op.	30:30	Ray Harlan	06-01-02

OUTDOOR FREE FLIGHT

Cargo

Age	Min/Sec	Held By	Date Set
Jr.	55:97	Darryl Stevens	07-10-77
Sr.	25:00	Melanie Sanford	07-02-88
Op.	100:21	Roman Ramirez	09-19-82

Coupe d'Hiver

Age	Min/Sec	Held By	Date Set
Jr.	11:54	Gregg Ferrer	01-26-75
Sr.	24:49	Jason Kendy	03-28-81
Op.	53:03	Robert P. White	03-28-81

P-30

Age	Min/Sec	Held By	Date Set
Jr.	11:01	David Ellis	01-01-95
Sr.	23:00	Don DeLoach	06-23-90
Op.	42:58	Stanley Buddenbohm	11-10-01

HL Glider

Age	Min/Sec	Held By	Date Set
Jr.	10:56	Darryl Stevens	02-20-77
Sr.	12:06	John Dowey	09-27-75
Op.	25:50	Bill Moody	10-04-97

A-2 Towline

Age	Min/Sec	Held By	Date Set
Jr.	38:31	Benjamin Coussens	10-06-01
Sr.	29:42	Dallas Parker	01-18-03
Op.	69:35	Tom Coussens	03-13-94

Wakefield

Age	Min/Sec	Held By	Date Set
Jr.	23:45	Taylor Gunder	06-22-02
Sr.	28:01	David Ellis	12-29-99
Op.	41:42	Walter Ghio	10-13-03

Age	Min/Sec	Held By	Date Set
F1C Power (FAI)			
Jr.	15:37	J. E. Troutman	10-13-91
Sr.	25:11	James Troutman	01-14-95
Op.	49:50	Ed Keck	03-14-98
F1H A-1 Towline			
Jr.	23:09	Ryan Archer	01-18-04
Sr.	47:21	Dallas Parker	03-10-04
Op.	58:42	Martyn Cowley	03-15-98
F1G Coupe (FAI)			
Jr.	16:56	Taylor Burrows	03-23-03
Sr.	12:09	Ben Thomson	07-31-03
Op.	38:24	John Sessums	03-10-02
F1J 1/2A Power (FAI)			
Jr.	20:20	Austin Gunder	08-02-00
Sr.	41:10	Austin Gunder	08-02-01
Op.	38:07	Douglas Galbreath	08-19-95
CO_2 Unlimited			
Jr.	05:30	Jeff Wittman	05-30-88
Sr.	26:05	Jeff Wittman	05-29-89
Op.	20:25	H. Graham Selick Jr	07-31-03
USA F1K CO_2			
Jr.		No record established	
Sr.		No record established	
Op.	25:48	Stanley Buddenbohm	09-27-03
Catapult Glider			
Jr.	06:38	Taylor Boe	05-01-04
Sr.	05:21	Sean Andrews	09-28-03
Op.	25:12	Stan Buddenbohm	03-09-03

OUTDOOR FREE FLIGHT CAT. I

Age	*Min/Sec*	*Held By*	*Date Set*
1/2A Gas			
Jr.	40:49	Jon-Mark Carman	05-26-90
Sr.	51:28	Paul Munana	05-25-80
Op.	73:01	Terry Thorkildsen	05-27-89
1/2A Gas ROW			
Jr.	10:20	Darryl Stevens	11-27-77
Sr.	12:49	Dana White	07-13-75
Op.	11:14	John Drobschoff	07-10-77
A Gas			
Jr.	25:00	Joey Foster	05-28-77
Sr.	47:02	Randy Weiler	05-29-76
Op.	129:25	Terry Thorkildsen	02-26-95
A Gas ROW			
Jr.	07:53	Darryl Stevens	11-27-77
Sr.	11:50	Dana White	07-12-75
Op.	13:44	Richard Myers	08-27-72
B Gas			
Jr.	25:57	Travis Hunter	05-30-83
Sr.	26:52	Bob Scully	03-07-76
Op.	105:00	Terry Kerger	02-25-95
B Gas /ROW			
Jr.	20:00	Steve Calhoun	11-16-75
Sr.	08:08	Dana White	07-12-75
Op.	12:03	H. Doering, Jr.	11-27-77
C Gas			
Jr.	33:46	Travis Hunter	05-28-83
Sr.	26:32	Keith Morgan	05-26-79
Op.	100:00	Douglas Galbreath	09-06-92
C Gas ROW			
Jr.		No record established	
Sr.		No record established	
Op.	10:43	Lyman Armstrong	07-27-80
D Gas			
Jr.	12:17	Mike Kelley	11-08-86
Sr.	29:50	Travis Hunter	01-15-84
Op.	69:04	Hulan Matthies	03-11-79
D Gas ROW			
Jr.		No record established	
Sr.		No record established	
Op.	05:51	John Drobshoff	07-04-81
Mulvihill Rubber			
Jr.	23:28	J. Cunnyngham	05-28-77
Sr.	28:10	Jason Kendy	04-19-80
Op.	88:24	Robert P. White	03-11-79
Mulvihill Rubber ROW			
Jr.	04:46	Darryl Stevens	11-27-77
Sr.	01:22	John Banck	09-19-76
Op.	12:42	Joe Bilgri	08-15-82
Ornithopter			
Jr.	01:01	Chris Scott	04-22-79
Sr.		No record established	
Op.	09:21	Roy White	10-13-85
Autogiro			
Jr.	02:36	Nancy Andrews	08-03-75
Sr.	01:35	Mike Johnson	01-17-76
Op.	09:46	Jean Andrews	05-29-89
Helicopter			
Jr.	13:24	Darryl Stevens	10-29-77
Sr.	07:16	Charles K. Martin	09-14-75
Op.	19:12	Stan Buddenbohm	03-10-04
Electric Power (Cl A)			
Jr.	13:01	Jason Villnave	04-27-86
Sr.	07:20	Allen Porter	09-04-94
Op.	40:04	B. Baron-Rawdon	05-27-85
Electric Power (Cl B)			
Jr.	08:51	Jason Villnave	10-26-86
Sr.	05:31	Allen Porter	09-05-94
Op.	41:31	John Oian	05-26-86
Class 101C (.000-.050 cubic inch)			
Jr.		No record established	
Sr.		No record established	
Op.	12:51	Joel Schwartzman	12-14-03
Class 102-103C (.051-.30 cubic inch)			
Jr.		No record established	
Sr.		No record established	
Op.	12:12	John Finn	12-14-03
Class 104-105C (.301-.67 cubic inch)			
Jr.		No record established	
Sr.		No record established	
Op.		No record established	

OUTDOOR FREE FLIGHT CAT. II

Age	*Min/Sec*	*Held By*	*Date Set*
1/2A Gas			
Jr.	16:01	Keith Morgan	01-30-77
Sr.	19:30	Kenneth Keegan	04-29-84
Op.	40:42	Michael Ryan	04-08-00
1/2A Gas ROW			
Jr.	4:53	Dana White	09-15-74
Sr.	7:52	Randy Archer	09-19-76
Op.	9:59	Joe W. Dodson	08-11-74
A Gas			
Jr.	21:45	Joey Foster	04-03-77
Sr.	26:46	Parker McQuown	05-24-80
Op.	83:40	Kenneth E. Oliver	09-28-91
A Gas ROW			
Jr.	04:35	Tom Regan	11-04-73
Sr.	06:54	Dana White	07-12-75
Op.	08:03	Jerald Murphy	10-23-83
B Gas			
Jr.	11:37	Travis Hunter	09-25-83
Sr.	19:11	Mark Woodrey	09-03-78
Op.	54:00	William Morgan	11-29-81
B Gas ROW			
Jr.	01:29	Timothy Young	07-13-75
Sr.	07:27	Dana White	07-12-75
Op.	08:24	Micheal Thompson	05-27-01
C Gas			
Jr.	23:13	Mike Keller	12-07-86
Sr.	20:52	Keith Morgan	11-11-79
Op.	68:18	Guy Menanno	09-27-03
C Gas ROW			
Jr.		No record established	
Sr.		No record established	
Op.	05:08	John Drobshoff	07-05-81
D Gas			
Jr.	22:22	Mike Keller	04-06-86
Sr.	14.28	Travis Hunter	04-29-84
Op.	120:00	Guy Menanno	09-28-03
D Gas ROW			
Jr.		No record established	
Sr.		No record established	
Op.	04:59	John Drobshoff	07-04-81
Payload			
Jr.	06:39	Steve Wittman	05-30-77
Sr.	07:11	Steve Wittman	05-29-78
Op.	39:00	Edward Eliot	08-29-76
Mulvihill Rubber			
Jr.	62:13	Randy Secor	11-05-72

Sr.	30:52	Joseph Kubina	08-05-78
Op.	22:21	Paul Andrade	01-17-04
Mulvihill Rubber ROW			
Jr.	03:21	Tim Young	09-21-75
Sr.	03:49	Gary Stevens	02-26-78
Op.	16:15	Henry Smith	01-15-00
Ornithopter			
Jr.	08:15	Jolyn Andrews	06-27-82
Sr.		No record established	
Op.	06:38	Roy White	10-21-84
Autogiro			
Jr.	02:38	Jeanine Andrews	06-08-75
Sr.	00:23	Mike Reagan	07-21-74
Op.	05:59	Jean F. Andrews	11-04-90
Helicopter			
Jr.	06:14	Jillian Anderson	08-02-00
Sr.	06:57	Charles K. Martin	09-14-75
Op.	16:56	Stanley Buddenbohm	05-23-98
Electric Power (Cl A)			
Jr.	07:12	Allen J. Porter	09-30-91
Sr.		No record established	
Op.	38:15	Donald L. Hughes	02-03-91
Electric Power (Cl B)			
Jr.		No record established	
Sr.		No record established	
Op.	26:32	William Rietdorf	5-16-93
Class 101C (.000-.050 cubic inch)			
Jr.		No record established	
Sr.		No record established	
Op.		No record established	
Class 102-103C (.051-.30 cubic inch)			
Jr.		No record established	
Sr.		No record established	
Op.	26:45	Don DeLoach	01-18-04
Class 104-105C (.301-.67 cubic inch)			
Jr.		No record established	
Sr.		No record established	
Op.	22:57	Don DeLoach	01-19-04

OUTDOOR FREE FLIGHT CAT. III

Age	*Min/Sec*	*Held By*	*Date Set*
1/2A Gas			
Jr.	11:00	Jason Poti	08-03-00
Sr.	07:52	Jason Greer	08-01-97
Op.	54:21	Charles Caton	10-22-00
1/2A Gas ROW			
Jr.		No record established	
Sr.		No record established	
Op.	20:01	Jerry Rocha	05-04-96
A Gas			
Jr.	09:33	James Troutman	07-08-90
Sr.	12:20	Mike Keller	05-27-90
Op	105:00	Ronnie Thompson	07-30-98
A Gas ROW			
Jr.		No record established	
Sr.		No record established	
Op.	05:09	Norman Peterson	04-23-94
B Gas			
Jr.	10:08	Mike Keller	10-18-86
Sr.	11:45	Justin Aronholt	05-18-97
Op.	90:00	Robert P. Johannes	07-22-96
B Gas ROW			
Jr.		No record established	
Sr.		No record established	
Op.	07:16	William Vanderbeek	05-06-00
C Gas			
Jr.	07:55	Mike Keller	10-26-86
Sr.	12:15	James P. Lebda	01-27-91
Op.	70:00	Edward Keck	07-29-98
C Gas ROW			
Jr.		No record established	
Sr.		No record established	
Op.	02:39	John Drobshoff	08-15-82
D Gas			
Jr.	11:00	Mike Keller	10-26-86
Sr.	07:19	James E. Troutman	05-29-94
Op.	60:00	Robert Johannes	08-30-98
D Gas ROW			
Jr.		No record established	
Sr.		No record established	
Op.	04:20	John Drobshoff	07-04-81
Payload			
Jr.	05:03	Tony Hutchins	08-09-86
Sr.	03:09	David Crofoot	07-28-97
Op.	10:58	Warren Kurth	07-30-00
Mulvihill Rubber			
Jr.	30:55	John Shailor	07-28-03
Sr.	27:56	Roderick T. Ioerger	07-06-94
Op.	43:00	Joseph Williams	06-28-02
Mulvihill Rubber ROW			
Jr.	02:12	Jason Thorton	11-24-85
Sr		No record established	
Op.	17:15	Henry Smith	05-01-99
Ornithopter			
Jr.	01:11	Jolyn Andrews	08-23-81
Sr.		No record established	
Op.	16:17	Roy White	10-13-85
Autogiro			
Jr.	01:43	Jolyn Andrews	07-23-78
Sr.		No record established	
Op.	05:50	Jean Andrews	09-14-97
Helicopter			
Jr.	05:31.9	Stephen Burhenn	07-14-99
Sr.	01:57	Don DeLoach	05-18-91
Op.	11:38	Stan Buddenbohm	06-19-94
Electric Power A			
Jr.	04:17	David Crofoot	06-30-96
Sr.	05:32	Melanie Sanford	09-03-88
Op.	30:00	Charles Groth	08-24-03
Electric Power B			
Jr.		No record established	
Sr.	07:14	David Crofoot	07-30-98
Op.	24:00	William Rietdorf	09-01-93
Moffett			
Jr.		No record established	
Sr.		No record established	
Op.	43:00	Robert Bienenstein	07-22-96
Class 101C (.000-.050 cubic inch)			
Jr.		No record established	
Sr.		No record established	
Op.	26:36	Jack Marsh	09-20-03
Class 102-103C (.051-.30 cubic inch)			
Jr.		No record established	
Sr.		No record established	
Op.	31:50	Ronnie Thompson	06-29-02
Class 104-105C (.301-.67 cubic inch)			
Jr.		No record established	
Sr.		No record established	
Op.	26:52	Marvin Mace	06-01-02

CONTROL LINE

Age	*MPH*	*Held By*	*Date Set*
1/2A Speed			
Jr.	146.72	Bobby Fogg III	10-09-94
Sr.	143.61	Bobby Fogg III	04-02-95
Op.	156.42	R. Fogg/C Aloise/ Brown/Shahan	12-11-94
A Speed			
Jr.	160.51	Bobby Fogg	09-04-94
Sr.	162.39	Bobby Fogg III	04-02-95
Op.	194.31	Jerry Rocha	12-02-00
B Speed			
Jr.	166.75	Bobby Fogg	12-11-94
Sr.	163.42	Bobby Fogg III	10-12-96
Op.	186.84	Team Fogg/Brown/ Mancini	10-19-02
D Speed			
Jr.		No record established	
Sr.		No record established	
Op.	199.92	R. Fogg/G. Brown	12-04-99
1/2A Profile Proto			
Jr.	103.76	Bobby Fogg	12-13-92
Sr.	104.73	Bobby Fogg III	07-14-97
Op.	116.08	Jerry Rocha	10-13-01

.21 Sport Speed

Jr.	143.83	Peter Brown	10-04-93
Sr.	145.93	Peter Brown	07-16-97
Op.	158.39	R. Fogg/G. Brown	12-06-98

Formula. 40 Speed

Jr.		No record established	
Sr.	146.16	Mike Wisniewski	07-11-96
Op.	161.87	James Rhoades	12-06-98

Jet

Jr.		No record established	
Sr.	123.41	Howell Pugh	09-02-95
Op.	204.93	L. Waltemath/ J. Mathison	12-07-02

F2A (FAI)

Jr.	162.62	Krystal King	06-17-00
Sr.	173.21	Krystal King	10-06-01
Op.	183.52	William Naemura	09-18-99

Rat Racing

Jr.		No record established	
Sr		No record established	
Op.	02:19.67	Team Fogg/ Shahan (70 Lap)	07-21-94
	04:50.87	Mike Shahan (140 Lap)	07-13-95

Slow Rat Racing

Jr.	05:16.20	Scott Matson (70 Lap)	07-10-00
	06:47.37	Scott Matson (140 Lap)	07-10-00
Sr.	04:29.63	Howell Pugh (70 Lap)	07-20-94
	10:58.47	Doug Short (140 Lap)	07-10-00
Op.	02:36.31	R. Ogez (70 Lap)	07-18-91
	05:24.95	Mike Greb (140 Lap)	07-19-90

1/2A Mouse Race I

Jr.	02:37.57	Scott Matson (50 lap)	07-15-99
	05:17.68	Scott Matson (100 lap)	07-15-99
Sr.	02:44.68	David Rolley Jr. (50 lap)	07-15-99
	05:20.11	D.J. Parr (100 lap)	07-16-98
Op.	2:14:35	Todd Ryan/ Michael MacCarthy (50 lap)	07-13-01
	04:22.00	Paul Gibeault/ Todd Ryan (100 lap)	07-15-99

Scale Racing

Jr.	02:50.65	Bobby Fogg III (70 lap)	07-16-91
	06:08.55	Bobby Fogg III (140 lap)	06-23-92
Sr.	03:15.12	Douglas Short (70 lap)	07-11-00
	05:40.05	Bobby Fogg III (140 lap)	07-11-95
Op.	02:39.38	Team Willoughby/ Oge (70 lap)	07-15-97
	05:33.04	Robert Fogg (140 lap)	07-16-91

F2C Team Race (FAI)

Jr.		No record established	
Sr.		No record established	
Op.	03:16.7	Aaron Ascher/ Lenard Ascher	07-19-02
	06:57.36	Lambert/Ballard (200 lap)	07-15-98

Carrier Class I

Jr.	386.0	Andy Westerheim	10-19-97
Sr.	435.8	Andy Westerheim	07-13-01
Op.	510.0	Kelvin Hite	02-18-01

Carrier Class II

Jr.	366.77	D. Silversmith	10-13-85
Sr.	416.5	Andy Westerheim	10-28-01
Op.	517.9	Peter O. Mazur	11-05-00

Profile Carrier

Jr.	279.50	Andy Westerheim	08-31-97
Sr.	342.1	Andy Westerheim	06-17-01
Op.	391.5	Peter Mazur	07-12-01

Endurance

Jr.	00:44.12	Mark E. Williams	10-06-90
Sr.	00:25.41	Mark Williams	11-27-94
Op.	02:28:02	Edwin Gifford	09-15-02

Electric Speed Class A

Jr.		No record established	
Sr.	64.35	Howell A. Pugh	06-20-92
Op.	90.42	H.Doering Jr.	10-10-93

Electric Speed Class B

Jr.		No record established	
Sr.		No record established	
Op.	119.63	William G. Stewart	06-05-99

PYLON RACING

Age	*Min/Sec*	*Held By*	*Date Set*
Q-40			
Jr.	01:02.24	Tanner Pacini (Long Course)	04-18-04
Sr.	01:00.73	Gino DelPonte (Long Course)	04-12-03
Op.	00:59.38	Thomas Scott (Long Course)	05-27-00
	01:03.13	Santiago Panzardi (Short Course)	10-22-94
F3D			
Jr.	01:08.69	Matthew Van Baren	11-15-98
Sr.	01:02.06	Matthew Van Baren	11-02-02
Op.	00:58.38	Richard Verano/ Robert Holik	08-30-03
Q-500			
Jr.	01:10.43	Bucky Miller (Short Course)	06-13-92
	01:09.92	Tanner Pacini (Long Course)	01-18-04
Sr.	01:01.36	David Wright (Short Course)	07-10-94
	01:05.83	Matthew Van Baren (Long Course)	06-16-01
Op.	00:56.49	Gordon Hyde (Short Course)	05-29-94
	01:03.54	Travis Flynn (Long Course)	02-03-02

RC SOARING

SLOPE DURATION

Age	*H/M/S*	*Held By*	*Date Set*
Class A			
Jr.	04:02:49	Matthew Collier	5-15-87
Sr.	02:58:09	Dieter Rozek	7-25-87
Op.	12:28:20	Alvin Battad	5-23-98
Class B			
Jr.	8:12:08	Patrick McIntire	04-21-01
Sr.	06:01:13	Christopher Ritter	5-31-93
Op.	12:08:04	Dale E. Collier	7-17-88
Class C			
Jr.	11:31:18	Caleb Lesher	04-21-01
Sr.	05:01:43	Ken Merenda	5-31-93
Op.	12:33:57	Jess Walls	5-05-92
Class D			
Jr.	04:02:49	Matthew Collier	5-15-88
Sr.	5:08:30	Joel Magnuson	04-21-01
Op.	13:32:11	Rex Coffman	04-21-01

THERMAL DURATION

Age	*H/M/S*	*Held By*	*Date Set*
Class A			
Jr.	00:30:27	Matthew Andren	6-12-93
Sr.	00:50:34	Andy Schuler	8-27-94
Op.	02:43:43	Bill Hoelcher	09-12-02
Class B			
Jr.	00:41:40	Matthew Andren	6-19-93
Sr.	01:27:04	Charles Vergo	5-31-92
Op.	05:00:55	Jerry Krainock	8-23-81
Class C			
Jr.	00:43:49	Matthew Andren	6-19-93
Sr.	01:54:09	Charles Vergo	6-14-92
Op.	08:03:00	Jay Siren	5-30-82
Class D			
Jr.	01:18:05	Christopher Ritter	5-11-91
Sr.	01:23:14	Rodney Cooper	5-09-92
Op.	09:47:00	Keith Kindrick	5-30-82

DECLARED DISTANCE

Age	Miles	Held By	Date Set
Class A			
Jr.	3.18	Christopher Ritter	5-4-91
Sr.	2.86	Rodney A. Cooper	5-4-91
Op.	5.58	David Hall	05-27-03
Class B			
Jr.	4.75	Christopher Ritter	4-7-91
Sr.	5.58	Christopher Ritter	5-16-92
Op.	14.76	Jerry Krainock	5-30-81
Class C			
Jr.	4.78	Christopher Ritter	5-24-91
Sr.	4.82	Kenneth Merenda	5-16-92
Op.	38.76	Jerry Krainock	5-30-82
Class D			
Jr.	44.63	Brian Isenhart	9-3-84
Sr.	16.95	Alex Bower	5-22-81
Op.	140.67	Joseph M. Wurts	5-28-88

OPEN DISTANCES

Age	Miles	Held By	Date Set
Class A			
Jr.	4.32	Christopher Ritter	4-7-91
Sr.	3.69	Rodney A. Cooper	5-4-91
Op.	6.60	Paul J. Wilson	7-7-88
Class B			
Jr.	4.75	Christopher Ritter	4-07-91
Sr.	5.58	Christopher Ritter	5-16-92
Op.	17.475	Jack R. Hiner	5-27-89
Class C			
Jr.	12.3	Brian Isenhart	5-27-84
Sr.	9.90	Alex Bower	5-30-81
Op.	38.76	Jerry Krainock	5-30-82
Class D			
Jr.	44.63	Brian Isenhart	09-03-84
Sr.	16.95	Alex Bower	5-22-81
Op.	140.67	Joseph M. Wurts	5-28-88

CLOSED COURSE DISTANCE

Age	Km.	Held By	Date Set
Class A			
Jr.	32.4	George Rodriguez	10-17-86
Sr.	69.80	Dieter Rozek	07-25-87
Op.	129.4	Arlie Stoner	06-25-89
Class B			
Jr.	57.0	Christopher Ritter	5-27-91
Sr.	69.80	Dieter Rozek	7-25-87
Op.	248.00	Dale E. Collier	7-17-88
Class C			
Jr.	57.0	Christopher Ritter	5-27-91
Sr.	69.80	Dieter Rozek	7-25-87
Op.	248.00	Dale E. Collier	7-17-88
Class D			
Jr.	57.0	Christopher Ritter	5-27-91
Sr.	69.80	Dieter Rozek	7-25-87
Op.	248.00	Dale E. Collier	7-17-88

ALTITUDE

Age	Feet	Held By	Date Set
Class A			
Jr.	1,887.5	Matthew Collier	05-7-89
Sr.		No record established	
Op.	2,220	Gary Fogel	10-13-02
Class B			
Jr.	1,887.5	Matthew Collier	05-7-89
Sr.	1,993	Alex Bower	10-4-81
Op.	3,100	Jack R. Hiner	9-14-85
Class C			
Jr.	309.7	Christopher Ritter	01-18-87
Sr.	3,045	Norman Timbs	10-04-81
Op.	3,600	Jack Hiner	07-27-85
Class D			
Jr.	4,010	Scott Christian	7-29-84
Sr.	3,045	Norman Timbs	10-4-81
Op.	6,025	Jack Hiner	6-11-82

SPEED

Age	MPH	Held By	Date Set
Class A			
Jr.	49.22	Christopher Ritter	5-18-91
Sr.	53.014	Christopher Ritter	5-09-92
Op.	45.747	Karl Cranford	5-31-92
Class B			
Jr.	49.22	Christopher Ritter	5-18-91
Sr.	47.006	Christopher Ritter	5-31-92
Op.	75.61	John Wyss	9-15-91
Class C			
Jr.	49.22	Christopher Ritter	5-18-91
Sr.	51.057	Christopher Ritter	5-9-92
Op.	75.61	John Wyss	9-15-91
Class D			
Jr.	49.22	Christopher Ritter	5-18-91
Sr.	57.23	Alex Bower	4-10-83
Op.	86.37	John H. Sasson Jr.	9-15-91

RC ELECTRIC

DURATION

Age	HM/S	Held By	Date Set
RC Class A Sailplane (UMR)			
Jr.		No record established	
Sr.		No record established	
Op.	2:16:58	Joachim Nave	06-22-02
RC Class A Sailplane (LMR)			
Jr.		No record established	
Sr.		No record established	
Op.	2:18:50	Art Chmielewski	07-08-01
RC Class B Sailplane (UMR)			
Jr.		No record established	
Sr.		No record established	
Op.	3:33:33	Joachim Nave	08-05-01
RC Class B Sailplane (LMR)			
Jr.		No record established	
Sr.		No record established	
Op.	1:47:30	Joachim Nave	05-27-00
Electric R/C Duration			
Jr.		No Record Established	
Sr.		No Record Established	
Op.	0:28:13	Raymond Harlan	08-31-03

DISTANCE

Age	Miles	Held By	Date Set
RC Class A Sailplane (UMR)			
Jr.		No Record Established	
Sr.		No Record Established	
Op.	9.53	Christopher Silva	09-13-03
RC Class A Sailplane (LMR)			
Jr.		No Record Established	
Sr.		No Record Established	
Op.	5.53	Christopher Silva	05-25-03
RC Class B Sailplane (UMR)			
Jr.		No Record Established	
Sr.		No Record Established	
Op.	9.54	Gary Fogel	09-13-03
RC Class B Sailplane (LMR)			
Jr.		No Record Established	
Sr.		No Record Established	
Op.	5.54	Gary Fogel	03-29-03

UNMANNED AERIAL VEHICLES

ABSOLUTE WORLD RECORDS - UNMANNED AERIAL VEHICLES

DISTANCE (USA)
Northrop Grumman Ryan Aeronautical Center
Northrop Grumman RQ-4A Global Hawk — 13,219.86 km / 8,214.43 mi
1 Rolls-Royce Allison AE-3007H, 7,600 lbs
Edwards AFB, CA
4/23/01

SPEED
(No record registered)

DURATION (USA)
Northrop Grumman Ryan Aeronautical Center
Northrop Grumman RQ-4A Global Hawk — 30 hrs 24 min 1 sec
1 Rolls-Royce Allison AE-3007H, 7,600 lbs
Edwards AFB, CA
3/21/01

ALTITUDE (USA)
Greg Kendall, Derek L. Lisoski, Rik D. Meininger, Wyatt C. Sadler — 29,524 m / 96,863 ft
AeroVironment Helios
14 AeroVironment Centurion, 2 hp
Kekaha, HI
8/14/01

U.S. NATIONAL AND WORLD CLASS RECORDS

CLASS U-1.C (50 < 500 KG / 110 < 1,102 LBS) GROUP 2 (ELECTRIC)

TRUE ALTITUDE*
Derek L. Lisoski — 80,201 ft
Wyatt C. Sadler
AeroVironment Pathfinder Plus
8 AeroVironment, 3 hp
Kekaha, HI
8/6/98

CLASS U-1.D (500 < 2,500 KG / 1,102 < 5,512 LBS) GROUP 2 (ELECTRIC)

TRUE ALTITUDE (USA) ★
Greg Kendall — 29,524 m / 96,863 ft
Derek L. Lisoski
Rik D. Meininger
Wyatt C. Sadler
AeroVironment Helios
14 AeroVironment Centurion, 2 hp
Kekaha, HI
8/14/01

CLASS U-2.F (5,000 < 10,000 KG / 11,023 < 22,046 LBS) GROUP 1 (INTERNAL COMBUSTION AND JET)

TRUE ALTITUDE*
The Boeing Company — 67,028 ft
Condor
2 Continental LTSIO-300, 175 bhp
Moses Lake, WA
2/26/89

CLASS U-2.G (10,000 < 20,000 KG / 22,046 < 44,092 LBS) GROUP 1 (INTERNAL COMBUSTION AND JET)

DISTANCE IN A STRAIGHT LINE (USA) ★
Northrop Grumman-
Ryan Aeronautical Center — 13,219.86 km / 8,214.43 mi
Northrop Grumman RQ-4A Global Hawk
1 Rolls-Royce Allison AE-3007H, 7,600 lbs
Edwards AFB, CA
4/23/01

DURATION (USA)
Northrop Grumman-
Ryan Aeronautical Center — 30 hrs 24 min 1 sec
Northrop Grumman RQ-4A Global Hawk
1 Rolls-Royce Allison AE-3007H, 7,600 lbs
Edwards AFB, CA
3/21/01

TRUE ALTITUDE (USA)
Northrop Grumman-
Ryan Aeronautical Center — 19,928 m / 65,380 ft
Northrop Grumman RQ-4A Global Hawk
1 Rolls-Royce Allison AE-3007H, 7,600 lbs
Edwards AFB, CA
3/21/01

NATIONAL RECORDS

Each national aero club has the authority to establish record categories of its own, and the National Aeronautic Association has chosen to do so for the U.S.A.

Some of these national records parallel FAI record categories and others are in "special" categories that have been created to recognize unique achievements.

SPECIAL CATEGORIES

ALTITUDE AND SPEED IN HORIZONTAL FLIGHT UNMANNED AIRCRAFT
Teledyne-Ryan Firebolt — 103,000 ft
Eglin AFB, FL — 4.1 Mach

COMBINED AIRCRAFT PERFORMANCE
(CAFE Triaviathon All-Time High Score) — 2,381.24
David W. Anders
RV-4
Santa Rosa, CA
9/27/97

CROSSING OF THE CONTINENTAL DIVIDE IN A HOT AIR BALLOON
Edward V. Sagon — Twice
Aerostar S77A — (during the same flight)
5/1/91

DISTANCE IN A CLOSED CIRCUIT BELOW SEA LEVEL
Russel J. Saunders — 680.35 mi
Cessna 140, N89965
Continental C-85, 85 hp
Death Valley, CA
3/29/86

FASTEST TIME TO FLY AROUND THE BORDER OF THE CONTINENTAL U.S. ★
Randolph M. Pentel — 1 day
Mark T. Anderson — 21 hrs 27 min
Cessna 560 Citation V Ultra
2 P&WC JT15D, 3,045 lbs
12/16-17/03

FASTEST TIME TO LAND AND TAKEOFF IN EACH OF THE CONTIGUOUS 48 STATES IN A SINGLE ENGINE AIRCRAFT
Joseph M. Dougherty — 5 days
Cessna 182E — 11 hrs 5 min
1 Continental O-470, 230 bhp
8/6-8/12/99

FASTEST TIME TO VISIT ALL SEVEN CONTINENTS ★
William E. Signs — 50 days
Ruth E. Jacobs — 1 hr 16 min
Cessna 210L — 24 sec
1 Continental IO-520, 300 bhp
Dallas, TX
12/31/95-2/18/96

FASTEST TIME TO VISIT ALL THE HARD SURFACE PUBLIC AIRPORTS IN FLORIDA
Benjamin D. Riecken — 1 day
Giangiacomo Palombo — 21 hrs 57 min
Cessna 172S — 9 sec
1 Lycoming IO-360, 180 bhp
8/16-18/03

FASTEST TIME TO VISIT ALL THE HARD SURFACE, PUBLIC AIRPORTS IN MICHIGAN (128) ★
Patrick J. Curley — 1 day
Juanita D. Curley — 12 hrs 5 min
Piper PA-28-181 — 3 sec
1 Lycoming O-360, 180 bhp
6/30/93

FASTEST TIME TO VISIT ALL THE HARD SURFACE PUBLIC AIRPORTS IN OREGON ★
Parker Johnstone — 1 day
James S. Murphy — 7 hrs 16 min
Beechcraft Bonanza 36 — 37 sec
1 Continental IO-550, 300 bhp
6/20/01

FASTEST TIME TO VISIT ALL THE HARD SURFACE, PUBLIC AIRPORTS IN TENNESSEE
James D. Howard, Pilot — 18 hrs
Keith Sewell, Copilot — 46 min
Mooney M20A — 36 sec
1 Lycoming O-360, 180 bhp
2/19/96

FASTEST TIME TO VISIT ALL THE STATE CAPITALS IN THE CONTIGUOUS U.S. AND THE DISTRICT OF COLUMBIA
Addison P. Farmer — 8 days
Harley E. Hice — 9 hrs 8 min
Piper Comanche PA-30 — 7 sec
2 Lycoming IO-320, 160 bhp
7/4-13/97

FASTEST TIME TO CONSECUTIVELY TAKEOFF AND LAND IN ALL 50 STATES AND THE DISTRICT OF COLUMBIA
Mike Hance — 334 hrs
Mooney 252, N252BA — 22 min
Departed: Honolulu, HI, 7/17/87
Arrived: Oshkosh, WI, 7/31/87

FLIGHT EFFICIENCY
(CAFE Challenge All-Time High Score) — 1,379,364
F. Gary Hertzler
Rutan VariEze
Santa Rosa, CA
6/4/94

GREATEST RECORDED WEIGHT AT WHICH ANY AIRPLANE HAS EVER FLOWN
Jesse T. Allen — 920,836 lbs
Lockheed C-5A
Georgia
12/17/84

HEAVIEST LOAD DROPPED FROM AN AIRCRAFT
Lockheed C-5B, Galaxy — 190,493 lbs
Pope AFB, NC
6/7/89

LARGEST GENERAL AVIATION FORMATION ★
Pilots in Command: 132 airplanes
Hal Aavang, Albert C. Ackerman, Bill Aikman, Jerry D. Anderson, Myron J. Babler, Raymond Balwierczak, David Biba, Mark Boe, Brent A. Bostwick, Dennis Boyet, Lionel G. Brown, Barry Brannan, Robert Breitbarth, Wayne Collins, Bruce A. Campbell, Roger D. Cannell, Michael L. Casto, Patrick Chavez, Christopher L. Coles, Terry A. Cone, Stanford Cook, Matt Dalton, Roland Davies, Raymond G. Davis, Domenic S. De Nardo, Joe Wagner Dimock, Grant Dorsett, Ted E. Dorsey, Frazier Eales, Michael L. Edwards, Adrian A. Eichhorn, Allan L. Fink, Louis J. Fox, Kevin Frank, Virgil R. Franz, David J. Friis, Frank E. Gibson, Mike B. Griffith, Frank Haile, Martin Hairabedian, Ralph Harter, Jr., David Hendricks, Jimmy W. Hively, Dean Holliday, Mitchell Januszewski, Michael D. Johnsen, Paul Joseph, Tom Keesling, Tom Kendall, F. Barry Knotts, Robert P. Landes, Donald M. Leach, M. Rudolph Lenich, Donald E. Lucy, II, Randy Lyons, Robert Mark, Jimmie Kent Martin, James E. McCormack, John S. Moffitt, Joseph A. Montineri, Arthur N. Moose, Thomas M. Neary, David Norgart, David Novak, Michael Oettinger, George H. Oswald, Jeffrey P. Parker, Alfred G. Peterson, Tim Pfeiffer, Bob B. Polehla, Marvin Polzien, Bruce Poulton, Jame W. Reagan, Jeff Renfrow, Leroy B. Roper, Harold J. Rucker, James L. Schenck, Elliott Schiffman, Gary Schlabaugh, Marshall Segal, James D. Sheen, Robert W. Siegfried, Richard D. Siegrist, John M. Slais, James E. Sok, Leet Sommerfeld, Richard I. Stearns, Stanley J. Stewart, Dick Strickland, Keith Stutts, Bill Tugaw, Edward O. Vernon, Mike Van Dinter, Garry C. Waite, Donald E. Walker, Jim Webber, Bill D. Whitman, John Wiebener, Terry Wilcox, Lyle R. Wilkinson, John T. Wilson, Glenn Wimbish, Carl N. Winans, Ken Wittekiend, Richard Wright
Rockford, IL to Oshkosh, WI
7/25/95

LONGEST HOVER OF A HUMAN POWERED HELICOPTER
Neal Saiki 6.8 sec
Kyle Naydo
Bill Patterson
Greg McNeil
Robert Faye
Scott Larwood
"Da Vinci III"
California State Polytechnic University
San Luis Obispo, CA
12/11/89

TIME TO CLIMB TO 51,000 FEET (Class C-1.g, Jet Engine)
Gene "Ed" Allen 25 min
Tim Brennan 36.5 sec
Falcon 900, N407FJ
Teterboro, NJ
2/21/87

TIME TO CLIMB TO 51,000 FEET (Class C-1.k, Jet Engine)
Gary M. Freeman 22 min
Edward D. Mendenhall 21.9 sec
William M. Osborne
Gulfstream V
2 BMW Rolls-Royce BR710, 14,845 lbs
Savannah, GA
9/14/97

TIME TO CLIMB TO 51,000 FEET WITH 1,000 KG PAYLOAD (Class C-1.k, Jet Engine)
Gary M. Freeman 22 min
Edward D. Mendenhall 21.9 sec
William M. Osborne
Gulfstream V
2 BMW Rolls-Royce BR710, 14,845 lbs
Savannah, GA
9/14/97

TIME TO CLIMB TO 51,000 FEET WITH 2,000 KG PAYLOAD (Class C-1.k, Jet Engine)
Gary M. Freeman 22 min
Edward D. Mendenhall 21.9 sec
William M. Osborne
Gulfstream V
2 BMW Rolls-Royce BR710, 14,845 lbs
Savannah, GA
9/14/97

INDEX

INDEX OF ADVERTISERS:

INDEX OF ENHANCED RECORD LISTINGS:

Back Cover: *First Flight of the Reproduction 1903 Flyer, November 20, 2003 at Wright Brothers National Memorial*

Photo by Paul Glenshaw, courtesy of The Wright Experience, Inc.